IFIP Advances in Information and Communication Technology

494

Editor-in-Chief

Kai Rannenberg, Goethe University Frankfurt, Germany

IFIP – The International Federation for Information Processing

IFIP was founded in 1960 under the auspices of UNESCO, following the first World Computer Congress held in Paris the previous year. A federation for societies working in information processing, IFIP's aim is two-fold: to support information processing in the countries of its members and to encourage technology transfer to developing nations. As its mission statement clearly states:

> IFIP is the global non-profit federation of societies of ICT professionals that aims at achieving a worldwide professional and socially responsible development and application of information and communication technologies.

IFIP is a non-profit-making organization, run almost solely by 2500 volunteers. It operates through a number of technical committees and working groups, which organize events and publications. IFIP's events range from large international open conferences to working conferences and local seminars.

The flagship event is the IFIP World Computer Congress, at which both invited and contributed papers are presented. Contributed papers are rigorously refereed and the rejection rate is high.

As with the Congress, participation in the open conferences is open to all and papers may be invited or submitted. Again, submitted papers are stringently refereed.

The working conferences are structured differently. They are usually run by a working group and attendance is generally smaller and occasionally by invitation only. Their purpose is to create an atmosphere conducive to innovation and development. Refereeing is also rigorous and papers are subjected to extensive group discussion.

Publications arising from IFIP events vary. The papers presented at the IFIP World Computer Congress and at open conferences are published as conference proceedings, while the results of the working conferences are often published as collections of selected and edited papers.

IFIP distinguishes three types of institutional membership: Country Representative Members, Members at Large, and Associate Members. The type of organization that can apply for membership is a wide variety and includes national or international societies of individual computer scientists/ICT professionals, associations or federations of such societies, government institutions/government related organizations, national or international research institutes or consortia, universities, academies of sciences, companies, national or international associations or federations of companies.

More information about this series at http://www.springer.com/series/6102

Lorena Bociu · Jean-Antoine Désidéri
Abderrahmane Habbal (Eds.)

System Modeling and Optimization

27th IFIP TC 7 Conference, CSMO 2015
Sophia Antipolis, France, June 29 – July 3, 2015
Revised Selected Papers

 Springer

Editors
Lorena Bociu
North Carolina State University
Raleigh, NC
USA

Abderrahmane Habbal
Université Côte d'Azur
Nice
France

Jean-Antoine Désidéri
Inria Sophia Antipolis
Sophia Antipolis
France

ISSN 1868-4238   ISSN 1868-422X   (electronic)
IFIP Advances in Information and Communication Technology
ISBN 978-3-319-85749-7   ISBN 978-3-319-55795-3   (eBook)
DOI 10.1007/978-3-319-55795-3

Printed on acid-free paper

This Springer imprint is published by Springer Nature
The registered company is Springer International Publishing AG
The registered company address is: Gewerbestrasse 11, 6330 Cham, Switzerland

Preface

This volume comprises selected contributions to the 27th Conference on System Modelling and Optimization that took place from June 29 to July 3, 2015, on the SophiaTech Campus, Sophia Antipolis, France.

These articles encompass broad aspects of system modelling and optimization, such as modelling and analysis of systems governed by partial differential equations (PDEs) or ordinary differential equations (ODEs), control of PDEs/ODEs, nonlinear optimization, stochastic optimization, multi-objective optimization, combinatorial optimization, industrial applications, and numerics of PDEs. These themes are the focus of the IFIP TC7 community.

The conference was co-organized by two local institutions, Inria Sophia Antipolis Méditerranée and Université Côte d'Azur jointly with North Carolina State University, (visit http://ifip2015.inria.fr).

This scientific event was attended by more than 250 participants, from about 30 different countries. The conference program was composed of eight plenary talks, and 26 invited mini-symposia, plus three sessions of refereed contributed papers, resulting in a total of 62 sessions, and altogether 230 presentations.

The 48 refereed contributions included in the present proceedings cover the latest progress in their respective areas, and give a flavor of the wide range and exciting topics discussed at the meeting.

We warmly thank the members of the Scientific Committee and our solicited reviewers for their valuable contributions.

January 2017

Lorena Bociu
Jean-Antoine Désidéri
Abderrahmane Habbal

Organization

Scientific Committee

Gregoire Allaire	École Polytechnique, France
A.V. Balakrishnan	University of California (UCLA), USA
Alfio Borzi	Universität Würzburg, Germany
Hector Cancela	University of the Republic of Uruguay
Christian Clason	Universität Duisburg-Essen, Germany
Gianni Di Pillo	Sapienza University of Rome, Italy
Yu.G. Evtushenko	Dorodnicyn Computing Centre of the Russian Academy of Sciences, Russia
Janusz Granat	Warsaw University of Technology, Poland
Jacques Henry	Inria, France
Dietmar Hömberg	TU Berlin, Germany
Peter Kall	University of Zurich, Switzerland
Alfred Kalliauer	TU Wien, Austria
Barbara Kaltenbacher	Universität Klagenfurt, Austria
Hisao Kameda	University of Tsukuba, Japan
Irena Lasiecka	University of Memphis, USA
Istvan Maros	Imperial College, London, UK
Kurt Marti	Federal Armed Forces University Munich, Germany
Jiri Outrata	Institute of Information Theory and Automation, Czech Republic
Andrzej Palczewski	University of Warsaw, Poland
Stefan Scholtes	University of Cambridge, UK
Volker Schulz	Universität Trier, Germany
Mark S. Squillante	IBM T.J. Watson Research Center, USA
Lukasz Stettner	Institute of Mathematics, Polish Academy of Sciences, Poland
Daniel Straub	Technische Universität München, Germany
Philippe Toint	University of Namur, Belgium
Fredi Tröltzsch	TU Berlin, Germany
José Manuel Valério de Carvalho	Universidade do Minho, Portugal
Lidija Zadnik-Stirn	University of Ljubljana, Slovenia
Jean-Paul Zolésio	CNRS, France

Organizing Committee

Lorena Bociu	NC State University, USA
Jean-Antoine Désidéri	Inria, France
Abderrahmane Habbal	Université Côte d'Azur, France
Agnes Cortell	Inria, France
Catherine Bonnet	Inria, France

Inria

Inria, the French National Institute for computer science and applied mathematics, promotes scientific excellence for technology transfer and society. Graduates from the world's top universities, Inria's 2,700 employees rise to the challenges of digital sciences. Research at Inria is organized in project teams that bring together researchers with complementary skills to focus on specific scientific projects. With this open, agile model, Inria is able to explore original approaches with its partners in industry and academia and provide an efficient response to the multidisciplinary and application challenges of the digital transformation. The source of many innovations that add value and create jobs, Inria transfers expertise and research results to companies (start-ups, SMEs, and major groups) in fields as diverse as health care, transport, energy, communications, security and privacy protection, smart cities, and the factory of the future.

NC State University

NC State is a pre-eminent research enterprise that excels in science, technology, engineering, math, design, textiles, and veterinary medicine, and consistently rates as one of the best values in higher education. NC State students, faculty, and staff take problems in hand and work with industry, government, and nonprofit partners to solve them. The institution's 34,000-plus high-performing students apply what they learn in the real world by conducting research, working in internships and co-ops, and performing acts of world-changing service.

UCA

The Mathematics Laboratory Jean-Alexandre Dieudonné LJAD is a mixed research unit — UMR No. 7351 — dependent on the two organizations: the National Center for Scientific Research (CNRS) and the Université Côte d'Azur (UCA). The laboratory is structured into six teams: (1) Algebra, Geometry and Topology, (2) Geometry, Analysis and Dynamic, (3) PDEs and Numerical Analysis, (4) Numerical Modeling and Fluid Dynamics, (5) Probability and Statistics, and, (6) Interfaces of Mathematics and Complex Systems. Bringing together 134 researchers and teacher-researchers, 13 administrative staff and support engineers to research, and 70 doctoral and post-doctoral students, the laboratory also has joint laboratories with Inria Sophia Antipolis and CEA (Commissariat of Atomic Energy) within the RSC Fusion.

Contents

Control Methods for the Optimization of Plasma Scenarios in a Tokamak

Jacques Blum[✉], Cédric Boulbe, Blaise Faugeras, and Holger Heumann

Université Côte d'Azur, Inria, CNRS, LJAD, Nice, France
jacques.blum@unice.fr

Abstract. This paper presents the modelling of the evolution of plasma equilibrium in the presence of external poloidal field circuits and passive structures. The optimization of plasma scenarios is formulated as an optimal control problem where the equations for the evolution of the plasma equilibrium are the constraints. The procedure determines the voltages applied to the external circuits that minimize a certain cost-function representing the distance to a desired plasma augmented by an energetic cost of the electrical system. A sequential quadratic programming method is used to solve the minimization of the cost-function and an application to the optimization of a discharge for ITER is shown.

Keywords: Plasma equilibrium · Optimization · Tokamak · Magnetohydrodynamics · Optimal control

1 Introduction

A tokamak is an experimental device whose purpose is to confine a plasma (ionized gas) in a magnetic field so as to control the nuclear fusion of atoms of low mass (deuterium, tritium,..) and to produce energy. The magnetic field has two components (see Fig. 1):

- a toroidal field created by toroidal field coils, that is necessary for the stability of the plasma,
- a poloidal field in the section of the torus created by poloidal field coils and by the plasma itself.

The plasma current is obtained by induction from currents in these poloidal field coils. The tokamak thus appears as a transformer whose plasma is the secondary. The currents in the external coils play another role, that of creating and controlling the equilibrium of the plasma. The goal of this paper is to provide a model for the evolution in time of the equilibrium of the plasma and to derive control methods in order to optimize a typical scenario of a discharge of the plasma in a tokamak.

Laboratoire J.-A. Dieudonné, Université de Nice Sophia-Antipolis, Parc Valrose, 06108 Nice Cedex 02, France.

L. Bociu et al. (Eds.): CSMO 2015, IFIP AICT 494, pp. 1–20, 2016.
DOI: 10.1007/978-3-319-55795-3_1

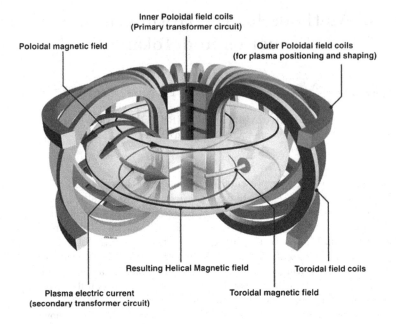

Fig. 1. Schematic representation of a tokamak

There are two approaches for simulating a plasma made of electrons and ions:

- the microscopic approach based on kinetic equations (Vlasov, Boltzmann, Fokker-Planck) that are 6D (3D in space and 3D in terms of the velocity) and 1D in time.
- the macroscopic approach based on magnetohydrodynamics (MHD) equations that are obtained by taking moments of the kinetic equations, and which are 3D in space and 1D in time. The validity of the MHD equations is clearly more restrictive than the one of the kinetic equations. We will present in Sect. 2 the way in which the MHD equations are obtained from the kinetic ones.

At the slow resistive diffusion time-scale, the plasma is in equilibrium at each instant (the kinetic pressure force balances at each point the Lorentz force due to the magnetic field) and hence the plasma follows the so-called quasi-static evolution of the equilibrium. The resistive diffusion in the external passive structures surrounding the plasma and the equations of the circuits of the poloidal field system enable to follow in time this quasi-static evolution. An axisymmetric hypothesis enables to reduce the problem to a 2D p.d.e. formulation, with the Grad-Shafranov equation for the equilibrium of the plasma. The plasma boundary is a free boundary, which is a particular poloidal flux line. It is either the outermost closed flux line inside the limiter, which prevents the plasma from touching the vacuum vessel, or a separatrix (with a hyperbolic X-point), as they are in presence of a poloidal divertor. This equilibrium model will be presented in Sect. 3 of the paper.

In order to solve numerically the set of equations for the poloidal flux, it is necessary to derive the weak formulation of this system and then a finite element method, coupled to Newton iterations for the treatment of the non-linearities, enables to solve the evolution of the equilibrium configuration in a tokamak. This is presented in Sect. 4 of this paper.

A typical discharge in a tokamak is made of several phases: ramp-up of total plasma current, plateau phase (stationary phase), ramp-down. The plasma shape can also move from a small circular plasma (at the beginning of the discharge) to a large elongated one with an X-point. The goal of this work is to determine, thanks to optimal control theory of systems governed by partial differential equations [1], the voltages applied to the poloidal field circuits that achieve at best the desired scenario, by minimizing a certain cost-function which represents the sum of the distance to the desired plasma and of the energetic cost of the electrical system. The introduction of an appropriate lagrangian taking as constraints the equilibrium system of the previous sections and the determination of the corresponding adjoint state enable the computation of the gradient of the cost-function in terms of the adjoint state. The minimization of this cost-function is performed thanks to a SQP (Sequential Quadratic Programming) method. An interesting test-case, solved by using these techniques, will be presented for the ITER (International Thermonuclear Experimental Reactor) tokamak. This is presented in Sect. 5 of this paper. This method has the purpose to replace the empirical methods used commonly to compute the pre-programmed voltages that enable to go from one snapshot to another one. This method can of course be extended to other type of optimization of the scenarios just by modifying the cost-function and the control variables (consumption of flux, desired profile of plasma current density,..).

2 The Magnetohydrodynamic Equations

A plasma is a ionized gas composed of ions and electrons. The kinetic equations describe the plasma thanks to a distribution function $f_\alpha(\mathbf{x}, \mathbf{v}, t)$ (with $\alpha = e$ for electrons and $\alpha = i$ for ions) where $\mathbf{x}$ is the point position and $\mathbf{v}$ the particles velocity. For a collisional plasma the kinetic equations are based on the Fokker-Planck equation

$$\frac{\partial f_\alpha}{\partial t} + (\mathbf{v}.\nabla_{\mathbf{x}})f_\alpha + \frac{\mathbf{F}_\alpha}{m_\alpha}.\nabla_{\mathbf{v}}f_\alpha = C_\alpha, \tag{1}$$

where m_α is the mass of the particles, $\mathbf{F}_\alpha$ the force applied to these particles and C_α the term due to collisions between particles. This microscopic approach requires the resolution of a partial differential equations in 6 dimensions (space and velocity) plus the time dimension. This is extremely difficult from a computational point of view. Therefore from this equation one derives a macroscopic representation based on the fluid equations in the following way. Let us define the density of particles by

$$n_\alpha(\boldsymbol{x}, t) = \int f_\alpha(\boldsymbol{x}, \boldsymbol{w}, t)d\boldsymbol{w},$$

the fluid velocity by

$$\boldsymbol{u}_\alpha(\boldsymbol{x},t) = \frac{1}{n_\alpha} \int f_\alpha(\boldsymbol{x},\boldsymbol{w},t)\boldsymbol{w}d\boldsymbol{w},$$

and the pressure tensor

$$P_\alpha(\boldsymbol{x},t) = m_\alpha \int f_\alpha(\boldsymbol{x},\boldsymbol{w},t)(\boldsymbol{w}-\boldsymbol{u}_\alpha)(\boldsymbol{w}-\boldsymbol{u}_\alpha)d\boldsymbol{w},$$

which under the isotropic assumption becomes

$$p_\alpha(\boldsymbol{x},t) = \frac{m_\alpha}{3} \int f_\alpha(\boldsymbol{x},\boldsymbol{w},t)(\boldsymbol{w}-\boldsymbol{u}_\alpha)^2 d\boldsymbol{w}.$$

Multiplying Eq. (1) by a test function $\phi(\boldsymbol{w})$ and integrating over the space of velocities leads to the fluid equations. The first moment (corresponding to $\phi = 1$) gives the equation for the density of particles:

$$\frac{\partial n_\alpha}{\partial t} + \nabla . \int f_\alpha \boldsymbol{w}d\boldsymbol{w} - \frac{1}{m_\alpha} \int \frac{\partial \boldsymbol{F}_\alpha}{\partial \boldsymbol{w}} f_\alpha d\boldsymbol{w} = 0.$$

Since for electromagnetic forces $\dfrac{\partial \boldsymbol{F}_\alpha}{\partial \boldsymbol{w}} = 0$, and since collisions do not change the number of particles one obtains:

$$\frac{\partial n_\alpha}{\partial t} + \nabla . (n_\alpha \boldsymbol{u}_\alpha) = 0.$$

The second moment is obtained by taking $\phi = m_\alpha \boldsymbol{w}$ which leads to the momentum equation

$$m_\alpha \frac{\partial}{\partial t}(n_\alpha \boldsymbol{u}_\alpha) + m_\alpha \nabla_x . \int f_\alpha \boldsymbol{w}\boldsymbol{w}d\boldsymbol{w} - \int \nabla_{\boldsymbol{w}} . (\boldsymbol{F}_\alpha . \boldsymbol{w}) f_\alpha d\boldsymbol{w} = \int m_\alpha \boldsymbol{w} C_\alpha d\boldsymbol{w},$$

where we have set $\boldsymbol{w} = (\boldsymbol{w}-\boldsymbol{u}_\alpha) + \boldsymbol{u}_\alpha$. Using the equation for conservation of the density one gets

$$m_\alpha n_\alpha (\frac{\partial \boldsymbol{u}_\alpha}{\partial t} + \boldsymbol{u}_\alpha . \nabla \boldsymbol{u}_\alpha) = -\nabla . P_\alpha + n_\alpha \overline{\boldsymbol{F}_\alpha} + \boldsymbol{R}_\alpha,$$

whith $\overline{\boldsymbol{F}_\alpha} = Ze(\boldsymbol{E} + \boldsymbol{u}_\alpha \times \boldsymbol{B})$ where Ze is the charge of particles and $\boldsymbol{R}_\alpha$ is the change rate of the momentum due to collisions.

The third moment gives the energy equation which needs to be complemented with closing relations on the heat flux. These latter come from a transport model. The single fluid magnetohydrodynamic equations are derived by defining the mass density

$$mn = m_e n_e + m_i n_i$$
$$= m_e Z n_i + m_i n_i \approx m_i n_i$$

the velocity of the fluid

$$u = \frac{m_e n_e u_e + m_i n_i u_i}{\rho} \approx u_i,$$

the current density

$$j = -e n_e u_e + Z e n_i u_i,$$
$$= e n_e (u_i - u_e),$$

and the scalar pressure

$$p = n_e k T_e + n_i k T_i,$$

where k is the Boltzmann constant. The Maxwell equations need to be added since we are in the presence of a magnetic field $\boldsymbol{B}$ and of an electric field $\boldsymbol{E}$. Finally the resistive MHD equations for a single fluid [2] read:

$$
\begin{cases}
\dfrac{\partial n}{\partial t} + \nabla.(n\boldsymbol{u}) = s & \text{(Conservation of particles)} \\[2mm]
mn(\dfrac{\partial \boldsymbol{u}}{\partial t} + \boldsymbol{u}.\nabla \boldsymbol{u}) + \nabla p = \boldsymbol{j} \times \boldsymbol{B} & \text{(Conservation of momentum)} \\[2mm]
\dfrac{3}{2}(\dfrac{\partial p}{\partial t} + \boldsymbol{u}.\nabla p) + \dfrac{5}{2}p\nabla.\boldsymbol{u} + \nabla Q = s' & \text{(Conservation of particle energy)} \\[2mm]
\nabla \times \boldsymbol{E} = -\dfrac{\partial \boldsymbol{B}}{\partial t} & \text{(Faraday's law)} \\[2mm]
\nabla.\boldsymbol{B} = 0 & \text{(Conservation of } \boldsymbol{B}) \\
\boldsymbol{E} + \boldsymbol{u} \times \boldsymbol{B} = \eta \boldsymbol{j} & \text{(Ohm's law)} \\
\nabla \times \boldsymbol{H} = \boldsymbol{j} & \text{(Ampere's law)} \\
\boldsymbol{B} = \mu \boldsymbol{H} & \text{(Magnetic permeability)} \\
p = nkT & \text{(Law of perfect gases)}
\end{cases}
\tag{2}
$$

where n denotes the density of the particles, m their mass, $\boldsymbol{u}$ their mean velocity, p their pressure, T their temperature, Q the heat flux, η the resistivity tensor, s and s' the source terms and k the Boltzmann constant.

3   Equilibrium of a Plasma in a Tokamak

In order to simplify system (2) some characteristic time constants of the plasma need to be defined. The Alfven time constant τ_A is

$$\tau_A = \frac{a(\mu_0 mn)^{1/2}}{B_0},$$

where a is the minor radius of the plasma and $\boldsymbol{B}_0$ is the toroidal magnetic field. It is of the order of a microsecond for present tokamaks.

The diffusion time constant of the particle density n is

$$\tau_n = \frac{a^2}{D},$$

where D is the particle diffusion coefficient. Likewise, the time constants for diffusion of heat of the electrons and of the ions are

$$\tau_e = \frac{n_e a^2}{K_e},$$

$$\tau_i = \frac{n_i a^2}{K_i},$$

where n_e, n_i are the density of electrons and ions, respectively, and K_e, K_i are their thermal conductivities. These constants τ_n, τ_e, τ_i are of the order of a millisecond on tokamaks currently operating.

Finally, the resistive time constant for the diffusion of the current density and magnetic field in the plasma is given by

$$\tau_r = \frac{\mu_0 a^2}{\eta},$$

and is of the order of a second.

If a global time constant for plasma diffusion is defined by

$$\tau_p = \inf(\tau_n, \tau_e, \tau_i, \tau_r),$$

we note that

$$\tau_A \ll \tau_p.$$

On the diffusion time-scale τ_p the term $(\frac{\partial u}{\partial t} + u\nabla u)$ is small compared with ∇p (see [3,4]) and the equilibrium equation

$$\nabla p = j \times B \tag{3}$$

is thus satisfied at every instant in the plasma.

Consequently the equations which govern the equilibrium of a plasma in the presence of a magnetic field in a tokamak are on the one hand Maxwell's equations satisfied in the whole of space (including the plasma):

$$\begin{cases} \nabla \cdot B = 0, \\ \nabla \times (\dfrac{B}{\mu}) = j, \end{cases} \tag{4}$$

and on the other hand the equilibrium Eq. (3) for the plasma itself.

Equation (3) means that the plasma is in equilibrium when the force ∇p due the kinetic pressure p is equal to the Lorentz force of the magnetic pressure $j \times B$. We deduce immediately from (3) that

$$B \cdot \nabla p = 0, \tag{5}$$

and

$$j \cdot \nabla p = 0. \tag{6}$$

Thus for a plasma in equilibrium the field lines and the current lines lie on isobaric surfaces ($p = const.$); these surfaces, generated by the field lines, are called magnetic surfaces. In order for them to remain within a bounded volume of space it is necessary that they have a toroidal topology. These surfaces form a family of nested tori. The innermost torus degenerates into a curve which is called the magnetic axis.

In a cylindrical coordinate system (r, ϕ, z) (where $r = 0$ is the major axis of the torus) the hypothesis of axial symmetry consists in assuming that the magnetic field B is independent of the toroidal angle ϕ. The magnetic field can be decomposed as $B = B_p + B_\phi$, where $B_p = (B_r, B_z)$ is the poloidal component and B_ϕ is the toroidal component. From Eq. (4) one can define the poloidal flux $\psi(r, z)$ such that

$$\begin{cases} B_r = -\dfrac{1}{r}\dfrac{\partial \psi}{\partial z}, \\ B_z = \dfrac{1}{r}\dfrac{\partial \psi}{\partial r}. \end{cases} \tag{7}$$

Concerning the toroidal component B_ϕ we define f by

$$B_\phi = \frac{f}{r} e_\phi, \tag{8}$$

where e_ϕ is the unit vector in the toroidal direction, and f is the diamagnetic function. The magnetic field can be written as:

$$\begin{cases} B = B_p + B_\phi, \\ B_p = \dfrac{1}{r}[\nabla \psi \times e_\phi], \\ B_\phi = \dfrac{f}{r} e_\phi. \end{cases} \tag{9}$$

According to (9), in an axisymmetric configuration the magnetic surfaces are generated by the rotation of the flux lines $\psi = const.$ around the axis $r = 0$ of the torus.

From (9) and the second relation of (4) we obtain the following expression for j:

$$\begin{cases} j = j_p + j_\phi, \\ j_p = \dfrac{1}{r}[\nabla(\dfrac{f}{\mu}) \times e_\phi], \\ j_\phi = (-\Delta^*\psi)e_\phi, \end{cases} \tag{10}$$

where j_p and j_ϕ are the poloidal and toroidal components respectively of j, and the operator Δ^* is defined by

$$\Delta^*\cdot = \partial_r\left(\frac{1}{\mu r}\partial_r \cdot\right) + \partial_z\left(\frac{1}{\mu r}\partial_z \cdot\right) = \nabla\left(\frac{1}{\mu r}\nabla\cdot\right). \tag{11}$$

Expressions (9) and (10) for B and j are valid in the whole of space since they involve only Maxwell's equations and the hypothesis of axisymmetry. Hence they can be reduced to one equation given in 2 space dimensions in the poloidal plane $(r, z) \in \Omega_\infty = (0, \infty) \times (-\infty, \infty)$ for the poloidal flux ψ:

$$-\Delta^* \psi = j_\phi. \tag{12}$$

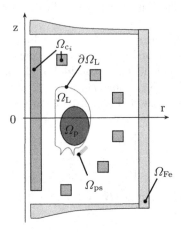

Fig. 2. Schematic representation of the poloidal plane of a tokamak. Ω_p is the plasma domain, Ω_L is the limiter domain accessible to the plasma, Ω_{c_i} represent poloidal field coils, Ω_{ps} the passive structures and Ω_{Fe} the ferromagnetic structures.

The toroidal component of the current density j_ϕ is zero everywhere outside the plasma domain, the poloidal field coils and the passive structures. The different sub-domains of the poloidal plane of a tokamak (see Fig. 2) as well as the corresponding expression for j_ϕ are described below:

- Ω_L is the domain accessible to the plasma. Its boundary is the limiter $\partial \Omega_L$.
- Ω_p is the plasma domain where relation (5) implies that ∇p and $\nabla \psi$ are co-linear, and therefore p is constant on each magnetic surface. This can be denoted by

$$p = p(\psi). \tag{13}$$

Relation (6) combined with the expression (10) implies that ∇f and ∇p are co-linear, and therefore f is likewise constant on each magnetic surface

$$f = f(\psi). \tag{14}$$

The equilibrium relation (3) combined with the expression (9) and (10) for B and j implies that:

$$\nabla p = -\frac{\Delta^* \psi}{r} \nabla \psi - \frac{f}{\mu_0 r^2} \nabla f, \tag{15}$$

which leads to the so-called Grad-Shafranov equilibrium equation:

$$-\Delta^* \psi = r p'(\psi) + \frac{1}{\mu_0 r}(f f')(\psi). \tag{16}$$

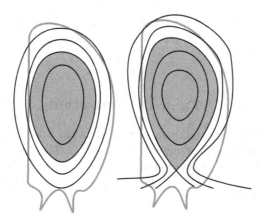

Fig. 3. Example of a plasma whose boundary is defined by the contact with limiter (left) or by the presence of an X-point (right).

Here μ is equal to the magnetic permeability μ_0 of the vacuum and Δ^* is a linear elliptic operator. From (10) it is clear that right-hand side of (16) represents the toroidal component of the plasma current density. It involves functions $p(\psi)$ and $f(\psi)$ which are not directly measured inside the plasma. The plasma domain is unknown, $\Omega_p = \Omega_p(\psi)$, and this is a free boundary problem. This domain is defined by its boundary which is the largest closed ψ iso-contour contained within the limiter Ω_L. The plasma can either be limited if this iso-contour is tangent to the limiter $\partial\Omega_L$ (see Fig. 3, left) or defined by the presence of a saddle-point also called X-point (see Fig. 3, right). In the later configuration which is obtained in presence of a divertor, the plasma does not touch any physical component and the performances and the confinement of the plasma are improved (see [5]).

- Ω_{Fe} represents the ferromagnetic structures. They do not carry any current, $j_\phi = 0$ but the magnetic permeability μ is not constant and depends on the magnetic field:

$$\mu = \mu_{Fe}\left(\frac{|\nabla\psi|^2}{r^2}\right). \tag{17}$$

- Domains Ω_{c_i} represent the poloidal field coils carrying currents. If we consider that the voltages V_i applied to these coils are given, using Faraday and Ohm laws the current density can be written as

$$j_\phi = \frac{n_i V_i}{R_i |\Omega_{c_i}|} - \frac{2\pi n_i^2}{R_i |\Omega_{c_i}|^2} \int_{\Omega_{C_i}} \dot\psi ds, \tag{18}$$

where n_i is the number of windings in the coil, $|\Omega_{c_i}|$ its section area, R_i its resistance and $\dot\psi$ is the time derivative of ψ,

- Ω_{ps} represents passive structures where the current density can be written as

$$j_\phi = -\frac{\sigma}{r}\dot\psi, \tag{19}$$

where σ is the conductivity.

In summary we are seeking for the poloidal flux $\psi(t)$ that is a solution of (12) with j_ϕ given by (16), (18) and (19) and verifies boundary conditions

$$\psi(0, z) = 0 \quad \text{and} \quad \lim_{\|(r,z)\| \to +\infty} \psi(r, z) = 0.$$

4 Weak Formulation and Discretization

We chose a semi-circle Γ of radius ρ_Γ surrounding the iron domain Ω_{Fe}, the coil domains Ω_{c_i} and the passive structures domain Ω_{ps}. The truncated domain, we use for our computations, is the domain Ω having the boundary $\partial\Omega = \Gamma \cup \Gamma_0$, where $\Gamma_0 := \{(0, z), z_{min} \le z \le z_{max}\}$. The weak formulation for $\psi(t)$ uses the following Sobolev space:

$$H := \left\{ \psi : \Omega \to \mathbb{R}, \|\psi\| < \infty, \left\|\frac{|\nabla\psi|}{r}\right\| < \infty, \psi_{|\Gamma_0} = 0 \right\} \cap C^0(\overline{\Omega}),$$

with

$$\|\psi\|^2 = \int_\Omega \psi^2\, r\, dr dz.$$

It reads as: Given $\mathbf{V}(t) = \{V_i(t)\}_{i=1}^N$ find $\psi(t) \in H$ such that for all $\xi \in H$

$$\mathsf{A}(\psi(t), \xi) - \mathsf{J}_\mathrm{p}(\psi(t), \xi) + \mathsf{j}^{\mathrm{ps}}(\dot\psi(t), \xi) + \mathsf{j}^{\mathrm{c}}(\dot\psi(t), \xi) + \mathsf{c}(\psi(t), \xi) = \ell(\mathbf{V}(t), \xi), \quad (20)$$

where

$$\mathsf{A}(\psi, \xi) := \int_\Omega \frac{1}{\mu(\psi)r} \nabla\psi \cdot \nabla\xi\, dr dz,$$

$$\mathsf{J}_\mathrm{p}(\psi, \xi) := \int_{\Omega_\mathrm{p}(\psi)} \left(r S_{p'}(\psi_\mathrm{N}) + \frac{1}{\mu_0 r} S_{ff'}(\psi_\mathrm{N}) \right) \xi\, dr dz,$$

$$\ell(\mathbf{V}(t), \xi) := \sum_{i=1}^N \frac{n_i}{R_i |\Omega_{c_i}|} V_i(t) \int_{\Omega_{c_i}} \xi\, dr dz, \qquad (21)$$

$$\mathsf{j}^{\mathrm{ps}}(\psi, \xi) := \int_{\Omega_\mathrm{ps}} \frac{\sigma}{r} \psi\xi\, dr dz,$$

$$\mathsf{j}^{\mathrm{c}}(\psi, \xi) := \sum_{i=1}^{N_i} \frac{2\pi n_i^2}{R_i |\Omega_{c_i}|^2} \int_{\Omega_{c_i}} \psi\, dr dz \int_{\Omega_{c_i}} \xi\, dr dz,$$

and

$$\mathsf{c}(\psi, \xi) := \frac{1}{\mu_0} \int_\Gamma \psi(\mathbf{P}_1) N(\mathbf{P}_1) \xi(\mathbf{P}_1) dS_1$$

$$+ \frac{1}{2\mu_0} \int_\Gamma \int_\Gamma (\psi(\mathbf{P}_1) - \psi(\mathbf{P}_2)) M(\mathbf{P}_1, \mathbf{P}_2)(\xi(\mathbf{P}_1) - \xi(\mathbf{P}_2)) dS_1 dS_2, \quad (22)$$

with

$$M(\mathbf{P}_1, \mathbf{P}_2) = \frac{k_{\mathbf{P}_1, \mathbf{P}_2}}{2\pi(r_1 r_2)^{\frac{3}{2}}} \left(\frac{2 - k_{\mathbf{P}_1, \mathbf{P}_2}^2}{2 - 2k_{\mathbf{P}_1, \mathbf{P}_2}^2} E(k_{\mathbf{P}_1, \mathbf{P}_2}) - K(k_{\mathbf{P}_1, \mathbf{P}_2}) \right),$$

$$N(\mathbf{P}_1) = \frac{1}{r_1} \left(\frac{1}{\delta_+} + \frac{1}{\delta_-} - \frac{1}{\rho_\Gamma} \right) \quad \text{and} \quad \delta_\pm = \sqrt{r_1^2 + (\rho_\Gamma \pm z_1)^2}.$$

where $\mathbf{P}_i = (r_i, z_i)$ and K and E the complete elliptic integrals of first and second kind, respectively and

$$k_{\mathbf{P}_j, \mathbf{P}_k} = \sqrt{\frac{4 r_j r_k}{(r_j + r_k)^2 + (z_j - z_k)^2}}.$$

The bilinear form $c : H \times H \to \mathbb{R}$ is accounting for the boundary conditions at infinity [6]. We refer to [7, Chap. 2.4] for the details of the derivation. The bilinear form $c(\cdot, \cdot)$ follows basically from the so-called *uncoupling procedure* in [8] for the usual coupling of boundary integral and finite element methods. As we focus here on the equilibrium problem the two functions p' and $f f'$ have to be supplied as data, called $S_{p'}$ and $S_{ff'}$ in the definition of $J_p(\psi, \xi)$. While the domain of p' and $f f'$ depends on the poloidal flux itself, it is more practical to supply those profiles $S_{p'}$ and $S_{ff'}$ as functions of the normalized poloidal flux $\psi_N(r, z)$:

$$\psi_N(r, z) = \frac{\psi(r, z) - \psi_{\mathrm{ax}}(\psi)}{\psi_{\mathrm{bd}}(\psi) - \psi_{\mathrm{ax}}(\psi)}, \tag{23}$$

where

$$\begin{aligned} \psi_{\mathrm{ax}}(\psi) &:= \psi(r_{\mathrm{ax}}(\psi), z_{\mathrm{ax}}(\psi)), \\ \psi_{\mathrm{bnd}}(\psi) &:= \psi(r_{\mathrm{bd}}(\psi), z_{\mathrm{bd}}(\psi)) \end{aligned} \tag{24}$$

with $(r_{\mathrm{ax}}(\psi), z_{\mathrm{ax}}(\psi))$ the magnetic axis, where ψ has its global maximum in Ω_L and $(r_{\mathrm{bnd}}(\psi), z_{\mathrm{bnd}}(\psi))$ the coordinates of the point that determines the plasma boundary. The point $(r_{\mathrm{bnd}}, z_{\mathrm{bnd}})$ is either an X-point of ψ or the contact point with the limiter $\partial \Omega_L$. $S_{p'}$ and $S_{ff'}$, have, independently of ψ, a fixed domain $[0, 1]$ and are usually given as (piecewise) polynomial functions. Another frequent a priori model is

$$S_{p'}(\psi_N) = \lambda \frac{\beta}{r_0} (1 - \psi_N^\alpha)^\gamma, \quad S_{ff'}(\psi_N) = \lambda (1 - \beta) \mu_0 r_0 (1 - \psi_N^\alpha)^\gamma \tag{25}$$

with r_0 the major radius of the vacuum chamber and $\alpha, \beta, \gamma \in \mathbb{R}$ given parameters. We refer to [9] for a physical interpretation of these parameters. The parameter β is related to the poloidal beta, whereas α and γ describe the peakage of the current profile and λ is a normalization factor.

Numerical Methods. It is straightforward to combine Galerkin methods in space and time-stepping schemes to get approximation schemes for solving (20) numerically. For the choice of the spatial discretization, the fine details of realistic tokamak sections (see Fig. 4) give here favor to finite element spaces based on triangular meshes. Since for many years now the piecewise affine approximations are the standard choice for the stationary free-boundary equilibrium problems [7,10,11], we stay also here with linear Lagrangian finite elements for the discretization in space. Higher order methods are likewise implementable.

In order to prohibit numerical instablities it is advisable to use implicit time-stepping methods such as implicit Euler, which leads to non-linear finite-dimensional problems. The Newton-type methods for solving such non-linear

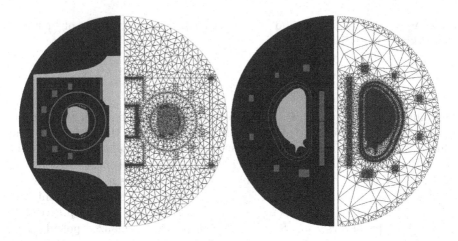

Fig. 4. The different subdomains of the geometry of the tokamak WEST (left) and ITER (right) and triangulations that resolve the geometric details.

problems can be based on the Gâteaux derivative

$$D_\psi \mathsf{A}(\psi, \xi)(\tilde{\psi}) = \int_\Omega \frac{1}{\mu(\psi)r} \nabla \tilde{\psi} \cdot \nabla \xi \, dr dz$$

$$- 2 \int_{\Omega_{\mathrm{Fe}}} \frac{\mu'_{\mathrm{Fe}}(\frac{|\nabla\psi|^2}{r^2})}{\mu^2_{\mathrm{Fe}}(\frac{|\nabla\psi|^2}{r^2})r^3} (\nabla \tilde{\psi} \cdot \nabla \psi)(\nabla \psi \cdot \nabla \xi) \, dr dz$$

of $\mathsf{A}(\psi, \xi)$ and the Gâteaux derivative

$$
\begin{aligned}
D_\psi \mathsf{J}_{\mathrm{p}}(\psi, \xi)(\tilde{\psi}) = & \int_{\Omega_{\mathrm{p}}(\psi)} \frac{\partial j_{\mathrm{p}}(r, \psi_{\mathrm{N}}(\psi))}{\partial \psi_{\mathrm{N}}} \frac{\partial \psi_{\mathrm{N}}(\psi)}{\partial \psi} \tilde{\psi} \xi \, dr dz \\
& - \int_{\Gamma_{\mathrm{p}}(\psi)} j_{\mathrm{p}}(r, 1)|\nabla\psi|^{-1}(\tilde{\psi} - \tilde{\psi}(r_{\mathrm{bd}}(\psi), z_{\mathrm{bd}}(\psi)))\xi \, d\Gamma \\
& + \int_{\Omega_{\mathrm{p}}(\psi)} \frac{\partial j_{\mathrm{p}}(r, \psi_{\mathrm{N}}(\psi))}{\partial \psi_{\mathrm{N}}} \frac{\partial \psi_{\mathrm{N}}(\psi)}{\partial \psi_{\mathrm{ax}}} \tilde{\psi}(r_{\mathrm{ax}}(\psi), z_{\mathrm{ax}}(\psi))\xi \, dr dz \\
& + \int_{\Omega_{\mathrm{p}}(\psi)} \frac{\partial j_{\mathrm{p}}(r, \psi_{\mathrm{N}}(\psi))}{\partial \psi_{\mathrm{N}}} \frac{\partial \psi_{\mathrm{N}}(\psi)}{\partial \psi_{\mathrm{bd}}} \tilde{\psi}(r_{\mathrm{bd}}(\psi), z_{\mathrm{bd}}(\psi))\xi \, dr dz
\end{aligned}
\tag{26}
$$

of $\mathsf{J}_{\mathrm{p}}(\psi, \xi)$, where Γ_{p} is the plasma boundary $\partial\Omega_{\mathrm{p}}$ and

$$j_{\mathrm{p}}(r, \psi_{\mathrm{N}}(\psi)) = rS_{p'}(\psi_{\mathrm{N}}(\psi)) + \frac{1}{\mu_0 r} S_{ff'}(\psi_{\mathrm{N}}(\psi)). \tag{27}$$

The derivation of the linearization $D_\psi \mathsf{J}_{\mathrm{p}}(\psi, \xi)(\tilde{\psi})$ requires to assume that $\nabla\psi \neq 0$ on $\partial\Omega_{\mathrm{p}}$ and involves shape calculus [12,13] and the non-trivial derivatives:

$$D_\psi \psi_{\mathrm{ax}}(\psi)(\tilde{\psi}) = \tilde{\psi}(r_{\mathrm{ax}}(\psi), z_{\mathrm{ax}}(\psi)) \text{ and } D_\psi \psi_{\mathrm{bd}}(\psi)(\tilde{\psi}) = \tilde{\psi}(r_{\mathrm{bd}}(\psi), z_{\mathrm{bd}}(\psi)).$$

Clearly, $\nabla \psi \neq 0$ on $\partial \Omega_p$ will not be true for the nowadays important X-point equilibria. Nevertheless this theoretical difficulty is not very essential for practical computations. In [14] it is pointed out that accurate Newton methods for discretized versions of the weak formulation (20) need to use accurate derivatives for the discretized non-linear operator, which is not necessarily equal to the discretization of the analytical derivatives. Here, the discretization and linearization of $J_p(\psi, \xi)$ needs special attention due to the ψ-dependent domain of integration. We refer to [14, Sect. 3.2] and [14, Sect. 3.3] for the technical details.

5   The Optimal Control Problem

We intend to determine the voltages $V_i(t)$ applied to the poloidal field circuits so that the plasma boundary Γ_p fit to a desired boundary Γ_{desi} during the whole discharge while minimizing a certain energetic cost.

Let $\Gamma_{desi}(t) \subset \Omega_L$ denote the evolution of a closed line, contained in the domain Ω_L that is either smooth and touches the limiter at one point or has at least one corner. The former case prescribes a desired plasma boundary that touches the limiter. The latter case aims at a plasma with X-point that is entirely in the interior of Ω_L. Further let $(r_{desi}(t), z_{desi}(t)) \in \Gamma_{desi}(t)$ and $(r_1(t), z_1(t)), \ldots, (r_{N_{desi}}(t), z_{N_{desi}}(t)) \in \Gamma_{desi}(t)$ be $N_{desi} + 1$ points on that line. We define a *quadratic* functional $K(\psi)$ that evaluates to zero if $\Gamma_{desi}(t)$ is an $\psi(t)$-isoline, i.e. if $\psi(t)$ is constant on $\Gamma_{desi}(t)$:

$$K(\psi, t) := \frac{1}{2} \left(\sum_{i=1}^{N_{desi}} \left(\psi(r_i(t), z_i(t)) - \psi(r_{desi}(t), z_{desi}(t)) \right)^2 \right). \qquad (28)$$

Another functional, that will serve as *regularization*, is

$$R(\mathbf{V}(t)) := \sum_{i=1}^{N} \frac{w_i}{2} \mathbf{V}_i^2 \qquad (29)$$

with *regularization weights* $w_i \geq 0$. The regularization functional penalizes the strength of the voltages V_i and represents the energetic cost in the coil system.

We consider the following minimization problem:

$$\min_{\psi(t), \mathbf{V}(t)} \int_0^T K(\psi(t), t) + R(\mathbf{V}(t)) \, dt \qquad (30)$$

subject to

$$\mathsf{A}(\psi(t), \xi) - \mathsf{J}_p(\psi(t), \xi) + \mathsf{j}^{ps}(\dot{\psi}(t), \xi) + \mathsf{j}^c(\dot{\psi}(t), \xi) + \mathsf{c}(\psi(t), \xi) = \ell(\mathbf{V}(t), \xi) \; \forall \xi \in H.$$

This minimization problem for transient axisymmetric equilibria extends the minimization problems for static axisysmmetric equilibria introduced in [15, Chap. II]. Hence, theoretical assertions for (30) such as the first order necessary conditions for optimality follow by similar arguments as those in [15, p. 80–84].

The Lagrangian for the optimization problem (30) with Lagrange multiplier ϕ is:

$$\mathcal{L}(\psi(t), \mathbf{V}(t), \phi(t)) = \int_0^T K(\psi(t), t) + R(\mathbf{V}(t)) \, dt$$
$$- \int_0^T A(\psi(t), \phi(t)) - J_p(\psi(t), \phi(t)) + c(\psi(t), \phi(t)) dt$$
$$- \int_0^T j^{ps}(\dot{\psi}(t), \phi(t)) + j^c(\dot{\psi}(t), \phi(t)) - \ell(\mathbf{V}(t), \phi(t)) dt.$$

We can state the first order necessary conditions for optimality under the following three assumptions in the limiter case:

1. $\sup_{\Omega_L} \psi$ is attained at one and only one point $\mathbf{M}_0 = (r_{bd}, z_{bd})$.
2. $\sup_{\Omega_p} \psi$ is attained at one and only one point $\mathbf{M}_1$, which is an interior point of Ω_p and $\mathbf{M}_1 = (r_{ax}, z_{ax})$. ψ is of class C^2 in a neighbourhood of $\mathbf{M}_1$ and the point $\mathbf{M}_1$ is a non-degenerated elliptic point.
3. $\nabla\psi$ vanishes nowhere on $\partial\Omega_p$.

Equivalent necessary conditions can be obtained in the X-point case.

Then necessary conditions for $(\psi(t), \mathbf{V}(t), \phi(t))$ to be a saddle point of $\mathcal{L}$ are obtained, after integrating by parts in time the Lagrangian:

– $\psi(t)$ and $\mathbf{V}(t)$ are solution of the direct problem (20)
– $\psi(t)$ and $\phi(t)$ are solution of the adjoint problem

$$D_\psi A(\psi(t), \phi(t))(\xi) - D_\psi J_p(\psi(t), \phi(t))(\xi) + c(\xi, \phi(t))$$
$$- j^{ps}(\xi, \dot{\phi}(t)) - j^c(\xi, \dot{\phi}(t)) = D_\psi K(\psi(t), t)(\xi) \quad \forall \xi \in H \quad (31)$$

with $\phi(T) = 0$ and

$$D_\psi K(\psi, t)(\xi) = \sum_{i=1}^{N_{desi}} \left(\psi(r_i(t), z_i(t)) - \psi(r_{desi}(t), z_{desi}(t)) \right) \cdot$$
$$\left(\xi(r_i(t), z_i(t)) - \xi(r_{desi}(t), z_{desi}(t)) \right).$$

– $\mathbf{V}(t)$ and $\phi(t)$ are solution to

$$w_i V_i(t) + \frac{n_i}{R_i |\Omega_{c_i}|} \int_{\Omega_{c_i}} \phi(t) \, dr dz = 0, \quad 1 \le i \le N. \quad (32)$$

The adjoint problem has the following strong formulation:

$$-\Delta^* \phi(t) + 1_{\Omega_{Fe}} \nabla \cdot \left(2 \frac{\mu'_{Fe}(\frac{|\nabla\psi|^2}{r^2})}{\mu^2_{Fe}(\frac{|\nabla\psi|^2}{r^2})r^3} (\nabla\phi(t) \cdot \nabla\psi)\nabla\psi \right)$$
$$- 1_{\Omega_p(\psi)} \frac{\partial j_p(r, \psi_N(\psi))}{\partial \psi_N} \frac{\partial \psi_N(\psi)}{\partial \psi} \phi(t)$$

$$-\delta_{\mathrm{bd}} \int_{\Gamma_{\mathrm{p}}(\psi)} \frac{j_{\mathrm{p}}(r,1)}{|\nabla\psi|} \phi(t)\, d\Gamma + \left(\delta_{\Gamma_{\mathrm{p}}}, \frac{j_{\mathrm{p}}(r,1)}{|\nabla\psi|} \phi(t)\right)$$

$$-\delta_{\mathrm{ax}} \int_{\Omega_{\mathrm{p}}(\psi)} \frac{\partial j_{\mathrm{p}}(r,\psi_{\mathrm{N}}(\psi))}{\partial \psi_{\mathrm{N}}} \frac{\partial \psi_{\mathrm{N}}(\psi)}{\partial \psi_{\mathrm{ax}}} \phi(t)\, dr\,dz$$

$$-\delta_{\mathrm{bd}} \int_{\Omega_{\mathrm{p}}(\psi)} \frac{\partial j_{\mathrm{p}}(r,\psi_{\mathrm{N}}(\psi))}{\partial \psi_{\mathrm{N}}} \frac{\partial \psi_{\mathrm{N}}(\psi)}{\partial \psi_{\mathrm{bd}}} \phi(t)\, dr\,dz$$

$$-1_{\Omega_{\mathrm{ps}}} \frac{\sigma}{r} \dot{\phi}(t) - \sum_{i=1}^{N_i} 1_{\Omega_{\mathrm{c}_i}} \frac{2\pi n_i^2}{R_i |\Omega_{\mathrm{c}_i}|^2} \int_{\Omega_{\mathrm{c}_i}} \dot{\phi}\, dr\,dz$$

$$= \left(\sum_{i=1}^{N_{\mathrm{desi}}} \big(\psi(r_i(t), z_i(t), t) - \psi(r_{\mathrm{desi}}(t), z_{\mathrm{desi}}(t), t)\big)\right) \big(\delta_{(r_i, z_i)} - \delta_{(r_{\mathrm{desi}}, z_{\mathrm{desi}})}\big)$$

with $\phi(T) = 0$, where δ_{ax} and δ_{bd} are the Dirac masses at the points $(r_{\mathrm{ax}}, r_{\mathrm{ax}})$ and $(r_{\mathrm{bd}}, r_{\mathrm{bd}})$, respectively. $\delta_{\Gamma_{\mathrm{p}}}$ is the Dirac mass of Γ_{p} with

$$\left(\delta_{\Gamma_{\mathrm{p}}}, \frac{j_{\mathrm{p}}(r,1)}{|\nabla\psi|} \phi(t)\xi\right) = \int_{\Gamma_{\mathrm{p}}} \frac{j_{\mathrm{p}}(r,1)}{|\nabla\psi|} \phi(t)\xi\, d\Gamma.$$

Equation (32) is the Euler equation for the minimization of (30). Equations (20), (31) and (32) constitute the optimality system for problem (30).

Numerical Methods. The discretization of our minimization problem (30) builds on the space-time discretization for (20) that we outlined in the previous section. Next, the discrete minimization problem can be recast as the following constrained optimization problem

$$\min_{\mathbf{u},\mathbf{y}} J(\mathbf{y},\mathbf{u}) \quad \text{s.t.} \quad \mathbf{B}(\mathbf{y}) = \mathbf{F}(\mathbf{u}), \tag{33}$$

where $\mathbf{y}$ and $\mathbf{u}$ are the so-called state and control variables. In our setting $\mathbf{y}$ will be the variable that describes the plasma and $\mathbf{u}$ will be the externally applied voltages. We think of $\mathbf{y}$ as the vector of degrees of freedoms describing the space and time evolution of the poloidal flux ψ, and $\mathbf{B}(\mathbf{y})$ and $\mathbf{F}(\mathbf{u})$ are the discretizations of the non-linear operators in the variational formulation (20). *Sequential Quadratic Programming (SQP)* is one of the most effective methods for non-linear constrained optimization with significant non-linearities in the constraints [16, Chap. 18]. SQP methods find a numerical solution by generating iteration steps that minimize quadratic cost functions subject to linear constraints. The Lagrange function formalism in combination with Newton-type iterations is one approach to derive the SQP-methods: the Lagrangian for (33) is

$$L(\mathbf{y},\mathbf{u},\mathbf{p}) = J(\mathbf{y},\mathbf{u}) + \langle \mathbf{p}, \mathbf{B}(\mathbf{y}) - \mathbf{F}(\mathbf{u})\rangle, \tag{34}$$

and the solution of (33) is a stationary point of this Lagrangian:

$$\begin{aligned} D_{\mathbf{y}} J(\mathbf{y},\mathbf{u}) + D_{\mathbf{y}} \mathbf{B}^T(\mathbf{y})\mathbf{p} &= 0, \\ D_{\mathbf{u}} J(\mathbf{y},\mathbf{u}) - D_{\mathbf{u}} \mathbf{F}^T(\mathbf{u})\mathbf{p} &= 0, \\ \mathbf{B}(\mathbf{y}) - \mathbf{F}(\mathbf{u}) &= 0. \end{aligned} \tag{35}$$

A Newton-type method for solving (35) are iterations of the type

$$
\begin{pmatrix}
\mathbf{H}^k_{\mathbf{y},\mathbf{y}} & \mathbf{H}^k_{\mathbf{y},\mathbf{u}} & D_{\mathbf{y}}\mathbf{B}^T(\mathbf{y}^k) \\
\mathbf{H}^k_{\mathbf{u},\mathbf{y}} & \mathbf{H}^k_{\mathbf{u},\mathbf{u}} & -D_{\mathbf{u}}\mathbf{F}^T(\mathbf{u}^k) \\
D_{\mathbf{y}}\mathbf{B}(\mathbf{y}^k) & -D_{\mathbf{u}}\mathbf{F}(\mathbf{u}^k) & 0
\end{pmatrix}
\begin{pmatrix}
\mathbf{y}^{k+1} - \mathbf{y}^k \\
\mathbf{u}^{k+1} - \mathbf{u}^k \\
\mathbf{p}^{k+1} - \mathbf{p}^k
\end{pmatrix}
$$
$$
= -\begin{pmatrix}
D_{\mathbf{y}}J(\mathbf{y}^k,\mathbf{u}^k) + D_{\mathbf{y}}\mathbf{B}^T(\mathbf{y}^k)\mathbf{p}^k \\
D_{\mathbf{u}}J(\mathbf{y}^k,\mathbf{u}^k) - D_{\mathbf{u}}\mathbf{F}^T(\mathbf{u}^k)\mathbf{p}^k \\
\mathbf{B}(\mathbf{y}^k) - \mathbf{F}(\mathbf{u}^k)
\end{pmatrix}
\tag{36}
$$

with

$$
\begin{pmatrix}
\mathbf{H}^k_{\mathbf{y},\mathbf{y}} & \mathbf{H}^k_{\mathbf{y},\mathbf{u}} \\
\mathbf{H}^k_{\mathbf{u},\mathbf{y}} & \mathbf{H}^k_{\mathbf{u},\mathbf{u}}
\end{pmatrix} =
\begin{pmatrix}
D_{\mathbf{y},\mathbf{y}}L(\mathbf{y}^k,\mathbf{u}^k,\mathbf{p}^k) & D_{\mathbf{y},\mathbf{u}}L(\mathbf{y}^k,\mathbf{u}^k,\mathbf{p}^k) \\
D_{\mathbf{u},\mathbf{y}}L(\mathbf{y}^k,\mathbf{u}^k,\mathbf{p}^k) & D_{\mathbf{u},\mathbf{u}}L(\mathbf{y}^k,\mathbf{u}^k,\mathbf{p}^k)
\end{pmatrix}.
$$

If the linear systems in (36) become too large, we are pursuing the null space approach to arrive at the SQP formulation with the reduced Hessian for the increment $\Delta\mathbf{u}^k := \mathbf{u}^{k+1} - \mathbf{u}^k$:

$$
\mathbf{M}(\mathbf{y}^k,\mathbf{u}^k)\Delta\mathbf{u}^k = -\mathbf{h}(\mathbf{y}^k,\mathbf{u}^k),
\tag{37}
$$

where

$$
\mathbf{M}(\mathbf{y}^k,\mathbf{u}^k) := \left(D_{\mathbf{u}}\mathbf{F}^T(\mathbf{u}^k)D_{\mathbf{y}}\mathbf{B}^{-T}(\mathbf{y}^k)\ Id\right)
\begin{pmatrix}
\mathbf{H}^k_{\mathbf{y},\mathbf{y}} & \mathbf{H}^k_{\mathbf{y},\mathbf{u}} \\
\mathbf{H}^k_{\mathbf{u},\mathbf{y}} & \mathbf{H}^k_{\mathbf{u},\mathbf{u}}
\end{pmatrix}
\begin{pmatrix}
D_{\mathbf{y}}\mathbf{B}^{-1}(\mathbf{y}^k)D_{\mathbf{u}}\mathbf{F}(\mathbf{u}^k) \\
Id
\end{pmatrix}
$$

and

$$
\begin{aligned}
\mathbf{h}(\mathbf{y}^k,\mathbf{u}^k) :=& D_{\mathbf{u}}J(\mathbf{y}^k,\mathbf{u}^k) + D_{\mathbf{u}}\mathbf{F}^T(\mathbf{u}^k)\lambda^k \\
& - \left(D_{\mathbf{u}}\mathbf{F}^T(\mathbf{u}^k)D_{\mathbf{y}}\mathbf{B}^{-T}(\mathbf{y}^k)\mathbf{H}^k_{\mathbf{y},\mathbf{y}} + \mathbf{H}^k_{\mathbf{u},\mathbf{y}}\right)D_{\mathbf{y}}\mathbf{B}^{-1}(\mathbf{y}^k)\mathbf{r}(\mathbf{y}^k,\mathbf{u}^k))
\end{aligned}
$$

with

$$
\lambda^k := D_{\mathbf{y}}\mathbf{B}^{-T}(\mathbf{y}^k)D_{\mathbf{y}}J(\mathbf{y}^k,\mathbf{u}^k), \qquad \mathbf{r}(\mathbf{y}^k,\mathbf{u}^k) := \mathbf{B}(\mathbf{y}^k) - \mathbf{F}(\mathbf{u}^k).
$$

We are using iterative methods, e.g. the conjugate gradient methods, to solve (37). Since in our case the number of control variables will be small we can expect convergence within very few iterations. Within each iteration step of the iterative method, we still have to solve the two linear systems corresponding to $D_{\mathbf{y}}\mathbf{B}(\mathbf{y}^k)$ and $D_{\mathbf{y}}\mathbf{B}^T(\mathbf{y}^k)$. Alternatively, if we have sufficient memory to store $\mathbf{M}(\cdot,\cdot)$, we can compute $\mathbf{M}(\cdot,\cdot)$ explicitly. Clearly, we never compute neither $D_{\mathbf{y}}\mathbf{B}^{-1}(\mathbf{y}^k)$ nor $D_{\mathbf{y}}\mathbf{B}^{-T}(\mathbf{y}^k)$ explicitly.

Once we know $\Delta\mathbf{u}^k$ we can compute $\mathbf{y}^{k+1}$ and $\mathbf{p}^{k+1}$ by:

$$
\begin{aligned}
\mathbf{y}^{k+1} - \mathbf{y}^k &= D_{\mathbf{y}}\mathbf{B}^{-1}(\mathbf{y}^k)D_{\mathbf{u}}\mathbf{F}(\mathbf{u}^k)\Delta\mathbf{u}^k - \mathbf{r}(\mathbf{u}^k,\mathbf{y}^k), \\
\mathbf{p}^{k+1} + \lambda^k &= -D_{\mathbf{y}}\mathbf{B}^{-T}(\mathbf{y}^k)(\mathbf{H}^k_{\mathbf{y},\mathbf{y}}(\mathbf{y}^{k+1}-\mathbf{y}^k) + \mathbf{H}^k_{\mathbf{y},\mathbf{u}}(\mathbf{u}^{k+1}-\mathbf{u}^k)).
\end{aligned}
$$

We would like to highlight that the SQP-method relies on proper derivatives of the non-linear operators $\mathbf{B}$ and $\mathbf{F}$. In our case $\mathbf{F}$ is affine, hence the derivative of $\mathbf{B}$ remains the most difficult part. On the other hand these are exactly the same terms that appear in the Newton iterations for the direct problem (20) and we can reuse the methodology presented at the end of Sect. 4. For practical purposes we do neglect all involved second order derivatives of $\mathbf{B}$.

It is very instrumental to compare the expression involved in the reduced formulation (37) of SQP to the gradient and the Hessian of the reduced cost function, that would appear when using algorithms for unconstrained optimization problems.

Let $\widehat{J}(\mathbf{u}) := J(\mathbf{y}(\mathbf{u}), \mathbf{u})$, with $\mathbf{B}(\mathbf{y}(\mathbf{u})) = \mathbf{F}(\mathbf{u})$ be the *reduced* cost function, then we have the following expressions for gradient

$$D_{\mathbf{u}}\widehat{J}(\mathbf{u}) = D_{\mathbf{u}}J(\mathbf{y}, \mathbf{u}) + D_{\mathbf{u}}\mathbf{F}^T(\mathbf{u}), \lambda$$

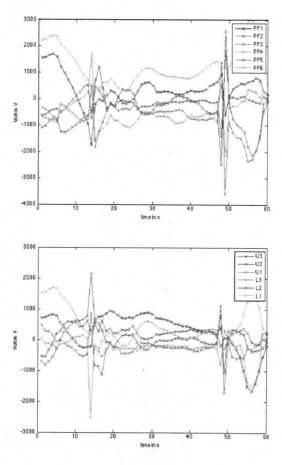

Fig. 5. The optimal voltages.

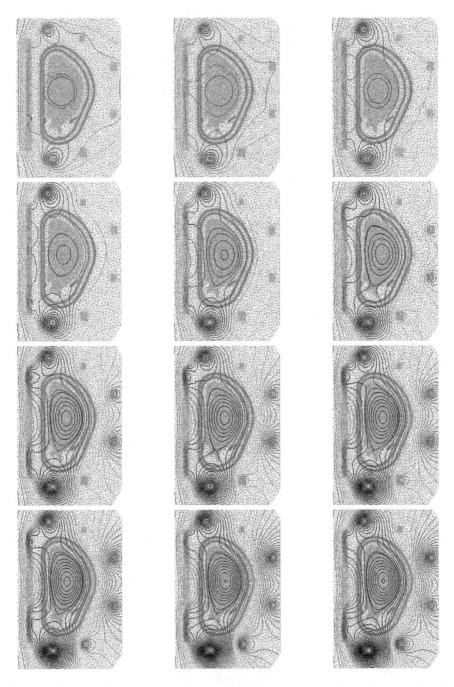

Fig. 6. Optimal control for a ramp-up scenario: the plasma boundary (green) follows the prescribed boundary (black points), snapshots at $t = 0, 2, 6, 10, 20, 30, 40, 45, 50, 54, 58, 60\,s$ (from left to right, top to down). (Color figure online)

and Hessian

$$D_{\mathbf{u},\mathbf{u}}\widehat{J}(\mathbf{u}) = \mathbf{Z}^T \begin{pmatrix} D_{\mathbf{y},\mathbf{y}}J(\mathbf{y},\mathbf{u}) & D_{\mathbf{y},\mathbf{u}}J(\mathbf{y},\mathbf{u}) \\ D_{\mathbf{u},\mathbf{y}}J(\mathbf{y},\mathbf{u}) & D_{\mathbf{u},\mathbf{u}}J(\mathbf{y},\mathbf{u}) \end{pmatrix} \mathbf{Z}$$
$$+ \mathbf{Z}^T \begin{pmatrix} -D_{\mathbf{y}}(D_{\mathbf{y}}\mathbf{B}^T(\mathbf{y})\lambda) & 0 \\ 0 & D_{\mathbf{u}}(D_{\mathbf{u}}\mathbf{F}^T(\mathbf{u})\lambda) \end{pmatrix} \mathbf{Z}$$

with

$$\lambda = D_{\mathbf{y}}\mathbf{B}^{-T}(\mathbf{y})D_{\mathbf{y}}J(\mathbf{y},\mathbf{u}) \quad \text{and} \quad \mathbf{Z} = \begin{pmatrix} D_{\mathbf{y}}\mathbf{B}^{-1}(\mathbf{y})D_{\mathbf{u}}\mathbf{F}(\mathbf{u}) \\ Id \end{pmatrix}$$

Hence, the reduced gradient $\mathbf{h}(\mathbf{y}^k, \mathbf{u}^k)$ is not the gradient of the reduced cost function, unless the state and control variable $\mathbf{y}^k$ and $\mathbf{u}^k$ verify the equation of state $\mathbf{B}(\mathbf{y}^k) = \mathbf{F}(\mathbf{u}^k)$.

Preliminary Example. Finally, we would like to show first results for a so-called ramp-up scenario in an ITER-like tokamak, where the plasma evolves from a small circular to a large elongated plasma. The optimal coil voltages are depicted in Fig. 5. Then, if we use those as data to solve the direct problem we verify that the plasma boundary follows indeed the prescribed trajectory (see Fig. 6).

Conclusion. The study and the optimization of scenarios is more and more important for the realization of objectives in magnetic confinement controlled fusion and will certainly be crucial for the ITER project. The first results presented in this paper are very encouraging and are the starting point of the development of new tools devoted to the preparation of scenarios of the future devices.

References

1. Lions, J.L.: Optimal Control of Systems Governed by Partial Differential Equations. Springer, Heildelberg (1971)
2. Brangiskii, S.I.: Reviews of Plasma Physics, vol. 1. Consultant Bureau, New York (1965)
3. Maschke, E.K., Sudano, J.P.: Etude analytique de l'évolution d'un plasma toroidal de type tokamak à section non circulaire (analytic study of the evolution of a torodal plasma of tokamak type with non-circular section). Technical report, EUR-CEA-FC-668 (1972)
4. Nelson, D.B., Grad, H.: Heating and transport in tokamaks of arbitrary shape and beta. Technical report, Oak Ridge Report ORNL/TM-6094 (1978)
5. Wesson, J.: Tokamaks. Oxford Science Publications, Oxford (2004)
6. Albanese, R., Blum, J., Barbieri, O.: On the solution of the magnetic flux equation in an infinite domain. In: EPS, 8th Europhysics Conference on Computing in Plasma Physics (1986)
7. Grandgirard, V.: Modélisation de l'équilibre d'un plasma de tokamak. Ph.D. thesis, l'Université de Franche-Comté (1999)

8. Gatica, G.N., Hsiao, G.C.: The uncoupling of boundary integral and finite element methods for nonlinear boundary value problems. J. Math. Anal. Appl. **189**(2), 442–461 (1995)
9. Luxon, J.L., Brown, B.B.: Magnetic analysis of non-circular cross-section tokamaks. Nucl. Fusion **22**(6), 813 (1982)
10. Blum, J., Le Foll, J., Thooris, B.: The self-consistent equilibrium and diffusion code SCED. Comput. Phys. Commun. **24**, 235–254 (1981)
11. Albanese, R., Blum, J., De Barbieri, O.: Numerical studies of the Next European Torus via the PROTEUS code. In: 12th Conference on Numerical Simulation of Plasmas, San Francisco (1987)
12. Murat, F., Simon, J.: Sur le contrôle par un domaine géométrique. Technical report 76015, Laboratoire d'Analyse Numérique, Université de Paris 6 (1976)
13. Delfour, M.C., Zolésio, J.-P.: Shapes and Geometries. Advances in Design and Control, vol. 22, 2nd edn. Society for Industrial and Applied Mathematics (SIAM), Philadelphia (2011)
14. Heumann, H., Blum, J., Boulbe, C., Faugeras, B., Selig, G., Ané, J.-M., Brémond, S., Grandgirard, V., Hertout, P., Nardon, E.: Quasi-static free-boundary equilibrium of toroidal plasma with CEDRES++: computational methods and applications. J. Plasma Phys. **81**, 1–35 (2015)
15. Blum, J.: Numerical Simulation and Optimal Control in Plasma Physics with Applications to Tokamaks. Series in Modern Applied Mathematics. Wiley Gauthier-Villars, Paris (1989)
16. Nocedal, J., Wright, S.J.: Numerical Optimization. Springer Series in Operations Research and Financial Engineering, 2nd edn. Springer, New York (2006)

Approximation of the Equations of the Humid Atmosphere with Saturation

Roger Temam[(⊠)] and Xiaoyan Wang

Institute for Scientific Computing and Applied Mathematics,
Indiana University, Rawles Hall, Bloomington, IN 47405, USA
{temam,wang264}@indiana.edu

Abstract. We investigate the numerical approximation of solutions to some variational inequalities modeling the humid atmosphere when the saturation of water vapor in the air is taken into account. Here we describe part of our work [31] and extend our former results to the case where the saturation q_s evolves with time.

Keywords: Atmosphere equation · Variational inequality · Penalization · Regularization · Uniform estimates · Fractional step method

1 Introduction

The rigorous mathematical theory of the equations of humid atmosphere has been initiated in [21,22] and has attracted the attention of a large number of researchers, see e.g., [1,3–8,13–16,24,30] and the references therein. These cited research works solved a large class of practical problems by investigating the system of partial differential equations based on different accuracies of the mathematical modelings [17,18,26]. However, in the modelings in [17,18,23,26], the saturation of vapor is not taken into account. As shown in [29,32], the resulted systems of partial differential equations are not physically correct in the extreme cases where the atmosphere is totally dry, $q = 0$, or when the atmosphere is totally humid, $q = 1$. To remedy this drawback, we have proposed in [32] a new formulation of the problem in the context of the variational inequalities [2,9,19,20,25]. This new variational *inequality* formulation also involves discontinuities due to phase changes. In this work, we describe the numerical approximation of the solutions to the variational inequalities derived from the humidity equations when the saturation of water vapor in the air is taken into account. As explained above, a striking feature of our work here is that the problems we study contain discontinuities and involve inequalities which come from the changes of phases and the extreme cases for the vapor concentration respectively (see e.g., [10–12]). In addition, we manage to extend our recent study [31] to the case that the saturation concentration q_s evolves according to the thermodynamical laws.

In [31], we proposed an implicit Euler scheme to approach the solutions to the system which involves a variational inequality. However, we can not simply proceed directly as usual due to the difficulties induced from the discontinuities and

L. Bociu et al. (Eds.): CSMO 2015, IFIP AICT 494, pp. 21–42, 2016.
DOI: 10.1007/978-3-319-55795-3_2

physical requirement of vapor concentration (i.e. $0 \le q \le 1$). To overcome the difficulty caused by the discontinuities in our current modeling, we use a regularization method. This regularization enables us to first study a system of partial differential equations and then discuss the approximation of the solutions to the original variational inequality. The constraint requirement $q \in \mathcal{K}$ for almost every $t \in [0, t_1]$ for the vapor concentration $q = q(t, \mathbf{x})$ brings us great technical challenges. Here $t_1 > 0$ is an arbitrary but fixed constant. See Sect. 2.1 for more details about this physical range requirement. The source of challenges in our study is that this range requirement can not be preserved in the discretization procedure in the implicit Euler scheme. To deal with these challenges, we devised a penalization technique in the regularized Euler scheme. Together with delicate energy estimates, the penalization technique can elegantly help us achieve the physical requirement on q. We point out here that the forms and the signs of the penalization terms encode very elegant structural propositions and are crucial for us to obtain the desired energy estimates. Finally, when we extend the study to the case where the saturation vapor concentration q_s evolves according to the thermodynamical laws, we emphasize that the discretization of the q_s-equation is of different nature from the discretization of the temperature equation on T and vapor concentration equations on q. See Remark 2 for more detailed comments.

The rest of the article is organized as follows. In Sect. 2, we give the formulation of the problem. In Sect. 3, we introduce the Euler scheme and derive various uniform estimates for the functions associated with the penalized and regularized scheme. In Sect. 4, we investigate the convergence of the Euler scheme. We devote Sect. 5 to the study of the implicit Euler scheme in the case where q_s depends on time.

2   The Problem

2.1   Formulation of the System

Let $\mathcal{M} = \mathcal{M}' \times (p_0, p_1)$ where $\mathcal{M}' \subset \mathbb{R}^2$ is a bounded domain with smooth boundary and p_0, p_1 are two real numbers with $0 < p_0 < p_1$. We will use $\mathbf{x} = (x, y, p)$ to denote a typical point in $\mathcal{M}$, and use n to denote the outward normal vector to $\partial\mathcal{M}$, the boundary of $\mathcal{M}$. Let $\mathcal{K}$ be the non-empty closed convex set in $H^1(\mathcal{M})$ defined as $\mathcal{K} = \{q \in H^1(\mathcal{M}); 0 \le q \le 1, a.e.\}$. Given a *fixed* $t_1 > 0$, we consider the following problem:

To find $T : (0, t_1) \to H^1(\mathcal{M})$, $q : (0, t_1) \to \mathcal{K}$ and $h_q \in \mathcal{H}(q - q_s)$ such that for $q^b \in \mathcal{K}$, there hold

$$\partial_t T + \mathcal{A}_T T + \mathbf{v} \cdot \nabla T + \omega \partial_p T - \frac{R\omega}{c_p p} T = \frac{1}{p}\omega^- h_q \varphi(T), \tag{1}$$

$$\langle \partial_t q, q^b - q \rangle + \left(\mathcal{A}_q q + \mathbf{v} \cdot \nabla q + \omega \partial_p q, q^b - q \right) \ge \left(-\frac{1}{p}\omega^- h_q F(T), q^b - q \right), \tag{2}$$

with initial and boundary conditions to be specified. Here $\mathcal{H}$ is the multivalued Heaviside function such that $\mathcal{H}(\tau) = 0$ for $\tau < 0$, $\mathcal{H}(0) = [0, 1]$, $\mathcal{H}(\tau) = 1$ for $\tau > 0$.

For the sake of simplicity, the velocity field of the fluid $\mathbf{u} :=$ $(\mathbf{v}(\mathbf{x},t), \omega(\mathbf{x},t)) \in \mathbb{R}^3$ is considered as a given data in this article. Throughout the presentation, we assume that the time-dependent velocity field $\mathbf{u}$ satisfies $\mathbf{u} \in L^r(0, t_1; V) \cap L^\infty(0, t_1; H)$ for some given $r \in (4, +\infty]$; $\nabla = (\partial_x, \partial_y)$ and $\Delta = \partial_x^2 + \partial_y^2$ are the horizontal gradient and horizontal Laplace operators respectively. In this way, the operators $\mathcal{A}_T$ and $\mathcal{A}_q$ are defined as

$$\mathcal{A}_T = -\mu_1 \Delta - \nu_1 \partial_p ((\frac{gp}{R\bar{T}})^2 \partial_p), \quad \mathcal{A}_q = -\mu_2 \Delta - \nu_2 \partial_p ((\frac{gp}{R\bar{T}})^2 \partial_p), \qquad (3)$$

where μ_i, ν_i, g, R, c_p are positive constants and $\bar{T} = \bar{T}(p)$ is the average temperature over the isobar with pressure p. We assume that $\bar{T}$ satisfies:

$$\bar{T}_* \le \bar{T}(p) \le \bar{T}^*, \ |\partial_p \bar{T}(p)| \le M, \ \text{for some postive constants } \bar{T}_*, \bar{T}^*, M \text{ and } p \in [p_0, p_1]. \quad (4)$$

Concerning the right hand sides of Eqs. (1)–(2), the functions F and φ both from $\mathbb{R}^1$ to $\mathbb{R}^1$ are defined as

$$F(\zeta) = q_s \zeta \frac{RL(\zeta) - c_p R_v \zeta}{c_p R_v \zeta^2 + q_s L(\zeta)^2}, \ \text{with } L(\zeta) = c_1 - c_2 \zeta; \ \varphi(\zeta) = \frac{1}{c_p} L(\zeta) F(\zeta). \quad (5)$$

Above, c_1, c_2, R_v, R_q are all strictly positive constants. It is easy to see that F is bounded and that both functions F and φ are globally Lipschitz; $\omega^+ :=$ $\max\{\omega, 0\}$ refers to the positive part of ω.

We partition the boundary of $\mathcal{M}$ as $\partial \mathcal{M} = \Gamma_i \cup \Gamma_u \cup \Gamma_l$ with Γ_i, Γ_u and Γ_l defined by

$$\Gamma_i = \{\mathbf{x} \in \overline{\mathcal{M}}; p = p_1\}, \ \Gamma_u = \{\mathbf{x} \in \overline{\mathcal{M}}; p = p_0\}, \ \Gamma_l = \{\mathbf{x} \in \overline{\mathcal{M}}; p_0 \le p \le p_1, (x,y) \in \partial \mathcal{M}'\}.$$

We supplement the system (1)–(2) with the boundary conditions

$$\begin{cases} \partial_p T = \alpha(T_* - T), \partial_p q = \beta(q_* - q) \text{ on } \Gamma_i, \\ \partial_p T = 0, \partial_p q = 0 \text{ on } \Gamma_u, \\ \partial_n T = 0, \partial_n q = 0 \text{ on } \Gamma_l, \end{cases} \qquad (6)$$

and initial conditions

$$T(\mathbf{x}, 0) = T_0(\mathbf{x}), q(\mathbf{x}, 0) = q_0(\mathbf{x}). \qquad (7)$$

If we allow q_s to evolve, the dependence of the nonlinear functions F and φ on q_s should be made explicit, i.e., $F = F(T, q_s)$, $\varphi = \varphi(T, q_s)$. In this case, we augment (1)–(2) with the following governing equation for the evolution of q_s (see [17, 18, 26]):

$$\frac{dq_s}{dt} = -\frac{\delta \omega^-}{p} h_q F(T, q_s). \qquad (8)$$

We will also impose a further initial condition

$$q_s(\mathbf{x}, 0) = q_{s,0}(\mathbf{x}). \qquad (9)$$

We shall always assume $q_0 \in L^2(\mathcal{M})$, $0 \le q_0 \le 1$ for a.e. $\mathbf{x} \in \mathcal{M}$, and $q_{s,0} \in L^2(\mathcal{M}) \cap L^\infty(\mathcal{M})$, $0 < q_{s,0} < 1$ and $0 \le q_* \le 1$ for a.e. $\mathbf{x} \in \mathcal{M}$, and assume the boundary datum T_* and q_* to satisfy $T_*, q_* \in L^2(0, t_1; L^2(\Gamma_i))$. For the convenience of fixing the ideas and of presentation, we shall first assume q_s is stationary during our study. In the last section, we will explain the case where q_s evolves according to the governing Eq. (8).

2.2   Functional Analytic Framework

We denote as usual $H = L^2(\mathcal{M})$, $V = H^1(\mathcal{M})$. We use $(\cdot, \cdot)_{L^2}$ (regarded the same as $(\cdot, \cdot)_H$) and $|\cdot|_{L^2}$ to denote the usual scalar product and induced norm in H. In the space V, we will use $((\cdot, \cdot))$ and $\|\cdot\|$ to denote the scalar product adapted to the problem under investigation

$$((\varphi, \phi)) := (\nabla\varphi, \nabla\phi) + (\partial_p\varphi, \partial_p\phi) + \int_{\Gamma_i} \varphi\phi \, d\Gamma_i,$$

and the induced norm. The symbol $\langle \cdot, \cdot \rangle$ will denote the duality pair between a Banach space E and its dual space E^*. We use the following standard function spaces for the vector field $\mathbf{u}$:

$$\mathbf{H} = \{\mathbf{u} \in H \times H \times H \mid div\,\mathbf{u} = 0 \text{ and } \mathbf{u} \cdot n = 0 \text{ on } \partial\mathcal{M}\},$$
$$\mathbf{V} = \{\mathbf{u} \in V \times V \times V \mid div\,\mathbf{u} = 0 \text{ and } \mathbf{u} \cdot n = 0 \text{ on } \partial\mathcal{M}\}.$$

For $T, T^b, q, q^b \in V$, we have the following specific forms for the duality pairs through integration by parts and in view of the Neumann boundary conditions:

$$\langle \mathcal{A}_T T, T^b \rangle = \mu_1(\nabla T, \nabla T^b)_H + \nu_1 \int_{\mathcal{M}} \left(\frac{gp}{R\overline{T}}\right)^2 \partial_p T \partial_p T^b \, d\mathcal{M} + \nu_1 \int_{\Gamma_i} \left(\frac{gp_1}{R\overline{T}}\right)^2 \alpha(T - T_*)T^b \, d\Gamma_i, \quad (10)$$

$$\langle \mathcal{A}_q q, q^b \rangle = \mu_2(\nabla q, \nabla q^b)_H + \nu_2 \int_{\mathcal{M}} \left(\frac{gp}{R\overline{T}}\right)^2 \partial_p q \partial_p q^b \, d\mathcal{M} + \nu_2 \int_{\Gamma_i} \left(\frac{gp_1}{R\overline{T}}\right)^2 \beta(q - q_*)q^b \, d\Gamma_i. \quad (11)$$

Consequently, we define the following bilinear forms

$$a_T(T, T^b) = \mu_1(\nabla T, \nabla T^b)_H + \nu_1 \int_{\mathcal{M}} \left(\frac{gp}{R\overline{T}}\right)^2 \partial_p T \partial_p T^b \, d\mathcal{M} + \nu_1\alpha \int_{\Gamma_i} \left(\frac{gp_1}{R\overline{T}}\right)^2 TT^b \, d\Gamma_i, \quad (12)$$

$$a_q(q, q^b) = \mu_2(\nabla q, \nabla q^b)_H + \nu_2 \int_{\mathcal{M}} \left(\frac{gp}{R\overline{T}}\right)^2 \partial_p q \partial_p q^b \, d\mathcal{M} + \nu_2\beta \int_{\Gamma_i} \left(\frac{gp_1}{R\overline{T}}\right)^2 qq^b \, d\Gamma_i. \quad (13)$$

Meanwhile, we set $U := (T, q)$, $U^b := (T^b, q^b)$ and introduce the bilinear form

$$a(U, U^b) := a_T(T, T^b) + a_q(q, q^b). \quad (14)$$

As for the Navier-Stokes equation, we define

$$b(\mathbf{u}, U, U^b) := \int_{\mathcal{M}} (\mathbf{u} \cdot \nabla_{x,y,p} U) \cdot U^b \, d\mathcal{M} = b_T(\mathbf{u}, T, T^b) + b_q(\mathbf{u}, q, q^b), \quad (15)$$

where $b_T(\mathbf{u}, T, T^b)$ and $b_q(\mathbf{u}, q, q^b)$ are given by

$$b_T(\mathbf{u}, T, T^b) = \int_{\mathcal{M}} (\mathbf{v} \cdot \nabla T + \omega \partial_p T) T^b \, d\mathcal{M}, \; b_q(\mathbf{u}, q, q^b) = \int_{\mathcal{M}} (\mathbf{v} \cdot \nabla q + \omega \partial_p q) q^b \, d\mathcal{M}. \quad (16)$$

In view of the last term in the left hand side of the (1), we introduce the following bilinear form

$$d(\omega, T, T^b) = \int_{\mathcal{M}} \frac{R\omega T T^b}{c_p p} \, d\mathcal{M}. \quad (17)$$

Similarly, in view of the last terms in (10) and (11), we define the linear functional

$$l(U^b) := l_T(T^b) + l_q(q^b) = \nu_1 \alpha \int_{\Gamma_i} \left(\frac{gp_1}{R\overline{T}}\right)^2 T_* T^b \, d\Gamma_i + \nu_2 \beta \int_{\Gamma_i} \left(\frac{gp_1}{R\overline{T}}\right)^2 q_* q^b \, d\Gamma_i. \quad (18)$$

Next, we consider the mapping relations related to the operators $\mathcal{A}_T$, $\mathcal{A}_q$ and the above defined functionals.

It is well-known that the linear operators A_T, $A_q : V \to V^*$ defined through the relations

$$\langle A_T u, v \rangle := a_T(u, v), \; \langle A_q u, v \rangle := a_q(u, v), \forall u, v \in V, \quad (19)$$

are both bounded linear operators.

Similarly, the operators $B(\mathbf{u}, U) = (B_T(\mathbf{u}, U), B_q(\mathbf{u}, q)) : \mathbf{V} \times V^2 \to (V^*)^2$ and $D(\mathbf{u}, u) : \mathbf{H} \times V \to V'$ defined by

$$\langle B(\mathbf{u}, U), U^b \rangle := (b_T(\mathbf{u}, T, T^b), b_q(\mathbf{u}, q, q^b)), \; \forall \mathbf{u} \in \mathbf{V}, U, U^b \in V^2, \quad (20)$$

and

$$\langle D(\mathbf{u}, u), v \rangle := d(\omega, u, v), \forall \mathbf{u} \in \mathbf{H}, u, v \in V, \quad (21)$$

are also bounded.

Due to the divergence free condition of $\mathbf{u}$, we easily see that for any $T, q \in V$,

$$b_T(\mathbf{u}, T, T) = 0, \; b_q(\mathbf{u}, q, q) = 0. \quad (22)$$

Concerning the boundedness of the above functionals, we have the following lemma.

Lemma 1 (Boundedness of the functionals). *Assume $U, U^b \in V^2$ and $\mathbf{u} \in \mathbf{V}$. There exist universal positive constants λ and K_i, $1 \leq i \leq 6$ such that*

$$|a_T(T, T^b)| \leq K_1 \|T\| \|T^b\|, \; a_T(T, T) \geq \lambda \|T\|^2; \quad (23)$$

$$|a_q(q, q^b)| \leq K_2 \|q\| \|q^b\|, \; a_q(q, q) \geq \lambda \|q\|^2; \quad (24)$$

$$|b(\mathbf{u}, U, U^b)| \leq K_3 \|\mathbf{u}\|_{\mathbf{v}} |U|_{L^2}^{\frac{1}{2}} \|U\|^{\frac{1}{2}} \|U^b\|; \quad (25)$$

$$|d(\omega, T, T^b)| \leq K_4 |\omega|_{L^2} |T|_{L^2}^{\frac{1}{4}} \|T\|^{\frac{3}{4}} |T^b|_{L^2}^{\frac{1}{4}} \|T^b\|^{\frac{3}{4}}; \quad (26)$$

$$|l_T(T^b)| \leq K_5 \|T^b\|, \; |l_q(q^b)| \leq K_6 \|q^b\|. \quad (27)$$

Definition 1. *Let* $(T_0, q_0) \in H \times H$ *be such that* $0 \le q_0 \le 1$ *a.e. in* $\mathcal{M}$ *and let* $t_1 > 0$ *be fixed. A vector* $U = (T, q) \in L^2(0, t_1; V \times V) \cap C([0, t_1]; H \times H)$ *with* $(\partial_t T, \partial_t q) \in L^2(0, t_1; V^* \times V^*)$ *is a solution to the initial and boundary value problem described by (1), (2), (6) and (7), if for almost every* $t \in [0, t_1]$ *and for every* $(T^b, q^b) \in V \times \mathcal{K}$, *we have*

$$\langle \partial_t T, T^b \rangle + a_T(T, T^b) + b_T(\mathbf{u}, T, T^b) - d(\omega, T, T^b) - l_T(T^b) = (\frac{1}{p}\omega^-(t)h_q\varphi(T), T^b),$$
(28)

$$\langle \partial_t q, q^b - q \rangle + a_q(q, q^b - q) + b_q(\mathbf{u}, q, q^b - q) - l_q(q^b - q) \ge (-\frac{1}{p}\omega^-(t)h_q F(T), q^b - q),$$
(29)

for some $h_q \in \mathcal{H}(q - q_s)$ *and*

$$U_0 = (T_0, q_0). \tag{30}$$

3   Time Discretization-The Euler Scheme

3.1   Time-Discretization

We assume that the velocity field $\mathbf{u}$ is given, time-dependent and satisfies $\mathbf{u} \in L^r(0, t_1; V) \cap L^\infty(0, t_1; H)$ for some given $r \in (4, +\infty]$.

Let N be an integer which will eventually go to $+\infty$ and set $\Delta t := k = t_1/N$. We will define recursively a family of elements of $V \times \mathcal{K}$, say (T^0, q^0), (T^1, q^1), $\cdots$, (T^N, q^N), where (T^m, q^m) will be in some sense an approximation of the functions (T, q) we are looking for, on the interval $[(m-1)k, mk]$.

First, we define $\mathbf{u}^m = \frac{1}{k} \int_{(m-1)k}^{mk} \mathbf{u}(t) \, dt$, $m = 1, 2, \cdots, N$. Our discretization is as follows:

We begin with $(T^0, q^0) := (T_0, q_0)$, i.e., the given initial datum. When (T^0, q^0), (T^1, q^1), $\cdots$, (T^{m-1}, q^{m-1}) are known, $T^m \in V$ and $q^m \in \mathcal{K}$ are determined by:

$$\langle \frac{T^m - T^{m-1}}{k}, T^b \rangle + a_T(T^m, T^b) + b_T(\mathbf{u}^m, T^m, T^b) - d(\omega^m, T^{m-1}, T^b) - l_T(T^b)$$
$$= (\frac{1}{p}[\omega^m]^- h_{Q^m} \varphi(T^{m-1}), T^b),$$
(31)

$$\langle \frac{q^m - q^{m-1}}{k}, q^b - q^m \rangle + a_q(q^m, q^b - q^m) + b_q(\mathbf{u}^m, q^m, q^b - q^m) - l_q(q^b - q^m)$$
$$\ge (-\frac{1}{p}[\omega^m]^- h_{Q^m} F(T^{m-1}), q^b - q^m),$$
(32)

for some $h_{Q^m} \in \mathcal{H}(Q^m - q_s)$ where Q^m is either q^{m-1} or q^m.

In the above construction of the discretization scheme (31)–(32), one shall pay special attention to the indices in the terms d and φ. Notice that we have $d(\omega^m, T^{m-1}, T^b)$ and $\varphi(T^{m-1})$. Obviously, this choice of indices will have influence on our search for T^m and q^m recursively. More importantly for us, it is crucial for us to obtain energy estimates later: the required estimates would not be true if we changed the indices $m-1$ to be m in d and φ. However, the choice of the indices in F and h_{Q^m} is not so sensitive.

Remark 1. In the above discretization, we have to deal with variational inequalities due to the q-equation (32). Meanwhile, we shall keep in mind that the physical constraint on the function q in our problem, $q^m \in \mathcal{K}$ is not preserved during the discretization (31)–(32). Finally, the problem we meet is nonlinear. The above three aspects form the main sources of difficulties for our study.

3.2   Regularization and Penalization

In view of Remark 1, we proceed our investigation by way of regularization and penalization. Let $\varepsilon = (\varepsilon_1, \varepsilon_2)$ and $\varepsilon_i > 0$ be small for $i = 1, 2$. For $\varepsilon_2 > 0$, we define as follows the regularization H_{ε_2} of $\mathcal{H}(\cdot) : \mathbb{R} \to [0,1]$: equal to 0 for $\eta \geq 0$, to 1 for $\eta \geq \varepsilon_2$, and linear continuous between 0 and ε_2. And consider the associated regularized and penalized problem:
   To find $T_\varepsilon^m, q_\varepsilon^m \in V$ such that

$$\langle \frac{T_\varepsilon^m - T_\varepsilon^{m-1}}{k}, T^b \rangle + a_T(T_\varepsilon^m, T^b) + b_T(\mathbf{u}^m, T_\varepsilon^m, T^b) - d(\omega^m, T_\varepsilon^{m-1}, T^b) - l_T(T^b)$$

$$= (\frac{1}{p}[\omega^m]^- H_{\varepsilon_2}(Q_\varepsilon^m - q_s)\varphi(T_\varepsilon^{m-1}), T^b),$$
(33)

$$\langle \frac{q_\varepsilon^m - q_\varepsilon^{m-1}}{k}, q^b \rangle + a_q(q_\varepsilon^m, q^b) + b_q(\mathbf{u}^m, q_\varepsilon^m, q^b) - l_q(q^b)$$

$$= (\frac{1}{\varepsilon_1}[q_\varepsilon^m]^-, q^b) - (\frac{1}{\varepsilon_1}[q_\varepsilon^m - 1]^+, q^b) - (\frac{1}{p}[\omega^m]^- H_{\varepsilon_2}(Q_\varepsilon^m - q_s)F(T_\varepsilon^{m-1}), q^b),$$
(34)

for all $T^b, q^b \in V$.
   Notice that we have two choices for Q_ε^m either $Q_\varepsilon^m = q_\varepsilon^{m-1}$ or $Q_\varepsilon^m = q_\varepsilon^m$. The introduction of penalization in the scheme (33)–(34) is designed to remedy the difficulty brought by the physical range requirement for the humidity q. The regularization process will overcome the difficulty caused by the variational inequality and the requirement on h_q. Here one may suspect that the two penalization terms in (34) may be potentially dangerous due to the blowing up factor $\frac{1}{\varepsilon_1}$. However, we point out that we could still obtain elegant estimates which do not depend on ε_1 (and ε_2, k) though we have a blowing up factor $\frac{1}{\varepsilon_1}$ when we pass to the limit $\varepsilon \to (0+, 0+)$. These estimates will yield that the limit functions q^m of q_ε^m satisfy the range requirement, i.e., $0 \leq q^m \leq 1$ for $m = 1, 2, \cdots N$ and a.e. $\mathbf{x} \in \mathcal{M}$.

3.3   Validity of Iteration

The scheme (33)–(34) yields elliptic system on $(T_\varepsilon^m, q_\varepsilon^m)$ when $(T_\varepsilon^{m-1}, q_\varepsilon^{m-1})$ is known. To carry out our program, the step of finding $(T_\varepsilon^m, q_\varepsilon^m)$ given

$(T_\varepsilon^{m-1}, q_\varepsilon^{m-1})$ is indispensable. To realize this iteration step, we need some surjective or existence theorems. Typically, we can use the Minty-Browder surjective
theorem, Lax-Milgram theorem or Galerkin method. Here we can realize the iteration step by different methods depending on the choices of Q_ε^m in our scheme
(33)–(34). When $Q_\varepsilon^m = q_\varepsilon^{m-1}$, the factor $H_{\varepsilon_2}(\cdot)$ is known when we proceed to
obtain T_ε^m and q_ε^m once T_ε^{m-1} and q_ε^{m-1} are known. We can apply the Minty-
Browder surjective theorem (see e.g., [20]) or Galerkin method (see e.g. [28])
to derive the existence of $(T_\varepsilon^m, q_\varepsilon^m)$. On the other hand, if $Q_\varepsilon^m = q_\varepsilon^m$, we only
can proceed by Galerkin method (see e.g., [27,28]) to derive the existence of
$(T_\varepsilon^m, q_\varepsilon^m)$, since the factor $H_{\varepsilon_2}(\cdot)$ is not known when we proceed to obtain T_ε^m
and q_ε^m even though T_ε^{m-1} and q_ε^{m-1} are known.

3.4 *A Priori* Estimate for $(T_\varepsilon^m, q_\varepsilon^m)$

The *a priori* estimates on $(T_\varepsilon^m, q_\varepsilon^m)$ independent of k and ε for the regularized
and penalized problem (33)–(34) will be crucial for the processes of passing to
the limits $\varepsilon \to (0+, 0+)$ and $k \to 0+$.

Lemma 2. *We have the estimates*

$$|U_\varepsilon^j|_{L^2}^2 \le C(\mathbf{u}, U_0, t_1), \ \forall\, 1 \le j \le N,$$

$$\sum_{m=1}^N |U_\varepsilon^m - U_\varepsilon^{m-1}|_{L^2}^2 \le C(\mathbf{u}, U_0, t_1),$$

$$k \sum_{m=1}^N \|U_\varepsilon^m\|^2 \le C(\mathbf{u}, U_0, t_1), \tag{35}$$

*where $C(\mathbf{u}, U_0, t_1)$ is a finite constant depending on the given datum $\mathbf{u}$, U_0 and
t_1, but independent of ε and k.*

In Lemma 2, the process to obtain the estimates on the T^m's is more involved
than that for the q^m's. The reason lies in the fact that the function F is bounded
while φ is not. Due to this reason, we need the following version of the so-called
discrete Gronwall lemma [33]:

Lemma 3 (Discrete Gronwall Lemma). *Let θ be any positive constant
and $N_0 > 1$ be an integer. Suppose the three nonnegative number sequences
$(X_m), (Y_m)$ and (Z_m) for $m = 0, 1, 2, \cdots, N_0$ satisfy the following relation*

$$X_m \le X_{m-1}(1 + \theta Y_m) + \theta Z_m. \tag{36}$$

Then for $m = 1, 2, 3, \cdots, N_0$, the following estimates hold

$$X_m \le X_0 \exp\left(\sum_{i=0}^{m-1} \theta Y_{i+1}\right) + \sum_{i=1}^{m-1} \theta Z_i \exp\left(\sum_{j=i}^{m-1} \theta Y_{j+1}\right) + \theta Z_n. \tag{37}$$

The iteration relation (36) in the Discrete Gronwall Lemma explains our
choice of index in $\varphi(T^{m-1})$ and $d(\omega^m, T^{m-1}, T^b)$ in the initial discretization.

We have the following *a priori* bound for the norm $k \sum_{m=1}^{N} \|\frac{U_\varepsilon^m - U_\varepsilon^{m-1}}{k}\|_{V^*}^2$, which will be used in our compactness argument.

Lemma 4. *For any $\varepsilon_1 > 0$ and any $\varepsilon_2 > 0$, the inequality*

$$k \sum_{m=1}^{N} \|\frac{U_\varepsilon^m - U_\varepsilon^{m-1}}{k}\|_{V^*}^2 \le C(\mathbf{u}, U_0, t_1) < +\infty, \tag{38}$$

holds for some constant $C(\mathbf{u}, U_0, t_1)$ depending on $U_0, \mathbf{u}, t_1$, but not on ε and k.

The main point of Lemma 4 is that the bound is independent of $\varepsilon = (\varepsilon_1, \varepsilon_2)$ and any k. As ε_2 comes into play through the regularization function H_{ε_2} and H_{ε_2} is bounded say by 1, it is easy to obtain the bound independent of ε_2. Therefore, the main issue here is to control the penalization terms which contain a blowing up factor $\frac{1}{\varepsilon_1}$ in the limit process $\varepsilon \to (0+, 0+)$. We have the following bounds for the penalization terms.

Lemma 5. *The following bounds hold:*

$$k \sum_{m=1}^{N} |\frac{[q_\varepsilon^m]^-}{\varepsilon_1}|_{L^2}^2 \le C|\omega|_{L^2(0,t_1;H)}^2, \quad k \sum_{m=1}^{N} |\frac{[q_\varepsilon^m - 1]^+}{\varepsilon_1}|_{L^2}^2 \le C|\omega|_{L^2(0,t_1;H)}^2. \tag{39}$$

The proof of the estimates on the two penalization terms is subtle. Let us briefly illustrate this point. To prove the two estimates in Lemma 5, we choose the test function $q^b = [q^m]^-$ and $[q^m - 1]^+$ in (34) respectively. Then the terms $\langle \frac{q^m - q^{m-1}}{k}, [q^m]^- \rangle$ and $\langle \frac{q^m - q^{m-1}}{k}, [q^m - 1]^+ \rangle$ will appear. However, neither term has a favorable sign. Actually, we think that the two kinds of terms may not have the same sign for different $k = 1, 2, \cdots, N$. Here we need more quantitative estimates. Interestingly, though for each *fixed* k, the above two kinds of terms may not have a definite sign, we have the following definite signs for their sums

$$\sum_{m=1}^{N} \langle \frac{q^m - q^{m-1}}{k}, [q^m]^- \rangle \le 0, \quad -\sum_{m=1}^{N} \langle \frac{q^m - q^{m-1}}{k}, [q^m - 1]^+ \rangle \le 0, \tag{40}$$

by writing $q^m = [q^m]^+ - [q^m]^-$ and the same for q^{m-1} directly. Due to the form of the estimates in Lemma 5, the two relations in (40) are sufficient to derive the proof of Lemma 5.

3.5 Passage to the Limit $\varepsilon \to (0+, 0+)$

Assume the time step $k > 0$ is *fixed*. Our goal, in this part, is to pass to the limit $\varepsilon \to (0+, 0+)$ in the scheme (33)–(34). The limit functions (T^m, q^m) of $(T_\varepsilon^m, q_\varepsilon^m)$ will be solutions to the time discretized scheme (31)–(32). These solutions will serve as building blocks for us to construct approximate solutions to our original problem.

After extracting a finite number of subsequences, $\varepsilon \to 0$, we infer from Lemma 2 that, for $m = 1, 2, \cdots, N$ there exist functions $U^m \in V$ such that, as $\varepsilon \to 0+$

$$U_\varepsilon^m \rightharpoonup U^m \text{ weakly in } V. \tag{41}$$

We still use ε as the index for the subsequence.

Since the inclusion $V \subset H$ is compact and U_ε^m is weakly convergent in V, it is strongly convergent in H, i.e., we also have

$$U_\varepsilon^m \to U^m \text{ strongly in } H. \tag{42}$$

By an additional extraction of subsequences we see that:

$$U_\varepsilon^m(x) \to U^m(x) \text{ a.e., } m = 1, 2, \cdots, N. \tag{43}$$

Meanwhile, we have $H_{\varepsilon_2}(Q_\varepsilon^m - q_s) \rightharpoonup h_{Q^m}$ weak-* in $L^\infty(\mathcal{M})$.

Concerning the limit functions q^m, the second component of U^m for $m = 1, 2, \cdots, N$, we know from Lemma 5 that

$$k \sum_{m=1}^{N} \left(|[q_\varepsilon^m]^-|_{L^2}^2 + |[q_\varepsilon^m - 1]^+|_{L^2}^2 \right) \le C\varepsilon_1^2 |\omega|_{L^2(0,t_1;H)}^2. \tag{44}$$

As the real functions $g_\pm(\theta) = \theta^\pm$ are both Lipschitz functions with Lipschitz constant 1 on $\mathbb{R}$, i.e., $|g_\pm(\theta_1) - g_\pm(\theta_2)| \le |\theta_1 - \theta_2|$, we have

$$||q_\varepsilon^m]^- - [q^m]^-|_{L^2} \le |q_\varepsilon^m - q^m|_{L^2}, \quad |[q_\varepsilon^m - 1]^+ - [q^m - 1]^+|_{L^2} \le |q_\varepsilon^m - q^m|_{L^2}.$$

Consequently, with (42) we have $[q_\varepsilon^m]^- \to [q^m]^-$ and $[q_\varepsilon^m - 1]^+ \to [q^m - 1]^+$ in H. As $k > 0$ is a *fixed* number, we can pass to the limit on ε in (44) to obtain that

$$\sum_{m=1}^{N} \left(|[q^m]^-|_{L^2}^2 + |[q^m - 1]^+|_{L^2}^2 \right) = 0,$$

which implies

$$0 \le q^m \le 1, \text{ a.e. in } \mathbf{x} \in \mathcal{M}, \text{ i.e., } q^m \in K. \tag{45}$$

With the above preparations, we could pass to the limit from the scheme (33)–(34) to the scheme (31)–(32) term by term. We only point out the following three subtle points. First, here we obtain the strong convergence of the functions U_ε^m to their limits U^m in H by the compact embedding $V \subset H$. The strong convergence will imply the a.e. convergence of U_ε^m in $\mathcal{M}$ up to subsequences. The strong convergence and the a.e. convergence of U_ε^m are crucial when we pass to the limit in the nonlinear terms in the scheme (33)–(34). The reason is well-known: nonlinear mappings do not preserve weak convergences. Second, for the two penalization terms, we have for $q^b \in K$ that

$$\left(\frac{1}{\varepsilon_1} [q_\varepsilon^m]^-, q^b - q_\varepsilon^m \right) \ge 0, \quad -\left(\frac{1}{\varepsilon_1} [q_\varepsilon^m - 1]^+, q^b - q_\varepsilon^m \right) \ge 0. \tag{46}$$

Due to weak convergence of q^m in V and the weak lower semi-continuity property of the norm, we also have

$$\limsup a_q(q^m_\varepsilon, q^b - q^m_\varepsilon) = \lim a_q(q^m_\varepsilon, q^b) - \liminf a_q(q^m_\varepsilon, q^m_\varepsilon)$$
$$\leq a_q(q^m, q^b) - a_q(q^m, q^m) \tag{47}$$
$$= a_q(q^m, q^b - q^m).$$

All the above three terms produce correct direction of inequalities during the limit process. Therefore, we obtain the desired variational inequalities in (32). Third, in order to show that $h_{Q^m} \in \mathcal{H}(Q^m - q_s)$, we shall use the idea of subdifferential for convex functions.

Summarizing the above arguments, we obtain from (33)–(34) via passing to the limit on ε the existence of a solution (T^m, q^m) to (31)–(32).

4   Convergence of the Euler Scheme

In this section, we want to prove the convergence of the solutions of the Euler scheme (31)–(32) to the solutions of the system (28)–(30). We shall use the same conventions on subsequences and indices as in the last section, that is, the limit process in this part is $N \to +\infty$ or equivalently $k \to 0+$ and up to subsequences.

Due to the weak lower semi-continuity property of the norms, we know that for the limit functions U^m which now have no dependence on ε, the bounds in Lemmas 2 and 4 are now valid with U^m_ε replaced by the limit functions U^m.

4.1   Construction of Approximations

For each *fixed* k (or N), we associate to the elements $U^0, U^1, U^2, \cdots, U^N$ the following approximate functions $U_k = (T_k, q_k)$, $\tilde{U}_k = (\tilde{T}_k, \tilde{q}_k)$ and $W_k = (\mathcal{T}_k, \mathcal{Q}_k)$ which are defined piecewise on $[0, t_1]$ and take values in the space V^2:

$$U_k(t) = U^m, \ \tilde{U}_k(t) = U^{m-1}, \text{ for } t \in [(m-1)k, mk), m = 1, 2, \cdots, N. \tag{48}$$

$$W_k(t) = \frac{U^m - U^{m-1}}{k}(t - (m-1)k) + U^{m-1}, \text{ for } t \in [(m-1)k, mk), m = 1, 2, \cdots, N. \tag{49}$$

4.2   Reinterpretation of *A Priori* Estimates

First, we give a lemma measuring the distance in $L^2(0, t_1; H)$ of the functions $U_k, \tilde{U}_k$ and W_k in the limit process $k \to 0+$.

Lemma 6. *For the functions U_k, W_k and $\tilde{U}_k$ defined above, there hold*

$$|U_k - W_k|_{L^2(0,t_1;H)} \leq C(\mathbf{u}, U_0, t_1)\sqrt{k}, \ |U_k - \tilde{U}_k|_{L^2(0,t_1;H)} \leq C(\mathbf{u}, U_0, t_1)\sqrt{k}.$$

Now, we state a result concerning the boundedness of the functions $U_k, \tilde{U}_k$ and W_k.

Lemma 7. *The functions U_k, $\tilde{U}_k$ and W_k remain in a bounded set of $L^2(0,t_1;V) \cap L^\infty(0,t_1;H)$ as $k \to 0+$. The functions $\partial_t W_k$ form a bounded set in $L^2(0,t_1;V^*)$ and $U_k - W_k \to 0$ in $L^2(0,t_1;H)$ strongly as $k \to 0+$.*

Lemmas 6 and 7 can be regarded as reinterpretations of Lemmas 2 and 4 in terms of the functions $U_k, W_k, \tilde{U}_k$.

Define $\mathbf{u}_k : [0,t_1] \to \mathbf{V}$ as follows:

$$\mathbf{u}_k(t) = \mathbf{u}^m, \text{ for } t \in [(m-1)k, mk), m = 1, 2, \cdots, N. \tag{50}$$

We have the following classical lemma:

Lemma 8 (Convergence of $\mathbf{u}_k$). *For the functions $\mathbf{u}_k$ defined above, there holds*

$$\mathbf{u}_k \to \mathbf{u}, \text{ in } L^r(0,t_1,\mathbf{V}) \text{ as } k \to 0+. \tag{51}$$

For later use, we also define the linear averaging map for the test functions $U^b = (T^b, q^b) \in L^2(0,t_1;V)$ that we will use below, that is, we define $U_k^b : [0,t_1] \to V^2$ piecewise by

$$U_k^b(t) = \frac{1}{k} \int_{(m-1)k}^{mk} U^b(t) \, dt \text{ on } [(m-1)k, mk).$$

Similarly as in Lemma 8, we conclude that $U_k^b \to U^b$ strongly in $L^2(0,t_1;V^2)$ as $k \to 0$. Moreover, if $q^b \in \mathcal{K}$ for a.e. $t \in [0,t_1]$, we have $q_k^b \in \mathcal{K}$ for all $t \in [0,t_1]$.

4.3   Passage to the Limit: $k \to 0+$

We first reinterpret as follows the scheme (31)–(32) in terms of the functions $U_k = (T_k, q_k), \tilde{U}_k = (\tilde{T}_k, \tilde{q}_k), W_k = (\mathcal{T}_k, \mathcal{Q}_k)$ and $U_k^b = (T_k^b, q_k^b)$:

$$\langle \partial_t \mathcal{T}_k, T_k^b \rangle + a_T(T_k, T_k^b) + b_T(\mathbf{u}_k, T_k, T_k^b) - d(\omega_k, \tilde{T}_k, T_k^b) - l_T(T_k^b)$$
$$= (\frac{1}{p}[\omega_k]^- h_{Q_k} \varphi(\tilde{T}_k), T_k^b), \tag{52}$$

$$\langle \partial_t \mathcal{Q}_k, q_k^b - q_k \rangle + a_q(q_k, q_k^b - q_k) + b_q(\mathbf{u}_k, q_k, q_k^b - q_k) - l_q(q_k^b - q_k)$$
$$\geq (-\frac{1}{p}[\omega_k]^- h_{Q_k} F(\tilde{T}_k), q_k^b - q_k), \tag{53}$$

where Q_k is either $\tilde{q}_k$ or q_k. Furthermore, h_{Q_k} is defined by $h_{Q_k}(t) = h_{Q^m}$ when $t \in [(m-1)k, mk)$. Here we emphasize that we require $q_k^b \in L^2(0,t_1;\mathcal{K})$.

Due to Lemma 7, we have, up to subsequences, in the limit $k \to 0+$, that

$$U_k \rightharpoonup U = (T,q), \text{ weakly in } L^2(0,t_1;V) \text{ and weak-} * \text{ in } L^\infty(0,t_1;H), \tag{54}$$

$$W_k \rightharpoonup W = (\mathcal{T}, \mathcal{Q}), \text{ weakly in } L^2(0,t_1;V) \text{ and weak-} * \text{ in } L^\infty(0,t_1;H), \tag{55}$$

and

$$\partial_t W_k \rightharpoonup \partial_t W = (\partial_t \mathcal{T}, \partial_t \mathcal{Q}), \text{ weakly in } L^2(0,t_1;V^*). \tag{56}$$

Obviously, $\tilde{U}_k = U(\cdot - k)$ converges also to U in $L^2(0, t_1; V)$ weakly and in $L^\infty(0, t_1; H)$ weak-$*$.

In view of Lemma 6, we know that

$$U = W. \tag{57}$$

Now, we consider the inclusions $V \subset H \subset V^*$ where the first inclusion is compact and the second inclusion is continuous. In view of (55) and (56), we conclude, by applying the Aubin-Lions compactness theorem, that

$$W_k \to W, \text{ strongly in } L^2(0, t_1; H). \tag{58}$$

By Lemma 6 again, we conclude that

$$U_k, \tilde{U}_k, W_k \to U, \text{ strongly in } L^2(0, t_1; H). \tag{59}$$

With the above preparations, we can now pass to the limit $k \to 0+$ from (52)–(53) to (28)–(30) term by term. Here we also point several subtle points. First, we obtain the strong convergence, i.e. (59), in this step by the Aubin-Lions compactness argument. For further details, see [28]. Second, the limit function q satisfies the required range condition. Indeed, regarded as a convex subset of $L^2(0, t_1; V)$, $L^2(0, t_1; \mathcal{K})$ is closed with respect to the strong topology induced by the $L^2(0, t_1; V)$-norm. Therefore, it is also closed with respect to the weak topology. Furthermore, in view of the fact that $q_k \in L^2(0, t_1; \mathcal{K})$ which is obvious from the definition and that q_k converge to q weakly in $L^2(0, t_1; V)$, we conclude that $q \in L^2(0, t_1; \mathcal{K})$. Third, the subtle point in this passage to the limit is to deal with the term $\int_0^{t_1} \langle \partial_t \mathcal{Q}_k, q_k^b - q_k \rangle \, dt$ which is the sum of $\int_0^{t_1} \langle \partial_t \mathcal{Q}_k, q_k^b - \mathcal{Q}_k \rangle \, dt$ and $\int_0^{t_1} \langle \partial_t \mathcal{Q}_k, \mathcal{Q}_k - q_k \rangle \, dt$. Using integration by parts, (56) and the lower semi-continuity of the norm, we write

$$
\begin{aligned}
\limsup \int_0^{t_1} \langle \partial_t \mathcal{Q}_k, q_k^b - \mathcal{Q}_k \rangle \, dt &= -\liminf \int_0^{t_1} \langle \partial_t \mathcal{Q}_k, \mathcal{Q}_k \rangle \, dt + \lim \int_0^{t_1} \langle \partial_t \mathcal{Q}_k, q_k^b \rangle \, dt \\
&= -\liminf \frac{1}{2} |\mathcal{Q}_k(t_1)|_{L^2}^2 + \frac{1}{2} |q_0|_{L^2}^2 + \int_0^{t_1} \langle \partial_t q, q^b \rangle \, dt \\
&\le -\frac{1}{2} |q(t_1)|_{L^2}^2 + \frac{1}{2} |q_0|_{L^2}^2 + \int_0^{t_1} \langle \partial_t q, q^b \rangle \, dt \\
&= -\int_0^{t_1} \langle \partial_t q, q \rangle \, dt + \int_0^{t_1} \langle \partial_t q, q^b \rangle \, dt \\
&= \int_0^{t_1} \langle \partial_t q, q^b - q \rangle \, dt.
\end{aligned}
\tag{60}
$$

where we have used, in the second equality of (60), the observation

$$\lim \int_0^{t_1} \langle \partial_t \mathcal{Q}_k, q_k^b \rangle \, dt \to \int_0^{t_1} \langle \partial_t q, q^b \rangle \, dt, \tag{61}$$

which is a simple consequence of (56) and the strong convergence of q_k^b to q^b in $L^2(0, t_1; V)$.

A subtle point is the treatment of $\int_0^{t_1} \langle \partial_t \mathcal{Q}_k, \mathcal{Q}_k - q_k \rangle \, dt$. Though we have (56) (which implies in particular that $\partial_t \mathcal{Q}_k$ is bounded in $L^2(0, t_1; V^*)$ and $\mathcal{Q}_k - q_k \rightharpoonup 0$ weakly in $L^2(0, t_1; V)$, we can not conclude that the limit of $\int_0^{t_1} \langle \partial_t \mathcal{Q}_k, \mathcal{Q}_k - q_k \rangle \, dt$ is 0. Rather, we show, by the specific forms of $\mathcal{Q}_k$ and q_k, that

$$\limsup \int_0^{t_1} \langle \partial_t \mathcal{Q}_k, \mathcal{Q}_k - q_k \rangle \, dt \leq 0. \tag{62}$$

Indeed, noticing that $\partial_t \mathcal{Q}_k = \frac{q^m - q^{m-1}}{k}$ and $\mathcal{Q}_k - q_k = \frac{q^m - q^{m-1}}{k}(t - mk)$ on the subinterval $[(m-1)k, mk)$ of $[0, t_1]$, we have:

$$
\begin{aligned}
\int_0^{t_1} \langle \partial_t \mathcal{Q}_k, \mathcal{Q}_k - q_k \rangle \, dt &= \sum_{m=1}^{N} \int_{(m-1)t}^{mt} \langle \partial_t \mathcal{Q}_k, \mathcal{Q}_k - q_k \rangle \, dt \\
&= \sum_{m=1}^{N} \int_{(m-1)t}^{mk} \langle \frac{q^m - q^{m-1}}{k}, \frac{q^m - q^{m-1}}{k}(t - mk) \rangle \, dt \\
&= \sum_{m=1}^{N} \int_{(m-1)t}^{mk} \frac{|q^m - q^{m-1}|_{L^2}^2}{k^2}(t - mk) \, dt \\
&\leq 0,
\end{aligned}
$$

which implies (62). From (60) and (62), we can conclude that

$$\limsup \int_0^{t_1} \langle \partial_t \mathcal{Q}_k, q_k^b - q_k \rangle \, dt \leq \int_0^{t_1} \langle \partial_t q, q^b - q \rangle \, dt. \tag{63}$$

To sum up, we have proved the following theorem when q_s is constant:

Theorem 1. *Given $T_0, q_0 \in H$ with $0 \leq q_0 \leq 1$ a.e. in $\mathcal{M}$, the Euler scheme (31)–(32) contains a subsequence which converges to a solution of the system (1)–(7).*

5   The Case Where q_s Depends on Time

In this part, we extend our former results to the case where q_s is not constant.

5.1   The Nonlinearities φ and F

When q_s evolves according to (8), the nonlinearities F and φ will also depend on the function q_s. The nonlinearities (see e.g., [17,18]) φ and $F \colon \mathbb{R} \times \mathbb{R} \to \mathbb{R}$ are defined as follows:

$$F(T, q_s) = q_s G(T, q_s) = q_s T \frac{RL(T) - c_p R_v T}{c_p R_v T^2 + q_s L(T)^2}, \tag{64}$$

$$\varphi(T, q_s) = \frac{L(T)}{c_p} F(T, q_s), \tag{65}$$

where

$$L(T) = c_1 - c_2 T, \quad G(T, q_s) = T \frac{RL(T) - c_p R_v T}{c_p R_v T^2 + q_s L(T)^2}. \tag{66}$$

In the above, c_1, c_2, R, c_p, R_v are all strictly positive constants. The additional dependence of F and φ on q_s will bring us technical complexities during the passages to the limits.

Notice that the functions F, G, and φ have a singularity at $(0,0)$; $F(T, q_s)$ is bounded but discontinuous at $(0,0)$ and $G(T, q_s)$ may blow up at $(0,0)$. To overcome this difficulty, we introduce the following regularized version φ_r, F_r and G_r for φ, F and G.

$$F_r(T, q_s) = q_s G_r(T, q_s) = q_s T \frac{RL(T) - c_p R_v T}{c_p R_v \max{(T, \gamma)}^2 + q_s L(T)^2}, \tag{67}$$

$$\varphi_r(T, q_s) = \frac{L(T)}{c_p} F_r(T, q_s), \tag{68}$$

$$G_r(T, q_s) = T \frac{RL(T) - c_p R_v T}{c_p R_v \max{(T, \gamma)}^2 + q_s L(T)^2}, \tag{69}$$

where $\gamma > 0$ is smaller than any temperature on Earth. Once arriving at $q_s \geq 0$, we can derive G_r is nonnegative by observing that $T \leq \frac{LR}{c_p R_v}$ for any temperature on Earth. It is easy to see the rational function F_r is bounded and globally Lipschitz on $\mathbb{R} \times \mathbb{R}$, i.e.,

$$|F_r(\zeta_1, \xi_1) - F_r(\zeta_2, \xi_2)| \leq C(|\zeta_1 - \zeta_2| + |\xi_1 - \xi_2|), \ \forall \ \zeta_1, \zeta_2 \in \mathbb{R}, \xi_1, \xi_2 \in [0, \infty), \tag{70}$$

and

$$|F_r(\zeta, \xi)| \leq C, \ \forall \ \zeta \in \mathbb{R}, \xi \in [0, \infty). \tag{71}$$

The function φ_r is also globally Lipschitz,

$$|\varphi_r(\zeta_1, \xi_1) - \varphi_r(\zeta_2, \xi_2)| \leq C(|\zeta_1 - \zeta_2| + |\xi_1 - \xi_2|), \ \forall \ \zeta_1, \zeta_2 \in \mathbb{R}, \xi_1, \xi_2 \in [0, \infty). \tag{72}$$

In addition, as $F_r(0,0) = 0$, we have $\varphi_r(0,0) = 0$. Hence the Lipschitz function φ_r also satisfies $|\varphi_r(\zeta, \xi)| \leq C(|\zeta| + |\xi|)$.

Definition 2. *Let $(T_0, q_0, q_{s,0}) \in H \times H \times H$ be such that $0 \leq q_0 \leq 1$, $0 < q_{s,0} < 1$ a.e. in $\mathcal{M}$ and let $t_1 > 0$ be fixed. A vector $(T, q, q_s) \in L^2(0, t_1; V \times V) \cap C([0, t_1]; H \times H) \times L^\infty(\mathcal{M} \times [0, t_1])$ with $0 < q_s < 1$, $(\partial_t T, \partial_t q, \partial_t q_s) \in L^2(0, t_1; V^* \times V^*) \times L^\infty(\mathcal{M} \times [0, t_1])$ is a solution to the initial boundary value problem described by (1), (2), (8), (6), (7) and (9), if for a.e. $t \in [0, t_1]$ and for every $(T^b, q^b) \in V \times \mathcal{K}$, we have*

$$\langle \partial_t T, T^b \rangle + a_T(T, T^b) + b_T(\mathbf{u}, T, T^b) - d(\omega, T, T^b) - l_T(T^b) = (\frac{1}{p} \omega^-(t) h_q \varphi(T, q_s), T^b), \tag{73}$$

$$\langle \partial_t q, q^b - q \rangle + a_q(q, q^b - q) + b_q(\mathbf{u}, q, q^b - q) - l_q(q^b - q) \geq (-\frac{1}{p}\omega^-(t)h_q F(T, q_s), q^b - q),$$

(74)

$$\frac{dq_s}{dt} = -\frac{1}{p}\omega^- h_q F(T, q_s) \quad \text{for a.e. } (t, \mathbf{x}) \in [0, t_1] \times \mathcal{M},$$

(75)

for some $h_q \in \mathcal{H}(q - q_s)$ and

$$U_0 = (T_0, q_0, q_{s,0}).$$

(76)

5.2   The Discretization Scheme

We begin with

$$(T^0, q^0, q_s^0) := (T_0, q_0, q_{s,0}), \quad \text{i.e., the given initial datum.}$$

(77)

When (T^0, q^0, q_s^0), (T^1, q^1, q_s^1), $\cdots$, $(T^{m-1}, q^{m-1}, q_s^{m-1})$ are known, $T^m \in V$ and $q^m \in \mathcal{K}$ are determined by:

$$\langle \frac{T^m - T^{m-1}}{k}, T^b \rangle + a_T(T^m, T^b) + b_T(\mathbf{u}^m, T^m, T^b) - d(\omega^m, T^{m-1}, T^b) - l_T(T^b)$$

$$= (\frac{1}{p}[\omega^m]^- h_{q^{m-1}}\varphi(T^{m-1}, q_s^{m-1}), T^b),$$

(78)

$$\langle \frac{q^m - q^{m-1}}{k}, q^b - q^m \rangle + a_q(q^m, q^b - q^m) + b_q(\mathbf{u}^m, q^m, q^b - q^m) - l_q(q^b - q^m)$$

$$\geq (-\frac{1}{p}[\omega^m]^- h_{q^{m-1}} F(T^{m-1}, q_s^{m-1}), q^b - q^m),$$

(79)

and

$$q_s^m := Z^m(mk),$$

(80)

where $Z^m(t)$ is the solution to the following initial value problem

$$\begin{cases} \frac{dZ^m(t)}{dt} - -\frac{1}{p}[\omega^m]^- h_{q^{m-1}} F(T^{m-1}, Z^m(t)), \\ Z^m((m-1)k) = q_s^{m-1}, \end{cases}$$

(81)

where $h_{q^{m-1}} \in H(q^{m-1} - q_s^{m-1})$.

Remark 2. We shall point out the distinct feature of the discretization for the q_s-equation. First, the q_s-equation shall be discretized once the (T, q)-equation is discretized as the (T, q)-equation depends on q_s through the nonlinear functions $F(T, q_s)$ and $\varphi(T, q_s)$ and we do not allow time dependence in the discretized equation on (T, q). Here we did not use the standard Euler algorithm for ordinary differential equations (ODEs). Rather, we define q_s^m by successively solving the ordinary differential equation (81) and make evaluations at specific time points.

The latter will make the q_s^m's satisfy the required range condition inherited from that of q_s. The Euler algorithm for ODEs cannot guarantee this range condition. Finally, we emphasize that we need to use $h_{q^{m-1}}$ in the scheme (78), (79) and (81). This choice makes the (T^m, q^m)-equation and the q_s^m-equation decouple.

To show the existence of a solution (T^m, q^m, Z^m) to (78)–(81), we consider the associated regularized and penalized problem:

To find $T_\varepsilon^m, q_\varepsilon^m \in V$ and $Z_\varepsilon^m \in H$ such that

$$\langle \frac{T_\varepsilon^m - T_\varepsilon^{m-1}}{k}, T^b \rangle + a_T(T_\varepsilon^m, T^b) + b_T(\mathbf{u}^m, T_\varepsilon^m, T^b) - d(\omega^m, T_\varepsilon^{m-1}, T^b) - l_T(T^b)$$

$$= (\frac{1}{p}[\omega^m]^- H_{\varepsilon_2}(q_\varepsilon^{m-1} - q_{s,\varepsilon}^{m-1})\varphi_r(T_\varepsilon^{m-1}, q_{s,\varepsilon}^{m-1}), T^b), \tag{82}$$

$$\langle \frac{q_\varepsilon^m - q_\varepsilon^{m-1}}{k}, q^b \rangle + a_q(q_\varepsilon^m, q^b) + b_q(\mathbf{u}^m, q_\varepsilon^m, q^b) - l_q(q^b)$$

$$= (\frac{1}{\varepsilon_1}[q_\varepsilon^m]^-, q^b) - (\frac{1}{\varepsilon_1}[q_\varepsilon^m - 1]^+, q^b) - (\frac{1}{p}[\omega^m]^- H_{\varepsilon_2}(q_\varepsilon^{m-1} - q_{s,\varepsilon}^{m-1})F_r(T_\varepsilon^{m-1}, q_{s,\varepsilon}^{m-1}), q^b), \tag{83}$$

and

$$q_{s,\varepsilon}^m := Z_\varepsilon^m(mk), \tag{84}$$

where Z_ε^m is the solution to the following initial value problem

$$\begin{cases} \frac{dZ_\varepsilon^m(t)}{dt} = -\frac{1}{p}[\omega^m]^- H_{\varepsilon_2}(q_\varepsilon^{m-1} - q_{s,\varepsilon}^{m-1})F_r(T_\varepsilon^{m-1}, Z_\varepsilon^m(t)), \\ Z_\varepsilon^m((m-1)k) = q_{s,\varepsilon}^{m-1}. \end{cases} \tag{85}$$

5.3   Validity of the Iteration

In order to have a valid scheme, we should be able to obtain $(T_\varepsilon^m, q_\varepsilon^m, q_{s,\varepsilon}^m)$ when $(T_\varepsilon^{m-1}, q_\varepsilon^{m-1}, q_{s,\varepsilon}^{m-1})$ are known. This is true. First, observe that the equation on $(T_\varepsilon^m, q_\varepsilon^m)$ and $q_{s,\varepsilon}^m$ are decoupled. Noticing that the arguments of F_r and φ_r have index $m-1$ in (82) and (83), we can obtain $(T_\varepsilon^m, q_\varepsilon^m)$ either by the Minty-Browder Theorem or by the Galerkin method when $(T_\varepsilon^{m-1}, q_\varepsilon^{m-1}, q_{s,\varepsilon}^{m-1})$ are known. The determination of $q_{s,\varepsilon}^m$ is easy when $(T_\varepsilon^{m-1}, q_\varepsilon^{m-1}, q_{s,\varepsilon}^{m-1})$ are known: we just need to use the initial value problem (85). The initial value problem (85) admits a unique solution $Z_\varepsilon^m(t)$ on $[(m-1)k, T^*)$. The maximum time of existence $T^* = +\infty$ due to the form of F_r. Actually, we have

$$F_r(T_\varepsilon^{m-1}, Z_\varepsilon^m(t)) = Z_\varepsilon^m(t)G_r(T_\varepsilon^{m-1}, Z_\varepsilon^m(t)).$$

Therefore, we have the following integral form of (85):

$$q_{s,\varepsilon}^m(t) = q_\varepsilon^{m-1} exp\left\{ -\int_{(m-1)k}^t \frac{1}{p}[\omega^m]^- H_{\varepsilon_2}(q_\varepsilon^{m-1} - q_{s,\varepsilon}^{m-1})G_r(T_\varepsilon^{m-1}, Z_\varepsilon^m(t)) \, d\tau \right\}. \tag{86}$$

Due to the above expression of $q_{s,\varepsilon}^m(t)$, no blow-up can happen as we always have

$$|Z_\varepsilon^m(t)|_{L^2} \le |q_{s,\varepsilon}^{m-1}|_{L^2}, \quad |Z_\varepsilon^m(t)|_{L^\infty} \le |q_{s,\varepsilon}^{m-1}|_{L^\infty} \text{ for } t \ge (m-1)k. \tag{87}$$

For our purpose, we just need $T^* \ge mk$ so that $q_{s,\varepsilon}^m$ is well-defined.

5.4 *A Priori* Estimates for $(T_\varepsilon^m, q_\varepsilon^m, q_{s,\varepsilon}^m)$

Integrating the first equation in (85) from $(m-1)k$ to mk, we find

$$q_{s,\varepsilon}^m = q_{s,\varepsilon}^{m-1} exp\Big\{ -\int_{(m-1)k}^{mk} \frac{1}{p}[\omega^m]^- H_{\varepsilon 2}(q_\varepsilon^{m-1} - q_{s,\varepsilon}^{m-1})G_r(T_\varepsilon^{m-1}, Z_\varepsilon^m(t))\, d\tau\Big\}. \quad (88)$$

As $0 < q_s^0 = q_{s,0} < 1$, we easily conclude that $0 < q_{s,\varepsilon}^m < 1$ for all $1 \le m \le N$ by repeatedly using (88) for $m = 1, 2, \cdots, N$. Actually, we have the following pointwise monotone relations

$$0 < q_{s,\varepsilon}^m(\mathbf{x}) \le q_{s,\varepsilon}^{m-1}(\mathbf{x}) < 1, \; m = 1, 2, \cdots, N. \quad (89)$$

Now we aim to obtain *a priori* estimates on $(T_\varepsilon^m, q_\varepsilon^m)$ independent of k and ε for the regularized and penalized problem (82)–(83). Due to the form of φ_r, F_r and our estimate on $q_{s,\varepsilon}^m$, we know that

$$|\varphi_r(T_\varepsilon^{m-1}, q_{s,\varepsilon}^{m-1})| \le C(|T_\varepsilon^{m-1}| + 1) \text{ and } |F_r(T_\varepsilon^{m-1}, q_{s,\varepsilon}^{m-1})| \le C, \quad (90)$$

where C is constant independent of ε and k. Therefore, Lemma 2 is still valid.

Now, we explain that Lemma 4 is still valid even when q_s is not constant. Since $|F_r(T_\varepsilon^{m-1}, q_{s,\varepsilon}^{m-1})|$ can be bounded by a universal constant, we still have, for the penalization terms, that Lemma 5 holds. Then estimating the duality pair $\langle \frac{U_\varepsilon^m - U_\varepsilon^{m-1}}{k}, U^b \rangle$ where $U^b \in V^2$, we can derive the conclusion of Lemma 4.

5.5 Passage to the Limit $\varepsilon \to 0+$ and $k \to 0+$

The passage to the limit can be proceeded essentially as before. By the compact Sobolev embedding theorem as before, we can derive that $(T_\varepsilon^m, q_\varepsilon^m) \to (T^m, q^m)$ strongly in H. In addition, by the above estimates on $q_{s,\varepsilon}^m$, we know that there exist $q_s^m \in H \cap L^\infty(\mathcal{M})$ such that, up to subsequences,

$$q_{s,\varepsilon}^m \rightharpoonup q_s^m \text{ weakly in } H \text{ and weak-* in } L^\infty(\mathcal{M}). \quad (91)$$

During the passage to the limit: $k \to 0+$, besides defining U_k, $\tilde{U}_k$, W_k, $h_{\tilde{q}_k}$ as before, we should also define $q_{s,k} : [0, t_1] \to L^\infty(\mathcal{M}) \cap L^2(\mathcal{M})$ as follows:

$$q_{s,k}(t) = \begin{cases} Z^1(t), & \text{when } t \in [0, k), \\ Z^2(t), & \text{when } t \in [k, 2k), \\ \cdots, \\ Z^m(t), & \text{when } t \in [(m-1)k, mk), \\ \cdots, \\ Z^N(t), & \text{when } t \in [(N-1)k, Nk). \end{cases} \quad (92)$$

By our definition of $q_{s,k}$, we know that $q_{s,k}(0) = q_{s,0}$ and $q_{s,k}$ is piecewise differentiable and satisfies (84) on the whole interval $[0, t_1]$. Then, we can reinterpret our scheme in terms of the above functions $U_k = (T_k, q_k)$, $\tilde{U}_k = (\tilde{T}_k, \tilde{q}_k)$,

$W_k = (\mathcal{T}_k, \mathcal{Q}_k)$ and $U_k^b = (T_k^b, q_k^b)$ as follows

$$\langle \partial_t \mathcal{T}_k, T_k^b \rangle + a_T(T_k, T_k^b) + b_T(\mathbf{u}_k, T_k, T_k^b) - d(\omega_k, \tilde{T}_k, T_k^b) - l_T(T_k^b)$$
$$= (\frac{1}{p}[\omega_k]^- h_{\tilde{q}_k} \varphi_r(\tilde{T}_k, \tilde{q}_{s,k}), T_k^b), \tag{93}$$

$$\langle \partial_t \mathcal{Q}_k, q_k^b - q_k \rangle + a_q(q_k, q_k^b - q_k) + b_q(\mathbf{u}_k, q_k, q_k^b - q_k) - l_q(q_k^b - q_k)$$
$$\geq (-\frac{1}{p}[\omega_k]^- h_{\tilde{q}_k} F_r(\tilde{T}_k, \tilde{q}_{s,k})), q_k^b - q_k), \tag{94}$$

$$\frac{dq_{s,k}}{dt} = -\frac{1}{p}[\omega_k]^- h_{\tilde{q}_k} F_r(\tilde{T}_k, q_{s,k}). \tag{95}$$

The two passages to the limit processes $\varepsilon \to 0+$ and $k \to 0+$ are parallel to those in the case that q_s is constant. However, we shall pay attention to the convergences of the terms involving $\varphi_r(T_\varepsilon^{m-1}, q_{s,\varepsilon}^{m-1})$, $F_r(T_\varepsilon^{m-1}, q_{s,\varepsilon}^{m-1})$ during the passage to the limit $\varepsilon \to (0+, +)$, and $\varphi_r(\tilde{T}_k, \tilde{q}_{s,k})$, $F_r(\tilde{T}_k, \tilde{q}_{s,k})$ during the passage to the limit $k \to 0+$ respectively. In both of the two limit processes, we could achieve strong convergences for the sequences $(T_\varepsilon^m)_\varepsilon$ and $(\tilde{T}_k)_k$ up to subsequences by compact Sobolev embedding and Aubin-Lions argument respectively. While for $(q_{s,\varepsilon}^m)_\varepsilon$ and $(\tilde{q}_{s,k})_k$, we could not obtain strong convergences. Fortunately, we could obtain a.e. convergences for them for $\mathbf{x} \in \mathcal{M}$ and $(t, \mathbf{x}) \in [0, t_1] \times \mathcal{M}$ respectively. These pointwise convergences, together with the Lebesgue Dominated Convergence Theorem, will enable us to pass to the limits for these terms. The a.e. convergences are guaranteed by the following lemma.

Lemma 9. *Consider the following equation*

$$\begin{cases} \frac{dq_s^j(t)}{dt} = F(T^j(t), q_s^j(t)), \\ q_s^j(0) = q_{s,0}, \end{cases} \tag{96}$$

where $F(\cdot, \cdot)$ is a real-valued bounded Lipschitz function. Suppose $T^j = T^j(x, t)$ converges to some $T = T(x, t)$ strongly in $L^2(0, l; H)$ as $j \to \infty$. Then up to subsequences, we have

$$q_s^j(x, t) \to q_s(x, t) \text{ for any } t \in [0, l] \text{ and a.e. } x \in \Omega \setminus \Omega_0, \tag{97}$$

where $q_s(x, t)$ is the solution to the initial value problem

$$\begin{cases} \frac{dq_s(t)}{dt} = F(T(t), q_s(t)), \\ q_s(0) = q_{s,0}, \end{cases} \tag{98}$$

and Ω_0 is a subset of $\mathcal{M}$ which has Lebesgue measure 0 and is independent of t.

Proof. By assumption, we have up to a subsequence that

$$T^j(x, t) \to T(x, t) \text{ a.e. } (x, t).$$

Writing the ODE on q_s^j in the integral equation form, we have

$$q_s^j(t) = q_s^j(0) + \int_0^t F(T^j(x,\tau), q_s^j(x,\tau)) \, d\tau. \tag{99}$$

By assumption, we have

$$\int_0^l \int_M |T^j(x,t) - T(x,t)|^2 \, dx dt = \int_M \left(\int_0^l |T^j(x,t) - T(x,t)|^2 \, dt \right) dx \to 0, \text{ as } j \to \infty.$$

By Fubini's theorem, we have for a.e. x that

$$\int_0^l |T^j(x,t) - T(x,t)|^2 \, dt \to 0.$$

Now take x_* such that

$$\int_0^l |T^j(x_*,t) - T(x_*,t)|^2 \, dt \to 0. \tag{100}$$

Notice that

$$\begin{cases} \frac{dq_s^j(x_*,t)}{dt} = F(T^j(x_*,t), q_s^j(x_*,t)), \\ q_s^j(x_*,0) = q_s(x_*,0). \end{cases} \tag{101}$$

As $|\frac{dq_s^j}{dt}|_{L^\infty(0,l)} \le C$, we have up to a subsequence and with x_* *fixed* that $q_s^{j_i} \to q_s$ uniformly on $[0, l]$.

Since $F(T^{j_i}(x_*,t))$ and $q_s^{j_i}(x_*,t)$ converge for a.e. $t \in [0,l]$, we can pass to the limit in the following equation

$$q_s^{j_i}(x_*,t) = q_s^j(x_*,0) + \int_0^t F(T^{j_i}(x_*,\tau), q_s^{j_i}(x_*,\tau)) \, d\tau,$$

and by the Lebesgue Dominated Convergence Theorem, we have

$$q_s(x_*,t) = q_s(x_*,0) + \int_0^t F(T(x_*,\tau), q_s(x_*,\tau)) \, d\tau. \tag{102}$$

By uniqueness of the solution to (99) and the fact that the limit solution is independent of the subsequence, we conclude that

$$q_s^j(\cdot, x_*) \to q_s(\cdot, x_*), \text{ uniformly in } t,$$

for any x_* such that (100) holds, i.e., for $x_* \in \Omega \setminus \Omega_0$ where $|\Omega_0| = 0$, Hence,

$$q_s^j(x,t) \to q_s(x,t) \text{ for any } t \text{ and for a.e. } x \in \Omega \setminus \Omega_0. \tag{103}$$

Remark 3. A particular case of the above lemma is that $T^j = T^j(x)$, i.e., T^j has no dependence on the time variable $t \in [0, l]$. In this case we can regard T^j as a constant function of $t \in [0, l]$ and consequently, we still have $T^j \to T$ in $L^2(0, l; H)$.

In summary, when q_s depends on time, we have the following theorem.

Theorem 2. *Given $T_0, q_0, q_{s,0} \in H$ with $0 \le q_0 \le 1$, and $0 < q_{s,0} < 1$ a.e. in $\mathcal{M}$, the scheme (78)–(81) contains a subsequence which converges to a solution of the system (1)–(9).*

Acknowledgments. The research was partially supported by the National Science Foundation under the grants NSF-DMS-1206438, NSF-DMS-1510249 and by the Research Fund of Indiana University.

References

1. Bousquet, A., Coti Zelati, M., Temam, R.: Phase transition models in atmospheric dynamics. Milan J. Math. **82**, 99–128 (2014)
2. Brézis, H.: Problèmes unilatéraux. J. Math. Pures. Appl. **9**, 1–168 (1972)
3. Cao, C., Titi, E.S.: Global well-posedness of the three dimensional viscous primitive equations of large scale ocean and atmosphere dynamics. Ann. Math. (2) **166**, 245–267 (2007)
4. Coti Zelati, M., Frémond, M., Temam, R., Tribbia, J.: The equations of the atmosphere with humidity and saturation: uniqueness and physical bounds. Physica D **264**, 49–65 (2013)
5. Coti Zelati, M., Huang, A., Kukavica, I., Temam, R., Ziane, M.: The primitive equations of the atmosphere in presence of vapor saturation. Nonlinearity (2015, in press). http://dx.doi.org/10.1088/0951-7715/28/3/625
6. Diaz, J.I.: Mathematical analysis of some diffusive energy balance models in climatology. In: Díaz, J.I., Lions, J.L. (Eds.) Mathematics, Climate and Environment. Research Notes in Applied Mathematics, vol. 27, Masson, Paris, pp. 28–56 (1993)
7. Diaz, J.I., Tello, L.: On a nonlinear parabolic problem on a Riemannian manifold without boundary arising in climatology. Collect. Math. L (Fascicle 1) **50**, 19–51 (1999)
8. Duvaut, G., Lions, J.-L.: Inequalities in Mechanics and Physics. Springer, Berlin (1976). (Translated from the French by C.W. John, 397 p.)
9. Ekeland, I., Temam, R.: Convex Analysis and Variational Problems. Classics in Applied Mathematics. North-Holland Publishing Company, Amsterdam (1987). (+416 p.)
10. Feireisl, E., Norbury, J.: Some existence, uniqueness and nonuniqueness theorems for solutions of parabolic equations with discontinuous nonlinearities. Proc. R. Soc. Edinb. Sect. A **119**, 1–17 (1991)
11. Frémond, M.: Phase Change in Mechanics. Springer, Heidelberg (2012)
12. Gianni, R., Hulshof, J.: The semilinear heat equation with a Heaviside source term. Eur. J. Appl. Math. **3**, 369–379 (1992)
13. Gill, A.E.: Atmosphere-Ocean Dynamics. International Geophysics Series, vol. 30. Academic Press, San Diego (1982)
14. Grabowski, W.W., Smolarkiewicz, P.K.: A multiscale anelastic model for meteorological research. Mon. Weather Rev. **130**, 939–956 (2002)
15. Guo, B., Huang, D.: Existence of weak solutions and trajectory attractors for the moist atmospheric equations in geophysics. J. Math. Phys. **47**, 083508 (2006)
16. Guo, B., Huang, D.: Existence of the universal attractor for the 3-D viscous primitive equations of large-scale moist atmosphere. J. Differ. Equ. **251**, 457–491 (2011)

17. Haltiner, G.J.: Numerical Weather Prediction. Wiley, New York (1971)
18. Haltiner, G.J., Williams, R.T.: Numerical Prediction and Dynamic Meteorology. Wiley, New York (1980)
19. Kinderlehrer, D., Stampacchia, G.: An Introduction to Variational Inequalities and Their Applications. Academic Press Inc., Harcourt Brace Jovanovich Publishers, New York (1980)
20. Lions, J.L.: Quelques méthodes de résolution des problèmes aux limites non linéaires, Dunod, Paris (1969, reprinted in 2002)
21. Lions, J., Temam, R., Wang, S.: New formulations of the primitive equations of atmosphere and applications. Nonlinearity 5(2), 237–288 (1992)
22. Lions, J., Temam, R., Wang, S.: Models for the coupled atmosphere and ocean (CAO1). Comput. Mech. Adv. 1(1), 5–119 (1993)
23. Pedlosky, J.: Geophysical Fluid Dynamics. Springer, New York (1987)
24. Petcu, M., Temam, R., Ziane, M.: Some mathematical problems in geophysical fluid dynamics, computational methods for the atmosphere and the oceans. In: Handbook of Numerical Analysis, Special vol. XIV. Elsevier, Amsterdam (2008)
25. Rockafellar, R.T.: Convex Analysis. Princeton University Press, Princeton (1997)
26. Rogers, R.R., Yau, M.K.: A Short Course in Cloud Physics. Pergamon Press, Oxford (1989)
27. Temam, R.: Navier-Stokes Equations and Nonlinear Functional Analysis. SIAM, Philadelphia (1995)
28. Temam, R.: Navier-Stokes Equations: Theory and Numerical Analysis. AMS Chelsea Publishing, Providence (2000). (+408 p.)
29. Temam, R., Tribbia, J.: Uniqueness of solutions for moist advection problems. Q. J. R. Meteorol. Soc. 140, 1315–1318 (2014)
30. Temam, R., Tribbia, J.: The equations of the moist advection: a unilateral problem. Q. J. R. Meteorol. Soc. 142, 1–4 (2015). doi:10.1002/qj.2638
31. Temam, R., Wang, X.: Numerical approximation of a variational inequality related to the humid atmosphere. SIAM J. Numer. Anal. 55, 217–239 (2017)
32. Temam, R., Wu, K.: Formulation of the equations of the humid atmosphere in the context of variational inequalities. J. Funct. Anal. 269, 2187–2221 (2015)
33. Tone, F., Wirosoetisno, D.: On the long-time stability of the implicit Euler scheme for the two-dimensional Navier-Stokes equations. SIAM J. Numer. Anal. 44, 29–40 (2006)

Parameter Estimation in a Size-Structured Population Model with Distributed States-at-Birth

Azmy S. Ackleh[1(✉)], Xinyu Li[1], and Baoling Ma[2]

[1] Department of Mathematics, University of Louisiana at Lafayette,
Lafayette, LA 70504, USA
{ackleh,xxl0154}@louisiana.edu
[2] Department of Mathematics,
Millersville University of Pennsylvania, Millersville, PA 17551, USA
baoling.ma@millersville.edu

Abstract. A least-squares method is developed for estimating parameters in a size-structured population model with distributed states-at-birth from field data. First and second order finite difference schemes for approximating the nonlinear-nonlocal partial differential equation model are utilized in the least-squares problem. Convergence results for the computed parameters are established. Numerical results demonstrating the efficiency of the technique are provided.

1 Introduction

It is often the case that direct observations of vital rates of individual organisms are not accessible and our knowledge of the vital rates is incomplete. Therefore, the inverse problem approach often plays an important role in deducing such information at the individual level from observation at the population level. In recent years substantial attention has been given to inverse problems governed by age/stage/size structured population models [1–3, 6–9, 12, 13]. Methodologies applied to solve such inverse problems include the least-squares approach [1, 3, 9] and the fixed point iterative technique [17]. The least-squares approach has been often used in inverse problems governed by size-structured models. For example, in [8, 9] the authors used least-squares method to estimate the growth rate distribution in a linear size-structured population model. A similar technique was applied to a semi-linear size-structured model in [14] where the mortality rate depends on the total population due to competition between individuals. Furthermore, such least-squares methodology has been applied for estimating parameters in general conservation laws [13]. Therein the author utilizes monotone finite-difference schemes to numerically solve the conservation law and present numerical results for estimation the flux function from numerically generated data. And in [3], the authors solved an inverse problem governed by structured juvenile-adult model. Therein, the least-squares approach was used

© IFIP International Federation for Information Processing 2016
Published by Springer International Publishing AG 2016. All Rights Reserved
L. Bociu et al. (Eds.): CSMO 2015, IFIP AICT 494, pp. 43–57, 2016.
DOI: 10.1007/978-3-319-55795-3_3

to estimate growth, mortality and reproduction rates in the adult stage from field data on an urban green tree frog population. The estimated parameters were then utilized to understand the long-term dynamics of this green tree frog population.

In this paper we consider the following nonlinear Gurtin-MacCamy type model with distributed states-at-birth:

$$
\begin{aligned}
\frac{\partial}{\partial t}p(x,t;\theta) + \frac{\partial}{\partial x}\left(g(x,t,Q(t;\theta))p(x,t;\theta)\right) &= -\mu(x,t,Q(t;\theta))p(x,t;\theta) \\
+ \int_{x_{min}}^{x_{max}} \beta(x,y,t,Q(t;\theta))p(y,t;\theta)dy, &\ (x,t) \in (x_{min},x_{max}) \times (0,T), \\
g(x_{min},t,Q(t;\theta))p(x_{min},t;\theta) = 0, &\qquad t \in [0,T], \\
p(x,0;\theta) = p^0(x), &\qquad x \in [x_{min},x_{max}].
\end{aligned}
$$
$$(1)$$

Here, $\theta = (g,\mu,\beta)$ is the vector of parameters to be identified. The function $p(x,t;\theta)$ is the parameter-dependent density of individuals of size x at time t. Therefore, $Q(t;\theta) = \int_{x_{min}}^{x_{max}} p(x,t;\theta)dx$ provides the total population at time t which depends on the vector of parameters $\theta = (g,\mu,\beta)$. The functions g and μ represent the individual growth and mortality rates, respectively. It is assumed that individuals may be recruited into the population at different sizes with $\beta(x,y,t,Q)$ representing the rate at which an individual of size y gives birth to an individual of size x. Henceforth, we will call the model (1) Distributed Size Structured Model and abbreviate it as DSSM.

The main goal of this paper is to develop a least-squares approach for estimating the parameter θ from population data and to provide convergence results for the parameter estimates. The paper is organized as follows. In Sect. 2, we set up a least-squares problem and present finite difference schemes for computing an approximate solution to this least-squares problem. In Sect. 3 we provide convergence results for the computed parameters. In Sect. 4, numerical examples showing the performance of the least-squares technique and an application to a set of field data on green tree frogs are presented.

2   The Least-Squares Problem and Approximation Schemes

Let $\mathbb{D}_1 = [x_{min},x_{max}] \times [0,T] \times [0,\infty)$ and $\mathbb{D}_2 = [x_{min},x_{max}] \times [x_{min},x_{max}] \times [0,T] \times [0,\infty)$ throughout the discussion. Let $\mathbf{B} = C_b^1(\mathbb{D}_1) \times C_b(\mathbb{D}_1) \times C_b(\mathbb{D}_2)$, where $C_b(\Omega)$ denotes the Banach space of bounded continuous functions on Ω endowed with the usual supremum norm and $C_b^1(\Omega)$ is the Banach space of bounded continuous functions with bounded continuous derivatives on Ω and endowed with the usual supremum norm. Clearly, $\mathbf{B}$ is a Banach space when endowed with the natural product topology. Let c be a sufficiently large positive constant and assume that the admissible parameter space Θ is any compact subset of $\mathbf{B}$ (endowed with the topology of $\mathbf{B}$) such that every $\theta = (g,\mu,\beta) \in \Theta$ satisfies (H1)–(H4) below.

(H1) $g \in C_b^1(\mathbb{D}_1)$ with $g_x(x, t, Q)$ and $g_Q(x, t, Q)$ being Lipschitz continuous in x with Lipschitz constant c, uniformly in t and Q. Moreover, $0 < g(x, t, Q) \le c$ for $x \in [x_{min}, x_{max})$ and $g(x_{max}, t, Q) = 0$.

(H2) $\mu \in C_b(\mathbb{D}_1)$ is Lipschitz continuous in x and Q with Lipschitz constant c, uniformly in t. Also, $0 \le \mu(x, t, Q) \le c$.

(H3) $\beta \in C_b(\mathbb{D}_2)$ is Lipschitz continuous in Q with Lipschitz constant c, uniformly in x, y and t. Also, $0 \le \beta(x, y, t, Q) \le c$ and for every partition $\{x_i\}_{i=1}^N$ of $[x_{min}, x_{max}]$, we have

$$\sup_{(y,t,Q) \in [x_{min}, x_{max}] \times [0,T] \times [0,\infty)} \sum_{i=1}^N |\beta(x_i, y, t, Q) - \beta(x_{i-1}, y, t, Q)| \le c.$$

(H4) $p^0 \in BV([x_{min}, x_{max}])$, the space of functions with bounded total variation on $[x_{min}, x_{max}]$, and $p^0(x) \ge 0$.

We now define a weak solution to the model (1).

Definition 21. *Given $\theta \in \Theta$, by a weak solution to problem (1) we mean a function $p(\cdot, \cdot; \theta) \in L^\infty([x_{min}, x_{max}] \times [0, T])$, $p(\cdot, t; \theta) \in BV([x_{min}, x_{max}])$ for $t \in [0, T]$, and satisfies*

$$\int_{x_{min}}^{x_{max}} p(x, t; \theta)\phi(x, t)dx - \int_{x_{min}}^{x_{max}} p^0(x)\phi(x, 0)dx$$
$$= \int_0^t \int_{x_{min}}^{x_{max}} p(x, \tau; \theta)[\phi_\tau(x, \tau) + g(x, \tau, Q(\tau; \theta))\phi_x(x, \tau) - \mu(x, \tau, Q(\tau; \theta))\phi(x, \tau)]dx d\tau$$
$$+ \int_0^t \int_{x_{min}}^{x_{max}} \int_{x_{min}}^{x_{max}} \beta(x, y, \tau, Q(\tau; \theta))p(y, \tau; \theta)\phi(x, \tau)dy dx d\tau.$$

$$(2)$$

for every test function $\phi \in C^1((x_{min}, x_{max}) \times (0, T))$ and $t \in [0, T]$.

We are interested in the following least-squares problem: given data X_s which corresponds to the number of individuals at time t_s, $s = 1, 2, \cdots, S$, find a parameter $\theta = (g, \mu, \beta) \in \Theta$ such that the weighted least-squares cost functional $F(\theta) = \sum_{s=1}^S |W(Q(t_s; \theta)) - W(X_s)|^2$ is minimized over the admissible parameter space Θ, i.e., find θ^* such that

$$\theta^* = \arg \min_{\theta \in \Theta} F(\theta) = \arg \min_{\theta \in \Theta} \sum_{s=1}^S |W(Q(t_s; \theta)) - W(X_s)|^2, \qquad (3)$$

where $W \in C([0, \infty))$ is a weight function.

In order to numerically approximate the solution to the minimization problem (3), we first need to approximate the solution of model (1). To this end, we utilize similar finite-difference approximation schemes as those developed in [5]. Suppose that the intervals $[x_{min}, x_{max}]$ and $[0, T]$ are divided into N and L subintervals, respectively. The following notations will be used throughout the pater: $\Delta x = (x_{max} - x_{min})/N$ and $\Delta t = T/L$. The discrete mesh points are given by $x_i = x_{min} + i\Delta x$, $t_k = k\Delta t$ for $i = 0, 1, \cdots, N$, $k = 0, 1, \cdots, L$. For ease of notation, we take a uniform mesh with constant sizes Δx and Δt. More general nonuniform meshes can be similarly considered. We shall denote by $p_i^k(\theta)$ and $Q^k(\theta)$ the finite difference approximation of $p(x_i, t_k; \theta)$ and $Q(t_k; \theta)$, respectively. For convenience we will also use the notation p_i^k and Q^k without explicitly stating their dependence on θ. We also let $g_i^k = g(x_i, t_k, Q^k)$, $\mu_i^k = \mu(x_i, t_k, Q^k)$, $\beta_{i,j}^k = \beta(x_i, y_j, t_k, Q^k)$. Here, $Q^k = \sum\limits_{i=1}^{N} p_i^k \Delta x$.

We define the ℓ^1, ℓ^∞ norms and TV (total variation) seminorm of the grid functions p^k by

$$\|p^k\|_1 = \sum_{i=1}^{N} |p_i^k| \Delta x, \quad \|p^k\|_\infty = \max_{0 \le i \le N} |p_i^k|, \quad TV(p^k) = \sum_{i=0}^{N-1} |p_{i+1}^k - p_i^k|,$$

and the finite difference operators by

$$\Delta_+ p_i^k = p_{i+1}^k - p_i^k, \; 0 \le i \le N-1, \qquad \Delta_- p_i^k = p_i^k - p_{i-1}^k, \; 1 \le i \le N.$$

Throughout the discussion, we impose the following CFL condition concerning Δx and Δt:

(H5) $\frac{\Delta t}{\Delta x} + \Delta t \le \frac{1}{c}$.

We discretize model (1) using the following first order explicit upwind finite difference scheme (FOEU):

$$\frac{p_i^{k+1} - p_i^k}{\Delta t} + \frac{g_i^k p_i^k - g_{i-1}^k p_{i-1}^k}{\Delta x} = -\mu_i^k p_i^k + \sum_{j=1}^{N} \beta_{i,j}^k p_j^k \Delta x, 1 \le i \le N, \quad 0 \le k \le M-1,$$

$$g_0^k p_0^k = 0, \qquad\qquad\qquad\qquad\qquad 0 \le k \le M,$$
$$p_i^0 = p^0(x_i), \qquad\qquad\qquad\qquad\qquad 0 \le i \le N.$$

$$(4)$$

We could write the first equation in (4) equivalently as

$$p_i^{k+1} = \frac{\Delta t}{\Delta x} g_{i-1}^k p_{i-1}^k + \left(1 - \frac{\Delta t}{\Delta x} g_i^k - \mu_i^k \Delta t\right) p_i^k + \left(\sum_{j=1}^{N} \beta_{i,j}^k p_j^k \Delta x\right) \Delta t,$$

$$1 \le i \le N, \quad 0 \le k \le M-1. \qquad (5)$$

It is easy to check that under assumptions (H1)–(H4) FOEU scheme converges to a unique weak solution of system (1) as proved in [5]. The above approximation can be extended to a family of functions $\{p_{\Delta x, \Delta t}(x, t; \theta)\}$ defined by $p_{\Delta x, \Delta t}(x, t; \theta) = p_i^k(\theta)$ for $(x, t) \in [x_{i-1}, x_i) \times [t_{k-1}, t_k)$, $i = 1, 2, \cdots, N, k = 1, 2, \cdots, M$.

Since the parameter set is infinite dimensional, a finite-dimensional approximation of the parameter space is necessary for computing minimizers. Thus, we consider the following finite-dimensional approximations of (3): Let

$Q_{\Delta x, \Delta t}(t; \theta) = \int_{x_{min}}^{x_{max}} p_{\Delta x, \Delta t}(x, t; \theta) dx$ denote the finite difference approximation of the total population and consider the finite dimensional minimization
problem

$$\arg \min_{\theta \in \Theta_m} F_{\Delta x, \Delta t}(\theta) = \arg \min_{\theta \in \Theta_m} \sum_{s=1}^{S} |W(Q_{\Delta x, \Delta t}(t_s; \theta)) - W(X_s)|^2. \qquad (6)$$

Here, $\Theta_m \subseteq \Theta$ is a sequence of compact finite-dimensional subsets that approximate the parameter space Θ, i.e., for each $\theta \in \Theta$, there exist a sequence of
$\theta_m \in \Theta_m$ such that $\theta_m \to \theta$ in the topology of $\mathbf{B}$ as $m \to \infty$.

Remark 22. *If the compact parameter space Θ is chosen to be finite dimensional, then the approximation space sequence can be taken to be $\Theta_m = \Theta$.*

Since the FOEU (4) scheme is first order it would require a large number of grid
points to achieve high accuracy. Thus, we next propose a second order minmod
finite difference scheme based on MUSCL schemes [5,15,18] to approximate the
solutions of the DSSM model (1) in the least-squares problem. We begin by using
the following second order approximations for the integrals:

$$Q^k = \sum_{i=0}^{N} {}^{\bigstar} p_i^k \Delta x = \frac{1}{2} p_0^k \Delta x + \sum_{i=1}^{N-1} p_i^k \Delta x + \frac{1}{2} p_N^k \Delta x$$

and

$$\sum_{j=0}^{N} {}^{\bigstar} \beta_{i,j}^k p_j^k \Delta x = \frac{1}{2} \beta_{i,0}^k p_0^k \Delta x + \sum_{j=1}^{N-1} \beta_{i,j}^k p_j^k \Delta x + \frac{1}{2} \beta_{i,N}^k p_N^k \Delta x.$$

Then we approximate the model (1) by

$$\frac{p_i^{k+1} - p_i^k}{\Delta t} + \frac{\hat{f}_{i+\frac{1}{2}}^k - \hat{f}_{i-\frac{1}{2}}^k}{\Delta x} = -\mu_i^k p_i^k + \sum_{j=0}^{N} {}^{\bigstar} \beta_{i,j}^k p_j^k \Delta x, \; i = 1, 2, \cdots, N, \quad k = 0, 1, \cdots, L-1,$$

$$g_0^k p_0^k = 0, \qquad\qquad\qquad\qquad k = 0, 1, \cdots, L,$$
$$p_i^0 = p^0(x_i), \qquad\qquad\qquad\qquad i = 0, 1, \cdots, N.$$

$$(7)$$

Here, the limiter is defined as

$$\hat{f}_{i+\frac{1}{2}}^k = \begin{cases} g_i^k p_i^k + \frac{1}{2}(g_{i+1}^k - g_i^k) p_i^k + \frac{1}{2} g_i^k mm(\Delta_+ p_i^k, \Delta_- p_i^k), & i = 2, \cdots, N-2, \\ g_i^k p_i^k, & i = 0, 1, N-1, N, \end{cases}$$

$$(8)$$

where $mm(a, b) = \frac{sign(a) + sign(b)}{2} \min(|a|, |b|)$.

3 Convergence Theory for the Parameter Estimation Problem Using FOEU

The results in this section pertain to the case when (1) is approximated by the
FOEU scheme (4). Our future efforts will focus on extending these theoretical

results to case when the model (1) is approximated by the higher order SOEM scheme (7). We establish the convergence results for the parameter estimation problem using an approach based on the abstract theory in [8]. To this end, we have the following theorem:

Theorem 31. *Let $\theta^r = (g^r, \mu^r, \beta^r) \in \Theta$. Suppose that $\theta^r \to \theta$ in Θ and $\Delta x_r, \Delta t_r \to 0$ as $r \to \infty$. Let $p_{\Delta x_r, \Delta t_r}(x, t; \theta^r)$ denote the solution of the finite difference scheme with parameter θ^r and initial condition p^0, and let $p(x, t; \theta)$ be the unique weak solution of the problem with initial condition $p^0(x)$ and parameter θ. Then $p_{\Delta x_r, \Delta t_r}(\cdot, t; \theta^r) \to p(\cdot, t; \theta)$ in $L^1(x_{min}, x_{max})$, uniformly in $t \in [0, T]$.*

Proof. Define $p_i^{k,r} = p_i^k(\theta^r)$. From the fact that Θ is compact and using similar arguments as in [5], there exist positive constants c_1, c_2, c_3 and c_4 such that $\|p^{k,r}\|_1 \le c_1$, $\|p^{k,r}\|_\infty \le c_2$, $TV(p^{k,r}) \le c_3$ and

$$\sum_{i=1}^{N} \left| \frac{p_i^{m,r} - p_i^{n,r}}{\Delta t_r} \right| \Delta x_r \le c_4(m - n),$$

where $m > n$. Thus, there exist $\hat{p} \in BV([x_{min}, x_{max}])$ such that $p_{\Delta x_r, \Delta t_r}(\cdot, t; \theta^r) \to \hat{p}(\cdot, t)$ in $L^1(x_{min}, x_{max})$ uniformly in t. Hence, from the uniqueness of bounded variation weak solutions which can be established using similar techniques as in [5], we just need to show that $\hat{p}(x, t)$ is the weak solution corresponding to the parameter θ.

In order to prove this, let $\phi \in C^1([x_{min}, x_{max}] \times [0, T])$ and denote the value of $\phi(x_i, t_k)$ by ϕ_i^k. Multiplying Eq. (5) by ϕ_i^{k+1} and rearranging some terms we have

$$p_i^{k+1,r}\phi_i^{k+1} - p_i^{k,r}\phi_i^k = p_i^{k,r}(\phi_i^{k+1} - \phi_i^k) + \frac{\Delta t}{\Delta x}[g_{i-1}^{k,r}p_{i-1}^{k,r}(\phi_i^{k+1} - \phi_{i-1}^{k+1})$$
$$-(g_{i-1}^{k,r}p_{i-1}^{k,r}\phi_{i-1}^{k+1} - g_i^{k,r}p_i^{k,r}\phi_i^{k+1})] \qquad (9)$$
$$-\mu_i^{k,r}p_i^{k,r}\phi_i^{k+1}\Delta t + \sum_{j=1}^{N}\beta_{i,j}^{k,r}p_j^{k,r}\phi_i^{k+1}\Delta x \Delta t.$$

Multiplying the above equation by Δx, summing over $i = 1, 2, \cdots, N$, $k = 0, 1, \cdots, M - 1$, and applying $p_0^k = 0$ and $g_N^k = 0$ we obtain,

$$\sum_{i=1}^{N}\left(p_i^{L,r}\phi_i^L - p_i^{0,r}\phi_i^0\right)\Delta x = \sum_{k=0}^{M-1}\sum_{i=1}^{N}p_i^{k,r}\frac{\phi_i^{k+1} - \phi_i^k}{\Delta t}\Delta x \Delta t$$
$$+ \sum_{k=0}^{M-1}\sum_{i=0}^{N-1}g_{i-1}^{k,r}p_{i-1}^{k,r}\frac{\phi_i^{k+1} - \phi_{i-1}^{k+1}}{\Delta x}\Delta x \Delta t$$
$$- \sum_{k=0}^{M-1}\sum_{i=1}^{N}\mu_i^{k,r}p_i^{k,r}\phi_i^{k+1}\Delta x \Delta t \qquad (10)$$
$$+ \sum_{k=1}^{M-1}\sum_{i=1}^{N}\sum_{j=1}^{N}\beta_{i,j}^{k,r}p_j^{k,r}\phi_i^{k+1}\Delta x \Delta t \Delta x.$$

Using the fact that $\theta^r \to \theta$ as $r \to \infty$ in Θ, passing to the limit in (10) we find that $\hat{p}(x, t)$ is the weak solution corresponding to the parameter θ.

Since W is a continuous on $[0, \infty)$, as an immediate consequence of Theorem 31, we obtain the following.

Corollary 32. *Let $p_{\Delta x_r, \Delta t_r}(x, t; \theta^r)$ denote the numerical solution of the finite difference scheme with parameter $\theta^r \to \theta$ in Θ and $\Delta x_r, \Delta t_r \to 0$ as $r \to \infty$. Then*

$$F_{\Delta x_r, \Delta t_r}(\theta^r) \to F(\theta), \quad as \ r \to \infty.$$

In the next theorem, we establish the continuity of the approximate cost functional in the parameter $\theta \in \Theta$ (a compact set), so that the computational problem of finding an approximate minimizer has a solution.

Theorem 33. *Let Δx and Δt be fixed. For each $\theta \in \Theta$, let $p_{\Delta x, \Delta t}(x, t; \theta)$ denote the solution of the finite difference scheme and $\theta^r \to \theta$ as $r \to \infty$ in Θ; then $p_{\Delta x, \Delta t}(\cdot, t; \theta^r) \to p_{\Delta x, \Delta t}(\cdot, t; \theta)$ as $r \to \infty$ in $L^1(x_{min}, x_{max})$ uniformly in $t \in [0, T]$.*

Proof. Fix Δx and Δt. Define p_i^{k, θ^r} and $p_i^{k, \theta}$ to be the solution of the finite difference scheme with parameter θ^r and θ, respectively. Let $v_i^{k, \theta} = p_i^{k, \theta^r} - p_i^{k, \theta}$. Then $v_i^{k, \theta}$ satisfy the following

$$
\begin{aligned}
v_i^{k+1, \theta} &= \frac{\Delta t}{\Delta x}\left(g_{i-1}^{k, \theta^r} p_{i-1}^{k, \theta^r} - g_{i-1}^{k, \theta} p_{i-1}^{k, \theta}\right) + \left(p_i^{k, \theta^r} - p_i^{k, \theta}\right) - \frac{\Delta t}{\Delta x}\left(g_i^{k, \theta^r} p_i^{k, \theta^r} - g_i^{k, \theta} p_i^{k, \theta}\right) \\
&\quad - \Delta t\left(\mu_i^{k, \theta^r} p_i^{k, \theta^r} - \mu_i^{k, \theta} p_i^{k, \theta}\right) + \sum_{j=1}^{N}\left(\beta_{i,j}^{k, \theta^r} p_j^{k, \theta^r} - \beta_{i,j}^{k, \theta} p_j^{k, \theta}\right) \Delta x \Delta t, \\
&\qquad\qquad\qquad\qquad\qquad 1 \le i \le N, \ 0 \le k \le M-1, \\
v_0^{k+1, \theta} &= p_0^{k+1, \theta^r} - p_0^{k+1, \theta} = 0, \qquad\qquad 0 \le k \le M-1.
\end{aligned}
$$
$$(11)$$

Here, $Q^{k, \theta^r} = \sum_{i=1}^{N} p_i^{k, \theta^r} \Delta x$, $g_i^{k, \theta^r} = g^{\theta^r}(x_i, t_k, Q^{k, \theta^r})$ and similar notations are used for μ_i^{k, θ^r} and $\beta_{i,j}^{k, \theta^r}$. Using the first equation in (11) and assumption (H5) we obtain

$$
\begin{aligned}
\sum_{i=1}^{N} |v_i^{k+1, \theta}| \Delta x &\le \sum_{i=1}^{N} \left[1 - \Delta t \mu_i^{k, \theta^r} + \left(\sum_{j=1}^{N} \beta_{i,j}^{k, \theta^r} \Delta x\right)\Delta t\right] |v_i^{k, \theta}| \Delta x - \Delta t \sum_{i=1}^{N}\left(g_i^{k, \theta^r} |v_i^{k, \theta}|\right. \\
&\quad \left. - g_{i-1}^{k, \theta^r} |v_{i-1}^{k, \theta}|\right) + \Delta t \sum_{i=1}^{N} |\left(g_{i-1}^{k, \theta^r} - g_{i-1}^{k, \theta}\right) p_{i-1}^{k, \theta} - \left(g_i^{k, \theta^r} - g_i^{k, \theta}\right) p_i^{k, \theta}| \\
&\quad + \Delta t \sum_{i=1}^{N} |\mu_i^{k, \theta^r} - \mu_i^{k, \theta}| p_i^{k, \theta} \Delta x + \Delta t \sum_{i=1}^{N}\sum_{j=1}^{N} |\beta_{i,j}^{k, \theta^r} - \beta_{i,j}^{k, \theta}| p_j^{k, \theta} \Delta x \Delta x.
\end{aligned}
$$
$$(12)$$

By assumption (H1), and (11)

$$\sum_{i=1}^{N} \left(g_i^{k, \theta^r} |v_i^{k, \theta}| - g_{i-1}^{k, \theta^r} |v_{i-1}^{k, \theta}|\right) = \left(g_N^{k, \theta^r} |v_N^{k, \theta}| - g_0^{k, \theta^r} |v_0^{k, \theta}|\right) = 0. \quad (13)$$

By assumptions (H2) and (H3) we have

$$\sum_{i=1}^{N} \left[1 - \mu_i^{k, \theta^r} \Delta t + \left(\sum_{j=1}^{N} \beta_{i,j}^{k, \theta^r} \Delta x\right)\Delta t\right] |u_i^{k, \theta}| \Delta x \le (1 + c(x_{max} - x_{min})\Delta t) \|v^{k, \theta}\|_1.$$
$$(14)$$

By assumption (H1),

$$\sum_{i=1}^{N} \left| \left(g_{i-1}^{k,\theta^r} - g_{i-1}^{k,\theta} \right) p_{i-1}^{k,\theta} - \left(g_i^{k,\theta^r} - g_i^{k,\theta} \right) p_i^{k,\theta} \right|$$
$$\leq \sup_i |g_{i-1}^{k,\theta^r} - g_{i-1}^{k,\theta}| \sum_{i=1}^{N} |p_i^{k,\theta} - p_{i-1}^{k,\theta}| + \sum_{i=1}^{N} \left| \frac{\left(g_i^{k,\theta^r} - g_{i-1}^{k,\theta^r} \right) - \left(g_i^{k,\theta} - g_{i-1}^{k,\theta} \right)}{\Delta x} \right| p_i^{k,\theta} \Delta x.$$

(15)

$$\left| \frac{\left(g_i^{k,\theta^r} - g_{i-1}^{k,\theta^r} \right) - \left(g_i^{k,\theta} - g_{i-1}^{k,\theta} \right)}{\Delta x} \right| \leq \int_0^1 \left| g_x^{\theta^r} \left(\tau x_{i-1} + (1-\tau)x_i, t_k, Q^{k,\theta^r} \right) \right.$$
$$\left. - g_x^{\theta^r} \left(\tau x_{i-1} + (1-\tau)x_i, t_k, Q^{k,\theta} \right) \right| d\tau$$
$$+ \int_0^1 \left| g_x^{\theta^r} \left(\tau x_{i-1} + (1-\tau)x_i, t_k, Q^{k,\theta} \right) \right.$$
$$\left. - g_x^{\theta} (\tau x_{i-1} + (1-\tau)x_i, t_k, Q^{k,\theta}) \right| d\tau.$$

(16)

Assumption (H1), (15) and (16) yield

$$\sum_{i=1}^{N} \left| \left(g_{i-1}^{k,\theta^r} - g_{i-1}^{k,\theta} \right) p_{i-1}^{k,\theta} - \left(g_i^{k,\theta^r} - g_i^{k,\theta} \right) p_i^{k,\theta} \right| \leq \sup_i |g_{i-1}^{k,\theta^r} - g_{i-1}^{k,\theta}| \sum_{i=1}^{N} |p_i^{k,\theta} - p_{i-1}^{k,\theta}|$$
$$+ \sum_{i=1}^{N} \left[\int_0^1 \left| g_x^{\theta^r} (\tau x_{i-1} + (1-\tau)x_i, t_k, Q^{k,\theta^r}) \right. \right.$$
$$\left. - g_x^{\theta^r} (\tau x_{i-1} + (1-\tau)x_i, t_k, Q^{k,\theta}) \right| d\tau$$
$$+ \int_0^1 \left| g_x^{\theta^r} (\tau x_{i-1} + (1-\tau)x_i, t_k, Q^{k,\theta}) \right.$$
$$\left. - g_x^{\theta} (\tau x_{i-1} + (1-\tau)x_i, t_k, Q^{k,\theta}) \right| d\tau \right] p_i^{k,\theta} \Delta x.$$

(17)

Note that

$$|Q^{k,\theta^r} - Q^{k,\theta}| = \left| \sum_{i=1}^{N} (p_j^{k,\theta^r} - p_j^{k,\theta}) \Delta x \right| \leq \sum_{i=1}^{N} |v_i^{k,\theta}| \Delta x = \|v^{k,\theta}\|_1.$$

(18)

By the assumption (H1) and the equation above, (17) yields

$$\sum_{i=1}^{N} \left| \left(g_{i-1}^{k,\theta^r} - g_{i-1}^{k,\theta} \right) p_{i-1}^{k,\theta} - \left(g_i^{k,\theta^r} - g_i^{k,\theta} \right) p_i^{k,\theta} \right|$$
$$\leq \sup_i |g_{i-1}^{k,\theta^r} - g_{i-1}^{k,\theta}| TV(p^{k,\theta}) + \left(c \|v^{k,\theta}\|_1 + \sup_i \int_0^1 \left| g_x^{\theta^r} (\bar{x}_i, t_k, Q^{k,\theta}) \right. \right.$$
$$\left. \left. - g_x^{\theta} (\bar{x}_i, t_k, Q^{k,\theta}) \right| dx \right) \|p^{k,\theta}\|_1,$$

(19)

where $\bar{x}_i = \tau x_{i-1} + (1-\tau)x_i$. By assumption (H2)

$$\sum_{i=1}^{N} |\mu_i^{k,\theta^r} - \mu_i^{k,\theta}| p_i^{k,\theta} \Delta x \leq \sup_i |\mu_i^{k,\theta^r} - \mu_i^{k,\theta}| \|p^{k,\theta}\|_1.$$

(20)

And from assumption (H3) we obtain

$$\sum_{i=1}^{N} \sum_{j=1}^{N} |\beta_{i,j}^{k,\theta^r} - \beta_{i,j}^{k,\theta}| p_j^{k,\theta} \Delta x \Delta x \leq \sup_{i,j} |\beta_{i,j}^{k,\theta^r} - \beta_{i,j}^{k,\theta}| \|p^{k,\theta}\|_1.$$

(21)

Using (13)–(21) we arrive at

$$
\begin{aligned}
\|v^{k+1,\theta}\|_1 \leq & (1 + c(x_{\max} - x_{\min})\Delta t)\|v^{k,\theta}\|_1 \\
& + \Delta t \Big[\sup_i |\mu_i^{k,\theta^r} - \mu_i^{k,\theta}| \|p^{k,\theta}\|_1 + \sup_{i,j} |\beta_{i,j}^{k,\theta^r} - \beta_{i,j}^{k,\theta}| \|p^{k,\theta}\|_1 \\
& + \sup_i |g_{i-1}^{k,\theta^r} - g_{i-1}^{k,\theta}| TV(p^{k,\theta}) + \Big(c\|v^{k,\theta}\|_1 + \sup_i \int_0^1 \big| g_x^{\theta^r}(\bar{x}_i, t_k, Q^{k,\theta}) \\
& - g_x^{\theta}(\bar{x}_i, t_k, Q^{k,\theta}) \big| \, dx \Big) \|p^{k,\theta}\|_1 \Big] .
\end{aligned}
$$
(22)

Note that

$$
\begin{aligned}
|\mu_i^{k,\theta^r} - \mu_i^{k,\theta}| &\leq |\mu^{\theta^r}(x_i, t_k, Q^{k,\theta^r}) - \mu^{\theta^r}(x_i, t_k, Q^{k,\theta})| + |\mu^{\theta^r}(x_i, t_k, Q^{k,\theta}) \\
& \quad - \mu^{\theta}(x_i, t_k, Q^{k,\theta})|, \\
|g_{i-1}^{k,\theta^r} - g_{i-1}^{k,\theta}| &\leq |g^{\theta^r}(x_{i-1}, t_k, Q^{k,\theta^r}) - g^{\theta^r}(x_{i-1}, t_k, Q^{k,\theta})| + |g^{\theta^r}(x_{i-1}, t_k, Q^{k,\theta}) \\
& \quad - g^{\theta}(x_{i-1}, t_k, Q^{k,\theta})|, \\
|\beta_{i,j}^{k,\theta^r} - \beta_{i,j}^{k,\theta}| &\leq |\beta^{\theta^r}(x_i, y_j, t_k, Q^{k,\theta^r}) - \beta^{\theta^r}(x_i, y_j, t_k, Q^{k,\theta})| \\
& \quad + |\beta^{\theta^r}(x_i, y_j, t_k, Q^{k,\theta}) - \beta^{\theta}(x_i, y_j, t_k, Q^{k,\theta})|.
\end{aligned}
$$
(23)

Thus, by assumptions (H1)–(H4) and the Eqs. (23) (18), we have

$$
\begin{aligned}
\sup_i |\mu_i^{k,\theta^r} - \mu_i^{k,\theta}| &\leq c\|v^{k,\theta}\|_1 + \sup_i |\mu^{\theta^r}(x_i, t_k, Q^{k,\theta}) - \mu^{\theta}(x_i, t_k, Q^{k,\theta})|, \\
\sup_i |g_{i-1}^{k,\theta^r} - g_{i-1}^{k,\theta}| &\leq c\|v^{k,\theta}\|_1 + \sup_i |g^{\theta^r}(x_{i-1}, t_k, Q^{k,\theta}) - g^{\theta}(x_{i-1}, t_k, Q^{k,\theta})|, \\
\sup_{i,j} |\beta_{i,j}^{k,\theta^r} - \beta_{i,j}^{k,\theta}| &\leq c\|v^{k,\theta}\|_1 + \sup_{i,j} |\beta^{\theta^r}(x_i, y_j, t_k, Q^{k,\theta}) - \beta^{\theta}(x_i, y_j, t_k, Q^{k,\theta})|.
\end{aligned}
$$

Set $\delta_k = 3c\|p^{k,\theta}\|_1 + cTV(p^{k,\theta})$ and

$$
\begin{aligned}
\rho_k^r = & \|p^{k,\theta}\|_1 \Big(\sup_i |\mu^{\theta^r}(x_i, t_k, Q^{k,\theta}) - \mu^{\theta}(x_i, t_k, Q^{k,\theta})| + \sup_{i,j} |\beta^{\theta^r}(x_i, y_j, t_k, Q^{k,\theta}) \\
& - \beta^{\theta}(x_i, y_j, t_k, Q^{k,\theta})| \\
& + \sup_i \int_0^1 \big| g_x^{\theta^r}(\bar{x}_i, t_k, Q^{k,\theta}) - g_x^{\theta}(\bar{x}_i, t_k, Q^{k,\theta}) \big| \, dx \Big) + \sup_i |g^{\theta^r}(x_{i-1}, t_k, Q^{k,\theta}) \\
& - g^{\theta}(x_{i-1}, t_k, Q^{k,\theta})| TV(p^{k,\theta}).
\end{aligned}
$$

Then we have

$$
\|v^{k+1,\theta}\|_1 \leq (1 + c(x_{\max} - x_{\min})\Delta t)\|v^{k,\theta}\|_1 + \Delta t(\rho_k^r + \delta_k)\|v^{k,\theta}\|_1.
$$
(24)

Since for each k, $\rho_k^r \to 0$ as $r \to \infty$, the result follows from (24).

Next, we establish subsequential convergence of minimizers of the finite dimensional problem (6) to a minimizer of the infinite dimensional problem (3).

Theorem 34. *Suppose that Θ_m is a sequence of compact subsets of Θ. Moreover, assume that for each $\theta \in \Theta$, there exist a sequence of $\theta_m \in \Theta_m$ such that $\theta_m \to \theta$ as $m \to \infty$. Then the function $F_{\Delta x, \Delta t}$ has a minimizer over Θ_m. Furthermore, if θ_m^r denotes a minimizer of $F_{\Delta x_r, \Delta t_r}$ over Θ_m and $\Delta x_r, \Delta t_r \to 0$, then any subsequence of θ_m^r has a further subsequence which convergence to a minimizer of F.*

Proof. The proof here is a direct application of the abstract theory in [10], base on the convergence of $F_{\Delta x_r, \Delta t_r}(\theta^r) \to F(\theta)$.

4   Numerical Results

In this section we present several numerical simulations to demonstrate the performance of the parameter estimation methodology. Although the theory presented here applies for the case of infinite dimensional parameter space Θ, for simplicity we restrict the unknown parameter space to finite-dimensional in the examples below.

4.1   Convergence of Parameter Estimates Computed by FOEU and SOEM in the Least-Squares Problem

In this example, we test the performance of the parameter-estimation technique using both FOEU and SOEM approximation schemes. As a first step in generating data, we choose $\Delta x = 0.0100, \Delta t = 0.0025, x_{min} = 0, x_{max} = 1, T = 1.0, g(x, t, Q) = (1-x)/2, \beta(x, y, t, Q) = 10\sin(4t)+10, \mu(x, t, Q) = 1/4\exp(Q)$, and the initial condition

$$p^0(x) = \begin{cases} 0.8, 0.25 \leq x \leq 0.45, \\ 2.5, 0.45 < x \leq 0.65, \\ 0.7, 0.65 < x < 0.8, \\ 0, \quad \text{else.} \end{cases}$$

We then solve system (1) with this choice of parameters in MATLAB using SOEM discretization and collect the resulting total population $Q(t^k) = \int_0^1 p(x, t^k)dx$ for $t^k = k/20, k = 1, \cdots, 20$. Observe that while $p(x, t)$ is discontinuous because $p^0(x)$ is, $Q(t)$ is a smooth function.

Assume all parameters are known except for $\mu = b\exp(Q)$ with b being an unknown parameter to be estimated. In our parameter estimation simulations, we fixed $\Delta x = 0.005, \Delta t = 0.0025$ for FOEU scheme. As for the SOEM scheme, the mesh sizes were chosen to be four times larger, that is, $\Delta x = 0.020$ and $\Delta t = 0.010$. We began with the above-mentioned data without noise in the least-squares problem described in Sect. 2. We then modified the data by adding normally distributed noise with mean zero and standard deviation $\sigma = 0.05, 0.10$, and 0.15, respectively, to the data. To solve the least-squares minimization problem we set $\theta = b$ and use the goal function

$$F(\theta) = \sum_{s=1}^{S} |Q(t_s; \theta) - X_s|^2,$$

i.e., $W = 1$. For each data set the least-squares minimization process was performed to estimate b using both numerical schemes. Our simulation results corroborate the convergence results of computed parameters. Figure 1 demonstrates the agreement of best fit model solutions obtained using FOEU and SOEM schemes in solving DSSM with the corresponding data sets with no noise as well as with different noise levels. A comparison of the two finite difference methods

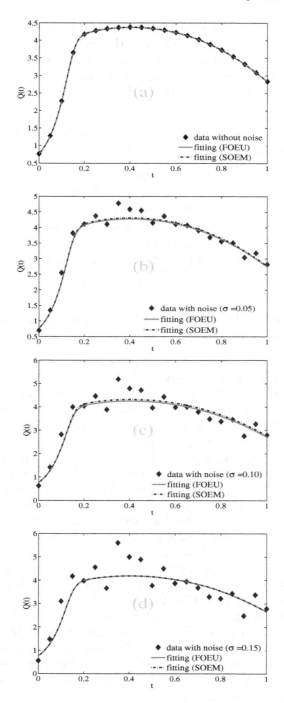

Fig. 1. (a) Comparison of data sets without noise and the corresponding best fit model solution using FOEU and SOEM schemes in solving DSSM. (b) (c) (d) Comparison of data at different noise levels ($\sigma = 0.05, 0.10, 0.15$) and the corresponding best fit model solutions using FOEU and SOEM schemes in solving DSSM.

performing in the least-squares parameter estimation process was provided in Table 1. It can be seen that using the SOEM scheme for approximating DSSM in the least-squares problem one could obtain a similar estimated value b (where the least-squares errors are at the same scale) with less than 10% of the CPU time compared to that using the FOEU scheme.

Table 1. Performance comparisons of FOEU and SOEM schemes in least-squares parameter estimates process

	Without noise		With noise					
			$\sigma = 0.05$		$\sigma = 0.10$		$\sigma = 0.15$	
	FOEU	SOEM	FOEU	SOEM	FOEU	SOEM	FOEU	SOEM
Estimated value of b	0.2507	0.2503	0.2674	0.2768	0.2853	0.2837	0.3033	0.3028
Least-squares error	6.33-04	3.85e-04	0.0838	0.1871	0.3112	0.2857	0.6583	0.6473
CPU time (in minutes)	285.48	20.67	264.59	16.45	367.49	22.47	368.71	15.33

4.2 Fitting DSSM Model to Green Tree Frog Population Estimates from Field Data

In this example, we fit DSSM to a set of green tree frog population estimates obtained from capture-mark-recapture (CMR) field data during years 2004–2009 (as shown in Fig. 2 (Left) [3]. The purpose is to estimate individual level vital rates for adult green tree frogs, and then use these parameter estimates to gain understanding of the dynamics of this population. In [3,4] the authors developed a model (referred to as JA model hereafter) to describe the dynamics of green tree frog population by dividing individuals into two stages, juveniles and adults, and set a least-squares problem to obtain the best fitted parameters to the CMR field data. With those estimated parameter values, they obtained the adult frog population curve (see Fig. 2 (Left)). Due to the relative short duration for the tadpole stage and the lack of other information regarding tadpoles, the authors in [3] simply assumed constant vital rates for tadpole stage. To circumvent this issue, we consider the short duration of juvenile stage as part of the reproduction process and adopt the DSSM to describe the adult frog dynamics. Here, $\beta(x, y, t, Q)$ in DSSM represents the rate at which an adult frog of size y gives birth to tadpoles that survive to metamorphose into frogs of size x.

As in [3], we take $t = 0$ to be the first week in January 2004. We chose $\Delta t = 1/52$. Since there are 52 weeks every year Δt represents one week. Let X_s, $s = 1, 2, \cdots, 136$, denote the observed number of frogs which was estimated statistically from CMR experiment data for 136 weeks during the breeding seasons in this six year experiment. The growth rate and mortality rates are assumed to take the same forms as in [3]:

$$g(x, t, Q) = \alpha_1(6 - x),$$

and

$$\mu(x,t,Q) = \begin{cases} \left(\alpha_2(1-\frac{t}{2})+\alpha_3\frac{t}{2}\right)(1+0.00343Q)\exp(\alpha_5 x), & 0 \le t \le 2, \\ \left(\alpha_3(2-\frac{t}{2})+\alpha_4(\frac{t}{2}-1)\right)(1+0.00343Q)\exp(\alpha_5 x), & 2 \le t \le 4, \\ \left(\alpha_4(3-\frac{t}{2})+3.093(\frac{t}{2}-2)\right)(1+0.00343Q)\exp(\alpha_5 x), & 4 \le t \le 6. \end{cases}$$

Here, the mortality rate was assumed to depend linearly on density as well as time since frogs hibernate during winter time. By monitoring program [16] the breeding season begins around the middle of April and ends in early August. Thus, similar to [3] the birth rate function was assumed to be

$$\beta(x,y,t,Q) = \begin{cases} \frac{\alpha_6}{0.3+\varepsilon}\gamma(x,\alpha_7,\alpha_8), & 0.3 \le t \le 0.6, 3 \le y \le 6, \\ \frac{\alpha_6(y-3+\varepsilon)}{\varepsilon(0.3+\varepsilon)}\gamma(x,\alpha_7,\alpha_8), & 0.3-\varepsilon \le t < 0.3, 2.7+t < y < 6, \\ \frac{\alpha_6(y-3+\varepsilon)}{\varepsilon(0.3+\varepsilon)}\gamma(x,\alpha_7,\alpha_8), & y-2.7 < t < 3.6-y, 0.3-\varepsilon < y < 0.3, \\ \frac{\alpha_6(0.6+\varepsilon-t)}{\varepsilon(0.3+\varepsilon)}\gamma(x,\alpha_7,\alpha_8), & 0.6 < t < 0.6+\varepsilon, 3.6-t \le y \le 6, \\ 0, & \text{else}. \end{cases}$$

Here, the gamma distribution density function with the shape parameter α_7 and scale parameter α_8, $\gamma(x,\alpha_7,\alpha_8)$, was chosen to model the size distribution of newly metamorphosed frogs. The constant ϵ is a positive small number that allows β to be extended to $(x,y,t) \in [1.5,6] \times [1.5,6] \times [0,1]$ and to satisfy the smoothness properties in (H4). We then also extend β periodically over one year intervals $[t,t+1]$, $t = 1,2,\cdots,5$.

We have $\alpha_1,\cdots,\alpha_8$ as unknown constants to be estimated (i.e., $\theta = (\alpha_1,\cdots,\alpha_8)$). We chose the initial condition in DSSM to be $p(x,0) = 615.96\exp(-0.75x)$ which implies $Q(0) = 257.2$ (cf. [3]). To solve the least-squares minimization problem, similar to [3], we set the goal function to be

$$F(\theta) = \sum_{s=1}^{S} |\log(Q(t_s;\theta)+1) - \log(X_s+1)|^2.$$

To guarantee that the estimated parameter values are biologically relevant, we set appropriate upper and lower bounds for each α_i. That is, $\underline{\alpha_i} \le \alpha_i \le \overline{\alpha_i}$ $i = 1,\cdots,8$. Using the vital rates determined by estimated α_i, $i = 1,\cdots,8$ given in Table 2, we simulated DSSM and compared the resulting adult frog population approximations to the data as well as the population estimates from the JA model in [3]. The comparison results are demonstrated in Fig. 2 (Left). It shows clearly that DSSM model output agrees with the population estimates resulting from field observations better than the JA model. Specifically, DSSM is more accurate in capturing the population dynamics when adult frog numbers are relatively low. Also, the DSSM fitting bears a smaller least-squares error of 33.5 compared to the JA model fitting which yielded an error of 37.8. Furthermore, the γ distribution that provided the best fit as presented in Fig. 2 (Right) indicates that the newly metamorphosed frogs have body length between 1.5 cm and 2 cm and approximately 99.6% of adult frogs give birth to tadpoles that eventually metamorphose into frogs of size between 1.5 cm and 2.0 cm.

Table 2. Parameter estimation values and corresponding standard deviation

	α_1	α_2	α_3	α_4	α_5	α_6	α_7	α_8
Estimated value	0.486	2.973	0.0085	2.376	0.000	47.061	6.740	1.849
Standard deviation	0.0183	0.7044	0.1164	1.6373	0.125	2.8295	0.8000	1.4763

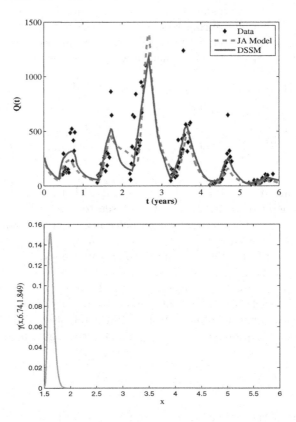

Fig. 2. Left: A comparison of the total population resulting from CMR field data to the total population resulting from model (1) and the JA model in [3]. Right: The probability density function that an adult frog gives birth to tadpoles that eventually metamorphose into frogs of size x.

We also applied a statistically based method to compute the variance in the estimated model parameters $\theta = (\alpha_1, \cdots, \alpha_8)$ similar to the work in [3] using standard regression formulations [12]. Table 2 provides the standard deviation for $\alpha_i's$ estimated above.

Acknowledgments. The work of A.S. Ackleh, X. Li, and B. Ma are partially supported by the National Science Foundation under grant # DMS-1312963.

References

1. Ackleh, A.S.: Parameter estimation in the nonlinear size-structured population model. Adv. Syst. Sci. Appl. **1**, 315–320 (1997). Special Issue
2. Ackleh, A.S.: Parameter identification in size-structured population models with nonlinear indivudual rates. Math. Comput. Model. **30**, 81–92 (1999)
3. Ackleh, A.S., Carter, J., Deng, K., Huang, Q., Pal, N., Yang, X.: Fitting a structured juvenile-adult model for green tree frogs to population estimate from capture-mark-recapture field data. Bull. Math. Biol. **74**, 641–665 (2012)
4. Ackleh, A.S., Deng, K.: A nonautonomous juvenile-adult model: well-posedness and long-time behavior via a comparison principle. SIAM J. Appl. Math. **69**, 1644–1661 (2009)
5. Ackleh, A.S., Farkas, J., Li, X., Ma, B.: A second order finite difference scheme for a size-structured population model with distributed states-at-birth. J. Biol. Dyn. **9**, 2–31 (2015)
6. Adams, B.M., Banks, H.T., Banks, J.E., Stark, J.D.: Population dynamics models in plant-insect herbivore-pesticide interactions. Math. Biosci. **196**, 39–64 (2005)
7. Banks, H.T., Botsford, L.W., Kappel, F., Wang, C.: Modeling and estimation in size structured population models. In: Mathematical Ecology, pp. 521–541. World Scientific Publishing, Singapore (1988)
8. Banks, H.T., Botsford, L.W., Kappel, F., Wang, C.: Estimation of growth and survival in size-structured cohort data. J. Math. Biol. **30**, 125–150 (1991)
9. Banks, H.T., Fitzpatrick, B.G.: Estimation of growth rate distributions in size structured population models. Quart. Appl. Math. **49**, 215–235 (1991)
10. Banks, H.T., Kunisch, K.: Estimation Techniques for Distributed Parameter Systems. Systems & Control: Foundations & Applications. Birkhäuser, Boston (1989)
11. Calsina, A., Saldana, J.: A model of physiologically structured population dynamics with a nonlinear individual growth rate. J. Math. Biol. **33**, 335–364 (1995)
12. Davidian, M., Giltinan, D.M.: Nonlinear models for repeated measurement data: an overview and update. J. Agric. Biol. Environ. Stat. **8**, 387–419 (2003)
13. Fitzpatrick, B.G.: Parameter estimation in conservation laws. J. Math. Syst. Estimation Control **3**, 413–425 (1993)
14. Huyer, W.: A size stuctured population model with dispersion. J. Math. Anal. Appl. **181**, 716–754 (1994)
15. LeVeque, R.J.: Numerical methods for conservation laws, birkhauser verlag, basel. J. Appl. Math. Mech.-Uss. **72**, 558 (1992)
16. Pham, L., Boudreaux, S., Karhbet, S., Price, B., Ackleh, A.S., Carter, J., Pal, N.: Population estimates of hyla cinerea (Schneider) (Green Tree Frog) in an urban environment. Southeast. Nat. **6**, 161–167 (2007)
17. Rundell, W.: Determing the death rate for an age-structured population from census data. SIAM J. Appl. Math. **53**, 1731–1746 (1993)
18. Shen, J., Shu, C.-W., Zhang, M.: High resolution schemes for a hierarchical size-structured model. SIAM J. Numer. Anal. **45**, 352–370 (2007)

On the Optimal Control of Opinion Dynamics on Evolving Networks

Giacomo Albi[1], Lorenzo Pareschi[2], and Mattia Zanella[2(✉)]

[1] Faculty of Mathematics, TU München, Boltzmanstraße 3,
85748 Garching (München), Germany
[2] Department of Mathematics and Computer Science,
University of Ferrara, Via N. Machiavelli 35, 44121 Ferrara, Italy
mattia.zanella@unife.it

Abstract. In this work we are interested in the modelling and control of opinion dynamics spreading on a time evolving network with scale-free asymptotic degree distribution. The mathematical model is formulated as a coupling of an opinion alignment system with a probabilistic description of the network. The optimal control problem aims at forcing consensus over the network, to this goal a control strategy based on the degree of connection of each agent has been designed. A numerical method based on a model predictive strategy is then developed and different numerical tests are reported. The results show that in this way it is possible to drive the overall opinion toward a desired state even if we control only a suitable fraction of the nodes.

Keywords: Multi-agent systems · Consensus dynamics · Scale-free networks · Collective behavior · Model predictive control

1 Introduction

Graph theory has emerged in recent years as one of the most active fields of research [1,7–10,24]. In fact, the study of technological and communication networks earned a special attention thanks to a huge amount of data coming from empirical observations and more recently from online platforms like Facebook, Twitter, Instagram and many others. This fact offered a real laboratory for testing on a large-scale the collective behavior of large populations of agents [16,17,20] and new challenges for the scientific research has emerged. In particular, the necessity to handle millions, and often billions, of vertices implied a substantial shift to large-scale statistical properties of graphs giving rise to the study of the so-called scale-free networks [8,18,24].

In this work, we will focus our attention on the modelling and control of opinion dynamics on a time evolving network. We consider a system of agents, each one belonging to a node of the network, interacting only if they are connected through the network. Each agent modifies his/her opinion through a compromise

© IFIP International Federation for Information Processing 2016
Published by Springer International Publishing AG 2016. All Rights Reserved
L. Bociu et al. (Eds.): CSMO 2015, IFIP AICT 494, pp. 58–67, 2016.
DOI: 10.1007/978-3-319-55795-3_4

function which depends both on opinions and the network [3–5,13,14,19,21,23]. At the same time new connections are created and removed from the network following a preferential attachment process. For simplicity here we restrict to non-growing network, that is a graph where the total number of nodes and the total number of edges are conserved in time. An optimal control problem is then introduced in order to drive the agents toward a desired opinion. The rest of the paper is organized as follows. In Sect. 2 we describe the alignment model for opinions spreading on a non-growing network. In order to control the trajectories of the model we introduce in Sect. 3 a general setting for a control technique weighted by a function on the number of connections. A numerical method based on model predictive control is then developed. Finally in Sect. 4 we perform numerical experiments showing the effectiveness of the present approach. Some conclusion are then reported in the last Section.

2 Modelling Opinion Dynamics on Networks

In the model each agent $i = 1, \ldots, N$ is characterized by two quantities $(w_i, c_i), i = 1, \ldots, N$, representing the opinion and the number of connections of the agent ith respectively. This latter term is strictly related to the architecture of the social graph where each agent shares its opinion and influences the interaction between individuals. Each agent is seen here as a node of a time evolving graph $\mathcal{G}^N = \mathcal{G}^N(t), t \in [t_0, t_f]$ whose nodes are connected through a given set of edges. In the following we will indicates the density of connectivity the constant $\gamma \geq 0$.

2.1 Network Evolution Without Nodes' Growth

In the sequel we will consider a graph with both a fixed number of nodes N and a fixed number of edges E. In order to describe the network's evolution we take into account a preferential attachment probabilistic process. This mechanism, known also as Yule process or Matthew effect, has been used in the modeling of several phenomena in biology, economics and sociology, and it is strictly connected to the generation of power law distributions [8,24]. The initial state of the network, $\mathcal{G}^N(0)$, is chosen randomly and, at each time step an edge is randomly selected and removed from the network. At the same time, a node is selected with probability

$$\Pi_\alpha(c_i) = \frac{c_i + \alpha}{\sum_{j=0}^{N}(c_j + \alpha)} = \frac{c_i + \alpha}{2E + N\alpha}, \qquad i = 1, \ldots, N, \tag{1}$$

among all possible nodes of $\mathcal{G}^N$, with $\alpha > 0$ an attraction coefficient. Based on the probability (1) another node is chosen at time t and connected with the formerly selected one. The described process is repeated at each time step. In this way both the number of nodes and the total number of edges remains constant in the reference time interval. Let $p(c, t)$ indicates the probability that a node is endowed of degree the c at time t. We have

$$\sum_{c} p(c,t) = 1, \qquad \sum_{c} c\, p(c,t) = \gamma. \tag{2}$$

The described process may be described by the following master equation [6]

$$\frac{d}{dt}p(c,t) = \frac{D}{E}[(c+1)p(c+1,t) - cp(c,t)]$$
$$+ \frac{2D}{2E+N\alpha}[(c-1+\alpha)p(c-1,t) - (c+\alpha)p(c,t)], \tag{3}$$

where $D > 0$ characterizes the relaxation velocity of the network toward an asymptotic degree distribution $p_\infty(c)$, the righthand side consists of four terms, the first and the third terms account the rate of gaining a node of degree c and respectively the second and fourth terms the rate of losing a node of degree c. The equation (3) holds in the interval $c \le E$, whereas for each $c > E$ we set $p(c,t) = 0$. While most the random graphs models with fixed number of nodes and vertices produces unrealistic degree distributions like the Watts and Strogatz generation model, called small-world model [22], the main advantage of the graph generated through the described rewiring process stands in the possibility to recover the scale-free properties. Indeed we can easily show that if $\gamma = 2E/N \ge 1$ with attraction coefficient $\alpha \ll 1$ then the stationary degree distribution $p_\infty(c)$ obeys a power-law of the following form

$$p_\infty(c) = \left(\frac{\alpha}{\gamma}\right)^\alpha \frac{\alpha}{c}. \tag{4}$$

When $\alpha \gg 1$ we loose the features of the preferential attachment mechanism, in fact high degree nodes are selected approximately with the same probability of the nodes with low degree of connection. Then the selection occurs in a non preferential way and the asymptotic degree distribution obeys the Poisson distribution

$$p_\infty(c) = \frac{e^{-\gamma}}{c!}\gamma^c. \tag{5}$$

A simple graph is sketched in Fig. 1 where we can observe how the initial degree of the nodes influences the evolution of the connections. In order to correctly observe the creation of the new links, that preferentially connect nodes with the highest connection degree, we marked each node with a number $i = 1, \ldots, 20$ and the nodes' diameters are proportional with their number of connections.

2.2   The Opinion Alignment Dynamics

The opinion of the ith agent ranges in the closed set $I = [-1, 1]$, that is $w_i = w_i(t) \in I$ for each $t \in [t_0, t_f]$, and its opinion changes over time according to the following differential system

$$\dot{w}_i = \frac{1}{|S_i|}\sum_{j \in S_i} P_{ij}(w_j - w_i), \qquad i = 1, \ldots, N \tag{6}$$

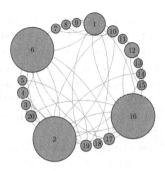 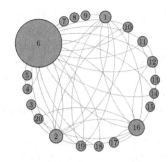

Fig. 1. Left: initial configuration of the sample network $\mathcal{G}^{20}$ with density of connectivity $\gamma = 5$. Right: a simulation of the network $\mathcal{G}^{20}$ after 10 time steps of the preferential attachment process. The diameter of each node is proportional to its degree of connection.

where S_i indicates the set of vertex connected with the ith agent and reflects the architecture of the chosen network, whereas $c_i = |S_i| < N$ stands for the cardinality of the set S_i, also known as degree of vertex i. Note that the number of connections c_i evolves in time accordingly to the process described in Sect. 2.1. Furthermore we introduced the interaction function $P_{ij} \in [0,1]$, depending on the opinions of the agents and the graph $\mathcal{G}^N$ which can be written as follows

$$P_{ij} = P(w_i, w_j; \mathcal{G}^N). \tag{7}$$

A possible choice for the interaction function is the following

$$P(w_i, w_j; \mathcal{G}^N) = H(w_i, w_j)K(\mathcal{G}^N), \tag{8}$$

where $H(\cdot, \cdot)$ represents the positive compromise propensity, and K a general function taking into account statistical properties of the graph $\mathcal{G}$. In what follows we will consider $K = K(c_i, c_j)$, a function depending on the vertices' connections.

3   Optimal Control Problem of the Alignment Model

In this section we introduce a control strategy which characterizes the action of an external agent with the aim of driving opinions toward a given target w_d. To this goal, we consider the evolution of the network $\mathcal{G}^N(t)$ and the opinion dynamics in the interval $[t_0, t_f]$. Therefore we introduce the following optimal control problem

$$\min_{u \in \mathcal{U}} J(\mathbf{w}, u) := \frac{1}{2} \int_{t_0}^{t_f} \left\{ \frac{1}{N} \sum_{j=1}^{N} (w_j(s) - w_d)^2 + \nu u(s)^2 \right\} ds, \tag{9}$$

subject to

$$\dot{w}_i = \frac{1}{|S_i|} \sum_{j \in S_i} P_{ij}(w_j - w_i) + u\chi(c_i \geq c^*), \quad w_i(0) = w_i^0, \tag{10}$$

where we indicated with $\mathcal{U}$ the set of admissible controls, with $\nu > 0$ a regularization parameter which expresses the strength of the control in the overall dynamics and $w_d \in [-1, 1]$ the target opinion. Note that the action of the control u is weighted by an indicator function $\chi(\cdot)$, which is active only for the nodes with degree $c_i \geq c^*$. In general this selective control approach models an a-priori strategy of a policy maker, possibly acting under limited resources or unable to influence the whole ensemble of agents. For example we can consider a varying horizon control acting on a fixed portion of connected agents.

The solution of this kind of control problems is in general a difficult task, given that their direct solution is prohibitively expensive for a large number of agents. Different strategies have been developed for alignment modeling in order to obtain feedback controls or more general numerical control techniques [2–5, 11, 12, 15]. To tackle numerically the described problem a standard strategy makes use of a model predictive control (MPC) approach, also referred as receding horizon strategy.

In general MPC strategies solves a finite horizon open-loop optimal control problem predicting the dynamic behavior over a predict horizon $t_p \leq t_f$, with initial state sampled at time t (initially $t = t_0$), and computing the control on a control horizon $t_c \leq t_p$. The optimization is computed introducing a new integral functional $J_p(\cdot, \cdot)$, which is an approximation of (9) on the time interval $[t, t+t_p]$, namely

$$J_p(\mathbf{w}, \bar{u}) := \frac{1}{2} \int_t^{t+t_p} \left\{ \frac{1}{N} \sum_{j=1}^N (w_j(s) - w_d)^2 + \nu_p \bar{u}(s)^2 \right\} ds \qquad (11)$$

where the control, $\bar{u} : [t, t+t_p] \to \mathcal{U}$, is supposed to be an admissible control in the set of admissible control $\mathcal{U}$, subset of $\mathbb{R}$, and ν_p a possibly different penalization parameter with respect to the full optimal control problem. Thus the computed optimal open-loop control $\bar{u}(\cdot)$ is applied feedback to the system dynamics until the next sampling time $t + t_s$ is evaluated, with $t_s \leq t_c$, thereafter the procedure is repeated taking as initial state of the dynamics at time $t + t_s$ and shifting forward the prediction and control horizons, until the final time t_f is reached. This process generates a sub-optimal solution with respect to the solution of the full optimal control problem (9)–(10).

Let us consider now the full discretize problem, defining the time sequence $[t_0, t_1, \ldots, t_M]$, where $t_n - t_{n-1} = t_s = \Delta t > 0$ and $t_M := M\Delta t = t_f$, for all $n = 1, \ldots, M$, assuming furthermore that $t_c = t_p = p\Delta t$, with $p > 0$. Hence the linear MPC method look for a piecewise control on the time frame $[t_0, t_M]$, defined as follows

$$\bar{u}(t) = \sum_{n=0}^{M-1} \bar{u}^n \chi_{[t_n, t_{n+1}]}(t). \qquad (12)$$

In order to discretize the evolution dynamics we consider a Runge-Kutta scheme, the full discretized optimal control problem on the time frame $[t_n, t_n + p\Delta t]$ reads

$$\min_{\bar{u} \in \mathcal{U}} J_p(\mathbf{w}, \bar{u}) := \frac{1}{2} \int_{t_n}^{t_n + p\Delta t} \left\{ \frac{1}{N} \sum_{j=1}^{N} (w_j(s) - w_d)^2 + \nu_p \bar{u}^2 \right\} ds \qquad (13)$$

subject to

$$W_{i,l}^{(n)} = w_i^n + \Delta t \sum_{k=1}^{s} a_{l,k} \left(F(t + \theta_k \Delta t, W_{i,k}^{(n)}) + \bar{U}_k^{(n)} Q_i(t + \theta_k \Delta t) \right),$$

$$w_i^{n+1} = w_i^n + \Delta t \sum_{l=1}^{s} b_l \left(F(t + \theta_l \Delta t, W_{i,l}^{(n)}) + \bar{U}_l^{(n)} Q_i(t + \theta_l \Delta t) \right), \qquad (15)$$

$$w_i^n = w_i(t_n),$$

for all $n = 1, \ldots, p - 1;\ l = 1, \ldots, s;\ i, \ldots, N$ and having defined the following functions

$$F(t, w_i) = \frac{1}{|S_i(t)|} \sum_{j \in S_i(t)} P_{ij}(w_j - w_i), \quad Q_i(t) = \chi(c_i(t) \geq c^*).$$

The coefficients $(a_{l,k})_{l,k}$, $(b_l)_l$ and $(\theta_l)_l$, with $l, k = 1, \ldots, s$, define the Runge-Kutta method and $(\bar{U}^{(n)})_l$, $(W_{i,l}^{(n)})_l$ are the internal stages associated to $\bar{u}(t), w_i(t)$ on time frame $[t_n, t_{n+1}]$.

3.1 Instantaneous Control

Let us restrict to the case of a single prediction horizon, $p = 1$, where we discretize the dynamics with an explicit Euler scheme ($a_{1,1} = \theta_1 = 0$ and $b_1 = 1$). Notice that since the control $\bar{u}$ is a constant value and assuming that the network, $\mathcal{G}^N$ remains fixed over the time interval $[t_n, t_n + \Delta t]$ the discrete optimal control problem (13) reduces to

$$\min_{\bar{u} \in \mathcal{U}} J_p(\mathbf{w}, \bar{u}^n) := \Delta t \left\{ \frac{1}{N} \sum_{j=1}^{N} (w_j^{n+1}(\bar{u}^n) - w_d)^2 + \nu_p (\bar{u}^n)^2 \right\} \qquad (16)$$

with

$$w_i^{n+1} = w_i^n + \Delta t \left(F(t_n, w_i^n) + \bar{u}^n Q_i^n \right), \quad w_i^n = w_i(t_n). \qquad (17)$$

In order to find the minima of (13) is sufficient to find the value $\bar{u}$ satisfying $\partial_{\bar{u}} J_p(\mathbf{w}, \bar{u}) = 0$, which can be computed by a straightforward calculation

$$\bar{u}^n = -\frac{1}{N\nu + \Delta t \sum_{j=1}^{N} (Q_j^n)^2} \left(\sum_{j=1}^{N} Q_j^n (w_j^n - w_d) + \Delta t \sum_{j=1}^{N} Q_j^n F(t_n, w_i^n) \right).$$

$$(18)$$

where we scaled the penalization parameter with $\nu_p = \Delta t \nu$.

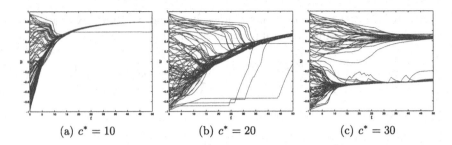

(a) $c^* = 10$ (b) $c^* = 20$ (c) $c^* = 30$

Fig. 2. Evolution of the constrained opinion dynamics with uniform initial distribution of opinions over the time interval $[0, 50]$ for different values of $c^* = 10, 15, 30$ with target opinion $w_d = 0.8$, control parameter $\kappa = 0.1$, $\Delta t = 10^{-3}$ and confidence bound $\Delta = 0.4$.

4   Numerical Results

In this section we present some numerical results in order to show the main features of the control introduced in the previous paragraphs. We considered a population of $N = 100$ agents, each of them representing a node of an undirected graph with density of connectivity $\gamma = 30$. The network $\mathcal{G}^{100}$ evolves in the time interval $[0, 50]$ with attraction coefficient $\alpha = 0.01$ and represents a single sample of the evolution of the master equation (3) with $D = 20$. The control problem is solved by the instantaneous control method described in Remark 3.1 with $\Delta t = 5 \; 10^{-2}$. In Fig. 3 we present the evolution over the reference time interval of the constrained opinion dynamics. The interaction terms have been chosen as follows

$$K(c_i, c_j) = e^{-\lambda c_i} \left(1 - e^{-\beta c_j}\right), \qquad H(w_i, w_j) = \chi(|w_i - w_j| \leq \Delta), \qquad (19)$$

where the function $H(\cdot, \cdot)$ is a bounded confidence function with $\Delta = 0.4$, while $K(\cdot, \cdot)$ defines the interactions between the agents i and j taking into account that agents with a large number of connections are more difficult to influence and at the same time they have more influence over other agents. The action of the control is characterized by a parameter $\kappa = 0.1$ and target opinion $w_d = 0.8$. We present the resulting opinion dynamics for a choice of constants $\lambda = 1/100, \beta = 1$ in Fig. 2. We report the evolution of the network and of the opinion in Fig. 3, here the diameter of each node is proportional with its degree of connection whereas the color indicates its opinion. As a measure of consensus over the agents we introduce the quantity

$$V_{w_d} = \frac{1}{N-1} \sum_{i=1}^{N} (w_i(t_f) - w_d)^2, \qquad (20)$$

where $w_i(t_f)$ is the opinion of the ith agent at the final time t_f. In Fig. 4 we compare different values of V_{w_d} as a function of c^*. Here we calculated the size of the controlled agents and the values of V_{w_d} both, starting from a given uniform

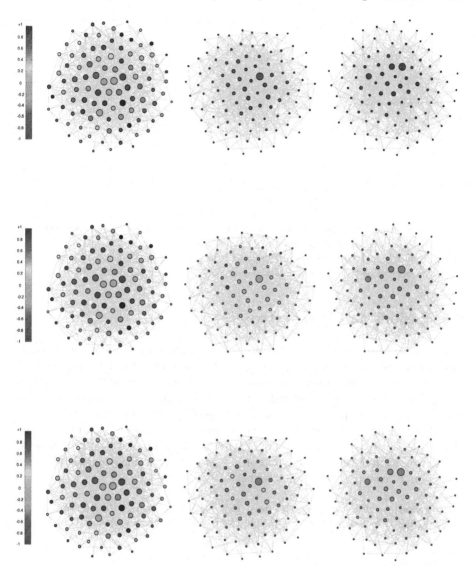

Fig. 3. Evolution of opinion and connection degree of each node of the previously evolved graph $\mathcal{G}^{100}$. From left to right: graph at times $t = 0, 25, 50$. From the top: opinion dynamics for threshold values $c^* = 10, 20, 30$. The target opinion is set $w_d = 0.8$ and the control parameter $\kappa = 0.1$. (Color figure online)

initial opinion and the same graph with initial uniform degree distribution. It can be observed how the control is capable to drive the overall dynamics toward the desired state acting only on a portion of the nodes.

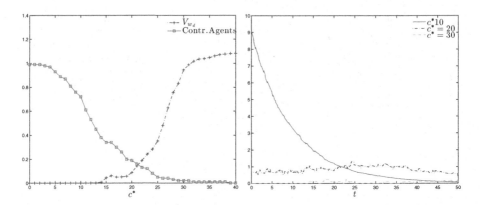

Fig. 4. Left: the red squared plot indicates the size of the set of controlled agent at the final time t_f in dependence on c^* whereas the blue line indicates the mean square displacement V_{w_d}. Right: values of the control u at each time step for $c^* = 10, 20, 30$. In the numerical test we assumed $\Delta = 0.4, \Delta t = 5 \ 10^{-2}, \kappa = 0.1$. (Color figure online)

Conclusions and Perspectives

In this short note we focus our attention on a control problem for the dynamic of opinion over a time evolving network. We show that the introduction of a suitable selective control depending on the connection degree of the agent's node is capable to drive the overall opinion toward consensus. In a related work we have considered this problem in a mean-field setting where the number of agents, and therefore nodes, is very large [6]. In future works we plan to concentrate on the model predictive control setting, where the evolution of the control is based on the evolution of the network, and on the case with varying prediction horizon acting on a given portion of the agents.

Acknowledgments. GA acknowledges the support of the ERC-Starting Grant project High-Dimensional Sparse Optimal Control (HDSPCONTR).

References

1. Albert, R., Barabási, A.-L.: Statistical mechanics of complex networks. Rev. Mod. phys. **74**(1), 47–97 (2002)
2. Albi, G., Bongini, M., Cristiani, E., Kalise, D.: Invisible control of self-organizing agents leaving unknown environments. SIAM J. Appl. Math **76**, 1683–1710 (2016). in press
3. Albi, G., Herty, M., Pareschi, L.: Kinetic description of optimal control problems and applications to opinion consensus. Commun. Math. Sci. **13**(6), 1407–1429 (2015)
4. Albi, G., Pareschi, L., Zanella, M.: Boltzmann-type control of opinion consensus through leaders. Philos. Trans. R. Soc. A **372**(2028), 20140138 (2014)

5. Albi, G., Pareschi, L., Zanella, M.: Uncertainty quantification in control problems for flocking models. Math. Prob. Eng. **2015**, 850124 (2015). 14 p
6. Albi, G., Pareschi, L., Zanella, M.: Opinion dynamics over complex networks: kinetic modeling and numerical methods. Kinet. Relat. Models **10**(1), 1–32 (2017)
7. Amaral, L.A.N., Scala, A., Barthélemy, M., Stanley, H.E.: Classes of small-world networks. Proc. Natl. Acad. Sci. U. S. Am. **97**(21), 11149–11152 (2000)
8. Barabási, A.-L., Albert, R., Jeong, H.: Mean-field theory for scale-free random networks. Phys. A: Stat. Meach. Appl. **272**(1), 173–187 (1999)
9. Barrat, A., Barthélemy, M., Vespignani, A.: Dynamical Processes on Complex Networks. Cambridge University Press, Cambridge (2008)
10. Benczik, I.J., Benczik, S.Z., Schmittmann, B., Zia, R.K.: Opinion dynamics on an adaptive random network. Phys. Rev. E **79**(4), 046104 (2009)
11. Bongini, M., Fornasier, M., Fröhlich, F., Haghverdi, L.: Sparse stabilization of dynamical systems driven by attraction and avoidance forces. Netw. Heterogen. Media **9**(1), 1–31 (2014)
12. Wongkaew, S., Caponigro, M., Borzì, A.: On the control through leadership of the Hegselmann-Krause opinion formation model. Math. Models Methods Appl. Sci. **25**(2), 255–282 (2015)
13. Chi, L.: Binary opinion dynamics with noise on random networks. Chin. Sci. Bull. **56**(34), 3630–3632 (2011)
14. Das, A., Gollapudi, S., Munagala, K.: Modeling opinion dynamics in social networks. In: Proceedings of the 7th ACM International Conference on Web Search and Data Mining, ACM (2014)
15. Herty, M., Zanella, M.: Performance bounds for the mean-field limit of constrained dynamics. Preprint (2015)
16. Jin, E.M., Girvan, M., Newman, M.E.J.: Structure of growing social networks. Phys. Rev. E **64**(4), 046132 (2001)
17. Kramer, A.D.I., Guillory, J.E., Hancock, J.T.: Experimental evidence of massive scale emotional contagion through social networks. Proc. Nat. Acad. Sci. **111**(24), 8788–8789 (2014)
18. Newman, M.E.J.: The structure and function on complex networks. SIAM Rev. **45**(2), 167–256 (2003)
19. Pareschi, L., Toscani, G.: Interacting Multiagent Systems. Kinetic Equations and Monte Carlo Methods. Oxford University Press, Oxford (2013)
20. Strogatz, S.H.: Exploring complex networks. Nature **410**(6825), 268–276 (2001)
21. Sznajd-Weron, K., Sznajd, J.: Opinion evolution in closed community. Int. J. Mod. Phys. C **11**(6), 1157–1165 (2000)
22. Watts, D.J., Strogatz, S.H.: Collective dynamics of small-world networks. Nature **393**, 440–442 (1998)
23. Weisbuch, G.: Bounded confidence and social networks. Eur. Phys. J. B-Condens. Matter Complex Syst. **38**(2), 339–343 (2004)
24. Xie, Y.-B., Zhou, T., Wang, B.-H.: Scale-free networks without growth. Phys. A **387**, 1683–1688 (2008)

Coupling MPC and DP Methods for an Efficient Solution of Optimal Control Problems

A. Alla[1], G. Fabrini[2,3](✉), and M. Falcone[4]

[1] Department of Scientific Computing, Florida State University, Tallahassee, USA
aalla@fsu.edu
[2] DIME, Università di Genova, Genova, Italy
fabrini@dime.unige.it
[3] Laboratoire Jacques-Louis Lions,
Sorbonne Universités UPMC Univ Paris 06, Paris, France
[4] Dipartimento di Matematica, La Sapienza Università di Roma, Roma, Italy
falcone@mat.uniroma1.it

Abstract. We study the approximation of optimal control problems via the solution of a Hamilton-Jacobi equation in a tube around a reference trajectory which is first obtained solving a Model Predictive Control problem. The coupling between the two methods is introduced to improve the initial local solution and to reduce the computational complexity of the Dynamic Programming algorithm. We present some features of the method and show some results obtained via this technique showing that it can produce an improvement with respect to the two uncoupled methods.

Keywords: Optimal control · Dynamic Programming · Model Predictive Control · Semi-Lagrangian schemes

1 Introduction

The numerical solution of partial differential equations obtained by applying the Dynamic Programming Principle (DPP) to nonlinear optimal control problems is a challenging topic that can have a great impact in many areas, e.g. robotics, aeronautics, electrical and aerospace engineering. Indeed, by means of the DPP one can characterize the value function of a fully–nonlinear control problem (including also state/control constraints) as the unique viscosity solution of a nonlinear Hamilton–Jacobi equation, and, even more important, from the solution of this equation one can derive the approximation of a feedback control. This result is the main motivation for the PDE approach to control problems and represents the main advantage over other methods, such as those based on the Pontryagin minimum principle. It is worth to mention that the characterization via the Pontryagin principle gives only necessary conditions for the optimal trajectory and optimal open-loop control. Although from the numerical point of view the control system can be solved via shooting methods for the associated

L. Bociu et al. (Eds.): CSMO 2015, IFIP AICT 494, pp. 68–77, 2016.
DOI: 10.1007/978-3-319-55795-3_5

two point boundary value problem, in real applications a good initial guess for the co-state is particularly difficult and often requires a long and tedious trial-and-error procedure to be found. In any case, it can be interesting to obtain a local version of the DP method around a reference trajectory to improve a sub-optimal strategy. The reference trajectory can be obtained via the Pontryagin principle (with open-loop controls), via a Model Predictive Control (MPC) approach (using feedback sub-optimal controls) or simply via the already known engineering experience. The application of DP in an appropriate neighborhood of the reference trajectory will not guarantee the global optimality of the new feedback controls but could improve the result within the given constraints.

In this paper we focus our attention on the coupling between the MPC approach and the DP method. Although this coupling can be applied to rather general nonlinear control problems governed by ordinary differential equations we present the main ideas of this approach using the *infinite horizon optimal control*, which is associated to the following Hamilton–Jacobi–Bellman equation:

$$\lambda v(x) + \max_{u \in U}\{-f(x,u) \cdot Dv(x) - \ell(x,u)\} = 0, \quad \text{for } x \in \mathbb{R}^d.$$

For numerical purposes, the equation is solved in a bounded domain $\Omega \subset \mathbb{R}^d$, so that also boundary conditions on $\partial\Omega$ are needed. A rather standard choice when one does not have additional information on the solution is to impose state constraints boundary conditions. It is clear that the domain Ω should be large enough in order to contain as much information as possible. It is, in general, computed without any information about the optimal trajectory. Here we construct the domain Ω around a reference trajectory obtained by a fast solution with a Model Predictive Control (MPC). MPC is a receding horizon method which allows to compute the optimal solution for a given initial condition by solving iteratively a finite horizon open-loop problem (see [5,7]).

2   A Local Version of DP via MPC Models

Let us present the method for the classical *infinite horizon problem*. Let the controlled dynamics be given by the solution of the following Cauchy problem:

$$\begin{cases} \dot{y}(t) = f(y(t), u(t)), & t > 0, \\ y(0) = x, \end{cases} \tag{1}$$

where $x, y \in \mathbb{R}^d$, $u \in \mathbb{R}^m$ and $u \in \mathcal{U} \equiv \{u : \mathbb{R}_+ \to U, \text{measurable}\}$. If f is Lipschitz continuous with respect to the state variable and continuous with respect to (x, u), the classical assumptions for the existence and uniqueness result for the Cauchy problem (1) are satisfied. To be more precise, the Carathéodory theorem (see [2]) implies that for any given control $u(\cdot) \in \mathcal{U}$ there exists a unique trajectory $y(\cdot; u)$ satisfying (1) almost everywhere. Changing the control policy the trajectory will change and we will have a family of infinitely many solutions of the controlled system (1) parametrized with respect to the control u.

Let us introduce the *cost functional* $J : \mathcal{U} \to \mathbb{R}$ which will be used to select the *optimal trajectory*. For the infinite horizon problem the cost functional is

$$J_x(u(\cdot)) = \int_0^\infty \ell(y(s), u(s))e^{-\lambda s}ds, \qquad (2)$$

where $\lambda > 0$ is a given parameter and ℓ is typically Lipschitz continuous function (although this is not strictly necessary to define the integral). We remark that for the numerical simulations we are working on a compact set and, in order to apply the error estimates for approximation (as in [3,4]), we will just need ℓ locally Lipschitz continuous in both arguments. The function ℓ represents the running cost and λ is the discount factor which allows to compare the costs at different times rescaling the costs at time 0. From the technical point of view, the presence of the discount factor guarantees that the integral is finite whenever ℓ is bounded, i.e. $||\ell||_\infty \le M_\ell$, where $||\ell||_\infty$ is defined as the supremum norm in $\mathbb{R}^d \times U$. In this section we will summarize the basic results for the two methods as they are the building blocks for our new method.

2.1 Hamilton–Jacobi–Bellman Equations

The essential features will be briefly sketched, and more details in the framework of viscosity solutions can be found in [2,4].

Let us define the value function of the problem as

$$v(x) = \inf_{u(\cdot) \in \mathcal{U}} J_x(u(\cdot)). \qquad (3)$$

It is well known that passing to the limit in the Dynamic Programming Principle one can obtain a characterization of the value function in terms of the following first order non linear Bellman equation

$$\lambda v(x) + \max_{u \in U}\{-f(x,u) \cdot Dv(x) - \ell(x,u)\} = 0, \qquad \text{for } x \in \mathbb{R}^d. \qquad (4)$$

Several approximation schemes on a fixed grid G have been proposed for (4). To simplify the presentation, let us consider a uniform structured grid with constant space step $k := \Delta x$. We will use a semi-Lagrangian method based on a Discrete Time Dynamic Programming Principle. A first discretization in time of the original control problem [2] leads to a characterization of the corresponding value function v^h (for the time step $h := \Delta t$) as

$$v^h(x) = \min_{u \in U}\{e^{-\lambda h}v_h(x + hf(x,u)) + h\ell(x,u)\}. \qquad (5)$$

Then, we have to project on the grid and reconstruct the value $v_h(x + hf(x,u))$ by interpolation (for example by a linear interpolation). Finally, we obtain the following fixed point formulation of the DP equation

$$w(x_i) = \min_{u \in U}\{e^{-\lambda h}w(x_i + hf(x_i,u)) + h\ell(x_i,u)\}, \quad \text{for } x_i \in G, \qquad (6)$$

where $w(x_i) = v^{h,k}(x_i)$ is the approximation of the value function at the node x_i (see [3, 4] for more details). Under appropriate assumptions, $v^{h,k}$ converges to $v(x)$ when $(\Delta t, \Delta x)$ goes to 0 (precise a-priori-estimates are available, e.g. [3] for more details). This method is referred in the literature as the *value iteration method* because, starting from an initial guess for the value function, it modifies the values on the grid according to the foot of the characteristics. It is well-known that the convergence of the value iteration can be very slow, since the contraction constant $e^{-\lambda \Delta t}$ is close to 1 when Δt is close to 0. This means that a higher accuracy will also require more iterations. Then, there is a need for an acceleration technique in order to cut the link between accuracy and complexity of the value iteration. One possible choice is the iteration in the policy space or the coupling between value iteration and the policy iteration in [1]. We refer the interested reader to the book [4] for a complete guide on the numerical approximation of the equation and the reference therein. One of the strength of this method is that it provides the feedback control once the value function is computed (and the feedback is computed at every node even in the fixed point iteration). In fact, we can characterize the optimal feedback control everywhere in Ω

$$u^*(x) = \arg\min_{u \in U}\{-f(x, u) \cdot Dv(x) - \ell(x, u)\}, \quad x \in \Omega,$$

where Dv is an approximation of the value function obtained by the values at the nodes.

2.2 Model Predictive Control

Nonlinear model predictive control (NMPC) is an optimization based method for the feedback control of nonlinear systems. It consists on solving iteratively a finite horizon open loop optimal control problem subject to system dynamics and constraints involving states and controls.

The infinite horizon problem, described at the beginning of Sect. 2, turns out to be computationally unfeasible for the open-loop approach. Therefore, we solve a sequence of finite horizon problems. In order to formulate the algorithm we need to introduce the finite horizon cost functional:

$$J_{y_0}^N(u(\cdot)) = \int_{t_0}^{t_0^N} \ell(y(s), u(s))e^{-\lambda s} ds$$

where N is a natural number, $t_0^N = t_0 + N\Delta t$ is the final time, $N\Delta t$ denotes the length of the prediction horizon for the chosen time step $\Delta t > 0$ and the state y solves $\dot{y}(t) = f(y(t), u(t))$, $y(t_0) = y_0$, $t \in [t_0, t_0^N)$ and is denoted by $y(\cdot, t_0; u(\cdot))$. We also note that $y_0 = x$ at $t = 0$ as in Eq. (1). The basic idea of NMPC algorithm is summarized at the end of sub-section.

The method works as follows: we store the optimal control on the first subinterval $[t_0, t_0 + \Delta t]$ together with the associated optimal trajectory. Then, we initialize a new finite horizon optimal control problem whose initial condition is given by the optimal trajectory $y(t) = y(t; t_0, u^N(t))$ at $t = t_0 + \Delta t$ using the sub-optimal control $u^N(t)$ for $t \in (t_0, t_0 + \Delta t]$. We iterate this process by setting $t_0 = t_0 + \Delta t$. Note that (7) is an open loop problem on a finite time horizon $[t_0, t_0 + N\Delta t]$ which can be treated by classical techniques, see e.g. [6]. The interested reader can find in [5] a detailed presentation of the method and a long list of references.

In general, one can obtain a better feedback approximation increasing the prediction horizon, but this will of course make the CPU time grow. Typically one is interested in short prediction horizons (or even horizon of minimal length) which can guarantee stabilization properties of the MPC scheme. The problem is that when the horizon N is too short we will lose these properties (see [5] Example 6.26). Estimates on the minimum value for N which ensures asympotitic stability are based on the relaxed dynamic programming principle and can be found in [5] and the references therein. The computation of this minimal horizon is related to a relaxed dynamic programming principle in terms of the value function for the finite horizon problem (7).

MPC Algorithm

Start: choose $\Delta t > 0$, $N \in \mathbb{N}$, $\lambda > 0$.
for $n = 0, 1, 2, \ldots$
Step 1: Compute the state $y(t_n)$ of the system at $t_n = n\Delta t$,
Step 2: Set $t_0 = t_n = n\Delta t$, $y_0 = y(t_n)$ and compute a global solution,

$$u^N(t) := \arg\min_{u \in \mathcal{U}} J_{y_0}^N(u(t_0)). \tag{7}$$

Step 3: Define the MPC feedback value $u^N(t)$, $t \in (t_0, t_0 + \Delta t]$
and use this control to compute the associated
state $y = y(t; t_0, u^N(t))$ by solving (1) in $[t_0, t_0 + \Delta t]$.
end for
end

2.3   Coupling MPC with Bellman Equation

The idea behind the coupling is to combine the advantages from both methods. The Dynamic Programming approach is global and gives information on the value function in a domain, provided we solve the Bellman equation. It gives the feedback synthesis in the whole domain. Model Predictive control is local and gives an approximate feedback control just for a single initial condition. Clearly MPC is faster but does not give the same amount of information.

In many real situations, we need a control to improve the solution around a reference trajectory starting at x, $\overline{y}_x(\cdot)$, so we can reduce the domain to a neighborhood of $\overline{y}_x(\cdot)$. Now let us assume that we are interested in the approximation of feedbacks for an optimal control problem given the initial condition x. First of all we have to select a (possibly small) domain where we are going to compute the approximate value function and to this end we need to compute a first guess that we will use as reference trajectory.

MPC can provide quickly a reasonable reference trajectory $\overline{y}_x(\cdot) := y^{MPC}(\cdot)$, but this trajectory is not guaranteed to be globally optimal (or have the required stabilization properties as we said in the previous section). In our approach, we can choose a rather *short* prediction horizon in order to have a fast approximation of the initial guess. This will not give the final feedback synthesis but will be just used to build the domain Ω_ρ where we are going to apply the DP approach. It is clear that MPC may provide inaccurate solutions if N is too short but its rough information about the trajectory y^{MPC} will be later compensated by the knowledge of the value function obtained by solving the Bellman equation. We construct Ω_ρ as a tube around y^{MPC} defining

$$\Omega_\rho := \{x \in \Omega : dist(x, y^{MPC}) \leq \rho\} \tag{8}$$

This tube can be actually computed via the Eikonal equation, i.e. solving the Dirichlet problem

$$|\nabla v(x)| = 1, \quad x \in \mathbb{R}^N \backslash \mathcal{T}, \quad \text{with } v(x) = 0, \quad x \in \mathcal{T}, \tag{9}$$

where the target is $\mathcal{T} := \{y^{MPC}(t), t \in [0, T]\}$. We just want to mention that for this problem several fast methods are available (e.g. Fast Marching [8] and Fast Sweeping [9]) so this step can be solved very efficiently. The interested reader can find in [4] many details on numerical approximation of the weak solutions to the eikonal equation.

Solving the eikonal Eq. (9) (in the viscosity sense) we obtain the distance function from the target. Then, we choose a radius $\rho > 0$ in order to build the tube Ω_ρ. In this way the domain of the HJB is not built by scratch but takes into account some information on the controlled system. To localize the solution in the tube we impose state constraints boundary conditions on $\partial\Omega_\rho$ penalizing in the scheme (6) the points outside the domain. It is clear that a larger ρ will allow for a more accurate approximation of the value function but at the same time enlarging ρ we will lose the localization around our trajectory increasing the number of nodes (and the CPU time). Finally, we compute the optimal feedback from the value function computed and the corresponding optimal trajectories in Ω_ρ The algorithm is summarized below:

Localized DP algorithm (LDP)

Start: Inizialization
Step 1: Solve MPC and compute y_x^{MPC} starting at x
Step 2: Compute the distance from y_x^{MPC} via the Eikonal equation
Step 3: Select the tube Ω_ρ of radius ρ centered at y_x^{MPC}
Step 4: Compute the constrained value function v^{tube} in Ω_ρ via HJB
Step 5: Compute the optimal feedbacks and trajectory using v^{tube}.
End

3   Numerical Tests

In this section we present two numerical tests for the infinite horizon problem to illustrate the performances of the proposed algorithm. However, the localization procedure can be applied to more general optimal control problems.

All the numerical simulations have been made on a MacBook Pro with 1 CPU Intel Core i5 2.4 GHz and 8 GB RAM. The codes used for the simulations are written in Matlab. The routine for the approximation of MPC is provided in [5].

Test 1: 2D Linear Dynamics. Let us consider the following controlled dynamics:

$$\begin{cases} \dot{y}(t) = u(t), & t \in [0,T], \\ y(0) = x \end{cases} \tag{10}$$

where $u = (u_1, u_2)$ is the control, $y : [0,T] \to \mathbb{R}^2$ is the dynamic and x is the initial condition. The cost functional we want to minimize is:

$$J_x(u) := \int_0^\infty \min\{|y(t;u) - P|^2, |y(t;u) - Q|^2 - 2\} \, e^{-\lambda t} \, dt \tag{11}$$

where $\lambda > 0$ is the discount factor.

In this example, the running cost has two local minima in P and Q. We set $P := (0,0)$ and $Q := (2,2)$ so that the value of the running cost is 0 at P and -2 at Q. Note that we have included a discount factor λ, which guarantees the integrability of the cost functional $J_x(u)$ and the existence and uniqueness of the viscosity solution. The main task of the discount factor is to penalize long prediction horizons. Since we want to make a comparison we introduce it also in the setting of MPC, although this is not a standard choice. As we mentioned, MPC will just provide a first guess which is used to define the domain where we are solving the HJB equation.

In this test the chosen parameters are: $u \in [-1,1]^2$, $\rho = 0.2$, $\Omega = [-4,6]^2$, $\Delta t_{MPC} = 0.05 = \Delta t_{HJB}$, $\Delta x_{HJB} = 0.025$, $\Delta \tau = 0.01$ (the time step to integrate the trajectories). In particular, we focus on $\lambda = 0.1$ and $\lambda = 1$. The number of controls are 21^2 for the value function and 3^2 for the trajectories. Note that the time step used in the HJB approach for the approximation of the trajectory

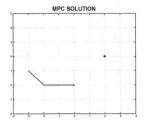
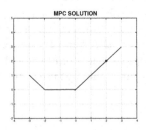

Fig. 1. Test 1: MPC solver with $\lambda = 0.1$ (left) and $\lambda = 1$ (right)

$(\Delta\tau)$ is smaller than the one used for MPC: this is because with MPC we want to have a rough and quick approximation of the solution. In Fig. 1, we show the results of MPC with $\lambda = 0.1$ on the left and $\lambda = 1$ on the right. As one can see, none of them is an accurate solution. In the first case, the solution goes to the local minimum $(0,0)$ and is trapped there, whereas when we increase λ the optimal solution does not stop at the global minimum y_2. On the other hand these two approximations help us to localize the behavior of the optimal solution in order to apply the Bellman equation in a reference domain Ω_ρ.

In Fig. 2, we show the contour lines of value function in the whole interval Ω for $\lambda = 1$ and the corresponding value function in Ω_ρ. Finally, the optimal trajectories for $\lambda = 1$ are shown in Fig. 3. On the right we propose the optimal solution obtained by the approximation of the value function in Ω whereas, on the left we can see the first approximation of the MPC solver (dotted line), the tube (solid lines) and the optimal solution via Bellman equation (dashed line). As you can see in the pictures, the solutions provided from the DP approach in Ω and Ω_ρ are able to reach the global desired minimum y_2. In Table 1, we present the CPU time and the evaluation of the cost functional for different tests. As far as the CPU time is concerned, in the fourth column we show the global time needed to get the approximation of the value function in the whole domain and the time to obtain the optimal trajectory, whereas the third column shows the global time needed to compute all the steps of our LDP algorithm: the trajectory obtained via MPC, to build the tube, to compute the value function in the reduced domain and to compute the optimal trajectory. As we expected, the value of the cost functional is lower when we compute the value function in the whole domain (just because $\Omega_\rho \subset \Omega$). It is important to note that the approximation in Ω_ρ guarantees a reduction of the CPU time of the 62.5%.

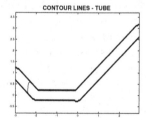
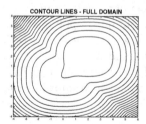

Fig. 2. Test 1: contour lines of the value function in the tube Ω_ρ (left) and in Ω (right).

Test 2: Van der Pol Dynamics. In this test we consider the two-dimensional nonlinear system dynamics given by the Van Der Pol oscillator:

$$\begin{cases} \dot{x}(t) = y(t) \\ \dot{y}(t) = (1 - x(t)^2)y(t) - x(t) + u(t) \\ x(0) = x_0,\ y(0) = y_0. \end{cases} \qquad (12)$$

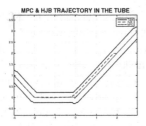 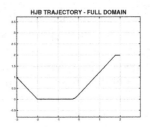

Fig. 3. Test 1: optimal trajectory via MPC (dotted line) and via HJB (dashed line) in the tube (solid lines) (left), optimal trajectory via HJB in Ω (right).

Table 1. A comparison of CPU time (seconds) and values of the cost functional.

$\lambda = 1$	MPC N $=5$	HJB in Ω_ρ	HJB in Ω
CPU	16 s	239 s	638 s
$J_x(u)$	5.41	5.33	5.3

The cost functional we want to minimize with respect to u is:

$$J_x(u) := \int_0^\infty (x^2 + y^2) e^{-\lambda t}\, dt. \tag{13}$$

We are dealing with a standard tracking problem where the state we want to reach is the origin. The chosen parameters are: $\lambda = \{0.1, 1\}$, $u \in [-1, 1]$, $\rho = 0.4$, $\Omega = [-6, 6]^2$, $\Delta t_{MPC} = 0.05 = \Delta t_{HJB}$, $\Delta x_{HJB} = 0.025$, $\Delta \tau = 0.01$, $x_0 = -3$, $y_0 = 2$. We took 21 controls for the approximation of the value function and 3 for the optimal trajectory. In Fig. 4, we present the optimal trajectory: on the right, the one obtained solving the HJB equation in the whole domain, on the left, the one obtained applying the algorithm we propose.

In Table 2 we present the CPU time and the evaluation of the cost functional with $\lambda = 0.1$ and $\lambda = 1$. In both cases we can observe that the algorithm we propose is faster than solving HJB in the whole domain and the cost functional provides a value which improves the one obtained with the MPC algorithm.

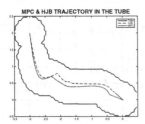 

Fig. 4. Test 2: optimal trajectory via MPC (dotted line) and via HJB (dashed line) in the tube Ω_ρ (left) and in Ω (right) for $\lambda = 0.1$.

Table 2. Test 2: a comparison of CPU time (seconds) and values of the cost functional for $\lambda = \{0.1, 1\}$.

$\lambda = 0.1$	MPC N = 10	HJB in Ω_ρ	HJB in Ω
CPU	79 s	155 s	228 s
$J_x(u)$	14.31	13.13	12.41
$\lambda = 1$	MPC N = 10	HJB in Ω_ρ	HJB in Ω
CPU	23 s	49 s	63 s
$J_x(u)$	6.45	6.09	6.07

4 Conclusions

We have proposed a local version of the dynamic programming approach for the solution of the infinite horizon problem showing that the coupling between MPC and DP methods can produce rather accurate results. The coupling improves the original guess obtained by the MPC method and allows to save memory allocations and CPU time with respect to the global solution computed via Hamilton-Jacobi equations. An extension of this approach to other classical control problems and more technical details on the choice of the parameters λ and ρ will be given in a future paper.

References

1. Alla, A., Falcone, M., Kalise, D.: An efficient policy iteration algorithm for dynamic programming equations. SIAM J. Sci. Comput. **37**(1), 181–200 (2015)
2. Bardi, M., Capuzzo Dolcetta, I.: Optimal Control and Viscosity Solutions of Hamilton-Jacobi-Bellman Equations. Birkhäuser, Basel (1997)
3. Falcone, M.: Numerical solution of dynamic programming equations. Appendix A in the volume. In: Bardi, M., Capuzzo Dolcetta, I. (eds.) Optimal Control and Viscosity Solutions of Hamilton-Jacobi-Bellman Equations, pp. 471–504. Birkhäuser, Boston (1997)
4. Falcone, M., Ferretti, R.: Semi-Lagrangian Approximation Schemes for Linear and Hamilton-Jacobi Equations. SIAM (2014)
5. Grüne, L., Pannek, J.: Nonlinear Model Predictive Control. Springer, London (2011)
6. Nocedal, J., Wright, S.J.: Numerical Optimization. Operation Research and Financial Engineering, 2nd edn. Springer, New York (2006)
7. Rawlings, J.B., Mayne, D.Q.: Model Predictive Control: Theory and Design. Nob Hill Publishing, LLC, Madison (2009)
8. Sethian, J.A.: Level Set Methods and Fast Marching Methods. Cambridge University Press, Cambridge (1999)
9. Zhao, H.: A fast sweeping method for Eikonal equations. Math. Comput. **74**, 603–627 (2005)

Real Options and Threshold Strategies

Vadim Arkin and Alexander Slastnikov[✉]

Central Economics and Mathematics Institute, Russian Academy of Sciences,
Nakhimovskii pr. 47, 117418 Moscow, Russia
slast@cemi.rssi.ru

Abstract. The paper deals an investment timing problem appearing in real options theory. The present values from an investment project are modeled by general diffusion process. We find necessary and sufficient conditions under which the optimal investment time is induced by a threshold strategy. We study also conditions for optimality of the threshold strategy (over all threshold strategies) and discuss the connection between the solutions to the investment timing problem and the free-boundary problem.

Keywords: Real options · Investment timing problem · Diffusion process · Optimal stopping · Threshold stopping time · Free-boundary problem

1 Introduction

One of the fundamental problems in real options theory concerns the determination of the optimal time for investment into a given project (see, e.g., the classical monograph [6]).

Let us think of an investment project, for example, a founding of a new firm in the real sector of economy. This project is characterized by a pair $(X_t, t \geq 0, I)$, where X_t is the present value of the firm founded at time t, and I is a cost of investment required to implement the project (for example, to found the firm). The input and the output production prices are assumed to be stochastic, so $X_t, t \geq 0$ is considered as a stochastic process, defined on a general filtered probability space $(\Omega, \mathcal{F}, \{\mathcal{F}_t, t \geq 0\}, \mathbf{P})$. This model assumes that:

- at any moment, a decision-maker (investor) can either *accept* the project and proceed with the investment or *delay* the decision until he obtains new information;
- investment are considered to be instantaneous and irreversible so that they cannot be withdrawn from the project any more and used for other purposes.

The investor's problem is to evaluate the project and determine an appropriate time for the investment (investment timing problem). In real options theory investment times are considered as stopping times (adapted to the flow of σ-algebras $\{\mathcal{F}_t, t \geq 0\}$).

© IFIP International Federation for Information Processing 2016
Published by Springer International Publishing AG 2016. All Rights Reserved
L. Bociu et al. (Eds.): CSMO 2015, IFIP AICT 494, pp. 78–88, 2016.
DOI: 10.1007/978-3-319-55795-3_6

In real options theory there are two different approaches to solving investment timing problem (see [6]).

The project value under the first approach is the maximum of the net present value from the implemented project over all stopping times (investment rules):

$$F = \max_{\tau} \mathbf{E}(X_\tau - I)e^{-\rho\tau}, \tag{1}$$

where ρ is a given discount rate. An optimal stopping time τ^* in (1) is viewed as the optimal investment time (investment rule).

Within the second approach an opportunity to invest is considered as an American call option – the right but not obligation to buy the asset on predetermined price. The exercise time is viewed as an investment time, and the option value is accepted as the investment project value. In this framework a project is spanned with some traded asset, whose price is completely correlated with the present value X_t of the project. In order to evaluate the (rational) value of this real option one can use methods of financial options pricing theory, especially, contingent claims analysis (see, e.g., [6]).

In this paper we follow the approach that the optimal investment timing decision can be mathematically determined as the solution of an optimal stopping problem (1). Such an approach originated in the well-known McDonald–Siegel model (see [6,11]), in which the underlying present value's dynamics is modeled by a geometric Brownian motion. The majority of results on this problem (optimal investment strategy) has a threshold structure: to invest when the present value of the project exceeds a certain level (threshold). At a heuristic level this is so for the cases of geometric Brownian motion, arithmetic Brownian motion, mean-reverting process and a few others (see [6]). However the following general question arises: For which underlying processes the optimal decision in the investment timing problem will have a threshold structure?

Another investment timing problem (with additional restrictions on stopping times) was considered by Alvarez [1], who established sufficient (but not necessary) conditions for optimality.

In this paper we focus our attention on finding of necessary and sufficient conditions for optimality of threshold strategies in the investment timing problem. Since this problem is a special case of the optimal stopping problem, a similar question may be addressed in the general optimal stopping problem: Under what conditions (on both process and payoff function) the optimal stopping time will have a threshold structure? Some results in this direction (in the form of necessary and sufficient conditions) were obtained in [2,3,5] under additional assumptions on underlying process and/or payoffs.

The paper is organized as follows. After a formal description of the investment timing problem and the assumptions on the underlying process (Sect. 2.1), we turn to study the threshold strategies in this problem. Since the investment timing problem by threshold strategies is reduced to one-dimensional maximization problem, then a related problem is to find the optimal threshold. In Sect. 2.2 we give necessary and sufficient conditions for the optimal threshold (over all thresholds). Solving a free-boundary problem (based on smooth-pasting principle) is the most commonly used method (but this is not the only method, see, e.g., [12]) that allows to find a solution to the optimal stopping problem. In Sect. 2.3 we discuss the connection between solutions to the investment timing problem and the free-boundary problem. Finally, in Sect. 2.4 we prove the main result on necessary and sufficient conditions under which the optimal investment time is generated by a threshold strategy.

2   Investment Timing Problem

Let I be the cost of investment required for implementing a project, and X_t the present value from the project started at time t. As usual the investment is supposed to be instantaneous and irreversible, and the project—infinitely-lived.

At any time a decision-maker (investor) can either *accept* the project and proceed with the investment or *delay* the decision until she/he obtains new information regarding its environment (prices of the product and resources, demand etc.). The goal of a decision-maker in this situation is to use the available information and find the optimal time for investing in the project (investment timing problem), i.e., find a time τ that maximizes the net present value from the project:

$$\mathbf{E}^x \left(X_\tau - I \right) e^{-\rho \tau} \mathbf{1}_{\{\tau < \infty\}} \to \max_{\tau \in \mathcal{M}}. \tag{2}$$

Here $\mathbf{E}^x$ is the expectation for the process X_t starting from the initial state x, $\mathbf{1}_A$ is indicator function of the set A, and the maximum is taken over all stopping times τ from a certain class $\mathcal{M}$ of stopping times[1].

We treat the interesting case $I < r$; otherwise the optimal time in (2) will be $+\infty$.

2.1   Mathematical Assumptions

Let $X_t, t > 0$ be a diffusion process with values in the interval $D \subseteq \mathbb{R}^1$ with boundary points l and r, where $-\infty \leq l < r \leq +\infty$, open or closed (i.e. it may be (l, r), $[l, r)$, $(l, r]$, or $[l, r]$), which is the solution to the stochastic differential equation:

$$dX_t = a(X_t)dt + \sigma(X_t)dw_t, \quad X_0 = x, \tag{3}$$

where w_t is a standard Wiener process, $a : D \mapsto \mathbb{R}^1$ and $\sigma : D \mapsto \mathbb{R}^1_+$ are the drift and the diffusion coefficients, respectively. Denote $\mathcal{I} = \mathrm{int}(D) = (l, r)$.

The process X_t is assumed to be regular; this means that, starting from an arbitrary point $x \in \mathcal{I}$, this process reaches any point $y \in \mathcal{I}$ in finite time with positive probability.

It is known that the following local integrability condition:

$$\int_{x-\varepsilon}^{x+\varepsilon} \frac{1 + |a(y)|}{\sigma^2(y)} dy < \infty \quad \text{for some } \varepsilon > 0, \tag{4}$$

---

[1] We consider stopping times which can take infinite values (with positive probability).

at any $x \in \mathcal{I}$ guarantees the existence of a weak solution of equation (3) and its regularity (see, e.g. [10]).

The process X_t is associated with the infinitesimal operator

$$\mathbb{L}f(x) = a(x)f'(x) + \frac{1}{2}\sigma^2(x)f''(x). \tag{5}$$

Under the condition (4) there exist (unique up to constant positive multipliers) increasing and decreasing functions $\psi(x)$ and $\varphi(x)$ with absolutely continuous derivatives, which are the fundamental solutions to the ODE

$$\mathbb{L}f(p) = \rho f(p) \tag{6}$$

almost sure (with respect to Lebesque measure) on the interval $\mathcal{I}$ (see, e.g. [10, Chap. 5, Lemma 5.26]). Moreover, $0 < \psi(p)$, $\varphi(p) < \infty$ for $p \in \mathcal{I}$. Note, if the functions $a(x)$, $\sigma(x)$ are continuous, then ψ, $\varphi \in C^2(\mathcal{I})$.

2.2   Optimality of Threshold Strategies

Let us define $\tau_p = \tau_p(x) = \inf\{t \geq 0 : X_t \geq p\}$—the first time when the process X_t, starting from x, *exceeds* level p. The time τ_p is said to be a threshold stopping time generated by a threshold strategy—to stop when the process exceeds threshold p. Let $\mathcal{M}_{\text{th}} = \{\tau_p, \ p \in \mathcal{I}\}$ be a class of all such threshold stopping times.

For the class $\mathcal{M}_{\text{th}}$ the investment timing problem (2) can be written as follows:

$$(p - I)\,\mathbf{E}^x e^{-\rho\tau_p} \rightarrow \max_{p \in (l,r)}. \tag{7}$$

Such a problem appeared in [7] as the heuristic method for solving a general investment timing problem (2) over the class of all stopping times.

We say that the threshold p^* is optimal for the investment timing problem (7) if the threshold stopping time τ_{p^*} is optimal in (7). The following result gives necessary and sufficient conditions for the optimal threshold.

Theorem 1. *Threshold $p^* \in \mathcal{I}$ is optimal in the problem (7) for all $x \in \mathcal{I}$, if and only if the following conditions hold:*

$$\frac{p - I}{\psi(p)} \leq \frac{p^* - I}{\psi(p^*)} \quad \text{if } p < p^*; \tag{8}$$

$$\frac{p - I}{\psi(p)} \quad \text{does not increase for } p \geq p^*, \tag{9}$$

where $\psi(p)$ is an increasing solution to the ODE (6).

Proof. Let us denote the left-hand side in (7) by $V(p; x)$. Obviously, $V(p; x) = x - I$ for $x \geq p$.

Along with the above stopping time τ_p let us define the first hitting time to the threshold p: $T_p = \inf\{t \geq 0 : X_t = p\}$, $p \in (l, r)$.

For $x < p$, clearly, $\tau_p = T_p$ and using known formula $\mathbf{E}^x e^{-\rho T_p} = \psi(x)/\psi(p)$ (see, e.g., [4,9]) we obtain:

$$V(p;x) = (p-I)\mathbf{E}^x e^{-\rho T_p} \mathbf{1}_{\{T_p < \infty\}} = (p-I)\mathbf{E}^x e^{-\rho T_p} = \frac{p-I}{\psi(p)}\psi(x). \qquad (10)$$

Denote $h(x) = (x-I)/\psi(x)$.

(i) *Necessity.* Let $p^* \in \mathcal{I}$ be an optimal threshold in the problem (7) for all $x \in \mathcal{I}$. Then for $p < p^*$ we have

$$V(p;p) = p - I \leq V(p^*;p) = \frac{p^*-I}{\psi(p^*)}\psi(p),$$

i.e. (8) holds. If $p^* \leq p_1 < p_2$, then

$$V(p_2;p_1) = h(p_2)\psi(p_1) \leq V(p^*;p_1) = p_1 - I = h(p_1)\psi(p_1),$$

and it follows that (9) is true.

(ii) *Sufficiency.* Now, suppose that conditions (8) and (9) hold.
Let $p < p^*$. If $x \geq p^*$, then $V(p;x) = x - I = V(p^*;x)$.
If $p \leq x < p^*$, then, due to (8), $V_p(x) = x - I = h(x)\psi(x) \leq h(p^*)\psi(x) = V(p^*;x)$.
Finally, if $x < p$, then, using (8) and (10), we obtain:
$V(p;x) = h(p)\psi(x) \leq h(p^*)\psi(x) = V(p^*;x)$.
Consider the case $p > p^*$. If $x \geq p$, then $V(p;x) = x - I = V(p^*;x)$.
Whenever $p^* \leq x < p$, then, due to (9), $V(p;x) = h(p)\psi(x) \leq h(x)\psi(x) = x - I = V(p^*;x)$.
When $x < p^*$, then $V(p;x) = h(p)\psi(x) \leq h(p^*)\psi(x) = V(p^*;x)$, since $h(p) \leq h(p^*)$.
Theorem is completely proved.

Remark 1. The condition (9) is equivalent to the inequality

$$(p-I)\psi'(p) \geq \psi(p) \quad \text{for } p \geq p^*.$$

This relation implies, in particular, that the optimal threshold p^* must be strictly greater than the cost I (because the values $\psi(p^*)$, $\psi'(p^*)$ are positive).

Remark 2. Assume that $\log \psi(x)$ is a convex function, i.e. $\psi'(x)/\psi(x)$ increases. In this case there exists a unique point p^* which satisfies the equation

$$(p^*-I)\psi'(p^*) = \psi(p^*). \qquad (11)$$

This value p^* constitutes the optimal threshold in the problem (7) for all $x \in \mathcal{I}$. Indeed, the sign of the derivative of the function $(p-I)/\psi(p)$ coincides with the sign of $\psi(p) - (p-I)\psi'(p)$. Therefore, in the considered case the conditions (8) and (9) in Theorem 1 are true automatically.

There are a number of cases of diffusion processes X_t which are more or less realistic for modeling the present values of a project. Some of them are listed below.

(1) *Geometric Brownian motion (GBM):*

$$dX_t = X_t(\alpha dt + \sigma dw_t). \tag{12}$$

In this case $\psi(x) = x^\beta$, where β is the positive root of the equation $\frac{1}{2}\sigma^2\beta(\beta - 1) + \alpha\beta - \rho = 0$.

(2) *Arithmetic Brownian motion (ABM):*

$$dX_t = x + \alpha dt + \sigma dw_t. \tag{13}$$

In this case $\psi(x) = e^{\beta x}$, where β is the positive root of the equation $\frac{1}{2}\sigma^2\beta^2 + \alpha\beta - \rho = 0$.

(3) *Mean-reverting process (or geometric Ornstein–Uhlenbeck process):*

$$dX_t = \alpha(\bar{x} - X_t)X_t dt + \sigma X_t\, dw_t. \tag{14}$$

In this case $\psi(x) = x^\beta {}_1F_1\left(\beta, 2\beta + \dfrac{2\alpha\bar{x}}{\sigma^2}; \dfrac{2\alpha}{\sigma^2}x\right)$, where β is the positive root of equation $\frac{1}{2}\sigma^2\beta(\beta - 1) + \alpha\bar{x}\beta - \rho = 0$, and ${}_1F_1(p, q; x)$ is the confluent hypergeometric function satisfying Kummer's equation $xf''(x) + (q - x)f'(x) - pf(x) = 0$.

(4) *Square-root mean-reverting process (or Cox–Ingersoll–Ross process):*

$$dX_t = \alpha(\bar{x} - X_t)dt + \sigma\sqrt{X_t}\, dw_t. \tag{15}$$

In this case $\psi(x) = {}_1F_1\left(\dfrac{\rho}{\alpha}, \dfrac{2\alpha\bar{x}}{\sigma^2}; \dfrac{2\alpha}{\sigma^2}x\right)$.

The above processes are well studied in the literature (in connection with real options and optimal stopping problems see, e.g., [6,8]).

For the first two processes, (12) and (13), Theorem 1 gives explicit formulas for the optimal threshold in the investment timing problem:

$$p^* = \frac{\beta}{\beta - 1}I \text{ for the GBM,} \quad \text{and } p^* = I + \frac{1}{\beta} \text{ for the ABM.}$$

On the contrary, for mean-reverting processes (14) and (15) the function $\psi(x)$ is represented as an infinite series, and the optimal threshold can be find only numerically.

So, Theorem 1 states that optimal threshold p^* is a point of maximum for the function $h(x) = (x - I)/\psi(x)$. This implies the first-order optimality condition $h'(p^*) = 0$, i.e. the equality (11), and smooth-pasting principle:

$$V_x'(p^*; x)\big|_{x=p^*} = 1.$$

In the next section we discuss smooth-pasting principle and appropriate free-boundary problem more closely.

2.3   Threshold Strategies and Free-Boundary Problem

There is a common opinion (especially among engineers and economists) that the solution to a free-boundary problem always gives a solution to an optimal stopping problem.

A free-boundary problem in the case of threshold strategies in the investment timing problem can be formulated as follows: find the threshold $p^* \in (l, r)$ and a twice differentiable function $H(x)$, $l < x < p^*$, such that

$$\mathbb{L}H(x) = \rho H(x), \quad l < x < p^*; \tag{16}$$

$$H(p^*-0) = p^* - I, \quad H'(p^*-0) = 1. \tag{17}$$

If $\psi(x)$ is a twice differentiable function, then the solution to the problem (16) and (17) has the form

$$H(x) = \frac{p^* - I}{\psi(p^*)}\psi(x), \quad l < x < p^*. \tag{18}$$

Here $\psi(x)$ is an increasing solution to the ODE (6) and p^* satisfies the smooth-pasting condition (11). We call such p^* a solution to a free-boundary problem.

According to Theorem 1 the optimal threshold in problem (7) must be the point of maximum of the function $h(x) = (x - I)/\psi(x)$. However the smooth-pasting condition (11) provides only a stationary point for $h(x)$. Thus, we can apply standard second-order optimality conditions to derive relations between the solutions to the investment timing problem and the free-boundary problem.

Let p^* be a solution to the free-boundary problem (16) and (17). If p^* is also an optimal threshold in the investment timing problem (7), then, of course, $h''(p^*) \leq 0$. This means that

$$\psi''(p^*) = -\frac{h''(p^*)\psi(p^*) + 2h'(p^*)\psi'(p^*)}{h(p^*)} = -\frac{h''(p^*)\psi(p^*)}{h(p^*)} \geq 0.$$

Thus, the inequality $\psi''(p^*) \geq 0$ may be viewed as a necessary condition for a solution of the free-boundary problem to be optimal in the investment timing problem. The inverse relation between solutions can be stated as follows.

Statement 1. *If p^* is the unique solution to the free-boundary problem (16) and (17), and $\psi''(p^*) > 0$, then p^* is an optimal threshold in the problem (7) for all $x \in \mathcal{I}$.*

Proof. Since $h'(p^*) = 0$ and $\psi''(p^*) > 0$ then $h''(p^*) = -h(p^*)\psi''(p^*)/\psi(p^*) < 0$. Therefore, $h'(p)$ strictly decreases at some neighborhood of p^*.

Then, it is easy to see that $h'(p) > 0$ for $p < p^*$ and $h'(p) < 0$ for $p > p^*$. Otherwise $h'(q) = 0$ for some $q \neq p^*$, that contradicts the uniqueness of the solution to the free-boundary problem (16) and (17). Therefore, conditions (8) and (9) hold and Theorem 1 gives the optimality of threshold p^*.

The following result concerns the general case when the free-boundary problem has several solutions.

Statement 2. *Let p^* and $\tilde{p}$ be two solutions to the free-boundary problem (16) and (17) such that $\psi''(p^*) > 0$ and $(x-I)/\psi(x) \leq (p^*-I)/\psi(p^*)$ for $l < x < p^*$. If $\tilde{p} > p^*$ is such that $\psi^{(k)}(\tilde{p}) = 0, k = 2, \ldots, n-1$ and $\psi^{(n)}(\tilde{p}) > 0$ for some $n > 2$, then p^* is an optimal threshold in the problem (7) for all $x \in \mathcal{I}$.*

Proof. Let us prove that $h'(p) \leq 0$ for all $p > p^*$. The inequality $\psi''(p^*) > 0$ implies (as above) that $h''(p^*) < 0$, and, therefore, $h'(p) < 0$ for all $p^* < p < p_1$ with some p_1. If we suppose that $h'(p_2) > 0$ for some $p_2 > p^*$, then there exists $p_0 \in (p_1, p_2)$ such that $h'(p_0) = 0$ and $h'(p) > 0$ for all $p_0 < p < p_2$. Therefore, p_0 would be another solution to the free-boundary problem (16)–(17). The conditions of the Statement imply that $h^{(k)}(p_0) = 0$, $k = 2, \ldots, n-1$, $h^{(n)}(p_0) < 0$ for some $n > 2$, which contradicts the positivity of $h'(p)$ for $p_0 < p < p_2$.

Hence, $h'(p) \leq 0$ for all $p > p^*$ and conditions (8) and (9) hold. Thus, according to Theorem 1, p^* is an optimal threshold in the problem (7).

2.4 Optimal Strategies in the Investment Timing Problem

Now, let us return to the 'general' investment timing problem (2).

A specific version of the investment timing problem (2) over the class $\mathcal{M}_0$ of stopping times τ such that $\tau < \tau(0) = \inf\{t \geq 0 : X_t \leq 0\}$ was considered by Alvarez [1]. He derived sufficient conditions under which an optimal investment time in (2) over the class $\mathcal{M}_0$ will be a threshold stopping time. However these conditions are not necessary.

In this section we give necessary and sufficient conditions (criterion) for optimality of the threshold stopping time in the investment timing problem (2) over the class of *all stopping times*.

To reduce some technical difficulties we assume below that the drift $a(x)$ and the diffusion $\sigma(x)$ of the underlying process X_t are continuous functions.

Theorem 2. *The threshold stopping time τ_{p^*}, $p^* \in \mathcal{I}$, is optimal in the investment timing problem (2) for all $x \in \mathcal{I}$ if and only if the following conditions hold:*

$$(p-I)\psi(p^*) \leq (p^*-I)\psi(p) \quad \text{for } p < p^*; \tag{19}$$

$$\psi(p^*) = (p^*-I)\psi'(p^*); \tag{20}$$

$$a(p) \leq \rho(p-I) \quad \text{for } p > p^*. \tag{21}$$

Here $\psi(x)$ is an increasing solution to the ODE (6) and $a(p)$ is the drift coefficient of the process X_t.

Proof. Define the value function for the problem (2) over the class $\mathcal{M}$ of all stopping times as follows:

$$V(x) = \sup_{\tau \in \mathcal{M}} \mathbf{E}^x \left(X_\tau - I \right) e^{-\rho\tau} \mathbf{1}_{\{\tau < \infty\}}.$$

(i) *Sufficiency.* Let conditions (19)–(21) hold. Take the function

$$\Phi(x) = V(p^*; x) = \begin{cases} \dfrac{p^* - I}{\psi(p^*)}\psi(x), & \text{for } x < p^*, \\ x - I, & \text{for } x \geq p^*. \end{cases}$$

Obviously, $\Phi(x) > 0$ (due to condition (20)) and $V(x) \geq \Phi(x)$ for all $x \in \mathcal{I}$. On the other hand, condition (19) implies

$$\frac{p^* - I}{\psi(p^*)}\psi(x) \geq \frac{x - I}{\psi(x)}\psi(x) = x - I.$$

Therefore $\Phi(x) \geq x - I$ for all $x \in (l, r)$, i.e. $\Phi(x)$ is a majorant of the specific payoff function $x - I$.

For any stopping time $\tau \in \mathcal{M}$ and a real number $N > 0$ put $\tilde{\tau} = \tau \wedge N$. From Itô–Tanaka–Meyer formula (see, e.g. [10]) we have:

$$\mathbf{E}^x\Phi(X_{\tilde{\tau}})e^{-\rho\tilde{\tau}} = \Phi(x) + \mathbf{E}^x \int_0^{\tilde{\tau}} (\mathbb{L}\Phi - \rho\Phi)(X_t)e^{-\rho t}dt$$
$$+ \frac{1}{2}\sigma^2(p^*)[\Phi'(p^*+0) - \Phi'(p^*-0)]\mathbf{E}^x \int_0^{\tilde{\tau}} e^{-\rho t}dL_t(p^*), \quad (22)$$

where $L_t(p^*)$ is the local time of the process X_t at the point p^*.

By definition and in view of condition (20) we have

$$\Phi'(p^* + 0) - \Phi'(p^* - 0) = 1 - \frac{p^* - I}{\psi(p^*)}\psi'(p^*) = 0.$$

Denote $T_1 = \{t : 0 \leq t \leq \tilde{\tau}, X_t < p^*\}$, $T_2 = \{t : 0 \leq t \leq \tilde{\tau}, X_t > p^*\}$. We have:

$$\mathbb{L}\Phi(X_t) - \rho\Phi(X_t) = \frac{p^* - I}{\psi(p^*)}\Big(\mathbb{L}\psi(X_t) - \rho\psi(X_t)\Big) = 0 \quad \text{for } t \in T_1,$$
$$\mathbb{L}\Phi(X_t) - \rho\Phi(X_t) = a(X_t) - \rho(X_t - I) \leq 0 \quad \text{for } t \in T_2.$$

These relations follow from the definition of the function $\psi(x)$ and in view of condition (21). Then

$$\mathbf{E}^x\Phi(X_{\tilde{\tau}})e^{-\rho\tilde{\tau}} \leq \Phi(x) + \mathbf{E}^x \left(\int_{T_1} (\mathbb{L}\Phi - \rho\Phi)(X_t)e^{-\rho t}dt + \int_{T_2} (\mathbb{L}\Phi - \rho\Phi)(X_t)e^{-\rho t}dt \right)$$
$$\leq \Phi(x).$$

Since $\Phi(X_{\tilde{\tau}})e^{-\rho\tilde{\tau}} \xrightarrow{\text{a.s.}} \Phi(X_\tau)e^{-\rho\tau}\mathbf{1}_{\{\tau<\infty\}}$ when $N \to \infty$, then due to Fatou's Lemma: $\mathbf{E}^x\Phi(X_\tau)e^{-\rho\tau}\mathbf{1}_{\{\tau<\infty\}} \leq \Phi(x)$ for all $\tau \in \mathcal{M}$ and $x \in \mathcal{I}$. Therefore, $\Phi(x)$ is a ρ-excessive function, which majorates the payoff function $x - I$. Since, by Dynkin's characterization, the value function $V(x)$ is the least ρ-excessive majorant, then $V(x) \leq \Phi(x)$.

Therefore, $V(x) = \Phi(x) = V(p^*; x)$, i.e. τ_{p^*} is the optimal stopping time in problem (2) for all x.

(ii) *Necessity.* Now, let τ_{p^*} be an optimal stopping time in the problem (2). Note, that τ_{p^*} will be also an optimal stopping time in the problem (7). Therefore, Theorem 1 implies conditions (19) and (20), since p^* is a point of maximum for the function $(x - I)/\psi(x)$.

Further, assume that inequality (21) is not true at some point $p_0 > p^*$, i.e. $a(p) > \rho(p - I)$ in some interval $J \subset (p^*, r)$ (by continuity). For some $\tilde{x} \in J$ define $\tau = \inf\{t \geq 0 : X_t \notin J\}$, where the process X_t starts from the point $\tilde{x}$. Then for any $N > 0$ from Dynkin's formula

$$\mathbf{E}^{\tilde{x}}(X_{\tau \wedge N} - I)e^{-\rho(\tau \wedge N)} = \tilde{x} - I + \mathbf{E}^{\tilde{x}} \int_0^{\tau \wedge N} [a(X_t) - \rho(X_t - I)]e^{-\rho t}dt > \tilde{x} - I.$$

Therefore, $V(\tilde{x}) > \tilde{x} - I$ which contradicts the relation $V(\tilde{x}) = V(p^*; \tilde{x}) = g(\tilde{x})$, since $\tilde{x} > p^*$.

Example 1. Let the process X_t be the geometric Brownian motion (12). Then Theorem 2 implies that the threshold stopping time τ_{p^*} will be optimal in the investment timing problem (2) over all investment times if and only if $p^* = I\beta/(\beta - 1)$, where β is the positive root of the equation $\frac{1}{2}\sigma^2\beta(\beta-1)+\alpha\beta-\rho = 0$.

Acknowledgments. The work was supported by Russian Foundation for Basic Researches (project 15-06-03723) and Russian Foundation for Humanities (project 14-02-00036).

References

1. Alvarez, L.H.R.: Reward functionals, salvage values, and optimal stopping. Math. Methods Oper. Res. **54**, 315–337 (2001)
2. Arkin, V.I.: Threshold strategies in optimal stopping problem for one-dimensional diffusion processes. Theory Probab. Appl. **59**, 311–319 (2015)
3. Arkin, V.I., Slastnikov, A.D.: Threshold stopping rules for diffusion processes and Stefan's problem. Dokl. Math. **86**, 626–629 (2012)
4. Borodin, A.N., Salminen, P.: Handbook of Brownian Motion - Facts and Formulae. Birkhäuser, Basel (2002)
5. Crocce, F., Mordecki, E.: Explicit solutions in one-sided optimal stopping problems for one-dimensional diffusions. Stochastics **86**, 491–509 (2014)
6. Dixit, A., Pindyck, R.S.: Investment Under Uncertainty. Princeton University Press, Princeton (1994)
7. Dixit, A., Pindyck, R.S., Sødal, S.: A markup interpretation of optimal investment rules. Econ. J. **109**, 179–189 (1999)
8. Johnson, T.C., Zervos, M.: A discretionary stopping problem with applications to the optimal timing of investment decisions (2005, preprint). https://vm171.newton.cam.ac.uk/files/preprints/ni05045.pdf
9. Ito, K., McKean, H.: Diffusion Processes and Their Sample Paths. Springer, Berlin (1974)

10. Karatzas, I., Shreve, S.E.: Brownian Motion and Stochastic Calculus, 2nd edn. Springer, Berlin (1991)
11. McDonald, R., Siegel, D.: The value of waiting to invest. Q. J. Econ. **101**, 707–728 (1986)
12. Peskir, G., Shiryaev, A.: Optimal Stopping and Free-Boundary Problems. Birkhäuser, Basel (2006)

Electrostatic Approximation of Vector Fields

Giles Auchmuty[(⊠)]

Department of Mathematics, University of Houston, Houston, TX 77204-3008, USA
auchmuty@uh.edu

Abstract. This paper provides expressions for the boundary potential that provides the best electrostatic potential approximation of a given L^2 vector field on a nice bounded region in $\mathbb{R}^N$. The permittivity of the region is assumed to be known and the potential is required to be zero on the conducting part of the boundary. The boundary potential is found by solving the minimization conditions and using a special basis of the trace space for the space of allowable potentials. The trace space is identified by its representation with respect to a basis of Σ-Steklov eigenfunctions.

Keywords: Boundary control · Trace spaces · Best approximation · Steklov eigenproblems

2010 Mathematics Subject Classification: Primary 35Q60 · Secondary 35J25 · 78A30

1 Introduction

Quite often in physical applications one wishes to produce electrostatic fields in a region of known permittivity $\epsilon(x)$ that approximate a prescribed (given) field **F**. Such fields are determined by their boundary values, so a natural question is what imposed potentials on the boundary provide good approximations, in an energy norm, to **F**? Very often the boundary includes surface patches that are conductors as well as patches where nonzero potentials may be imposed. This may be regarded as a problem of stationary control or approximation.

The difficulty with such problems has been how to work with the control space of allowable boundary conditions as it will be a trace space of allowable H^1 functions. The standard Lions-Magenes description of trace spaces is not amenable to nice constructions of solutions for problems of this type. This problem is treated here using methods based on the spectral characterization of trace spaces as described in Auchmuty [3,4] which provides constructive methods and explicit bases for the traces. The allowable trace space is characterized as being isomorphic to a class of weak solutions of a linear elliptic equation and a basis of Σ-Steklov eigenfunctions of this space is identified.

G. Auchmuty—This research has been supported in part by NSF award DMS 11008754.

L. Bociu et al. (Eds.): CSMO 2015, IFIP AICT 494, pp. 89–94, 2016.
DOI: 10.1007/978-3-319-55795-3_7

Here an explicit expression for the best approximation in terms of the data is found. Moreover boundary data that is close in a boundary norm to this optimal solution will provide good approximations to the field $\mathbf{F}$ in an energy norm on Ω. The analysis described here is described for quite general N-dimensional regions since the results are essentially independent of the dimension $N \geq 2$ and the methods may be of interest for other approximation questions.

2 Definitions and Requirements

To analyze this problem, standard definitions, terminology and assumptions will be used as in Attouch et al. [1]. All functions in this paper will take values in $\overline{\mathbb{R}} := [-\infty, \infty]$, derivatives should be taken in a weak sense and $N \geq 2$ throughout.

A *region* is a non-empty, connected, open subset of $\mathbb{R}^N$. Its closure is denoted $\overline{\Omega}$ and its boundary is $\partial\Omega := \overline{\Omega} \setminus \Omega$. Let $L^p(\Omega), H^1(\Omega)$ be the usual real Lebesgue and Sobolev spaces of functions on Ω. The norm on $L^p(\Omega)$ is denoted $\|.\|_p$ and the inner product on $L^2(\Omega)$ by $\langle .,. \rangle$. The basic requirement on Ω is

(B1): Ω *is a bounded region in* $\mathbb{R}^N$ *whose boundary* $\partial\Omega$ *is the union of a finite number of disjoint closed Lipschitz surfaces; each surface having finite surface area.*

The region Ω is said to satisfy *Rellich's theorem* provided the imbedding of $H^1(\Omega)$ into $L^p(\Omega)$ is compact for $1 \leq p < p_S$ where $p_S(N) := 2N/(N-2)$ when $N \geq 3$, or $p_S(2) = \infty$ when $N = 2$.

The *trace map* is the linear extension of the map restricting Lipschitz continuous functions on $\overline{\Omega}$ to $\partial\Omega$. When (B1) holds, this map has an extension to $W^{1,1}(\Omega)$ and then the trace of u on $\partial\Omega$ will be Lebesgue integrable with respect to σ, see [5], Sect. 4.2 for details. The region Ω is said to satisfy the *compact trace theorem* provided the trace mapping $\gamma : H^1(\Omega) \to L^2(\partial\Omega, d\sigma)$ is compact. We will use the inner product

$$[u,v]_\partial := \int_\Omega \nabla u(x) \cdot \nabla v(x) \, dx + \int_{\partial\Omega} \gamma(u)\,\gamma(v) \, d\sigma \qquad (2.1)$$

on $H^1(\Omega)$ and the associated norm is denoted $\|u\|_\partial$. This is an equivalent inner product to the usual inner product when Ω obeys (B1) - see [2] for a proof. Here $\nabla u := (D_1 u, \ldots, D_N u)$ is the gradient of the function u.

Our interest is in a problem that arises in electrostatics where part of the boundary Σ is a conductor and a potential can be imposed on the complementary part of the boundary $\tilde{\Sigma} := \partial\Omega \setminus \overline{\Sigma}$. Mathematically our requirements are

(B2): Σ *is an nonempty open subset of* $\partial\Omega$, Σ *and* $\tilde{\Sigma}$ *have strictly positive surface measure and* $\sigma(\partial\Sigma) = 0$.

A function $u \in H^1(\Omega)$ is said to be in $H^1_{\Sigma 0}(\Omega)$ provided $\gamma(u) = 0$, $\sigma a.e.$ on Σ. This is equivalent to requiring that

$$\int_{\partial\Omega} \gamma(u)\,\gamma(v) \, d\sigma = 0 \quad \text{for all} \quad v \in X_\Sigma \qquad (2.2)$$

Let X be the space $H^1(\Omega) \cap C(\overline{\Omega})$ and X_Σ be the subspace of functions in X with $supp\, v \cap \partial\Omega \subset \Sigma$. The space $H^1_{\Sigma 0}(\Omega)$ is a closed subspace of $H^1(\Omega)$ that contains $H^1_0(\Omega)$.

3   The Boundary Control Problem

The problem to be studied here is given a vector field $\mathbf{F}$ on Ω to find the potential φ that provides the best L^2-approximation when the region Ω has known permittivity tensor $\epsilon(.)$ and part of the boundary Σ is a conductor held at zero potential. That is we want to find the function $\tilde{\varphi}$ that minimizes

$$\| \epsilon\nabla\varphi - \mathbf{F} \|_2^2 := \int_\Omega |\epsilon\nabla\varphi - \mathbf{F}|^2 \, dx \qquad \text{over all} \quad \varphi \in H^1_{\Sigma 0}(\Omega). \qquad (3.3)$$

with $|.|$ is the Euclidean norm on $\mathbb{R}^N$. Since $\mathbf{F}$ is known this reduces to minimizing the functional

$$\mathcal{E}(\varphi) := \int_\Omega \left[(A(x)\nabla\varphi) \cdot \nabla\varphi - 2\, \mathbf{G} \cdot \nabla\varphi \right] dx \qquad \text{over all} \quad \varphi \in H^1_{\Sigma 0}(\Omega). \quad (3.4)$$

where $A(x) := \epsilon(x)^T \epsilon(x)$ is real symmetric, $\mathbf{G}(\mathbf{x}) := \epsilon(\mathbf{x})^{\mathbf{T}}\mathbf{F}(\mathbf{x})$ on Ω and the superscript T denotes the vector transpose. The following will be assumed.

(A1): $A(x) := (a_{jk}(x))$ is a real symmetric matrix whose components are bounded Lebesgue-measurable functions on Ω and there exist constants c_0, c_1 such that

$$c_0 |\xi|^2 \leq \xi^T A(x)\xi \leq c_1 |\xi|^2 \qquad \text{for all} \quad \xi \in \mathbb{R}^N, \; x \in \Omega. \qquad (3.5)$$

Existence uniqueness and extremality conditions for this problem may be obtained using standard methods. The problem is a convex quadratic minimization problem on a Hilbert space so the existence may be stated as follows.

Theorem 3.1. *Assume that (A1), (B1), (B2) hold and $\mathbf{F} \in L^2(\Omega : \mathbb{R}^N)$ is given, then there is a unique minimizer $\tilde{\varphi}$ of $\mathcal{E}$ on $H^1_{\Sigma 0}(\Omega)$.*

The functional $\mathcal{E}$ also is G-differentiable so the minimizers satisfy the following.

Theorem 3.2. *Assume that (A1), (B1), (B2) hold and $\mathbf{F} \in L^2(\Omega : \mathbb{R}^N)$ is given, then the minimizer $\tilde{\varphi}$ of $\mathcal{E}$ on $H^1_{\Sigma 0}(\Omega)$ satisfies the equation*

$$\int_\Omega (A(x)\nabla\varphi - \mathbf{G}) \cdot \nabla\psi \; \mathrm{dx} = 0 \quad \text{for all} \quad \psi \in H^1_{\Sigma 0}(\Omega). \qquad (3.6)$$

To obtain further results about this problem a decomposition of the space $H^1_{\Sigma 0}(\Omega)$ will be used. Consider the bilinear form $a : H^1_{\Sigma 0}(\Omega) \to \mathbb{R}$ defined by

$$a(u,v) := [u,v]_a := \int_\Omega (A\nabla u) \cdot \nabla v \; d^N x + \int_{\tilde{\Sigma}} \gamma(u)\,\gamma(v) \; d\sigma \qquad (3.7)$$

This bilinear form defines the a-inner product on $H^1_{\Sigma 0}(\Omega)$ and is equivalent to the ∂-norm (2.1) when A satisfies (A1).

Observe that a function $w \in H^1_{\Sigma 0}(\Omega)$ is a-orthogonal to $H^1_0(\Omega)$ if and only if

$$\int_\Omega (A\nabla w) \cdot \nabla v \, dx = 0 \quad \text{for all} \quad v \in H^1_0(\Omega). \tag{3.8}$$

That is $\gamma(w)$ is zero on Σ and $\mathcal{L}_A w(x) := \operatorname{div}(A\nabla w)(x) = 0$ in a weak sense on Ω. The space of all such functions will be denoted $N(\mathcal{L}_A, \Sigma)$ so we have the orthogonal decomposition

$$H^1_{\Sigma 0}(\Omega) = H^1_0(\Omega) \oplus_a N(\mathcal{L}_A, \Sigma) \tag{3.9}$$

where $\oplus_a$ indicates that the a-inner product is used.

In light of this result, the minimizer $\tilde{\varphi}$ has a decomposition of the form $\tilde{\varphi} = \varphi_0 + \varphi_b$ where φ_0 is the minimizer of $\mathcal{E}$ on $H^1_0(\Omega)$ and $\varphi_b \in N(\mathcal{L}_A, \Sigma)$ is the solution of

$$\int_\Omega (A\nabla\varphi) \cdot \nabla\psi \, dx = \int_\Omega (\mathbf{G} - A\nabla\varphi_0) \cdot \nabla\psi \, \mathbf{dx} \quad \text{for all} \quad \psi \in \mathbf{N}(\mathcal{L_A}, \boldsymbol{\Sigma}). \tag{3.10}$$

The fact that φ_0 is a solution of the extremality condition on $H^1_0(\Omega)$ implies that $\operatorname{div}(A\nabla\varphi_0 - \mathbf{G}) = \mathbf{0}$ on Ω in a weak sense so this last equation may be written

$$\int_\Omega (A\nabla\varphi) \cdot \nabla\psi \, dx = \int_{\tilde{\Sigma}} \psi\,(\mathbf{G} - A\nabla\varphi_0) \cdot \nu \, \mathbf{d\sigma} \quad \text{for all} \quad \psi \in \mathbf{N}(\mathcal{L_A}, \boldsymbol{\Sigma}). \tag{3.11}$$

This is the weak form of the equation $\mathcal{L}_A \varphi = 0$ on Ω subject to the boundary conditions

$$\varphi(z) = 0 \quad \text{on} \quad \Sigma \quad \text{and} \quad A\nabla\varphi \cdot \nu = (\mathbf{G} - A\nabla\varphi_0) \cdot \nu \quad \text{on} \quad \tilde{\Sigma}. \tag{3.12}$$

The solution of the problem for φ_0 is a standard Dirichlet boundary value problem and it is worth noting that the value of $\mathcal{E}(\varphi_0) = 0$ if and only if $\operatorname{div}\mathbf{G} = 0$ on Ω in a weak sense. In this case the general problem reduces to that of solving (3.10) or (3.11) alone.

Our interest is in the problem of finding an expression for the boundary trace of $\tilde{\varphi}$ or φ_b on $\tilde{\Sigma}$. That is what boundary data gives the best approximating potential for the given field $\mathbf{F}$ on Ω?

4 Bases and Representations of $N(\mathcal{L}_A, \Sigma)$

To find the boundary data that provides the best L^2-approximation to the field $\mathbf{F}$, an orthogonal basis of the space consisting of certain Steklov-type eigenfunctions of $\mathcal{L}_A$ is constructed and used. This will yield a spectral representation of φ_b as described in Theorem 5.1 below.

An a-orthonormal basis of $N(\mathcal{L}_A, \Sigma)$ may be found using the algorithm described in Auchmuty [4]. In the notation of that paper take $V = N(\mathcal{L}_A, \Sigma)$, a as above and

$$m(u, v) := \int_{\tilde{\Sigma}} \gamma(u) \gamma(v) \, d\sigma. \tag{4.13}$$

A function $\chi \in N(\mathcal{L}_A, \Sigma)$ is said to be a Σ-Steklov eigenfunction of $\mathcal{L}_A$ on Ω provided it is a solution of

$$\int_{\Omega} (A\nabla\chi) \cdot \nabla\psi \, dx = \lambda \int_{\tilde{\Sigma}} \gamma(\chi) \gamma(\psi) \, d\sigma \quad \text{for all} \quad \psi \in H^1_{\Sigma 0}(\Omega). \tag{4.14}$$

This problem has the form of Eq. 2.1 of [4] and the bilinear forms a, m satisfy conditions (A1)–(A4) of that paper. Moreover condition (A5) there holds with $H = L^2(\tilde{\Sigma}, d\sigma)$.

Define $\mathcal{A}(\chi) = a(\chi, \chi)$, $\mathcal{M}(\chi) := m(\chi, \chi)$ and C_1 to be the closed unit ball in $H^1_{\Sigma 0}(\Omega)$ with respect to the a-norm. Consider the variational problem of maximizing $\mathcal{M}$ on C_1. This problem has maximizers $\pm\chi_1$ that have a-norm 1 and are solutions of (4.14) associated with an eigenvalue $\lambda_1 > 0$. Moreover one has the coercivity inequality

$$\mathcal{A}(\chi) \geq (\lambda_1 + 1) \int_{\tilde{\Sigma}} \gamma(\chi)^2 \, d\sigma \quad \text{for all} \quad \chi \in H^1_{\Sigma 0}(\Omega). \tag{4.15}$$

Using the construction of Sect. 4 of [4], a countably infinite a-orthonormal basis $\mathcal{B} := \{\chi_j : j \geq 1\}$ of $N(\mathcal{L}_A, \Sigma)$ may be constructed using a sequence of constrained maximization problems for $\mathcal{M}$.

These eigenfunctions also are m-orthogonal so that $m(\chi_j, \chi_k) = 0$ when $j \neq k$. Define $\tilde{\chi}_j := \chi_j / \sqrt{\lambda_j + 1}$ then $\tilde{\mathcal{B}} := \{\gamma(\tilde{\chi}_j) : j \geq 1\}$ will be an m-orthonormal basis of $L^2(\tilde{\Sigma}, d\sigma)$ from theorem 4.6. Since these functions constitute bases of the various Hilbert spaces, there is a spectral representation of functions $\varphi \in N(\mathcal{L}_A, \Sigma)$ in terms of their boundary values. Namely when $\varphi \in N(\mathcal{L}_A, \Sigma)$ then,

$$\varphi(x) = \sum_{j=1}^{\infty} c_j \, \tilde{\chi}_j(x) \quad \text{on} \quad \Omega \quad \text{with} \quad c_j := m(\varphi, \tilde{\chi}_j) \tag{4.16}$$

and this series converges strongly in $H^1_{\Sigma 0}(\Omega)$. Thus these constructions yield the following result.

Theorem 4.1. *Assume that (A1), (B1), (B2) hold and $\mathcal{B}, \tilde{\mathcal{B}}$ are defined as above. If $\varphi \in N(\mathcal{L}_A, \Sigma)$ then (4.16) holds and the series converges in a-norm. Moreover $a(\varphi, \varphi) = \sum_{j=1}^{\infty} (1 + \lambda_j) c_j^2$.*

Let $H^{1/2}(\tilde{\Sigma})$ be the subspace of $L^2(\tilde{\Sigma}, d\sigma)$ of all functions with $\sum_{j=1}^{\infty} (1 + \lambda_j) c_j^2 < \infty$. It will be a Hilbert space with respect to the inner product

$$\langle \varphi, \psi \rangle_{1/2, \tilde{\Sigma}} := \sum_{j=1}^{\infty} (1 + \lambda_j) \, m(\phi, \chi_j) \, m(\psi, \chi_j) \tag{4.17}$$

In particular this yields an isomorphism between functions in $N(\mathcal{L}_A, \Sigma)$ and functions in $H^{1/2}(\tilde{\Sigma})$. Thus $H^{1/2}(\tilde{\Sigma})$ may be regarded as the boundary trace subspace (on $\tilde{\Sigma}$) of functions in $N(\mathcal{L}_A, \Sigma)$ and this is an isometry with $a(\varphi, \varphi) = \langle \varphi, \varphi \rangle^2_{1/2, \tilde{\Sigma}}$.

5   The Best Approximating Potential

We are now in a position to specify the boundary data for the potential that minimizes (3.3). The preceding analysis enables the derivation of an explicit representation of the solution φ_b of (3.10) or (3.11). Equation (3.10) implies that φ_b satisfies

$$a(\varphi_b, \tilde{\chi}_j) = g_j := \int_\Omega (\mathbf{G} - A\nabla\varphi_0) \cdot \nabla\tilde{\chi}_j \; \mathbf{dx} \quad \text{for} \quad j \geq 1.$$

Note that the g_j depend onlds only on the data, the eigenfunction $\tilde{\chi}_j$ and the solution φ_0 of the zero-Dirichlet variational problem. Then the eigenfunction equation (4.14) yields that the solution is

$$\varphi_b(x) = \sum_{j=1}^{\infty} \frac{g_j}{(1 + \lambda_j)} \, \tilde{\chi}_j(x). \tag{5.18}$$

Thus the boundary trace on $\tilde{\Sigma}$ of the best approximation is given by the boundary trace of this right hand side. That is imposing Dirichlet boundary data $\gamma(\varphi_b)$ on $\tilde{\Sigma}$ given by (5.18) yields the minimizing potential φ_b of $\mathcal{E}$.

Theorem 5.1. *Assume that (A1), (B1), (B2) hold and $\tilde{\mathcal{B}}, g_j$ are defined as above. Then the potential $\tilde{\varphi}$ that minimizes the norm in (3.3) or $\mathcal{E}$ on $H^1_{\Sigma 0}(\Omega)$ is given by $\tilde{\varphi} = \varphi_0 + \varphi_b$ where φ_0 minimizes $\mathcal{E}$ on $H^1_0(\Omega)$ and φ_b is given by (5.18). When $\mathrm{div}\, \mathbf{G} = \mathbf{0}$ on Ω, then $\mathbf{F} = \epsilon\nabla\tilde{\varphi}$ on Ω.*

Proof. This is a restatement of the preceding results. Note that when $\mathrm{div}\, \mathbf{G} = \mathbf{0}$ on Ω, then $\varphi_0 = 0$ so the last sentence holds.

Moreover when the boundary potential φ is a good approximation of this φ_b in the norm of $H^{1/2}(\tilde{\Sigma})$ then the fact that the a-norm and the norm on $H^{1/2}(\tilde{\Sigma})$ are isometric implies that such potentials φ will provide a good approximation of $\mathbf{F}$ on Ω in any equivalent norm on $H^1_{\Sigma 0}(\Omega)$.

References

1. Attouch, H., Buttazzo, G., Michaille, G.: Variational Analysis in Sobolev and BV Spaces. SIAM Publications, Philadelphia (2006)
2. Auchmuty, G.: Steklov eigenproblems and the representation of solutions of elliptic boundary value problems. Numer. Funct. Anal. Optim. **25**, 321–348 (2004)
3. Auchmuty, G.: Spectral characterizations of the trace spaces $H^s(\partial\Omega)$. SIAM J. Math. Anal. **38**, 894–905 (2006)
4. Auchmuty, G.: Bases and comparison results for linear elliptic eigenproblems. J. Math. Anal. Appl. **390**(1), 394–406 (2012)
5. Evans, L.C., Gariepy, R.F.: Measure Theory and Fine Properties of Functions. CRC Press, Boca Raton (1992)

Modelling Pesticide Treatment Effects on *Lygus hesperus* in Cotton Fields

H.T. Banks[1]($\boxtimes$), J.E. Banks[1,2], Neha Murad[1], J.A. Rosenheim[3], and K. Tillman[1]

[1] Center for Research in Scientific Computation, North Carolina State University, Raleigh, NC 27695-8212, USA
htbanks@ncsu.edu
[2] Undergraduate Research Opportunities Center, California State University, Monterey Bay, Seaside, CA 93955, USA
[3] Department of Entomology and Nematology, Center for Population Biology, University of California, Davis, Davis, CA 95616, USA

Abstract. We continue our efforts on modeling of the population dynamics of herbivorous insects in order to develop and implement effective pest control protocols. In the context of inverse problems, we explore the dynamic effects of pesticide treatments on *Lygus hesperus*, a common pest of cotton in the western United States. Fitting models to field data, we consider model selection for an appropriate mathematical model and corresponding statistical models, and use techniques to compare models. We address the question of whether data, as it is currently collected, can support time-dependent (as opposed to constant) parameter estimates.

Keywords: Inverse problems · Generalized least squares · Model selection · Information content · Residual plots · Piecewise linear splines · Hemiptera · Herbivory · Pesticide

1 Introduction

When addressing questions in fields ranging from conservation science to agricultural production, ecologists frequently collect time-series data in order to better understand how populations are affected when subjected to abiotic or biotic disturbance [12,13,22]. Fitting models to data, which generally requires a broad understanding of both statistics and mathematics, is an important component of understanding pattern and process in population studies. In agricultural ecology, pesticide disturbance may disrupt predator-prey interactions [27,28] as well as impose both acute and chronic effects on arthropod populations. In the past several decades, the focus of many studies of pesticide effects on pests and their natural enemies has shifted away from static measures such as the LC50, as authors have emphasized population metrics/outcomes [17–19,26,29]. Simple mathematical models, parameterized with field data, are often used to then predict the consequences of increasing or decreasing pesticide exposure in the field. Accuracy in parameter estimation and quantification of uncertainty in fitting

© IFIP International Federation for Information Processing 2016
Published by Springer International Publishing AG 2016. All Rights Reserved
L. Bociu et al. (Eds.): CSMO 2015, IFIP AICT 494, pp. 95–106, 2016.
DOI: 10.1007/978-3-319-55795-3_8

data to models, which has recently received increasing attention in ecological circles [20,21], depends critically on the appropriate model selection. In most cases, this includes selection of both **statistical and mathematical models** in fits-to-data – something that is not always fully explicitly addressed in the ecological literature. We first addressed this gap in [1,3] using data from pest population counts of *Lygus hesperus* Knight (Hemiptera: Meridae) feeding on pesticide-treated cotton fields in the San Joaquin Valley of California [23].

In particular, in [1,3] we investigated the effect of pesticide treatments on the growth dynamics of *Lygus hesperus*. This was done by constructing mathematical models and then fitting these models to field data so as to estimate growth rate parameters of *Lygus hesperus* both in the absence and in the presence of pesticide application. Overall, compelling evidence was found for the untreated fields, using model comparison tests, that it may be reasonable to ignore nymph mortality (i.e., just count total number of *L. hesperus* and not distinguish between nymphs and adults). This would greatly simplify the models, *as well as the data collection process.*

In the present effort we further examine the importance of model selection and demonstrate how optimal selection of both statistical and mathematical models is crucial for accuracy in parameter estimation and uncertainty quantification in fitting data to models. This report further investigates these issues by testing different data sets from the same database as in [1,3] but with a varied number of pesticide applications in treated fields.

2   Methods

The data used came from a database consisting of approximately 1500 replicates of L. hesperus density counts, using sweep counts, in over 500 Pima or Acala cotton fields in 1997–2008 in the San Joaquin Valley of California. This data is described more fully in [3]. We selected subsets to analyze using the following criteria:

- In each replicate (corresponding to data collected during one season at one field) we considered data that was collected by pest control advisors (PCAs) between June 1 and August 30.
- We considered data which had the pesticide applications that targeted beet armyworms, aphids, mites as well as Lygus.
- We used only replicates where adult and nymph counts were combined into a total insect count.
- No counts were made on the days of pesticide applications.
- Superposition of pesticide applications has not been incorporated in the algorithm, so we chose samples with at least a week gap between consecutive pesticide applications.

We consider inverse or parameter estimation problems in the context of a parameterized (with vector parameter $q \in \Omega^{\kappa_q} \subset \mathbb{R}^{\kappa_q}$) N-dimensional vector dynamical system or **mathematical model** given by

$$\frac{d\boldsymbol{x}}{dt}(t) = \boldsymbol{g}(t, \boldsymbol{x}(t), \boldsymbol{q}), \tag{1}$$

$$\boldsymbol{x}(t_0) = \boldsymbol{x}_0, \tag{2}$$

with <u>scalar</u> observation process

$$f(t; \boldsymbol{\theta}) = \mathcal{C}\boldsymbol{x}(t; \boldsymbol{\theta}), \tag{3}$$

where $\boldsymbol{\theta} = (\boldsymbol{q}^\mathsf{T}, \tilde{\boldsymbol{x}}_0^\mathsf{T})^\mathsf{T} \in \boldsymbol{\Omega}^{\kappa_\theta} \subset \mathbb{R}^{\kappa_q + \tilde{N}} = \mathbb{R}^{\kappa_\theta}, \tilde{N} \leq N$, and the observation operator $\mathcal{C}$ maps $\mathbb{R}^N$ to $\mathbb{R}^1$. The sets $\boldsymbol{\Omega}^{\kappa_q}$ and $\boldsymbol{\Omega}^{\kappa_\theta}$ are assumed known restraint sets for the parameters.

We make some standard statistical assumptions (see [7,8,16,24]) underlying our inverse problem formulations.

- (A1) Assume $\mathcal{E}_i$ are independent identically distributed *i.i.d.* with $\mathbb{E}(\mathcal{E}_i) = 0$ and $\text{cov}(\mathcal{E}_i, \mathcal{E}_i) = \sigma_0^2$, where $i = 1, ..., n$ and n is the number of observations or data points in the given data set taken from a time interval $[0, T]$.
- (A2) Assume that there exists a *true or nominal set* of parameters $\boldsymbol{\theta}_0 \in \boldsymbol{\Omega} \equiv \boldsymbol{\Omega}^{\kappa_\theta}$.
- (A3) $\boldsymbol{\Omega}$ is a compact subset of Euclidian space of $\mathbb{R}^{\kappa_\theta}$ and $f(t, \boldsymbol{\theta})$ is continuous on $[0, T] \times \boldsymbol{\Omega}$.

Denote as $\hat{\boldsymbol{\theta}}$ the estimated parameter for $\boldsymbol{\theta}_0 \in \boldsymbol{\Omega}$. The inverse problem is based on statistical assumptions on the observation error in the data. If one assumes some type of *generalized relative error data model*, then the error is proportional in some sense to the measured observation. This can be represented by a **statistical model** with observations of the form

$$\boldsymbol{Y}_i = f(t_i; \boldsymbol{\theta}_0) + f(t_i; \boldsymbol{\theta}_0)^\gamma \mathcal{E}_i, \quad \gamma \in [0, 2], \tag{4}$$

with corresponding realizations

$$y_i = f(t_i; \boldsymbol{\theta}_0) + f(t_i; \boldsymbol{\theta}_0)^\gamma \epsilon_i, \quad \gamma \in [0, 2], \tag{5}$$

where the ϵ_i are realizations of the $\mathcal{E}_i$, $i = 1, ..., n$.

For relative error models one should use inverse problem formulations with *Generalized Least Squares (GLS)* cost functional

$$J^n(\boldsymbol{Y}; \boldsymbol{\theta}) = \sum_{i=1}^{n} \left(\frac{\boldsymbol{Y}_i - f(t_i; \boldsymbol{\theta})}{f(t_i; \boldsymbol{\theta})^\gamma} \right)^2. \tag{6}$$

The corresponding estimator and estimates are respectively defined by

$$\boldsymbol{\Theta}_{GLS} = \operatorname*{argmin}_{\boldsymbol{\theta} \in \boldsymbol{\Omega}} \sum_{i=1}^{n} \left(\frac{\boldsymbol{Y}_i - f(t_i; \boldsymbol{\theta})}{f(t_i; \boldsymbol{\theta})^\gamma} \right)^2, \quad \gamma \in [0, 2], \tag{7}$$

with realizations

$$\hat{\boldsymbol{\theta}}_{GLS} = \underset{\boldsymbol{\theta} \in \Omega}{\text{argmin}} \sum_{i=1}^{n} \left(\frac{y_i - f(t_i; \boldsymbol{\theta})}{f(t_i; \boldsymbol{\theta})^{\gamma}} \right)^2, \quad \gamma \in [0, 2]. \tag{8}$$

GLS estimates $\hat{\boldsymbol{\theta}}^n$ and weights $\{\omega_j\}_{j=1}^{n}$ are found using an iterative method as defined below (see [7]). For the sake of notation, we will suppress the superscript n (i.e., $\hat{\boldsymbol{\theta}}_{GLS} := \hat{\boldsymbol{\theta}}_{GLS}^n$).

1. Estimate $\hat{\boldsymbol{\theta}}_{GLS}$ by $\hat{\boldsymbol{\theta}}^{(0)}$ using the OLS method ((8) with $\gamma = 0$). Set $k = 0$.
2. Compute weights $\hat{\omega}_j = f^{-2\gamma}(t_j, \hat{\boldsymbol{\theta}}^{(k)})$.
3. Obtain the $k + 1$ estimate for $\hat{\boldsymbol{\theta}}_{GLS}$ by $\hat{\boldsymbol{\theta}}^{(k+1)} := \text{argmin} \sum_{j=1}^{n} \hat{\omega}_j [y_j - f(t_j, \boldsymbol{\theta})]^2$.
4. Set $k := k + 1$ and return to step 2. Terminate when the two successive estimates for $\hat{\boldsymbol{\theta}}_{GLS}$ are sufficiently close.

3   Mathematical Models

Our focus here is on the comparison of two different models for insect (*L. hesperus*) population growth/mortality in pesticide-treated fields. The simplest model (which we denote as model B) is for constant reduced growth due to effects of pesticides versus an added time-varying mortality (denoted by model A) to reflect this decreased total population growth rate. Model B is given by

$$\frac{dx}{dt} = \eta x \tag{9}$$
$$x(t_1) = x_0,$$

where x_0 is defined as initial population count at time t_1 of initial observation and η is the reduced population growth rate in the presence of pesticides.

Model A is given by

$$\frac{dx}{dt} = k(t)x \tag{10}$$
$$x(t_1) = x_0,$$

where t_1 is again the time of the first data point, and $k(t)$ is a time dependent growth rate

$$k(t) = \begin{cases} \eta + p(t) & t \in P_j, j \in \{1, 2, 3, 4\} \\ \eta & \text{otherwise.} \end{cases}$$

Here $p(t)$ is composed of piecewise linear splines as described below, and $P_j = [t_{p_j}, t_{p_j} + 1/4], j = 1, \ldots j^*$ with t_{p_j} as the time point of the jth pesticide application. Observe that these t_{p_j} are *not* the same as the observation or data points t_j. Also note that $|P_j| = 1/4$ which is approximately the length of time of one week when t is measured in months. This reflects the general assumption that pesticides are most active during the 7 days immediately following treatment. Clearly, η is a reduced **constant** growth rate of the total population in the presence of pesticides. In addition, t = 0 refers to June 1 (as no data is present before June 1 in our database).

Piecewise linear splines [25] were used to approximate $p(t)$ as follows. Consider m linear splines

$$p(t) = \sum_{i=1}^{m} \lambda_i l_i(t),$$

where

$$l_i(t) = 1/h \begin{cases} t - t_{i-1} & t_{i-1} \leq t < t_i \\ t_{i+1} - t & t_i \leq t \leq t_{i+1} \\ 0 & \text{otherwise,} \end{cases}$$

where h is the step size, $h = \frac{|P_j|}{(m+1)}$. Piecewise linear spline representations are simple, yet flexible in that they allow the modeler to avoid assuming a certain shape to the curve being approximated. Incorporating a time-dependent component such as $p(t)$ is useful when modeling a system with discontinuous perturbations (such as the removal of a predator, or the application of an insecticide). The addition of more splines ($m > 3$) provides a finer approximation, but demands more terms in the parameter estimates. We assume here that it is likely that $m = 3$ is sufficient. (Our subsequent findings suggest that perhaps even $m = 2$ is sufficient!). In our analysis, we first estimated the initial condition x_0 using model B (as this data point precedes any pesticide applications and provides a good estimate for x_0), and then fixed this parameter in all subsequent parameter estimates. Therefore, the parameters to be estimated in model A are $\boldsymbol{\theta} = \boldsymbol{q} = \{\eta, \lambda_1, \lambda_2, \lambda_3\}$ whereas for model B we must only estimate $\boldsymbol{\theta} = \boldsymbol{q} = \{\eta, 0, 0, 0\}$ since model A reduces to model B when applying the constraint $p(t) \equiv 0$, i.e. $\lambda_i = 0$ for $i = 1, 2, 3$.

4   Parameter Estimation

Using the model information provided in [3] we try to estimate parameters for new data sets and determine whether the fit-to-data provided by model A does provide a statistically significantly better fit than the fit provided by model B. A big part of the parameter estimation process is the minimization of the respective cost functions for both model A and model B. The constrained nonlinear optimization solver in Matlab, fmincon was initially being used for minimization of GLS cost functionals in model A while fminsearch was being used for minimization of cost functionals for model B. We later switched to lsqnonlin which gave faster and better results. Since model A is stiff in nature, Matlab solver ode15s was used whereas for model B ode45 was used. Both fmincon and fminsearch require an initial guess of parameters. While for model B fminsearch was able to find a minimum fairly quickly, the initial guess of $\boldsymbol{\theta} = \{\eta, \lambda_1, \lambda_2, \lambda_3\}$ for model A involved a fairly detailed process given below:

1. Create a trial file of data selected based on a specific set of rules.
2. Choose a parameter space $\boldsymbol{\Omega} = [\epsilon, K] \times [-K, -\epsilon]^3$.

3. Choose a constant $\gamma \in [0, 1.5]$. (Note that best way to choose (see [7,8]) gamma is to consider the plot of residuals using both residual vs time and residual vs model plots to ascertain whether the scatter of the error appears to violate the statistical assumption of being $i.i.d.$).
4. Choose an initial condition x_0 to the exponential model by considering the plot of the data and model and visually estimating x_0. We observe that this did not produce acceptable results so we eventually solved a separate inverse problem to estimate intrinsic growth rate ignoring the effects of pesticides and initial condition.

5   Model Comparsion: Nested Restraint Sets

Here we summarize the use of *statistically based model comparison tests*. These residual sum of squares model comparison tests as developed in [5], described in [7,8] and extended in [9] to GLS problems is used in the same manner as used in [1,3]. This test is used to determine which of several nested models is the best fit to the data; therefore, this test can be applied to the comparison of models A and B. In these examples below we are interested in questions related to whether the data will support a more detailed or sophisticated model to describe it. In the next section we recall the fundamental statistical tests to be employed here.

5.1   Statistical Comparison Tests

In general, assume we have an inverse problem for the model observations $f(t, \boldsymbol{\theta})$ and are given n observations. As in (6), we define

$$J^n(\boldsymbol{Y}; \boldsymbol{\theta}) = \sum_{i=1}^{n} \left(\frac{\boldsymbol{Y}_i - f(t_i; \boldsymbol{\theta})}{f(t_i; \boldsymbol{\theta})^\gamma} \right)^2,$$

where our statistical model has the form (4). Here, as before, $\boldsymbol{\theta}_0$, is the nominal value of $\boldsymbol{\theta}$ which we assume to exist. We use $\boldsymbol{\Omega}$ to represent the set of all the admissible parameters $\boldsymbol{\theta}$. We make some further assumptions.

- (A4) Observations are taken at $\{t_j\}_{j=1}^n$ in $[0, T]$. There exists some finite measure μ on $[0, T]$ such that

$$\frac{1}{n} \sum_{j=1}^{n} h(t_j) \longrightarrow \int_0^T h(t) d\mu(t)$$

as $n \to \infty$, for all continuous functions h.
- (A5) $J_0(\boldsymbol{\theta}) = \int_0^T (f(t; \boldsymbol{\theta}_0) - f(t; \boldsymbol{\theta}))^2 d\mu(t) = \sigma^2$ has a unique minimizer in $\boldsymbol{\Omega}$ at $\boldsymbol{\theta}_0$.

Let $\boldsymbol{\Theta}^n = \boldsymbol{\Theta}_{GLS}^n(\boldsymbol{Y})$ be the GLS estimator for J^n as defined in (7) so that

$$\boldsymbol{\Theta}_{GLS}^n(\boldsymbol{Y}) = \operatorname*{argmin}_{\boldsymbol{\theta} \in \boldsymbol{\Omega}} J^n(\boldsymbol{Y}; \boldsymbol{\theta})$$

and

$$\hat{\boldsymbol{\theta}}_{GLS}^n = \operatorname*{argmin}_{\boldsymbol{\theta} \in \Omega} J^n(\boldsymbol{y}; \boldsymbol{\theta}),$$

where as above $\boldsymbol{y}$ is a realization for $\boldsymbol{Y}$.

One can then establish a series of useful results (see [5,7,9] for detailed proofs).

Result 1: Under (A1) to (A5), $\frac{1}{n}\boldsymbol{\Theta}^n = \frac{1}{n}\boldsymbol{\Theta}_{GLS}^n(\boldsymbol{Y}) \longrightarrow \boldsymbol{\theta}_0$ as $n \to \infty$ with probability 1.

We will need further assumptions to proceed (these will be denoted by (A7)–(A11) to facilitate reference to [5,7]). These include:

- (A7) Ω is finite dimensional in R^p and $\boldsymbol{\theta}_0 \in \operatorname{int}\Omega$.
- (A8) $f : \Omega \to C[0,T]$ is a C^2 function.
- (A10) $\mathcal{J} = \frac{\partial^2 J_0}{\partial \theta^2}(\boldsymbol{\theta}_0)$ is positive definite.
- (A11) $\Omega_H = \{\boldsymbol{\theta} \in \Omega | H\boldsymbol{\theta} = c\}$ where H is an $r \times p$ matrix of full rank, and c is a known constant.

In many instances, including the examples discussed here, one is interested in using data to question whether the "nominal" parameter $\boldsymbol{\theta}_0$ can be found in a subset $\Omega_H \subset \Omega$ which we assume for discussions here is defined by the constraints of assumption (A11). Thus, we want to test the *null hypothesis H_0*: $\boldsymbol{\theta}_0 \in \Omega_H$, i.e., that the constrained model provides an adequate fit to the data.

Define then

$$\boldsymbol{\Theta}_H^n(\boldsymbol{Y}) = \operatorname*{argmin}_{\boldsymbol{\theta} \in \Omega_H} J^n(\boldsymbol{Y}; \boldsymbol{\theta})$$

and

$$\hat{\boldsymbol{\theta}}_H^n = \operatorname*{argmin}_{\boldsymbol{\theta} \in \Omega_H} J^n(\boldsymbol{y}; \boldsymbol{\theta}).$$

Observe that $J^n(\boldsymbol{y}; \hat{\boldsymbol{\theta}}_H^n) \geq J^n(\boldsymbol{y}; \hat{\boldsymbol{\theta}}^n)$. We define the related non-negative test statistics and their realizations, respectively, by

$$T_n(\boldsymbol{Y}) = J^n(\boldsymbol{Y}; \boldsymbol{\theta}_H^n) - J^n(\boldsymbol{Y}; \boldsymbol{\theta}^n)$$

and

$$\hat{T}_n = T_n(\boldsymbol{y}) = J^n(\boldsymbol{y}; \hat{\boldsymbol{\theta}}_H^n) - J^n(\boldsymbol{y}; \hat{\boldsymbol{\theta}}^n).$$

One can establish asymptotic convergence results for the test statistics $T_n(\boldsymbol{Y})$–see [5]. These results can, in turn, be used to establish a fundamental result about much more useful statistics for model comparison. We define these statistics by

$$U_n(\boldsymbol{Y}) = \frac{nT_n(\boldsymbol{Y})}{J^n(\boldsymbol{Y}; \boldsymbol{\theta}^n)}, \tag{11}$$

with corresponding realizations

$$\hat{u}_n = U_n(\boldsymbol{y}).$$

We then have the asymptotic result that is the basis of our analysis-of-variance–type tests.

Results 2: Under the assumptions (A1)–(A5) and (A7)–(A11) above and assuming the null hypothesis H_0 is true, then $U_n(Y)$ converges in distribution (as $n \to \infty$) to a random variable $U(r)$, i.e.,

$$U_n \xrightarrow{\;D\;} U(r),$$

with $U(r)$ having a chi-square distribution $\chi^2(r)$ with r degrees of freedom.

In any graph of a χ^2 density there are two parameters (τ, α) of interest. For a given value τ, the value α is simply the probability that the random variable U will take on a value greater than τ. That is, $\mathrm{Prob}\{U > \tau\} = \alpha$ where in hypothesis testing, α is the *significance level* and τ is the *threshold*.

We then wish to use this distribution $U_n \sim \chi^2(r)$ to test the null hypothesis, H_0, that the restricted model provides an adequate fit to represent the data. If the test statistic, $\hat{u}_n > \tau$, then we *reject H_0 as false* with confidence level $(1 - \alpha)100\%$. Otherwise, we *do not reject H_0*. For our examples below, we use a $\chi^2(3)$ table, which can be found in any elementary statistics text, online or the partial summary below. Typical confidence levels of interest are $90\%, 95\%, 99\%, 99.9\%$, with corresponding (α, τ) values given in Table 1 below.

To test the null hypothesis H_0, we choose a significance level α and use χ^2 tables to obtain the corresponding threshold $\tau = \tau(\alpha)$ so that $\mathrm{Prob}\{\chi^2(r) > \tau\} = \alpha$. We next compute $\hat{u}_n = \overline{\tau}$ and compare it to τ. If $\hat{u}_n > \tau$, then we reject H_0 as false; otherwise, we do not reject the null hypothesis H_0.

We use a $\chi^2(3)$ for our comparison tests as summarized here.

Table 1. Chi-square table: $\chi^2(3)$

α	.10	.05	.01	.001
τ	6.251	7.815	11.345	16.266

We can then formulate the null and alternative hypotheses:

H_0: The fit provided by model A *does not* provide a statistically significantly better fit to the data than the fit provided by model B.

H_A: The fit provided by model A *does* provide a statistically significantly better fit to the data than the fit provided by model B.

We considered such comparison tests for a number of data sets with a varying no. of pesticides applications among the fields. These included

- Replicate number 296 (1 pesticide application at t = 0.5 months)
- Replicate number 350 (2 pesticide applications at times t = 1 and t = 1.7 months)
- Replicate number 277 (3 pesticide applications at t = .4, .7, 1.9 months)
- Replicate number 178 (4 pesticide applications at t = .13, .77, 1.9, 2.27 months)
- Replicate number 174 (4 pesticide applications at t = .13, .77, 1.83, 2.27 months)

We carried out multiple inverse problems with varying values of γ for these data sets (see [2]). We visually examined the resulting residual plots (residual vs time and residual vs observed output) and determined whether the scatter of the error appear to be i.i.d. On examining the plots for a wide range of γ we observed that the statistical i.i.d. assumptions were approximately satisfied for γ values ranging around 0.7 to 0.8. We therefore used these values of γ in the results reported here and in [2].

We note here that one could, as an alternative to use of residual plots, include the parameter γ as a parameter to be estimated along with θ as often done in statistical formulations [16] for the joint estimation of $\beta = (\theta, \gamma)$, or one could attempt to estimate the form of the statistical model (i.e., the value of γ directly from the data itself as suggested in [6]. Both of these methods offer some advantages but include more complex inverse problem analysis. We have therefore chosen here to use the simpler but less sophisticated analysis of residuals in our approach.

6 Results

We examine the importance of model selection and how optimal selection of both statistical and mathematical models is crucial for accuracy in parameter estimation and fitting data to models testing different data sets from the same database with varied number of pesticide applications. Tables 2, 3 and 4 contain summaries of the results for the investigated replicates.

Table 2. Parameter estimates for models A and B

Replicate no.	No. of pest. apps	Par. Est. for model A $(\eta, \lambda_1, \lambda_2, \lambda_3)$	Par. Est. for model B $(\eta, \lambda_1, \lambda_2, \lambda_3)$
296 (n = 19)	1	$(1.3551, -7.8782, -7.1632, -9.9904)$	$(0.3548, 0, 0, 0)$
350 (n = 19)	2	$(3.0150, -16.7393, -17.9356, -8.580)$	$(0.7683, 0, 0, 0)$
277 (n = 19)	3	$(3.1255, -3.5165, -6.2870, -7.1868)$	$(1.7284, 0, 0, 0)$
178 (n = 22)	4	$(1.8201, -6.6352, -2.5786, -4.1605)$	$(0.6335, 0, 0, 0)$
174 (n = 20)	4	$(2.8600, -7.9867, -13.2780, -4.6992)$	$(0.4110, 0, 0, 0)$

Table 3. Cost functional values for models A and B

Replicate no.	Cost functional model A	Cost functional model B	Range of γ values	x_0
296	3.9577	7.3663	0.5 to 1.0	0.39
350	7.4100	18.223	0.5 to 0.8	0.28
277	5.5290	8.075	0.7 to 0.9	0.05
178	5.7596	8.371	0.7 to 1.0	0.37
174	6.7480	9.390	0.0 to 0.8	0.34

Table 4. Statistical comparison test results for degrees of freedom $r = 3$

Replicate no.	Test statistic $\hat{u}_n$	Threshold τ	Signif. level α	Null hyp. (H_0) accept or reject	Confid. level
296	16.36	16.26	0.001	Rejected	99.9%
350	27.73	16.26	0.001	Rejected	99.9%
277	8.7458	7.815	0.05	Rejected	95%
178	9.9715	7.815	0.05	Rejected	95%
174	7.8281	7.815	0.05	Rejected	95%

7   Concluding Remarks

The above results strongly support the notion that time varying reduced growth/mortality rates as opposed to constant rates provide substantially better models at the population levels for the description of the effects of pesticides on the growth rates. It is interesting to note that our findings hold consistently across the differing levels of pesticide applications, even in the case of only one pesticide application. Another interesting observation is that this is consistent even when one uses total pest counts as opposed to individual nymph and adult counts. This, of course, has significant implications for data collection procedures. The model comparison techniques employed here are just one of several tools that one can use to determine aspects of information content in support of model sophistication/complexity. Of note are the use of the Akiake Information Criterion (AIC) and its variations [1, 4, 10, 11, 14, 15] for model comparison.

Acknowledgements. This research was supported by the Air Force Office of Scientific Research under grant number AFOSR FA9550-12-1-0188 and grant number AFOSR FA9550-15-1-0298. J.E. Banks is grateful to J.A. Rosenheim for hosting him during his sabbatical, and also extends thanks to the University of Washington, Tacoma for providing support for him during his sabbatical leave.

References

1. Banks, H.T., Banks, J.E., Link, K., Rosenheim, J.A., Ross, C., Tillman, K.A.: Model comparison tests to determine data information content. Appl. Math. Lett. **43**, 10–18 (2015). CRSC-TR14-13, North Carolina State University, Raleigh, NC, October 2014
2. Banks, H.T., Banks, J.E., Murad, N., Rosenheim, J.A., Tillman, K.: Modelling pesticide treatment effects on Lygus hesperus in cotton fields. CRSC-TR15-09, Center for Research in Scientific Computation, North Carolina State University, Raleigh, NC, September 2015
3. Banks, H.T., Banks, J.E., Rosenheim, J., Tillman, K.: Modeling populations of Lygus hesperus on cotton fields in the San Joaquin Valley of California: the importance of statistical and mathematical model choice. J. Biol. Dyn. **11**, 25–39 (2017)
4. Banks, H.T., Doumic, M., Kruse, C., Prigent, S., Rezaei, H.: Information content in data sets for a nucleated-polymerization model. J. Biol. Dyn. **9**, 172–197 (2015). doi:10.1080/17513758.2015.1050465

5. Banks, H.T., Fitzpatrick, B.G.: Statistical methods for model comparison in parameter estimation problems for distributed systems. J. Math. Biol. **28**, 501–527 (1990)
6. Banks, H.T., Catenacci, J., Hu, S.: Applications of difference-based variance estimation methods. J. Inverse Ill-Posed Prob. **24**, 413–433 (2016)
7. Banks, H.T., Hu, S., Thompson, W.C.: Modeling and Inverse Problems in the Presence of Uncertainty. Taylor/Francis-Chapman/Hall-CRC Press, Boca Raton (2014)
8. Banks, H.T., Tran, H.T.: Mathematical and Experimental Modeling of Physical and Biological Processes. CRC Press, Boca Raton (2009)
9. Banks, H.T., Kenz, Z.R., Thompson, W.C.: An extension of RSS-based model comparison tests for weighted least squares. Int. J. Pure Appl. Math. **79**, 155–183 (2012)
10. Bozdogan, H.: Model selection and Akaike's information criterion (AIC): the general theory and its analytical extensions. Psychometrika **52**, 345–370 (1987)
11. Bozdogan, H.: Akaike's information criterion and recent developments in information complexity. J. Math. Psychol. **44**, 62–91 (2000)
12. Brabec, M., Honěk, A., Pekár, S., Martinková, Z.: Population dynamics of aphids on cereals: digging in the time-series data to reveal population regulation caused by temperature. PLoS ONE **9**, e106228 (2014)
13. Brawn, J.D., Robinson, S.K., Thompson, F.R.: The role of disturbance in the ecology and conservation of birds. Annu. Rev. Ecol. Syst. **32**(2001), 251–276 (2001)
14. Burnham, K.P., Anderson, D.R.: Model Selection and Inference: A Practical Information-Theoretical Approach, 2nd edn. Springer, New York (2002)
15. Burnham, K.P., Anderson, D.R.: Multimodel inference: understanding AIC and BIC in model selection. Sociol. Methods Res. **33**, 261–304 (2004)
16. Davidian, M., Giltinan, D.M.: Nonlinear Models for Repeated Measurement Data. Chapman and Hall, London (2000)
17. Forbes, V.E., Calow, P.: Is the per capita rate of increase a good measure of population-level effects in ecotoxicology? Environ. Toxicol. Chem. **18**, 1544–1556 (1999)
18. Forbes, V.E., Calow, P., Sibly, R.M.: The extrapolation problem and how population modeling can help. Environ. Toxicol. Chem. **27**, 1987–1994 (2008)
19. Forbes, V.E., Calow, P., Grimm, V., Hayashi, T., Jager, T., Palmpvist, A., Pastorok, R., Salvito, D., Sibly, R., Spromberg, J., Stark, J., Stillman, R.A.: Integrating population modeling into ecological risk assessment. Integr. Environ. Assess. Manag. **6**, 191–193 (2010)
20. Hilborn, R.: The Ecological Detective: Confronting Models with Data, vol. 28. Princeton University Press, Princeton (1997)
21. Motulsky, H., Christopoulos, A.: Fitting Models to Biological Data Using Linear and Nonlinear Regression: A Practical Guide to Curve Fitting. Oxford University Press, Oxford (2004)
22. Pickett, S., White, P.: Natural disturbance and patch dynamics: an introduction. In: Pickett, S.T.A., White, P.S. (eds.) The Ecology of Natural Disturbance and Patch Dynamics, pp. 3–13. Academic Press, Orlando (1985)
23. Rosenheim, J.A., Steinmann, K., Langelloto, G.A., Link, A.G.: Estimating the impact of Lygus hesperus on cotton: the insect, plant and human observer as sources of variability. Environ. Entomol. **35**, 1141–1153 (2006)
24. Seber, G.A.F., Wild, C.J.: Nonlinear Regression. Wiley, Hoboken (2003)
25. Schultz, M.H.: Spline Analysis. Prentice-Hall, Englewood Cliffs (1973)

26. Stark, J.D., Banks, J.E.: Population-level effects of pesticides and other toxicants on arthropods. Annu. Rev. Entomol. **48**, 505–519 (2003)
27. Stark, J.D., Banks, J.E., Vargas, R.I.: How risky is risk assessment? The role that life history strategies play in susceptibility of species to pesticides and other toxicants. Proc. Nat. Acad. Sci. **101**, 732–736 (2004)
28. Stark, J.D., Vargas, R., Banks, J.E.: Incorporating ecologically relevant measures of pesticide effect for estimating the compatibility of pesticides and biocontrol agents. J. Econ. Entomol. **100**, 1027–1032 (2007)
29. Stark, J.D., Vargas, R., Banks, J.E.: Incorporating variability in point estimates in risk assessment: bridging the gap between LC50 and population endpoints. Environ. Toxicol. Chem. **34**, 1683–1688 (2015)

Extension of p-Laplace Operator
for Image Denoising

George Baravdish[1(✉)], Yuanji Cheng[2], Olof Svensson[1], and Freddie Åström[3]

[1] Linköping University, Linköping, Sweden
{george.baravdish,olof.svensson}@liu.se
[2] Malmö University, Malmö, Sweden
yuanji.cheng@mah.se
[3] Heidelberg Collaboratory for Image Processing,
Heidelberg University, Heidelberg, Germany
freddie.astroem@iwr.uni-heidelberg.de

Abstract. In this work we introduce a novel operator $\Delta_{(p,q)}$ as an extended family of operators that generalize the p-Laplace operator. The operator is derived with an emphasis on image processing applications, and particularly, with a focus on image denoising applications. We propose a non-linear transition function, coupling p and q, which yields a non-linear filtering scheme analogous to adaptive spatially dependent total variation and linear filtering. Well-posedness of the final parabolic PDE is established via pertubation theory and connection to classical results in functional analysis. Numerical results demonstrates the applicability of the novel operator $\Delta_{(p,q)}$.

Keywords: p-Laplace operator · Parabolic equations · Image denoising · Anisotropic diffusion · Inverse problems

1 Introduction

A well known inverse problem in image processing is image denoising [4]. In the last decades the energy functional approach together with its corresponding Euler-Lagrange (E-L) equation has attracted great attention in solving inverse problems applied to image reconstruction. One important case of E-L equations is the one which involves the p-Laplace operator

$$\Delta_p u = \operatorname{div}(|\nabla u|^{p-2}\nabla u), \quad p \geq 1, \tag{1}$$

associated with the evolution equation of p-Laplacian

$$\begin{cases} \partial_t u - \Delta_p u = 0, & \text{in } \Omega \times (0,T) \\ u(0) = u_0, \text{in } \Omega \\ \partial_n u = 0, & \text{on } \partial\Omega \times (0,T) \end{cases} \tag{2}$$

© IFIP International Federation for Information Processing 2016
Published by Springer International Publishing AG 2016. All Rights Reserved
L. Bociu et al. (Eds.): CSMO 2015, IFIP AICT 494, pp. 107–116, 2016.
DOI: 10.1007/978-3-319-55795-3_9

where Ω is a bounded domain in $\mathbf{R}^2$ and $u_0 : \Omega \to \mathbf{R}$ is a given degraded image [5,7,13] and ∇u is the gradient. The degenerate parabolic Eq. in (2) has been studied by many authors and we limit ourselves here to refer the reader to [10]. It is well known that the case $p = 2$ gives the linear Gaussian filter, which however, impose strong spatial regularity and therefore image details such as lines and edges are oversmoothed. $p = 1$ is often refereed to as the method of total variation [13] and $p = 0$ is an instance of the so called balanced forward backward evolution [9].

In this work we study a decoupled form of the p-Laplace operator, expressed as a non-linear combination of the Δ_1 and Δ_∞ operators, introduced below. We call our new operator $\Delta_{(p,q)}$. Via established existence theory we show that the corresponding perturbed parabolic equation is well-posed and close to the original operator.

In Sect. 2 we review some of the properties of the p-Laplace operator applied in image denoising and compare it with Perona-Malik models. Our main contribution is in Sect. 3 where we extend the p-Laplace operator to a new operator $\Delta_{(p,q)}$ with focus on image denoising. We consider the corresponding variable version $\Delta_{(p(x),q(x))}$ in Sect. 4. Finally, in Sect. 5 we demonstrates the applicability of $\Delta_{(p,q)}$ by numerical results.

2 p-Laplacian for Image Denosing

An important feature in any evolution process for image denoising is preservation of certain geometrical features of the underlying image. In the case of image restoration these features include edges and corners. It is straight-forward to express the p-Laplace operator (1) as

$$\Delta_p u = |\nabla u|^{p-1} \Delta_1 u + (p-1)|\nabla u|^{p-2} \Delta_\infty u, \tag{3}$$

where $\Delta_1 u = \operatorname{div}\left(\dfrac{\nabla u}{|\nabla u|}\right)$, $\Delta_\infty u = \dfrac{\nabla u}{|\nabla u|} \cdot (D^2 u)\dfrac{\nabla u}{|\nabla u|}$ and $D^2 u$ is the Hessian of u. However, an intuitive way to represent Δ_p, giving direct interpretation of the diffusivity directions is to express Δ_p by using Gauge coordinates $(x, y) \to (\eta, \xi)$:

$$\Delta_p u = |\nabla u|^{p-2}(u_{\xi\xi} + (p-1)u_{\eta\eta}) \tag{4}$$

where

$$\eta = \frac{\nabla u}{|\nabla u|}, \qquad \xi = \frac{\nabla^\perp u}{|\nabla u|}, \tag{5}$$

and

$$u_{\xi\xi} = |\nabla u|\Delta_1 u, \qquad u_{\eta\eta} = \Delta_\infty u. \tag{6}$$

From (4) it is now clear that Δ_p imposes the same diffusivity strength in both directions ξ and η independent of the magnitude of the gradient. In an attempt to resolve this drawback, Perona and Malik (P-M) [12] proposed to replace $|\nabla u|^{p-2}$ with $g(|\nabla u|^2)$ in (1). The idea is that the weight should satisfy $g(s) \to 0$, $s \to \infty$

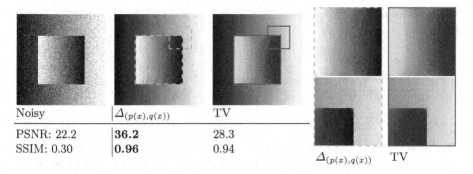

Noisy	$\Delta_{(p(x),q(x))}$	TV
PSNR: 22.2	**36.2**	28.3
SSIM: 0.30	**0.96**	0.94

$$\Delta_{(p(x),q(x))} \qquad \text{TV}$$

Fig. 1. Synthetic test image with 20 standard deviations of noise (left) and the obtained results for TV (green/thick) and the $\Delta_{(p(x),q(x))}$-operator (red/dashed). We see that both error measures improve with our operator and we get less staircasing artifacts while preserving corner points and edges as shown in the detailed images. (Color figure online)

and $g(s) \to 1$, $s \to 0^+$. P-M studied weights like $g = k/(k+s)$ and $g = e^{-ks}$ and demonstrated the advantages of these weight functions for edge preservations. Rewriting the operator given by the P-M method in Gauge coordinates, we obtain

$$\text{div}\big(g(|\nabla u|^2)\nabla u\big) = g(|\nabla u|^2)u_{\xi\xi} + \phi(|\nabla u|^2)u_{\eta\eta}, \tag{7}$$

where $\phi(s) = (sg(s))'$. Thus the diffusion in the direction η differs from the diffusion in the direction ξ. Since $\phi(s)$ is negative for large s, the evolution will be of backward diffusion effect near edges. This backward evolution cause problem for the well-posedness of the model and could also lead to staircasing problem [17].

The Perona-Malik PDE is a forward-backward type equation, and the diffusion is forward in the region $\{|\nabla u| < k\}$ and is backward and hence ill-posed in the region $\{|\nabla u| > k\}$. From evolution point of view, the forward-backward type equation may have infinitely many solutions and from variational minimization perspective the energy functional may have infinitely many minima [17]. To overcome the ill-posedness in the P-M model the authors in [17] introduced a regularization and proposed the following model

$$\partial_t u = \text{div}(g(|\nabla G_\sigma * u|)\nabla u),$$

where G_σ is a Gaussian function with standard deviation σ. A similar but time dependent variance $\sigma(t)$ was used in [16]. However, it's rather tricky to choose $\sigma(t)$ since it should neither decay too fast nor too slow during the evolution.

3 Extending the p-Laplace Operator

3.1 Anisotropic Decomposition via Constant Coefficients

The p-Laplace operator in (1), is an isotropic operator and expanding the divergence we obtain the equivalent form

$$\Delta_p u = \partial_x([(\partial_x u)^2 + (\partial_y u)^2]^{\frac{p-2}{2}}\partial_x u) + \partial_y([(\partial_x u)^2 + (\partial_y u)^2]^{\frac{p-2}{2}}\partial_y u). \tag{8}$$

An anisotropic behavior of the Δ_p-operator is induced by suppressing mixed derivatives, i.e., we define

$$L_{(p,p)}u = \partial_x(|\partial_x u|^{p-2}\partial_x u) + \partial_y(|\partial_y u|^{p-2}\partial_y u), \qquad 1 \le p < \infty. \qquad (9)$$

In the case $p = 1$, (8) is known as isotropic TV and (9) is anisotropic TV [8]. By decoupling the exponents in (9) one obtains the operator $L_{(p_1,p_2)}$

$$L_{(p_1,p_2)}u = \partial_x(|\partial_x u|^{p_1-2}\partial_x u) + \partial_y(|\partial_y u|^{p_2-2}\partial_y u), \quad 1 \le p_1, p_2 < \infty, \qquad (10)$$

which has previously appeared in fluid mechanics and we refer to [2,3].

Next, to see how the diffusion appears in the operator (10) we reformulate it in Gauge coordinates (5) by making the following definition.

Definition 1. *The $L_{(p_1,p_2)}$-operator is given by*

$$L_{(p_1,p_2)}u = \partial_\xi(|\partial_\xi u|^{p_1-2}\partial_\xi u) + \partial_\eta(|\partial_\eta u|^{p_2-2}\partial_\eta u), \qquad (11)$$

where $1 \le p_1, p_2 < \infty$.

The above operator $L_{(p_1,p_2)}$ is in fact a generalization of several known operators. We have

1. **The case $p_1 = p_2 = 2$.** The operator $L_{(2,2)}$ in (11) is the Laplacian, by now well studied. Due to the Laplacian's rotation invariance property we get

$$L_{(2,2)}u = \partial_\xi(\partial_\xi u) + \partial_\eta(\partial_\eta u) = \Delta u = u_{xx} + u_{yy}. \qquad (12)$$

2. **The case $p_1 = 2$ and $p_2 = 1$.** The operator $L_{(p_1,p_2)}$ in (11) is then given by

$$L_{(2,1)}u = \partial_\xi(\partial_\xi u) + \partial_\eta(|\partial_\eta u|^{-1}\partial_\eta u). \qquad (13)$$

Since we have $|\partial_\eta u|^{-1}\partial_\eta u = 1$, it follows that $L_{(2,1)}u = u_{\xi\xi}$. In Cartesian coordinates this corresponds to

$$L_{(2,1)}u = |\nabla u|\Delta_1 u, \qquad (14)$$

i.e. the mean curvature equation (see e.g., [6]). The corresponding regularized (weighted) mean curvature equation is given by

$$\partial_t u = g(|\nabla G_\sigma * u|)|\nabla u|\Delta_1 u \qquad (15)$$

previously studied in the context of image analysis, see e.g., [1].

3. **The case $p_1 = 2$ and $p_2 = p \in (1,2)$.** It follows from (11) that

$$L_{(2,p)}u = \partial_{\xi\xi}u + \partial_\eta(|\partial_\eta u|^{p-2}\partial_\eta u) = \partial_{\xi\xi}u + (p-1)u_\eta^{p-2}u_{\eta\eta} \qquad (16a)$$

$$= |\nabla u|\Delta_1 u + (p-1)|\nabla u|^{p-2}\Delta_\infty u \qquad (16b)$$

i.e. a mean curvature operator (14) with a second order term corresponding to (3). Note that the second order term induce invariant smoothing of the image data. Since (16) is merely a special case of (3), we further relax the mean curvature term next to better reflect the trade-off between edge-preservation and obtained smoothness. Although this modification appears straight-forward, its implications are non-trivial, however.

The anisotropic $L_{(2,p)}$ operator is given by

Definition 2. *The operator $\Delta_{(p,q)}$ is*

$$\Delta_{(p,q)}u = |\nabla u|^q \Delta_1 u + (p-1)|\nabla u|^{p-2}\Delta_\infty u, \quad p \in [1,2], \quad q \geq 0. \tag{17}$$

Remark 1. Definition 2 is a straight-forward relaxation of the exponents in (3), motivated by the discussion in above point 3. The introduction of q in (17), defines an additional degree of freedom, allowing us to control the trade-off between Δ_1 and Δ_∞, i.e., a trade-off between edge preservation and smoothness.

The corresponding evolution problem of the operator $\Delta_{(p,q)}$ is

Definition 3. *The evolution problem of $\Delta_{(p,q)}$ is given by*

$$\begin{cases} \partial_t u - \Delta_{(p,q)}u = 0, & p \in [1,2], \quad q \geq 0 \\ \quad\quad\quad u(0) = u_0 \\ \quad\quad\quad \partial_n u = 0 \end{cases} \tag{18}$$

We point out that $q = 0, p = 1$ yields the familiar isotropic TV regularizer and $q = 1, p = 2$ results in the heat equation, i.e., isotropic filtering. In the next section, we propose to couple p and q via a smooth non-linear transition function such that $\Delta_{(p,q)}$ can be thought of a spatially variant TV and isotropic regularizer.

4   Variable Coefficients

4.1   Coefficient Coupling

In the previous section we motivated the $\Delta_{(p,q)}$ operator and advocated to introduce variable coefficients, depending on the image data. The behavior of the operator $\Delta_{(p(x),q(x))}$ that we seek, is the edge-preserving effect of TV leading to $\Delta_{(p(x),q(x))} \rightarrow \Delta_{(1,0)}$ (the case of TV) as $|\nabla u| \rightarrow \infty$. In regions with small gradients we define the $(p(x), q(x))$-coefficients such that the operator is a linear filter leading to $\Delta_{(p(x),q(x))} \rightarrow \Delta_{(2,1)} = \Delta$ (the case of isotropic diffusion) as $|\nabla u| \rightarrow 0$. To see how the diffusion appears in $\Delta_{(p(x),q(x))}$ we rewrite the operator in Gauge coordinates.

Lemma 1. *The operator $\Delta_{(p,q)}$ in (17) can be written as*

$$\Delta_{(p,q)}u = |\nabla u|^{q-1}\left(\xi^\top(D^2 u)\xi + (p-1)|\nabla u|^{p-q-1}\eta^\top(D^2 u)\eta\right). \tag{19}$$

Proof. Using the relations in (6) and by rotation invariance of Laplacian, we obtain

$$\Delta u = u_{\xi\xi} + u_{\eta\eta} = \Delta_\infty u + |\nabla u|\Delta_1 u. \tag{20}$$

Now we observe that the operator $\Delta_1 u$ can be expanded as

$$\Delta_1 u = \frac{u_y^2 u_{xx} - 2u_x u_y u_{xy} + u_x^2 u_{yy}}{|\nabla u|^3} = \frac{(\nabla^\perp u)^\top(D^2 u)\nabla^\perp u}{|\nabla u|^3}, \tag{21}$$

and using this in (20) gives

$$\Delta_\infty u = |\nabla u|^2 \Delta u - (\nabla^\perp u)^\top (D^2 u) \nabla^\perp u = \nabla^\top u (D^2 u) \nabla u. \tag{22}$$

Thus, the $\Delta_{(p,q)}$ operator in (17) reformulates to

$$\begin{aligned}\Delta_{(p,q)} u &= |\nabla u|^{q-3} (\nabla^\perp u)^\top (D^2 u) \nabla^\perp u + (p-1)|\nabla u|^{p-4} \nabla^\top u (D^2 u) \nabla u \\ &= |\nabla u|^{q-1} \xi^\top (D^2 u) \xi + (p-1)|\nabla u|^{p-2} \eta^\top (D^2 u) \eta,\end{aligned}$$

which shows the result.   □

Next, we couple p and q via the relation

$$p(|\nabla u|) = 1 + q(|\nabla u|) \tag{23}$$

and from which we derive the following result.

Lemma 2. *If $p(x) = 1 + q(x)$, then*

$$\Delta_{(1+q(x),q(x))} u = |\nabla u|^{q(x)-1} \left(\xi^\top (D^2 u) \xi + q(x) \eta^\top (D^2 u) \eta \right). \tag{24}$$

Proof. The proof follows immediately from Lemma 1.   □

In this study, we couple p and q via the negative exponential function (although other selections are possible), i.e., we set

$$q(|\nabla u|) = k_2 \exp\left(-|\nabla u|/k_1\right), \tag{25a}$$
$$p(|\nabla u|) = 1 + q(|\nabla u|), \tag{25b}$$

where $k_1 > 0$ and $0 < k_2 < 1$. Under this selection we form the following parabolic PDE

$$\partial_t u - |\nabla u|^{q(|\nabla u|)-1} \left(\xi^\top (D^2 u) \xi + q(|\nabla u|) \eta^\top (D^2 u) \eta \right) = 0. \tag{26}$$

One easily checks that (26) describes a smooth transition between total variation and linear filtering for the selection of p, q in (25). By using $|\nabla u| \Delta_1 u = \Delta u - \Delta_\infty u$, the operator $\Delta_{(1+q,q)}$ becomes

$$\Delta_{(1+q,q)} u = |\nabla u|^{q(|\nabla u|)-1} \left(Tr(D^2 u) + \frac{q(|\nabla u|)-1}{|\nabla u|^2} \nabla u^\top (D^2 u) \nabla u \right). \tag{27}$$

Remark 2. The operator $\Delta_{(1+q,q)}$ is non-linear and have unbounded coefficients. In this first study, we perturb the $\Delta_{(1+q,q)}$-operator to obtain a regularized version of the evolution problem (18). This regularization enables us to pose the necessary conditions for well-posedness, introduced next.

4.2 Regularity of Solutions

In order to set up the framework for numerical calculation, we define

Definition 4. *Let $0 < \varepsilon, \delta < 1$ and q as in (25). Then*

$$\Delta^{\varepsilon,\delta}_{(1+q,q)} u =$$

$$(|\nabla u|^2 + \varepsilon^2)^{(q(|\nabla u|)-1)/2}\left((1+\delta)Tr(D^2u) + \frac{q(|\nabla u|)-1}{|\nabla u|^2 + \varepsilon^2}\nabla u^\top(D^2u)\nabla u\right). \quad (28)$$

Thus we study the regularized evolution equation

$$u_t = \Delta^{\varepsilon,\delta}_{(1+q,q)} u = \sum_{i,j=1} a^{\varepsilon,\delta}_{ij}(\nabla u)u_{ij} \quad (29)$$

where

$$a^{\varepsilon,\delta}_{ij}(\zeta) = (|\zeta|^2 + \varepsilon^2)^{(q(|\zeta|)-1)/2}\left[(1+\delta)\delta_{ij} + \frac{q(|\zeta|)-1}{|\zeta|^2 + \varepsilon^2}\zeta_i\zeta_j\right], \quad (30)$$

and $\zeta = \nabla u$. Given $u^{(k)}$, we obtain the next update $u^{(k+1)}$ by solving the following, equivalent (see [15]), initial value problem iteratively

$$\begin{cases} u_t^{(k+1)} = \sum a^{\varepsilon,\delta}_{ij}(\nabla u^{(k)})u_{ij}^{(k+1)}, & \text{in } \Omega \times (0,T) \\ u^{(k+1)}(0) = u_0, & \text{in } \Omega \\ \partial_n u^{(k+1)} = 0, & \text{on } \partial\Omega \times (0,T) \end{cases} \quad (31)$$

Proposition 1. *If the initial data $u^{(0)} = 0$ then the solution $u^{(k)}$ to (31) exists and is in $C^\infty(\Omega \times (0,T))$.*

Proof. If the initial guess $u^{(0)} = 0$, then $\sum a^{\varepsilon,\delta}_{ij}(0)u_{ij} = (1+\delta)\varepsilon^{q(0)-1}\Delta u$ and we deduce from Theorem 4.31 in [10] that $u^{(1)}$ exists and is $C^\infty(\Omega \times (0,T))$. Given $u^{(k)} \in C^\infty$, then $||u^{(k)}||_{C^1(\Omega\times(0,T))}$ is bounded, $(|\zeta|^2 + \varepsilon^2)^{(q(|\zeta|)-1)/2}$ is bounded from below and $a^{\varepsilon,\delta}_{ij}$ are also $C^\infty(\Omega\times(0,T))$. Hence, there are constants $c = c(\varepsilon,\delta,k)$ and $C = C(\varepsilon,\delta,k) > 0$, depending only on $\varepsilon, \delta, ||u^{(k)}||_{C^1}$, such that

$$c|\zeta|^2 \leq a^{\varepsilon,\delta}_{ij}(\nabla u^{(k)})\zeta_i\zeta_j \leq C|\zeta|^2.$$

It follows once again from Theorem 4.31 in [10] that $u^{(k+1)}$ exists and also is $C^\infty(\Omega \times (0,T))$. $\qquad\square$

5 Evaluation

5.1 Implementation

The parabolic PDE is discretized by using finite differences and a simple forward Euler scheme. We used a state of the art Split Bregman (SB) implementation

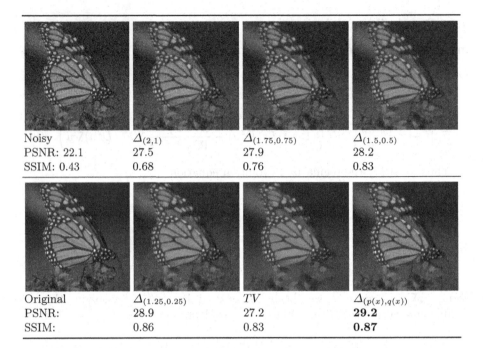

Fig. 2. Example results for the $\Delta_{(p,q)}$-operator where we stopped the filtering process at the maximum SSIM value. "TV"-was obtained using the Split Bregman method [8]. We set $k_1 = k_2 = 0.1$ in $\Delta_{(p(x),q(x))}$. See text for details.

of total variation. For details on SB see [8]. The regularization parameter for the SB was optimized in the range $[0.01, 1]$ in increasing steps of 0.11. The regularization parameter of the Bregman-variables of SB was set to 1 and the scheme was terminated as $10^{-3} > ||u^{(k)} - u^{(k-1)}||_2/||u^{(k-1)}||_2$ and we choose the regularization parameter that produces highest SSIM value [14]. We also report the peak signal-to-noise value (PSNR). For the $(p(x), q(x))$-operator we found that $k_1 = k_2 = 0.1$ and the update stepsize as $\alpha = 10^{-5}$ and $\tau = 0.5$ works well for the considered noise level of 20 standard deviations of additive Gaussian noise. These values are ad-hoc and future work include methods for parameter estimation.

5.2   Results

First we test our algorithm on a synthetic test image seen in Fig. 1. We see that the result of the proposed operator appears smoother than the result of TV in the center region, but yet preserves the corner point well. In Fig. 2 (cropped 256×256 pixels of image 35049.jpg [11]) we compare the visual quality and the SSIM-values for a range of p, q-values. As expected for $\Delta_{(2,1)}$ (isotropic filtering) performs the worst whereas the operator produce improved result w.r.t. SSIM as well as perceptual appearance. In the case of non-adaptive parameters, TV

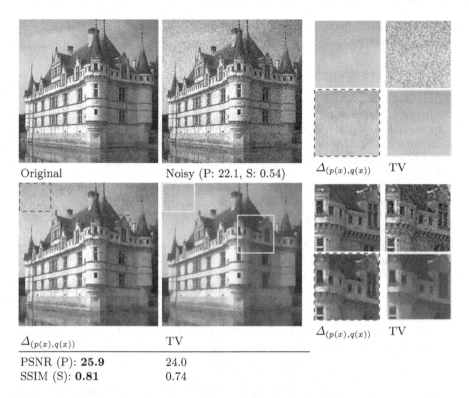

$\Delta_{(p(x),q(x))}$	TV
PSNR (P): **25.9**	24.0
SSIM (S): **0.81**	0.74

Fig. 3. Example denoising a grayscale image with 20 standard deviations of noise. In the close-up images to the right it can be seen that TV (thick/green) produces the characteristic staircasing effect while the operator $\Delta_{(p,q)}$ (dashed/red) shows good visual similarity with the noise free patch. TV shows good result in the sky (up right), but oversmooths, e.g., the window tiles seen in the close-up down right. (Color figure online)

performs the best. However, the operator with adaptive coefficients improve both SSIM, PSNR values and produces less oversmoothing (down-right figure). We also include the result from a grayscale image "Castle" (cropped 256×256 pixels of image 102061.jpg [11]) in Fig. 3. In this image TV performs very well in the sky (detail up right with green/thick frame) whereas the result from the $\Delta_{(p,q)}$ operator appears less noisy and looks visually more crisp. In both examples the proposed operator shows an improvement in PSNR as well as SSIM values.

6   Conclusion

In this paper we introduced a new family of operators, $\Delta_{(p,q)}$. Preliminary numerical results indicate that there could be a relationship between $p(x)$ and $q(x)$ that further improves the restoration effect. In forthcoming works we will investigate the operator $\Delta_{(p,q)}$ further regarding both regularity and different areas of applications.

116   G. Baravdish et al.

Acknowledgment. Support by the German Science Foundation and the Research Training Group (GRK 1653) is gratefully acknowledged.

References

1. Alvarez, L., Lions, P.L., Morel, J.M.: Image selective smoothing and edge detection by nonlinear diffusion. SIAM J. Numer. Anal. **29**(3), 845–866 (1992)
2. Antontsev, S.N., Shmarev, S.I.: Existence and uniqueness of solutions of degenerate parabolic equations with variable exponents of nonlinearity. Fundam. Prikl. Mat. **12**(4), 3–19 (2006)
3. Antontsev, S., Shmarev, S.: Elliptic equations with anisotropic nonlinearity and nonstandard growth conditions. In: Handbook of Differential Equations. Stationary Partial Differential Equations, vol. 3, pp. 1–100. Elsevier/North Holland, Amsterdam (2006)
4. Aubert, G., Kornprobst, P.: Mathematical Problems in Image Processing. Applied Mathematical Sciences, vol. 147, 2nd edn. Springer, New York (2006)
5. Baravdish, G., Svensson, O., Åström, F.: On backward $p(x)$-parabolic equations for image enhancement. Numer. Funct. Anal. Optim. **36**(2), 147–168 (2015)
6. Chen, Y.G., Giga, Y., Goto, S.: Uniqueness and existence of viscosity solutions of generalized mean curvature flow equations. J. Differ. Geom. **33**(3), 749–786 (1991)
7. Does, K.: An evolution equation involving the normalized p-laplacian. Commun. Pure Appl. Anal. **10**(1), 361–396 (2011)
8. Goldstein, T., Osher, S.: The split Bregman method for l1-regularized problems. SIAM J. Imaging Sci. **2**(2), 323–343 (2009)
9. Keeling, S.L., Stollberger, R.: Nonlinear anisotropic diffusion filtering for multiscale edge enhancement. Inverse Prob. **18**(1), 175–190 (2002)
10. Lieberman, G.M.: Second Order Parabolic Differential Equations. World Scientific Publishing Co., Inc., River Edge (1996)
11. Martin, D., Fowlkes, C., Tal, D., Malik, J.: A database of human segmented natural images and its application to evaluating segmentation algorithms and measuring ecological statistics. In: ICCV, vol. 2, pp. 416–423 (2001)
12. Perona, P., Malik, J.: Scale-space and edge detection using anisotropic diffusion. IEEE Trans. PAMI **12**, 629–639 (1990)
13. Rudin, L.I., Osher, S., Fatemi, E.: Nonlinear total variation based noise removal algorithms. Physica D **60**(1–4), 259–268 (1992)
14. Wang, Z., Bovik, A., Sheikh, H., Simoncelli, E.: Image quality assessment: from error visibility to structural similarity. IEEE Trans. Image Process. **13**(4), 600–612 (2004)
15. Vogel, C.R., Oman, M.E.: Iterative methods for total variation denoising. SIAM J. Sci. Comput. **17**, 227–238 (1996)
16. Whitaker, R.T., Pizer, S.M.: A multi-scale approach to nonuniform diffusion. CVGIP: Image Underst. **57**(1), 99–110 (1993)
17. You, Y.L., Xu, W., Tannenbaum, A., Kaveh, M.: Behavioral analysis of anisotropic diffusion in image processing. IEEE Trans. Image Process. **5**(11), 1539–1553 (1996)

Preconditioned ADMM with Nonlinear Operator Constraint

Martin Benning[1](✉), Florian Knoll[2], Carola-Bibiane Schönlieb[1], and Tuomo Valkonen[3]

[1] Department of Applied Mathematics and Theoretical Physics, University of Cambridge, Wilberforce Road, Cambridge CB3 0WA, UK
{mb941,cbs31}@cam.ac.uk
[2] Center for Advanced Imaging, Innovation and Research, New York University, 4th Floor 660 First Avenue, New York, NY 10016, USA
florian.knoll@nyumc.org
[3] Department of Mathematical Sciences, University of Liverpool, Mathematical Sciences Building, Liverpool L69 7ZL, UK
tuomo.valkonen@liverpool.ac.uk

Abstract. We are presenting a modification of the well-known Alternating Direction Method of Multipliers (ADMM) algorithm with additional preconditioning that aims at solving convex optimisation problems with nonlinear operator constraints. Connections to the recently developed Nonlinear Primal-Dual Hybrid Gradient Method (NL-PDHGM) are presented, and the algorithm is demonstrated to handle the nonlinear inverse problem of parallel Magnetic Resonance Imaging (MRI).

Keywords: ADMM · Primal-dual · Nonlinear inverse problems · Parallel MRI · Proximal point method · Operator splitting · Iterative Bregman method

1 Introduction

Non-smooth regularisation methods are popular tools in the imaging sciences. They allow to promote sparsity of inverse problem solutions with respect to specific representations; they can implicitly restrict the null-space of the forward operator while guaranteeing noise suppression at the same time. The most prominent representatives of this class are total variation regularisation [19] and ℓ^1-norm regularisation as in the broader context of compressed sensing [8,10].

In order to solve convex, non-smooth regularisation methods with linear operator constraints computationally, first-order operator splitting methods have gained increasing interest over the last decade, see [3,9,11,12] to name just a few. Despite some recent extensions to certain types of non-convex problems [7,14–16] there has to our knowledge only been made little progress for nonlinear operators constraints [2,22].

L. Bociu et al. (Eds.): CSMO 2015, IFIP AICT 494, pp. 117–126, 2016.
DOI: 10.1007/978-3-319-55795-3_10

In this paper we are particularly interested in minimising non-smooth, convex functionals with nonlinear operator constraints. This model covers many interesting applications; one particular application that we are going to address is the joint reconstruction of the spin-proton-density and coil sensitivity maps in parallel MRI [13,21].

The paper is structured as follows: we will introduce the generic problem formulation, then address its numerical minimisation via a generalised ADMM method with linearised operator constraints. Subsequently we will show connections to the recently proposed NL-PDHGM method (indicating a local convergence result of the proposed algorithm) and conclude with the joint spin-proton-density and coil sensitivity map estimation as a numerical example.

2   Problem Formulation

We consider the following generic constrained minimisation problem:

$$(\hat{u}, \hat{v}) = \arg\min_{u,v} \left\{ H(u) + J(v) \text{ subject to } F(u,v) = c \right\}. \tag{1}$$

Here H and J denote proper, convex and lower semi-continuous functionals, F is a nonlinear operator and c a given function. Note that for nonlinear operators of the form $F(u,v) = G(u) - v$ and $c = 0$ problem (1) can be written as

$$\hat{u} = \arg\min_{u} \left\{ H(u) + J(G(u)) \right\}. \tag{2}$$

In the following we want to propose a strategy for solving (1) that is based on simultaneous linearisation of the nonlinear operator constraint and the solution of an inexact ADMM problem.

3   Alternating Direction Method of Multipliers

We solve (1) by alternating optimisation of the augmented Lagrange function

$$\mathcal{L}_\delta(u,v;\mu) = H(u) + J(v) + \langle \mu, F(u,v) - c \rangle + \frac{\delta}{2} \|F(u,v) - c\|_2^2. \tag{3}$$

Alternating minimisation of (3) in u, v and subsequent maximisation of μ via a step of gradient ascent yields this nonlinear version of ADMM [11]:

$$u^{k+1} \in \arg\min_{u} \left\{ \frac{\delta}{2} \|F(u,v^k) - c\|_2^2 + \langle \mu^k, F(u,v^k) \rangle + H(u) \right\}, \tag{4}$$

$$v^{k+1} \in \arg\min_{v} \left\{ \frac{\delta}{2} \|F(u^{k+1},v) - c\|_2^2 + \langle \mu^k, F(u^{k+1},v) \rangle + J(v) \right\}, \tag{5}$$

$$\mu^{k+1} = \mu^k + \delta \left(F(u^{k+1}, v^{k+1}) - c \right). \tag{6}$$

Not having to deal with nonlinear subproblems, we replace $F(u,v^k)$ and $F(u^{k+1},v)$ by their Taylor linearisations around u^k and v^k, which yields $F(u,v^k) \approx F(u^k,v^k) + \partial_u F(u^k,v^k) (u - u^k)$ and $F(u^{k+1},v) \approx F(u^{k+1},v^k) + \partial_v F(u^{k+1},v^k) (v - v^k)$, respectively. The updates (4) and (5) modify to

$$u^{k+1} \in \arg\min_{u} \left\{ \frac{\delta}{2} \left\| A^k u - c_1^k \right\|_2^2 + \langle \mu^k, A^k u \rangle + H(u) \right\}, \tag{7}$$

$$v^{k+1} \in \arg\min_{v} \left\{ \frac{\delta}{2} \left\| B^k v - c_2^k \right\|_2^2 + \langle \mu^k, B^k v \rangle + J(v) \right\}, \tag{8}$$

with $A^k := \partial_u F(u^k, v^k)$, $B^k := \partial_v F(u^{k+1}, v^k)$, $c_1^k := c + A^k u^k - F(u^k, v^k)$ and $c_2^k := c + B^k v^k - F(u^{k+1}, v^k)$. Note that the updates (7) and (8) are still implicit, regardless of H and J. In the following, we want to modify the updates such that they become simple proximity operations.

4   Preconditioned ADMM

Based on [23], we modify (7) and (8) by adding the surrogate terms $\|u^{k+1} - u^k\|_{Q_1^k}^2 / 2$ and $\|v^{k+1} - v^k\|_{Q_2^k}^2 / 2$, with $\|w\|_Q := \sqrt{\langle Qw, w \rangle}$ (note that if Q is chosen to be positive definite, $\| \cdot \|_Q$ becomes a norm). We then obtain

$$u^{k+1} \in \arg\min_{u} \left\{ \frac{\delta}{2} \left\| A^k u - c_1^k \right\|_2^2 + \langle \mu^k, A^k u \rangle + H(u) + \frac{1}{2} \|u - u^k\|_{Q_1^k}^2 \right\},$$

$$v^{k+1} \in \arg\min_{v} \left\{ \frac{\delta}{2} \left\| B^k v - c_2^k \right\|_2^2 + \langle \mu^k, B^k v \rangle + J(v) + \frac{1}{2} \|v - v^k\|_{Q_2^k}^2 \right\}.$$

If we choose $Q_1^k := \tau_1^k I - \delta A^{k*} A^k$ with $\tau_1^k \delta < 1/\|A^k\|^2$ and $Q_2^k := \tau_2^k I - \delta B^{k*} B^k$ with $\tau_2^k \delta < 1/\|B^k\|^2$ and if we define $\overline{\mu}^k := 2\mu^k - \mu^{k-1}$ we obtain

$$u^{k+1} = \left(I + \tau_1^k \partial H \right)^{-1} \left(u^k - \tau_1^k A^{k*} \overline{\mu}^k \right), \tag{9}$$

$$v^{k+1} = \left(I + \tau_2^k \partial J \right)^{-1} \left(v^k - \tau_2^k B^{k*} \left(\mu^k + \delta \left(F(u^{k+1}, v^k) - c \right) \right) \right), \tag{10}$$

with $(I + \alpha \partial E)^{-1}(w)$ denoting the proximity or resolvent operator

$$(I + \alpha \partial E)^{-1}(w) := \arg\min_{u} \left\{ \frac{1}{2} \|u - w\|_2^2 + \alpha E(u) \right\}.$$

The entire proposed algorithm with updates (9), (10) and (6) reads as

---
Algorithm 1. Preconditioned ADMM with nonlinear operator constraint

---
Parameters: H, J, F, c
Initialization: u^0, v^0, μ^0, δ
$\overline{\mu}^0 = \mu^0$
while convergence criterion is not met **do**
    $A^k = \partial_u F(u^k, v^k)$
    Set τ_1^k such that $\tau_1^k \delta < 1/\|A^k\|^2$
    $u^{k+1} = \left(I + \tau_1^k \partial H \right)^{-1} \left(u^k - \tau_1^k A^{k*} \overline{\mu}^k \right)$
    $B^k = \partial_v F(u^{k+1}, v^k)$
    Set τ_2^k such that $\tau_2^k \delta < 1/\|B^k\|^2$
    $v^{k+1} = \left(I + \tau_2^k \partial J \right)^{-1} \left(v^k - \tau_2^k B^{k*} \left(\mu^k + \delta \left(F(u^{k+1}, v^k) - c \right) \right) \right)$
    $\mu^{k+1} = \mu^k + \delta \left(F(u^{k+1}, v^{k+1}) - c \right)$
    $\overline{\mu}^{k+1} = 2\mu^{k+1} - \mu^k$
end while
return u^k, v^k, μ^k, $\overline{\mu}^k$

---

5   Connection to NL-PDHGM

In the following we want to show how the algorithm simplifies in case the non-linear operator constraint is only nonlinear in one variable, which is sufficient for problems of the form (2). Without loss of generality we consider constraints of the form $F(u, v) = G(u) - v$, where G represents a nonlinear operator in u. Then we have $A^k = \mathcal{J}G(u^k)$ (with $\mathcal{J}G(u^k)$ denoting the Jacobi matrix of G at u^k), $B^k = -I$ and if we further choose $\tau_2^k = 1/\delta$ for all k, update (10) reads

$$v^{k+1} = \left(I + \frac{1}{\delta}\partial J\right)^{-1} \left(G(u^{k+1}) + \frac{1}{\delta}\mu^k\right).$$

Applying Moreau's identity [18] $b = \left(I + \frac{1}{\delta}\partial J\right)^{-1}(b) + \frac{1}{\delta}(I + \delta\partial J^*)^{-1}(\delta b)$ yields

$$\mu^{k+1} = (I + \delta\partial J^*)^{-1}\left(\mu^k + \delta G(u^{k+1})\right).$$

If we further change the order of the updates, starting with the update for μ, the whole algorithm reads

$$\mu^{k+1} = (I + \delta\partial J^*)^{-1}\left(\mu^k + \delta G(u^k)\right),$$
$$\overline{\mu}^{k+1} = 2\mu^{k+1} - \mu^k,$$
$$u^{k+1} = \left(I + \tau_1^k\partial H\right)^{-1}\left(u^k - \tau_1^k \, \mathcal{J}G(u^k)^*\overline{\mu}^{k+1}\right).$$

Note that this algorithm is almost the same as NL-PDHGM proposed in [22] for $\theta = 1$, except that the extrapolation step is carried out on the dual variable μ instead of the primal variable u. In the following we want to briefly sketch how to prove convergence for this algorithm in analogy to [22]. We define

$$N(\mu^{k+1}, u^{k+1}) := \begin{pmatrix} \partial J^*(\mu^{k+1}) - \nabla G(u^k)u^{k+1} - c^k \\ \partial H(u^{k+1}) + \mathcal{J}G(u^k)^*\mu^{k+1} \end{pmatrix},$$

$$L^k := \begin{pmatrix} \frac{1}{\delta}I & \mathcal{J}G(u^k) \\ \mathcal{J}G(u^k)^* & \frac{1}{\tau_1^k}I \end{pmatrix},$$

with $c^k := G(u^k) - \mathcal{J}G(u^k)u^k$. Now the algorithm is: find (μ^{k+1}, u^{k+1}) such that

$$N(\mu^{k+1}, u^{k+1}) + L^k(\mu^{k+1} - \mu^k, u^{k+1} - u^k) \ni 0.$$

If we exchange the order of μ and u here, i.e., reorder the rows of N, and the rows and columns of L^k, we obtain *almost* the "linearised" NL-PDHGM of [22]. The difference is that the sign of $\mathcal{J}G$ in L^k is inverted. The only points in [22] where the exact structure of L^k (M_{x^k} therein) is used, are Lemma 3.1, Lemma 3.6 and Lemma 3.10. The first two go through exactly as before with the negated structure. Reproducing Lemma 3.10 demands bounding actual step lengths $\|u^k - u^{k+1}\|$ and $\|\mu^k - \mu^{k+1}\|$ from below, near a solution for arbitrary $\epsilon > 0$. A proof would go beyond the page limit of this proceeding. Let us just point out that this can be done, implying that the convergence results of [22] apply for this algorithm as well. This means that under somewhat technical regularity conditions, which for TV type problems amount to Huber regularisation, local convergence in a neighbourhood of the true solution can be guaranteed.

6   Joint Estimation of the Spin-Proton Density and Coil Sensitivities in Parallel MRI

We want to demonstrate the numerical capabilities of Algorithm 1 by applying it to the nonlinear problem of joint estimation of the spin-proton density and the coil sensitivities in parallel MRI. The discrete problem of joint reconstruction from sub-sampled k-space data on a rectangular grid reads

$$
\begin{pmatrix} \hat{u} \\ \hat{c}_1 \\ \vdots \\ \hat{c}_2 \end{pmatrix} \in \operatorname*{arg\,min}_{\mathbf{v}=(u,c_1,\ldots,c_n)} \left\{ \frac{1}{2} \sum_{j=1}^{n} \| S\mathcal{F}(G(\mathbf{v}))_j - f_j \|_2^2 + \alpha_0 R_0(u) + \sum_{j=1}^{n} \alpha_j R_j(c_j) \right\},
$$

where $\mathcal{F}$ is the 2D discrete Fourier transform, f_j are the k-space measurements for each of the n coils, S is the sub-sampling operator and R_j denote appropriate regularisation functionals. The nonlinear operator G maps the unknown spin-proton density u and the different coil sensitivities c_j as follows [21]:

$$
G(u, c_1, \ldots, c_n) = (uc_1, uc_2, \ldots, uc_n)^T. \tag{11}
$$

In order to compensate for sub-sampling artefacts in sub-sampled MRI it is common practice to use total variation as a regulariser [6,17]. Coil sensitivities are assumed to be smooth, cf. Fig. 1, motivating a reconstruction model similar to the one proposed in [13]. We therefore choose the discrete isotropic total variation, $R_0(u) = \|\nabla u\|_{2,1}$, and the smooth 2-norm of the discretised gradient, i.e. $R_j(c_j) := \|\nabla c_j\|_{2,2}$, for all $j > 0$, following the notation in [4]. We further introduce regularisation parameters λ_j in front of the data fidelities and rescale all regularisation parameters such that $\alpha_0 + \frac{1}{n}\left(\sum_{j=1}^{n}\lambda_j + \sum_{j=1}^{n}\alpha_j\right) = 1$. In order to realise this model via Algorithm 1 we consider the following operator splitting strategy. We define $F(u_0, \ldots, u_n, v_0, \ldots, v_{2n})$ as

$$
F(u_0, \ldots, u_n, v_1, \ldots, v_n) := \begin{pmatrix} G(u_0, \ldots, u_n) \\ \nabla u_0 & 0 & \cdots & 0 \\ 0 & \nabla u_1 & \ddots & \vdots \\ \vdots & \ddots & \ddots & 0 \\ 0 & \cdots & 0 & \nabla u_n \end{pmatrix} \begin{pmatrix} v_0 \\ \vdots \\ v_n \\ \vdots \\ v_{2n} \end{pmatrix},
$$

set $H(u_0, \ldots, u_n) \equiv 0$, and $J(v_0, \ldots, v_{2n}) = \sum_{j=0}^{2n} J_j(v_j)$ with $J_j(v_j) := \frac{\lambda_j}{2}\|S\mathcal{F}v_j - f_j\|_2^2$ for $j \in \{0, \ldots, n-1\}$, $J_n(v_n) = \alpha_0\|v_n\|_{2,1}$ and $J_j(v_j) = \alpha_{j-n}\|v_j\|_{2,2}$ for $j \in \{n+1, \ldots, 2n\}$. Note that with these choices of functions, all the resolvent operations can be carried out easily. In particular, we obtain

$$(I + \tau_1^k \partial H)^{-1}(w) = w,$$

$$(I + \tau_2^k \partial J_j)^{-1}(w) = \mathcal{F}^{-1}\left(\frac{\mathcal{F}w_j + \tau_2^k \lambda_j S^T f_j}{1 + \tau_2^k \lambda_j \operatorname{diag}(S^T S)}\right) \text{ for } j \in \{0, \ldots, n-1\},$$

$$(I + \tau_2^k \partial J_n)^{-1}(w) = \frac{w_n}{\|w_n\|_2} \max\left(\|w_n\|_2 - \alpha_0 \tau_2^k, 0\right),$$

$$(I + \tau_2^k \partial J_j)^{-1}(w) = \frac{w_j}{\|w_j\|_{2,2}} \max\left(\|w_j\|_{2,2} - \alpha_{j-n} \tau_2^k, 0\right) \text{ for } j \in \{n+1, \ldots, 2n\}.$$

Moreover, as $B_k = -I$ (and thus, $\|B^k\| = 1$) for all k, we can simply eliminate τ_2^k by replacing it with $1/\delta$, similar to Sect. 5.

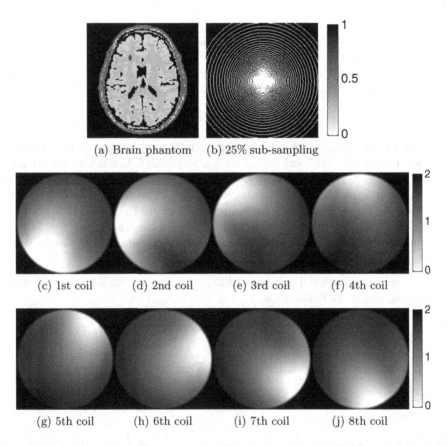

(a) Brain phantom   (b) 25% sub-sampling

(c) 1st coil   (d) 2nd coil   (e) 3rd coil   (f) 4th coil

(g) 5th coil   (h) 6th coil   (i) 7th coil   (j) 8th coil

Fig. 1. (a) shows the brain phantom as described in Sect. 6.1. (c)–(j) show visualisations of the measured coil sensitivities of a water bottle. (b) shows the simulated, spiral-shaped sub-sampling scheme used to sub-sample the k-space data.

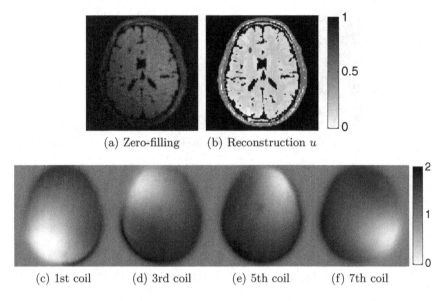

(a) Zero-filling   (b) Reconstruction u

(c) 1st coil   (d) 3rd coil   (e) 5th coil   (f) 7th coil

Fig. 2. Reconstructions for noise with low noise level $\sigma = 0.05$. Despite the sub-sampling, features of the brain phantom are very well preserved. In addition, the coil sensitivities seem to correspond well to the original ones, despite a slight loss of contrast. Note that coil sensitivities remain the initial value where the signal is zero.

6.1   Experimental Setup

We now want to discuss the experimental setup. We want to reconstruct the synthetic brain phantom in Fig. 1a from sub-sampled k-space measurements. The numerical phantom is based on the design in [1] with a matrix size of 190×190. It consists of several different tissue types like cerebrospinal fluid (CSF), gray matter (GM), white matter (WM) and cortical bone. Each pixel is assigned a set of MR tissue properties: Relaxation times $T_1(x,y)$ and $T_2(x,y)$ and spin density $\rho(x,y)$. These parameters were also selected according to [1]. The MR signal $s(x,y)$ in each pixel was then calculated by using the signal equation of a fluid attenuation inversion recovery (FLAIR) sequence [5]:

$$s(x,y) = \rho(x,y)(1 - 2\ e^{-TI/T_1(x,y)})(1 - e^{-TR/T_1(x,y)})\ e^{-TE/T_2(x,y)}.$$

The sequence parameters were selected: $TR = 10000$ ms, $TE = 90$ ms. TI was set to 1781 ms to achieve signal nulling of CSF ($T_1^{csf} \log(2)$ with $T_1^{csf} = 2569$ ms).

In order to generate artificial k-space measurements for each coil, we proceed as follows. First, we produce 8 images of the brain phantom multiplied by the measured coil sensitivity maps shown in Fig. 1c–j. The coil sensitivity maps were generated from the measurements of a water bottle with an 8-channel head coil array. Then we produce artificial k-space data by applying the 2D discrete Fourier-transform to each of those individual images. Subsequently, we sub-sample only approx. 25% of each of the k-space datasets via the spiral shown in Fig. 1b. Finally, we add Gaußian noise with standard deviation σ to the sub-sampled data.

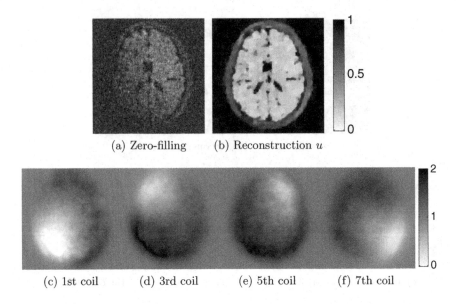

(a) Zero-filling   (b) Reconstruction u

(c) 1st coil   (d) 3rd coil   (e) 5th coil   (f) 7th coil

Fig. 3. Reconstructions for noise with high noise level $\sigma = 0.95$. Due to the large amount of noise, higher regularisation parameters are necessary. As a consequence, fine structures are smoothed out and in contrast to the case of little noise, compensation of sub-sampling artefacts is less successful.

6.2   Computations

For the actual computations we use two noisy versions f_j of the simulated k-space data; one with small noise ($\sigma = 0.05$) and one with a high amount of noise ($\sigma = 0.95$). As stopping criterion we simply choose a fixed number of iterations; for both the low noise level as well as the high noise level dataset we have fixed the number of iterations to 1500. The initial values used for the algorithm are $u_j^0 = 1$ with $1 \in \mathbb{R}^{l \times 1}$ being the constant one-vector, for all $j \in \{0, \dots, n\}$. All other initial variables (v^0, μ^0, $\overline{\mu}^0$) are set to zero.

Low Noise Level. We have computed reconstructions from the noisy data with noise level $\sigma = 0.05$ via Algorithm 1, with regularisation parameters set to $\lambda_j = 0.0621$, $\alpha_0 = 0.0062$ and $\alpha_j = 0.9317$ for $j \in \{1, \dots, n\}$. We have further created a naïve reconstruction by averaging the individual inverse Fourier-transformed images obtained from zero-filling the k-space data. The modulus images of the results are visualised in Fig. 2. The PSNR values for the averaged zero-filled reconstruction is 10.2185, whereas the PSNR of the reconstruction with the proposed method is 24.5572.

High Noise Level. We proceeded as in the previous section, but for noisy data with noise level $\sigma = 0.95$. The regularisation parameters were set to $\lambda_j =$

0.0149, $\alpha_0 = 0.0135$ and $\alpha_j = 0.9716$ for $j \in \{1, \ldots, n\}$. The modulus images of the results are visualised in Fig. 3. The PSNR values for the averaged zero-filled reconstruction is 9.9621, whereas the PSNR of the reconstruction with the proposed method is 16.672.

7   Conclusions and Outlook

We have presented a novel algorithm that allows to compute minimisers of a sum of convex functionals with nonlinear operator constraint. We have shown the connection to the recently proposed NL-PDHGM algorithm which implies local convergence results in analogy to those derived in [22]. Subsequently we have demonstrated the computational capabilities of the algorithm by applying it to a nonlinear joint reconstruction problem in parallel MRI.

For future work, the convergence of the algorithm in the general setting has to be verified, and possible extensions to guarantee global convergence have to be studied. Generalisation of stopping criteria such as a linearised primal-dual gap will also be of interest as well. With respect to the presented parallel MRI application, exact conditions for the convergence (like the exact norm of the bounds) have to be verified. The impact of the algorithm - as well as the regularisation-parameters on the reconstruction has to be analysed, and a rigorous study with artificial and real data would also be desirable. Moreover, future research will focus on alternative regularisation functions, e.g. based on spherical harmonics motivated by [20]. Last but not least, other applications that can be modelled via (1) should be considered in future research.

Acknowledgments. MB, CS and TV acknowledge EPSRC grant EP/M00483X/1. FK ackowledges National Institutes of Health grant NIH P41 EB017183.

EPSRC Data Statement: the corresponding code and data are available for download at https://www.repository.cam.ac.uk/handle/1810/256221.

References

1. Aubert-Broche, B., Evans, A.C., Collins, L.: A new improved version of the realistic digital brain phantom. NeuroImage **32**(1), 138–145 (2006)
2. Bachmayr, M., Burger, M.: Iterative total variation schemes for nonlinear inverse problems. Inverse Prob. **25**(10), 105004 (2009)
3. Beck, A., Teboulle, M.: Fast gradient-based algorithms for constrained total variation image denoising and deblurring problems. IEEE Trans. Image Process. **18**(11), 2419–2434 (2009)
4. Benning, M., Gladden, L., Holland, D., Schönlieb, C.-B., Valkonen, T.: Phase reconstruction from velocity-encoded MRI measurements-a survey of sparsity-promoting variational approaches. J. Magn. Reson. **238**, 26–43 (2014)
5. Bernstein, M.A., King, K.F., Zhou, X.J.: Handbook of MRI Pulse Sequences. Elsevier, Amsterdam (2004)

6. Block, K.T., Uecker, M., Frahm, J.: Undersampled radial MRI with multiple coils. Iterative image reconstruction using a total variation constraint. Magn. Reson. Med. **57**(6), 1086–1098 (2007)
7. Bonettini, S., Loris, I., Porta, F., Prato, M.: Variable metric inexact line-search based methods for nonsmooth optimization. Siam J. Optim. **26**, 891–921 (2015)
8. Candes, E.J., et al.: Compressive sampling. In: Proceedings of the International Congress of Mathematicians, vol. 3, Madrid, Spain, pp. 1433–1452 (2006)
9. Chambolle, A., Pock, T.: A first-order primal-dual algorithm for convex problems with applications to imaging. J. Math. Imaging Vis. **40**(1), 120–145 (2011)
10. Donoho, D.L.: Compressed sensing. IEEE Trans. Inf. Theory **52**(4), 1289–1306 (2006)
11. Gabay, D.: Applications of the method of multipliers to variational inequalities. Stud. Math. Appl. **15**, 299–331 (1983)
12. Goldstein, T., Osher, S.: The split Bregman method for L1-regularized problems. SIAM J. Imaging Sci. **2**(2), 323–343 (2009)
13. Knoll, F., Clason, C., Bredies, K., Uecker, M., Stollberger, R.: Parallel imaging with nonlinear reconstruction using variational penalties. Magn. Reson. Med. **67**(1), 34–41 (2012)
14. Möllenhoff, T., Strekalovskiy, E., Möller, M., Cremers, D.: The primal-dual hybrid gradient method for semiconvex splittings. SIAM J. Imaging Sci. **8**(2), 827–857 (2015)
15. Möller, M., Benning, M., Schönlieb, C., Cremers, D.: Variational depth from focus reconstruction. IEEE Trans. Image Process. **24**(12), 5369–5378 (2015)
16. Ochs, P., Chen, Y., Brox, T., Pock, T.: iPiano: inertial proximal algorithm for nonconvex optimization. SIAM J. Imaging Sci. **7**(2), 1388–1419 (2014)
17. Ramani, S., Fessler, J., et al.: Parallel MR image reconstruction using augmented Lagrangian methods. IEEE Trans. Med. Imaging **30**(3), 694–706 (2011)
18. Rockafellar, R.T.: Convex Analysis. Princeton Mathematical Series, 46:49. Princeton University Press, Princeton (1970)
19. Rudin, L.I., Osher, S., Fatemi, E.: Nonlinear total variation based noise removal algorithms. Phys. D: Nonlinear Phenom. **60**(1), 259–268 (1992)
20. Sbrizzi, A., Hoogduin, H., Lagendijk, J.J., Luijten, P., den Berg, C.A.T.: Robust reconstruction of B1+ maps by projection into a spherical functions space. Magn. Reson. Med. **71**(1), 394–401 (2014)
21. Uecker, M., Hohage, T., Block, K.T., Frahm, J.: Image reconstruction by regularized nonlinear inversion joint estimation of coil sensitivities and image content. Magn. Reson. Med. **60**(3), 674–682 (2008)
22. Valkonen, T.: A primal-dual hybrid gradient method for nonlinear operators with applications to MRI. Inverse Prob. **30**(5), 055012 (2014)
23. Zhang, X., Burger, M., Osher, S.: A unified primal-dual algorithm framework based on Bregman iteration. J. Sci. Comput. **46**(1), 20–46 (2011)

Regularized Optimal Design Problem for a Viscoelastic Plate Vibrating Against a Rigid Obstacle

Igor Bock[✉]

Institute of Computer Science and Mathematics FEI,
Slovak University of Technology, Ilkovičova 3, 81219 Bratislava, Slovak Republic
igor.bock@stuba.sk
http://www.stuba.sk

Abstract. We deal with a regularized optimal control problem governed by a nonlinear hyperbolic initial-boundary value problem describing behaviour of a viscoelastic plate vibrating against a rigid obstacle. A variable thickness of a plate plays the role of a control variable. The original problem for the deflection is regularized in order to have the uniqueness of a solution to the state problem and only the existence of an optimal thickness but also necessary optimality conditions.

Keywords: Viscoelastic anisotropic plate · Variable thickness · Rigid foundation · Regularization · Optimal control · Optimality conditions

1 Introduction

Shape design optimization problems belong to frequently solved problems with many engineering applications. We deal here with a regularized optimal design problem for a viscoelastic anisotropic plate vibrating against a rigid foundation. A variable thickness of a plate plays the role of a control variable. The corresponding state initial-boundary value contact problem represents one of the most natural problem of mechanics not frequently solved because of the hyperbolic character of the presented evolutional variational inequality. We deal here with a plate made of short memory viscoelastic material. It characterizes constructions made of concrete for example [7]. The dynamic contact for a viscoelastic bridge in a contact with a fixed road has been solved in [3]. The similar optimal control problems for the beams in a boundary contact are investigated in [1,4] respectively.

Due to the variable thickness e and the contact between a bottom of the plate and the obstacle represented by a function Φ the equation for the movement u of the middle surface and the complementarity conditions have the form

I. Bock—This work was financially supported by the Ministry of Education of Slovak Republic under grant VEGA-1/0159/16 and by grant of Science and Technology Assistance Agency no. APVV-0246-12.

L. Bociu et al. (Eds.): CSMO 2015, IFIP AICT 494, pp. 127–136, 2016.
DOI: 10.1007/978-3-319-55795-3_11

$$\frac{1}{2}\rho e(x)u_{tt} - \frac{1}{12}\mathrm{div}\left[e^3(x)\mathrm{grad}\,u_{tt}\right] + \left[e^3(x)(A_{ijk\ell}u_{t,x_ix_j} + B_{ijk\ell}u_{x_ix_j})\right]_{x_kx_\ell}$$
$$= F + G, \qquad 0 \le G \perp u - \tfrac{1}{2}e - \Phi \ge 0 \qquad \text{in } (0,T] \times \Omega,$$

where F and G express a perpendicular force acting on the plate and an unknown contact force respectively. In order to derive not only the existence of optimal variable thickness e but also the necessary optimality conditions we regularize the contact condition using the function

$$\omega \mapsto g_\delta(\omega), \quad g_\delta(\omega) = \begin{cases} 0 \text{ for } w \le 0 \\ \frac{6}{\delta^3}\omega^3 - \frac{8}{\delta^4}\omega^4 + \frac{3}{\delta^5}\omega^5 \text{ for } 0 < \omega < \delta \\ \frac{1}{\delta}\omega \text{ for } \omega \ge \delta. \end{cases}$$

in an analogous way as in [5], where the control problem for an elastic beam vibrating against an elastic foundation of Winkler's type was considered. We remark that instead of the function g_δ we can use any not negative nondecreasing function $g \in C^2(\mathbb{R})$ of the variable ω vanishing for $\omega \le 0$ and equaled to $\frac{1}{\delta}\omega$ for $\omega \ge \delta$.

Solving the state problem we apply the Galerkin method in the same way as in [1], where the rigid obstacle acting against a beam is considered or in [2] where the problem for a viscoelastic von Kármán plate vibrating against a rigid obstacle has been solved. The compactness method will be used in solving the minimum problem for a cost functional. We apply the approach from [5] in deriving the optimality conditions.

2   Solving the State Problem

2.1   Setting of the State Problem

We consider an anisotropic plate short memory viscoelastic plate with the middle surface $\Omega \subset \mathbb{R}^2$. The variable thickness of the plate is expressed by a positive function $x \mapsto e(x)$, $x \in \bar\Omega$, the positive constant ρ is the density of the material, $A_{ijk\ell}$, $B_{ijk\ell}$ are the symmetric and positively definite tensor expressing the viscoelastic and elastic properties of the material. The plate is clamped on its boundary. Let $F : (0,T] \times \Omega \mapsto \mathbb{R}$ be a perpendicular load per a square unit acting on the plate. Let u_0, $v_0 : \Omega \mapsto \mathbb{R}$ be the initial displacement and velocity, and

$$a = \frac{1}{6\rho}, \; a_{ijk\ell} = \frac{2}{\rho}A_{ijk\ell}, \; b_{ijk\ell} = \frac{2}{\rho}B_{ijk\ell}, \; f = \frac{2F}{\rho}$$

be the new mechanical and material characteristics. Then the vertical displacement $u : (0,T] \times \Omega \mapsto \mathbb{R}$ is a solution of the following regularized hyperbolic initial-boundary value problem

$$e(x)u_{tt} - a\frac{1}{12}\text{div}\,[e^3\text{grad}\,u_{tt}] + [e^3(x)(a_{ijk\ell}u_{t,x_ix_j} + b_{ijk\ell}u_{x_ix_j})]_{x_kx_\ell}$$
$$- g_\delta(\tfrac{1}{2}e(x) + \varPhi(x) - u) = f(t,x),\ \text{in}\ (0,T] \times \varOmega. \tag{1}$$

$$u(t,\xi) = \frac{\partial u}{\partial \boldsymbol{n}}(t,\xi) = 0,\ t \in (0,T],\ \xi \in \partial\varOmega \tag{2}$$

$$u(0,x) = u_0(x),\ u_t(0,x) = v_0(x),\ x \in \varOmega. \tag{3}$$

We introduce the Hilbert spaces

$$H \equiv L_2(\varOmega),\ H^k(\varOmega) = \{y \in H:\ D^\alpha y \in H,\ |\alpha| \le k\},\ k \in \mathbb{N}$$

with the standard inner products $(\cdot,\cdot)$, $(\cdot,\cdot)_k$ and the norms $|\cdot|_0$, $\|\cdot\|_k$,

$$\mathring{H}^1(\varOmega) = \{y \in H^1(\varOmega):\ y(\xi) = 0,\ \xi \in \partial\varOmega\ (\text{in the sense of traces})\}$$

and

$$V \equiv \mathring{H}^2(\varOmega) = \{y \in H^2(\varOmega):\ y(\xi) = \frac{\partial y}{\partial \boldsymbol{n}}(\xi) = 0,\ \xi \in \partial\varOmega\ (\text{in the sense of traces})\}$$

with the inner product and the norm

$$((y,z)) = \int_\varOmega y_{x_ix_j}(x)z_{x_ix_j}(x)\,dx,\ \|y\| = ((y,y))^{1/2},\ y,\,z \in V.$$

We denote by V^* the dual space of linear bounded functionals over V with duality pairing $\langle F,y\rangle_* = F(y)$, $F \in V^*$, $y \in V$. It is a Banach space with a norm $\|\cdot\|_*$.

The spaces V, H, V^* form the Gelfand triple meaning the dense and compact embeddings

$$V \hookrightarrow\hookrightarrow H \hookrightarrow\hookrightarrow V^*.$$

We set $I = (0,T)$, $Q = I \times \varOmega$. For a Banach space X we denote by $L_p(I;X)$ the Banach space of all functions $y : I \mapsto X$ such that $\|y(\cdot)\|_X \in L_p(0,T)$, $p \ge 1$, by $L_\infty(I;X)$ the space of essentially bounded functions with values in X, by $C(\bar I;X)$ the space of continuous functions $y : \bar I \mapsto X$, $\bar I = [0,T]$. For $k \in \mathbb{N}$ we denote by $C^k(\bar I;X)$ the spaces of k-times continuously differentiable functions defined on $\bar I$ with values in X. If X is a Hilbert space we set

$$H^k(I;X) = \{v \in C^{k-1}(\bar I;X):\ \frac{d^k v}{dt^k} \in L_2(I;X)\}$$

the Hilbert spaces with the inner products

$$(u,v)_{H^k(I,X)} = \int_I \Big[(u,v)_X + \sum_{j=1}^k (u^j,v^j)_X\Big]dt,\ k \in \mathbb{N}.$$

We denote by $\dot w$, $\ddot w$ and $\dddot w$ the first, the second and the third time derivative of a function $w : I \to X$. In order to derive necessary optimality conditions in the next chapter we assume stronger regularity of data:

$$e \in E_{ad} := \left\{ e \in H^2(\Omega) : \; 0 < e_{min} \le e(x) \le e_{max} \; \forall x \in \bar{\Omega}, \; \|e\|_2 \le \hat{e} \right\};$$

$$u_0 \in V \cap H^4(\Omega), \; u_0(x) \ge \tfrac{1}{2} e_{max} + \Phi(x) \, \forall x \in \Omega; \tag{4}$$

$$\Phi \in C(\bar{\Omega}), \; \Phi(\xi) \le 0 \; \forall \xi \in \partial\Omega;, \; v_0 \in V, \; f \in H^1(I; H).$$

The symmetric and positively definite fourth-order tensors $a_{ijk\ell}$, $b_{ijk\ell}$ fulfil

$$a_{ijk\ell} = a_{k\ell ij} = a_{jik\ell}, \; b_{ijk\ell} = b_{k\ell ij} = b_{jik\ell},$$

$$\alpha_0 > 0, \; \alpha_0 \varepsilon_{ij}\varepsilon_{ij} \le a_{ijk\ell}\,\varepsilon_{ij}\varepsilon_{k\ell} \le \alpha_1 \varepsilon_{ij}\varepsilon_{ij} \; \forall \; \{\varepsilon_{ij}\} \in \mathbb{R}^{2\times 2}_{sym}, \tag{5}$$

$$\beta_0 > 0, \; \beta_0 \varepsilon_{ij}\varepsilon_{ij} \le b_{ijk\ell}\,\varepsilon_{ij}\varepsilon_{k\ell} \le \beta_1 \varepsilon_{ij}\varepsilon_{ij} \; \forall \; \{\varepsilon_{ij}\} \in \mathbb{R}^{2\times 2}_{sym},$$

where the Einstein summation convention is employed and $\mathbb{R}^{2\times 2}_{sym}$ is the set of all second-order symmetric tensors. For $e, u, y \in H^2(\Omega)$ we define bilinear forms $A(e)$, $B(e)$ by

$$A(e)(u,y) = e^3 a_{ijk\ell} u_{ij} y_{k\ell}, \; B(e)(u,y) = e^3 b_{ijk\ell} u_{ij} y_{k\ell}.$$

Definition 1. *A function u is a weak solution of the problem (1)–(3) if $\ddot{u} \in L_2(I; \mathring{H}^1(\Omega))$, $\dot{u} \in L_2(I; V)$, there hold the identity*

$$\int_Q \left[e(x)\ddot{u}y + ae^3(x)\nabla\ddot{u}\cdot\nabla y + A(e;\dot{u},y) + B(e;u,y) \right] dx\,dt$$

$$= \int_Q \left[g_\delta(\tfrac{1}{2}e(x) + \Phi(x) - u) + f(t,x) \right] y \, dx\,dt \; \forall\, y \in L_2(I; V) \tag{6}$$

and the initial conditions

$$u(0) = u_0, \; \dot{u}(0) = v_0. \tag{7}$$

2.2   Existence and Uniqueness of the State Problem

We verify the existence and uniqueness of a weak solution.

Theorem 1. *There exists a unique solution u of the problem (6) and (7) such that $u \in C^1(\bar{I}; V)$, $\ddot{u} \in L_2(I; V) \cap L_\infty(I; \mathring{H}^1(\Omega)) \cap C^1(\bar{I}; H)$, $\dddot{u} \in L_2(I; H)$ and there hold the estimates*

$$\|u\|_{C(\bar{I},V)}\| + \|\dot{u}\|_{L_2(I,V)} + \|\dot{u}\|_{C(\bar{I},\mathring{H}^1(\Omega))}$$

$$\le C_0(\alpha_0, \alpha_1, \beta_0, \beta_1, e_{min}, e_{max}, \hat{e}, u_0, v_0, f), \tag{8}$$

$$\|\dot{u}\|_{C(\bar{I},V)} + \|\ddot{u}\|_{L_2(I,V)} + \|\ddot{u}\|_{L_\infty(I,\mathring{H}^1(\Omega))} + \|\dddot{u}\|_{L_2(I,H)}$$

$$\le C_1(\delta, \alpha_0, \alpha_1, \beta_0, \beta_1, e_{min}, e_{max}, \hat{e}, u_0, v_0, f). \tag{9}$$

Proof. Let $\{w_i \in V \cap H^4(\Omega); \; i \in \mathbb{N}\}$ be a basis of V. We introduce the Galerkin approximation u_m of a solution in a form

$$u_m(t) = \sum_{i=1}^{m} \alpha_i(t)w_i, \; \alpha_i(t) \in \mathbb{R}, \; i = 1, \ldots, m, \; m \in \mathbb{N},$$

$$\int_{\Omega} \left[e(x)\ddot{u}_m w_i + ae^3(x)\nabla\ddot{u}_m \cdot \nabla w_i + A(e)(\dot{u}_m, w_i) + B(e)(u_m, w_i) \right] dx =$$

$$\int_{\Omega} \left[g_\delta(\tfrac{1}{2}e(x) + \Phi(x) - u_m) + f(t) \right] w_i \, dx, \; i = 1, \ldots, m;$$

$$u_m(0) = u_{0m}, \; \dot{u}_m(0) = v_{0m}; \; u_{0m} \to u_0 \text{ in } H^4(\Omega), \; v_{0m} \to v_0 \text{ in } V.$$

A solution originally existing only locally can be prolonged to the whole time interval I with the *a priori* estimates

$$\|u_m\|_{C(\bar{I},V)} + \|\dot{u}_m\|_{L_2(I,V)} + \|\dot{u}_m\|_{C(\bar{I},\mathring{H}^1(\Omega))}$$
$$\leq C_2(\alpha_0, \alpha_1, \beta_0, \beta_1, e_{\min}, e_{\max}, \hat{e}, u_0, v_0, f). \tag{10}$$

Better estimates can be achieved after differentiating the Galerkin equation with respect to t:

$$\|\dot{u}_m\|_{C(\bar{I},V)} + \|\ddot{u}_m\|_{L_2(I,V)} + \|\ddot{u}_m\|_{C(\bar{I},\mathring{H}^1(\Omega))}$$
$$\leq C_3(\delta, \alpha_0, \alpha_1, \beta_0, \beta_1, e_{\min}, e_{\max}, \hat{e}, u_0, v_0, f). \tag{11}$$

We proceed with the convergence of the Galerkin approximation. Applying the estimates (10) and (11), the Aubin-Lions compact imbedding theorem [9], Sobolev imbedding theorems and the interpolation theorems in Sobolev spaces [8] we obtain for a subsequence of $\{u_m\}$ (denoted again by $\{u_m\}$) a function $u \in C(\bar{I}, V)$ with $\dot{u} \in L_\infty(I, V)$, $\ddot{u} \in L_\infty(I, \mathring{H}^1(\Omega))$ and the convergences

$$
\begin{aligned}
\ddot{u}_m &\rightharpoonup^* \ddot{u} && \text{in } L_\infty(I, \mathring{H}^1(\Omega)), \\
\ddot{u}_m &\rightharpoonup \ddot{u} && \text{in } L_2(I, V), \\
\dot{u}_m &\rightharpoonup^* \dot{u} && \text{in } L_\infty(I; V), \\
u_m &\to u && \text{in } C(\bar{I}; V), \\
u_m &\to u && \text{in } C^1(\bar{I}; H^{2-\varepsilon}(\Omega)) \; \forall \, \varepsilon > 0, \\
u_m &\to u && \text{in } C^1(\bar{I}; C(\bar{\Omega})).
\end{aligned}
\tag{12}
$$

The convergence process (12) implies that a function u fulfils for a.e. $t \in I$

$$\int_{\Omega} \left[e\ddot{u}w + ae^3(x)\nabla\ddot{u} \cdot \nabla w + A(e; \dot{u}, w) + B(e; u, w) \right] dx$$
$$= \int_{\Omega} \left[g_\delta(\tfrac{1}{2}e(x) + \Phi(x) - u) + f \right] w \, dx, \; \forall \, w \in V. \tag{13}$$

The identity (6) follows directly after setting $w \equiv y(t, \cdot)$, $y \in L_2(I; V)$ in (13). The estimate (10) together with the convergences (12) implies the estimate (8).

Due to the differentiability of g_δ, f we obtain the third time derivative $\dddot{u} \in L_2(I; H)$ fulfilling

$$
\int_Q \left[\dddot{u} \left(e(x)y - \operatorname{div}(e^3(x)\nabla y) \right) + A(e; \ddot{u}, y) + B(e; \dot{u}, y) \right] dx\, dt
$$
$$
= \int_Q \left[-g_\delta'(\tfrac{1}{2}e(x) + \Phi(x) - u)\dot{u} + \dot{f}(t, x) \right] y\, dx\, dt \ \forall y \in L_2(I; V). \tag{14}
$$

The estimate (9) is then the consequence of (11) together with the convergences (12) and the relation (14). The proof of the uniqueness can be performed in a standard way using the Gronwall lemma.

Remark 1. The constant $C_0(\alpha, \beta, e_{\min}, e_{\max}, \hat{e}, u_0, v_0, f, q)$ in the estimate (8) does not depend on δ for $\delta \in (0, \delta_0)$. It is possible to derive the existence of a variational solution u of the original problem with the rigid obstacle in a similar way as in [2], where the method of penalization was applied. We can use the limit of a subsequence of solutions $\{u_{\delta_n}\}$, $\delta_n \to 0+$ to the problem (6), and (7) for $\delta \equiv \delta_n$, $n \in \mathbb{N}$ instead of the sequence of penalized solutions.

Let $\mathcal{K}$ be a closed convex set in $L_2(I; V)$ of the form

$$
\mathcal{K} := \{ y \in L_2(I; V); \ \dot{y} \in L_2(I; \mathring{H}^1(\Omega)), \ y \geq \tfrac{1}{2}e + \Phi \}. \tag{15}
$$

A function $u \in \mathcal{K}$ such that $\dot{u} \in L_2(I; V)$ and $u(0, \cdot) = u_0$ solves the initial value problem for a nonstationary variational inequality

$$
\int_Q \left(A(e; \dot{u}, y - u) + B(e; u, y - u) - ae^3 \nabla \dot{u} \cdot \nabla(\dot{y} - \dot{u}) - e\dot{u}(\dot{y} - \dot{u}) \right) dx\, dt
$$
$$
+ \int_\Omega \left(ae^3 \nabla \dot{u} \cdot \nabla(y - u) + e\dot{u}(y - u) \right)(T, \cdot)\, dx
$$
$$
\geq \int_\Omega \left(a\nabla v_0 \cdot (\nabla y(0, \cdot) - \nabla u_0) + v_0(y(0, \cdot) - u_0) \right) dx \tag{16}
$$
$$
+ \int_Q f(y - u)\, dx\, dt \quad \forall y \in \mathcal{K}.
$$

3   Optimal Control Problem

3.1   The Existence of an Optimal Thickness

We consider a cost functional $J : L_2(I; V) \times H^2(\Omega) \mapsto \mathbb{R}$ fulfilling the assumption

$$
u_n \rightharpoonup u \text{ in } L_2(I; V), \ e_n \rightharpoonup e \text{ in } H^2(\Omega) \Rightarrow J(u, e) \leq \liminf_{n \to \infty} J(u_n, e_n) \tag{17}
$$

and formulate

Optimal control problem $\mathcal{P}$: To find a control $e_* \in E_{ad}$ such that

$$
J(u(e_*), e_*) \leq J(u(e), e) \ \forall e \in E_{ad}, \tag{18}
$$

where $u(e)$ is a (unique) weak solution of the Problem (1)–(3).

Theorem 2. *There exists a solution of the Optimal control problem* $\mathcal{P}$.

Proof. We use the weak lower semicontinuity property of the functional J and the compactness of the admissible set E_{ad} of thicknesses in the space $C(\bar{\Omega})$. Let $\{e_n\} \subset E_{ad}$ be a minimizing sequence for (18). i.e.

$$\lim_{n\to\infty} J(u(e_n); e_n) = \inf_{e\in E_{ad}} J(u(e), e). \tag{19}$$

The set E_{ad} is convex and closed and hence a weakly closed in $H^2(\Omega)$ as the closed convex set. Then there exists a subsequence of $\{e_n\}$ (denoted again by $\{e_n\}$) and an element $e_* \in E_{ad}$ such that

$$e_n \rightharpoonup e_* \text{ in } H^2(\Omega), \ e_n \to e_* \text{ in } C(\bar{\Omega}). \tag{20}$$

The *a priori* estimates (8), Sobolev imbedding theorems and the Ascoli theorem on uniform convergence on $\bar{I}$ imply the existence of a function $u^* \in C(\bar{I}; V)$ such that $\dot{u} \in L_\infty(I; V) \cap C(\bar{I}; \overset{\circ}{H}{}^1(\Omega))$, $\ddot{u} \in L_\infty(I; \overset{\circ}{H}{}^1(\Omega))$ and the convergences

$$\ddot{u}(e_n) \rightharpoonup^* \ddot{u}^* \text{ in } L_\infty(I; \overset{\circ}{H}{}^1(\Omega)),$$
$$\dot{u}(e_n) \rightharpoonup^* \dot{u}^* \text{ in } L_\infty(I; V), \ \dot{u}(e_n) \to \dot{u}^* \text{ in } C(\bar{I}; \overset{\circ}{H}{}^1(\Omega)), \tag{21}$$
$$u(e_n) \rightharpoonup^* u^* \text{ in } L_\infty(I; V), \ u(e_n) \to u^* \text{ in } C(\bar{I}; C^1(\bar{\Omega}))$$

for a chosen subsequence. Functions $u_n \equiv u(e_n)$ solve the initial value state problem (6), and (7) for $e \equiv e_n$. We verify that u_* solves the problem (6), and (7) with $e \equiv e_*$. The previous convergences together with the Lipschitz continuity of g_δ imply

$$e_n \ddot{u}_n \rightharpoonup e_* \ddot{u}_* \text{ in } L_2(Q), \ e_n^3 \ddot{u}_{n,i} \rightharpoonup e_*^3 \ddot{u}_{*i} \text{ in } L_2(Q), \ i = 1, 2;$$
$$e_n^3 \dot{u}_{n,ij} \rightharpoonup e_*^3 \dot{u}_{*ij} \text{ in } L_2(Q), \ e_n^3 u_{n,ij} \rightharpoonup e_*^3 u_{*ij} \text{ in } L_2(Q), \ i, j \in \{1, 2\};$$
$$g_\delta(\tfrac{1}{2}(e_n + \Phi - u_n)) \rightharpoonup g_\delta(\tfrac{1}{2}(e_* + \Phi - u_*)) \text{ in } L_2(Q).$$

Then $u_* \equiv u(e_*)$ and hence

$$u(e_n) \rightharpoonup u(e_*) \text{ in } L_2(I; V), \quad e_n \rightharpoonup e \text{ in } H^2(\Omega).$$

Property (17) together with (19) then imply that

$$J(u(e_*), e_*) = \min_{e\in E_{ad}} J(u(e), e)$$

and the proof is complete.

3.2 Necessary Optimality Conditions

We introduce the Banach space $\mathcal{W} = \{w \in H^1(I; V) : \ddot{w} \in L_2(I; V^*)\}$ with a norm

$$\|w\|_{\mathcal{W}} = \|w\|_{H^1(I;V)} + \|\ddot{w}\|_{L_2(I;V^*)}$$

and operators $\mathcal{A}(e) : W \to L_2(I; V^*)$, $\mathcal{B}(e) : H^2(\Omega) \to L_2(I; V^*)$ by

$$
\langle\langle \mathcal{A}(e)z, y \rangle\rangle = \int_Q \ddot{z}[ey - a\,\mathrm{div}\,(e^3 \nabla y)]\,dx\,dt +
$$
$$
\int_Q [A(e)(\dot{z}, y) + B(e)(z, y) + g'_\delta(\omega(e))zy]\,dx\,dt, \tag{22}
$$

$$
\langle\langle \mathcal{B}(e)h, y \rangle\rangle = \int_Q h\left[\ddot{u}(e)y - 3a\,\mathrm{div}\,(e^2 \nabla y)\right]\,dx\,dt +
$$
$$
\int_Q h\left[A'(e)(\dot{z}, y) + B'(e)(z, y) - \frac{1}{2}g'_\delta(\omega(e))y\right]\,dx\,dt, \tag{23}
$$
$$
\omega(e) = \tfrac{1}{2}e + \varPhi - u(e), \; y \in L_2(I; V).
$$

In a similar way as in [5] or [6] the following theorem about Fréchet differentiability of the mapping $e \mapsto u(e)$ can be verified.

Theorem 3. *The mapping* $u(\cdot) : E_{ad} \to W$ *is Fréchet differentiable and its derivative* $z \equiv z(h) = u'(e)h \in W$, $h \in H^2(\Omega)$ *fulfils for every* $e \in E_{ad}$ *uniquely the operator equation*

$$
\mathcal{A}(e)z = -\mathcal{B}(e)h, \; z(0) = \dot{z}(0) = 0 \tag{24}
$$

Proof. The existence of a solution z to the Eq. (24) can be verified using the standard Galerkin method and its uniqueness by the Gronwall lemma.

We proceed with the differentiability of $e \mapsto z(e)$:

Let $h \in H^2(\Omega)$ with $e + h \in E_{ad}$ and

$$
r(h) = u(e + h) - u(e) - z(h)
$$

We have

$$
\mathcal{A}(e)r(h) = \mathcal{A}(e)[u(e + h) - u(e)] + \mathcal{B}(e)h,
$$

and verify

$$
r(h) = o(h) \text{ i.e. } \lim_{\|h\|_2 \to 0} \frac{r(h)}{\|h\|_2} = 0 \; \Rightarrow \; z(h) = u'(e)h.
$$

In order to derive necessary optimality conditions we assume that the cost functional $J(\cdot, \cdot) : L_2(I; V) \times H^2(\Omega) \to \mathbb{R}$ is Fréchet differentiable.

The optimal control problem can be expressed in a form

$$
j(e_*) = \min_{e \in E_{ad}} j(e), \; j(e) = J(u(e), e). \tag{25}
$$

The functional j in (25) is Fréchet differentiable and its derivative in $e_* \in E_{ad}$ has the form

$$
\langle j'(e_*), h \rangle = \langle\langle J_u(u(e_*), e_*), u'(e_*)h \rangle\rangle + \langle J_e(u(e_*), e_*), h \rangle_{-2}, \; h \in H^2(\Omega) \tag{26}
$$

with the duality pairings $\langle\langle \cdot, \cdot \rangle\rangle$, $\langle \cdot, \cdot \rangle_{-2}$ between $L_2(I; V)^*$ and $L_2(I; V)$, $(H^2(\Omega))^*$ and $H^2(\Omega)$ respectively.

The optimal thickness $e_* \in E_{ad}$ fulfils the variational inequality

$$\langle j'(e_*), e - e_* \rangle_{-2} \geq 0 \ \forall e \in E_{ad}. \tag{27}$$

which can be expressed in a form

$$\langle \langle J_u(u(e_*), e_*), u'(e_*)(e - e_*) \rangle \rangle + \langle J_e(u(e_*), e_*), e - e_* \rangle_{-2} \geq 0 \ \forall e \in E_{ad}. \tag{28}$$

Applying Theorem 3 we obtain necessary optimality conditions in a form of a system with an adjoint state p:

Theorem 4. *The optimal thickness e_*, the corresponding state (deflection) $u^* \equiv u(e_*)$ and the adjoint state $p^* \equiv p(e_*)$ are solutions of the initial value problem*

$$\int_Q \left[e_* u_{tt}^* y + a e^3 \nabla u_{tt}^* \cdot \nabla y + A(e)(u_t^*, y) + B(e)(u^*, y) \right] dx \, dt$$

$$= \int_Q \left[g_\delta(\tfrac{1}{2}e + \Phi - u^*) + f(t, x) \right] y \, dx \, dt \ \forall \, y \in L_2(I; V),$$

$$u^*(0) = u_0, \ u_t^*(0) = v_0,$$

$$A(e_*)p_* = -J_u(u^*, e_*); \ p^*(T) = p_t^*(T) = 0,$$

$$\langle \langle B(e_*)(e - e_*), p^* \rangle \rangle + \langle J_e(u^*, e_*), e - e_* \rangle_{-2} \geq 0 \ \forall e \in E_{ad}.$$

Remark 2. If the partial derivative $e \mapsto J_e(u(e), e)$ is strongly monotone i.e.

$$\langle J_e(u(e_1), e_1) - J_e(u(e_2), e_2), e_1 - e_2 \rangle_2 \geq N \|e\|_2^2 \ \forall e_1, e_2 \in H^2(\Omega), \ N > 0,$$

then it is possible after using the variational inequality (28) to obtain for sufficiently large N the uniqueness of the Optimal control e_*.

Remark 3. We have mentioned in Remark 1 that there is a sequence $\{\delta_n\}$, $\delta_n \to 0+$ such that a corresponding sequence of regularized solutions u_{δ_n} of (6), and (7) converges to a solution u of the original problem. If $e_* \equiv e_*(\delta_n)$ is a sequence of optimal thicknesses tending to some $\tilde{e}_* \in E_{ad}$ then it is an open question if $\tilde{e}_* \in E_{ad}$ is a solution of the corresponding Optimal control problem connected with the Problem (16). In this case there is no uniqueness of solutions and hence this Optimal control problem has the form

Optimal control problem $\tilde{\mathcal{P}}$: To find a couple $\{\tilde{u}_*, \tilde{e}_*\} \in \mathcal{U} \times E_{ad}$ such that

$$J(\tilde{u}_*, \tilde{e}_*) \leq J(u, e) \ \forall \{u, e\} \in \mathcal{U} \times E_{ad},$$

where

$$\mathcal{U} = \{u \in \mathcal{K}; \ \dot{u} \in L_2(I; V), \ u(\cdot, 0) = u_0 \text{ and } (16) \text{ holds}\}.$$

References

1. Bock, I.: An optimization of a Mindlin-Timoshenko beam with a dynamic boundary contact on the boundary. Proc. Appl. Math. Mech. **11**, 791–792 (2011)
2. Bock, I., Jarušek, J.: Unilateral dynamic contact of viscoelastic von Kármán plates. Adv. Math. Sci. Appl. **16**, 175–187 (2006)
3. Bock, I., Jarušek, J.: Dynamic contact problem for a bridge modeled by a viscoelastic von Kármán system. Z. Angev. Math. Phys. **61**, 865–876 (2010)
4. Bock, I., Kečkemétyová, M.: An optimal design with respect to a variable thickness of a viscoelastic beam in a dynamic boundary contact. Tatra Mt. Math. Publ. **48**, 15–24 (2011)
5. Bock, I., Kečkemétyová, M.: Regularized optimal control problem for a beam vibrating against an elastic foundation. Tatra Mt. Math. Publ. **63**, 53–71 (2015)
6. Bock, I., Lovíšek, J.: Optimal control of a viscoelastic plate bending with respect to a thickness. Math. Nachrichten **125**, 135–151 (1986)
7. Christensen, R.M.: Theory of Viscoelasticity. An Introduction Academic Press, New York (1982)
8. Eck, C., Jarušek, J., Krbec, M.: Unilateral Contact Problems in Mechanics. Variational Methods and Existence Theorems. Monographs and Textbooks in Pure and Applied Mathematics, no. 270. Chapman & Hall/CRC (Taylor & Francis Group), Boca Raton (2005). ISBN 1-57444-629-0
9. Lions, J.L.: Quelques méthodes de résolutions sep problémes aux limites non linéaires. Dunod, Paris (1969)

Optimal Abort Landing in the Presence of Severe Windshears

Nikolai Botkin and Varvara Turova[✉]

Zentrum Mathematik, Technische Universität München,
Boltzmannstr. 3, 85748 Garching bei München, Germany
{botkin,turova}@ma.tum.de
http://www-m6.ma.tum.de/Lehrstuhl/NikolaiBotkin,
http://www-m6.ma.tum.de/Lehrstuhl/VarvaraTurova

Abstract. In this paper, the abort landing problem is considered with reference to a point-mass aircraft model describing flight in a vertical plane. It is assumed that the pilot linearly increases the power setting to maximum upon sensing the presence of a windshear. This option is accounted for in the aircraft model and is not considered as a control. The only control is the angle of attack, which is assumed to lie between minimum and maximum values. The aim of this paper is to construct a feedback strategy that ensures a safe abort landing. An algorithm for solving nonlinear differential games is used for the design of such a strategy. The feedback strategy obtained is discontinuous in time and space so that realizations of control may have a bang-bang structure. To be realistic, outputs of the feedback strategy are being smoothed in time, and this signal is used as control.

Keywords: Aircraft model · Penetration landing · Abort landing · Differential game · Hamilton-Jacobi equation · Grid method · Feedback strategy · Optimal trajectories

1 Introduction

Many aircraft accidents are caused by severe windshears such as e.g. downbursts. A downburst appears when a descending column of air hits the ground and then spreads horizontally. This phenomenon is especially dangerous for aircrafts during landing or take-off, because a headwind can be followed by a downdraft and then by a tailwind at relatively low altitudes.

There are a large number of works devoted to the problem of aircraft control in the presence of severe windshears. In particular, papers [1–8] address the problem of aircraft control during take-off in the presence of windshears. In works [1,2], the wind velocity field is assumed to be known. It is shown that open loop controls obtained as solutions of appropriate optimization problems provide satisfactory results for rather severe wind disturbances. Nevertheless, it is clear that the spatial distribution of wind velocity cannot be measured with

© IFIP International Federation for Information Processing 2016
Published by Springer International Publishing AG 2016. All Rights Reserved
L. Bociu et al. (Eds.): CSMO 2015, IFIP AICT 494, pp. 137–146, 2016.
DOI: 10.1007/978-3-319-55795-3_12

an appropriate accuracy, and therefore feedback principles of control design are more realistic. Different types of feedback controls are proposed in papers [3–6]. In [3], the design of a feedback robust control is based on the construction of an appropriate Lyapunov function. Robust control theory is used in [4] to develop feedback controls stabilizing the relative path inclination and, in [5,6], for the design of feedback controls stabilizing the climb rate. In papers [7,8], feedback controls, which are effective against downbursts, are designed using differential game approach (see e.g. [9]). The value function, which is a viscosity solution (see. e.g. [10,11]) of an appropriate Hamilton-Jacobi equation, are computed using dynamic programming techniques described in [12,13]. Both the case of known wind velocity field and the case of unknown wind disturbance are considered.

An approach based on differential game theory is used in paper [14] in concern with the problem of landing. A full nonlinear system of model equations is linearized and reduced to a two-dimensional differential game using a transformation of variables. The resulting differential game is numerically solved, and optimal feedback controls are constructed.

Paper [15] considers the penetration landing problem with reference to flight in a vertical plane. The model is governed either by one control (the angle of attack, if the power setting is predetermined) or two controls (the angle of attack and the power setting). The wind field is simulated by a downburst, and an near-optimal open-loop control is computed.

Works [16–20] refer to the abort landing problem. In paper [16], the optimization problem, a Chebysbev problem of optimal control, is converted into a Bolza problem through suitable transformations. The Bolza problem is then solved employing the dual sequential gradient-restoration algorithm for optimal control problems. Numerical results are obtained for several combinations of windshear intensities, initial altitudes, and power setting rates. Papers [17–19] are also concerned with a Chebysbev problem of optimal control. They utilize a multiple shooting method to compute a near-optimal control maximizing a performance index and providing necessary state constraints. Paper [20] deals with the application of differential games theory to take-off and abort landing problems. The same as in [1,16], nonlinear aircraft model describing flight in a vertical plane is considered, the dynamics equations are linearized about some reference trajectory, and the resulting differential game is reduced to a two-dimensional one under the assumption that the performance index is being computed at a fixed termination time and depends on two state variables. Feedback strategies are constructed in the form of switch lines that divide the reduced two-dimensional state space into components where certain constant values of control are prescribed. A careful tuning of this method, which includes the use of a "sliding" termination time, allows the author to obtain trajectories comparable with those from work [16].

The current paper concerns with the abort landing problem considered in [20] in the framework of differential game theory. The difference consists in the application of numerical methods described in [12,13] to the original nonlinear model reported in [16]. Moreover, a performance index of Chebysbev type is used in the current paper. The optimal trajectories are comparable with these obtained in [16,20]. It should be noted that the method described in the current paper does not require fine tuning of parameters.

2   Model Equations

We use a simplified aircraft model describing flight in a vertical plane, see papers [1,2,16]. Hence, the following system of four ordinary differential equations governing the horizontal distance x, the altitude h, the aircraft relative velocity V, and the relative path inclination γ is considered:

$$
\begin{aligned}
\dot{x} &= V \cos\gamma + W_x, \\
\dot{h} &= V \sin\gamma + W_h, \\
m\dot{V} &= T \cos(\alpha + \delta) - D - mg \sin\gamma - m\dot{W}_x \cos\gamma - m\dot{W}_h \sin\gamma, \\
mV\dot{\gamma} &= T \sin(\alpha + \delta) + L - mg \cos\gamma + m\dot{W}_x \sin\gamma - m\dot{W}_h \cos\gamma.
\end{aligned}
\tag{1}
$$

Here, α is the angle of attack; W_x and W_h are the longitudinal and vertical components of the wind velocity, respectively; g is the acceleration of gravity; m the aircraft mass; δ the thrust inclination; $T, D,$ and L are the thrust, drag, and lift forces, respectively. The following definitions hold:

$$
T = \beta(t)(A_0 + A_1 V + A_2 V^2), \quad \beta(t) = \begin{cases} \beta_0 + \dot{\beta}_0\, t, & t \in [0, t_0] \\ 1, & t \in [t_0, t_f] \end{cases},
$$

$$
D = \frac{1}{2} C_D \rho S V^2, \quad C_D = B_0 + B_1 \alpha + B_2 \alpha^2, \quad L = \frac{1}{2} C_L \rho S V^2,
$$

$$
C_L = \begin{cases} C_0 + C_1 \alpha, & \alpha \le \alpha_{**} \\ C_0 + C_1 \alpha + C_2 (\alpha - \alpha_{**})^2, & \alpha \in [\alpha_{**}, \alpha_*], \end{cases}
$$

where $\beta(\cdot)$ is a function that changes the power setting at time t_0, upon sensing the presence of a windshear; t_f is the end time; and $\beta_0, \dot{\beta}_0, \alpha_*, \alpha_{**}$ are given constants. The attack angle α is considered as the control parameter constrained by the inequalities $0 \le \alpha \le \alpha_*$.

If the components $W_x(x, h)$ and $W_h(x, h)$ of the velocity field are known, the derivatives $\dot{W}_x$ and $\dot{W}_h$ in model (1) are computed using the first two equations.

Two wind velocity field models are used in our simulations.

Wind Model 1. The first, downburst, model is borrowed from [3]:

$$
W_x = \begin{cases} -k, & x \le a \\ -k + 2k\dfrac{(x-a)}{(b-a)}, & a \le x \le b\, ; \\ k, & x \ge b. \end{cases}
\quad
W_h = \begin{cases} 0, & x \le a \\ -k(h/h_*)\dfrac{(x-a)}{(c-a)}, & a \le x \le c \\ -k(h/h_*)\dfrac{(b-x)}{(b-c)}, & c \le x \le b \\ 0, & x \ge b, \end{cases}
$$

where a, b, c, and k are parameters defining the location and the strength of the downburst. It was set $k = 50$, which corresponds to strong-to-severe windshears.

Wind Model 2. The second, double vortex model, is taken from [4]. It has two cores of radius R located symmetric about the vertical line $x = 1500$. The vortex motion of air about the centres of the cores occurs as follows. Inside of each core, the tangential speed, W_θ, of wind increases linearly from zero (at the center) to a maximum value W_0 (at the core boundary). Outside of the core, W_θ decreases in inverse proportion to the distance from the core. In the polar coordinate system with the origin at the core center, the tangential speed of wind is given by the formula

$$W_\theta = \begin{cases} W_0\, r/R, & 0 \le r \le R, \\ W_0\, R/r, & r > R. \end{cases}$$

It was chosen $W_0 = 100$, which corresponds to strong-to-severe windshears.

The conflict control problem is stated in the same way as in [20]:

$$\begin{aligned}
m\dot{V} &= T\cos(\alpha + \delta) - D - mg\sin\gamma - m\dot{W}_x\cos\gamma - m\dot{W}_h\sin\gamma \\
mV\dot{\gamma} &= T\sin(\alpha + \delta) + L - mg\cos\gamma + m\dot{W}_x\sin\gamma - m\dot{W}_h\cos\gamma \\
\dot{W}_x &= -\kappa\,(W_x - v_1) \\
\dot{W}_h &= -\kappa\,(W_h - v_2).
\end{aligned} \tag{2}$$

Here, v_1 and v_2 are artificial disturbances that may have instantaneous jumps. The wind components W_x and W_h smoothly track v_1 and v_2, respectively, with a time lag depending on the parameter κ (set $\kappa = 0.2$). The following constraints are imposed (cf. [20]):

$$\alpha \in [0, 16]\ \text{deg}, \quad |v_1| \le 50\ \text{ft/s} \quad |v_2| \le 20\ \text{ft/s}. \tag{3}$$

3   Problem Statement

Two problem statements will be considered. In both cases, the wind velocity field is supposed to be unknown. Feedback controls will be constructed from the corresponding conflict control problems.

P1. The objective of the control α in system (2) is to maximize a payoff functional defined below, i.e.

$$J = \min_{\tau \in [0, t_f]} \left(V(\tau)\sin\gamma(\tau) + W_h(\tau)\right) \to \max_{\alpha[\cdot]}\ \min_{v_1(\cdot), v_2(\cdot)}. \tag{4}$$

It is easily seen that the expression in the parentheses of (4) is the climb rate $\dot{h}(t)$. Besides, the maximum in (4) is taken over all feedback strategies $\alpha[\cdot]$.

In this variant, the full four-dimensional differential game (2)–(4) will be numerically solved.

P2. In the second variant, the idea to derive an equation for the climb rate (see [6]) is used. Computing $\ddot{h}$ from system (2) yields the formulae

$$\ddot{h} = \frac{T}{m}[\cos(\alpha + \delta)\sin\gamma + \sin(\alpha + \delta)\cos\gamma] - \frac{D}{m}\sin\gamma + \frac{L}{m}\cos\gamma - g,$$

$$\sin\gamma = \frac{\dot{h} - W_h}{V}, \quad \cos\gamma = \sqrt{1 - (\dot{h} - W_h)^2/V^2}.$$

Thus, the following system arises:

$$\dot{h} = z, \quad \dot{z} = \mathcal{Z}(z, \alpha, V, W_h),$$

where the function $\mathcal{Z}$ is defined by the above formulae. Moreover, the payoff functional is chosen the same as in P1:

$$J = \min_{\tau \in [0, t_f]} z(t) \; \rightarrow \; \max_{\alpha[\cdot]} \; \min_{V(\cdot), W_h(\cdot)}$$

It should be noted that the vertical wind velocity, W_h, and the relative velocity, V, are considered as disturbances. The same constraints as in [7,8] are imposed:

$$0 \leq \alpha \leq 16\,\mathrm{deg}, \quad V \in [V_{\mathrm{ref}} - 20, V_{\mathrm{ref}}], \quad W_h \in [-100, 0], \tag{5}$$

where $V_{\mathrm{ref}} = 276\,\mathrm{ft/s}$ is a reference value. It is worth to mention that only negative deviations of V and W_h from their reference values are taken in (5), because negative deviations are more dangerous.

4   Numerical Method

Let us shortly outline the solution method for problems P1 and P2. The description will be given in terms of general nonlinear differential games, which is similar to that presented in [7,8].

4.1   Differential Game and Value Function

Consider a differential game defined as follows:

$$\dot{x} = f(t, x, \alpha, \beta), \; t \in [0, t_f], \; x \in R^n, \; \alpha \in A \subset R^p, \; \beta \in B \subset R^q \tag{6}$$

where x is the state vector, α and β are control parameters of the first and second player, respectively. The sets A and B are given compacts. The game starts at $t_0 \in [0, t_f]$ and finishes at t_f. The objective of the first player (control α) is to minimize the functional

$$J(x(\cdot)) = \min_{\tau \in [t_0, t_f]} \sigma(x(\tau)). \tag{7}$$

It is assumed that the first player uses pure feedback strategies, i.e. functions of the form:

$$\mathcal{A} : [0, t_f] \times R^n \rightarrow A.$$

The second player (wind) uses feedback counter strategies:

$$\mathcal{B}^c : [0, t_f] \times R^n \times A \to B.$$

Thus, it is assumed that second player can measure the current value of the attack angle ("future" values are not available), which meets the concept of guaranteeing control.

For any initial position $(t_0, x_0) \in [0, t_f] \times R^n$ and any strategies $\mathcal{A}$ and $\mathcal{B}^c$, two functional sets $\mathcal{X}_1(t_0, x_0, \mathcal{A})$ and $\mathcal{X}_2(t_0, x_0, \mathcal{B}^c)$ are defined (see [9]). It is proven in [9] that the differential game (6), (7) has a value function defined by:

$$\mathcal{V}(t_0, x_0) = \max_{\mathcal{A}} \min_{x(\cdot) \in \mathcal{X}_1(t_0,x_0,\mathcal{A})} J(x(\cdot)) = \min_{\mathcal{B}^c} \max_{x(\cdot) \in \mathcal{X}_2(t_0,x_0,\mathcal{B}^c)} J(x(\cdot)).$$

It is known (see [10–12]) that the value function is a viscosity solution of the Hamilton-Jacobi equation:

$$\mathcal{V}_t + H(t, x, \mathcal{V}_x) = 0, \text{ where } H(t, x, p) = \max_{\alpha \in A} \min_{\beta \in B} \langle p, f(t, x, \alpha, \beta) \rangle. \tag{8}$$

4.2   Grid Method for Computing the Value Function

To compute viscosity solutions of (8), the following finite difference scheme can be used (see [7,8,12,13]).

Let $h_1, ..., h_n$ and τ be space and time discretization steps, and F an operator defined on continuous functions by the relation

$$F(\mathcal{V}; t, \tau)(x) = \max_{\alpha \in A} \min_{\beta \in B} \mathcal{V}(x + \tau f(t, x, \alpha, \beta)). \tag{9}$$

Set $\Lambda = t_f/\tau$, $t_\ell = \ell\tau$, $\ell = 0, ..., \Lambda$, and introduce the following notation:

$$\mathcal{V}^\ell(x_{i_1}, ..., x_{i_n}) = \mathcal{V}(t_\ell, i_1 h_1, ..., i_n h_n), \quad \sigma^h(x_{i_1}, ..., x_{i_n}) = \sigma(i_1 h_1, ..., i_n h_n).$$

The following backward in time finite-difference scheme yields an approximate solution:

$$\mathcal{V}^{\ell-1} = \max\left\{F\big(\mathcal{L}_h[\mathcal{V}^\ell]; t_\ell, \tau\big), \sigma^h\right\}, \quad \mathcal{V}^\Lambda = \sigma^h, \quad \ell = \Lambda, \Lambda - 1, ..., 1. \tag{10}$$

Here, $\mathcal{L}_h$ is an interpolation operator that maps grid functions to continuous functions.

4.3   Control Design

During the performance of the algorithm (10), the optimal grid values of the attack angle,

$$\alpha^\ell_{i_1 i_2 ... i_n} = \arg \max_\alpha \min_{\alpha \in A} \min_{\beta \in B} \mathcal{L}_h[\mathcal{V}^\ell](x_{i_1 i_2 ... i_n} + \tau f(t_\ell, x_{i_1 i_2 ... i_n}, \alpha, \beta)),$$

are stored on a hard disk for all ℓ. The control at a time instant t_s and the current state $x(t_s)$ is computed as $\mathcal{L}_h[\alpha^s](x(t_s))$, where α^s denotes the grid function $\alpha^s_{i_1 i_2 ... i_n}$.

5   Simulation Results

This section describes numerical results of simulations where optimal controls obtained in problems P1 and P2 work against wind models 1 and 2. Numerical values of parameters appearing in our considerations are the same as in [3,16,20]. The parameters correspond to Boeing-727.

In all simulations, $t \in [0, 40]$ s, and the initial values of the state variables are chosen the same as in [16,20]: $x(0) = 0$, $h(0) = 600$ ft, $V(0) = 239.7$ feet/s, and $\gamma(0) = -2.249$ deg.

In all figures, the horizontal axes measure either the traveled distance (from 0 to 10700 ft) or the time of flight (from 0 to 40 s).

The calculations are performed on a Linux SMP-computer with 8xQuad-Core AMD Opteron processors (Model 8384, 2.7 GHz) and shared 64 GB memory. The programming language C with OpenMP (Open Multiprocessing) support is used. The efficiency of the parallelization is up to 80%.

Simulation 1. An optimal feedback strategy $\alpha[\cdot]$ computed from problem P1 works up against wind models 1 and 2 in the simulation of model (1). When solving problem P1, a $100 \times 10 \times 40 \times 40$ grid in the state space (V, γ, W_x, W_h) is used. Another variant concerns the application of sparse grid techniques (see e.g. [21]). Namely, the grid functions $\mathcal{V}^\ell$ are stored on a sparse grid, and the operator $\mathcal{L}_h$ is implemented as interpolation on this grid.

Simulation 2. An optimal feedback strategy $\alpha[\cdot]$ computed from problem P2 works up against wind models 1 and 2 in the simulation of model (1). When solving problem P2, a 400×200 grid in the state space $(h, \dot{h})$ is used.

Figure 2A shows that the maximal attack angle guidance fails against wind model 1. The reason is that the aircraft relative velocity drops just in the beginning of the trajectory because of the large attack angle.

Figure 1A shows Simulation 1 in the case of wind model 1. It is seen that the angle of attack is close to zero in the beginning of the trajectory. The aircraft drifts down for a while and gains the relative velocity, which enables a safe abort landing. Figure 1B shows the same but for wind model 2.

Figure 2B shows Simulation 1 for wind model 2 in the case where the differential game (2) is solved using sparse grid techniques.

Figure 3 shows the change of results if the output, $\bar{\alpha}$, of an optimal feedback strategy found from the differential game (2) is being smoothed using the filter $\dot{\alpha} = -(\alpha - \bar{\alpha})$ when computing trajectories in model (1). Wind model 1 is used in this simulation. It should be noted that a comparable divergence of trajectories occurs when using wind model 2.

Figure 4 shows the difference of results in Simulations 1 and 2. Wind model 1 is used in both simulations. The solid line corresponds to Simulations 1, and the dashed one stands for Simulation 2.

Note that our simulation results are in a good agreement with those of paper [16] where an open loop control is designed for the aircraft dynamics given by (1). Besides, our results are in conformity with those of paper [20] where a control

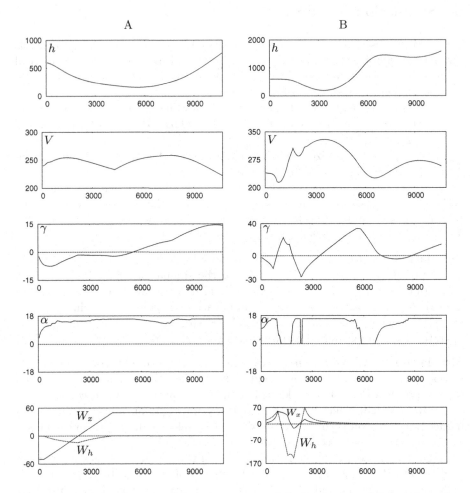

Fig. 1. Simulation of system (1) with an optimal feedback control found from the differential game (2). (A) wind model 1; (B) wind model 2.

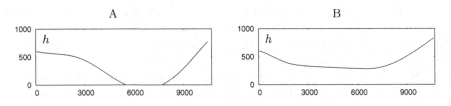

Fig. 2. (A) Simulation of system (1) with wind model 1 and $\alpha \equiv 16$ deg. (B) Simulation of system (1) with wind model 2 and an optimal feedback control found from the differential game (2) using sparse grid techniques.

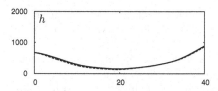
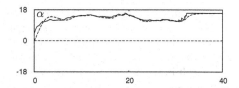

Fig. 3. The change of results if the output of an optimal strategy found from the differential game (2) is being smoothed with a filter. Wind model 1 is used. The dashed line shows the case of smoothing.

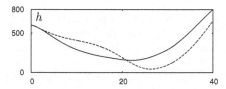
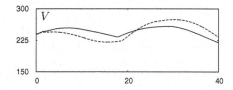

Fig. 4. The difference of results in Simulations 1 and 2. Wind model 1 is used in both simulations. The solid line corresponds to Simulation 1, and the dashed line stands for Simulation 2.

based on the computation of switch lines in an appropriate two-dimensional linear differential game is constructed.

6   Conclusion

The current investigation shows that methods based on the theory of differential games can be successfully applied to nonlinear conflict control problems concerned with aircraft maneuvers under windshear conditions. The paper demonstrates that the approach based on the solution of nonlinear differential games yields feedback controls that can work against strong-to-severe wind disturbances in abort landing. It should be noted that this approach does not require special adaptation of the controller to the problem, and, on the other hand, it is competitive with other approaches based, e.g. on robust control theory.

Acknowledgements. This work was supported in part by the DFG grant TU427/2-1.

References

1. Miele, A., Wang, T., Melvin, W.W.: Optimal take-off trajectories in the presence of windshear. J. Optim. Theory Appl. **49**, 1–45 (1986)
2. Miele, A., Wang, T., Melvin, W.W.: Guidance strategies for near-optimum take-off performance in windshear. J. Optim. Theory Appl. **50**(1), 1–47 (1986)
3. Chen, Y.H., Pandey, S.: Robust control strategy for take-off performance in a windshear. Optim. Control Appl. Methods **10**(1), 65–79 (1989)

4. Leitmann, G., Pandey, S.: Aircraft control under conditions of windshear. In: Leondes, C.T. (ed.) Control and Dynamic Systems, Part 1, vol. 34, pp. 1–79. Academic Press, New York (1990)
5. Leitmann, G., Pandey, S.: Aircraft control for flight in an uncertain environment: takeoff in windshear. J. Optim. Theory Appl. **70**(1), 25–55 (1991)
6. Leitmann, G., Pandey, S., Ryan, E.: Adaptive control of aircraft in windshear. Int. J. Robust Nonlinear **3**, 133–153 (1993)
7. Botkin, N.D., Turova, V.L.: Application of dynamic programming approach to aircraft take-off in a windshear. In: Simos, T.E., Psihoyios, G., Tsitouras, C., Zacharias, A. (eds.) AIP Conference Proceedings, ICNAAM-2012, vol. 1479, pp. 1226–1229. AIP, Melville (2012)
8. Botkin, N.D., Turova, V.L.: Dynamic programming approach to aircraft control in a windshear. In: Křivan, V., Zaccour, G. (eds.) Advances in Dynamic Games: Theory, Applications, and Numerical Methods. Annals of the International Society of Dynamic Games, vol. 13, pp. 53–69 (2013)
9. Krasovskii, N.N., Subbotin, A.I.: Game-Theoretical Control Problems. Springer, New York (1988)
10. Crandall, M.G., Lions, P.L.: Viscosity solutions of Hamilton-Jacobi equations. Trans. Am. Math. Soc. **277**, 1–47 (1983)
11. Subbotin, A.I.: Generalized Solutions of First Order PDEs: The Dynamical Optimization Perspective. Birkhäuser, Boston (1995)
12. Botkin, N.D., Hoffmann, K.-H., Mayer, N., Turova, V.L.: Approximation schemes for solving disturbed control problems with non-terminal time and state constraints. Analysis **31**, 355–379 (2011)
13. Botkin, N.D., Hoffmann, K.-H., Turova, V.L.: Stable numerical schemes for solving Hamilton-Jacobi-Bellman-Isaacs equations. SIAM J. Sci. Comput. **33**(2), 992–1007 (2011)
14. Patsko, V.S., Botkin, N.D., Kein, V.M., Turova, V.L., Zarkh, M.A.: Control of an aircraft landing in windshear. J. Optim. Theory Appl. **83**(2), 237–267 (1994)
15. Miele, A., Wang, T., Wang, H., Melvin, W.W.: Optimal penetration landing trajectories in the presence of windshear. J. Optim. Theory Appl. **57**(1), 1–40 (1988)
16. Miele, A., Wang, T., Tzeng, C.Y., Melvin, W.W.: Optimal abort landing trajectories in the presence of windshear. J. Optim. Theory Appl. **55**(2), 165–202 (1987)
17. Bulirsch, R., Montrone, F., Pesch, H.J.: Abort landing in the presence of a windshear as a minimax optimal control problem, part 1: necessary conditions. J. Optim. Theory Appl. **70**(1), 1–23 (1991)
18. Bulirsch, R., Montrone, F., Pesch, H.J.: Abort landing in the presence of a windshear as a minimax optimal control problem, part 2: multiple shooting and homotopy. J. Optim. Theory Appl. **70**(2), 223–254 (1991)
19. Berkmann, P., Pesch, H.J.: Abort landing in windshear: optimal control problem with third-order state constraint and varied switching structure. J. Optim. Theory Appl. **85**(1), 21–57 (1995)
20. Turova, V.L.: Application of numerical methods of the theory of differential games to the problems of take-off and abort landing. In: Osipov, Y. (ed.) Proceedings of the Institute of Mathematics and Mechanics, Ross. Akad. Nauk Ural. Otdel., Inst. Mat. Mekh., Ekaterinburg, vol. 2., pp. 188–201 (1992). (in Russian)
21. Pflüger, D.: Spatially adaptive sparse grids for higher-dimensional problems. Dissertation, Verlag Dr. Hut, München (2010)

Aircraft Runway Acceleration in the Presence of Severe Wind Gusts

Nikolai Botkin$^{(\boxtimes)}$ and Varvara Turova

Zentrum Mathematik, Technische Universität München,
Boltzmannstr. 3, 85748 Garching bei München, Germany
{botkin,turova}@ma.tum.de
http://www-m6.ma.tum.de/Lehrstuhl/NikolaiBotkin
http://www-m6.ma.tum.de/Lehrstuhl/VarvaraTurova

Abstract. This paper concerns the problem of aircraft control during the takeoff roll in the presence of severe wind gusts. It is assumed that the aircraft moves on the runway with a constant axial acceleration from a stationary position up to a specific speed at which the aircraft can go into flight. The lateral motion is controlled by the steering wheel and the rudder and affected by side wind. The aim of control is to prevent rolling out of the aircraft from the runway strip. Additionally, the lateral deviation, lateral speed, yaw angle, and yaw rate should remain in certain thresholds during the whole takeoff roll. The problem is stated as a differential game with state constraints. A grid method for computing the value function and optimal feedback strategies for the control and disturbance is used. The paper deals both with a nonlinear and linearized models of an aircraft on the ground. Simulations of the trajectories are presented.

Keywords: Aircraft runway · Lateral runway model · Differential game · Grid method

1 Introduction

Control of aircraft on the ground is a very complicated problem because of nonlinear effects playing a significant role in the dynamics of aircraft. Moreover, severe wind gusts may lead to rolling out from the runway, especially during high-speed roll.

The following investigations are devoted to the enhancement of aircraft-on-ground models and to the development of controllers providing safe ground operations, including taxing and takeoff run.

In the report [1], a detailed explanation of essential requirements and basic assumptions for aircraft modeling is given, including a description of various elements needed in the model structure. The main focus lies on the description of the interface between the aircraft and the runway pavement.

© IFIP International Federation for Information Processing 2016
Published by Springer International Publishing AG 2016. All Rights Reserved
L. Bociu et al. (Eds.): CSMO 2015, IFIP AICT 494, pp. 147–158, 2016.
DOI: 10.1007/978-3-319-55795-3_13

In paper [2] a bifurcation analysis of steady-state solutions and a transient analysis are applied to the study of the behavior of aircraft on the ground. A general approach to assess an aircraft's performance during taxiway manoeuvres is introduced. This allows to the author to find maximal loads during taxiway manoeuvres, which is important for assessing existing regulations for the certification of aircraft.

The work [3] presents results and interpretations from the analytical analysis aimed to uncover the dominant directional characteristics of the aircraft. Three mathematical models, of growing complexity, of the aircraft on the ground are used. Some fundamental dynamic characteristics such as e.g. the yaw rate to steering command transfer function are determined.

Paper [4] presents the study of a yaw rate control of the aircraft on the ground. A highly nonlinear realistic model of the aircraft is used, and the control design is based on the feedback linearization technique aimed to design a non-linear controller that forces the system output to follow a linear reference behavior. This approach supposes that the linear reference model perfectly corresponds to the real system. It should be noted that wind disturbances are not included into the study.

Paper [5] uses a simplified LFT (Linear Fractional Transformation) model of an aircraft on the ground. In particular, the nonlinear lateral ground forces are reduced to saturation-type nonlinearities. A robust anti-windup control technique is applied to the simplified model to improve lateral control laws to exclude oversteer when working against lateral wind step inputs.

The works [6,7] are devoted to modeling of the takeoff and landing phases for an unmanned aerial vehicle. The investigation is aimed to the development of an automatic takeoff and landing control system reducing effects of human pilot errors. The main attention is concentrated on the takeoff phase and, in particular, on aircraft's lateral motion during the takeoff roll. The authors apply transfer function techniques to a linearized model of the aircraft on the ground to design a controller. This approach does not provide safety against worst-case disturbances.

Paper [8] concerns the application of differential game theory (see e.g. [9]) to the aircraft takeoff roll. A linearized model of aircraft's lateral motion on the runway is considered there, and a conflict control problem, differential game, is formulated. It is assumed that the first player, autopilot, uses feedback strategies to minimize the objective functional of the form $J = \sigma\big(y(T), \dot{y}(T)\big)$, where y is the lateral deviation, and T is a fixed termination time. The second player, side wind, strives to maximize the objective functional using all possible constrained non-anticipative strategies. Thus, the lateral position and velocity of the aircraft are evaluated only at the termination time T, which is insufficient from the technical point of view. The reason to use such a simplified functional is that the authors could solve only two-dimensional games that time, and this simplification allowed the authors to reduce the original differential game to a two-dimensional one using a variable transformation. The main result of this paper is the construction of optimal feedback strategies of the autopilot in the form of switch lines that divide the reduced two-dimensional state space into components where certain constant values of control are prescribed. A similar representation of optimal strategies of side wind is also found.

The following limitations of this investigation should be mentioned. First, the transformation reducing the original differential game to a two-dimensional one is of course not invertible, and therefore imposing state constraints in the original problem is impossible. Second, the strategies found from the linearized model were not tested in the original nonlinear system. All these reasons give rise to the motivation to investigate the problem with modern tools for solving nonlinear state constrained differential games.

The current paper deals with the problem of aircraft control during the take-off roll and enhances the work [8]. The modification consists in the application of modern grid methods for solving nonlinear differential games (see [10,11]) to a nonlinear lateral motion runway model derived in [6,7]. These methods allow us to solve nonlinear differential games of a relative high dimension with accounting for state constraints. Speaking more certainly, it is now possible to consider the objective functional of the form $J = \max_{\tau \in [0,T]} \sigma\big(x_1(\tau), ..., x_n(\tau)\big)$, $n = 4$ or 5, and therefore to constrain all state variables for all time instants. This allows us to develop a control law that prevents rolling out of the airplane from the runway.

The model parameters are fitted to the characteristics of Boeing-727.

2   Model Equations

Consider an aircraft during the takeoff roll (see Fig. 1).

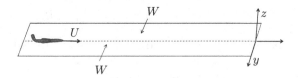

Fig. 1. Aircraft during the takeoff roll under wind gusts.

Let the state variables be defined as follows: y is the lateral deviation, V the lateral velocity, ψ the yaw angle, and R the yaw rate. The model derived in [6] reads:

$$\begin{aligned}
\dot{y} &= V, \\
\dot{V} &= -UR + (F_u + F_a)/m, \\
\dot{\psi} &= R, \\
\dot{R} &= (M_u + M_a)/I_z.
\end{aligned} \tag{1}$$

Here, $U = at$ is the axial velocity increasing linearly with time t according to the acceleration a; F_u and M_u are the undercarriage forces and moments, respectively; F_a and M_a are aerodynamic forces and moments, respectively; m is the aircraft mass, and I_z is the z-axis moment of inertia. The expressions for the forces and moments are given by the formulas

$$
\begin{aligned}
F_u &= N_s C_{\alpha\alpha} \left[\arctan \frac{V + l_s R}{U} - \delta_s \right] \cos \delta_s - N_s \mu_f \sin \delta_s \\
&\quad + N_l C_{\alpha\alpha} \arctan \frac{V - l_m R}{U + l_w/2\,R} + N_r C_{\alpha\alpha} \arctan \frac{V - l_m R}{U - l_w/2\,R},
\end{aligned}
$$

$$
\begin{aligned}
M_u &= l_s N_s C_{\alpha\alpha} \left[\arctan \frac{V + l_s R}{U} - \delta_s \right] \cos \delta_s - l_s N_s \mu_f \sin \delta_s \\
&\quad - l_m N_l C_{\alpha\alpha} \arctan \frac{V - l_m R}{U + l_w/2\,R} - l_m N_r C_{\alpha\alpha} \arctan \frac{V - l_m R}{U - l_w/2\,R},
\end{aligned}
\tag{2}
$$

$$
F_a = q \cdot S \cdot (C_{y\beta}\beta + b/(2V_a)C_{yr}R + C_{y\delta_r}\delta_r),
$$

$$
M_a = b \cdot q \cdot S \cdot (C_{n\beta}\beta + b/(2V_a)C_{nr}R + C_{n\delta_r}\delta_r).
\tag{3}
$$

Here, $V_a = \sqrt{U^2 + (V - W)^2}$ is the air speed, W the velocity of side wind; $q = 1/2\rho V_a^2$ the dynamic pressure; $\beta = \arcsin\left((V - W)/V_a\right)$ the sideslip angle; δ_s the steering wheel deflection; and δ_r the rudder deflection. It is assumed that $\delta_s = 1/3\delta_r$ for balanced manoeuvres. The control variable, u, and the disturbance, v, are introduced as follows:

$$
u := \delta_r \in [-25, 25]\,\text{deg}, \quad v := W \in [-17, 17]\,\text{m/s}.
\tag{4}
$$

The following notation for the components of the state vector is used below:

$$
x_1 := y, \quad x_2 := V, \quad x_3 := \psi, \quad x_4 := R.
\tag{5}
$$

The coefficients appearing in (1), (2), and (3) are listed in Table 1. The model is considered on the time interval $t \in [0, T]$, where $T = 34$ s.

The linearized, non-stationary, model reads:

$$
\begin{aligned}
\dot{x}_1 &= x_2, \\
\dot{x}_2 &= a_{22}(t)x_2 + a_{23}(t)x_3 + a_{24}(t)x_4 + a_{25}(t)u + c_2(t)v, \\
\dot{x}_3 &= x_4, \\
\dot{x}_4 &= a_{42}(t)x_2 + a_{43}(t)x_3 + a_{44}(t)x_4 + a_{45}(t)u + c_4(t)v, \\
\dot{u} &= -k(u - \bar{u}).
\end{aligned}
\tag{6}
$$

Here, an artificial control $\bar{u}$ that may have instantaneous jumps is introduced. The physical control u ($= \delta_r$) smoothly tracks $\bar{u}$ with a time lag depending on the parameter k. The artificial control is constrained just as u in (4).

Table 1. Model coefficients approximately corresponding to Boeing-727.

Notation	Name	Value	Units
CG	Center of gravity	-	-
μ_f	Coefficient of kinetic friction	0.5	-
ρ	Air density	1.207	kg/m^3
m	Aircraft mass	288773	kg
S	Wing area	511	m^2
b	Wing span	60	m
I_z	z-axis moment of inertia	67.38e6	$kg \cdot m^2$
l_s	Distance from CG to steering wheel along x	28.36	m
l_m	Distance from CG to main wheels along x	1.64	m
l_l, l_r	Distance from CG to left/right main wheel along y	6	m
l_w	Distance between main wheels ($l_w = l_l + l_r$)	12	m
l_L	Distance from steering to main wheels ($l_L = l_s + l_m$)	30	m
N_s	Normal reactions at steering wheel	154.863	kN
N_l, N_r	Normal reactions at main wheels	1338.99	kN
$C_{\alpha\alpha}$	Tire cornering coefficient	0.25	$1/rad$
$C_{y\beta}$	Output of y-force due to sideslip angle	−0.9	$1/rad$
C_{yr}	Output of y-force due to yaw rate	0	$1/rad$
$C_{y\delta_r}$	Output of y-force due to rudder deflection	0.120	$1/rad$
$C_{n\beta}$	Output of yawing moment due to sideslip angle	0	$1/rad$
C_{nr}	Output of yawing moment due to yaw rate	−0.280	$1/rad$
$C_{n\delta_r}$	Output of yawing moment due to rudder deflection	−0.1	$1/rad$

The coefficients appearing in (6) are defined by the formulas

$$a_{22}(t) = 0.229(1 - 100/\xi) - 0.345 \cdot 10^{-2}\xi,$$
$$a_{23}(t) = 0.12 \cdot 10^{-3}\,\xi^2 - 0.8(1 - 0.01\xi),$$
$$a_{24}(t) = -0.138 \cdot 10^{-2}(1 - 100/\xi),$$
$$a_{25}(t) = -0.2 \cdot 10^{-4}\,\xi^2 + 0.32 \cdot 10^{-1}(1 - 0.01\xi),$$
$$a_{42}(t) = -0.132 \cdot 10^{-1}\xi, \quad a_{43}(t) = -0.464 \cdot 10^{-3}\,\xi^2,$$
$$a_{44}(t) = 0.715 \cdot 10^{-1}(1 - 100/\xi),$$
$$a_{45}(t) = -0.164 \cdot 10^{-3}\,\xi^2 - 0.3(1 - 0.01\xi),$$
$$c_2(t) = 0.345 \cdot 10^{-2}\xi, \quad c_4(t) = 0.132 \cdot 10^{-1}\xi, \; \xi := t + 1, \; k = 4.$$

(7)

The model is considered on the time interval $t \in [0, T]$, where $T = 34\,\mathrm{s}$.

3   Numerical Method

Let us shortly outline the solution method that is applicable both to linear and nonlinear problems. The description will be given in terms of general nonlinear differential games, see [9,12,13].

3.1   Differential Game and Value Function

Consider the differential game

$$\dot{x} = f(t, x, u, v), \quad x \in R^n, \quad u \in P \subset R^p, \quad v \in Q \subset R^q, \tag{8}$$

where u and v are control parameters of the first and second player, respectively. The sets P and Q are given compacts. The game starts at $t_0 \in [0, T]$ and finishes at T. The aim of the first (resp. second) player is to minimize (resp. maximize) an objective functional of the form:

$$J\big(x(\cdot)\big) = \max \left\{ \sigma_0\big(x(T)\big), \max_{\tau \in [t_0, T]} \sigma\big(x(\tau)\big) \right\}, \tag{9}$$

where σ_0 and $\sigma : R^n \to R$ are given functions.

The value function, $\mathcal{W}$, is informally defined by the relation

$$\mathcal{W}(t, x) = \max_{\mathcal{V}^c} \min_{\mathcal{U}} J(x(\cdot)) = \min_{\mathcal{U}} \max_{\mathcal{V}^c} J(x(\cdot)),$$

where the minimum is taken over all admissible feedback strategies of the first player, and the maximum is computed over the so-called feedback counter-strategies of the second player (see [9]). This means that the second player (e.g. wind) can measure the current choice of the first player (e.g. the ruder deflection), which makes the second player more dangerous.

It should be noted that the strong definition of the value function (see [9]) is more complicated than that, because the strategies are in general discontinuous functions of x, and therefore cannot be directly substituted into (8) in place of u and v.

The value function plays a very important role, representing the guaranteed result of the players. For example, let the game starts from a position (t_0, x_0), and $\mathcal{W}(t_0, x_0) \le 0$. Then, there exists a feedback strategy $\mathcal{U}$ such that, for all trajectories $x(\cdot)$ generated by $\mathcal{U}$ and any $\mathcal{V}^c$, the inequalities $\sigma_0(x(T)) \le 0$ and $\sigma(x(t)) \le 0$, $t \in [t_0, T]$, hold. This can be interpreted as obtaining a guaranteed gain at the termination time T and keeping the object inside of prescribed state constraints at any time instant. Moreover, as the Subsect. 3.3 shows, optimal strategies of the players can be constructed in the course of computing the value function. Besides, an optimal feedback counter-strategy of the second player is directly derived from the value function.

It is established, see [10, 14, 15], that the value function is a viscosity solution of the following Hamilton-Jacobi equation:

$$\mathcal{W}_t + H(t, x, \mathcal{W}_x) = 0, \text{ where } H(t, x, p) = \min_{u \in P} \max_{v \in Q} \langle f(t, x, u, v), p \rangle. \quad (10)$$

This correspondence has given rise to numerical methods for computing value functions. The next subsection describes a grid method developed for computing viscosity solutions of (10) and, therefore, value functions in the differential game (8)–(9).

3.2 Grid Method for Computing the Value Function

To compute the value function, the following finite difference scheme is used, see [10–13].

Let $h_1, ..., h_n$, and τ be space and time discretization step lengths. Set $L = T/\tau$, $t_\ell = \ell\tau$, $\ell = 0, 1, ..., L$, and denote

$$\mathcal{W}^\ell(x_{i_1}, \ldots, x_{i_n}) = \mathcal{W}(\ell\tau, i_1 h_1, \ldots, i_n h_n),$$

$$\sigma_0^h(x_{i_1}, \ldots, x_{i_n}) := \sigma_0(i_1 h_1, \ldots, i_n h_n), \quad \sigma^h(x_{i_1}, \ldots, x_{i_n}) := \sigma(i_1 h_1, \ldots, i_n h_n).$$

Let c be a grid function. Assume that the variable x runs over all grid nodes and define the following upwind operator:

$$F(c; t, \tau, h_1, ..., h_n)(x) = c(x) + \tau \min_{u \in P} \max_{v \in Q} \sum_{i=1}^{n} (p_i^R f_i^+ + p_i^L f_i^-),$$

where $f_i = f_i(t, x, u, v)$ are the right hand sides of the control system, and

$$a^+ = \max\{a, 0\}, \quad a^- = \min\{a, 0\},$$

$$p_i^R = [c(x_1, ..., x_i + h_i, ..., x_n) - c(x_1, ..., x_i, ..., x_n)]/h_i,$$

$$p_i^L = [c(x_1, ..., x_i, ..., x_n) - c(x_1, ..., x_i - h_i, ..., x_n)]/h_i.$$

An approximate solution is the output of the following backward in time finite-difference scheme:

$$\mathcal{W}^{\ell-1} = \max\left\{ F(\mathcal{W}^\ell; t_\ell, \tau, h_1, ..., h_n), \sigma^\ell \right\}, \quad \mathcal{W}^L = \sigma_0^h, \quad \ell = L, L-1, ..., 0. \quad (11)$$

This algorithm is proposed and analyzed in [10–13]. It was stated there that its convergence rate is of order $\sqrt{\tau}$ if $\tau/h_i = c, i = 1...n$, where c is a small enough constant. This convergence rate is not improvable when applying grid methods to Hamilton-Jacobi equations arising from differential games.

3.3   Control Design

When running the algorithm (11), the minimizing grid values of the control,

$$u^\ell_{i_1 i_2 \dots i_n} = \arg\min_u \max_{u \in P} \max_{v \in Q} \sum_{i=1}^{n} (p_i^R f_i^+ + p_i^L f_i^-),$$

are stored on a hard disk for each grid multi index $i_1 i_2 \dots i_n$ and each time sampling index ℓ. The control at a time instant t_ℓ and the current state $x(t_\ell)$ is computed as $\mathcal{L}_h[u^\ell](x(t_\ell))$, where u^ℓ denotes the grid function $u^\ell_{i_1 i_2 \dots i_n}$, and $\mathcal{L}_h$ is an interpolation operator.

A counter-strategy of the second player is defined as follows. Let $(t_\ell, x(t_\ell))$ be the current position of the game, and a control u of the first player is chosen. Then the second player chooses its control as

$$v = \arg\max_v \max_{v \in Q} \mathcal{L}_h[\mathcal{W}^\ell]\big(x(t_\ell) + \tau f(t_\ell, x(t_\ell), u, v)\big),$$

where $\mathcal{W}^\ell$ is the grid approximation of the value function at the time instant t_ℓ, computed by formula (11).

4   Simulation Results

This section describes simulation results for the models (1)–(4) and (6)–(7). In both cases, the objective functional of the form (9) with the functions

$$\sigma_0(x) = \max\left\{ \frac{|x_1|}{10}, \frac{|x_2|}{5}, \frac{|x_3|}{10}, \frac{|x_4|}{5} \right\} - 1, \ \sigma(x) = \max\left\{ \frac{|x_1|}{15}, \frac{|x_2|}{5}, \frac{|x_3|}{15}, \frac{|x_4|}{5} \right\} - 1$$

is used. Thus, the controls u and $\bar{u}$, see (4) and the last equation of (6), strive to satisfy the conditions

$$\sigma_0(x(T)) \leq 0 \ \text{and} \ \sigma(x(t)) \leq 0, \ t \in [0, T],$$

for any realization of the disturbance v constrained as in (4). In other words, u (resp. $\bar{u}$) strives to satisfy the conditions

$$|y(T)| \leq 10\,\mathrm{m}, \ |V(T)| \leq 5\,\mathrm{m/s}, \ |\psi(T)| \leq 10\,\mathrm{deg}, \ |R(T)| \leq 5\,\mathrm{deg/s}$$

at the termination time $T = 34\,\mathrm{s}$ and to keep the state constraints

$$|y(t)| \leq 15\,\mathrm{m}, \ |V(t)| \leq 5\,\mathrm{m/s}, \ |\psi(t)| \leq 15\,\mathrm{deg}, \ |R(t)| \leq 5\,\mathrm{deg/s}$$

for all time instants. According to the problem statement, this is possible if the value function, see Sect. 3.1 and the explanation there, is non-positive at the initial state $\{t = 0, y = 0, V = 0, \psi = 0, R = 0\}$.

Differential games (1)–(4) and (6)–(7) are solved using numerical methods outlined in Sect. 3. The calculations are performed on a Linux SMP-computer with 8xQuad-Core AMD Opteron processors (Model 8384, 2.7 GHz) and shared 64 Gb memory. The programming language C with OpenMP (Open Multi-processing) support is used. The efficiency of the parallelization is up to 80%.

When solving the differential game related to the linear model (6)–(7), a rectangular $40 \times 20 \times 40 \times 20 \times 30$ grid is chosen. In the case of the nonlinear model (1)–(4), a rectangular $40 \times 20 \times 40 \times 20$ is used.

Figure 2 shows the simulation of the linear model (6)–(7) with an optimal feedback control strategy and the corresponding optimal feedback counter-strategy for wind. The horizontal axes measure the traveled distance in meter, the vertical axes measure the lateral deviation y (meter), the yaw angle ψ (degree), the rudder deflection δ_r (degree), and the velocity of side wind (meter/sec), respectively. The vertical bold bars, drawn to the right in the first two graphs, show the admissible interval for the terminal values of y and ψ, respectively. It is seen that the terminal and state constraints are satisfied for y and ψ. It should also be noted that the other two variables, V and R (not shown here), satisfy their terminal and state constraints too.

Figure 3 presents the simulation of the nonlinear model (1)–(4) using the optimal feedback control strategy found for the linear model (6)–(7), whereas the disturbance is formed using the optimal feedback counter-strategy for wind taken from the nonlinear model. It is seen that the terminal and state constraints are

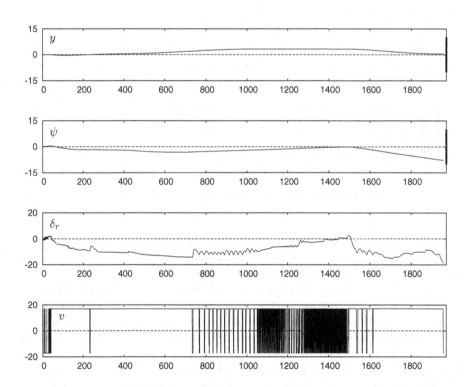

Fig. 2. Simulation of the linear model (6)–(7).

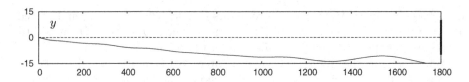

Fig. 3. Simulation of the nonlinear model (1)–(4) with the optimal feedback control strategy found for the linear model (6)–(7), whereas the optimal feedback counter-strategy for wind is taken from the nonlinear model.

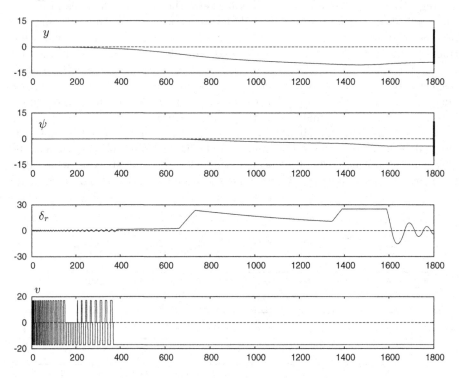

Fig. 4. Simulation of the nonlinear model (1)–(4).

violated. This means that the linearized model (6)–(7) does not properly reflect the dynamical properties of the real nonlinear plant. Thus, the construction of controllers based on linearized models is questionable.

Figure 4 shows the simulation of the nonlinear model (1)–(4) with the optimal feedback control strategy and the corresponding optimal feedback counter-strategy for wind, found from the four-dimensional nonlinear differential game (1)–(4). During the simulation of trajectories, the output, u, of the optimal control strategy is smoothed with the filter $\dot{\delta}_r = -4\,(\delta_r - u)$. It is seen that the terminal and state constraints are satisfied for y and ψ. The other two variables, V and R (not shown here), satisfy the terminal and state constraints too.

The simulation results show that the control strategy found from the linear differential game associated with the models (6)–(7) works perfectly in the linear model, and does not work in the nonlinear one.

The control strategy found from the nonlinear differential game (1)–(4) works perfectly in the real nonlinear model against very severe wind disturbances comparable with hurricane. It should be noted that none conventional control system cannot apparently keep the aircraft on the runway in the presence of smart wind gusts obtained from the nonlinear differential game. However, our control strategies ensure the desired terminal and state constraints (see Figs. 2 and 4). Moreover, the strategies work stable in a wide range of discretization parameters such as time sampling and spatial steps in the algorithm (11), which is checked in numerous test runs. Finally, these strategies can be physically implemented on board, because all state variables used in them are available for measurements.

5   Conclusion

The current investigation shows that methods based on the theory of differential games can be successfully applied to nonlinear conflict control problems related to aircraft's takeoff roll under severe wind gusts. The paper demonstrates the following advantages: A very detailed nonlinear model of aircraft's takeoff roll is used. The corresponding highly nonlinear differential games are solved using a novel grid method, and optimal control strategies ensuring a safe takeoff roll are designed. It is planned to test them on a flight simulator providing a fully realistic model of an aircraft.

Acknowledgements. This work is supported by the DFG grant TU427/2-1.

References

1. Barnes, A.G., Yager, T.J.: Enhancement of aircraft ground handling simulation capability. Technical report AGARDograph 333. AGARD (1998)
2. Rankin, J.: Bifurcation analysis of nonlinear ground handling of aircraft. Dissertation, University of Bristol (2010)
3. Klyde, D.H., Myers, T.T., Magdaleno, R.E., Reinsberg, J.G.: Identification of the dominant ground handling characteristics of a navy jet trainer. J. Guid. Control Dyn. **25**(3), 546–552 (2002)
4. Duprez, J., Mora-Camino, F., Villaume, F.: Control of the aircraft-on-ground lateral motion during low speed roll and manoeuvers. In: Proceedings of the 2004 IEEE Aerospace Conference, vol. 4, pp. 2656–2666 (2004)
5. Roos, C., Biannic, J.M.: Aircraft-on-ground lateral control by an adaptive LFT-based anti-windup approach. In: 2006 IEEE Conference on Computer Aided Control System Design, 2006 IEEE International Conference on Control Applications, 2006 IEEE International Symposium on Intelligent Control, pp. 2207–2212 (2006)
6. De Hart, R.D.: Advanced take-off and flight control algorithms for fixed wing unmanned aerial vehicles. Masters Dissertation, University of Stellenbosch (2010)

7. Essuri, M., Alkurmaji, K., Ghmmam, A.: Developing a dynamic model for unmanned aerial vehicle motion on ground during takeoff phase. Appl. Mech. Mater. **232**, 561–567 (2012)
8. Botkin, N.D., Krasov, A.I.: Positional control in a model problem of aircraft take-off while on the runway. In: Subbotin, A.I., Tarasiev, A.M. (eds.) Positional Control with Guaranteed Result, vol. 113, pp. 22–32. Akad Nauk SSSR, Ural. Otdel., Inst. Matem. i Mekhan., Sverdlovsk (1988). (in Russian)
9. Krasovskii, N.N., Subbotin, A.I.: Game-Theoretical Control Problems. Springer, New York (1988)
10. Botkin, N.D., Hoffmann, K.-H., Mayer, N., Turova, V.L.: Approximation schemes for solving disturbed control problems with non-terminal time and state constraints. Analysis **31**, 355–379 (2011)
11. Botkin, N.D., Hoffmann, K.-H., Turova, V.L.: Stable numerical schemes for solving Hamilton-Jacobi-Bellman-Isaacs equations. SIAM J. Sci. Comput. **33**(2), 992–1007 (2011)
12. Botkin, N.D., Turova, V.L.: Application of dynamic programming approach to aircraft take-off in a windshear. In: Simos, T.E., Psihoyios, G., Tsitouras, C., Zacharias, A. (eds.) ICNAAM-2012. AIP Conference Proceedings, vol. 1479, pp. 1226–1229. The American Institute of Physics, Melville (2012)
13. Botkin, N.D., Turova, V.L.: Dynamic programming approach to aircraft control in a windshear. In: Křivan, V., Zaccour, G. (eds.) Advances in Dynamic Games: Theory, Applications, and Numerical methods. Annals of the International Society of Dynamic Games, vol. 13, pp. 53–69. Birkäuser, Boston (2013)
14. Crandall, M.G., Lions, P.L.: Viscosity solutions of Hamilton-Jacobi equations. Trans. Am. Math. Soc. **277**, 1–47 (1983)
15. Subbotin, A.I.: Generalized Solutions of First Order PDEs: The Dynamical Optimization Perspective. Birkhäuser, Boston (1995)

Dynamic Programming Approach for Discrete-Valued Time Discrete Optimal Control Problems with Dwell Time Constraints

Michael Burger[1], Matthias Gerdts[2(✉)], Simone Göttlich[3], and Michael Herty[4]

[1] Abteilung Mathematische Methoden in Dynamik und Festigkeit MDF,
Fraunhofer-Institut für Techno- und Wirtschaftsmathematik ITWM,
Fraunhofer-Platz 1, 67663 Kaiserslautern, Germany
[2] Institut für Mathematik und Rechneranwendung (LRT), Universität der
Bundeswehr München, Werner-Heisenberg-Weg 39, 85577 Neubiberg, Germany
matthias.gerdts@unibw.de
[3] Department of Mathematics, University of Mannheim, 68131 Mannheim, Germany
[4] Department of Mathematics, RWTH Aachen,
Templergraben 55, 52056 Aachen, Germany

Abstract. The article discusses a numerical approach to solve optimal control problems in discrete time that involve continuous and discrete controls. Special attention is drawn to the modeling and treatment of dwell time constraints. For the solution of the optimal control problem in discrete time, a dynamic programming approach is employed. A numerical example is included that illustrates the impact of dwell time constraints in mixed integer optimal control.

Keywords: Dynamic programming · Mixed-integer optimization · Dwell time constraints

1 Introduction

Mixed-integer optimal control is a field of increasing importance as practical applications often include discrete decisions in addition to continuous-valued control variables. Examples of such problems can be found, e.g., in [1,2,4,5,8,14,15,18]. One way to approach mixed-integer optimal control problems is by solving necessary optimality conditions provided by the well-known maximum principle. These necessary conditions are valid even for discrete control sets. In [6] a graph-based solution method exploiting the maximum principle was developed, which is however limited to single-state problems. The drawback of maximum principle based methods is that a very good initial guess of the switching structure is needed which is often not available for practical applications. Direct discretization methods based on variable time transformations as in [1,2,9–11,18,19], or direct discretization methods based on relaxations and sum-up-rounding strategies as in [13,16,17] have shown their ability to solve difficult

L. Bociu et al. (Eds.): CSMO 2015, IFIP AICT 494, pp. 159–168, 2016.
DOI: 10.1007/978-3-319-55795-3_14

real-world examples. Extensions towards mixed control-state constraints, which depend on the discrete variable and lead to vanishing constraints, can be found in [7, Chap. 5] and [12]. The dynamic programming approach in [3, Chap. 7] allows to consider switching costs to avoid frequent switches of the discrete-valued controls. This paper particularly addresses so-called dwell time constraints in mixed-integer optimal control problems in discrete time. Dwell times are an important aspect in many engineering applications, especially in the context of switched systems. After a switch has occured, the physical process typically requires a certain time period to recover and to return to normal operation. During this dwell time the process is limited in its operation and follows specified dynamics. An example for such a behavior is a truck with gear shifts. As the clutch is engaged, the motor torque does not arrive at the wheels for a short time period. Only after the clutch is released again, the motor torque is distributed at the wheels. Throughout the paper, we focus on optimal control problems in discrete time, which exist in their own rights but often they are obtained as discretizations of optimal control problems in continuous time. The purpose of the paper is twofold. Firstly, a model taking into account dwell time is suggested. Herein, we will use delays in the dynamics. Secondly, it is shown that a dynamic programming principle applies and can be used for numerical computations, if the state dimension is low. The paper is organized as follows. Section 2 defines the mixed-integer optimal control problem in discrete time with dwell time constraints. Section 3 discusses a dynamic programming approach to solve the problems. Section 4 discusses an illustrative numerical example.

2   Modeling Dwell Time Constraints in Mixed-Integer Optimal Control in Discrete Time

Let a fixed grid $\mathbb{G}_N = \{t_0 < t_1 < \ldots < t_N\}$ with $N \in \mathbb{N}$ be given. For $i = 0, \ldots, N$ and $x \in \mathbb{R}^{n_x}$ let $\emptyset \neq \mathcal{X}(t_i) \subseteq \mathbb{R}^{n_x}$ and $\emptyset \neq \mathcal{U}(t_i, x) \subseteq \mathbb{R}^{n_u}$ be closed connected sets and

$$\mathcal{V} = \{v^1, \ldots, v^M\}$$

a discrete set of vectors $v^j \in \mathbb{R}^{n_v}$, $j = 1, \ldots, M$. On the grid $\mathbb{G}_N$, grid functions are introduced. The grid function $x : \mathbb{G}_N \longrightarrow \mathbb{R}^{n_x}$ is called state and it is restricted by the state constraints $x(t_i) \in \mathcal{X}(t_i)$, $i = 0, \ldots, N$. The grid function $u : \mathbb{G}_N \longrightarrow \mathbb{R}^{n_u}$ is called real-valued control and it is restricted by the control constraints $u(t_i) \in \mathcal{U}$, $i = 0, \ldots, N$. The grid function $v : \mathbb{G}_N \rightarrow \mathbb{R}^{n_v}$ is called discrete-valued control and it is restricted by the constraints $v(t_i) \in \mathcal{V}$, $i = 0, \ldots, N$. We say the discrete-valued control switches at time point $t_i \in \mathbb{G}_N$ with $i \in \{0, \ldots, N-1\}$, if $v(t_i) \neq v(t_{i+1})$. Switches in the grid function v can be measured with the help of the discrete variation $dv : \mathbb{G}_N \rightarrow \mathbb{R}^{n_v}$ defined by

$$dv(t_i) = \begin{cases} v(t_{i+1}) - v(t_i), & \text{for } i = 0, \ldots, N-1, \\ v(t_N) - v(t_{N-1}), & \text{for } i = N. \end{cases}$$

A jump of v at t_i occurs if and only if $dv(t_i) \neq 0$. In order to define dwell time constraints let $L \in \mathbb{N}$ denote a number of steps and $\bar{u} \in \mathcal{U}$ a fixed control vector.

Definition 1. *A dwell time constraint of time horizon length $L \in \mathbb{N}$ and control $\bar{u}$ applies, if and only if the following conditions are met:*

(a) v switches at t_i for some $i \in \{0, \ldots, N-1\}$ and
(b) $u(t_{i+\ell}) = \bar{u}$ for $\ell = 1, \ldots, L$.

In other words, a dwell time constraint is a logical implication of type

$$v(t_i) \neq v(t_{i+1}) \quad \Longrightarrow \quad u(t_{i+\ell}) = \bar{u} \text{ for } \ell = 1, \ldots, L,$$

i.e. if the discrete control switches, then control u is fixed to $\bar{u}$ for a defined future time horizon of length L.

Remark 1. For notational simplicity we fixed the full control vector u in Definition 1 (b) to a specified value $\bar{u}$ for all time points $t_{i+\ell}$, $\ell = 1, \ldots, L$. This can be generalized such that only some components of u are fixed to specified values which may even vary with the index ℓ. This more general setting would cause a more technical notation, but otherwise can be treated with the same techniques as below.

Let the state evolve according to the discrete dynamics

$$x(t_{i+1}) = f(t_i, x(t_i), u(t_i), v(t_i)), \qquad i = 0, 1, \ldots, N-1.$$

Imposing a dwell time constraint of length L and value $\bar{u}$ implies that the future state depends not just on the current control value but on the history of the control v, i.e.

$$
\begin{aligned}
u(t_i) &\in \bar{U}(v(t_i), v(t_{i-1}), \ldots, v(t_{i-L})) \\
&:= \begin{cases} \mathcal{U}, & \text{if } v(t_{i-\ell}) = v(t_{i-\ell-1}) \text{ for } \ell = 0, 1, \ldots, L-1, \\ \{\bar{u}\}, & \text{otherwise.} \end{cases}
\end{aligned}
$$

Herein and throughout the paper we use the convention

$$v(t_{-\ell}) := v(t_0) \qquad \text{for } \ell = 1, \ldots, L. \tag{1}$$

We arrived at the following mixed-integer optimal control problem in discrete time with a dwell time constraint of length L with value $\bar{u}$:

Probelm 1 (DMIOCP).
Minimize

$$F(x, u, v) := \varphi(x(t_N)) + \sum_{i=0}^{N-1} f_0(t_i, x(t_i), u(t_i), v(t_i))$$

subject to the constraints

$$
\begin{aligned}
x(t_{i+1}) &= f(t_i, x(t_i), u(t_i), v(t_i)), & i &= 0, 1, \ldots, N-1, \\
x(t_0) &= x_0, & & \\
x(t_i) &\in \mathcal{X}(t_i), & i &= 0, 1, \ldots, N, \\
u(t_i) &\in \bar{U}(v(t_i), v(t_{i-1}), \ldots, v(t_{i-L})), & i &= 0, 1, \ldots, N-1, \\
v(t_i) &\in \mathcal{V}, & i &= 0, 1, \ldots, N-1.
\end{aligned}
$$

Note that the control $v(t_N)$ does not have any influence in the problem. Consequently, only switches at the time points $t_0, \ldots, t_{N-1}$ are taken into account. In the sequel we will make use of the convention $v_N = v_{N-1}$ whenever useful.

3   A Dynamic Programming Approach

DMIOCP is embedded into a family of perturbed problems as follows. To this end we use the notation $v_{(k)}^L := (v_{k-1}, \ldots, v_{k-L})$ for $L \in \mathbb{N}$ and some index k with values $v_{k-\ell} \in \mathcal{V}$ for $\ell = 1, \ldots, L$. We denote with $\mathcal{V}^L$ the cartesian product $\mathcal{V} \times \cdots \times \mathcal{V}$ (L times).

Probelm 2 (DMIOCP($t_k, x_k, v_{(k)}^L$)).
For a given index $t_k \in \mathbb{G}_N$, $x_k \in \mathbb{R}^{n_x}$, and $v_{(k)}^L \in \mathcal{V}^L$, minimize

$$F_k(x, u, v) := \varphi(x(t_N)) + \sum_{i=k}^{N-1} f_0(t_i, x(t_i), u(t_i), v(t_i))$$

subject to

$$\left.\begin{array}{ll}
x(t_{i+1}) = f(t_i, x(t_i), u(t_i), v(t_i)), & i = k, \ldots, N-1, \\
x(t_k) = x_k, & \\
x(t_i) \in \mathcal{X}(t_i), & i = k, \ldots, N, \\
u(t_i) \in \bar{\mathcal{U}}(v(t_i), v(t_{i-1}), \ldots, v(t_{i-L})), & i = k, \ldots, N-1, \\
v(t_i) \in \mathcal{V}, & i = k, \ldots, N-1, \\
v(t_{k-\ell}) = v_{k-\ell}, & \ell = 1, \ldots, L.
\end{array}\right\} \quad (2)$$

Definition 2. *The function $W : \mathbb{G}_N \times \mathbb{R}^{n_x} \times \mathcal{V}^L \longrightarrow \mathbb{R}$ that assigns to every $(t_k, x_k, v_{(k)}^L) \in \mathbb{G}_N \times \mathbb{R}^{n_x} \times \mathcal{V}^L$ the optimal value of DMIOCP($t_k, x_k, v_{(k)}^L$) is called* optimal value function of DMIOCP, *i.e.*

$$W(t_k, x_k, v_{(k)}^L) := \begin{cases} \inf\limits_{x,u,v \text{ with } (2)} F_k(x, u, v), & \text{if DMIOCP}(t_k, x_k, v_{(k)}^L) \text{ is feasible,} \\ \infty, & \text{otherwise.} \end{cases}$$

The next theorem establishes Bellman's optimality principle.

Theorem 1. *Let $(\hat{x}, \hat{u}, \hat{v})$ be optimal grid functions for DMIOCP and let $\mathbb{G}_N^k := \mathbb{G}_N \setminus \{t_0, \ldots, t_{k-1}\}$. Then, the restrictions on $\mathbb{G}_N^k$ given by $\hat{x}|_{\mathbb{G}_N^k}$, $\hat{u}|_{\mathbb{G}_N^k}$, and $\hat{v}|_{\mathbb{G}_N^k}$ are optimal for DMIOCP($t_k, \hat{x}(t_k), \hat{v}_{(k)}^L$) with $\hat{v}_{(k)}^L = (\hat{v}(t_{k-1}), \ldots, \hat{v}(t_{k-L}))$ for all $k = 0, \ldots, N$ subject to the convention (1).*

Proof. Assume that the restrictions $\hat{x}|_{\mathbb{G}_N^k}$, $\hat{u}|_{\mathbb{G}_N^k}$, and $\hat{v}|_{\mathbb{G}_N^k}$ are not optimal for DMIOCP($t_k, \hat{x}(t_k), \hat{v}_{(k)}^L$) for some $k \in \{0, \ldots, N-1\}$. Then there exist feasible grid functions $\tilde{x} : \mathbb{G}_N^k \longrightarrow \mathbb{R}^{n_x}$, $\tilde{u} : \mathbb{G}_N^k \longrightarrow \mathbb{R}^{n_u}$, and $\tilde{v} : \mathbb{G}_N^k \longrightarrow \mathcal{V}$ for DMIOCP($t_k, \hat{x}(t_k), \hat{v}_{(k)}^L$) with

$$F_k(\tilde{x}, \tilde{u}, \tilde{v}) < F_k(\hat{x}, \hat{u}, \hat{v}), \quad (3)$$

$\tilde{x}(t_k) = \hat{x}(t_k)$, and $\tilde{v}(t_{k-\ell}) = \hat{v}(t_{k-\ell})$ for $\ell = 1, \ldots, L$. Hence, the trajectories $x : \mathbb{G}_N \longrightarrow \mathbb{R}^{n_x}$, $u : \mathbb{G}_N \longrightarrow \mathbb{R}^{n_u}$, and $v : \mathbb{G}_N \longrightarrow \mathcal{V}$ with

$$(x(t_i), u(t_i), v(t_i)) := \begin{cases} (\hat{x}(t_i), \hat{u}(t_i), \hat{v}(t_i)) \text{ for } i = 0, 1, \ldots, k-1, \\ (\tilde{x}(t_i), \tilde{u}(t_i), \tilde{v}(t_i)) \text{ for } i = k, k+1, \ldots, N, \end{cases}$$

are feasible for DMIOCP and satisfy

$$F(x, u, v) = \varphi(x(t_N)) + \sum_{i=0}^{N-1} f_0(t_i, x(t_i), u(t_i), v(t_i))$$

$$= \varphi(\tilde{x}(t_N)) + \sum_{i=0}^{k-1} f_0(t_i, \hat{x}(t_i), \hat{u}(t_i), \hat{v}(t_i)) + \sum_{i=k}^{N-1} f_0(t_i, \tilde{x}(t_i), \tilde{u}(t_i), \tilde{v}(t_i))$$

$$< \varphi(\hat{x}(t_N)) + \sum_{i=0}^{k-1} f_0(t_i, \hat{x}(t_i), \hat{u}(t_i), \hat{v}(t_i)) + \sum_{i=k}^{N-1} f_0(t_i, \hat{x}(t_i), \hat{u}(t_i), \hat{v}(t_i))$$

$$= \varphi(\hat{x}(t_N)) + \sum_{i=0}^{N-1} f_0(t_i, \hat{x}(t_i), \hat{u}(t_i), \hat{v}(t_i))$$

$$= F(\hat{x}, \hat{u}, \hat{v}),$$

where (3) is exploited. This contradicts the optimality of $\hat{x}(\cdot)$, $\hat{u}(\cdot)$, and $\hat{v}(\cdot)$. $\square$

From the definition of the optimal value function one immediately obtains

$$W(t_N, x_N, v_{(N)}^L) = \begin{cases} \varphi(x_N), \text{ if } x_N \in \mathcal{X}(t_N), \\ \infty, \quad \text{otherwise.} \end{cases} \tag{4}$$

Bellman's optimality principle allows to derive the following result.

Theorem 2. *For all* $(t_k, x_k, v_{(k)}^L) \in \mathbb{G}_N \times \mathcal{X}(t_k) \times \mathcal{V}^L$ *and* $k = N-1, \ldots, 0$, *the optimal value function in Definition 2 satisfies the recursion*

$$W(t_k, x_k, v_{(k)}^L) = \inf_{u \in \bar{\mathcal{U}}(v, v_{k-1}, \ldots, v_{k-L}), v \in \mathcal{V}} \Big\{ f_0(t_k, x_k, u, v) $$

$$+ W(t_{k+1}, f(t_k, x_k, u, v), v_{(k+1)}^L) \Big\}$$

where $v_{(k)}^L = (v_{k-1}, \ldots, v_{k-L})$ *and* $v_{(k+1)}^L = (v, v_{k-1}, \ldots, v_{k-L+1})$ *and convention (1) is used. Herein, W at $t = t_N$ is given by (4) and the convention* $W(t_k, x_k, v_{(k)}^L) = \infty$ *is used whenever* $(x_k, v_{(k)}^L) \notin \mathcal{X}(t_k) \times \mathcal{V}^L$.

Proof. Let $(t_k, x_k, v_{(k)}^L) \in \mathbb{G}_N \times \mathcal{X}(t_k) \times \mathcal{V}^L$ and $k \in \{0, \ldots, N-1\}$ be given. If $(x_k, v_{(k)}^L) \notin \mathcal{X}(t_k) \times \mathcal{V}^L$, then $W(t_k, x_k, v_{(k)}^L) = \infty$ by definition.

If $f(t_k, x_k, u, v) \notin \mathcal{X}(t_{k+1})$ for all $u \in \bar{\mathcal{U}}(v, v_{k-1}, \ldots, v_{k-L})$ and all $v \in \mathcal{V}$, then DMIOCP$(t_k, x_k, v_{(k)}^L)$ and DMIOCP$(t_{k+1}, f(t_k, x_k, u, v), v_{(k+1)}^L)$ are infeasible for every $u \in \bar{\mathcal{U}}(v, v_{k-1}, \ldots, v_{k-L})$ and every $v \in \mathcal{V}$ and hence

$W(t_k, x_k, v_{(k)}^L) = \infty$. For arbitrary $u \in \bar{\mathcal{U}}(v, v_{k-1}, \ldots, v_{k-L})$ and $v \in \mathcal{V}$ with $f(t_k, x_k, u, v) \in \mathcal{X}(t_{k+1})$ the definition of the optimal value function yields

$$W(t_k, x_k, v_{(k)}^L) \leq f_0(t_k, x_k, u, v) + W(t_{k+1}, f(t_k, x_k, u, v), v_{(k+1)}^L).$$

Taking the infimum over all $u \in \bar{\mathcal{U}}(v, v_{k-1}, \ldots, v_{k-L})$ and $v \in \mathcal{V}$ yields the first part of the assertion. Now let $\varepsilon > 0$ and feasible grid functions $\tilde{x}, \tilde{u}, \tilde{v}$ with $\tilde{x}(t_k) = x_k$ and

$$F_k(\tilde{x}, \tilde{u}, \tilde{v}) \leq W(t_k, x_k, \tilde{v}_{(k)}^L) + \varepsilon$$

with $\tilde{v}_{(k)}^L = (\tilde{v}(t_{k-1}), \ldots, \tilde{v}(t_{k-L}))$ be given. Then,

$$
\begin{aligned}
W(t_k, x_k, \tilde{v}_{(k)}^L) &\geq F_k(\tilde{x}, \tilde{u}, \tilde{v}) - \varepsilon \\
&\geq f_0(t_k, x_k, \tilde{u}(t_k), \tilde{v}(t_k)) \\
&\quad + W(t_{k+1}, f(t_k, x_k, \tilde{u}(t_k), \tilde{v}(t_k)), \tilde{v}_{(k+1)}^L) - \varepsilon \\
&\geq \inf_{u \in \bar{\mathcal{U}}(v, \tilde{v}(t_{k-1}), \ldots, \tilde{v}(t_{k-L})), v \in \mathcal{V}} \Big\{ f_0(t_k, x_k, u, v) \\
&\quad + W(t_{k+1}, f(t_k, x_k, u, v), v_{(k+1)}^L) \Big\} - \varepsilon,
\end{aligned}
$$

where $\tilde{v}_{(k+1)}^L = (\tilde{v}(t_k), \ldots, \tilde{v}(t_{k-L+1}))$ and $v_{(k+1)}^L = (v, \tilde{v}(t_{k-1}), \ldots, \tilde{v}(t_{k-L+1}))$. As $\varepsilon > 0$ was arbitrary, the assertion follows. $\square$

Theorem 2 allows to deduce the following dynamic programming algorithm. Note that owing to the presence of $v_{(k)}^L$ in DMIOCP, the recursive formula for the optimal value function is not of standard type. The vector $v_{(k)}^L$ contains the history of the discrete-valued control, which influences the decision at stage k. The vector $v_{(k)}^L$ can be interpreted as additional states in the argument of W and leads to an additional runtime factor of M^L when compared to standard dynamic programming. In general this is a prohibitively large factor, but often M and L are moderate numbers. For instance, a binary control ($M = 2$) and $L = 3$ leads to a factor of 8.

Algorithm: Dynamic Programming

Init: Let $\mathbb{G}_N = \{t_0 < t_1 < \ldots < t_N\}$ be given. Set

$$W(t_N, x_N, v_{(N)}^L) = \begin{cases} \varphi(x_N), & \text{if } x_N \in \mathcal{X}(t_N), \\ \infty, & \text{otherwise} \end{cases}$$

for all $x_N \in \mathbb{R}^{n_x}$ and all $v_{(N)}^L = (v_{N-1}, \ldots, v_{N-L}) \in \mathcal{V}^L$.

Phase 1 (Backward solution): For $k = N - 1, \ldots, 0$ compute

$$
\begin{aligned}
W(t_k, x_k, v_{(k)}^L) &= \inf_{u \in \bar{\mathcal{U}}(v, v_{k-1}, \ldots, v_{k-L}), v \in \mathcal{V}} \Big\{ f_0(t_k, x_k, u, v) \\
&\quad + W(t_{k+1}, f(t_k, x_k, u, v), v_{(k+1)}^L) \Big\}
\end{aligned}
\tag{5}
$$

with $v^L_{(k+1)} = (v, v_{k-1}, \ldots, v_{k-L+1})$ for all $x_k \in \mathbb{R}^{n_x}$ and all $v^L_{(k)} \in \mathcal{V}^L$.

Phase 2 (Forward solution): Set $\hat{x}(t_0) = x_0$. For $k = 0, \ldots, N-1$ find

$$
(\hat{u}(t_k), \hat{v}(t_k)) = \arg \min_{u \in \bar{\mathcal{U}}(v, \hat{v}(t_{k-1}), \ldots, \hat{v}(t_{k-L})), v \in \mathcal{V}} \Big\{ f_0(t_k, \hat{x}(t_k), u, v)
$$
$$
+ W(t_{k+1}, f(t_k, \hat{x}(t_k), u, v), \hat{v}^L_{(k+1)}) \Big\} \tag{6}
$$

with $\hat{v}^L_{(k+1)} = (v, \hat{v}(t_{k-1}), \ldots, \hat{v}(t_{k-L+1}))$ and set

$$
\hat{x}(t_{k+1}) = f(t_k, \hat{x}(t_k), \hat{u}(t_k), \hat{v}(t_k)).
$$

3.1 Practical Issues in the Dynamic Programming Algorithm

The implementation of the dynamic programming approach works on a compact state space $\Omega = \{x \in \mathbb{R}^{n_x} \mid x_\ell \leq x \leq x_u\} \subset \mathbb{R}^{n_x}$ with lower and upper bounds $x_\ell, x_u \in \mathbb{R}^{n_x}$ and $-\infty < x_\ell < x_u < \infty$. The bounds should be chosen such that all realistically relevant trajectories are contained in this set. The state space Ω is discretized with an equidistant partition

$$
\Omega_{N_x} = \left\{ (x_1, \ldots, x_{n_x})^\top \in \Omega \;\middle|\; \begin{array}{l} x_j = x_{\ell,j} + i \frac{x_{u,j} - x_{\ell,j}}{N_x}, \\ j = 1, \ldots, n_x, \\ i \in \{0, \ldots, N_x\} \end{array} \right\}
$$

with $N_x \in \mathbb{N}$. During the backward solution and forward solution phases of the dynamic programming algorithm values of the value function at points $x = f(t_k, x_k, u, v)$ not in Ω_{N_x} are approximated by polynomial interpolation of the value function on a cell of Ω_{N_x} that contains x.

The minimization in (5) and (6) is usually done by a sufficiently dense discretization of the control set $\bar{\mathcal{U}}$ and complete enumeration of all possible values. The v-component is not crucial as $\mathcal{V}$ is supposed to be a finite discrete set. For a fixed v the minimization w.r.t. the component u could be carried out using methods from nonsmooth optimization, e.g. Bundle methods or subgradient methods. However, such methods may result in a local minimum and therefore the previously mentioned enumeration technique is preferred. The general drawback of the dynamic programming approach is the curse of dimensions. Since the value function W depends on the state x and the control history $v^L_{(k)}$ of length L, the approach is computationally feasible merely for low state dimensions and moderate values of L and M, where M is the number of elements in $\mathcal{V}$.

Remark 2. Please note that a more efficient coding of the switching history is possible. To this end, it would be sufficient to replace $v^L_{(k)}$ in W by v_{k-1} and an additional state with values in $\{0, \ldots, L\}$, which encodes how long the last switch is ago (with 0 meaning longer than L).

4   Numerical Results

We consider a car of mass m driving on a road with a given slope profile $\gamma(\cdot)$ and aim to optimize the gear shift control on a given time horizon $[0,T]$. The car model is adapted from [1] with the state (x,v) (position and velocity) and controls $\phi \in \mathcal{U} := [0,1]$ (gas pedal position) and $\mu \in \mathcal{V} := \{1,\dots,5\}$ (gear shift). The optimal control problem in discrete time with step-size $h = T/N$, $N \in \mathbb{N}$, reads as follows:

Minimize

$$-\alpha x(t_N) + h \sum_{k=0}^{N-1} f_0(v(t_k), \phi(t_k), \mu(t_k))$$

subject to the constraints $x(0) = 1470$, $v(0) = 23$, *and for* $k = 0,1,\dots,N-1$,

$$x(t_{k+1}) = x(t_k) + hv(t_k),$$

$$v(t_{k+1}) = v(t_k) + \frac{h}{m}\left(\frac{i_g(\mu(t_k))i_t M_{mot}(\phi(t_k),v(t_k),\mu(t_k))}{R}\right.$$

$$\left. -F_R(v(t_k))\cos(\gamma(x(t_k))) - F_L(v(t_k)) - mg\sin(\gamma(x(t_k)))\right),$$

$$\mu(t_k) \in \mathcal{V}, \quad \phi(t_k) \in \mathcal{U}, \quad v(t_k) \in [22,30].$$

Herein, $\alpha \geq 0$ is a weight factor, i_g is the gear transmission coefficient, i_t the motor torque transmission, R the wheel radius, F_R the friction force of the rolling wheel, $F_L(v) = \frac{1}{2}c_w\rho Av^2$ the drag, and

$$f_0(v,\phi,\mu) = \beta_1 \omega_{mot}(v,\mu) + \beta_2 |M_{mot}(\phi,v,\mu)| + \beta_3 \omega_{mot}(v,\mu)|M_{mot}(\phi,v,\mu)|$$

models the fuel costs, where ω_{mot} denotes the rotary frequency of the motor and M_{mot} the motor torque, for details on ω_{mot}, M_{mot}, and F_R please refer to [1].

Figure 1 shows the results of the dynamic programming algorithm for both, the problem with dwell time constraint (with $L = 3$ and $\bar{u} = \bar{\phi} = 0$, i.e. ϕ is set to zero, if μ switches) and the problem without dwell time constraint. The parameters $\alpha = 1/10$, $\beta_1 = 1/120$, $\beta_2 = 10/258$, $\beta_3 = 1/10320$, $c_w = 0.3$, $\rho = 1.249512$, $A = 2$, $R = 0.302$, $i_t = 3.91$, $g = 9.81$, $m = 2000$, $T = 10$, $N = 25$, $N_x = (100,45)$, $N_u = 100$, $i_g(1) = 3.91$, $i_g(2) = 2.002$, $i_g(3) = 1.33$, $i_g(4) = 1$, $i_g(5) = 0.805$ were used. The optimal value for the problem with dwell time constraint is -79.7656, while the optimal value without dwell time constraint is -88.0430. In the presence of the dwell time constraint it can be nicely seen that ϕ is set to zero after μ switches at approximately $t = 4$. The CPU time on a PC with 2.3 GHz is 1 min 47.146 s without dwell time constraint and 14 min 38.9 s with dwell time constraint.

5   Conclusions

The paper discusses a dynamic programming principle that allows to consider dwell times in a time discrete optimal control problem with discrete-valued controls. Dwell times have to be taken into account in many applications where

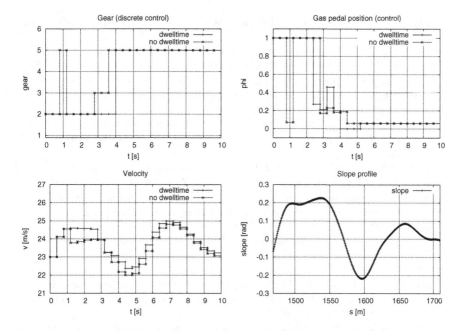

Fig. 1. Optimal gear sequence (top left), gas pedal position (top right), velocity (bottom left), and slope profile (bottom right).

a switch in a discrete-valued control fixes the dynamics of the process for a specified time period before the process can be controlled again. A particular example is the optimization of gear shifts in a car, where a shift in gears implies that the motor torque is reduced for a given time period. A dynamic programming approach is suggested to solve such problems and a numerical example shows the applicability of the method. However, the curse of dimension associated to dynamic programming approaches limits the method to low dimensional systems with few states. Hence, it is desirable to develop further methods for this problem class and extend it to optimal control problems in continuous time. To this end, a formulation of the dwell time constraint in a continuous time setting becomes necessary, which is subject to future research.

References

1. Gerdts, M.: Solving mixed-integer optimal control problems by branch & bound: a case study from automobile test-driving with gear shift. Optim. Control Appl. Methods **26**(1), 1–18 (2005)
2. Gerdts, M.: A variable time transformation method for mixed-integer optimal control problems. Optim. Control Appl. Methods **27**(3), 169–182 (2006)
3. Gerdts, M.: Optimal Control of ODEs and DAEs. DeGruyter, Berlin (2011)
4. Göttlich, S., Herty, M., Ziegler, U.: Modeling and optimizing traffic light settings on road networks. Comput. Oper. Res. **55**, 36–51 (2015)

5. Howlett, P.: Optimal strategies for the control of a train. Automatica **32**(4), 519–532 (1996)
6. Khmelnitsky, E.: A combinatorial, graph-based solution method for a class of continuous-time optimal control problems. Math. Oper. Res. **27**(2), 312–325 (2002)
7. Kirches, C.: Fast numerical methods for mixed-integer nonlinear model-predictive control. Ph.D. thesis, Naturwissenschaftlich-Mathematische Gesamtfakultät, Universität Heidelberg, Heidelberg, Germany (2010)
8. Kirches, C., Sager, S., Bock, H.G., Schlöder, J.P.: Time-optimal control of automobile test drives with gear shifts. Optim. Control Appl. Methods **31**, 137–153 (2010). doi:10.1002/oca.892
9. Lee, H.W.J., Teo, K.L., Cai, X.Q.: An optimal control approach to nonlinear mixed integer programming problems. Comput. Math. Appl. **36**(3), 87–105 (1998)
10. Lee, H.W.J., Teo, K.L., Rehbock, V., Jennings, L.S.: Control parameterization enhancing technique for time optimal control problems. Dyn. Syst. Appl. **6**(2), 243–262 (1997)
11. Lee, H.W.J., Teo, K.L., Rehbock, V., Jennings, L.S.: Control parametrization enhancing technique for optimal discrete-valued control problems. Automatica **35**(8), 1401–1407 (1999)
12. Palagachev, K., Gerdts, M.: Mathematical programs with blocks of vanishing constraints arising in discretized mixed-integer optimal control problems. Set-Valued Var. Anal. **23**(1), 149–167 (2015)
13. Sager, S.: Reformulations and algorithms for the optimization of switching decisions in nonlinear optimal control. J. Process Control **19**(8), 1238–1247 (2009)
14. Sager, S., Bock, H.G., Diehl, M., Reinelt, G., Schlöder, J.P.: Numerical methods for optimal control with binary control functions applied to a Lotka-Volterra type fishing problem. In: Seeger, A. (ed.) Recent Advances in Optimization (Proceedings of the 12th French-German-Spanish Conference on Optimization). Lectures Notes in Economics and Mathematical Systems, vol. 563, pp. 269–289. Springer, Heidelberg (2006)
15. Sager, S., Kirches, C., Bock, H.G.: Fast solution of periodic optimal control problems in automobile test-driving with gear shifts. In: Proceedings of the 47th IEEE Conference on Decision and Control (CDC 2008), Cancun, Mexico, pp. 1563–1568 (2008). ISBN: 978-1-4244-3124-3
16. Sager, S.: Numerical methods for mixed-integer optimal control problems. Ph.D. thesis, Heidelberg: Univ. Heidelberg, Naturwissenschaftlich-Mathematische Gesamtfakultät (Diss.). viii, 219 p. (2006)
17. Sager, S., Bock, H.G., Reinelt, G.: Direct methods with maximal lower bound for mixed-integer optimal control problems. Math. Program. (A) **118**(1), 109–149 (2009)
18. Siburian, A.: Numerical Methods for Robust, Singular and Discrete Valued Optimal Control Problems. Ph.D. thesis, Curtin University of Technology, Perth, Australia (2004)
19. Teo, K.L., Jennings, L.S., Lee, V., Rehbock, H.W.J.: The control parameterization enhancing transform for constrained optimal control problems. J. Aust. Math. Soc. **40**(3), 314–335 (1999)

Infimal Convolution Regularisation Functionals of BV and L^p Spaces. The Case $p = \infty$

Martin Burger[1], Konstantinos Papafitsoros[2($\boxtimes$)], Evangelos Papoutsellis[3], and Carola-Bibiane Schönlieb[4]

[1] Institute for Computational and Applied Mathematics,
University of Münster, Münster, Germany
martin.burger@wwu.de

[2] Weierstrass Institute for Applied Analysis and Stochastics, Berlin, Germany
kostas.papafitsoros@wias-berlin.de

[3] Laboratoire MAPMO, Fédération Denis Poisson,
Université d'Orléans, Orléans, France
evangelos.papoutsellis@univ-orleans.fr

[4] Department of Applied Mathematics and Theoretical Physics,
University of Cambridge, Cambridge, UK
cbs31@cam.ac.uk

Abstract. In this paper we analyse an infimal convolution type regularisation functional called TVL$^\infty$, based on the total variation (TV) and the L$^\infty$ norm of the gradient. The functional belongs to a more general family of TVL^p functionals ($1 < p \leq \infty$) introduced in [5]. There, the case $1 < p < \infty$ is examined while here we focus on the $p = \infty$ case. We show via analytical and numerical results that the minimisation of the TVL$^\infty$ functional promotes piecewise affine structures in the reconstructed images similar to the state of the art total generalised variation (TGV) but improving upon preservation of hat–like structures. We also propose a spatially adapted version of our model that produces results comparable to TGV and allows space for further improvement.

Keywords: Total variation · Infimal convolution · L$^\infty$ norm · Denoising · Staircasing

1 Introduction

In the variational setting for imaging, given image data $f \in$ L$^s(\Omega)$, $\Omega \subseteq \mathbb{R}^2$, ones aims to reconstruct an image u by minimising a functional of the type

$$\min_{u \in X} \frac{1}{s} \|f - Tu\|^s_{\mathrm{L}^s(\Omega)} + \Psi(u), \tag{1.1}$$

over a suitable function space X. Here T denotes a linear and bounded operator that encodes the transformation or degradation that the original image has

© IFIP International Federation for Information Processing 2016
Published by Springer International Publishing AG 2016. All Rights Reserved
L. Bociu et al. (Eds.): CSMO 2015, IFIP AICT 494, pp. 169–179, 2016.
DOI: 10.1007/978-3-319-55795-3_15

gone through. Random noise is also usually present in the degraded image, the statistics of which determine the norm in the first term of (1.1), the *fidelity term*. The presence of Ψ, the *regulariser*, makes the minimisation (1.1) a well–posed problem and its choice is crucial for the overall quality of the reconstruction. A classical regulariser in imaging is the total variation functional weighted with a parameter $\alpha > 0$, αTV [10], where

$$\mathrm{TV}(u) := \sup \left\{ \int_\Omega u \operatorname{div}\phi \, dx : \phi \in C_c^1(\Omega, \mathbb{R}^2), \|\phi\|_\infty \leq 1 \right\}. \tag{1.2}$$

While TV is able to preserve edges in the reconstructed image, it also promotes piecewise constant structures leading to undesirable staircasing artefacts. Several regularisers that incorporate higher order derivatives have been introduced in order to resolve this issue. The most prominent one has been the second order total generalised variation (TGV) [3] which can be interpreted as a special type of infimal convolution of first and second order derivatives. Its definition reads

$$\mathrm{TGV}_{\alpha,\beta}^2(u) := \min_{w \in \mathrm{BD}(\Omega)} \alpha\|Du - w\|_{\mathcal{M}} + \beta\|\mathcal{E}w\|_{\mathcal{M}}. \tag{1.3}$$

Here α, β are positive parameters, $\mathrm{BD}(\Omega)$ is the space of functions of bounded deformation, $\mathcal{E}w$ is the distributional symmetrised gradient and $\|\cdot\|_{\mathcal{M}}$ denotes the Radon norm of a finite Radon measure, i.e.,

$$\|\mu\|_{\mathcal{M}} = \sup \left\{ \int_\Omega \phi \, d\mu : \phi \in C_c^\infty(\Omega, \mathbb{R}^2), \|\phi\|_\infty \leq 1 \right\}.$$

TGV regularisation typically produces piecewise smooth reconstructions eliminating the staircasing effect. A plausible question is whether results of similar quality can be achieved using simpler, first order regularisers. For instance, it is known that Huber TV can reduce the staircasing up to an extent [6].

In [5], a family of first order infimal convolution type regularisation functionals is introduced, that reads

$$\mathrm{TVL}_{\alpha,\beta}^p(u) := \min_{w \in \mathrm{L}^p(\Omega)} \alpha\|Du - w\|_{\mathcal{M}} + \beta\|w\|_{\mathrm{L}^p(\Omega)}, \tag{1.4}$$

where $1 < p \leq \infty$. While in [5], basic properties of (1.4) are shown for the general case $1 < p \leq \infty$, see Proposition 1, the main focus remains on the finite p case. There, the TVL^p regulariser is successfully applied to image denoising and decomposition, reducing significantly the staircasing effect and producing piecewise smooth results that are very similar to the solutions obtained by TGV. Exact solutions of the L^2 fidelity denoising problem are also computed there for simple one dimensional data.

1.1   Contribution of the Present Work

The purpose of the present paper is to examine more thoroughly the case $p = \infty$, i.e.,

$$\mathrm{TVL}_{\alpha,\beta}^\infty(u) := \min_{w \in \mathrm{L}^\infty(\Omega)} \alpha\|Du - w\|_{\mathcal{M}} + \beta\|w\|_{\mathrm{L}^\infty(\Omega)}, \tag{1.5}$$

and the use of the TVL^∞ functional in L^2 fidelity denoising

$$\min_u \frac{1}{2}\|f - u\|^2_{L^2(\Omega)} + \mathrm{TVL}^\infty_{\alpha,\beta}(u). \tag{1.6}$$

We study thoroughly the one dimensional version of (1.6), by computing exact solutions for data f a piecewise constant and a piecewise affine step function. We show that the solutions are piecewise affine and we depict some similarities and differences to TGV solutions. The functional $\mathrm{TVL}^\infty_{\alpha,\beta}$ is further tested for Gaussian denoising. We show that $\mathrm{TVL}^\infty_{\alpha,\beta}$, unlike TGV, is able to recover hat–like structures, a property that is already present in the $\mathrm{TVL}^p_{\alpha,\beta}$ regulariser for large values of p, see [5], and it is enhanced here. After explaining some limitations of our model, we propose an extension where the parameter β is spatially varying, i.e., $\beta = \beta(x)$, and discuss a rule for selecting its values. The resulting denoised images are comparable to the TGV reconstructions and indeed the model has the potential to produce much better results.

2 Properties of the $\mathrm{TVL}^\infty_{\alpha,\beta}$ Functional

The following properties of the $\mathrm{TVL}^\infty_{\alpha,\beta}$ functional are shown in [5]. We refer the reader to [5,9] for the corresponding proofs and to [1] for an introduction to the space of functions of bounded variation $\mathrm{BV}(\Omega)$.

Proposition 1 [5]. *Let $\alpha, \beta > 0$, $d \geq 1$, let $\Omega \subseteq \mathbb{R}^d$ be an open, connected domain with Lipschitz boundary and define for $u \in L^1(\Omega)$*

$$\mathrm{TVL}^\infty_{\alpha,\beta}(u) := \min_{w \in L^\infty(\Omega)} \alpha\|Du - w\|_{\mathcal{M}} + \beta\|w\|_{L^\infty(\Omega)}. \tag{2.1}$$

Then we have the following:

(i) $\mathrm{TVL}^\infty_{\alpha,\beta}(u) < \infty$ if and only if $u \in \mathrm{BV}(\Omega)$.
(ii) If $u \in \mathrm{BV}(\Omega)$ then the minimum in (2.1) is attained.
(iii) $\mathrm{TVL}^\infty_{\alpha,\beta}(u)$ can equivalently be defined as

$$\mathrm{TVL}^\infty_{\alpha,\beta}(u) = \sup\left\{ \int_\Omega u\,\mathrm{div}\phi\,dx : \; \phi \in C^1_c(\Omega, \mathbb{R}^2), \; \|\phi\|_\infty \leq \alpha, \; \|\phi\|_{L^1(\Omega)} \leq \beta \right\},$$

and $\mathrm{TVL}^\infty_{\alpha,\beta}$ is lower semicontinuous w.r.t. the strong L^1 convergence.
(iv) There exist constants $0 < C_1 < C_2 < \infty$ such that

$$C_2 \mathrm{TV}(u) \leq \mathrm{TVL}^\infty_{\alpha,\beta}(u) \leq C_1 \mathrm{TV}(u), \quad \text{for all } u \in \mathrm{BV}(\Omega).$$

(v) If $f \in L^2(\Omega)$, then the minimisation problem

$$\min_{u \in \mathrm{BV}(\Omega)} \frac{1}{s}\|f - u\|^2_{L^2(\Omega)} + \mathrm{TVL}^\infty_{\alpha,\beta}(u),$$

has a unique solution.

3   The One Dimensional L^2–$TVL^\infty_{\alpha,\beta}$ Denoising Problem

In order to get an intuition about the underlying regularising mechanism of the $TVL^\infty_{\alpha,\beta}$ regulariser, we study here the one dimensional L^2 denoising problem

$$\min_{\substack{u \in BV(\Omega) \\ w \in L^\infty(\Omega)}} \frac{1}{2}\|f - u\|^2_{L^2(\Omega)} + \alpha\|Du - w\|_{\mathcal{M}} + \beta\|w\|_{L^\infty(\Omega)}. \qquad (3.1)$$

In particular, we present exact solutions for simple data functions. In order to do so, we use the following theorem:

Theorem 2 (Optimality conditions). *Let* $f \in L^2(\Omega)$. *A pair* $(u,w) \in BV$ $(\Omega) \times L^\infty(\Omega)$ *is a solution of* (3.1) *if and only if there exists a unique* $\phi \in H^1_0(\Omega)$ *such that*

$$\phi' = u - f, \qquad (3.2)$$

$$\phi \in \alpha \mathrm{Sgn}(Du - w), \qquad (3.3)$$

$$\phi \in \begin{cases} \{\psi \in L^1(\Omega) : \|\psi\|_{L^1(\Omega)} \leq \beta\}, & \text{if } w = 0, \\ \{\psi \in L^1(\Omega) : \langle\psi, w\rangle = \beta\|w\|_{L^\infty(\Omega)}, \|\psi\|_{L^1(\Omega)} \leq \beta\}, & \text{if } w \neq 0. \end{cases} \qquad (3.4)$$

Recall that for a finite Radon measure μ, $\mathrm{Sgn}(\mu)$ is defined as

$$\mathrm{Sgn}(\mu) = \left\{\phi \in L^\infty(\Omega) \cap L^\infty(\Omega, \mu) : \|\phi\|_{L^\infty(\Omega)} \leq 1, \ \phi = \frac{d\mu}{d|\mu|}, \ |\mu| - \text{a.e.}\right\}.$$

As it is shown in [4], $\mathrm{Sgn}(\mu) \cap C_0(\Omega) = \partial\|\cdot\|_{\mathcal{M}}(\mu) \cap C_0(\Omega)$.

Proof. The proof of Theorem 2 is based on Fenchel–Rockafellar duality theory and follows closely the corresponding proof of the finite p case. We thus omit it and we refer the reader to [5,9] for further details, see also [4,8] for the analogue optimality conditions for the one dimensional L^1–TGV and L^2–TGV problems. □

The following proposition states that the solution u of (3.1) is essentially piecewise affine.

Proposition 3 (Affine structures). *Let* (u,w) *be an optimal solution pair for* (3.1) *and* ϕ *be the corresponding dual function. Then* $|w| = \|w\|_{L^\infty(\Omega)}$ *a.e. in the set* $\{\phi \neq 0\}$. *Moreover,* $|u'| = \|w\|_{L^\infty(\Omega)}$ *whenever* $u > f$ *or* $u < f$.

Proof. Suppose that there exists a $U \subseteq \{\phi \neq 0\}$ of positive measure such that $|w(x)| < \|w\|_{L^\infty(\Omega)}$ for every $x \in U$. Then

$$\int_\Omega \phi w\, dx \leq \int_{\Omega \setminus U} |\phi||w|\, dx + \int_U |\phi||w|\, dx < \|w\|_{L^\infty(\Omega)} \left(\int_{\Omega \setminus U} |\phi|\, dx + \int_U |\phi|\, dx\right)$$

$$= \|w\|_{L^\infty(\Omega)}\|\phi\|_{L^1(\Omega)} = \beta\|w\|_{L^\infty(\Omega)},$$

where we used the fact that $\|\phi\|_{L^1(\Omega)} \leq \beta$ from (3.4). However this contradicts the fact that $\langle \phi, w \rangle = \beta \|w\|_{L^\infty(\Omega)}$ also from (3.4). Note also that from (3.2) we have that $\{u > f\} \cup \{u < f\} \subseteq \{\phi \neq 0\}$ up to null sets. Thus, the last statement of the proposition follows from the fact that whenever $u > f$ or $u < f$ then $u' = w$ there. This last fact can be shown exactly as in the corresponding TGV problems, see [4, Prop. 4.2]. □

Piecewise affinity is typically a characteristic of higher order regularisation models, e.g. TGV. Indeed, as the next proposition shows, TGV and TVL^∞ regularisation coincide in some simple special cases.

Proposition 4. *The one dimensional functionals* $TGV^2_{\alpha,\beta}$ *and* $TVL^\infty_{\alpha,2\beta}$ *coincide in the class of those* BV *functions* u, *for which an optimal* w *in both definitions of* TGV *and* TVL^∞ *is odd and monotone.*

Proof. Note first that for every odd and monotone bounded function w we have $\|Dw\|_{\mathcal{M}} = 2\|w\|_{L^\infty(\Omega)}$ and denote this set of functions by $\mathcal{A} \subseteq BV(\Omega)$. For a BV function u as in the statement of the proposition we have

$$TGV^2_{\alpha,\beta}(u) = \underset{w \in \mathcal{A}}{\operatorname{argmin}} \, \alpha \|Du - w\|_{\mathcal{M}} + \beta \|Dw\|_{\mathcal{M}}$$

$$= \underset{w \in \mathcal{A}}{\operatorname{argmin}} \, \alpha \|Du - w\|_{\mathcal{M}} + 2\beta \|w\|_{L^\infty(\Omega)} = TVL^\infty_{\alpha,2\beta}(u).$$

□

Exact Solutions: We present exact solutions for the minimisation problem (3.1), for two simple functions $f, g : (-L, L) \to \mathbb{R}$ as data, where $f(x) = h\mathcal{X}_{(0,L)}(x)$ and $g(x) = f(x) + \lambda x$, with $\lambda, h > 0$. Here $\mathcal{X}_C(x) = 1$ for $x \in C$ and 0 otherwise.

With the help of the optimality conditions (3.2)–(3.4) we are able to compute all possible solutions of (3.1) for data f and g and for all values of α and β. These solutions are depicted in Fig. 1. Every coloured region corresponds to a different type of solution. Note that there are regions where the solutions coincide with the corresponding solutions of TV minimisation, see the blue and red regions in Fig. 1a and the blue, purple and red regions in Fig. 1b. This is not surprising since as it is shown in [5] for all dimensions, $TVL^\infty_{\alpha,\beta} = \alpha TV$, whenever $\beta/\alpha \geq |\Omega|^{1/q}$ with $1/p + 1/q = 1$ and $p \in (1, \infty]$. Notice also the presence of affine structures in all solutions, predicted by Proposition 3. For demonstration purposes, we present the computation of the exact solution that corresponds to the yellow region of Fig. 1b and refer to [9] for the rest.

Since we require a piecewise affine solution, from symmetry and (3.2) we have that $\phi(x) = (c_1 - \lambda)\frac{x^2}{2} - c_2|x| + c_3$. Since we require u to have a discontinuity at 0, (3.3) implies $\phi(0) = a$ while from the fact that $\phi \in H^1_0(\Omega)$ and from (3.4) we must have $\phi(-L) = 0$ and $\langle \phi, w \rangle = \beta \|w\|_{L^\infty(\Omega)}$. These conditions give

$$c_1 = \frac{6(\alpha L - \beta)}{L^3} + \lambda, \quad c_2 = \frac{4\alpha L - 3\beta}{L^2}, \quad c_3 = \alpha.$$

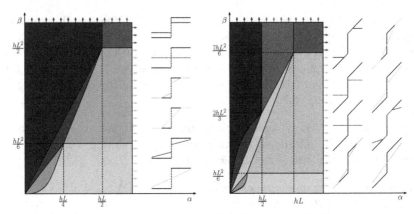

(a) Piecewise constant data function f (b) Piecewise affine data function g

Fig. 1. Exact solutions for the L^2–$TVL^\infty_{\alpha,\beta}$ one dimensional denoising problem (Color figure online)

We also have $c_1 = u' = w$ and thus we require $c_1 > 0$. Since we have a jump at $x = 0$ we also require $g(0) < u(0) < h$, i.e., $0 < c_2 < \frac{h}{2}$ and $u(-L) > g(-L)$ i.e., $\phi'(-L) > 0$. These last inequalities are translated to the following conditions

$$\left\{\beta < \alpha L + \frac{\lambda L^3}{6},\ \beta > \frac{4\alpha L}{3} - \frac{hL^2}{6},\ \beta > \frac{2\alpha L}{3},\ \beta < \frac{4\alpha L}{3}\right\},$$

which define the yellow area in Fig. 1b. We can easily compute u now:

$$u(x) = \begin{cases} \left(\frac{6(\alpha L - \beta)}{L^3} + \lambda\right)x + h - \frac{4\alpha L - 3\beta}{L^2}, & x \in (0, L), \\ \left(\frac{6(\alpha L - \beta)}{L^3} + \lambda\right)x + \frac{4\alpha L - 3\beta}{L^2}, & x \in (-L, 0). \end{cases}$$

Observe that when $\beta = \alpha L$, apart from the discontinuity, we can also recover the slope of the data $g' = \lambda$, something that neither TV nor TGV regularisation can give for this example, see [8, Sect. 5.2].

4   Numerical Experiments

In this section we present our numerical experiments for the discretised version of L^2–$TVL^\infty_{\alpha,\beta}$ denoising. We solve (3.1) using the split Bregman algorithm, see [9, Chap. 4] for more details.

First we present some one dimensional examples that verify numerically our analytical results. Figure 2a shows the TVL^∞ result for the function $g(x) = h\mathcal{X}_{(0,L)}(x) + \lambda x$ where α, β belong to the yellow region of Fig. 1b. Note that the numerical and the analytical results coincide. We have also computed the TGV solution where the parameters are selected so that $\|f - u_{TGV}\|_2 = \|f - u_{TVL^\infty}\|_2$. Figure 2b shows a numerical verification of Proposition 4. There, the

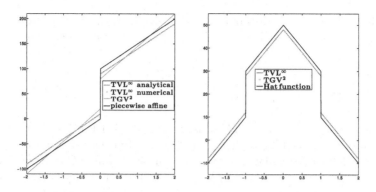

Fig. 2. One dimensional numerical examples

TVL$^\infty$ parameters are $\alpha = 2$ and $\beta = 4$, while the TGV ones are $\alpha = 2$ and $\beta = 2$. Both solutions coincide since they satisfy the symmetry properties of the proposition.

We proceed now to two dimensional examples, starting from Fig. 3. There we used a synthetic image corrupted by additive Gaussian noise of $\sigma = 0.01$, cf. Figs. 3a and b. We observe that TVL$^\infty$ denoises the image in a staircasing–free way in Fig. 3c. In order to do so however, one has to use large values of α and β

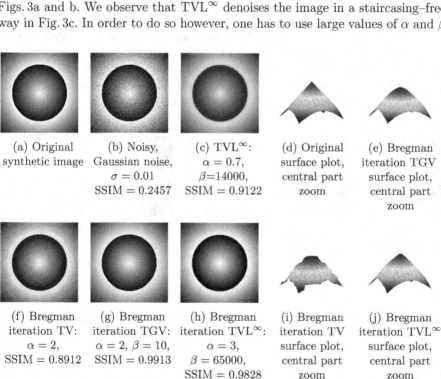

(a) Original synthetic image

(b) Noisy, Gaussian noise, $\sigma = 0.01$ SSIM = 0.2457

(c) TVL$^\infty$: $\alpha = 0.7$, β=14000, SSIM = 0.9122

(d) Original surface plot, central part zoom

(e) Bregman iteration TGV surface plot, central part zoom

(f) Bregman iteration TV: $\alpha = 2$, SSIM = 0.8912

(g) Bregman iteration TGV: $\alpha = 2$, $\beta = 10$, SSIM = 0.9913

(h) Bregman iteration TVL$^\infty$: $\alpha = 3$, $\beta = 65000$, SSIM = 0.9828

(i) Bregman iteration TV surface plot, central part zoom

(j) Bregman iteration TVL$^\infty$ surface plot, central part zoom

Fig. 3. TVL$^\infty$ based denoising and comparison with the corresponding TV and TGV results. All the parameters have been optimised for best SSIM.

something that leads to a loss of contrast. This can easily be treated by solving the *Bregman iteration* version of L^2–$TVL^\infty_{\alpha,\beta}$ minimisation, that is

$$u^{k+1} = \operatorname*{argmin}_{u,w} \frac{1}{2}\|f - v^k - u\|_2^2 + \alpha\|\nabla u - w\|_1 + \beta\|w\|_\infty,$$
$$v^{k+1} = v^k + f - u^{k+1}. \tag{4.1}$$

Bregman iteration has been widely used to deal with the loss of contrast in these type of regularisation methods, see [2,7] among others. For fair comparison we also employ the Bregman iteration version of TV and TGV regularisations. The Bregman iteration version of TVL^∞ regularisation produces visually a very similar result to the Bregman iteration version of TGV, even though it has a slightly smaller SSIM value, cf. Figs. 3g and h. However, TVL^∞ is able to reconstruct better the sharp spike at the middle of the figure, cf. Figs. 3d, e and j.

The reason for being able to obtain good reconstruction results with this particular example is due to the fact that the modulus of the gradient is essentially constant apart from the jump points. This is favoured by the TVL^∞ regularisation which promotes constant gradients as it is proved rigorously in dimension one in Proposition 3. We expect that a similar analytic result holds in higher dimensions and we leave that for future work. However, this is restrictive when gradients of different magnitude exist, see Fig. 4. There we see that in order to get a staircasing–free result with TVL^∞ we also get a loss of geometric information, Fig. 4b, as the model tries to fit an image with constant gradient, see the middle row profiles in Fig. 4c. While improved results can be achieved with the Bregman iteration version, Fig. 4d, the result is not yet fully satisfactory as an *affine staircasing* is now present in the image, Fig. 4e.

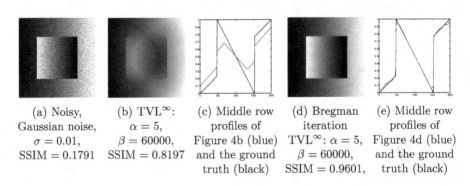

| (a) Noisy, Gaussian noise, $\sigma = 0.01$, SSIM = 0.1791 | (b) TVL^∞: $\alpha = 5$, $\beta = 60000$, SSIM = 0.8197 | (c) Middle row profiles of Figure 4b (blue) and the ground truth (black) | (d) Bregman iteration TVL^∞: $\alpha = 5$, $\beta = 60000$, SSIM = 0.9601 | (e) Middle row profiles of Figure 4d (blue) and the ground truth (black) |

Fig. 4. Illustration of the fact that TVL^∞ regularisation favours gradients of fixed modulus

Spatially Adapted TVL^∞: One way to allow for different gradient values in the reconstruction, or in other words allow the variable w to take different values, is to treat β as a spatially varying parameter, i.e., $\beta = \beta(x)$. This leads to the spatially adapted version of TVL^∞:

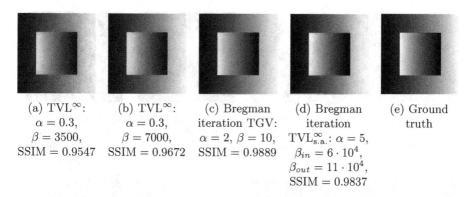

(a) TVL$^\infty$:
$\alpha = 0.3$,
$\beta = 3500$,
SSIM $= 0.9547$

(b) TVL$^\infty$:
$\alpha = 0.3$,
$\beta = 7000$,
SSIM $= 0.9672$

(c) Bregman
iteration TGV:
$\alpha = 2$, $\beta = 10$,
SSIM $= 0.9889$

(d) Bregman
iteration
TVL$^\infty_{\text{s.a.}}$: $\alpha = 5$,
$\beta_{in} = 6 \cdot 10^4$,
$\beta_{out} = 11 \cdot 10^4$,
SSIM $= 0.9837$

(e) Ground
truth

Fig. 5. TVL$^\infty$ reconstructions for different values of β. The best result is obtained by spatially varying β, setting it inversely proportional to the gradient, where we obtain a similar result to the TGV one

$$\text{TVL}^\infty_{\text{s.a.}}(u) = \min_{w \in L^\infty(\Omega)} \alpha \|Du - w\|_{\mathcal{M}} + \|\beta w\|_{L^\infty(\Omega)}. \qquad (4.2)$$

The idea is to choose β large in areas where the gradient is expected to be small and vice versa, see Fig. 5 for a simple illustration. In this example the slope inside the inner square is roughly twice the slope outside. We can achieve a perfect reconstruction inside the square by setting $\beta = 3500$, with artefacts outside, see Fig. 5a and a perfect reconstruction outside by setting the value of β twice as large, i.e., $\beta = 7000$, Fig. 5b. In that case, artefacts appear inside the square. By setting a spatially varying β with a ratio $\beta_{out}/\beta_{in} \simeq 2$ and using the Bregman iteration version, we achieve an almost perfect result, visually very similar to the TGV one, Figs. 5c and d. This example suggests that ideally β should be inversely proportional to the gradient of the ground truth. Since this information is not available in practice we use a pre-filtered version of the noisy image and we set

$$\beta(x) = \frac{c}{|\nabla f_\sigma(x)| + \epsilon}. \qquad (4.3)$$

Here c is a positive constant to be tuned, $\epsilon > 0$ is a small parameter and f_σ denotes a smoothing of the data f with a Gaussian kernel. We have applied the spatially adapted TVL$^\infty$ (non-Bregman) with the rule (4.3) in the natural image "Ladybug" in Fig. 6. There we pre-smooth the noisy data with a discrete Gaussian kernel of $\sigma = 2$, Fig. 6d, and then apply TVL$^\infty_{\text{s.a.}}$ with the rule (4.3), Fig. 6g. Comparing to the best TGV result in Fig. 6f, the SSIM value is slightly smaller but there is a better recovery of the image details (objective). Let us note here that we do not claim that our rule for choosing β is the optimal one. For demonstration purposes, we show a reconstruction where we have computed β using the gradient of ground truth $u_{\text{g.t.}}$, Fig. 6c, as $\beta(x) = c/(|\nabla u_{\text{g.t.}}(x)| + \epsilon)$, with excellent results, Fig. 6h. This is of course impractical, since the gradient of the ground truth is typically not available but it shows that there is plenty of room for improvement regarding the choice of β. One could also think of

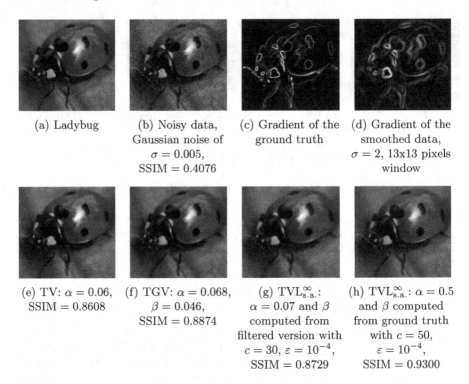

(a) Ladybug

(b) Noisy data, Gaussian noise of $\sigma = 0.005$, SSIM = 0.4076

(c) Gradient of the ground truth

(d) Gradient of the smoothed data, $\sigma = 2$, 13x13 pixels window

(e) TV: $\alpha = 0.06$, SSIM = 0.8608

(f) TGV: $\alpha = 0.068$, $\beta = 0.046$, SSIM = 0.8874

(g) TVL$^\infty_{s.a.}$: $\alpha = 0.07$ and β computed from filtered version with $c = 30$, $\varepsilon = 10^{-4}$, SSIM = 0.8729

(h) TVL$^\infty_{s.a.}$: $\alpha = 0.5$ and β computed from ground truth with $c = 50$, $\varepsilon = 10^{-4}$, SSIM = 0.9300

Fig. 6. Best reconstruction of the "Ladybug" in terms of SSIM using TV, TVG and spatially adapted TVL$^\infty$ regularisation. The β for the latter is computed both from the filtered (Fig. 6g) and the ground truth image (Fig. 6h)

reconstruction tasks where a good quality version of the gradient of the image is available, along with a noisy version of the image itself. Since the purpose of the present paper is to demonstrate the capabilities of the TVL$^\infty$ regulariser, we leave that for future work.

Acknowledgments. The authors acknowledge support of the Royal Society International Exchange Award Nr. IE110314. This work is further supported by the King Abdullah University for Science and Technology (KAUST) Award No. KUK-I1-007-43, the EPSRC first grant Nr. EP/J009539/1 and the EPSRC grant Nr. EP/M00483X/1. MB acknowledges further support by ERC via Grant EU FP 7-ERC Consolidator Grant 615216 LifeInverse. KP acknowledges the financial support of EPSRC and the Alexander von Humboldt Foundation while in UK and Germany respectively. EP acknowledges support by Jesus College, Cambridge and Embiricos Trust.

References

1. Ambrosio, L., Fusco, N., Pallara, D.: Functions of Bounded Variation and Free Discontinuity Problems. Oxford University Press, USA (2000)

2. Benning, M., Brune, C., Burger, M., Müller, J.: Higher-order TV methods - enhancement via Bregman iteration. J. Sci. Comput. **54**(2–3), 269–310 (2013)
3. Bredies, K., Kunisch, K., Pock, T.: Total generalized variation. SIAM J. Imaging Sci. **3**(3), 492–526 (2010)
4. Bredies, K., Kunisch, K., Valkonen, T.: Properties of L^1-TGV^2: the one-dimensional case. J. Math. Anal. Appl. **398**(1), 438–454 (2013)
5. Burger, M., Papafitsoros, K., Papoutsellis, E., Schönlieb, C.-B.: Infimal convolution regularisation functionals of BV and L^p spaces. Part I: the finite p case. arXiv:1504.01956 (submitted)
6. Huber, P.: Robust regression: asymptotics, conjectures and monte carlo. Ann. Stat. **1**, 799–821 (1973)
7. Osher, S., Burger, M., Goldfarb, D., Xu, J., Yin, W.: An iterative regularization method for total variation-based image restoration. Multiscale Model. Simul. **4**, 460–489 (2005)
8. Papafitsoros, K., Bredies, K.: A study of the one dimensional total generalised variation regularisation problem. Inverse Prob. Imaging **9**(2), 511–550 (2015)
9. Papoutsellis, E.: First-order gradient regularisation methods for image restoration. Reconstruction of tomographic images with thin structures and denoising piecewise affine images. Ph.D. thesis. University of Cambridge (2015)
10. Rudin, L., Osher, S., Fatemi, E.: Nonlinear total variation based noise removal algorithms. Phys. D **60**(1–4), 259–268 (1992)

Successive Approximation of Nonlinear Confidence Regions (SANCR)

Thomas Carraro$^{(\boxtimes)}$ and Vladislav Olkhovskiy

Institute for Applied Mathematics, Heidelberg University,
Im Neuenheimer Feld 205, 69120 Heidelberg, Germany
thomas.carraro@iwr.uni-heidelberg.de

Abstract. In parameter estimation problems an important issue is the approximation of the confidence region of the estimated parameters. Especially for models based on differential equations, the needed computational costs require particular attention. For this reason, in many cases only linearized confidence regions are used. However, despite the low computational cost of the linearized confidence regions, their accuracy is often limited. To combine high accuracy and low computational costs, we have developed a method that uses only successive linearizations in the vicinity of an estimator. To accelerate the process, a principal axis decomposition of the covariance matrix of the parameters is employed. A numerical example illustrates the method.

Keywords: Parameter estimation · Nonlinear confidence region · Covariance matrix · Differential equations

1 Introduction

To simplify the notation, we consider a nonlinear model $f(t, \theta)$, with $\theta \in \mathbb{R}^n$ and $t \in \mathbb{R}$, which does not depend on an additional (dynamical) system. We assume that f is differentiable with respect to θ and continuous with respect to t.

We consider the approximation of a confidence region about parameter values estimated by nonlinear least squares. The parameters are estimated by using experimental data y_i in some given points t_i with $i = 1, \ldots, m$. The observed values contain unknown errors e_i that we assume additive, so the response variable can be modeled by

$$y_i = f(t_i, \theta_{\text{true}}) + e_i, \tag{1}$$

where θ_{true} is the unknown true value of the parameters. Therefore, the least squares estimator $\hat{\theta}$ is the value that solves the following problem

$$\hat{\theta} = \operatorname{argmin} \frac{1}{2} S(\theta), \tag{2}$$

where $S(\theta)$ is the residual sum of squares

© IFIP International Federation for Information Processing 2016
Published by Springer International Publishing AG 2016. All Rights Reserved
L. Bociu et al. (Eds.): CSMO 2015, IFIP AICT 494, pp. 180–188, 2016.
DOI: 10.1007/978-3-319-55795-3_16

$$S(\theta) = \sum_{i=1}^{m} (y_i - f(t_i, \theta))^2. \tag{3}$$

We assume that the model is correct and that the errors are normal, independent and identically distributed (iid) random variables with zero mean and variance σ^2, i.e. $e_i \sim N(0, \sigma^2)$.

The confidence regions are here interpreted (from the frequentistic perspective [14]) as the regions in the parameter space covering the true value of the parameters θ_{true}, in large samples, with probability approximately $1 - \alpha$.

The use of linearized confidence regions with nonlinear algebraic models has been extensively treated in literature, see for example [1,2,6,8,11,16]. In particular, it has been shown that confidence regions derived for the linear case can be used in linearized form also for nonlinear models, but in many cases with limited accuracy [18]. Furthermore, there are approximation techniques for nonlinear models that are not based on linearizations [3,10,17,19].

To simplify the exposition, in this work we consider an algebraic model, but the method can be used for more complex models. In fact, the problem to approximate nonlinear confidence regions for implicit models, i.e. models based on a system of (differential) equations has been considered from different points of view and for different kind of applications by several authors. To cite only few of them, see the work [18] and the references therein for the design under uncertainty, [20] for an application to ground water flow, [13] for ecological systems, and [15] for additional examples. Newly, it has been presented a method based on second-order sensitivity for the approximation of nonlinear confidence regions applied to ODE based models [12]. It has been shown that higher order sensitivities give a higher accurate approximation of the confidence regions than methods using only the first order sensitivities.

With this work we show that the approximation using only linearized confidence regions can be substantially improved by a systematic successive application of linearizations, in the following called *Successive Approximation of Nonlinear Confidence Regions (SANCR)* method. We show results for the case with only two model parameters. An extension to more than two parameters is technically straightforward and could be partially parallelized, but the effect of successive linearizations in more than two (parameter space) dimensions has yet to be studied in this framework.

This paper is organized as follows (i) In Sect. 2 we report the two methods on which our approach is based; (ii) In Sect. 3 we describe the new method; (iii) In Sect. 4 we show a numerical realization of the SANCR method.

2 Linearized Confidence Region and Likelihood Ratio Test

As explained above, there are several methods to approximate (nonlinear) confidence regions. Our method is based on the following two approaches [19].

For a given estimator $\hat{\theta}$ of the parameter θ, we consider:

(i) The method derived from the likelihood ratio test (LR)

$$-2 \, log\big(L(\theta)/L(\hat{\theta})\big) \leq \gamma^2, \tag{4}$$

from which it follows

$$S(\theta) - S(\hat{\theta}) \leq \gamma^2. \tag{5}$$

where L is the likelihood function and γ^2 is the confidence level.

(ii) The method based on the Wald test that leads to the linearized confidence regions (CL):

$$(\theta - \hat{\theta})^T Cov^{-1}(\theta - \hat{\theta}) \leq \gamma^2, \tag{6}$$

where Cov is the estimated covariance matrix of the parameters. There are several approximations of Cov [18], we use the one based on the Jacobian J of f:

$$Cov = s^2 (J^T J)^{-1}, \tag{7}$$

where

$$J_{i,j} = \frac{\partial f(t_i, \theta)}{\partial \theta_j}. \tag{8}$$

The level $\gamma^2 = \chi^2_{1-\alpha,n}$ is given by the $1 - \alpha$ percentile of the chi-square distribution with n degrees of freedom in case σ^2 is known, and it is $\gamma^2 = s^2 n F_{(1-\alpha,n,m-n)}$ in case σ^2 is unknown, but approximated by $s^2 = S(\hat{\theta})/(m-n)$. It has been proved [7] that these two confidence regions are asymptotically equivalent, but far from the asymptotic behavior, i.e. in case of a small number of data, they perform differently as presented in [18]. Additionally, our method show the limitation of linearized confidence regions based only on (6).

One of the major goals in defining the confidence regions is the reduction of the costs associated to their computation. From the perspective of the computational costs, the method CL is cheap since it needs only one evaluation of the covariance matrix at the parameter value $\hat{\theta}$, while the method LR is much more expensive because it is based on the evaluation of the functional S in an adequately high number of points θ in the vicinity of $\hat{\theta}$ to produce a contour. In addition, the extension of the confidence region is not known a priori. In practice, the number of function evaluations needed for the method LR is in the order of several thousands, for example in our case with two parameters we use a grid of 10^4 points for the method LR.

On the contrary, as indicated in the expression (7), the covariance matrix can be evaluated at the cost of building the Jacobian J. Therefore, the major computational costs for the method CL are given by the computation of the derivatives of the model f with respect to the parameters. Thus, we have few computations of a linearized model for the method CL while many thousand computations of a nonlinear model are needed for the method LR. Unfortunately, the accuracy of these two methods is inversely related to their computational costs, with the CL method being much more inaccurate if the model is highly nonlinear. We remind that both methods are only asymptotically exact for linear models and their quality decreases far from the asymptotic behavior.

Therefore, a compromise between computational costs and precision is highly required for many practical applications especially in case the model is based on differential equations. To this aim we established a new method combining low computational costs and high accuracy.

3   Successive Linearizations of Nonlinear Confidence Regions

The SANCR method is based on the use of successive linearizations of the confidence region, starting from the estimated parameter value $\hat{\theta}$ (see expression (2)) combined with the likelihood ratio test (5) as explained below examplarily for a model with two parameters.

The likelihood ratio test is used to check whether a point belongs or not to the approximate nonlinear confidence region. Instead of testing all points in the vicinity of $\hat{\theta}$ we use an educated guess, i.e. the likelihood ratio test is performed only on few points lying on the contour of the *linearized* confidence regions. In fact, linearized confidence regions are ellipsoids in the parameter space and the directions of the semi-axis are defined by the eigenvectors of the covariance matrix as can be deduced by the quadratic form (6). Note that the covariance matrix has dimension $n \times n$, where n is the number of parameters to estimate. Therefore, starting from $\hat{\theta}$ we determine the directions of the principal axes and their length which is given by

$$\ell_i = \gamma\sqrt{\lambda_i},$$

where λ_i is the eigenvalue corresponding to the ith eigenvector. We perform the likelihood ratio test for the extreme points of the semi-axes, see points θ_A, θ_B, θ_C, θ_D in Fig. 1.

Let be θ_A the first point to be processed. If this point passes the test, i.e. if the following condition is fulfilled

$$S(\theta_A) - S(\hat{\theta}) \le \gamma^2,$$

it is considered for the construction of the confidence region and the procedure continues along the second axis. On the contrary, if the point θ_A does not pass the test, it is discarded and a new candidate in the same direction $\hat{\theta}\theta_A$ is chosen.

A new point θ'_A along the selected semi-axis is taken by scaling ℓ_1 by a factor $\alpha < 1$ as shown in Fig. 2(a). This procedure is repeated with a new likelihood ratio test and possibly a rescaling (reducing α) until a point that satisfies the test

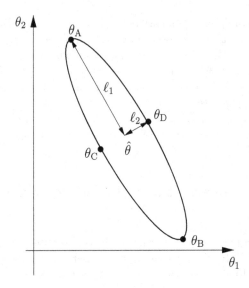

Fig. 1. Definition of the points to perform the likelihood ratio test for two parameters.

$$S(\theta'_A) - S(\hat{\theta}) \leq \gamma^2$$

is found. Once this point, say θ_{new}, has been found, we linearize the confidence region around this new point. To this aim we calculate the Jacobian $J(\theta_{new})$ (see (8)) and the covariance $Cov(\theta_{new})$ (see (7)).

After performing the eigendecomposition of the new covariance matrix, the principal axes might have changed direction due to the nonlinearity of the model, see Fig. 2(b). Following the new principal directions, we can analogously find the next candidate points belonging to the confidence region, i.e. the points $\theta_{new,A}, \theta_{new,C}$ and $\theta_{new,D}$, see Fig. 2(b). The point $\theta_{new,B}$ is not considered because it is the opposite extremal point of the same principal axis. In fact, instead of taking $\theta_{new,B}$, we perform the same procedure starting from θ_B to approximate the confidence region in the direction $\hat{\theta}\theta_B$. Therefore, this procedure is repeated along all principal axes considering both directions.

Stopping Criterion. The search along one principal axis is stopped if the distance of the next accepted point, let's say $\theta'_{new,A}$, to the previous one is less than a given tolerance

$$|\theta'_{new,A} - \theta_{new,A}| < TOL, \tag{9}$$

then the point $\theta_{new,A}$ is retained to define the nonlinear confidence region, see Fig. 3.

Contour Approximation. The countour of the nonlinear confidence region is approximated by connecting all retained points, in our case $\theta_{new,A}, \theta_{new,C}, \theta_{new,D}, \theta_C, \theta_D$ and θ_B. These points are linearly connected as shown in Fig. 3(b).

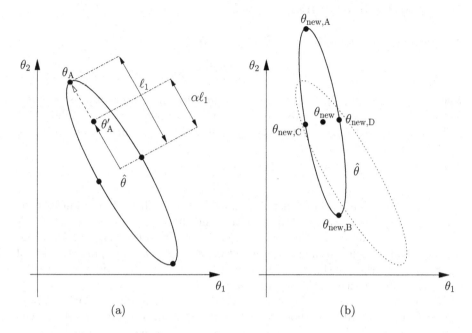

Fig. 2. Scaling the semi-axes (a) and linearize at the new point (b).

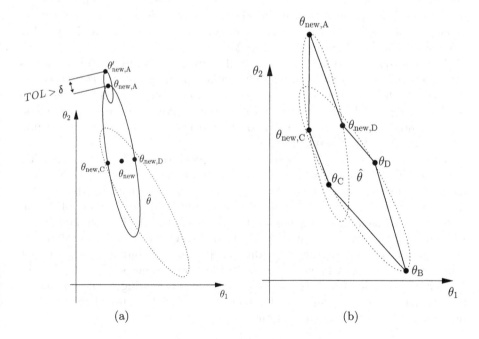

Fig. 3. Stopping criterion (a) and interpolation (b).

4   Numerical Results

As an example the following model is considered

$$y = \theta_1 t^{\theta_2},$$

where the parameter θ_1 and θ_2 are estimated by the nonlinear least squares method. To simulate the parameter estimation process we have applied perturbed data generated using the "true" values of the parameters according to the following model response:

$$y_i = f(t_i, \theta_{\text{true}}) + e_i, \tag{11}$$

where e_i is a random variable distributed as $N(0, \sigma^2)$. The Table 1 indicates the values θ_{true} and σ^2 used in the calculations, and the least squares estimated values $\hat{\theta}$ found by minimizing $S(\theta)$ for a realization of the observations y_i. Additionally, the Table 2 includes the measurement positions t_i. One stopping criterion of the SANCR method is that the distance of two successive candidates is smaller than a given tolerance TOL, see (9). We have used $TOL = 0.15$.

To evaluate the results of our approach we compare it with a Markov Chain Monte Carlo (MCMC) method described in [9] using the associated MCMC toolbox for Matlab. In fact, an alternative way to perform a statistical analysis of nonlinear models is the use of the Bayes's theorem [4]. Bayesian inference is not the focus of our work, therefore we refer for example to [5] for a presentation of the Bayesian approach. Since the MCMC method does not allow to easily define a stopping criterion to assure convergence, we have set to $5 \cdot 10^6$ the number of model evaluations in the MCMC code.

In Fig. 4 the approximations of the confidence region using the four methods can be qualitatively compared. The blue dots (for the colors see the electronic version) are the points of the MCMC method. The cyan ellipse is the linearized confidence region of the method CL. The green curve is the confidence region approximated by the method LR and the red curve is the confidence region approximated by the SANCR method.

One can observe that the linearized confidence region CL is much smaller than the MCMC approximation and that it is not centered in it. The SANCR method is an approximation of the confidence region defined by the method LR obtained at a much lower computational cost than the method LR itself. The computational costs are reported in Tables 3 and 4. The method CL is very cheap with only one evaluation of the nonlinear model and the evaluations of the sensitivities with respect to the two parameters, but its quality is not satisfactory. The SANCR method uses 59 function evaluations and 42 ellipses. The latter correspond to 84 sensitivity evaluations according to the number of two parameters. The LR and the MCMC methods have been used here with 10^4, respectively $5 \cdot 10^6$, model evaluations.

Table 1. Parameters and variance

σ^2	θ_{true}	$\hat{\theta}$
0.1	(0.725, 4)	(0.7279, 3.9974)

Table 2. Position of measurement points

x_1	x_2	x_3	x_4	x_5	x_6
1.309	1.471	1.490	1.565	1.611	1.680

Table 3. Model evaluations of the four methods

SANCR	CL	LR	MCMC
59	1	10^4	$5 \cdot 10^6$

Table 4. Derivatives computations of the four methods

SANCR	CL	LR	MCMC
84	2	0	0

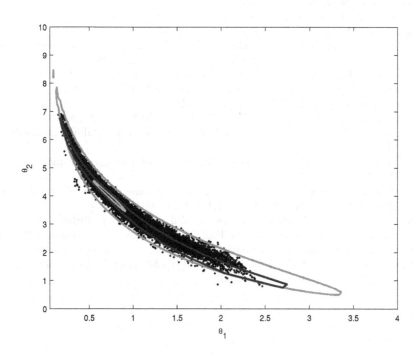

Fig. 4. Confidence region approximated by the four methods.

References

1. Bard, Y.: Nonlinear Parameter Estimation. Academic Press, a subsidiary of Harcourt Brace Jovanovich, Publishers, New York (1974)
2. Bates, D.M., Watts, D.G.: Relative curvature measures of nonlinearity. J. Royal Stat. Soc. Ser. B **42**, 1–25 (1980)
3. Bates, D.M., Watts, D.G.: Nonlinear regression analysis and its applications. Wiley series in probability and mathematical statistics. Applied probability and statistics. Wiley, Chichester (1988)
4. Box, G.E.P., Tiao, G.C.: Bayesian Inference in Statistical Analysis. Wiley, New York (1973)
5. Conrad, P., Marzouk, Y.M., Pillai, N., Smith, A.: Accelerating asymptotically exact MCMC for computationally intensive models via local approximations. Journal of the American Statistical Association (2015, in press)
6. Cook, R., Witmer, J.: A note on parameter-effects curvature. J. Am. Stat. Assoc. **80**, 872–878 (1985)
7. Cox, D., Hinkley, D.: Theoretical Statistics. Chapman and Hall, London (1974)
8. Gallant, A.R.: Confidence regions for the parameters of a nonlinear regression model. Mimeograph Series No. 1077, pp. 633–643. North Carolina University, Institute of Statistics (1976)
9. Haario, H., Laine, M., Mira, A., Saksman, E.: Dram: efficient adaptive MCMC. Stat. Comput. **16**(4), 339–354 (2006)
10. Hamilton, D.C., Watts, D.G., Bates, D.M.: Accounting for intrinsic nonlinearity in nonlinear regression parameter inference regions. Ann. Stat. **10**(2), 386–393 (1982)
11. Jennrich, R.I.: Asymptotic properties of non-linear least squares estimators. Ann. Math. Stat. **40**(2), 633–643 (1969)
12. Kostina, E., Nattermann, M.: Second-order sensitivity analysis of parameter estimation problems. Int. J. Uncertain. Quantif. **5**(3), 209–231 (2015)
13. Marsili-Libelli, S., Guerrizio, S., Checchi, N.: Confidence regions of estimated parameters for ecological systems. Ecol. Model. **165**(23), 127–146 (2003)
14. Meeker, W.Q., Escobar, L.A.: Teaching about approximate confidence regions based on maximum likelihood estimation. Am. Stat. **49**(1), 48–53 (1995)
15. Oberkampf, W.L., Barone, M.F.: Measures of agreement between computation and experiment: validation metrics. J. Comput. Phys. **217**(1), 5–36 (2006)
16. Pázman, A.: Nonlinear Statistical Models. Kluwer Academic Publishers, Dordrecht (1994)
17. Potocký, R., To, V.B.: Confidence regions in nonlinear regression models. Appl. Math. **37**(1), 29–39 (1992)
18. Rooney, W.C., Biegler, L.T.: Design for model parameter uncertainty using nonlinear confidence regions. AIChE J. **47**(8), 1794–1804 (2001)
19. Seber, G., Wild, C.: Nonlinear Regression. Wiley, New York (1989)
20. Vugrin, K.W., Swiler, L.P., Roberts, R.M., Stucky-Mack, N.J., Sullivan, S.P.: Confidence region estimation techniques for nonlinear regression in groundwater flow: three case studies. Water Resour. Res. **43**(3) (2007). http://onlinelibrary.wiley.com/doi/10.1029/2005WR004804/full

Two-Phase Multi-criteria Server Selection for Lightweight Video Distribution Systems

Octavian Catrina[1]([⊠]), Eugen Borcoci[1], and Piotr Krawiec[2]

[1] University Politehnica of Bucharest, Bucharest, Romania
octavian.catrina@elcom.pub.ro
[2] Warsaw University of Technology and National Institute of Telecommunications,
Warsaw, Poland

Abstract. Video streaming services need a server selection algorithm that allocates efficiently network and server resources. Solving this optimization problem requires information about current resources. A video streaming system that relies entirely on the service provider for this task needs an expensive monitoring infrastructure. In this paper, we consider a two-phase approach that reduces the monitoring requirements by involving the clients in the selection process: the provider recommends several servers based on limited information about the system's resources, and the clients make the final decision, using information obtained by interacting with these servers. We implemented these selection methods in a simulator and compared their performance. The results show that the two-phase selection is effective, improving substantially the performance of lightweight service providers, with limited monitoring capabilities.

Keywords: Content networks · Video streaming · Server selection · Multi-criteria decision algorithms

1 Introduction

An essential task of a video streaming service provider is to select a suitable content server. This task can be formalized as a multi-criteria optimization problem, that takes into account various static and dynamic attributes of the system's components. The goal is to allocate efficiently the network and server resources, while providing a suitable Quality of Experience (QoE) to the service users.

These problems are NP-complete, in general, but practical server selection algorithms are available. As these algorithms need information about network and server resources, a service provider has to deploy a complex monitoring infrastructure in order to use them. Lightweight service providers, without monitoring capabilities, can only make suboptimal decisions.

In this paper, we consider a more flexible approach, that splits the selection process between the service provider and the clients. The provider recommends a short list of content servers, using limited information (e.g., the lengths of

© IFIP International Federation for Information Processing 2016
Published by Springer International Publishing AG 2016. All Rights Reserved
L. Bociu et al. (Eds.): CSMO 2015, IFIP AICT 494, pp. 189–199, 2016.
DOI: 10.1007/978-3-319-55795-3_17

the paths between servers and clients). The client refines the selection using additional information, obtained by interacting directly with these servers. The provider and the clients can use various combinations of algorithms, corresponding to different capabilities to obtain information about the system, and different tradeoffs between performance and complexity. In particular, this approach can improve the performance for lightweight service providers.

The paper has the following structure. Section 2 introduces a generic video streaming model and Sect. 3 describes the server selection methods. We implemented these methods in a simulator, presented in Sect. 4. The results of the simulations are discussed in Sect. 5, followed by a summary in Sect. 6.

2   Video Streaming Framework

We introduce in the following the generic model of online video streaming used in this paper. Figure 1 shows the main actors, the interactions between them, and the infrastructure used to deliver the service. We separate the management of the service, assigned to a service provider (SP) entity, from the delivery of the video content, performed by a collection of content servers (CS). An essential function of the SP is to select a content server when a client initiates a session.

We assume over-the-top delivery, using video streaming over HTTP or a similar technology [6]. The video content is stored on a collection of servers deployed in the Internet, each video file being available from several servers. These servers usually operate in a Content Delivery Network (CDN) [5]. We expect efficient CDN operation, with server placement and content replication adapted to the video streaming traffic. The video streaming system can include a CDN, or use the services of one or more CDN providers [1].

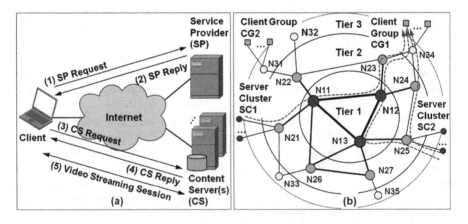

Fig. 1. Video streaming model: (a) Actors and interactions. (b) Infrastructure.

Table 1. Information used by service providers for content server selection.

Static or quasi-static information	Dynamic information
Lists of content servers and video files	Available capacity of the content servers
Mapping of video content to servers and of clients and servers to network nodes	Available capacity of the paths between servers and clients
Length of server-client network paths	Admission control for servers and paths

A video streaming session starts with the following interactions (Fig. 1a):

– The client sends a request to the SP, specifying the desired video content.
– The SP runs a server selection algorithm and replies with a list of servers that could deliver the requested video content (or rejects the request). Table 1 lists the information used by the SP in the server selection process.
– In a single-phase selection process, the client simply asks one of the servers indicated by the provider to deliver the desired video content.
– In a two-phase selection process, the client chooses a server from the provider's list by running its own selection algorithm, with additional information.

For example, if the service is implemented using the DASH standard [6], the provider can answer with a Media Presentation Description (MPD) file, containing information about the encoding of the video content (including available resolutions and data rates) and the URLs of the servers that can deliver it.

We assume that the main bottlenecks of the transport network are the links between the autonomous systems (AS) of the network service providers. Therefore, the network is modeled as a collection of interconnected nodes, corresponding to an AS-level, multi-tier network topology, similar to the Internet topology (Fig. 1b). The clients and the content servers are attached to these network nodes. The Border Gateway Protocol (BGP), responsible for inter-domain routing in the Internet, determines a single best path between ASes, based on network service provider policies and path length. Load balancing on multiple paths would help to allocate more efficiently the network resources to the video streaming sessions, and could be achieved using application-level overlays. We consider, therefore, this additional capability for some of the server selection algorithms.

3   Content Server Selection

Multi-Criteria Decision Algorithm (MCDA). Selection of a server and a path in content delivery systems can be formulated as a multi-criteria optimization problem. These problems are NP-complete in general, and are solved in practice using heuristics methods. Practical and efficient methods have been proposed for server and path selection based on Multi-Criteria Decision Algorithms (MCDA) [3]. The variant of MCDA used in this paper is described below:

- The algorithm determines a set of candidate solutions, $S = \{S_i\}_{i \in 1,...,n}$, from information about the request and the delivery system. A candidate solution is a vector of decision variables, $S_i = (v_{i,j})_{j \in 1,...,m}$, representing characteristics of a server and a path that could deliver the requested content.
- For each candidate solution S_i and variable $v_{i,j}$, the algorithm calculates the component achievement function $R_{i,j} = \frac{r_{i,j} - v_{i,j}}{r_{i,j} - a_{i,j}}$, where $v_{i,j}$ is the value of the variable, while $r_{i,j}$ and $a_{i,j}$ are the reservation level and aspiration level for this variable, respectively (the algorithm can be extended with more complex functions). We use two decision variables: the available (unused) capacity of the server and the available capacity (unused bandwidth) of the path. For these variables, the reservation level is a lower bound for suitable solutions, and the aspiration level is an upper bound beyond which the preference of the solutions is similar. We set $a_{i,j}$ to the maximum capacity of the server and of the path, respectively, and $r_{i,j}$ to the amount of server capacity and path capacity consumed by a session, respectively.
- The algorithm calculates the rank of each solution, $R_i = \min_j\{R_{i,j}\}$, and then determines the index of the best solution, $\text{argmin}_i\{R_i\}$.

Single-Phase Selection. The single-phase selection relies entirely on the service provider to find suitable servers for the clients' sessions. In its simplest variant, the provider answers a client's request by indicating a single server [3]. We consider a more general variant, in which the provider answers with an ordered list of servers. The client can start the session by connecting to one of these servers and use the others as back-up [1]. By delivering a list of servers, the provider enables the client to improve the initial server selection and/or use certain adaptive streaming techniques [2]. The selection proceeds as follows:

- The service provider maintains a resource database with the information about network and content servers that is necessary for server selection. The database contains static information and some of the dynamic information listed in Table 1, depending on the algorithm being used.
- When a service request arrives, the provider determines an ordered list of suitable servers and delivers it to the client. The selection is based on information received in the service request (e.g., client and video ids) and information available in the resource database. If the provider does not find any suitable server the request is rejected.
  Candidate solutions consist of a server that has a copy of the requested video file and a path from that server to the client. The selection algorithm takes as input a set of candidate solutions that (optionally) satisfy additional requirements. The provider can use a range of server selection algorithms, that offer different tradeoffs between the complexity of the system and the optimality of the solution (e.g., different requirements for the information collected by the provider about network and servers). The algorithms used in our simulation experiments are listed in Table 2.
- If the request is accepted, the client starts the video session by connecting to the first server in the list. In case of failure, the client can try the next servers, without having to ask the provider to recommend another server.

Table 2. Server selection algorithms used by the service provider.

Algorithm	Input	Output (list of servers)
Random servers	Set of feasible solutions	Randomly chosen servers; shortest or random path
Closest servers	Set of feasible solutions. Length of the paths from servers to client	Servers with the shortest paths to the client
MCDA	Set of feasible solutions. Available capacity of the servers and the paths	Servers in the solutions with the highest MCDA rank

This approach simplifies the functionality of the client. On the other hand, the service provider has, essentially, two main options: algorithms that use only static information and offer poor performance and algorithms that can achieve much better performance, but require an expensive monitoring infrastructure.

Two-Phase Selection. We consider now a more flexible, two-phase approach, that involves the clients in the selection process:

- The provider's resource database contains only the information about network and servers that is necessary for a preliminary selection: static and quasi-static information (Table 1) and, possibly, dynamic information that is easier to collect (e.g., server load).
- When a request arrives, the service provider makes a preliminary server selection and delivers to the client a short, ordered list of recommended servers. The procedure is essentially the same as for single-phase selection, using algorithms that match the limited information collected by the provider: e.g., random servers or closest servers.
- If the request is accepted, the client performs the second phase of the selection, choosing the server that will be used by the video streaming session. The input of the client's selection algorithm is the list of servers received from the provider and additional information obtained by the client:
  1. Local information: Client capabilities, processor and link load.
  2. Current state of the recommended servers: Reports from the servers, indicating if they can handle the request and/or the available capacity.
  3. Current state of the paths between the servers and the client: Throughput of the connection with each server.
  The client can use MCDA based on available server and path capacity, or choose the least loaded server, depending on the information it can obtain.

Admission Control. The video streaming sessions have to ensure continuous playout. This requires timely data delivery, and hence a certain lower bound for the end-to-end data rate. An important issue, therefore, is to add some form of admission control to the server selection process. This can be achieved by removing the solutions with fully loaded servers or paths from the set of candidate solutions given as input to the selection algorithms.

Admission control is simpler for content servers. The available capacity can be estimated by a local monitoring application and reported by the servers to the provider or the clients. Moreover, a fully loaded server can reject additional requests. This offers basic admission control, independent of the selection process. However, it is more difficult to determine if the path from a server to a client can handle additional sessions. The service providers need for this purpose a network monitoring infrastructure that measures the throughput between network nodes. The clients can measure the throughput of the connections with the recommended servers. These end-to-end measurements are more relevant, but increase the delay and overhead of session establishment. Also, they may fail to detect initial congestion and avoid the QoE degradation of current sessions.

4   Simulation Software

We analyzed the performance of the server selection methods described in Sect. 3 by simulation. The simulation software developed for this purpose is written in C++ and uses the discrete event simulator OMNeT++ [8] as basic simulation engine. Figure 2 shows the main components of the simulation software.

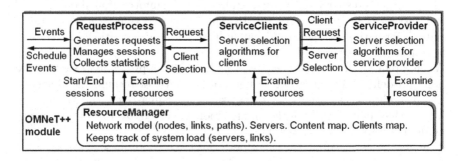

Fig. 2. High-level architecture of the simulation software.

The class ResourceManager provides a resource database with information about the entire system infrastructure: network topology (nodes and links, link capacity and load), network paths, list of content servers (server capacity and load), mapping of video content to servers, mapping of clients and servers to network nodes. The simulations are driven by the RequestProcess, which generates the service requests, manages the video streaming sessions, and collects statistics. When a session starts or ends, the RequestProcess notifies the ResourceManager, which updates accordingly the load of the server and the load of the links on the path from the server to the client.

The server selection process is implemented by the classes ServiceProvider and ServiceClients, and obtains information about system resources from the ResourceManager. The class ServiceProvider handles the clients' service requests and implements the server selection algorithms run by the service provider for single-phase and two-phase selection (Table 2). The class ServiceClients implements the server selection algorithms run by the clients in two-phase selection. The network model uses random topologies generated using the tool aSHIIP [7] and the network nodes are the autonomous systems (AS) of the network service providers (Fig. 1b). The topologies have a multi-tier structure similar to the Internet. We assume that this structure corresponds to the usual business relations between network service providers: peering relations between nodes in the same tier, and customer-provider relations between nodes in different tiers, with the customer in the lower tier. The paths used in the simulations are similar to the BGP paths, according to the relations between network providers [4].

The clients and the content servers are attached to network nodes. Servers are deployed all over the network, in well-connected nodes, close to clients, taking into account the number of links of the node and their bandwidth.

The resources consumed by the video streaming sessions are measured using a metric for server capacity and a metric for link capacity. To simplify the analysis, we assume that a session consumes 1 unit of server capacity and 1 unit of link capacity. The system is provisioned by assigning a fixed maximum capacity to servers and links, which is available for video streaming. The servers have the same capacity and are deployed in clusters; better connected nodes have larger clusters. The network is provisioned by assigning capacity to links according to its hierarchical structure, with higher capacity in upper tiers.

The main performance metric used in the analysis of the server selection process is the success rate, defined as the ratio between the number of successful service requests and the total number of service requests. A request is successful if the system has sufficient server and network resources to deliver the requested video content to the client, with the desired quality, for its entire duration.

Failures can occur in the server selection phase or during the video streaming session. If admission control is available, a request is rejected when the selection process does not find a server and a path with sufficient unused capacity; the load of the system remains unchanged. Otherwise, a new session is created, which may overload the server and/or some links on the path from server to client. The RequestProcess monitors the current sessions and updates their state. If a session has been successful so far and does not have sufficient resources any more, its state changes from success to fail.

5   Performance Evaluation

We discuss in this section the results of simulations with single-phase and two-phase selection, for different selection algorithms. The experiments use a network with 1000 nodes, structured into 4 tiers, and (up to) 3 paths for any pair of nodes. The network is provisioned so that the number of concurrent sessions can reach the total capacity of the servers, 61500 sessions, for a suitable server selection. Service requests are generated at random time intervals with exponential distribution. We measure the session success rate as a function of the request rate.

Table 3. Configurations used for server selection by the service provider.

Name	Server and path selection algorithm	Admission control (AC)
RandRand	Random servers and paths	Server reject
RandShrt	Random servers, shortest paths	Server reject
NearSrv	Servers with shortest paths to client	Server reject
MCDA	MCDA for available server and path capacity	AC for servers and paths

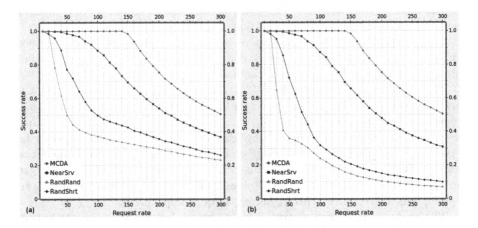

Fig. 3. Single-phase selection: (a) with session abort; (b) without session abort.

Analysis of Single-Phase Selection. Figure 3 shows the success rate for single-phase selection and the configurations in Table 3. Service providers with network monitoring capability can use MCDA and admission control for servers and paths. The success rate of MCDA is equal to 1 for request rates up to 150 requests/sec and about 60000 concurrent sessions, close to the total server capacity (the mean session duration is 400 s). The performance drops drastically if the providers have only static information and select random servers or closest servers, without admission control. The degradation is due to sessions rejected by fully loaded servers and, especially, congestion of network links.

Congested sessions may be aborted due to poor QoE. The freed resources can then be used to start other sessions, which may be successful (e.g., their paths do not include congested links). Figure 3 shows the simulations results when the congested sessions are aborted and when they are allowed to continue. Important performance differences appear for randomly chosen servers and/or paths, because the allocation of network resources is less efficient.

Intuitively, the performance should improve when the load is distributed on multiple network paths. We observe, however, that using the shortest path is better than the random choice of a path (e.g., RandShrt versus RandRand). In a multi-tier network with valley-free paths, the shortest path may avoid the core of the network, while the other paths are more likely to cross it (e.g., the paths between SC2 and CG1 in Fig. 1). Thus, the random choice of the path concentrates the traffic in the network core. Therefore, if multiple paths are available, the choice should take into account the path load (e.g., using MCDA).

For similar reasons, the performance improves when the provider selects the servers with the shortest paths to the client (NearSrv), instead of choosing them randomly (e.g., RandRand, RandShrt): the sessions use less network resources and the paths are more likely to avoid the core. However, the performance of NearSrv selection is very sensitive to server placement.

Analysis of Two-Phase Selection. In the two-phase selection process, the provider and the clients can use various combinations of algorithms. Our main goal is to see the effects of this approach on the performance of the video streaming services offered by lightweight providers. We assume, therefore, that the provider has global, static information about the entire system, and the clients can obtain dynamic information for a particular request and several candidate servers.

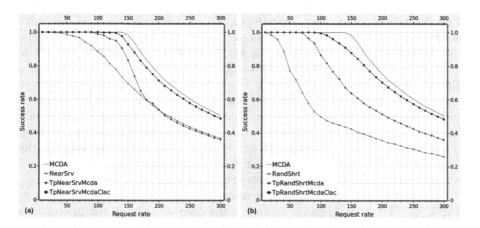

Fig. 4. Two-phase versus single-phase selection for: (a) NearSrv; (b) RandShrt.

Figure 4 shows the simulation results for the following configurations:

- TpNearSrvMcda/TpRandShrtMcda: Provider: NearSrv/RandShrt. Client: MCDA for available server and path capacity. Congested sessions are aborted.
- TpNearSrvMcdaClac/TpRandShrtMcdaClac: Similar to the previous configuration, with client admission control for servers and paths.
- The provider sends a list of 3 recommended servers, out of 8 candidates.
- The results for single-phase selection using NearSrv, RandShrt, and MCDA are added to Fig. 4 to facilitate the performance comparison.

The two-phase selection clearly improves the success rate with respect to single-phase selection when the provider chooses the recommended servers using

NearSrv or RandShrt and the clients use MCDA. NearSrv can achieve better performance than RandShrt, for a suitable placement of the servers in the network, but the performance varies significantly depending on server placement.

However, the two-phase selection cannot reach the performance of the single-phase selection using MCDA (with admission control): the provider runs MCDA with the entire set of candidate solutions, while the client has a short list of servers, chosen without information about system load. The size of the list is a trade-off between system performance and the delay and overhead of the interactions between client and servers, that collect additional information for the final selection. The results suggest that a short list of 3 servers is sufficient.

The performance improves significantly when the clients add admission control to MCDA, becoming close to the performance of a provider that is able to run MCDA itself. The improvement is more important when the provider uses RandShrt, which is more vulnerable to network congestion.

6 Conclusions

A service provider that offers over-the-top video streaming needs an expensive monitoring infrastructure for efficient server selection. Lightweight providers, without such capabilities, can only use simple selection methods, based on static information. Simulation results show a large performance gap between these two approaches. We explored, therefore, a two-phase selection process, that aims at improving the performance for lightweight providers, by involving the clients.

We studied the performance of two-phase selection by simulation. The results show that the approach is effective, improving substantially the success rate. In the ideal case, when the clients can reliably apply admission control, the performance of the two-phase selection becomes close to the performance of a provider that runs MCDA using a network monitoring infrastructure.

Future work will focus on implementing the two-phase selection, using, e.g., DASH [6]. The critical issue is to enable the clients to obtain the information needed for the final selection with minimum delay and overhead, and sufficient accuracy. Ideally, this information could be collected during the initial buffering phase. The DASH standard offers several features that could be exploited for this purpose, e.g., chunks of video content can be downloaded from several servers.

Acknowledgments. This work was partially supported by the Research Project DISEDAN, No.3-CHIST-ERA C3N, 2014–2015.

References

1. Adhikari, V.K., Guo, Y., Hao, F., Varvello, M., Hilt, V., Steiner, M., Zhang, Z.L.: Unreeling netflix: understanding and improving multi-CDN movie delivery. In: Proceedings of 31st IEEE International Conference on Computer Communications (IEEE INFOCOM 2012), pp. 1620–1628. IEEE (2012)

2. Batalla, J.M., Janikowski, S., Krawiec, P., Belter, B., Lopatowski, L., Gutkowski, J.: Dynamic seamless handover of video servers. In: Proceedings of 12th International Conference on Telecommunications (ConTEL 2013), pp. 139–146. IEEE (2013)
3. Beben, A., Batalla, J.M., Chai, W., Sliwinski, J.: Multi-criteria decision algorithms for efficient content delivery in content networks. Ann. Telecommun. (Annales des telecommunications) **68**(3–4), 1–13 (2013)
4. Gao, L.: On inferring autonomous system relationships in the internet. IEEE/ACM Trans. Netw. **9**(6), 733–745 (2001)
5. Hofmann, M., Beaumont, L.: Content Networking: Architecture, Protocols, and Practice. Morgan Kaufmann, Burlington (2005)
6. Sodagar, I.: The MPEG-DASH standard for multimedia streaming over the internet. IEEE MultiMed. **18**(4), 62–67 (2011)
7. Tommasik, J., Weisse, M.A.: The inter-domain hierarchy in measured and randomly generated AS-level topologies. In: Proceedings of IEEE International Conference on Communications (ICC 2012), pp. 1448–1453. IEEE (2012)
8. Wehrle, K., Gross, J., Gnes, M.: Modeling and Tools for Network Simulation. Springer, Heidelberg (2010)

Hamilton-Jacobi-Bellman Equation for a Time-Optimal Control Problem in the Space of Probability Measures

Giulia Cavagnari[1], Antonio Marigonda[2($\boxtimes$)], and Giandomenico Orlandi[2]

[1] Department of Mathematics, University of Trento,
Via Sommarive 14, 38123 Povo (TN), Italy
giulia.cavagnari@unitn.it
[2] Department of Computer Sciences, University of Verona,
Strada Le Grazie 15, 37134 Verona, Italy
{antonio.marigonda,giandomenico.orlandi}@univr.it

Abstract. In this paper we formulate a time-optimal control problem in the space of probability measures endowed with the Wasserstein metric as a natural generalization of the correspondent classical problem in $\mathbb{R}^d$ where the controlled dynamics is given by a differential inclusion. The main motivation is to model situations in which we have only a probabilistic knowledge of the initial state. In particular we prove first a Dynamic Programming Principle and then we give an Hamilton-Jacobi-Bellman equation in the space of probability measures which is solved by a generalization of the minimum time function in a suitable viscosity sense.

Keywords: Optimal transport · Differential inclusions · Time optimal control

1 Introduction

The controlled dynamics of a classical time-optimal control problem in finite-dimension can be presented by mean of a differential inclusion as follows:

$$\begin{cases} \dot{x}(t) \in F(x(t)), & \text{for a.e. } t > 0, \\ x(0) = x_0 \in \mathbb{R}^d, \end{cases} \tag{1}$$

where F is a set-valued map from $\mathbb{R}^d$ to $\mathbb{R}^d$. The problem in this setting is to minimize the time needed to steer x_0 to a given closed *target set* $S \subseteq \mathbb{R}^d$, $S \neq \emptyset$, defining the *minimum time function* $T : \mathbb{R}^d \to [0, +\infty]$ by

$$T(x_0) := \inf\{T > 0 : \exists x(\cdot) \text{ solving (1) such that } x(T) \in S\}. \tag{2}$$

Acknowledgements. The first two authors have been supported by INdAM - GNAMPA Project 2015: *Set-valued Analysis and Optimal Transportation Theory Methods in Deterministic and Stochastics Models of Financial Markets with Transaction Costs.*

L. Bociu et al. (Eds.): CSMO 2015, IFIP AICT 494, pp. 200–208, 2016.
DOI: 10.1007/978-3-319-55795-3_18

The main motivation of this work is to model situations in which the knowledge of the starting position x_0 is only probabilistic (for example in the case of affection by noise) and this can happen even if the evolution of the system is deterministic.

We thus consider as state space the space of Borel probability measures with finite p-*moment* endowed with the p-*Wasserstein metric* $W_p(\cdot,\cdot)$, $(\mathscr{P}_p(\mathbb{R}^d), W_p)$. In [2] the reader can find a detailed treatment about Wasserstein distance.

Following this idea we choose to describe the initial state by a probability measure $\mu_0 \in \mathscr{P}_p(\mathbb{R}^d)$ and for its evolution in time we take a time-depending probability measure on $\mathbb{R}^d$, $\boldsymbol{\mu} := \{\mu_t\}_{t\in[0,T]} \subseteq \mathscr{P}_p(\mathbb{R}^d)$, $\mu_{|t=0} = \mu_0$. In order to preserve the total mass $\mu_0(\mathbb{R}^d)$ during the evolution, the process will be described by a (controlled) *continuity* equation

$$\begin{cases} \partial_t \mu_t + \operatorname{div}(v_t \mu_t) = 0, & 0 < t < T, \\ \mu_{|t=0} = \mu_0, \end{cases} \tag{3}$$

where the time-depending Borel velocity field $v_t : \mathbb{R}^d \to \mathbb{R}^d$ has to be chosen in the set of $L^2_{\mu_t}$-selections of F in order to respect also the classical underlying control problem (1) which is the characteristic system of (3) in the smooth case.

It is well known that if $v_t(\cdot)$ is sufficiently regular then the solution of the continuity equation is characterized by the push-forward of μ_0 through the unique solution of the characteristic system.

In Theorem 8.2.1 in [2] and Theorem 5.8 in [4], the so called *Superposition Principle* states that, if we conversely require much milder assumptions on v_t, the solution μ_t of the continuity equation can be characterized by the push-forward $e_t \sharp \eta$, where $e_t : \mathbb{R}^d \times \Gamma_T \to \mathbb{R}^d, (x, \gamma) \mapsto \gamma(t)$, $\Gamma_T := C^0([0,T]; \mathbb{R}^d)$ and η is a probability measure in the infinite-dimensional space $\mathbb{R}^d \times \Gamma_T$ concentrated on those pairs $(x, \gamma) \in \mathbb{R}^d \times \Gamma_T$ such that γ is an integral solution of the underlying characteristic system, i.e. of an ODE of the form $\dot{\gamma}(t) = v_t(\gamma(t))$, with $\gamma(0) = x$. We refer the reader to the survey [1] and the references therein for a deep analysis of this approach that is at the basis of the present work.

Pursuing the goal of facing control systems involving measures, we define a generalization of the target set S by duality. We consider an observer that is interested in measuring some quantities $\phi(\cdot) \in \Phi$; the results of this measurements are the average of these quantities w.r.t. the state of the system. The elements of the generalized target set $\tilde{S}_p^\phi$ are the states for which the results of all these measurements are below a fixed threshold.

Once defined the admissible trajectories in this framework, the definition of the generalized minimum time function follows in a straightforward way the classical one.

Since classical minimum time function can be characterized as unique viscosity solution of a Hamilton-Jacobi-Bellman equation, the problem to study a similar formulation for the generalized setting would be quite interesting. Several authors have treated a similar problem in the space of probability measures or in a general metric space, giving different definitions of sub-/super differentials and viscosity solutions (see e.g. [2,3,8–10]). For example, the theory presented in [10] is quite complete: indeed there are proved also results on time-dependent problems, comparison principles granting uniqueness of the viscosity solutions under very reasonable assumptions.

However, when we consider as metric space the space $\mathscr{P}_2(\mathbb{R}^d)$, we notice that the class of equations that can be solved is quite small: the general structure of metric space of [10] allows only to rely on the metric gradient, while $\mathscr{P}_2(\mathbb{R}^d)$ enjoys a much more richer structure in the tangent space (which is a subset of L^2).

Dealing with the definition of sub-/superdifferential given in [8], the major bond is that the "perturbed" measure is assumed to be of the form $(\mathrm{Id}_{\mathbb{R}^d} + \phi)\,\sharp\mu$ in which a (rescaled) transport plan is used. It is well known that, by Brenier's Theorem, if $\mu \ll \mathscr{L}^d$ in this way we can describe *all* the measures near to μ. However in general this is not true. Thus if the set of admissible trajectories contains curves whose points are not all a.c. w.r.t. Lebesgue measure (as in our case), the definition in [8] cannot be used.

In order to fully exploit the richer structure of the tangent space of $\mathscr{P}_2(\mathbb{R}^d)$, recalling that AC curves in $\mathscr{P}_2(\mathbb{R}^d)$ are characterized to be weak solutions of the continuity equation (Theorem 8.3.1 in [2]), we considered a different definition than the one presented in [8] using the *Superposition Principle*.

The paper is structured as follows: in Sect. 2 we give the definitions of the generalized objects together with the proof of a Dynamic Programming Principle in this setting. In Sect. 3 we focus on the main result of this work, namely we outline a Hamilton-Jacobi-Bellman equation in $\mathscr{P}_2(\mathbb{R}^d)$ and we solve it in a suitable viscosity sense by the generalized minimum time function, assuming some regularity on the velocity field. Finally, in Sect. 4 we illustrate future research lines on the subject.

2 Generalized Minimum Time Function

Definition 1 (Standing Assumptions). *We will say that a set-valued function $F : \mathbb{R}^d \rightrightarrows \mathbb{R}^d$ satisfies the assumption (F_j), $j = 0, 1, 2$ if the following hold true*

(F_0) $F(x) \neq \emptyset$ *is compact and convex for every $x \in \mathbb{R}^d$, moreover $F(\cdot)$ is continuous with respect to the Hausdorff metric, i.e. given $x \in X$, for every $\varepsilon > 0$ there exists $\delta > 0$ such that $|y - x| \leq \delta$ implies $F(y) \subseteq F(x) + B(0, \varepsilon)$ and $F(x) \subseteq F(y) + B(0, \varepsilon)$.*
(F_1) $F(\cdot)$ *has linear growth, i.e. there exist nonnegative constants L_1 and L_2 such that $F(x) \subseteq \overline{B(0, L_1|x| + L_2)}$ for every $x \in \mathbb{R}^d$,*
(F_2) $F(\cdot)$ *is bounded, i.e. there exist $M > 0$ such that $\|y\| \leq M$ for all $x \in \mathbb{R}^d$, $y \in F(x)$.*

Definition 2 (Generalized target). *Let $p \geq 1$, $\Phi \subseteq C^0(\mathbb{R}^d, \mathbb{R})$ such that the following property holds*

(T_E) *there exists $x_0 \in \mathbb{R}^d$ with $\phi(x_0) \leq 0$ for all $\phi \in \Phi$.*

We define the generalized target $\tilde{S}_p^\Phi$ *as follows*

$$\tilde{S}_p^\Phi := \left\{ \mu \in \mathscr{P}_p(\mathbb{R}^d) : \int_{\mathbb{R}^d} \phi(x)\, d\mu(x) \le 0 \text{ for all } \phi \in \Phi \right\}.$$

For an analysis of the properties of the generalized target see [5] or [6] for deeper results.

Definition 3 (Admissible curves). *Let $F : \mathbb{R}^d \rightrightarrows \mathbb{R}^d$ be a set-valued function, $I = [a, b]$ a compact interval of $\mathbb{R}$, $\alpha, \beta \in \mathscr{P}_p(\mathbb{R}^d)$. We say that a Borel family of probability measures $\boldsymbol{\mu} = \{\mu_t\}_{t \in I} \subseteq \mathscr{P}_p(\mathbb{R}^d)$ is an admissible trajectory (curve) defined in I for the system Σ_F joining α and β, if there exists a family of Borel vector fields $v = \{v_t(\cdot)\}_{t \in I}$ such that*

1. $\boldsymbol{\mu}$ *is a narrowly continuous solution in the distributional sense of the continuity equation $\partial_t \mu_t + \operatorname{div}(v_t \mu_t) = 0$, with $\mu_{|t=a} = \alpha$ and $\mu_{|t=b} = \beta$.*
2. $J_F(\boldsymbol{\mu}, v) < +\infty$, *where $J_F(\cdot)$ is defined as*

$$J_F(\boldsymbol{\mu}, v) := \begin{cases} \displaystyle\int_I \int_{\mathbb{R}^d} \left(1 + I_{F(x)}\left(v_t(x) \right) \right) d\mu_t(x)\, dt, & \text{if } \|v_t\|_{L^1_{\mu_t}} \in L^1([0, T]), \\[2mm] +\infty, & \text{otherwise,} \end{cases} \tag{4}$$

where $I_{F(x)}$ is the indicator function of the set $F(x)$, i.e., $I_{F(x)}(\xi) = 0$ for all $\xi \in F(x)$ and $I_{F(x)}(\xi) = +\infty$ for all $\xi \notin F(x)$.

In this case, we will also shortly say that $\boldsymbol{\mu}$ is driven by v.

When $J_F(\cdot)$ is finite, this value expresses the time needed by the system to steer α to β along the trajectory $\boldsymbol{\mu}$ with family of velocity vector fields v.

Definition 4 (Generalized minimum time). *Given $p \ge 1$, let $\Phi \in C^0(\mathbb{R}^d; \mathbb{R})$ and $\tilde{S}_p^\Phi$ be the corresponding generalized target defined in Definition 2. In analogy with the classical case, we define the* generalized minimum time function $\tilde{T}_p^\Phi : \mathscr{P}_p(\mathbb{R}^d) \to [0, +\infty]$ *by setting*

$$\tilde{T}_p^\Phi(\mu_0) := \inf \left\{ J_F(\boldsymbol{\mu}, v) : \boldsymbol{\mu} \text{ is an admissible curve in } [0, T], \right. \tag{5}$$

$$\left. \text{driven by } v, \text{ with } \mu_{|t=0} = \mu_0, \, \mu_{|t=T} \in \tilde{S}_p^\Phi \right\},$$

where, by convention, $\inf \emptyset = +\infty$.

Given $\mu_0 \in \mathscr{P}_p(\mathbb{R}^d)$, *an admissible curve $\boldsymbol{\mu} = \{\mu_t\}_{t \in [0, \tilde{T}_p^\Phi(\mu_0)]} \subseteq \mathscr{P}_p(\mathbb{R}^d)$, driven by a time depending Borel vector-field $v = \{v_t\}_{t \in [0, \tilde{T}_p^\Phi(\mu_0)]}$ and satisfying $\mu_{|t=0} = \mu_0$ and $\mu_{|t=\tilde{T}_p^\Phi(\mu_0)} \in \tilde{S}_p^\Phi$ is* optimal *for μ_0 if*

$$\tilde{T}_p^\Phi(\mu_0) = J_F(\boldsymbol{\mu}, v).$$

Some interesting results concerning the generalized minimum time function together with comparisons with the classical definition are proved in the proceedings [5] and in the forthcoming paper [6].

Here we will focus our attention on the problem of finding an Hamilton-Jacobi-Bellman equation for our time-optimal control problem.

First of all we need to state and prove a Dynamic Programming Principle and, for this aim, the gluing result for solutions of the continuity equation stated in Lemma 4.4 in [7] will be used.

Theorem 1 (Dynamic programming principle). *Let $p \geq 1$, $0 \leq s \leq \tau$, let $F : \mathbb{R}^d \rightrightarrows \mathbb{R}^d$ be a set-valued function, let $\boldsymbol{\mu} = \{\mu_t\}_{t \in [0,\tau]}$ be an admissible curve for Σ_F. Then we have*

$$\tilde{T}_p^\Phi(\mu_0) \leq s + \tilde{T}_p^\Phi(\mu_s).$$

Moreover, if $\tilde{T}_p^\Phi(\mu_0) < +\infty$, equality holds for all $s \in [0, \tilde{T}_p^\Phi(\mu_0)]$ if and only if $\boldsymbol{\mu}$ is optimal for $\mu_0 = \mu_{|t=0}$.

Proof. The proof is based on the fact that, by Lemma 4.4 in [7], the juxtaposition of admissible curves is an admissible curve. Thus, for every $\varepsilon > 0$ we consider the curve obtained by following $\boldsymbol{\mu}$ up to time s, and then following an admissible curve steering μ_s to the generalized target in time $\tilde{T}_p^\Phi(\mu_s) + \varepsilon$. We obtain an admissible curve steering μ_0 to the generalized target in time $s + \tilde{T}_p^\Phi(\mu_s) + \varepsilon$, and so, by letting $\varepsilon \to 0^+$, we have $\tilde{T}_p^\Phi(\mu_0) \leq s + \tilde{T}_p^\Phi(\mu_s)$.

Assume now that $\tilde{T}_p^\Phi(\mu_0) < +\infty$ and equality holds for all $s \in [0, \tilde{T}_p^\Phi(\mu_0)]$. By taking $s = \tilde{T}_p^\Phi(\mu_0)$ we get $T_p^\Phi(\mu_{\tilde{T}_p^\Phi(\mu_0)}) = 0$, i.e., $\mu_{\tilde{T}_p^\Phi(\mu_0)} \in \tilde{S}_p^\Phi$. In particular, $\boldsymbol{\mu}$ steers μ_0 to $\tilde{S}_p^\Phi$ in time $\tilde{T}_p^\Phi(\mu_0)$, which is the infimum among all admissible trajectories steering μ_0 to the generalized target. So $\boldsymbol{\mu}$ is optimal.

Finally, assume that $\boldsymbol{\mu}$ is optimal for μ_0 and $\tilde{T}_p^\Phi(\mu_0) < +\infty$. Starting from μ_0, we follow $\boldsymbol{\mu}$ up to time s. Since $\boldsymbol{\mu}$ is still an admissible curve steering μ_s to the generalized target in time $\tilde{T}_p^\Phi(\mu_0) - s$, we must have $\tilde{T}_p^\Phi(\mu_s) \leq \tilde{T}_p^\Phi(\mu_0) - s$, and so $\tilde{T}_p^\Phi(\mu_s) + s = \tilde{T}_p^\Phi(\mu_0)$, since the reverse inequality always holds true. $\square$

3   Hamilton-Jacobi-Bellman Equation

In this section we will prove that, under some assumptions, the generalized minimum time functional $\tilde{T}_2^\Phi$ is a viscosity solution, in a sense we will precise, of a suitable Hamilton-Jacobi-Bellman equation on $\mathscr{P}_2(\mathbb{R}^d)$. In this paper we assume the velocity field to be continuous for simplicity. In the forthcoming paper [6] we prove a result of approximation of L_μ^2-selections of F with continuous and bounded ones in L_μ^2-norm that allows us to treat a more general case.

We recall that, given $T \in]0, +\infty]$, the *evaluation operator* $e_t : \mathbb{R}^d \times \Gamma_T \to \mathbb{R}^d$ is defined as $e_t(x, \gamma) = \gamma(t)$ for all $0 \leq t < T$. We set

$$\mathscr{T}_F(\mu_0) := \{\boldsymbol{\eta} \in \mathscr{P}(\mathbb{R}^d \times \Gamma_T) : T > 0, \boldsymbol{\eta} \text{ concentrated on trajectories of}$$
$$\dot{\gamma}(t) = v(\gamma(t)), \text{ with } v \in C^0(\mathbb{R}^d; \mathbb{R}^d), v(x) \in F(x) \, \forall x \in \mathbb{R}^d$$
$$\text{and satisfies } \gamma(0)\sharp\boldsymbol{\eta} = \mu_0\},$$

where $\mu_0 \in \mathscr{P}_2(\mathbb{R}^d)$.

It is not hard to prove the following result.

Lemma 1 (Properties of the evaluation operator). *Assume (F_0) and (F_1), and let $L_1, L_2 > 0$ be the constants as in (F_1). For any $\mu_0 \in \mathscr{P}_2(\mathbb{R}^d)$, $T \in \,]0,1]$, $\eta \in \mathscr{T}_F(\mu_0)$, we have:*

(i) $|e_t(x,\gamma)| \le (|e_0(x,\gamma)| + L_2)\, e^{L_1}$ *for all $t \in [0,T]$ and η-a.e. $(x,\gamma) \in \mathbb{R}^d \times \Gamma_T$;*
(ii) $e_t \in L^2_\eta(\mathbb{R}^d \times \Gamma_T; \mathbb{R}^d)$ *for all $t \in [0,T]$;*
(iii) *there exists $C > 0$ depending only on L_1, L_2 such that for all $t \in [0,T]$ we have*

$$\left\| \frac{e_t - e_0}{t} \right\|^2_{L^2_\eta} \le C\,(\mathrm{m}_2(\mu_0) + 1).$$

In the case we are considering, where the trajectory $t \mapsto e_t \sharp \eta$ is driven by a sufficiently smooth velocity field, we recover as initial velocity what we expected.

Lemma 2 (Regular driving vector fields). *Let $\boldsymbol{\mu} = \{\mu_t\}_{t \in [0,T]}$ be an absolutely continuous solution of*

$$\begin{cases} \partial_t \mu_t + \mathrm{div}(v\mu_t) = 0, \ t \in \,]0,T[\\[2mm] \mu_{|t=0} = \mu_0 \in \mathscr{P}_2(\mathbb{R}^d), \end{cases}$$

where $v \in C^0_b(\mathbb{R}^d; \mathbb{R}^d)$ satisfies $v(x) \in F(x)$ for all $x \in \mathbb{R}^d$. Then if $\eta \in \mathscr{T}_F(\mu_0)$ satisfies $\mu_t = e_t \sharp \eta$ for all $t \in [0,T]$, we have that

$$\lim_{t \to 0} \left\| \frac{e_t - e_0}{t} - v \circ e_0 \right\|_{L^2_\eta} = 0.$$

The proof is based on the boundedness of v and on the fact that, by hypothesis, $\gamma \in C^1$, $\dot{\gamma}(t) = v(\gamma(t))$. The conclusion comes applying Lebesgue's Dominated Convergence Theorem.

We give now the definitions of viscosity sub-/superdifferential and viscosity solutions that suit our problem. As presented in the Introduction, these concepts are different from the ones treated in [2,3,8–10], due mainly to the structure of $\mathscr{P}_2(\mathbb{R}^d)$.

Definition 5 (Sub-/Super-differential in $\mathscr{P}_2(\mathbb{R}^d)$). *Let $V : \mathscr{P}_2(\mathbb{R}^d) \to \mathbb{R}$ be a function. Fix $\mu \in \mathscr{P}_2(\mathbb{R}^d)$ and $\delta > 0$. We say that $p_\mu \in L^2_\mu(\mathbb{R}^d; \mathbb{R}^d)$ belongs to the δ-superdifferential $D^+_\delta V(\mu)$ at μ if for all $T > 0$ and $\eta \in \mathscr{P}(\mathbb{R}^d \times \Gamma_T)$ such that $t \mapsto e_t \sharp \eta$ is an absolutely continuous curve in $\mathscr{P}_2(\mathbb{R}^d)$ defined in $[0,T]$ with $e_0 \sharp \eta = \mu$ we have*

$$\limsup_{t \to 0^+} \frac{V(e_t \sharp \eta) - V(e_0 \sharp \eta) - \displaystyle\int_{\mathbb{R}^d \times \Gamma_T} \langle p_\mu \circ e_0(x,\gamma), e_t(x,\gamma) - e_0(x,\gamma) \rangle \, d\eta(x,\gamma)}{\|e_t - e_0\|_{L^2_\eta}} \le \delta.$$

$$(6)$$

In the same way, $q_\mu \in L^2_\mu(\mathbb{R}^d; \mathbb{R}^d)$ belongs to the δ-subdifferential $D^-_\delta V(\mu)$ at μ if $-q_\mu \in D^+_\delta[-V](\mu)$. Moreover, $D^\pm_\delta[V](\mu)$ is the closure in L^2_μ of $D^\pm_\delta[V](\mu) \cap C^0_b(\mathbb{R}^d; \mathbb{R}^d)$.

Definition 6 (Viscosity solutions). *Let $V : \mathscr{P}_2(\mathbb{R}^d) \to \mathbb{R}$ be a function and $\mathscr{H} : \mathscr{P}_2(\mathbb{R}^d) \times C^0_b(\mathbb{R}^d; \mathbb{R}^d) \to \mathbb{R}$. We say that V is a*

1. *viscosity supersolution of $\mathscr{H}(\mu, DV(\mu)) = 0$ if there exists $C > 0$ depending only on $\mathscr{H}$ such that $\mathscr{H}(\mu, q_\mu) \geq -C\delta$ for all $q_\mu \in D^-_\delta V(\mu) \cap C^0_b$, $\delta > 0$ and $\mu \in \mathscr{P}_2(\mathbb{R}^d)$.*
2. *viscosity subsolution of $\mathscr{H}(\mu, DV(\mu)) = 0$ if there exists $C > 0$ depending only on $\mathscr{H}$ such that $\mathscr{H}(\mu, p_\mu) \leq C\delta$ for all $p_\mu \in D^+_\delta V(\mu) \cap C^0_b$, $\delta > 0$ and $\mu \in \mathscr{P}_2(\mathbb{R}^d)$.*
3. *viscosity solution of $\mathscr{H}(\mu, DV(\mu)) = 0$ if it is both a viscosity subsolution and a viscosity supersolution.*

Definition 7 (Hamiltonian Function). *Given $\mu \in \mathscr{P}_2(\mathbb{R}^d)$, we define the map $\mathscr{H}_F : \mathscr{P}_2(\mathbb{R}^d) \times C^0_b(\mathbb{R}^d; \mathbb{R}^d) \to \mathbb{R}$ by setting*

$$\mathscr{H}_F(\mu, \psi) := -\left[1 + \inf_{\eta \in \mathscr{T}_F(\mu)} \int_{\mathbb{R}^d} \langle p(x), v(x) \rangle \, d\mu(x)\right].$$

Theorem 2 (Viscosity solution). *Assume (F_0) and (F_2). Then $\tilde{T}^\Phi_2(\cdot)$ is a viscosity solution of $\mathscr{H}_F(\mu, D\tilde{T}^\Phi_2(\mu)) = 0$, with $\mathscr{H}_F$ defined as in Definition 7.*

Proof. The proof is splitted in two claims.

Claim 1. $\tilde{T}^\Phi_2(\cdot)$ is a subsolution of $\mathscr{H}_F(\mu, D\tilde{T}^\Phi_2(\mu)) = 0$.

Proof of Claim 1. Given $\eta \in \mathscr{T}_F(\mu_0)$ and set $\mu_t = e_t \sharp \eta$ for all t by the Dynamic Programming Principle we have $\tilde{T}^\Phi_2(\mu_0) \leq \tilde{T}^\Phi_2(\mu_s) + s$ for all $0 < s \leq \tilde{T}^\Phi_2(\mu_0)$. Without loss of generality, we can assume $0 < s < 1$. Given any $p_{\mu_0} \in D^+_\delta \tilde{T}^\Phi_2(\mu_0) \cap C^0_b$, and set

$$A(s, p_{\mu_0}, \eta) := -s - \int_{\mathbb{R}^d \times \Gamma_T} \langle p_{\mu_0} \circ e_0(x, \gamma), e_s(x, \gamma) - e_0(x, \gamma) \rangle \, d\eta,$$

$$B(s, p_{\mu_0}, \eta) := \tilde{T}^\Phi_2(\mu_s) - \tilde{T}^\Phi_2(\mu_0) - \int_{\mathbb{R}^d \times \Gamma_T} \langle p_{\mu_0} \circ e_0(x, \gamma), e_s(x, \gamma) - e_0(x, \gamma) \rangle d\eta,$$

we have $A(s, p_{\mu_0}, \eta) \leq B(s, p_{\mu_0}, \eta)$.

We recall that since by definition $p_{\mu_0} \in L^2_{\mu_0}$, we have that $p_{\mu_0} \circ e_0 \in L^2_\eta$. Dividing by $s > 0$ the left hand side, we observe that we can use Lemma 2, indeed the velocity field $v(\cdot)$ associated to $\eta \in \mathscr{T}_F(\mu_0)$ satisfies all the hypothesis (the boundedness comes from (F_2)) and so we have

$$\limsup_{s \to 0^+} \frac{A(s, p_{\mu_0}, \eta)}{s} = -1 - \int_{\mathbb{R}^d \times \Gamma_T} \langle p_{\mu_0} \circ e_0(x, \gamma), v \circ e_0(x, \gamma) \rangle \, d\eta(x, \gamma).$$

Recalling that $p_{\mu_0} \in D_\delta^+ \tilde{T}_2^\Phi(\mu_0)$ and using Lemma 1(iii), we have

$$\limsup_{s \to 0^+} \frac{B(s, p_{\mu_0}, \boldsymbol{\eta})}{s} = \limsup_{s \to 0^+} \frac{B(s, p_{\mu_0}, \boldsymbol{\eta})}{\|e_s - e_0\|_{L_\eta^2}} \cdot \left\| \frac{e_s - e_0}{s} \right\|_{L_\eta^2} \leq C\delta,$$

where $C > 0$ is a suitable constant (we can take twice the upper bound on F given by (F_2)).

We thus obtain for all $\boldsymbol{\eta} \in \mathscr{T}_F(\mu_0)$ that

$$1 + \int_{\mathbb{R}^d \times \Gamma_T} \langle p_{\mu_0} \circ e_0(x, \gamma), v \circ e_0(x, \gamma) \rangle \, d\boldsymbol{\eta}(x, \gamma) \geq -C\delta.$$

By passing to the infimum on $\boldsymbol{\eta} \in \mathscr{T}_F(\mu_0)$ we have

$$-C\delta \leq 1 + \inf_{\boldsymbol{\eta} \in \mathscr{T}_F(\mu_0)} \int_{\mathbb{R}^d \times \Gamma_T} \langle p_{\mu_0} \circ e_0(x, \gamma), v \circ e_0(x, \gamma) \rangle \, d\boldsymbol{\eta}(x, \gamma)$$

$$= 1 + \inf_{\boldsymbol{\eta} \in \mathscr{T}_F(\mu_0)} \int_{\mathbb{R}^d} \langle p_{\mu_0}(x), v(x) \rangle \, d\mu_0(x) = -\mathscr{H}_F(\mu_0, p_{\mu_0}),$$

so $\tilde{T}_2^\Phi(\cdot)$ is a subsolution, thus confirming Claim 1.   ◇

Claim 2. $\tilde{T}_2^\Phi(\cdot)$ is a supersolution of $\mathscr{H}_F(\mu, D\tilde{T}_2^\Phi(\mu)) = 0$.

Proof of Claim 2. Given $\boldsymbol{\eta} \in \mathscr{T}_F(\mu_0)$, let us define the admissible trajectory $\boldsymbol{\mu} = \{\mu_t\}_{t \in [0,T]} = \{e_t \sharp \boldsymbol{\eta}\}_{t \in [0,T]}$. Given $q_{\mu_0} \in D_\delta^- \tilde{T}_2^\Phi(\mu_0) \cap C_b^0$, we have

$$\int_{\mathbb{R}^d \times \Gamma_T} \langle q_{\mu_0} \circ e_0(x, \gamma), \frac{e_s(x, \gamma) - e_0(x, \gamma)}{s} \rangle \, d\boldsymbol{\eta}(x, \gamma)$$

$$\leq 2\delta \left\| \frac{e_s - e_0}{s} \right\|_{L_\eta^2} - \frac{\tilde{T}_2^\Phi(\mu_0) - \tilde{T}_2^\Phi(\mu_s)}{s}.$$

Thus, using Lemma 2 and Lemma 1, we have

$$\int_{\mathbb{R}^d \times \Gamma_T} \langle q_{\mu_0} \circ e_0(x, \gamma), v \circ e_0(x, \gamma) \rangle \, d\boldsymbol{\eta}(x, \gamma) \leq 3C\delta - \frac{\tilde{T}_2^\Phi(\mu_0) - \tilde{T}_2^\Phi(\mu_s)}{s},$$

for all $s > 0$.

Then, by passing to the infimum on all admissible trajectories, we obtain

$$-\mathscr{H}_F(\mu_0, q_{\mu_0}) - 1 = \inf_{\boldsymbol{\eta} \in \mathscr{T}_F(\mu_0)} \int_{\mathbb{R}^d \times \Gamma_T} \langle q_{\mu_0} \circ e_0(x, \gamma), v \circ e_0(x, \gamma) \rangle \, d\boldsymbol{\eta}(x, \gamma)$$

$$\leq 3C\delta - \sup_{\boldsymbol{\eta} \in \mathscr{T}_F(\mu_0)} \frac{\tilde{T}_2^\Phi(\mu_0) - \tilde{T}_2^\Phi(\mu_s)}{s}.$$

Thus

$$\mathscr{H}_F(\mu_0, q_{\mu_0}) \geq -3C\delta + \sup_{\boldsymbol{\eta} \in \mathscr{T}_F(\mu_0)} \left[\frac{\tilde{T}_2^\Phi(\mu_0) - \tilde{T}_2^\Phi(\mu_s)}{s} - 1 \right].$$

By the Dynamic Programming Principle, recalling that $\dfrac{\tilde{T}_2^\Phi(\mu_0) - \tilde{T}_2^\Phi(\mu_s)}{s} - 1 \leq 0$ with equality holding if and only if $\boldsymbol{\mu}$ is optimal, we obtain $\mathscr{H}_F(\mu_0, q_{\mu_0}) \geq -C'\delta$, which proves that $\tilde{T}_2^\Phi(\cdot)$ is a supersolution, thus confirming Claim 2.   □

4 Conclusion

In this work we have studied a Hamilton-Jacobi-Bellman equation solved by a generalized minimum time function in a regular case. In the forthcoming paper [6] an existence result is proved for optimal trajectories as well as attainability properties in the space of probability measures. Furthermore, a suitable approximation result allows to give a sense to a Hamilton-Jacobi-Bellman equation in a more general case.

We plan to study if it is possible to prove a comparison principle for an Hamilton-Jacobi equation solved by the generalized minimum time function, as well as to give a Pontryagin maximum principle for our problem.

References

1. Ambrosio, L.: The flow associated to weakly differentiable vector fields: recent results and open problems. In: Bressan, A., Chen, G.Q., Lewicka, M., Wang, D. (eds.) Nonlinear Conservation Laws and Applications. The IMA Volumes in Mathematics and its Applications, vol. 153, pp. 181–193. Springer, Boston (2011). doi:10.1007/978-1-4419-9554-4_7
2. Ambrosio, L., Gigli, N., Savaré, G.: Gradient Flows in Metric Spaces and in the Space of Probability Measures. Lectures in Mathematics ETH Zürich, 2nd edn. Birkhäuser Verlag, Basel (2008)
3. Ambrosio, L., Feng, J.: On a class of first order Hamilton-Jacobi equations in metric spaces. J. Differ. Equ. **256**(7), 2194–2245 (2014)
4. Bernard, P.: Young measures, superpositions and transport. Indiana Univ. Math. J. **57**(1), 247–276 (2008)
5. Cavagnari, G., Marigonda, A.: Time-optimal control problem in the space of probability measures. In: Lirkov, I., Margenov, S.D., Waśniewski, J. (eds.) LSSC 2015. LNCS, vol. 9374, pp. 109–116. Springer, Cham (2015). doi:10.1007/978-3-319-26520-9_11
6. Cavagnari, G., Marigonda, A., Nguyen, K.T., Priuli, F.S.: Generalized control systems in the space of probability measures (submitted)
7. Dolbeault, J., Nazaret, B., Savaré, G.: A new class of transport distances between measures. Calc. Var. Partial Differ. Equ. **34**(2), 193–231 (2009)
8. Cardaliaguet, P., Quincampoix, M.: Deterministic differential games under probability knowledge of initial condition. Int. Game Theory Rev. **10**(1), 1–16 (2008)
9. Gangbo, W., Nguyen, T., Tudorascu, A.: Hamilton-Jacobi equations in the Wasserstein space. Methods Appl. Anal. **15**(2), 155–184 (2008)
10. Gangbo, W., Święch, A.: Optimal transport and large number of particles. Discret. Contin. Dyn. Syst. **34**(4), 1397–1441 (2014)

Strong Optimal Controls in a Steady-State Problem of Complex Heat Transfer

Alexander Yu. Chebotarev[1,2], Andrey E. Kovtanyuk[1,2], Nikolai D. Botkin[3(✉)],
and Karl-Heinz Hoffmann[3]

[1] Far Eastern Federal University, Sukhanova Street 8, 690950 Vladivostok, Russia
[2] Institute for Applied Mathematics FEB RAS,
Radio Street 7, 690041 Vladivostok, Russia
cheb@iam.dvo.ru, kovtanyuk.ae@dvfu.ru
[3] Technische Universität München, Zentrum Mathematik,
Boltzmannstr. 3, 85748 Garching bei München, Germany
{botkin,hoffmann}@ma.tum.de
http://www-m6.ma.tum.de/Lehrstuhl/NikolaiBotkin

Abstract. An optimal control problem of steady-state complex heat transfer with monotone objective functionals is under consideration. A coefficient function appearing in boundary conditions and reciprocally corresponding to the reflection index of the domain surface is considered as control. The concept of strong maximizing (resp. strong minimizing) optimal controls, i.e. controls that are optimal for all monotone objective functionals, is introduced. The existence of strong optimal controls is proven, and optimality conditions for such controls are derived. An iterative algorithm for computing strong optimal controls is proposed.

Keywords: Conductive-convective-radiative heat transfer · Diffusion approximation · Control problem · Strong optimal controls · Optimality condition

1 Introduction

The interest in studying problems of complex heat transfer (where the radiative, convective, and conductive contributions are simultaneously taken into account) is motivated by their importance for many engineering applications. The common feature of such processes is the radiative heat transfer dominating at high temperatures. The radiative heat transfer equation (RTE) is a first order integro-differential equation governing the radiation intensity. The radiation traveling along a path is attenuated as a result of absorption and scattering. The precise derivation and analysis of such models can be found in the monograph [1].

Solutions to the RTE can be represented in the form of the Neumann series whose terms are powers of an integral operator applied to a certain start function. The terms can be calculated using a Monte Carlo method, which may be

© IFIP International Federation for Information Processing 2016
Published by Springer International Publishing AG 2016. All Rights Reserved
L. Bociu et al. (Eds.): CSMO 2015, IFIP AICT 494, pp. 209–219, 2016.
DOI: 10.1007/978-3-319-55795-3_19

interpreted as tracking the history of energy bundles from emission to adsorption on the boundary or within the participating medium. The method assumes that the bundles start from random points, propagate in random directions, and show the energy exchange due to random scattering (see e.g. [2]).

A way to avoid solving the integro-differential RTE is the use of expansions of the local intensity in terms of spherical harmonics, with truncation to N terms in the series, and substitution into the moments of the differential form of the RTE (see e.g. [1]). This approach leads to the P_N approximations, where N is the approximation order. Especially interesting is the P_1 (diffusion) approximation because it does not require high computational efforts. Using the diffusion model instead of the integro-differential RTE becomes popular, and this is substantiated in various applications (e.g., image reconstruction [3] and modeling the radiative transfer in biological tissues [4]). In this connection, the work [5] can also be mentioned: It is shown there that the diffusion approximation yields a good accuracy for temperatures up to 1200°C. Thus, the diffuse approximation can successfully be applied to various heat transfer problems where very high accuracy is not required.

Optimal control problems of complex heat transfer draw the interest of researchers working in applied fields, e.g. glass manufacturing [6–8], laser thermotherapy [9], the design of cooling systems [10,11], etc. A considerable number of works is devoted to control problems related to evolutionary systems describing radiative heat transfer (see e.g. [6–10,12–14]). In the above-mentioned works, the transfer of radiation is described by an integral-differential equation or by its approximations. The temperature field is simulated by the conventional evolutionary heat transfer equation with additional source terms accounting for the contribution of radiation.

As for steady-state problems of complex heat transfer, there are few results in this direction. It is worth to mention the work [11], where an optimal boundary multiplicative control problem for a steady-state complex heat transfer model is considered. The problem is formulated as the maximization of the energy outflow from the model domain by controlling reflection properties of the boundary. The solvability of this problem is proven based on new a priori estimates for solutions of the model equations. Moreover, an analogue of the bang-bang principle arising in control theory for ordinary differential equations is proven. Notice that the optimization of energy in/out flow, which improves heating/cooling systems, is a quite popular problem in many engineering applications. In [15–18], similar problems are considered in the context of shape optimization.

In the current work, a conductive-convective-radiative heat transfer control problem with monotone objective functionals is under consideration. The definition of monotone functionals looks as follows. Let θ and φ be the state variables of the model, and $J(\theta, \varphi)$ the objective functional. This functional is called monotone increasing (resp. decreasing) if the relations $\theta_1 \leq \theta_2$ and $\varphi_1 \leq \varphi_2$, a. e., imply the relation $J(\theta_1, \varphi_1) \leq J(\theta_2, \varphi_2)$ (resp. $J(\theta_1, \varphi_1) \geq J(\theta_2, \varphi_2)$). It should be noticed that such functionals appear very often in applications. For example, the objective functional in the problem of maximum energy outflow is a monotone one.

Moreover, the concept of strong minimizing (resp. maximizing) optimal controls, i.e. controls that yield minimal (resp. maximal) state functions is introduced. Sufficient optimality conditions that do not involve solutions of adjoint equations are derived.

An iterative algorithm for finding strong optimal controls is proposed, and its convergence is proven. This, in particular, proves the existence of strong optimal controls.

2   Problem Setting

The normalized diffusion model, P_1 approximation, of radiative, conductive, and convective heat transfer in a bounded domain $\Omega \subset \mathbb{R}^3$ looks as follow (see [1, 19–21]):

$$- a\Delta\theta + \mathbf{v} \cdot \nabla\theta + b\kappa_a(|\theta|\theta^3 - \varphi) = 0, \tag{1}$$

$$- \alpha\Delta\varphi + \kappa_a(\varphi - |\theta|\theta^3) = 0. \tag{2}$$

Here, θ is the normalized temperature, φ the normalized intensity of radiation averaged over all direction, κ_a the absorbtion coefficient, and $\mathbf{v}$ a prescribed velocity field. The constants a, b, and α are defined by the formulas

$$a = \frac{k}{\rho c_p}, \quad b = \frac{4\sigma n^2 T_{max}^3}{\rho c_p}, \quad \alpha = \frac{1}{3\kappa - A\kappa_s},$$

where k is the thermal conductivity, c_p the specific heat capacity, ρ the density, σ the Stefan-Boltzmann constant, n the refractive index, T_{max} the maximum temperature in unnormalized model, $\kappa := \kappa_s + \kappa_a$ the extinction coefficient, κ_s the scattering coefficient. The coefficient $A \in [-1, 1]$ describes the anisotropy of scattering. The case $A = 0$ corresponds to isotropic scattering.

The following boundary conditions on $\Gamma := \partial\Omega$ are imposed:

$$a\partial_n\theta + \beta(\theta - \theta_b) = 0, \qquad \alpha\partial_n\varphi + u(\varphi - \theta_b^4) = 0. \tag{3}$$

Here, ∂_n denotes the derivative in the direction of the outward normal $\mathbf{n}$; $\theta_b = \theta_b(x)$, $x \in \Gamma$, and $\beta = \beta(x)$, $x \in \Gamma$, are given non-negative functions describing the normalized external temperature and the normalized overall heat transfer coefficient, respectively. The function $u = u(x)$, $x \in \Gamma$, describing the reflection properties of the boundary is considered as control input.

The minimum (resp. maximum) control problem is formulated as finding a control $\widehat{u} \in L^\infty(\Gamma)$, $\widehat{u}(x) \in [u_1(x), u_2(x)]$, a.e. on Γ, such that for any admissible control u the following relation holds: $y(\widehat{u}) \le y(u)$ (resp. $y(\widehat{u}) \ge y(u)$), a.e. in Ω, where $y(\widehat{u})$ and $y(u)$ are solution pairs satisfying relations (1)–(3) with $\widehat{u}$ and u, respectively. Here, u_1, u_2 are given non-negative functions defining inequality constraints imposed on the control.

It is clear that the control $\widehat{u}$ is optimal in the sense of minimization (resp. maximization) of monotone functionals outlined in Sect. 1.

3   Problem Formalization

Assume that Ω be a bounded Lipschitz domain. Let L^p, $p \in [1, \infty]$, denotes the Lebesgue space, and $H^s(\Omega)$ the Sobolev space $W_2^s(\Omega)$. Moreover, let the following conditions be fulfilled:

(i) $\mathbf{v} \in H^1(\Omega)$, div $\mathbf{v} = 0$;
(ii) $\beta, u_1, u_2, \theta_b \in L^\infty(\Gamma)$, $0 < \beta_0 \le \beta$, $0 < u_0 \le u_1 \le u_2$, $\theta_b \ge 0$, β_0, u_0 are const;
(iii) $\beta + \mathbf{v} \cdot \mathbf{n} \ge 0$, if $\mathbf{v} \cdot \mathbf{n} < 0$.

Denote $H = L^2(\Omega)$, $V = H^1(\Omega)$, and define the norms, $\|\cdot\|$ and $\|\cdot\|_V$, in H and V, respectively, as follows:

$$\|f\|^2 = (f, f), \quad \|f\|_V^2 = \|f\|^2 + \|\nabla f\|^2, \quad (f, g) = \int_\Omega f(x)g(x)dx.$$

Definition 1. *A pair $\{\theta, \varphi\} \in V \times V$ is called a (weak) solution of the problem (1)–(3) if*

$$a(\nabla\theta, \nabla\eta) + (\mathbf{v} \cdot \nabla\theta + b\kappa_a(|\theta|\theta^3 - \varphi), \eta) + \int_\Gamma \beta(\theta - \theta_b)\eta d\Gamma = 0 \quad \forall \eta \in V, \quad (4)$$

$$\alpha(\nabla\varphi, \nabla\psi) + \kappa_a(\varphi - |\theta|\theta^3, \psi) + \int_\Gamma u(\varphi - \theta_b^4)\psi d\Gamma = 0 \quad \forall \psi \in V. \quad (5)$$

Theorem 1 (see [21]). *Let the conditions (i)–(iii) be true. Then the problem (1)–(3) is uniquely solvable, a weak solution $\{\theta, \varphi\}$ belongs to $\left(L^\infty(\Omega)\right)^2$ and satisfies the inequalities $0 \le \theta \le M$ and $0 \le \varphi \le M^4$, where $M = \|\theta_b\|_{L^\infty(\Gamma)}$, and the following estimate is true:*

$$\|\theta\|_V + \|\varphi\|_V \le C, \quad (6)$$

where C depends only on Ω, M, $\|\beta\|_{L^\infty(\Gamma)}$, $\|u\|_{L^\infty(\Gamma)}$, $\|\mathbf{v}\|_V$, a, α, b, and κ_a.

Now, the conception of strong optimal controls will be introduced. Denote by $U_{ad} = \{u \in L^\infty(\Gamma) : u_1 \le u \le u_2\}$ the set of admissible controls.

Definition 2. *A function $\widehat{u} \in U_{ad}$ is called strong minimizing (resp. maximizing) optimal control if $\widehat{\theta} \le \theta$ and $\widehat{\varphi} \le \varphi$ (resp. $\widehat{\theta} \ge \theta$ and $\widehat{\varphi} \ge \varphi$), a.e. in Ω, for all $u \in U_{ad}$, where $\{\widehat{\theta}, \widehat{\varphi}\}$ and $\{\theta, \varphi\}$ are solution pairs corresponding to $\widehat{u}$ and u, respectively.*

Definition 3. *A functional $J : [V \cap L^\infty(\Omega)]^2 \to \mathbb{R}$ is called monotone if the relations $0 \le \theta_1 \le \theta_2$ and $0 \le \varphi_1 \le \varphi_2$, a.e. in Ω, imply the inequality $J(\theta_1, \varphi_1) \le J(\theta_2, \varphi_2)$, where $\theta_1, \theta_2, \varphi_1$, and φ_2 are arbitrary functions from $V \cap L^\infty(\Omega)$.*

Consider examples of monotone functionals.

1. The sum of weighted norms of θ and φ:

$$J(\theta, \varphi) = \int_{\Omega} \left(r_1\theta^2 + r_2\varphi^2\right) dx, \tag{7}$$

where r_1 and $r_2 \in L^{\infty}(\Omega)$ are non-negative given functions.

2. Energy flow through a part of the boundary. Let $\Gamma_1 \subset \Gamma$ be an outflow boundary part, i.e. $\mathbf{v} \cdot \mathbf{n} \geq 0$ on Γ_1. The density of the energy flow is defined by the formula

$$\mathbf{q} = -a\nabla\theta + \theta\mathbf{v} - \alpha b\nabla\varphi,$$

and therefore, the energy outflow through Γ_1 is given by the functional

$$J(\theta, \varphi) = \int_{\Gamma_1} \mathbf{q} \cdot \mathbf{n} \, d\Gamma = \int_{\Gamma_1} (\beta(\theta - \theta_b) + \theta\mathbf{v} \cdot \mathbf{n} + b\gamma(\varphi - \theta_b^4))d\Gamma. \tag{8}$$

Here, γ is a constant that replaces the function u in the boundary condition for φ on the opening Γ_1. If the Marshak boundary condition [22] is used, then $\gamma = 0.5$.

Say that a triple $\{\theta, \varphi, u\}$ is admissible if $u \in U_{ad}$ and $\{\theta, \varphi\} \in V \times V$ is a solution of the problem (1)–(3) corresponding to the control u. Denote the set of all admissible triples by $\mathcal{U}$.

Let J be a monotone functional (see Definition 3). Consider the following optimization problems:

Problem 1:

$$J(\theta, \varphi) \to \min, \qquad \{\theta, \varphi, u\} \in \mathcal{U}.$$

Problem 2:

$$J(\theta, \varphi) \to \max, \qquad \{\theta, \varphi, u\} \in \mathcal{U}.$$

A solution $\{\theta, \varphi, u\}$ of Problem 1 or 2 will be called *optimal triple*, and its components $\{\theta, \varphi\}$ and u will be referred as *optimal state* and *optimal control*, respectively.

The following proposition is an obvious corollary of Definitions 2 and 3, accounting for that the objective functionals of Problems 1 and 2 are monotone.

Proposition 1. *A strong minimizing (resp. maximizing) optimal control solves Problem 1 (resp. Problem 2).*

The next section describes the derivation of sufficient optimality conditions characterizing strong optimal controls and discusses the question of uniqueness. These considerations give rise to an iterative numerical method that always converges to a strong optimal control, which, in particular, proves the existence of such controls.

4   Conditions of Optimality

Similar to [21], introduce nonlinear operators $F_1 : L^\infty(\Omega) \times U_{ad} \to L^\infty(\Omega) \cap H^1(\Omega)$ and $F_2 : L^\infty(\Omega) \to L^\infty(\Omega) \cap H^1(\Omega)$ as follows: $\varphi = F_1(\theta, u)$ if

$$\alpha(\nabla\varphi, \nabla v) + \int_\Gamma u\,(\varphi - \theta_b^4)v d\Gamma + \kappa_a(\varphi, v) = \kappa_a(|\theta|\theta^3, v) \quad \forall v \in V, \qquad (9)$$

and $\theta = F_2(\varphi)$ if

$$a(\nabla\theta, \nabla v) + \int_\Gamma \beta(\theta - \theta_b)v d\Gamma + (\mathbf{v}\nabla\theta, v) + b\kappa_a(|\theta|\theta^3, v) = b\kappa_a(\varphi, v) \quad \forall v \in V. \quad (10)$$

Notice that $\{\widehat{\theta}, \widehat{\varphi}\}$ is a weak solution of the problem (1)–(3) with a control $\widehat{u}$ if and only if $\widehat{\theta} = F_2(F_1(\widehat{\theta}, \widehat{u}))$, $\widehat{\varphi} = F_1(F_2(\widehat{\varphi}), \widehat{u})$. The operators F_1 and F_2 have the following properties (see [21]):

1. If $u \in U_{ad}$, $M = \|\theta_b\|_{L^\infty(\Gamma)}$, $0 \le \theta \le M$, and $0 \le \varphi \le M^4$, a.e. in Ω, then $0 \le F_1(\theta, u) \le M^4$ and $0 \le F_2(\varphi) \le M$, a.e. in Ω.
2. If $u \in U_{ad}$, $\theta_1 \le \theta_2$, and $\varphi_1 \le \varphi_2$, a.e. in Ω, then $F_1(\theta_1, u) \le F_1(\theta_2, u)$ and $F_2(\varphi_1) \le F_2(\varphi_2)$, a.e. in Ω.

Define an operator $U \colon L^\infty(\Gamma) \to L^\infty(\Gamma)$ as follows:

$$U(\eta)(s) = \begin{cases} u_1(s), & \eta(s) - \theta_b^4(s) < 0, \\ u_2(s), & \eta(s) - \theta_b^4(s) > 0. \end{cases}$$

Lemma 1. *If $\widetilde{u} \in U_{ad}$, $\theta \le \widetilde{\theta}$ a.e. in Ω, $\varphi = F_1(\theta, u)$, $\widetilde{\varphi} = F_1(\widetilde{\theta}, \widetilde{u})$, where $u = U(\varphi)$ or $u = U(\widetilde{\varphi})$, then $\varphi \le \widetilde{\varphi}$ a.e. in Ω.*

Proof. Set $\overline{\theta} = \theta - \widetilde{\theta}$ and $\overline{\varphi} = \varphi - \widetilde{\varphi}$. Observe that Eq. (9) implies the equation

$$\alpha(\nabla\overline{\varphi}, \nabla v) + \int_\Gamma [u(\varphi - \theta_b^4) - \widetilde{u}(\widetilde{\varphi} - \theta_b^4)]v d\Gamma + \kappa_a(\overline{\varphi}, v) = \kappa_a(|\theta|\theta^3 - |\widetilde{\theta}|\widetilde{\theta}^3, v) \quad \forall v \in V.$$
$$(11)$$

Denote $\psi = \max(\overline{\varphi}, 0)$ and put $v = \psi$ into (11) to obtain the estimate

$$\alpha\|\nabla\psi\|^2 + \int_\Gamma [u(\varphi - \theta_b^4) - \widetilde{u}(\widetilde{\varphi} - \theta_b^4)]\psi d\Gamma + \kappa_a\|\psi\|^2 = \kappa_a(|\theta|\theta^3 - |\widetilde{\theta}|\widetilde{\theta}^3, \psi) \le 0.$$

Notice that the equality

$$u(\varphi - \theta_b^4) - \widetilde{u}(\widetilde{\varphi} - \theta_b^4) = \widetilde{u}\overline{\varphi} + (u - \widetilde{u})(\varphi - \theta_b^4) = u\overline{\varphi} + (u - \widetilde{u})(\widetilde{\varphi} - \theta_b^4)$$

implies the non-negativity of the boundary integral in the estimate provided that $u = U(\varphi)$ or $u = U(\widetilde{\varphi})$. Therefore, $\psi = 0$, i.e. $\varphi \le \widetilde{\varphi}$ a.e. in Ω.

Theorem 2. *In order for a function $u \in U_{ad}$ to be a strong minimizing optimal control, it is sufficient that $u = U(\varphi)$, where $\varphi = \varphi(u)$ is a solution of system (1)–(3) with the control u.*

Proof. Assume that u satisfies the conditions of Theorem 2. Let $\widetilde{u} \in U_{ad}$ be an arbitrary control, $\theta = \theta(u)$, $\varphi = \varphi(u)$, $\widetilde{\theta} = \theta(\widetilde{u})$, and $\widetilde{\varphi} = \varphi(\widetilde{u})$. Lemma 1 implies the inequality $\varphi = F_1(\theta, u) \leq F_1(\theta, \widetilde{u}) = \widetilde{\varphi}_1$. Let

$$\widetilde{\theta}_k = F_2(\widetilde{\varphi}_k), \quad \widetilde{\varphi}_{k+1} = F_1(\widetilde{\theta}_k, \widetilde{u}), \quad k = 1, 2, \ldots. \tag{12}$$

Since $\{\theta, \varphi\}$ is a solution pair, the equation $\theta = F_2(\varphi)$ holds, and therefore, due to above mentioned properties 1 and 2 of the operators F_1 and F_2, the following inequalities are true:

$$0 \leq \theta \leq \widetilde{\theta}_k \leq M, \quad 0 \leq \varphi \leq \widetilde{\varphi}_k \leq M^4, \quad k = 1, 2, \ldots.$$

Notice that the sequences $\{\widetilde{\theta}_k\}$ and $\{\widetilde{\varphi}_k\}$ are bounded in V. Therefore, there exist functions $\theta_*, \varphi_* \in L^\infty(\Omega) \cap H^1(\Omega)$ such that

$$\widetilde{\theta}_k \to \theta_*, \quad \widetilde{\varphi}_k \to \varphi_* \text{ a.e. in } \Omega, \quad \text{weakly in } H^1(\Omega), \quad \text{and strongly in } L^2(\Omega)$$

up to subsequences.

The above convergence allows us to pass to the limit in (12) as $k \to \infty$. Therefore, $\{\theta_*, \varphi_*\}$ is a weak solution of the problem (1)–(3) with the control $\widetilde{u}$. Moreover, $\theta \leq \theta_*$ and $\varphi \leq \varphi_*$. By the uniqueness of solutions of the problem (1)–(3), it holds that $\widetilde{\theta} = \theta_*$ and $\widetilde{\varphi} = \varphi_*$, and therefore $\theta \leq \widetilde{\theta}$ and $\varphi \leq \widetilde{\varphi}$, a.e. in Ω. This proves the theorem.

The proof of the next theorem is similar to that of Theorem 2.

Theorem 3. *In order for a function $u \in U_{ad}$ to be a strong maximizing optimal control, it is sufficient that*

$$u(s) = \begin{cases} u_1(s), & \text{if } \varphi(s) - \theta_b^4(s) > 0, \\ u_2(s), & \text{if } \varphi(s) - \theta_b^4(s) < 0, \end{cases}$$

where $\varphi = \varphi(u)$.

Theorem 4. *Let u and $\widetilde{u}$ be two strong optimal controls. Then $u = \widetilde{u}$ on Γ where $\varphi \neq \theta_b^4$.*

Proof. Let u and $\widetilde{u}$ be two strong optimal controls. From the definition of strong optimal controls, it follows that $\varphi(u) = \varphi(\widetilde{u}) = \varphi$ and $\theta(u) = \theta(\widetilde{u}) = \theta$, a.e. in Ω. Using Eq. (11) yields the relation

$$\int_\Gamma [u(\varphi - \theta_b^4) - \widetilde{u}(\varphi - \theta_b^4)]v \, d\Gamma = \int_\Gamma (u - \widetilde{u})(\varphi - \theta_b^4)v = 0 \quad \forall v \in V,$$

which yields that $(u - \widetilde{u})(\varphi - \theta_b^4) = 0$ a.e. in Γ. Therefore, $u = \widetilde{u}$ at points of Γ where $\varphi \neq \theta_b^4$.

Corollary 1. *If a strong optimal control $\widehat{u}$ is arbitrarily changed at points of Γ where $\varphi(\widehat{u}) = \theta_b^4$, then it remains to be a strong optimal control.*

Proof. Let $\widehat{u}$ be a strong optimal control, and $\widehat{\theta}$ and $\widehat{\varphi}$ the corresponding solutions satisfying Eqs. (4) and (5). Assume that $\widehat{u}$ is changed on the set $\{s \in \Gamma : \widehat{\varphi} - \theta_b^4 = 0\}$ to obtain a new control $\widehat{u}_{\mathrm{new}}$. Equations (4) and (5) imply that the pair $\{\widehat{\theta}, \widehat{\varphi}\}$ is also a weak solution corresponding to the modified control $\widehat{u}_{\mathrm{new}}$ because the last integral of Eq. (5) remains unchanged. Thus, the control $\widehat{u}_{\mathrm{new}}$ is a strong optimal control.

5   Iterative Algorithm for Finding the Optimal Control

By Theorem 2, a function $u \in U_{ad}$ such that $u = U(\varphi(u))$ is a strong minimizing optimal control. This gives rise to the idea to use an iterative procedure for finding strong minimizing optimal controls. Below, such a procedure will be proposed, and its convergence will be proven. This additionally proves the existence of strong minimizing optimal controls. The case of strong maximizing optimal controls is treated analogously.

Let $w \in U_{ad}$, $\varphi_0 = \varphi(w)$, $\theta_0 = \theta(w)$, i.e. $\varphi_0 = F_1(\theta_0, w)$, $\theta_0 = F_2(\varphi_0)$. Define the sequences

$$u_{k+1} = U(\varphi_k), \quad \theta_{k+1} = F_2(\varphi_k), \quad \varphi_{k+1} = F_1(\theta_{k+1}, u_{k+1}), \quad k = 0, 1, 2, \dots. \quad (13)$$

Using properties 1 and 2 of the operators F_1 and F_2 yields the following estimates:

$$0 \le \theta_k \le M, \quad 0 \le \varphi_k \le M^4, \quad k = 0, 1, 2, \dots,$$

and, additionally, the sequences $\{\theta_k\}$ and $\{\varphi_k\}$ are bounded in V. Observe that $\theta_1 = F_2(\varphi_0) = \theta_0$, and hence $\varphi_1 = F_1(\theta_1, U(\varphi_0)) \le \varphi_0 = F_1(\theta_0, w)$ by Lemma 1. Then, the monotonicity of F_2 yields the inequality $\theta_2 \le \theta_1$.

Now, inductive arguments yield the following relations:

$$\varphi_k \le \varphi_{k-1}, \quad \theta_{k+1} \le \theta_k, \quad u_{k+1} \le u_k, \quad \text{a.e. pointwise, for all } k \ge 1.$$

Indeed, if $\varphi_k \le \varphi_{k-1}$ and $\theta_{k+1} \le \theta_k$ for some $k \ge 1$, then, by the Lemma 1,

$$\varphi_{k+1} = F_1(\theta_{k+1}, U(\varphi_k)) \le \varphi_k = F_1(\theta_k, u_k),$$

and therefore $\theta_{k+2} \le \theta_{k+1}$. Moreover, the monotonicity of the mapping U with respect to the a.e. pointwise order yields the inequality $u_{k+1} = U(\varphi_k) \le U(\varphi_{k-1}) = u_k$.

Similar to the arguments used in the proof of Theorem 2, the properties of boundedness and monotonicity of the sequences $\{u_k\}, \{\theta_k\}$, and $\{\varphi_k\}$ allow us to clime the existence of functions $\widehat{u} \in L^\infty(\Gamma)$ and $\widehat{\theta}, \widehat{\varphi} \in L^\infty(\Omega) \cap H^1(\Omega)$ such that

$$u_k \to \widehat{u} \text{ a.e. in } \Gamma, \quad \theta_k \to \widehat{\theta}, \quad \varphi_k \to \widehat{\varphi} \text{ a.e. in } \Omega,$$

$$\text{weakly in } H^1(\Omega), \quad \text{and strongly in } L^2(\Omega). \quad (14)$$

The convergence (14), taking into account the upper semi-continuity of the mapping U, allows us to pass to the limit in (13) to obtain the equations

$$\widehat{u} = U(\widehat{\varphi}), \quad \widehat{\theta} = F_2(\widehat{\varphi}), \quad \widehat{\varphi} = F_1(\widehat{\theta}, \widehat{u}).$$

Therefore, $\widehat{u} = U(\varphi(\widehat{u}))$, $\widehat{\theta} \leq \theta(w)$, and $\widehat{\varphi} \leq \varphi(w)$, i.e. $\widehat{u}$ is a strong minimizing optimal control.

Thus, the following statement is true:

Theorem 5. *There exists a strong minimizing (resp. maximizing) control u uniquely defined on the set $\Gamma \setminus \{\eta \in \Gamma : \varphi(u) = \theta_b^4\}$. The modification of such a control on the set $\{\eta \in \Gamma : \varphi(u) = \theta_b^4\}$ does not violate its strong optimality.*

6   Numerical Experiment

The following data are used in the numerical experiments. The region Ω being a channel of the following form (the units are centimeters):

$$\Omega = \{r = (x_1, x_2, x_3) : 0 < x_2 < 50, \ 0 < x_{1,3} < 10\}.$$

The boundary parts at $x_2 = 0$ and $x_2 = 50$ are inflow and outflow regions, respectively. The side faces, parallel to the x_2 axis, are solid walls of the channel. The velocity field is specified as $\mathbf{v} = (0, 9, 0)$ [cm/s]. The function θ_b is defined as follows:

$$\theta_b(x_1, 0, x_3) = 0.5, \quad \theta_b(x_1, 50, x_3) = 1,$$

$$\theta_b(0, x_2, x_3) = \theta_b(10, x_2, x_3) = \theta_b(x_1, x_2, 0) = \theta_b(x_1, x_2, 10) = 0.5 + 0.01x_2.$$

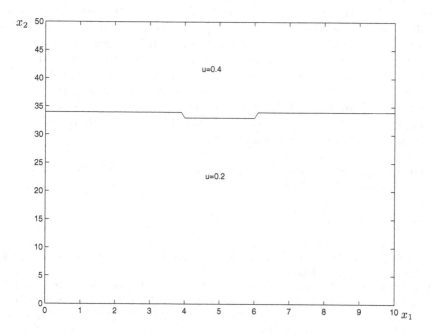

Fig. 1. Distribution of the control on the top face, $x_3 = 10$, of the channel.

The thermodynamical characteristics of the medium inside the channel correspond to air at the normal atmospheric pressure and the temperature of 400 °C. The maximum temperature in unnormalized model is chosen as $T_{max} = 500$ °C. The extinction coefficient κ is equal to 0.1 [cm^{-1}], $\alpha = 3.3(3)$, the absorption coefficient κ_a equals 0.01 [cm^{-1}], the anisotropy coefficient A equals 0, and the coefficient γ equals 10.

It is assumed that the control u is variable on the upper face, $x_3 = 10$, and constrained by the inequalities $0.2 \leq u \leq 0.4$. On the other faces, the control assumes prescribed constant values as follows:

$$u(x_1, 0, x_3) = u(x_1, 50, x_3) = 0.5,$$

$$u(0, x_2, x_3) = u(10, x_2, x_3) = u(x_1, x_2, 0) = 0.3.$$

The iterative algorithm requires only two steps to deliver a strong maximizing optimal control. The distribution of this control on the upper face of the channel is shown in Fig. 1.

7   Conclusion

The current paper deals with a nonstandard problem of optimal control and proposes its complete solution. The notion of strong optimal controls seems to be a little bit unrealistic for common control problems. Nevertheless, the model considered in this work does have such solutions. They are unique in some sense and can be easily computed. Another surprising point is that the intuition fails when predicts that e.g. a strong maximizing optimal control should assume possibly maximal admissible values. In contrast to that, the example presented shows the opposite. Some analysis shows that a strong maximizing optimal control assumes minimal admissible values on a part of the surface where the absorption of thermal radiation occurs. Thus, the structure of strong optimal controls may be rather complicated, and therefore some practical heuristic solutions can be improved using the study presented. It would be also interesting to find other problems permitting strong optimal controls.

Acknowledgements. The research was supported by the Russian Scientific Foundation (Project No. 14-11-00079).

References

1. Modest, M.F.: Radiative Heat Transfer. Academic Press, An Imprint of Elsevier Science, San Diego (2003)
2. Kovtanyuk, A., Nefedev, K., Prokhorov, I.: Advanced computing method for solving of the polarized-radiation transfer equation. In: Hsu, C.-H., Malyshkin, V. (eds.) MTPP 2010. LNCS, vol. 6083, pp. 268–276. Springer, Heidelberg (2010). doi:10.1007/978-3-642-14822-4_30

3. Qi, H., Qiao, Y., Sun, S., Yao, Y., Ruan, L.: Image reconstruction of two-dimensional highly scattering inhomogeneous medium using MAP-based estimation. Math. Probl. Eng. **2015**, 412315 (2015)
4. Klose, A.D., Larsen, E.W.: Light transport in biological tissue based on the simplifed spherical harmonics equations. J. Comput. Phys. **220**, 441–470 (2006)
5. Kovtanyuk, A.E., Botkin, N.D., Hoffmann, K.-H.: Numerical simulations of a coupled radiative-conductive heat transfer model using a modified Monte Carlo method. Int. J. Heat Mass Transf. **55**, 649–654 (2012)
6. Thömes, G., Pinnau, R., Seaïd, M., Götz, T., Klar, A.: Numerical methods and optimal control for glass cooling processes. Transp. Theor. Stat. **31**(4–6), 513–529 (2002)
7. Frank, M., Klar, A., Pinnau, R.: Optimal control of glass cooling using simplified P_N theory. Transp. Theor. Stat. **39**(2–4), 282–311 (2010)
8. Clever, D., Lang, J.: Optimal control of radiative heat transfer in glass cooling with restrictions on the temperature gradient. Optim. Control. Appl. Methods **33**(2), 157–175 (2012)
9. Tse, O., Pinnau, R., Siedow, N.: Identification of temperature dependent parameters in a simplified radiative heat transfer. Aust. J. Basic Appl. Sci. **5**(1), 7–14 (2011)
10. Tse, O., Pinnau, R.: Optimal control of a simplified natural convection-radiation model. Commun. Math. Sci. **11**(3), 679–707 (2013)
11. Kovtanyuk, A.E., Chebotarev, A.Y., Botkin, N.D., Hoffmann, K.-H.: Theoretical analysis of an optimal control problem of conductive-convective-radiative heat transfer. J. Math. Anal. Appl. **412**, 520–528 (2014)
12. Pinnau, R.: Analysis of optimal boundary control for radiative heat transfer modeled by the SPN system. Commun. Math. Sci. **5**(4), 951–969 (2007)
13. Herty, M., Pinnau, R., Thömes, G.: Asymptotic and discrete concepts for optimal control in radiative transfer. Z. Angew. Math. Mech. **87**(5), 333–347 (2007)
14. Grenkin, G.V., Chebotarev, A.Y., Kovtanyuk, A.E., Botkin, N.D., Hoffmann, K.-H.: Boundary optimal control problem of complex heat transfer model. J. Math. Anal. Appl. **433**, 1243–1260 (2016)
15. Bobaru, F., Rachakonda, S.: Optimal shape profiles for cooling fins of high and low conductivity. Int. J. Heat Mass Transf. **47**(23), 4953–4966 (2004)
16. Belinskiy, B.P., Hiestand, J.W., McCarthy, M.L.: Optimal design of a bar with an attached mass for maximizing the heat transfer. Electron. J. Diff. Eqn. **13**(181), 1–13 (2012)
17. Huang, C.-H., Wuchiu, C.-T.: A shape design problem in determining the interfacial surface of two bodies based on the desired system heat flux. Int. J. Heat Mass Transf. **54**(11–12), 2514–2524 (2011)
18. Marck, G., Nadin, G., Privat, Y.: What is the optimal shape of a fin for one-dimensional heat conduction? SIAM J. Appl. Math. **74**(4), 1194–1218 (2014)
19. Kovtanyuk, A.E., Chebotarev, A.Y.: Steady-state problem of complex heat transfer. Comp. Math. Math. Phys. **54**(4), 719–726 (2014)
20. Kovtanyuk, A.E., Chebotarev, A.Y., Botkin, N.D., Hoffmann, K.-H.: The unique solvability of a complex 3D heat transfer problem. J. Math. Anal. Appl. **409**, 808–815 (2014)
21. Kovtanyuk, A.E., Chebotarev, A.Y., Botkin, N.D., Hoffmann, K.-H.: Unique solvability of a steady-state complex heat transfer model. Commun. Nonlinear Sci. Numer. Simul. **20**, 776–784 (2015)
22. Marshak, R.: Note on the spherical harmonic method as applied to the milne problem for sphere. Phys. Rev. **71**(7), 443–446 (1947)

A Non-autonomous Stochastic Discrete Time System with Uniform Disturbances

Ioannis K. Dassios[1,2] and Krzysztof J. Szajowski[3(✉)]

[1] MACSI, Department of Mathematics and Statistics, University of Limerick,
Limerick, Ireland
[2] ERC, Electricity Research Centre, University College Dublin, Dublin, Ireland
jdasios@math.uoa.gr
[3] Faculty of Pure and Applied Mathematics, Wrocław University of Technology,
Wybrzeże Wyspiańskiego 27, 50-370 Wrocław, Poland
Krzysztof.Szajowski@pwr.edu.pl
https://ioannisdassios.wordpress.com/research-visits-invited-talks/,
http://neyman.im.pwr.edu.pl/~szajow/

Abstract. The main objective of this article is to present Bayesian optimal control over a class of non-autonomous linear stochastic discrete time systems with disturbances belonging to a family of the one parameter uniform distributions. It is proved that the Bayes control for the Pareto priors is the solution of a linear system of algebraic equations. For the case that this linear system is singular, we apply optimization techniques to gain the Bayesian optimal control. These results are extended to generalized linear stochastic systems of difference equations and provide the Bayesian optimal control for the case where the coefficients of these type of systems are non-square matrices. The paper extends the results of the authors developed for system with disturbances belonging to the exponential family.

Keywords: Bayes control · Optimal · Singular system · Disturbances · Pareto distribution

1 Introduction

Linear stochastic discrete time systems (or linear matrix stochastic difference equations), are systems in which the variables take their value at instantaneous time points. The horizon of control depends on the problem. The state at instance n depends on random disturbance and the chosen controls. Discrete time systems differ from continuous time ones in that their signals are in the form of sampled data. With the development of the digital computer, the stochastic discrete time system theory plays an important role in the control theory. In real systems, the discrete time system often appears when it is the result of sampling the continuous-time system or when only discrete data are available for use. The

L. Bociu et al. (Eds.): CSMO 2015, IFIP AICT 494, pp. 220–229, 2016.
DOI: 10.1007/978-3-319-55795-3_20

investigation aims are, when such system is under consideration, determining the control goals, performance measures and the information available at moments of controls' specification. The small deviations of the parameters can be treated as disturbances. As the random disturbance is admitted the performance measure will be the mean value of the deviation of the states from the required behavior of the system. When all the parameters of the system are known and the distribution of disturbances is well defined then the optimal control can be determined at least for the finite horizon case. The extension of the model to the adaptive one means that the disturbances are not precisely described. Adaptive control is the control method used by a controller which must adapt to a controlled system with parameters which vary, or are initially uncertain (c.f. Black et al. [2] or Tesfatsion [18] for the history of the adaptive control). Under some unification the model of adaptive control of the linear system is formulated as a control of the discrete time Markov process (cf. [5]).

It is assumed that the disturbance has a fixed probabilistic description which is determined by the assumption. In this paper it is assumed that the distribution function is known to be an accuracy of parameters and the disturbances additionally change the state of the system. It resembles the statistical problem of estimation. It was seminal paper by Wald [21] where the background of the modern decision theory was established (cf. [22, Chap. 7]). The decision theory approach to the control problems were immediately applied (see books by Sworder [16], Aoki [1], Sage and Melsa [14]). The new class of control systems under uncertainty was called *adaptive* (cf. [2,18]). In these adaptive control problems the important role have Bayesian systems. In this class of control models it is assumed that the preliminary knowledge of the disturbances is given by *a priori* distributions of their parameters. The aim is to construct the controls in a close form. The construction of the Bayes control is also auxiliary for the construction of minimax controls (see Szajowski and Trybuła [17], Porosiński and Szajowski [11], Grzybowski [8], González-Trejo et al. [7]). It is observed the interested in various models of disturbance structure (cf. Duncan and Pasik-Duncan [6]) and the disturbance distributions (cf. Walczak [19,20]). Stochastic discrete time systems have many applications which we have described in [3] where the Bayes control of the linear system with quadratic cost function and the disturbances having the distribution belonging to the exponential family with conjugate priors is solved.

The paper is organized as follows: the description of the stochastic discrete time systems is subject of the Sect. 1.1 and some remarks on disturbances are given in the Sect. 1.2. In the Sect. 2 we determine the Bayes control for the conjugate prior distribution π of the parameter $\overline{\lambda}$ as the solution of a singular linear system and provide optimal Bayesian control. We close the paper by studying the Bayes control of a class of generalized linear stochastic discrete time systems.

1.1 Stochastic Discrete Time Systems

Let $\overline{x}_n \in \mathbb{R}^m$ be the state of the system, $\overline{u}_n \in \mathbb{R}^m$ be the control. Assume that $\overline{v}_n \in V \subset \mathbb{R}^m$, with $\overline{v}_n = (v_n^1, v_n^2, \ldots, v_n^k, 0, \ldots, 0)^T$, is the disturbance at time n and α_n, b_n, $c_n \in \mathbb{R}^{m \times m}$. Consider a stochastic discrete time system (cf. Kushner [9])

$$\overline{x}_{n+1} = \alpha_n \overline{x}_n + b_n \overline{u}_n + c_n \overline{v}_n, \quad \forall n = 0, 1, \dots, N - 1. \tag{1}$$

The horizon N of the control, the time up to which the system is controlled, is a random variable, independent of the disturbances $\overline{v}_0, \overline{v}_1, \dots$, and has the following known distribution

$$P\{N = k\} = p_k, \quad \forall k = 0, 1, \dots, M, \quad \sum_{i=0}^{M} p_k = 1, \quad p_M \neq 0. \tag{2}$$

In the authors paper [3] it was considered the family of the exponentially distributed disturbances. Let us assume here that the disturbances v_n^i have the uniform distributions on $[0, \lambda_i]$ with parameter $\lambda_i \in \Re^+$, $i = 1, 2, \dots, k$ and

$$X_n = (\overline{x}_0, \overline{x}_1, \dots, \overline{x}_n), \quad U_n = (\overline{u}_0, \overline{u}_1, \dots, \overline{u}_n), \quad \overline{\lambda} = (\lambda_1, \lambda_2, \dots, \lambda_k, 0, \dots, 0)^T.$$

For convenience U_M will be denoted by U and called a control policy.

Definition 1. *The control cost for a given policy U (the loss function) is the following*

$$L(U, X_N) = \sum_{i=0}^{N} (\overline{y}_i^T s_i \overline{y}_i + \overline{u}_i^T k_i \overline{u}_i), \tag{3}$$

where $k_i \in \mathbb{R}^{m \times m} \geq 0_{m,m}$, are symmetric matrices, $s_i \in \mathbb{R}^{2m \times 2m} \geq 0_{2m,2m}$ and

$$\overline{y}_i = \begin{pmatrix} \overline{x}_i \\ \dots \\ \overline{\lambda} \end{pmatrix} \in \mathbb{R}^{2m}, \forall i = 0, 1, \dots, M. \text{ With } 0_{i,j} \text{ we will denote the zero matrix}$$

$i \times j$.

Let the prior distribution π of the parameter $\overline{\lambda}$ be given. It is considered the Pareto priors (see [4, Chap. 9.7], [10]) with parameters $r_i > 0$, $\beta_i > 2$

$$g(\lambda_i | \beta_i, r_i) = \frac{\beta_i r_i^{\beta_i}}{\lambda_i^{\beta_i + 1}} \mathbb{I}_{[r_i, \infty)}(\lambda_i). \tag{4}$$

Denote $E_N, E_{\overline{\lambda}}$ the expectations with respect to the distributions of N and random vectors $\overline{v}_0, \overline{v}_1, \dots$ (when $\overline{\lambda}$ is the parameter), E_π and E are the expectations with respect to the distribution π and to the joint distribution $\overline{v}_n$ and $\overline{\lambda}$, respectively.

Definition 2 *(see [9, 12, 15, 19, 20]). Let $L(\cdot, \cdot)$ be the loss function given by (3).*

(a) *The risk connected with the control policy U, when the parameter $\overline{\lambda}$ is given, is defined as follows*

$$R(\overline{\lambda}, U) = E_N \left[E_{\overline{\lambda}}[L(U, X_N) \mid X_0] \right] = E_N \left[E_{\overline{\lambda}}[\sum_{i=0}^{N} \overline{y}_i^T s_i \overline{y}_i + \overline{u}_i^T k_i \overline{u}_i \mid X_0] \right].$$

(b) *The expected risk r, associated with π and the control policy U, is equal to*

$$r(\pi, U) = E_\pi[R(\overline{\lambda}, U)] = E_N\left[E[\sum_{i=0}^{N} \overline{y}_i^T s_i \overline{y}_i + \overline{u}_i^T k_i \overline{u}_i \mid X_0]\right].$$

(c) *The expected risk r, associated with π and the control policy U, is equal to*

$$r_n(\pi, U^n) = E_N[E_\pi[R(\overline{\lambda}, U)]] = E_N\left[E[\sum_{i=n}^{N} \overline{y}_i^T s_i \overline{y}_i + \overline{u}_i^T k_i \overline{u}_i \mid X_0]\right].$$

Let the initial state $\overline{x}_0$ and the distribution π of the parameter $\overline{\lambda}$ be given.

Definition 3. *A control policy U^* is called the Bayes policy when $r(\pi, U^*) = \inf_{U \in \wp_\pi} r(\pi, U)$, where $\wp_\pi$ is the class of the control policies U for which exists $r(\pi, U)$.*

1.2   Filtering

Let us assume that the random variables $\overline{v}_n$ have the density $p(\overline{v}_n, \overline{\lambda})$ with respect to a σ-finite measure μ on $\mathbb{R}$. The consideration is focused on the special case when each coordinate has the uniform distribution, i.e. the density $p(\overline{v}_n, \overline{\lambda})$ has the following representation:

$$p(\overline{v}_n, \overline{\lambda}) = \prod_{i=1}^{k} p(v_n^i, \lambda_i), \tag{5}$$

where $p(v_n^i, \lambda_i) = \frac{1}{\lambda_i} \mathbb{I}_{[0,\lambda_i]}(v_n^i)$, for all $i = 1, 2, ..., k$. V_i^* is the set of the random variables v_n^i. We have:

$$E_{\lambda_i}[v_n^i] = \frac{\lambda_i}{2} = q_i \lambda_i \text{ and, } E_{\lambda_i}[(v_n^i)^2] = \frac{\lambda_i^2}{3} = q_{1,i} \lambda_i^2,$$

where $q_i, q_{1,i}$ are constants. Let $\overline{\lambda}$ have the a priori distribution π with density

$$g(\overline{\lambda} \mid \overline{\beta}, \overline{r}) = \prod_{i=1}^{k} g_i(\lambda_i; \beta^i, r^i), \tag{6}$$

where $g_i(\lambda_i \mid \beta^i, r^i)$ is given by (4) where $\overline{\beta} \in S_k^\beta \subset \mathbb{R}^m$, $\overline{r} \in S_k^r \subset \mathbb{R}^m$ with

$$\overline{\beta} = (\beta^1, \beta^2, \ldots, \beta^k, 0, \ldots, 0)^T,$$

and $\overline{r} = (r^1, r^2, \ldots, r^k, 0, \ldots, 0)^T$. When such the a priori density is assigned to λ_i and then the object of filtering, to determine the Bayes control, is to produce a posteriori density for λ_i after any new observations of the state of the system. We change the control after obtaining the new data. Hence, to determine the Bayes control, a posteriori density for $\overline{\lambda}$ must be obtained after any new observations. This is possible if for $n = 0, 1, ...N - 1$ and a given $\overline{x}_0$, we can derive $\overline{v}_n$ from (1), i.e. the equations

$$\bar{v}_n = c_n^{-1}[\bar{x}_{n+1} - \alpha_n \bar{x}_n - b_n \bar{u}_n].$$

If for a value of n, the matrix c_n is singular, we will have to compute the Moore-Penrose Pseudoinverse $c_n^\dagger$ and then use the following expression

$$\bar{v}_n = c_n^\dagger[\bar{x}_{n+1} - \alpha_n \bar{x}_n - b_n \bar{u}_n].$$

The Moore-Penrose pseudo-inverse can be calculated via the singular value decomposition of c_n (see [13]). In these cases *a posteriori* density $f(\bar{\lambda} \mid X_n, U_{n-1})$ of the parameter $\bar{\lambda}$, after having observed X_n and chosen U_{n-1}, has the same form as (6) i.e.

$$f(\bar{\lambda} \mid X_n, U_{n-1}) = f(\bar{\lambda} \mid V_{n-1}) = g(\bar{\lambda} \mid \bar{\beta}_n, \bar{r}_n),$$

where $V_{n-1} = (\bar{v}_0, \bar{v}_1, ..., \bar{v}_{n-1})$, $\bar{\beta}_n = \bar{\beta}_{n-1} + 2\bar{q}$, $\bar{q} = (q_1, q_2, ..., q_k, 0, ..., 0)^T \in Q_i^* \subset \mathbb{R}^m$ and $\bar{r}_n = \bar{r}_{n-1} \vee \bar{v}_n$ ($\bar{r}_0 = \bar{r}$). Under these denotations we have $E(\lambda_i \mid X_n, U_{n-1}) = T^{n,i} r_n^i = \frac{\beta_n}{\beta_n - 1} r_n^i$ and $E(\lambda_i^2 \mid X_n, U_{n-1}) = T_1^{n,i}(r_n^i)^2$. For known X_n and U_{n-1}, the conditional distribution of $\bar{v}_n$ has the density

$$h(\bar{v}_n \mid X_n, U_{n-1}) = \prod_{i=1}^k h_i(v_n^i \mid X_n, U_{n-1}),$$

where

$$h_i(v_n^i \mid X_n, U_{n-1}) = \int_0^\infty p(v_n^i, \lambda) g(\lambda \mid \beta_n^i, r_n^i) d\lambda$$

$$= \frac{\beta_n^i (r_n^i)^{\beta_n^i}}{\beta_{n+1}^i} \frac{1}{(r_{n+1}^i)^{\beta_{n+1}}} \mathbb{I}_{[0,\infty)}(v),$$

for $n = 0, 1, ..., M-1$, $i = 1, 2, ..., k$. In addition (see [19,20]) by direct calculation we get

Lemma 1. *The following equations are fulfilled:*

$$E(v_n^i \mid X_n, U_{n-1}) = \frac{1}{2} \frac{\beta_n^i}{\beta_{n+1}^i} r_n^i = Q^{n,i} r_n \tag{7}$$

$$E((v_n^i)^2 \mid X_n, U_{n-1}) = Q_1^{n,i}(r_n^i)^2 \text{ where } Q_1^{n,i} = \frac{\beta_n^i}{3(\beta_n - 2)}. \tag{8}$$

$$E(r_{n+1}^i \mid X_n, U_{n-1}) = Q_2^{n,i} r_n^i \text{ where } Q_2^{n,i} = \frac{(\beta_n^i)^2}{(\beta_n^i)^2 - 1}, \tag{9}$$

$$E((r_{n+1}^i)^2 \mid X_n, U_{n-1}) = Q_3^{n,i}(r_n^i)^2 \text{ where } Q_3^{n,i} = \frac{\beta_n^i(\beta_n^i - 1)}{(\beta_n + 1)(\beta_n + 2)}. \tag{10}$$

$$E(x_{n+1}^i \mid X_n, U_{n-1}) = \alpha_n x_n + u_n + \gamma_n Q^{n,i} r_n^i, \tag{11}$$

$$E((x_{n+1}^i)^2 \mid X_n, U_{n-1}) = (\alpha_n x_n + u_n)^2 + 2(\alpha_n x_n + u_n)\gamma_n^i Q^{n,i} r_n^i$$
$$+ \gamma_n^2 Q_1^{n,i}(r_n^i)^2, \tag{12}$$

$$E(x_{n+1} r_{n+1}^i \mid X_n, U_{n-1}) = (\alpha_n x_n + u_n) Q_2^{n,i} r_n^i + \gamma_n^i Q_4^{n,i} r_n^i, \tag{13}$$

where $Q_4^{n,i} = \frac{(\beta_n^i)^2}{(\beta_n + 1)(\beta_n - 2)}.$

2   The Bayes Control

Suppose the initial state $\overline{x}_0$ is given, the disturbances have the distribution with the density given by (5) and the prior distribution π of the parameter $\overline{\lambda}$ is given by (6). Let the distribution of the random horizon N be given by (2). Consider the problem of the Bayes control for the system (1) with the starting point at the moment n, when X_n, U_{n-1} are given. The expected risk is then given by (c.f Defintion 2 (c); see [3, 12])

$$r_n(\pi, U^n) = E\left[\sum_{i=n}^{M}(\overline{y}_i^T s_i \overline{y}_i + \overline{u}_i^T k_i \overline{u}_i) \mid X_n, U_{n-1}\right]. \tag{14}$$

Let us denote $\varphi_k = \sum_{i=k}^{M} p_i$. We have

$$r_n = E[\sum_{i=n}^{M}\frac{\varphi_i}{\varphi_n}(\overline{y}_i^T s_i \overline{y}_i + \overline{u}_i^T k_i \overline{u}_i) \mid X_n, U_{n-1}].$$

For the above truncated problem we provide the following definitions:

Definition 4. *The Bayes risk is defined as*

$$W_n = \inf_{U^n} r_n(\pi, U^n), \tag{15}$$

where $r(\pi, U^n)$ is the expected risk defined in the Definition 2 (c) and the formulae (14).

Definition 5. *If there exists $U^{n^*} = (\overline{u}_n^*, \overline{u}_{n+1}^*, .., \overline{u}_N^*)$ such that $W_n = r(\pi, U^{n^*})$, then U^{n^*} will be called the Bayes policy and $\overline{u}_i^*, i = n, n+1, ..., N$ the Bayes controls for truncated control problem.*

Obviously, $r(\pi, U^0) = r(\pi, U), W_0 = r(\pi, U^*)$. For the solution of the Bayes control problem we derive the Bayes controls $\overline{u}_n^*$ for $n = N, N-1, ..., 1, 0$ recursively. Then U^{0^*} is the solution of the problem. From the Bellman's dynamic programming optimality principle we obtain the following Lemma, see [12].

Lemma 2. *Assume the stochastic discrete time system (1). Then the Bayes risk W_n has the form*

$$W_n = \overline{x}_n^T A_n \overline{x}_n + 2\overline{r}_n^T B_n \overline{x}_n + 2\overline{r}_n^T C_n \overline{r}_n, \tag{16}$$

where $A_n, B_n, C_n \in \mathbb{R}^{m \times m}, \overline{D}_n \in \mathbb{R}^m$ with $A_n = f_1(s_n), B_n = f_2(Q^n, Q_2^n, s_n),$ $C_n = f_3(Q^n, Q_1^n, Q_3^n, Q_4^n, s_n)$. The functions $f_j, j = 1, 2, 3$ are strictly monotonic, differentiable. The constants $Q^{n,i}, Q_j^{n,i}, j = 1, 2, 3, 4, n = 0, 1, ..., N$ are given by (7), (8) and s_n is defined in (3).

2.1   Bayesian Optimal Control for Stochastic Discrete Time Systems

We can now prove the following theorem

Theorem 1. *Assume the stochastic discrete time system* (1). *Then, the Bayes control $\bar{u}_n^*$ is given by the solution of the linear system*

$$K_n \bar{u}_n^* = L_n, \tag{17}$$

where

$$K_n = k_n + b_n^T A_{n+1} b_n \tag{18}$$

and

$$L_n = -b_n^T [A_{n+1} \alpha_n \bar{x}_n + (A_{n+1} c_n Q^n + B_{n+1} Q_2^n) \bar{r}_n]. \tag{19}$$

The matrices k_n, A_n, Q^n, are defined in (3), (16), *the Lemma 1, respectively and $\bar{e} = \sum_{j=0}^{n-1} Q^j \bar{r}_j$.*

PROOF. From (15), the Bayes risk is given by $W_n = \inf_{U^n} r(\pi, U^n)$. It is, equivalently,

$$W_n = \min_{U^n} E \left[\sum_{i=n}^{M} (\bar{y}_i^T s_i \bar{y}_i + \bar{u}_i^T k_i \bar{u}_i) \mid X_n, U_{n-1} \right].$$

We have

$$W_n = \min_{\bar{u}_n} \left\{ \bar{u}_n^T k_n \bar{u}_n + E \left[\bar{y}_n^T s_n \bar{y}_n \mid X_n, U_{n-1} \right] \right.$$

$$\left. + \min_{U^{n+1}} E \left[E \left[\sum_{i=n+1}^{k} (\bar{y}_i^T s_i \bar{y}_i + \bar{u}_i^T k_i \bar{u}_i) \right] \mid X_n, U_{n-1} \right] \right\}.$$

It means $W_n = \min_{\bar{u}_n} \{ \bar{u}_n^T k_n \bar{u}_n + E [\bar{y}_n^T s_n \bar{y}_n \mid X_n, U_{n-1}] + E [W_{n+1} \mid X_n, U_{n-1}] \}$. Hence, the Bayes control $\bar{u}_n^*$ satisfies the equation (∇ is the gradient):

$$\nabla_{\bar{u}_n} \left\{ \bar{u}_n^T k_n \bar{u}_n + E \left[\bar{y}_n^T s_n \bar{y}_n \mid X_n, U_{n-1} \right] + E \left[W_{n+1} \mid X_n, U_{n-1} \right] \right\}_{\bar{u}_n = \bar{u}_n^*} = 0_{m,1}.$$

By using (16) we get

$$k_n \bar{u}_n^* + b_n^T A_{n+1} (\alpha_n \bar{x}_n + b_n \bar{u}_n$$

$$+ c_n E(\bar{v}_n \mid X_n, U_{n-1})) + E \left[\left\{ b_n^T B_{n+1} \bar{r}_{n+1} \right\}_{\bar{u}_n = \bar{u}_n^*} \mid X_n, U_{n-1} \right] = 0_{m,1}.$$

By the properties of conjugate priors for the uniform distribution (see the Lemma 1 we have

$$k_n \bar{u}_n^* + b_n^T A_{n+1} (\alpha_n \bar{x}_n + b_n \bar{u}_n + c_n Q^n \bar{r}_n) + E \left[\left\{ b_n^T B_{n+1} \bar{r}_{n+1} \right\}_{\bar{u}_n = \bar{u}_n^*} \mid X_n, U_{n-1} \right] = 0_{m,1},$$

and at the end $(k_n + b_n^T A_{n+1} b_n) \bar{u}_n^* = -b_n^T [A_{n+1} \alpha_n \bar{x}_n + (A_{n+1} c_n Q^n + B_{n+1} Q_2^n) \bar{r}_n]$.
The proof is completed. ∎

Similarly like for the system with the disturbances belonging to the exponential family (see [3]) we get

Theorem 2. *Consider the system* (1) *and the matrices* K_n, L_n *as defined in* (18), (19) *respectively. Then*

(a) $\forall n$ *such that* K_n *is full rank, the Bayes control* $\bar{u}_n^*$, *is given by*

$$\bar{u}_n^* = K_n^{-1} L_n. \tag{20}$$

(b) $\forall n$ *such that* K_n *is rank deficient, the Bayesian optimal control* $\hat{u}_n^*$ *is given by*

$$\hat{u}_n^* = (K_n^T K_n + E^T E)^{-1} K_n^T L_n. \tag{21}$$

Where E *is a matrix such that* $K_n^T K_n + E^T E$ *is invertible and* $\|E\|_2 = \theta$, $0 < \theta \ll 1$. *Where* $\|\cdot\|_2$ *is the Euclidean norm.*

2.2 Bayesian Optimal Control for Generalized Stochastic Discrete Time Systems

In this subsection we will expand the results of the Sect. 2.1 by studying Bayesian optimal control for a class of linear stochastic discrete time systems with non-square coefficients. We consider the following non-autonomous linear stochastic discrete time system

$$I_{r,m} \bar{x}_{n+1} = \alpha_n \bar{x}_n + b_n \bar{u}_n + c_n \bar{v}_n, \quad \forall n = 0, 1, ..., N - 1. \tag{22}$$

Where $\bar{x}_n \in \mathbb{R}^m$ is the state of the system, $\bar{u}_n \in \mathbb{R}^m$ is the control, $\bar{v}_n \in V \subset \mathbb{R}^m$, with $\bar{v}_n = (v_n^1, v_n^2, ..., v_n^k, 0, ..., 0)^T$, is the disturbance at time n and $\alpha_n, b_n, c_n \in \mathbb{R}^{r \times m}$. The horizon N of the control is fixed and independent of the disturbances $\bar{v}_n$, $n \geq 0$. If $r = m$, then $I_{r,m} = I_m$. If $r > m$, then $I_{r,m} = \begin{bmatrix} I_m \\ 0_{r-m,m} \end{bmatrix}$ and if $r < m$, then $I_{r,m} = \begin{bmatrix} I_r & 0_{r,m-r} \end{bmatrix}$ with I_m, I_r identity matrices.

Definition 6. *We will refer to system* (22) *as a generalized stochastic linear discrete time system.*

In the above definition we use the term *"generalized"* because the coefficients in the system (22) can be either square or non-square matrices.

Theorem 3. *Consider the system* (22) *for* $r \neq m$ *and assume the matrices* K_n, L_n *as defined in* (18), (19) *respectively. Then,* $\forall n$ *such that*

(a) $m < r$, $rank(K_n) = m$ *and* $L_n \in colspan K_n$, *the Bayes control* $\bar{u}_n^*$, *is given by*

$$\bar{u}_n^* = K_n^{-1} L_n. \tag{23}$$

(b) $m < r$, $rank(K_n) = m$ *and* $L_n \notin colspan(K_n)$, *a Bayesian optimal control is given by*

$$\hat{u}_n^* = (K_n^T K_n)^{-1} K_n^T L_n. \tag{24}$$

(c) $L_n \notin colspan K_n$ and K_n is rank deficient, a Bayesian optimal control is given by

$$\hat{u}_n^* = (K_n^T K_n + E^T E)^{-1} K_n^T L_n. \tag{25}$$

Where E is a matrix such that $K_n^T K + E^T E$ is invertible and $\|E\|_2 = \theta$, $0 < \theta \ll 1$.

(d) $m > r$, K_n is full rank, a Bayesian optimal control is given by

$$\hat{u}_n^* = K_n^T (K_n K_n^T)^{-1} L_n. \tag{26}$$

(e) $L_n \in colspan K_n$ and K_n is rank deficient, a Bayesian optimal control is given by (25).

The proof is based on ideas similar to those used in prove [3, Theorem 3] and is omitted here.

3 Conclusions

In this article we focused on developing the Bayesian optimal control for a class of non-autonomous linear stochastic discrete time systems of type (1). Firstly, we proved that the Bayes control of these type of systems is the solution of a linear system of algebraic equations which can also be singular. For this case we used optimization techniques to derive the optimal Bayes control for (1). In addition, we used these methods to obtain the Bayesian optimal control of the non-autonomous linear stochastic discrete time system of type (2), where the coefficients of this system are non-square matrices.

The further extension of this paper is to study to Bayes control problem of stochastic fractional discrete time systems. The fractional nabla operator is a very interesting tool when applied to systems of difference equations and has many applications especially in macroeconomics, since it succeeds to provide information from a specific year in the past until the current year. For all these there is some research in progress.

Acknowledgments. I. Dassios is supported by Science Foundation Ireland (award 09/SRC/E1780).

References

1. Aoki, M.: Optimization of Stochastic Systems. Topics in Discrete-Time Systems. Mathematics in Science and Engineering, vol. 32. Academic Press, New York-London (1967)
2. Black, W.S., Haghi, P., Ariyur, K.B.: Adaptive systems: history, techniques, problems, and perspectives. Systems **2**, 606–660 (2014). http://dx.doi.org/10.3390/systems2040606
3. Dassios, I.K., Szajowski, K.J.: Bayesian optimal control for a non-autonomous stochastic discrete time system. Technical report, Faculty of Pure and Applied Mathematicas, Wrocław, Poland (2015, to appear)

4. DeGroot, M.: Optimal Statistical Decision. McGraw Hill Book Co., New York (1970)
5. Duncan, T.E., Pasik-Duncan, B., Stettner, L.: Adaptive control of a partially observed discrete time Markov process. Appl. Math. Optim. **37**(3), 269–293 (1998). http://dx.doi.org/10.1007/s002459900077
6. Duncan, T.E., Pasik-Duncan, B.: Discrete time linear quadratic control with arbitrary correlated noise. IEEE Trans. Autom. Control **58**(5), 1290–1293 (2013). http://dx.doi.org/10.1109/TAC.2012.2220444
7. González-Trejo, J.I., Hernández-Lerma, O., Hoyos-Reyes, L.F.: Minimax control of discrete-time stochastic systems. SIAM J. Control Optim. **41**(5), 1626–1659 (2002). (electronic). http://dx.doi.org/10.1137/S0363012901383837
8. Grzybowski, A.: Minimax control of a system with actuation errors. Zastos. Mat. **21**(2), 235–252 (1991)
9. Kushner, H.: Introduction to Stochastic Control. Holt, Rinehart and Winston, Inc., New York-Montreal, Quebec-London (1971)
10. Philbrick, S.: A practical guide to the single parameter pareto distribution. In: Proceedings of the Casualty Actuarial Society, vol. LXXII, pp. 44–84. Casualty Actuarial Society, Arlington, USA (1985)
11. Porosiński, Z., Szajowski, K.: A minimax control of linear systems. In: Zabczyk, J. (ed.) Stochastic Systems and Optimization. LNCIS, vol. 136, pp. 344–355. Springer, Berlin (1989). doi:10.1007/BFb0002665. MR1180792; Zbl:0711.93095
12. Porosiński, Z., Szajowski, K., Trybuła, S.: Bayes control for a multidimensional stochastic system. Syst. Sci. **11**(2), 51–64 (1985, 1987)
13. Rugh, W.J.: Linear System Theory. Prentice Hall Information and System Sciences Series. Prentice Hall, Inc., Englewood Cliffs (1993)
14. Sage, A.P., Melsa, J.L.: Estimation Theory with Applications to Communications and Control. McGraw-Hill Series in Systems Science. McGraw-Hill Book Co., New York-Düsseldorf-London (1971)
15. Sawitzki, G.: Exact filtering in exponential families: discrete time. Math. Operationsforsch. Stat. Ser. Stat. **12**(3), 393–401 (1981). http://dx.doi.org/10.1080/02331888108801598
16. Sworder, D.: Optimal Adaptive Control Systems. Mathematics in Science and Engineering, vol. 25. Academic Press, New York-London (1966)
17. Szajowski, K., Trybuła, S.: Minimax control of a stochastic system with the loss function dependent on parameter of disturbances. Statistics **18**(1), 151–165 (1987). http://dx.doi.org/10.1080/02331888708802005
18. Tesfatsion, L.: A dual approach to Bayesian inference and adaptive control. Theory Decis. **14**(2), 177–194 (1982). http://dx.doi.org/10.1007/BF00133976
19. Walczak, D.: Bayes and minimax control of discrete time linear dynamical systems. Technical report, TU Wrocław, Master thesis (in Polish) (1986)
20. Walczak, D.: Bayesian control of a discrete-time linear system with uniformly distributed disturbances. Math. Appl. **43**(2), 173–186 (2016). http://dx.doi.org/10.1016/j.amc.2013.10.090
21. Wald, A.: Contributions to the theory of statistical estimation and testing hypotheses. Ann. Math. Stat. **10**, 299–326 (1939)
22. Wald, A.: Statistical Decision Functions. Wiley, Chapman and Hall, Ltd., New York, London (1950)

Differentials and Semidifferentials for Metric Spaces of Shapes and Geometries

Michel C. Delfour$^{(\boxtimes)}$

Centre de recherches mathématiques and Département de
mathématiques et de statistique, Université de Montréal,
CP 6128, succ. Centre-ville, Montréal, QC H3C 3J7, Canada
delfour@crm.umontreal.ca
http://dms.umontreal.ca/~delfour/

Abstract. The *Hadamard semidifferential* retains the advantages of the differential calculus such as the *chain rule* and semiconvex functions are Hadamard semidifferentiable. The *semidifferential calculus* extends to subsets of $\mathbb{R}^n$ without Euclidean smooth structure. This set-up is an ideal tool to study the semidifferentiability of objective functions with respect to families of sets which are non-linear non-convex complete metric spaces. *Shape derivatives* are differentials for spaces endowed with *Courant metrics*. *Topological derivatives* are shown to be *semidifferentials* on the group of Lebesgue measurable characteristic functions.

Keywords: Semidifferential · Shape and topological derivatives

1 Introduction

In the past decades, direct constructions of complete *metric spaces of shapes and geometries* (cf., for instance, Delfour and Zolésio [7]) and, additional new ones (cf., Delfour [4,6]) have been given without appealing to the classical notions of atlases or smooth manifolds encountered in classical Differential Geometry. Since, at best, such spaces are groups, the issue of making sense of tangent spaces and differentials naturally arises not only for "differentiable" functions but also for large classes of "non-differentiable" functions.

In that context, the geometrical definition of a differentiable function of Hadamard [5,11] is especially interesting since it implicitly involves the construction of trajectories (or paths) and tangent vectors to trajectories living in the space under investigation. His definition was relaxed by Fréchet [10] in 1937 by dropping the requirement that the differential be linear with respect to the direction or tangent vector while preserving two important properties of the differential calculus: the continuity of the function and the chain rule. A vast litterature on differentials on topological spaces followed (cf., for instance, the survey papers of Averbuh and Smoljanov [3] in 1988 and the 207-page paper of Nashed [12] for a rather complete account until 1971). The definition of Fréchet

M.C. Delfour—This research has been supported by a Discovery Grant from the Natural Sciences and Engineering Research Council of Canada.

L. Bociu et al. (Eds.): CSMO 2015, IFIP AICT 494, pp. 230–239, 2016.
DOI: 10.1007/978-3-319-55795-3_21

can be further relaxed to the one of semidifferential which handles convex and semiconvex functions while preserving the two properties.

Since the semidifferential is not required to be linear, they have far reaching consequences for a function $f : A \to B$ between arbitrary sets A and B. De facto, this relaxes the requirement that the tangent spaces in each points of the sets A and B be linear spaces. It is sufficient to work with tangent cones to A and B such as Bouligand's tangent cone to make sense of semidifferentials. Shortcircuiting the requirement of a smooth manifold makes it possible to directly study the tangent spaces to non-convex metric spaces of shapes and geometries.

We show that the metric group of Lebesgue measurable characteristic functions has semi-tangents and that the notion of *topological derivative* of Sokołowski and Żochowski [14] is in fact a semidifferential obtained by dilatation of a point creating a hole. By extending this construction via *dilatations*, we also show that the tangent space contains distributions creating topological perturbations along curves and surfaces that can break the connectivity of the set. In the same spirit *dilatations* of d-rectifiable and some H_d-rectifiable compact sets (H_d, d-dimensional Hausdorff measure) of Ambrosio et al. [1] also generate semi-tangents. *Orthogonal dilatations* of closed subsets of the boundary of a set of positive reach can also be used via Steiner formula (see Federer [8]).

2 Hadamard Differential and Semidifferential

Hadamard Differential. In 1923 Hadamard [11] gave a *geometrical definition* by using an *auxiliary function* $t \mapsto x(t) : \mathbb{R} \to \mathbb{R}^N$ such that

$$x(0) = a \quad \text{and} \quad x'(0) \stackrel{\text{def}}{=} \lim_{t \to 0} \frac{x(t) - a}{t} \text{ exists in } \mathbb{R}^N,$$

where $\mathbb{R}$ is the field of real numbers. It defines a path that induces a perturbation or a variation of the point a. We shall use the terminology *time* for the auxiliary variable t and *admissible trajectory* for the auxiliary function x. Note that x need not be continuous or differentiable at $t \neq 0$. The *vector $x'(0)$ is the tangent to the trajectory x at the point $x(0) = a$*. Scaling t by an arbitrary non-zero real number generates a whole line tangent to x at a

Definition 1. A function $f : \mathbb{R}^N \to \mathbb{R}^K$ *is Hadamard differentiable at $a \in \mathbb{R}^N$ if*

(i) for all *admissible trajectories* x at a the limit

$$(f \circ x)'(0) \stackrel{\text{def}}{=} \lim_{t \to 0} \frac{f(x(t)) - f(a)}{t} \text{ exists in } \mathbb{R}^K$$

(ii) and there exists a *linear function* $Df(a) : \mathbb{R}^N \to \mathbb{R}^K$ such that for all admissible trajectories x at a

$$(f \circ x)'(0) = Df(a)\,(x'(0)).$$

$Df(a)$ is the *differential of f at a.* □

The definition of *Hadamard differentiability* is equivalent to the one of Fréchet differentiability in finite dimension. In Banach and Fréchet spaces, a *Hadamard differentiable function* at a point a is *continuous* at a and the *chain rule is applicable*. In 1937, Fréchet [10] *insisted* on the fact that the *definition of Hadamard is more general than his* since it extends to functions $f : X \to \mathbb{R}^K$ defined on *topological vector spaces* X that are not *normed vector spaces*. Furthermore, in *Banach spaces of functions*, we can consider the set of *tangent vectors (functions)* $x'(0)$ as *weak limits* ... and even as *distributions*.

In his 1937 paper, Fréchet [10] observed that, in *function spaces*, the Hadamard differentiability is *not only* a notion more general than the one he introduced in 1911 but that the *linearity in part (ii) is not necessary* to preserve the continuity of the function and the chain rule. He gives the following example:

$$f(x_1, x_2) \overset{\text{def}}{=} x \sqrt{\frac{x_1^2}{x_1^2 + x_2^2}} \quad \text{for } (x_1, x_2) \neq (0,0) \quad \text{and} \quad f(0,0) \overset{\text{def}}{=} 0.$$

([10, p. 239]). Indeed, it is readily checked that for any trajectory $x : \mathbb{R}^2 \to \mathbb{R}$ such that $x(0) = (0,0)$ and $x'(0)$ exists

$$(f \circ x)'(0) \overset{\text{def}}{=} \lim_{t \to 0} \frac{f(x(t)) - f(0,0)}{t} = f(x'(0)).$$

Hadamard always insisted on the linearity and this new notion was criticized by P. Lévy. Yet, his example shows that such *nondifferentiable functions* exist.

Hadamard Semidifferential. By relaxing the linearity, we can deal with some families of non-differentiable functions. Unfortunately, some *convex continuous functions* and, in particular, the *norm* $\|x\|$ in $a = 0$, are *not differentiable* in this relaxed sense. To get around this, we need the notion of *semidifferential*.

For instance, in the case of the Euclidean norm $x \mapsto f(x) = \|x\| : \mathbb{R}^N \to \mathbb{R}$ at $x = 0$, consider a *semi-trajectory* $x : [0, +\infty) \to \mathbb{R}^N$ through the origin $x(0) = 0$ for which the right-hand limit $x'(0^+)$ exists. We get at $a = 0$

$$(f \circ x)'(0^+) \overset{\text{def}}{=} \lim_{t \searrow 0} \frac{f(x(t)) - f(0)}{t} = \lim_{t \searrow 0} \left\| \frac{x(t) - x(0)}{t} \right\| = \|x'(0^+)\|,$$

where the notation $t \searrow 0$ means that t goes to 0 by strictly positive values. We have a similar result for *convex and semiconvex* continuous functions. When $(f \circ x)'(0^+)$ is not a linear function of the semi-tangent $x'(0^+)$, we say that the function is *semidifferentiable*.

From Linear to Non-convex Spaces. The *hypothesis of linearity of the differential* is also a severe restriction to define a differential for a function $f : A \subset \mathbb{R}^N \to B \subset \mathbb{R}^K$ since it requires that the *tangent space to A at a* and the *tangent space to B at $f(a)$* be *linear subspaces*. This necessitates that the sets A and B be *sufficiently smooth* in the sense that, at each point of A and of B, the tangent spaces be linear subspaces of $\mathbb{R}^N$ and $\mathbb{R}^K$.

Since the Hadamard semidifferential does not require the linearity of the tangent space, the *a priori smoothness assumption* of the sets A and B can be de facto dropped since the semidifferential only needs to be defined on a tangent cone. Several tangent cones are available in the literature, but the following one is especially well suited for semidifferentials.

Definition 2. The *Bouligand tangent cone* to a set A at a point $a \in \overline{A}$ is

$$T_a A \overset{\text{def}}{=} \left\{ v \in \mathbb{R}^N : \exists \{x_n\} \subset A \text{ and } \{t_n \searrow 0\} \text{ such that } \lim_{n \to \infty} \frac{x_n - a}{t_n} = v \right\}.$$

□

When the boundary ∂A of A is smooth, $T_a A$ is a linear subspace of $\mathbb{R}^N$. However, the linearity of $T_a A$ puts a severe restriction on the sets A. For instance, the requirement that $T_a A$ be linear rules out a curve in $\mathbb{R}^2$ with kinks.

$x'(0) \overset{\text{def}}{=} \lim_{t \to 0} \frac{x(t)-a}{t}$ exists

- tangent linear subspace $T_a(A)$

a

path or trajectory $x(t)$

A

- tangent (non convex) cone $T_a(A)$

$x'(0^+) \overset{\text{def}}{=} \lim_{t \searrow 0} \frac{x(t)-a}{t}$ exists

a

half or semi-trajectory $x(t)$

A

This naturally leads to the following notions of admissible trajectory.

Definition 3. Given $A \subset \mathbb{R}^N$, an *admissible semi-trajectory* in A at $a \in \overline{A}$ is a function $x : [0, \tau] \to A$, $\tau > 0$, such that the *semi-tangent* at a

$$x'(0^+) \overset{\text{def}}{=} \lim_{t \searrow 0} \frac{x(t) - a}{t}$$

exists. When the limit $x'(0^+)$ exists, it follows that $x(t) \to a$ as $t \searrow 0$.   □

An equivalent characterization of the Bouligand's tangent cone is obtained.

Theorem 1. $T_a A = \{x'(0^+) : x \text{ is an admissible semi-trajectory in } A \text{ at } a\}$.

Following Fréchet, we now relax the linearity and formalize the notion of semidifferential for functions $f : A \to B$.

Definition 4 (Geometrical definition). Given $A \subset \mathbb{R}^N$ and $B \subset \mathbb{R}^K$, the function $f : A \to B$ is *Hadamard semidifferentiable* at $a \in A$ if

(i) for each *admissible semi-trajectory* x in A at a, the limit

$$(f \circ x)'(0^+) \stackrel{\text{def}}{=} \lim_{t \searrow 0} \frac{f(x(t)) - f(a)}{t} \text{ exists}$$

(ii) and there exists a (positively homogeneous) function $v \mapsto d_A f(a; v) : T_a A \to T_{f(a)} B$ such that for all *admissible semi-trajectories* x in A at a

$$(f \circ x)'(0^+) = d_A f(a; x'(0^+)).$$

The function $v \mapsto d_A f(a)(v) = d_A f(a; v)$ is referred to as the *(tangential) semi-differential of f* at $a \in A$. It can be shown that $d_A f(a)$ is continuous on $T_a A$.   □

This definition has an equivalent analytical counterpart.

Theorem 2 (Analytical definition). *Given $A \subset \mathbb{R}^N$ and $B \subset \mathbb{R}^K$, the function $f : A \to B$ is Hadamard semidifferentiable at $a \in A$ if and only if there exists a (positively homogeneous) function $v \mapsto d_A f(a; v) : T_a A \to T_{f(a)} B$ such that for all $v \in T_a A$ and all sequences $\{x_n\} \subset A$ and $\{t_n \searrow 0\}$ such that $(x_n - a)/t_n \to v$*

$$\lim_{n \to \infty} \frac{f(x_n) - f(a)}{t_n} = d_A f(a; v).$$

With the above definitions, the two important properties are preserved: continuity of f at a and the *chain rule*. The previous definitions extend to subsets A of topological vector spaces X, but we have to be careful and retain the abstract notions that are really meaningful. For shapes and geometries, the subset A will be a *complete metric space* with or without a group structure in a *surrounding Banach or Fréchet space*. We consider *Courant metrics* and the metric space of characteristic functions. Oriented distance functions can also be considered.

3   Metric Group of Characteristic Functions

Consider the metric Abelian group of characteristic functions on $\mathbb{R}^N$

$$X(\mathbb{R}^N) = \{\chi_\Omega : \Omega \subset \mathbb{R}^N \text{ Lebesgue measurable}\} \subset L^\infty(\mathbb{R}^N).$$

It is a *closed subset without interior* of the Banach space $L^\infty(\mathbb{R}^N)$ and of the Fréchet spaces $L^p_{\text{loc}}(\mathbb{R}^N)$, $1 \le p < \infty$. The analog would be the sphere in $\mathbb{R}^3$.

3.1   Velocity Method

For the *velocity method*, consider the following *continuous trajectory* in $X(\mathbb{R}^N)$

$$t \mapsto \chi_{T_t(V)(\Omega)} : [0,1] \to X(\mathbb{R}^N), \quad \frac{dT_t(V)}{dt} = V(t) \circ T_t(V), \quad T_0(V) = I.$$

The *semitangent* at χ_Ω is obtained by considering the limit of the differential quotient $(\chi_{T_t(V)(\Omega)} - \chi_\Omega)/t \in L^\infty(\mathbb{R}^N)$ which does not exist in $L^\infty(\mathbb{R}^N)$, but also not in $L^p_{\text{loc}}(\mathbb{R}^N)$, $1 \le p < \infty$.

To get a derivative consider the *distribution* associated with $\chi_{T_t(V)(\Omega)}$

$$\phi \mapsto \int_{\mathbb{R}^N} \chi_{T_t(V)(\Omega)} \, \phi \, dx = \int_{T_t(V)(\Omega)} \phi \, dx = \int_{\Omega} \phi \circ T_t \, \det DT_t \, dx : \mathcal{D}(\mathbb{R}^N) \to \mathbb{R}$$

If $V \in C^{0,1}(\overline{\mathbb{R}^N}, \mathbb{R}^N)$, then

$$\frac{d}{dt}\Big|_{t=0^+} \int_{\Omega} \phi \circ T_t \, \det DT_t \, dx = \int_{\Omega} \operatorname{div}\,(V(0)\,\phi)\, dx = \int_{\mathbb{R}^N} \chi_{\Omega} \operatorname{div}\,(V(0)\,\phi)\, dx$$

(see, for instance, [7, Theorem 4.1, Chap. 9, p. 483]). The bilinear function

$$(\phi, V) \mapsto \int_{\mathbb{R}^N} \chi_{\Omega} \operatorname{div}\,(V(0)\,\phi)\, dx : H_0^1(\mathbb{R}^N) \times C^{0,1}(\overline{\mathbb{R}^N}, \mathbb{R}^N) \to \mathbb{R}$$

is continuous. This generates the continuous linear mapping $V \mapsto \nabla\chi_{\Omega} \cdot V :$ $C^{0,1}(\overline{\mathbb{R}^N}, \mathbb{R}^N) \to H^{-1}(\mathbb{R}^N)$

$$(\nabla\chi_{\Omega} \cdot V)\,\phi \stackrel{\text{def}}{=} \int_{\mathbb{R}^N} \chi_{\Omega} \operatorname{div}\,(V(0)\,\phi)\, dx,$$

where $\nabla\chi_{\Omega}$ is the distributional gradient of χ_{Ω}. The support of $\nabla\chi_{\Omega} \cdot V$ is in Γ, the boundary of Ω. So, the tangent space to $X(\mathbb{R}^N)$ (considered as a subset of the space of distributions) at χ_{Ω} contains the linear subspace

$$\left\{ \nabla\chi_{\Omega} \cdot V : V \in C^{0,1}(\overline{\mathbb{R}^N}, \mathbb{R}^N) \right\} \subset H^{-1}(\mathbb{R}^N) \subset \mathcal{D}(\mathbb{R}^N)'$$

of functions in $H^{-1}(\mathbb{R}^N)$. When Ω is an open set with Lipschitz boundary

$$\frac{d}{dt}\Big|_{t=0^+} \int_{\mathbb{R}^N} \chi_{T_t(V)(\Omega)} \, \phi \, dx = \int_{\Gamma} V(0) \cdot n_{\Gamma}\, \phi \, dH_{N-1}$$

is a bounded measure, where H_d is the d-dimensional Hausdorff measure.

3.2   Topological Derivative via Dilatations

The rigorous introduction of the *topological derivative* in 1999 by Sokołowski and Zóchowski [14]) (see also the book by Novotny-Sokołowski [13]) opened a broader spectrum of notions of "differential" with respect to a set. The set Ω is topologically perturbed by introducing a *small hole* around a point $a \in \Omega$, that is, a dilatation of a. This idea can be readily extended to some families of closed subsets E of Ω of dimension d, $1 \le d \le N - 1$, for which $H_d(E)$ is finite.

Given $\Omega \subset \mathbb{R}^N$ open, we consider several examples where m_N denotes the Lebesgue measure in $\mathbb{R}^N$. The *distance function* $d_E(x)$ of x to a subset $E \subset \mathbb{R}^N$ and the *r-dilatation of E* are defined as

$$d_E(x) \stackrel{\text{def}}{=} \inf_{y \in E} |x - y|, \quad E_r \stackrel{\text{def}}{=} \{x \in \mathbb{R}^N : d_E(x) \le r\}. \tag{3.1}$$

Example 1. $E = \{a\}$, $a \in \mathbb{R}^3$, $\dim E = 0$. The *r-dilatation* of E is $\bar{B}_r(a)$,

$$t \stackrel{\text{def}}{=} m_3(\bar{B}_r(a)) = \alpha_3 r^3, \quad \alpha_3 = 4\pi/3 = \text{ volume of unit ball in } \mathbb{R}^3$$
$$\phi \mapsto \phi(a) : \mathcal{D}(\mathbb{R}^3) \to \mathbb{R} \text{ is a distribution.}$$

Assuming that $\bar{B}_r(a) \subset \Omega$ for some $r > 0$, the perturbed sets will be

$$t \mapsto \Omega_t \stackrel{\text{def}}{=} \Omega \setminus E_r = \Omega \setminus \bar{B}_{\sqrt[3]{t/\alpha_3}}(a).$$

Given $\phi \in \mathcal{D}(\mathbb{R}^3)$, the weak limit of the differential quotient $(\chi_{\Omega_t} - \chi_\Omega)/t$ is

$$\frac{1}{t}\left[\int_{\Omega_t} \phi\, dx - \int_\Omega \phi\, dx\right] = -\frac{1}{m_3(\bar{B}_{\sqrt[3]{t/\alpha_3}}(a))} \int_{\bar{B}_{\sqrt[3]{t/\alpha_3}}(a)} \chi_\Omega\, \phi\, dx$$

$$= -\frac{1}{m_3(\bar{B}_r(a))} \int_{\bar{B}_r(a)} \chi_\Omega\, \phi\, dx \to -\phi(a).$$

This distribution is a *half tangent* since for all $\rho > 0$

$$\frac{1}{t}\left[\int_{\Omega_{\rho t}} \phi\, dx - \int_\Omega \phi\, dx\right] \to -\rho\phi(a).$$

$\square$

Example 2. Let $A \subset \mathbb{R}^3$ be an open set of class $C^{1,1}$, ∂A compact, $\dim \partial A = 2$, and $b_A(x) \stackrel{\text{def}}{=} d_A(x) - d_{\mathbb{R}^3 \setminus A}(x)$ be the *oriented distance function*, then

$$\exists \varepsilon > 0 \text{ such that } b_A \in C^{1,1}(U_\varepsilon(\partial A)), \quad U_\varepsilon(\partial A) \stackrel{\text{def}}{=} \{x \in \mathbb{R}^3 : |b_A(x)| < \varepsilon\}$$
$$\text{projection onto } \partial A : p_{\partial A}(x) = x - b_A(x)\, \nabla b_A(x), \quad H_2(\partial A) < \infty.$$

Consider the *shell or sandwich of thickness* $t = 2r$ *around* $E = \partial A$ and, for $0 < r < \varepsilon$, the *r-dilatation* $E_r \stackrel{\text{def}}{=} \{x \in \mathbb{R}^3 : |b_A(x)| \leq r\} = \{x \in \mathbb{R}^3 : d_{\partial A}(x) \leq r\}$,

$$t = 2r = \alpha_1 r, \quad \alpha_1 = 2 = \text{ volume of the unit ball in } \mathbb{R}^1$$
$$\phi \mapsto \int_E \phi\, dH_2 : \mathcal{D}(\mathbb{R}^3) \to \mathbb{R} \text{ is a distribution.}$$

Assuming that $U_\varepsilon(\partial A) \subset \Omega$, the perturbed sets for $0 < t < \varepsilon$ are

$$t \mapsto \Omega_t \stackrel{\text{def}}{=} \Omega \setminus E_r = \Omega \setminus E_{t/2}.$$

Given $\phi \in \mathcal{D}(\mathbb{R}^3)$, the weak limit of the differential quotient $(\chi_{\Omega_t} - \chi_\Omega)/t$ is

$$\frac{1}{t}\left[\int_{\Omega_t} \phi\, dx - \int_\Omega \phi\, dx\right] = -\frac{1}{t}\int_{E_{t/2}} \chi_\Omega\, \phi\, dx$$

$$= -\frac{1}{\alpha_1 r}\int_{E_r} \chi_\Omega\, \phi\, dx \to -\int_E \phi\, dH_2.$$

This distribution is a *half tangent* since for all $\rho > 0$

$$\frac{1}{t}\left[\int_{\Omega_{\rho t}} \phi \, dx - \int_{\Omega} \phi \, dx\right] \rightarrow -\rho \int_E \phi \, dH_2.$$

When $E = \partial A$, we create a new connected component (cf. Fig. 1). This construction extends to a set of class $C^{1,1}$ with compact boundary in $\mathbb{R}^N$. $\square$

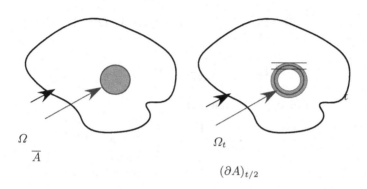

$$\Omega$$
$$\overline{A}$$

$$\Omega_t$$

$$(\partial A)_{t/2}$$

Fig. 1. For $E = \partial A$, Ω_t has two connected components

Example 3. Let $A = \partial A \subset \mathbb{R}^N$, $N \geq 3$, be a compact C^2-submanifold for which there exists $\varepsilon > 0$ such that $d_A^2 \in C^2(U_\varepsilon(A))$, then

$$\text{projection onto } \partial A : p_A(x) = x - \frac{1}{2}\nabla d_A^2(x), \quad Dp_A(x) = I - \frac{1}{2}D^2 d_A^2(x),$$

$$\text{Im } Dp_A(x) = \text{tangent space at } x \in A, \quad \dim A(x) = \dim(\text{Im } Dp_A(x)).$$

Let $\dim A = d$ and $H_d(A) < \infty$ for some d, $0 < d < N - 1$. Given $E = A$ and $0 < r < \varepsilon$, consider the r-*dilatation* E_r of E,

$$t = \alpha_{N-d} r^{N-d}, \quad \alpha_{N-d} = \text{ volume of the unit ball in } \mathbb{R}^{N-d}$$

$$\phi \mapsto \int_E \phi \, dH_d : \mathcal{D}(\mathbb{R}^N) \rightarrow \mathbb{R} \text{ is a distribution.}$$

Assuming that $U_\varepsilon(A) \subset \Omega$, the perturbed set for $0 < r < \varepsilon$ will be

$$t \mapsto \Omega_t \overset{\text{def}}{=} \Omega \backslash E_r = \Omega \backslash E_{N-\sqrt[d]{t/\alpha_{N-d}}}.$$

Given $\phi \in \mathcal{D}(\mathbb{R}^N)$, the weak limit of the differential quotient $(\chi_{\Omega_t} - \chi_\Omega)/t$ is

$$\frac{1}{t}\left[\int_{\Omega_t} \phi \, dm_N - \int_\Omega \phi \, dm_N\right] = -\frac{1}{t}\int_{E_{N-\sqrt[d]{t/\alpha_{N-d}}}} \chi_\Omega \, \phi \, dm_N$$

$$= -\frac{1}{\alpha_{N-d} r^{N-d}}\int_{E_r} \chi_\Omega \, \phi \, dm_N \rightarrow -\int_E \phi \, dH_d.$$

This distribution is a *half tangent* since for all $\rho > 0$

$$\frac{1}{t} \left[\int_{\Omega_{\rho t}} \phi \, dm_N - \int_{\Omega} \phi \, dm_N \right] \to -\rho \int_E \phi \, dH_d.$$

$\square$

4   Generalization and Concluding Remarks

In Sect. 3.2 we considered the *Minkowski content* $M^d(E)$ of closed subsets E of $\mathbb{R}^N$ (of *positive reach*) such that

$$M^d(E) \overset{\text{def}}{=} \lim_{r \searrow 0} \frac{m_N(E_r)}{\alpha_{N-d} \, r^{N-d}} = H_d(E), \quad 0 \le d \le N, \tag{4.1}$$

and the associated distribution (measure)

$$\phi \mapsto \int_E \phi \, dH_d = \lim_{r \searrow 0} \frac{1}{\alpha_{N-d} \, r^{N-d}} \int_{E_r} \phi \, dm_N : \mathcal{D}(\mathbb{R}^N) \to \mathbb{R}. \tag{4.2}$$

Choosing the volume $t = \alpha_{N-d} r^{N-d}$ of the ball of radius r in $\mathbb{R}^{N-d}$ as the auxiliary variable, that is, $r = (t/\alpha_{N-d})^{1/(N-d)}$,

$$\phi \mapsto \int_E \phi \, dH_d = \lim_{t \searrow 0} \frac{1}{t} \int_{E_{(t/\alpha_{N-d})^{1/(N-d)}}} \phi \, dm_N : \mathcal{D}(\mathbb{R}^N) \to \mathbb{R}. \tag{4.3}$$

Given a Lebesgue measurable $\Omega \subset \mathbb{R}^N$, we considered the perturbation

$$\Omega_t = \Omega \backslash E_r \tag{4.4}$$

and obtained a continuous trajectory $t \mapsto \chi_{\Omega_t}$ in $X(\mathbb{R}^N)$ such that

$$\chi_{\Omega_t} \to \chi_{\Omega \backslash E} \text{ in } L^p_{\text{loc}}(\mathbb{R}^N), 1 \le p < \infty.$$

If $m_N(E) = 0$, then $\chi_{\Omega_t} \to \chi_{\Omega}$ in $L^p_{\text{loc}}(\mathbb{R}^N)$, $1 \le p < \infty$.

Such a construction extends to *dilatations* of d-rectifiable compact sets (see Federer [9]) and to H_d-rectifiable sets E verifying a certain *density condition* (see Ambrosio et al. [2, Definition 2.57, p. 80] and [1, pp. 730–731]).

Another family of closed sets is provided by the extension of the *Steiner formula* by Federer [8, Theorem 5.6, p. 455] to closed sets A of *positive reach*. Given $E \subset \partial A$ closed and $0 \le r < \text{reach}(A)$, define the *orthogonal r-dilatation* of E: $E_r^A \overset{\text{def}}{=} \{x \in \mathbb{R}^N : d_A(x) \le r \text{ and } p_A(x) \in E\}$, where p_A is the projection onto A. Then $\lim_{r \searrow 0} m_N(E_r^A)/(\alpha_{N-d} r^{N-d})$ is a Radon measure for some d, $0 \le d \le N$.

The emerging point of view is to consider the elements of the group $X(\mathbb{R}^N)$ of characteristic functions χ_{Ω} of Lebesgue measurable subsets $\Omega \subset \mathbb{R}^N$ as a subset of measures in the space of distributions $\mathcal{D}(\mathbb{R}^N)'$:

$$\phi \mapsto \int_{\mathbb{R}^N} \chi_{\Omega} \, \phi \, dx = \int_{\Omega} \phi \, dx : \mathcal{D}(\mathbb{R}^N) \to \mathbb{R}, \quad X(\mathbb{R}^N) \subset \mathcal{D}(\mathbb{R}^N)'. \tag{4.5}$$

It is conjectured that the tangent cone $T_{\chi_\Omega} X(\mathbb{R}^N)$ is contained in $\mathcal{D}(\mathbb{R}^N)'$. In Sect. 3.1 the velocities generate *tangents* that are distributions in $H^1(\mathbb{R}^N)'$; in Sect. 3.2 the compact subsets E generate *semi-tangents* that are bounded measures. As a result, $T_{\chi_\Omega} X(\mathbb{R}^N)$ is not a linear space and it does not only contain measures, but we don't know how big it is. We could also attempt to characterize the tangent space to a family of measures.

References

1. Ambrosio, L., Colesanti, A., Villa, E.: Outer Minkowski content for some classes of closed sets. Math. Ann. **342**(4), 727–748 (2008)
2. Ambrosio, L., Fusco, N., Pallara, D.: Functions of Bounded Variation and Free Discontinuity Problems. The Clarendon Press/Oxford University Press, New York (2000)
3. Averbuh, V.I., Smoljanov, O.G.: The various definitions of the derivative in linear topological spaces. (Russian) Uspehi Mat. Nauk **23**(4(142)), 67–113 (1968). (English Translation, Russian Math. Surveys)
4. Delfour, M.C.: Groups of transformations for geometrical identification problems: metrics, geodesics. In: Hintermüller, M., Leugering, G., Sokolowski, J. (eds.) Mini-Workshop: Geometries, Shapes and Topologies in PDE-based Applications. Organizers, Mathematisches Forschungsinstitut Oberwolfach, pp. 3403–3406 (2012). (Report No. 57/2012, doi:10.4171/OWR/2012/57)
5. Delfour, M.C.: Introduction to Optimization and Semidifferential Calculus. MOS-SIAM Series on Optimization. Society for Industrial and Applied Mathematics, Philadelphia (2012)
6. Delfour, M.C.: Metrics spaces of shapes and geometries from set parametrized functions. In: Pratelli, A., Leugering, G. (eds.) New Trends in Shape Optimization, vol. 166, pp. 57–101, International Series of Numerical Mathematics, Birkhäuser Basel (2015)
7. Delfour, M.C., Zolésio, J.-P.: Shapes and Geometries, Metrics, Analysis, Differential Calculus, and Optimization. SIAM Series on Advances in Design and Control, 2nd edn. SIAM, Philadelphia (2011)
8. Federer, H.: Curvature measures. Trans. Am. Math. Soc. **93**, 418–419 (1959)
9. Federer, H.: Geometric Measure Theory. Die Grundlehren der Mathematischen Wissenschaften, vol. 153. Springer, New York (1969)
10. Fréchet, M.: Sur la notion de différentielle. J. de Mathématiques Pures et Appliquées **16**, 233–250 (1937)
11. Hadamard, J.: La notion de différentielle dans l'enseignement. Scripta Univ. Ab. Bib. Hierosolymitanarum Jerusalem (1923). Reprinted in the Mathematical Gazette **19**(236), 341–342 (1935)
12. Nashed, M.Z.: Differentiability and related properties of nonlinear operators: some aspects of the role of differentials in nonlinear functional analysis. In: Rail, L.B. (ed.) Nonlinear Functional Analysis and Applications, pp. 103–309. Academic Press, New York (1971)
13. Novotny, A.A., Sokołowski, J.: Topological Derivatives in Shape Optimization. Interaction of Mechanics and Mathematics. Springer, Heidelberg (2013)
14. Sokolowski, J., Zochowski, A.: On the topological derivative in shape optimization. SIAM J. Control Optim. **37**(4), 1251–1272 (1999)

Necessary Optimality Conditions in a Problem with Integral Equations on a Nonfixed Time Interval Subject to Mixed and State Constraints

Andrei Dmitruk[1,2(✉)] and Nikolai Osmolovskii[3,4]

[1] CEMI, Russian Academy of Sciences, Moscow, Russia
[2] Moscow Institute of Physics and Technology, Dolgoprudny, Russia
dmitruk@member.ams.org
[3] University of Technology and Humanities in Radom, Radom, Poland
[4] Moscow State University of Civil Engineering, Moscow, Russia
osmolovski@uph.edu.pl

Abstract. We consider an optimal control problem with Volterra-type integral equations on a nonfixed time interval subject to endpoint constraints, mixed state-control constraints of equality and inequality type, and pure state inequality constraints. The main assumption is the linear–positive independence of the gradients of active mixed constraints with respect to the control. We formulate first order necessary optimality conditions for an extended weak minimum, the notion of which is a natural generalization of the notion of weak minimum with account of variations of the time. The presented conditions generalize the local maximum principle in optimal control problems with ordinary differential equations.

Keywords: Volterra-type equation · Extended weak minimum · Local maximum principle · State-control constraints · Adjoint equation · Transversality conditions · Change of time variable · Linear–positive independence

1 Introduction

The results presented in this paper generalize the results obtained in our previous two papers [6,7]. Paper [6] was devoted to the first order necessary conditions for a weak minimum in a general optimal control problem with Volterra-type integral equations, considered on a *fixed* time interval, subject to endpoint constraints of equality and inequality type, mixed state-control constraints of inequality and equality type, and pure state constraints of inequality type. Paper [7] studied first order necessary conditions for an extended weak minimum in an optimal control problem with Volterra-type integral equations considered on a *non-fixed* time interval, subject to endpoint constraints of equality and inequality type, but without mixed state-control constraints and pure state constraints. Here we

L. Bociu et al. (Eds.): CSMO 2015, IFIP AICT 494, pp. 240–249, 2016.
DOI: 10.1007/978-3-319-55795-3_22

consider a problem generalizing both problems of [6,7]. We formulate first order necessary conditions for an extended weak minimum in this general problem. Following the tradition, we call them *stationarity conditions*, or conditions of the *local maximum principle*. They are presented in Theorem 1. As far as we know, such conditions for problems with integral equations on a variable time interval were not obtained up to now. Their novelty, as compared with those for problems on a fixed time interval is that the costate equation and transversality condition with respect to t involve nonstandard terms that are absent in problems with ODEs. More remarks concerning the existing literature on the problems with integral equations can be found in papers [1–4,6,7].

As was already mentioned in [6], the stationarity conditions in optimal control problems constitute an important stage in obtaining any further necessary optimality condition, including maximum principle or higher order conditions, and thus, they deserve a separate thorough study for each specific class of problems.

The paper is organized as follows. In Sect. 2 we formulate a general optimal control problem with integral equations on a variable time interval which we call Problem A. We also define in this section the notion of the extended weak minimum. Section 3 is devoted to formulation of the main result of the paper – the local maximum principle in Problem A, which is the first order necessary condition for an extended weak minimum (Theorem 1). A short discussions of its proof is given in Sect. 4.

2   General Optimal Control Problem with Integral Equations on a Variable Time Interval (Problem A)

Consider the following control system of Volterra-type integral equations on a variable time interval $[t_0, t_1]$:

$$x(t) = x(t_0) + \int_{t_0}^{t} f(t, s, x(s), u(s)) \, ds, \tag{1}$$

where $x(\cdot)$ is a continuous $n-$ dimensional and $u(\cdot)$ is a measurable essentially bounded $r-$ dimensional vector-functions on $[t_0, t_1]$. As usual, we call $x(\cdot)$ the *state* variable and $u(\cdot)$ the *control* variable (or simply the *control*). A pair $w(t) = (x(t), u(t))$ defined on its own interval $[t_0, t_1]$ and satisfying (1) for a.e. $t \in [t_0, t_1]$ is called a *process*. We assume that the function f is defined and *twice* continuously differentiable on an open set $\mathcal{R} \subset \mathbf{R}^{2+n+r}$.

The problem is to minimize the endpoint functional

$$J = \varphi_0(t_0, x(t_0), t_1, x(t_1)) \to \min \tag{2}$$

on the set of all processes (solutions of system (1)) satisfying the *endpoint constraints*

$$\eta_j(t_0, x(t_0), t_1, x(t_1)) = 0, \qquad j = 1, \ldots, k, \tag{3}$$

$$\varphi_i(t_0, x(t_0), t_1, x(t_1)) \leqslant 0, \qquad i = 1, \ldots, \nu, \tag{4}$$

the *mixed state-control constraints*

$$F_i(t, x(t), u(t)) \leqslant 0 \quad \text{for a.e. } t \in [t_0, t_1], \qquad i = 1, \ldots, d(F), \tag{5}$$

$$G_j(t, x(t), u(t)) = 0 \quad \text{for a.e. } t \in [t_0, t_1], \qquad j = 1, \ldots, d(G), \tag{6}$$

and the *state constraints*

$$\Phi_k(t, x(t)) \leqslant 0 \quad \text{for all } t \in [t_0, t_1], \qquad k = 1, \ldots, d(\Phi). \tag{7}$$

The functions $\varphi_0, \varphi_i, \eta_j$ are assumed to be defined and continuously differentiable on an open set $\mathcal{P} \subset \mathbf{R}^{2n+2}$, and the functions F_i, G_j, and Φ_k are assumed to be defined and continuously differentiable on an open set $\mathcal{Q} \subset \mathbf{R}^{1+n+r}$ (the smoothness assumptions). The notation $d(F), d(G)$, and $d(\Phi)$ stand for the numbers of these functions.

Moreover, we assume that the mixed constraints (5) and (6) are *regular* in the following sense: at any point $(t, x, u) \in \mathcal{Q}$ satisfying relations $F_i \leqslant 0 \; \forall i$ and $G_j = 0 \; \forall j$, the system of vectors

$$F_{iu}(t, x, u), \quad i \in I(t, x, u), \qquad G_{ju}(t, x, u), \quad j = 1, \ldots, d(G),$$

is *positively–linearly independent*, where $I(t, x, u) = \{i \mid F_i(t, x, u) = 0\}$ is the set of active indices of mixed inequality constraints at the given point. Here and in the sequel we denote by F_{iu} the partial derivative (gradient) of the function F_i with respect to the variable u. Similar notation is used for other functions an variables.

Recall that a system consisting of two tuples of vectors $p_1, \ldots, p_m$ and $q_1, \ldots, q_k$ in the space $\mathbf{R}^r$ is said to be *positively-linearly independent* if there does not exist a nontrivial tuple of multipliers $\alpha_1, \ldots, \alpha_m, \beta_1, \ldots \beta_k$ with all $\alpha_i \geqslant 0$ such that

$$\sum_i \alpha_i \, p_i + \sum_j \beta_j \, q_j = 0.$$

The problem (1)–(7) will be called *Problem A*, and the relations (2)–(4) its *endpoint block*.

Note that the function f explicitly depends on two time variables, t and s, the roles of which are essentially different. Conventionally, the variable s will be called *inner*, while t will be called *outer* time variable, and one should carefully distinguish between them in further considerations. Among the four arguments of the function f and its derivatives, the first argument will always be the outer and the second one be the inner time variables, no matter by which letters they will be denoted.

As in [6,7], we mention an important particular case of system (1): if f does not depend on the outer time variable t, i.e., $f = f(s, x(s), u(s))$, then the integral equation (1) is equivalent to the differential equation $\dot{x}(t) = f(t, x(t), u(t))$, hence Problem A becomes an optimal control problem of ordinary differential equations on a nonfixed time interval.

Obviously, each process under consideration must "lie" in the domain $\mathcal{R}$ of the function $f(t, s, x, u)$, i.e.

$$(t, s, x(s), u(s)) \in \mathcal{R} \quad \text{for a.e.} \quad (t, s) \in \Delta[t_0, t_1],$$

where $\Delta[t_0, t_1] = \{(t, s) : \ t_0 \leqslant s \leqslant t \leqslant t_1\}$. We will need even a stronger condition.

Definition. A process $w(t) = (x(t), u(t))$ defined on an interval $t \in [t_0, t_1]$ (with continuous $x(t)$ and measurable and essentially bounded $u(t)$) will be called *admissible with respect to $\mathcal{R}$* if its "extended graph"

$$\Gamma(w) = \{(t, s, x(s), u(s)) \,|\, (t, s) \in \Delta[t_0, t_1]\}$$

lies in the set $\mathcal{R}$ with some "margin", i.e.,

$$\text{dist} \left((t, s, x(s), u(s)), \partial\mathcal{R}\right) \geqslant \text{const} > 0 \qquad \text{for a.a. } (t, s) \in \Delta[t_0, t_1].$$

A process is called *admissible in problem A* if it is admissible with respect to $\mathcal{R}$ and satisfies all the constraints (1) and (3)–(7) of the problem.

Like in any problem on a *nonfixed* time interval, the notion of weak minimum in Problem A needs a modification.

Definition. We will say that an admissible process $w^0(t) = (x^0(t), u^0(t))$, $t \in [\hat{t}_0, \hat{t}_1]$, provides *the extended weak minimum* if there exists an $\varepsilon > 0$ such that for any Lipschitz continuous bijective mapping $\rho : [t_0, t_1] \to [\hat{t}_0, \hat{t}_1]$ satisfying the conditions $|\rho(t) - t| < \varepsilon$ and $|\dot{\rho}(t) - 1| < \varepsilon$, and for any admissible process $w(t) = (x(t), u(t))$, $t \in [t_0, t_1]$, satisfying the conditions

$$|x(t) - x^0(\rho(t))| \leqslant \varepsilon \ \ \forall t, \quad \text{and} \quad |u(t) - u^0(\rho(t))| \leqslant \varepsilon \ \ (\forall)t, \qquad (8)$$

we have $J(w) \geqslant J(w^0)$. (Notation $(\forall)$, as usual, means "for almost all".)

The conditions on ρ imply $\rho(t_0) = \hat{t}_0$ and $\rho(t_1) = \hat{t}_1$ with $|\hat{t}_0 - t_0| < \varepsilon$ and $|\hat{t}_1 - t_1| < \varepsilon$. If the interval $[t_0, t_1]$ is fixed and we take $\rho(t) = t$, then relations (8) describe the usual uniform closeness between the processes w^0 and w both in the state and control variables. However, for an arbitrary $\rho(t)$, relations (8) extend the set of "competing" processes, and thus, even for a fixed time interval, the extended weak minimum is stronger than the usual weak minimum.

3   Local Maximum Principle in Problem A

Let a process

$$w^0(t) = (x^0(t), u^0(t)), \quad t \in [\hat{t}_0, \hat{t}_1] \qquad (9)$$

provide an extended weak minimum in Problem A. We assume that the endpoints of the reference state $x^0(t)$ do not lie on the boundary of state constraints; moreover, that

$$\Phi_k(\hat{t}_0, x^0(\hat{t}_0)) < 0, \quad \Phi_k(\hat{t}_1, x^0(\hat{t}_1)) < 0, \qquad k = 1, \ldots, d(\Phi). \qquad (10)$$

For process (9), let us introduce a tuple of Lagrange multipliers

$$\lambda = (\alpha_0, \alpha, \beta, \psi_x(t), \psi_t(t), h_i(t), m_j(t), \mu_k(t)). \tag{11}$$

Here $\alpha_0 \in \mathbf{R}$ corresponds to the cost φ_0, the components of vectors $\alpha = (\alpha_1, \ldots, \alpha_\nu) \in \mathbf{R}^\nu$ and $\beta = (\beta_1, \ldots, \beta_k) \in \mathbf{R}^k$ correspond to endpoint constraints $\varphi_i \leqslant 0$, and $\eta_j = 0$, respectively, adjoint variables $\psi_x(t) : [\hat{t}_0, \hat{t}_1] \to \mathbf{R}^n$ and $\psi_t(t) : [\hat{t}_0, \hat{t}_1] \to \mathbf{R}$ correspond to the control system, $\mu_k(t) : [\hat{t}_0, \hat{t}_1] \to \mathbf{R}$, $k = 1, \ldots, d(\Phi)$, refer to the state constraints $\Phi_k(t, x) \leqslant 0$; moreover, ψ_x, ψ_t, and μ_k are functions of bounded variation, continuous at $\hat{t}_0$ and $\hat{t}_1$; multipliers $h_i(t) : [\hat{t}_0, \hat{t}_1] \to \mathbf{R}$, $i = 1, \ldots, d(F)$, and $m_j(t) : [\hat{t}_0, \hat{t}_1] \to \mathbf{R}$, $j = 1, \ldots, d(G)$ corresponding to the mixed constraints $F_i(t, x, u) \leqslant 0$, $G_j(t, x, u) = 0$, are measurable bounded functions.

Note that here ψ_x and ψ_t are not the partial derivatives with respect to x and t, but simply the adjoint variables, which refer to x and t, respectively. This notation was proposed by Dubovitskii and Milyutin and turned out to be highly convenient, especially in problems with many state variables. We hope it will not cause confusion.

We denote by $d\psi_x$, $d\psi_t$, $d\mu_k$ the Lebesgue-Stieltjes measures which correspond to the functions of bounded variation ψ_x, ψ_t, μ_k, respectively. These measures have no atoms at the points $\hat{t}_0$ and $\hat{t}_1$, and moreover, $d\mu_k \geqslant 0, k = 1, \ldots, d(\Phi)$, since it corresponds to the inequality constraint. Hence each μ_k is a monotone nondecreasing function. By

$$\frac{d\psi_x}{dt} = \dot{\psi}_x(t), \qquad \frac{d\psi_t}{dt} = \dot{\psi}_t(t), \qquad \frac{d\mu_k}{dt} = \dot{\mu}_k(t)$$

we denote the generalized derivatives of these functions with respect to t. Consequently, the following relations hold:

$$\dot{\psi}_x(t)\,dt = d\psi_x(t), \qquad \dot{\psi}_t(t)\,dt = d\psi_t(t), \qquad \dot{\mu}_k(t)\,dt = d\mu_k(t).$$

In what follows, all pointwise relations involving continuous functions hold for all t, while those involving measurable functions hold for almost all t.

In order to present optimality conditions in Problem A, introduce, for a tuple λ of (11), the *modified Pontryagin function*

$$H(t, s, x, u) = \psi_x(t-)\, f(t, s, x, u) + \int_t^{\hat{t}_1} \psi_x(\tau)\, f_t(\tau, s, x, u)\, d\tau. \tag{12}$$

Here $\psi_x(t-)$ means the left hand value of the function ψ_x at a point t, and f_t means the partial derivative of the function $f(t, s, x, u)$ with respect to the first, outer variable t.

Also, for w^0 and λ, let us introduce the *augmented modified Pontryagin function*

$$\overline{H}(t, s, x, u) = H(t, s, x, u)$$
$$- \sum_i h_i(t)\, F_i(s, x, u) - \sum_j m_j(t)\, G_j(s, x, u) - \sum_k \dot{\mu}_k(t)\, \Phi_k(s, x), \tag{13}$$

the *endpoint Lagrange function*

$$l(t_0, x_0, t_1, x_1) = \left(\sum_{i=0}^{\nu} \alpha_i \varphi_i + \sum_{j=1}^{k} \beta_j \eta_j \right)(t_0, x_0, t_1, x_1),$$

and a special auxiliary function

$$R(t) = \int_{\hat{t}_0}^{t} f_t(t, s, x^0(s), u^0(s))\, ds.$$

The main result of the paper is the following

Theorem 1 (local maximum principle)**.** *If a process* $w^0(t) = (x^0(t), u^0(t))$, $t \in [\hat{t}_0, \hat{t}_1]$ *provides the extended weak minimum in Problem A and satisfies assumption (10), then there exists a tuple of multipliers (11) satisfying the specified above properties and such that the following conditions hold true:*

(a) nonnegativity conditions

$$\alpha_0 \geqslant 0, \quad \alpha \geqslant 0, \quad h_i(t) \geqslant 0, \quad i = 1, \dots, d(F), \quad d\mu_k \geqslant 0, \quad k = 1, \dots, d(\Phi),$$

(b) nontrivality condition

$$\alpha_0 + |\alpha| + |\beta| + \sum_k (\mu_k(\hat{t}_1) - \mu_k(\hat{t}_0)) + \sum_i \int_{\hat{t}_0}^{\hat{t}_1} h_i(t)\, dt > 0,$$

(c) endpoint complementary slackness conditions

$$\alpha_i \, \varphi_i(\hat{t}_0, x^0(\hat{t}_0), \hat{t}_1, x^0(\hat{t}_1)) = 0, \qquad i = 1, \dots, \nu,$$

(d) pointwise complementary slackness conditions

$$d\mu_k(t)\, \Phi_k(t, x^0(t)) \equiv 0, \qquad k = 1, \dots, d(\Phi),$$
$$h_i(t)\, F_i(t, x^0(t), u^0(t)) = 0 \qquad a.e.\ on \quad [\hat{t}_0, \hat{t}_1],$$

(e) adjoint equation in x

$$-d\psi_x(t) = \overline{H}_x(t, t, x^0(t), u^0(t))dt$$
$$= \left(\psi_x(t-)f_x(t, t, x^0(t), u^0(t)) + \int_t^{\hat{t}_1} \psi_x(\tau)\, f_{tx}(\tau, t, x^0(t), u^0(t))\, d\tau \right) dt$$
$$- \left(\sum_i h_i(t)\, F_{ix}(t, x^0(t), u^0(t)) + \sum_j m_j(t)\, G_{jx}(t, x^0(t), u^0(t)) \right) dt$$
$$- \sum_k d\mu_k(t)\, \Phi_{kx}(t, x^0(t)),$$

(f) adjoint equation in t

$$-d\psi_t(t) = H_s(t, t, x^0(t), u^0(t))\, dt$$
$$-\left(\sum_i h_i(t)\, F_{it}(t, x^0(t), u^0(t)) + \sum_j m_j(t)\, G_{jt}(t, x^0(t), u^0(t))\right) dt$$
$$-\sum_k d\mu_k(t)\, \Phi_{kt}(t, x^0(t)) - d\psi_x(t) R(t), \tag{14}$$

with

$$H_s(t, t, x^0(t), u^0(t)) = \psi_x(t-)\, f_s(t, t, x^0(t), u^0(t))$$
$$+ \int_t^{\hat{t}_1} \psi_x(\tau)\, f_{ts}(\tau, t, x^0(t), u^0(t))\, d\tau,$$

where f_s is the partial derivative of the function $f(t, s, x, u)$ with respect to the second, inner variable s, and f_{ts} is its second partial derivative,

(g) transversality conditions in x,

$$\psi_x(\hat{t}_0) = l_{x_0}, \qquad -\psi_x(\hat{t}_1) = l_{x_1},$$

(h) transversality conditions in t,

$$\psi_t(\hat{t}_0) = l_{t_0}, \qquad -\psi_t(\hat{t}_1) = l_{t_1} - \psi_x(\hat{t}_1) R(\hat{t}_1), \tag{15}$$

(i) stationarity condition with respect to the control

$$\overline{H}_u(t, t, x^0(t), u^0(t)) = 0 \qquad a.e.\, on \quad [\hat{t}_0, \hat{t}_1],$$

i.e.,

$$\psi_x(t-) f_u(t, t, x^0(t), u^0(t)) + \int_t^{\hat{t}_1} \psi_x(\tau)\, f_{tu}(\tau, t, x^0(t), u^0(t))\, d\tau$$
$$-\sum_i h_i(t)\, F_{iu}(t, x^0(t), u^0(t)) + \sum_j m_j(t)\, G_{ju}(t, x^0(t), u^0(t)) = 0,$$

(k) the "energy evolution law"

$$H(t, t, x^0(t), u^0(t)) + \psi_t(t) = 0 \qquad a.e.\, on \quad [\hat{t}_0, \hat{t}_1].$$

The last condition is called in such a way, since together with (14) it gives the equation for evolution of the function $H(t, t, x^0(t), u^0(t))$, which is often (especially in mechanical problems) regarded as the total energy of the system:

$$\dot{H} = \overline{H}_s - \dot{\psi}_x R.$$

If the state and mixed constraints are absent and the dynamics does not explicitly depend on time: $f = f(x, u)$, then $\overline{H} = H, R = 0$, and we get "the energy conservation law": $H = const$ along the optimal process.

Note that both the adjoint equation in t and the right transversality condition in t involve additional terms, $d\psi_x(t)R(t)$ and $\psi_x(\hat{t}_1)R(\hat{t}_1)$, respectively. Both these terms are generated by the dependence of $f(t, s, x, u)$ on the outer time variable t, which was absent in problems with ODEs. (Indeed, in those problems $f_t = 0$, whence $F = 0$, so this term disappears.) In our opinion, this novelty in optimality conditions for problems on a variable time interval needs further study.

Using generalized derivatives of functions of bounded variation, we can represent the adjoint equation in x and t in the easy-to-remember form:

$$-\frac{d\psi_x(t)}{dt} = \overline{H}_x(t, t, x^0(t), u^0(t))$$

$$= H_x(t, t, x^0(t), u^0(t)) - \sum_i h_i(t) F_{ix}(t, x^0(t), u^0(t))$$

$$- \sum_j m_j(t) G_{jx}(t, x^0(t), u^0(t)) - \sum_k \frac{d\mu_k(t)}{dt} \Phi_{kx}(t, x^0(t)), \quad (16)$$

and

$$-\frac{d\psi_t(t)}{dt} = H_s(t, t, x^0(t), u^0(t))$$

$$- \sum_i h_i(t) F_{it}(t, x^0(t), u^0(t)) - \sum_j m_j(t) G_{jt}(t, x^0(t), u^0(t))$$

$$- \sum_k \frac{d\mu_k(t)}{dt} \Phi_{kt}(t, x^0(t)) - \frac{d\psi_x(t)}{dt} R(t). \quad (17)$$

4   About the Proof of Theorem 1

Like in our paper [7], in order to prove Theorem 1, we reduce Problem A to an auxiliary problem on a fixed time interval by using the change of time variable $t = t(\tau)$, where $dt/d\tau = v(\tau)$ and $v(\tau) > 0$. Setting $\tilde{x}(\tau) = x(t(\tau))$ and $\tilde{u}(\tau) = u(t(\tau))$, we come to the following system of integral equations:

$$\tilde{x}(\tau) = x(\tau_0) + \int_{\tau_0}^\tau f(t(\tau), t(\sigma), \tilde{x}(\sigma), \tilde{u}(\sigma)) \, v(\sigma) \, d\sigma, \quad (18)$$

$$t(\tau) = t(\tau_0) + \int_{\tau_0}^\tau v(\sigma) \, d\sigma, \quad (19)$$

where τ is a new time, $t(\tau)$ an additional state variable, $v(\tau)$ an additional control variable, and σ a new time of integration instead of s.

We see that here the integrand of the first equation involves the value $t(\tau)$ of a state variable, which was not allowed in (1). Abstracting from the specific form of the second equation and changing the notation $t(\tau)$ to a more general

$y(\tau)$ (and changing also τ to a more convenient t), we come to a system of the following form on a fixed interval $[t_0, t_1]$:

$$x(t) = x(t_0) + \int_{t_0}^{t} g(t, s, y(t), x(s), y(s), u(s))\, ds, \qquad (20)$$

$$y(t) = y(t_0) + \int_{t_0}^{t} h(t, s, y(s), u(s))\, ds, \qquad (21)$$

where $x(t)$ and $y(t)$ are continuous functions of dimensions n and m respectively, $u(t)$ is a measurable and essentially bounded function on $[t_0, t_1]$. We still denote the time by t. The data functions g and h, as before, are assumed to be twice continuously differentiable on an open set $\widetilde{\mathcal{R}} \subset \mathbf{R}^{2+2m+n+r}$.

This system does not fall into the framework of Eq. (1), since the integrand of the first equation depends on $y(t)$ (which can be regarded as the *outer state variable*). Thus, we have to study *a new, broader than* (1), *class of integral control systems*.

Adding to the obtained system the mixed constraints, the state constraints and the terminal block, we obtain the following

4.1   Problem B on a Fixed Time Interval

On the set of solutions $w = (x, y, u)$ to system (20)–(21) satisfying the constraints

$$F_i(t, x(t), y(t), u(t)) \leqslant 0 \quad \text{for a.e. } t \in [t_0, t_1], \qquad i = 1, \ldots, d(F), \qquad (22)$$

$$G_j(t, x(t), y(t), u(t)) = 0 \quad \text{for a.e. } t \in [t_0, t_1], \qquad j = 1, \ldots, d(G), \qquad (23)$$

$$\Phi_k(t, x(t), y(t)) \leqslant 0 \quad \text{for all } t \in [t_0, t_1], \qquad k = 1, \ldots, d(\Phi), \qquad (24)$$

$$\eta_j(x(t_0), y(t_0), x(t_1), y(t_1)) = 0, \qquad j = 1, \ldots, k, \qquad (25)$$

$$\varphi_i(x(t_0), y(t_0), x(t_1), y(t_1)) \leqslant 0, \qquad i = 1, \ldots, \nu, \qquad (26)$$

to minimize the endpoint functional

$$J = \varphi_0(x(t_0), y(t_0), x(t_1), y(t_1)) \to \min. \qquad (27)$$

Like before, the functions η_j, φ_i, and φ_0 are assumed to be continuously differentiable on an open set $\widetilde{\mathcal{P}} \subset \mathbf{R}^{2n+2m}$, the functions F_i, G_j, and Φ_k continuously differentiable on an open set $\widetilde{\mathcal{Q}} \subset \mathbf{R}^{1+m+n+r}$. We also assume that the mixed constraints (22)–(23) are regular in the same sense as in Problem A.

To derive optimality conditions in Problem B, we consider it as a particular case of an abstract nonsmooth problem in a Banach space, hence we can apply the well known abstract Lagrange multipliers rule for nonsmooth problems (see, e.g. [5,6]). Let us formulate it.

4.2 Lagrange Multipliers Rule for an Abstract Nonsmooth Problem

Let X, Y, Z_i, $i = 1, \ldots, \nu$ be Banach spaces, $\mathcal{D} \subset X$ an open set, $K_i \subset Z_i$, $i = 1, \ldots, \nu$ closed convex cones with nonempty interiors, $f_0 : \mathcal{D} \to \mathbf{R}$, $b_i : \mathcal{D} \to Z_i$, $i = 1, \ldots, \nu$, and $g : \mathcal{D} \to Y$ given mappings. Consider the following problem

$$f_0(x) \to \min, \qquad b_i(x) \in K_i, \quad i = 1, \ldots, \nu, \qquad g(x) = 0. \tag{28}$$

We study the local minimality of an admissible point $x^0 \in \mathcal{D}$. Assume that the cost f_0 and the mappings b_i are Frechet differentiable at x^0, the operator g is strictly differentiable at x^0, and the image of $g'(x^0)$ is closed. Let K_i^0 be the polar cone to K_i, $i = 1, \ldots, \nu$.

Theorem 2. *Let x^0 provide a local minimum in problem (28). Then there exist Lagrange multipliers $\alpha_0 \geqslant 0, \zeta_i^* \in K_i^0, i = 1, \ldots, \nu$, and $y^* \in Y^*$, not all equal to zero, satisfying the complementary slackness conditions $\langle \zeta_i^*, b_i(x^0) \rangle = 0$, $i = 1, \ldots, \nu$, and such that the Lagrange function*

$$L(x) = \alpha_0 f_0(x) + \sum_{i=1}^{\nu} \langle \zeta_i^*, b_i(x) \rangle + \langle y^*, g(x) \rangle$$

is stationary at x^0 : $L'(x^0) = 0$.

Applying Theorem 2 to Problem B, we perform some analysis of the obtained conditions and represent them in the form of local maximum principle for Problem B. The latter is then applied to the auxiliary problem with system (18)–(19), and finally, we rewrite the results in terms of the original Problem A.

References

1. De La Vega, C.: Necessary conditions for optimal terminal time control problems governed by a Volterra integral equation. JOTA **130**(1), 79–93 (2006)
2. Bonnans, J.F., De La Vega, C.: Optimal control of state constrained integral equations. Set-Valued Var. Anal. **18**(3–4), 307–326 (2010)
3. Bonnans, J.F., Dupuis, X., De La Vega, C.: First and second order optimality conditions for optimal control problems of state constrained integral equations. JOTA **159**(1), 1–40 (2013)
4. Filatova, D., Grzywaczewski, M., Osmolovskii, N.: Optimal control problems with integral equation as control object. Nonlin. Anal. **72**, 1235–1246 (2010)
5. Milyutin, A.A., Dmitruk, A.V., Osmolovskii, N.P.: Maximum principle in optimal control. Moscow State University, Faculty of Mechanics and Mathematics, Moscow (2004) (in Russian)
6. Dmitruk, A.V., Osmolovskii, N.P.: Necessary conditions for a weak minimum in optimal control problems with integral equations subject to state and mixed constraints. SIAM J. Control Optim. **52**, 3437–3462 (2014)
7. Dmitruk, A.V., Osmolovskii, N.P.: Necessary conditions for a weak minimum in optimal control problems with integral equations on a variable time interval. Discret. Continuous Dyn. Syst. Ser. A **35**(9), 4323–4343 (2015)

A Criterion for Robust Stability with Respect to Parametric Uncertainties Modeled by Multiplicative White Noise with Unknown Intensity, with Applications to Stability of Neural Networks

Vasile Dragan[1], Adrian-Mihail Stoica[2]([✉]), and Toader Morozan[1]

[1] Institute of Mathematics of the Romanian Academy,
P.O. Box 1-764, 70700 Bucharest, Romania
{vasile.dragan,teoader.morozan}@imar.ro
[2] Faculty of Aerospace Engineering,
University "Politehnica" of Bucharest, Bucharest, Romania
adrian.stoica@upb.ro

Abstract. In the present paper a robust stabilization problem of continuous-time linear dynamic systems with Markov jumps and corrupted with multiplicative (state-dependent) white noise perturbations is considered. The robustness analysis is performed with respect to the intensity of the white noises. It is proved that the robustness radius depends on the solution of an algebraic system of coupled Lyapunov matrix equations.

Keywords: Stochastic systems · Markovian jumps · Multiplicative white noise · Robust stability · Lyapunov operators

1 Introduction

The stochastic systems subject both to Markovian jumps and to multiplicative white noise perturbations received a considerable attention over the last years. Relevant results include the stability of such systems, optimal control and filtering (see e.g. [4–6,8] and their references). In the present paper a robust stabilization problem of continuous-time linear dynamic systems with Markov jumps and corrupted with multiplicative (state-dependent) white noise perturbations is considered. The robustness analysis is performed with respect to the intensity of the white noise terms. It is proved that the robustness radius depends on the solution of an algebraic system of coupled Lyapunov matrix equations. The derived results are a generalization of the ones proved in [10] for the case without Markovian jumps. The paper is organized as follows: in the next section the problem statement is presented. The third section includes some preliminary results concerning the Lyapunov operators associated to the considered class of

L. Bociu et al. (Eds.): CSMO 2015, IFIP AICT 494, pp. 250–260, 2016.
DOI: 10.1007/978-3-319-55795-3_23

stochastic systems. The main result is presented and proved in Sect. 4. In the last section the stability radius is determined for two relevant particular cases and a numerical example illustrates the theoretical developments.

2   The Problem Statement

Consider the system of stochastic linear differential equations:

$$dx(t) = A(\eta_t)x(t)dt + \sum_{l=1}^{r} \mu_l b_l(\eta_t) c_l^T(\eta_t)x(t)dw_l(t) \tag{1}$$

where $\{w_l(t)\}_{t\geq0}$, $1 \leq l \leq r$, are one-dimensional independent standard Wiener processes defined on a given probability space $(\Omega, \mathcal{F}, \mathcal{P})$; $\{\eta_t\}_{t\geq0}$ is a homogeneous standard right continuous Markov process defined on the same probability space $(\Omega, \mathcal{F}, \mathcal{P})$ and taking value in the finite set $\mathbb{N} = \{1, 2, ..., N\}$ and having the transition semigroup $P(t) = e^{Qt}$, $t \geq 0$, where $Q \in \mathbb{R}^{N \times N}$ is a matrix whose elements q_{ij} satisfy the condition

$$\begin{cases} q_{ij} \geq 0 \, if \, i \neq j, \, i, j \in \mathbb{N} \\ \sum_{j=1}^{N} q_{ij} = 0, \forall i \in \mathbb{N}. \end{cases} \tag{2}$$

For more details we refer to [1,3,9,12]. We also assume that $\{\eta_t\}_{t\geq0}$, $\{w_l(t)\}_{t\geq0}$, $1 \leq l \leq r$, are independent stochastic processes. In (1) the matrices $A(i) \in \mathbb{R}^{n \times n}$, $b_l(i)$, $c_l(i) \in \mathbb{R}^{n \times 1}$, $1 \leq l \leq r$, $1 \leq i \leq N$ are known, while the scalars $\mu_l \in \mathbb{R}$ are unknown. The system (1) can be regarded as a perturbation of the so called nominal system

$$\dot{x}(t) = A(\eta_t)x(t). \tag{3}$$

The perturbed system (1) emphasizes the fact that the coefficients of the nominal system are affected by parametric uncertainties modeled by state multiplicative white noise perturbations with unknown intensity μ_l. Often when we refer to the perturbed system (1) we shall say that it corresponds to the vector of parameters $\mu = (\mu_1, \mu_2, ..., \mu_r)$. Assuming that the nominal system (3) is exponentially stable in mean square (ESMS) we want to find necessary and sufficient conditions which will be satisfied by the parameters $\mu_l, 1 \leq l \leq r$ such that the perturbed system (1) to be also ESMS. In the special case $\mathbb{N} = \{1\}$ (no Markov jumps) the conditions derived in this note recover those derived in [10]. The concept of exponential stability in mean square of the linear stochastic systems of type (1) and (3) may be found in [2] and [8], respectively.

3   Some Preliminaries

Let $\mathcal{S}_n^N = \mathcal{S}_n \otimes \mathcal{S}_n \otimes ... \otimes \mathcal{S}_n$, where $\mathcal{S}_n \subset \mathbb{R}^{n \times n}$ is the subspace of symmetric matrices. Let $\mathcal{S}_{n+}^N$ be the convex cone defined by $\mathcal{S}_{n+}^N = \{\mathbf{X} = (X(1),$

$X(2), ..., X(N)) \in \mathcal{S}_n^N | X(i) \geq 0, 1 \leq i \leq N \}$. Here, $X(i) \geq 0$ means that $X(i)$ is positive semidefinite, $\mathcal{S}_{n+}^N$ is a closed convex cone with non empty interior. Its interior is $Int\mathcal{S}_{n+}^N = \{ \mathbf{X} = (X(1), X(2), ..., X(N)) \in \mathcal{S}_{n+}^N | X(i) > 0, 1 \leq i \leq N \}$. Applying Theorem 3.3.2 and Theorem 3.3.3 from [8] in the case of system (1) one obtains the following result.

Proposition 3.1. *The following are equivalent*

(i) *The system (4) is ESMS;*

(ii) *For any* $\mathbf{H} = (H(1), H(2), ..., H(N)) \in Int\mathcal{S}_{n+}^N$ *there exists* $\mathbf{X} = (X(1), X(2), ..., X(N)) \in Int\mathcal{S}_{n+}^N$ *solving the following equation on* $\mathcal{S}_n^N$

$$A^T(i)X(i) + X(i)A(i) + \sum_{j=1}^{N} q_{ij}X(j) + \sum_{l=1}^{r} \mu_l^2 c_l(i)b_l^T(i)X(i)b_l(i)c_l^T(i) \tag{4}$$
$$+H(i) = 0; 1 \leq i \leq N;$$

(iii) *For any* $\mathbf{H} \in Int\mathcal{S}_{n+}^N$, *there exists* $\mathbf{Y} = (Y(1), y(2), ..., Y(N)) \in Int\mathcal{S}_{n+}^N$ *solving the following equation on* $\mathcal{S}_n^N$

$$A(i)Y(i) + Y(i)A^T(i) + \sum_{j=1}^{N} q_{ji}Y(j) + \sum_{l=1}^{r} \mu_l^2 b_l(i)c_l^T(i)Y(i)c_l(i)b_l^T(i) \tag{5}$$
$$+H(i) = 0, 1 \leq i \leq N;$$

(iv) *There exists* $\mathbf{X} = (X(1), ..., X(N)) \in Int\mathcal{S}_{n+}^N$ *satisfying the following system of LMIs*

$$A^T(i)X(i) + X(i)A(i) + \sum_{j=1}^{N} q_{ij}X(j) + \sum_{l=1}^{r} \mu_l^2 c_l(i)b_l^T(i)X(i)b_l(i)c_l^T(i) < 0,$$

where $1 \leq i \leq N$.

Then one associates the following Lyapunov operators to the nominal system (3) $\mathcal{L} : \mathcal{S}_n^N \to \mathcal{S}_n^N$, $\mathfrak{L} : \mathcal{S}_n^N \to \mathcal{S}_n^N$ defined by

$$\mathcal{L}[\mathbf{X}](i) = A^T(i)X(i) + X(i)A(i) + \sum_{j=1}^{N} q_{ij}X(j), 1 \leq i \leq N, \tag{6}$$

and

$$\mathfrak{L}[\mathbf{X}](i) = A(i)X(i) + X(i)A^T(i) + \sum_{j=1}^{N} q_{ji}X(j), 1 \leq i \leq N \tag{7}$$

for all $\mathbf{X} = (X(1), ..., X(N)) \in \mathcal{S}_n^N$. It is easy to check that $\mathcal{L}$ is the adjoint operator of $\mathfrak{L}$ with respect to the usual inner product on $\mathcal{S}_n^N$:

$$< \mathbf{X}, \mathbf{Y} > = \sum_{j=1}^{N} Tr[X(i)Y(i)]. \tag{8}$$

Invoking Proposition 3.20 and Theorem 3.21 from [2] (see also Theorem 3.2.2 and Theorem 3.2.4 from [8]) one deduces that the nominal system (3) is ESMS if and only if the eigenvalues of the linear operator $\mathcal{L}$ are located in the half plane $\mathbb{C}_-$. This allows us to obtain the following result.

Corollary 3.2. *If the nominal system (3) is ESMS then for each* $\mathbf{H} \in \mathcal{S}_n^N$ *the equations*

$$\mathcal{L}[\mathbf{X}] + \mathbf{H} = 0 \tag{9}$$

and

$$\mathfrak{L}[\mathbf{Y}] + \mathbf{H} = 0 \tag{10}$$

have unique solutions given by $\mathbf{X} = -\mathcal{L}^{-1}[\mathbf{H}] = (-\mathcal{L}^{-1}[\mathbf{H}](1), ..., -\mathcal{L}^{-1}[\mathbf{H}](N)) \in \mathcal{S}_n^N$ *and* $\mathbf{Y} = -\mathfrak{L}^{-1}[\mathbf{H}] = (-\mathfrak{L}^{-1}[\mathbf{H}](1), ..., -\mathfrak{L}[\mathbf{H}](N)) \in \mathcal{S}_n^N$, *respectively. If* $\mathbf{H} \in \mathcal{S}_{n+}^N$ *then* $\mathbf{X} \in \mathcal{S}_{n+}^N$, $\mathbf{Y} \in \mathcal{S}_{n+}^N$. *Moreover, if* $\mathbf{H} \in Int\mathcal{S}_{n+}^N$ *then the unique solutions of (9) and (10), respectively are in* $Int\mathcal{S}_{n+}^N$ *i.e.*

$$-\mathcal{L}^{-1}[\mathbf{H}](i) > 0 \tag{11}$$

and

$$-\mathfrak{L}^{-1}[\mathbf{H}](i) > 0 \tag{12}$$

for all $1 \leq i \leq N$.

Further, let us consider the ordered space $(\mathbb{R}^d, \mathbb{R}_+^d)$ where the order relation is induced by the convex cone $\mathbb{R}_+^d = \{x = (x_1, x_2, ..., x_d)^T \in \mathbb{R}^d | x_i \geq 0, 1 \leq i \leq d\}$. The interior $Int\mathbb{R}_+^d$ of the convex cone $\mathbb{R}_+^d$ consists of the set $Int\mathbb{R}_+^d = \{x = (x_1, x_2, ..., x_d)^T \in \mathbb{R}_+^d | x_i > 0, 1 \leq i \leq d\}$. If $D = (d_{ij})_{ij} \in \mathbb{R}^{d \times d}$ is the matrix of a linear operator $\mathcal{D} : \mathbb{R}^d \to \mathbb{R}^d$ then $\mathcal{D}\mathbb{R}_+^d \subset \mathbb{R}_+^d$ if and only if $d_{ij} \geq 0, 1 \leq i, j \leq d$. In this case D will be called positive matrix. Applying Theorems 2.6 and 2.7 from [7] in the special case of the ordered linear space $(\mathbb{R}^d, \mathbb{R}_+^d)$ one obtains the following result.

Proposition 3.3. *For a positive matrix* $D \in \mathbb{R}^{d \times d}$ *the following are equivalent:*

(i) $\rho(D) < 1$, $\rho(\cdot)$ *being the spectral radius;*
(ii) There exists $\psi \in Int\mathbb{R}_+^d$ *such that the equation*

$$(I_d - D)\zeta = \psi \tag{13}$$

has a solution $\zeta \in Int\mathbb{R}_+^d$.

4   Main Results

The equivalence $(i) \leftrightarrow (iv)$ from Proposition 3.1 allows us to deduce that if the perturbed system (1) is ESMS for a value $\mu = (\mu_1, \mu_2, ..., \mu_r)$ of the intensities of the white noises, then this system is ESMS for every value $\mu^{'} = (\mu^{'}_1, ..., \mu^{'}_r)$ which are satisfying $|\mu^{'}_l| \le |\mu_l|$ for all $1 \le l \le r$. In this section we shall derive a set of necessary and sufficient conditions which guarantee the exponential stability in mean square of a perturbed system (1) corresponding to a set of unknown vector of intensities $\mu = (\mu_1, ..., \mu_r)$. Using Proposition 3.1 one notices that the system (1) is ESMS if and only if for any $\mathbf{H} \in IntS^N_{n+}$ the Eq. (4) has a solution $\mathbf{X} \in IntS^N_{n+}$. Using (6) we may rewrite (4) in the form

$$\mathcal{L}[\mathbf{X}] + \tilde{\mathbf{H}} = 0 \qquad (14)$$

where $\tilde{\mathbf{H}} = (\tilde{H}(1), ..., \tilde{H}(N))$,

$$\tilde{H}(i) = \sum_{l=1}^{r} \mu_l^2 b_l^T(i)X(i)b_l(i)c_l(i)c_l^T(i) + H(i), 1 \le i \le N. \qquad (15)$$

Further, we rewrite $\tilde{\mathbf{H}}$ in the form:

$$\tilde{\mathbf{H}} = \sum_{l=1}^{r}\sum_{j=1}^{N} \mu_l^2 b_l^T(j)X(j)b_l(j)\Xi_{lj} + \mathbf{H} \qquad (16)$$

where $\Xi_{lj} = (\Xi_{lj}(1), ..., \Xi_{lj}(N)) \in \mathcal{S}^N_{n+}$ with

$$\Xi_{lj}(i) = \begin{cases} c_l(j)c_l^T(j), & if\ i = j \\ 0, & otherwise. \end{cases} \qquad (17)$$

Since the nominal system (3) is necessarily ESMS if the perturbed system (1) is ESMS, we deduce via Corollary 3.2 and (16) that the solution of the equation (14) satisfies

$$\mathbf{X} = -\sum_{l=1}^{r}\sum_{j=1}^{N} b_l^T(j)X(j)b_l(j)\mu_l^2\mathcal{L}^{-1}[\Xi_{lj}] - \mathcal{L}^{-1}[\mathbf{H}].$$

The ith component of this solution is

$$X(i) = -\sum_{l=1}^{r}\sum_{j=1}^{N} b_l^T(j)X(j)b_l(j)\mu_l^2\mathcal{L}^{-1}[\Xi_{lj}](i) - \mathcal{L}^{-1}[\mathbf{H}](i), 1 \le i \le N. \qquad (18)$$

Multiplying (18) on the left by $b_k^T(i)$ and on the right by $b_k(i)$, one obtains

$$b_k^T(i)X(i)b_k(i) = \sum_{l=1}^{r}\sum_{j=1}^{N} b_l^T(j)X(j)b_l(j)m_{ki,lj} + \nu_{ki} \qquad (19)$$

$1 \le k \le r$, $1 \le i \le N$, where

$$m_{ki,lj} = -\mu_l^2 b_k^T(i)\mathcal{L}^{-1}[\Xi_{lj}](i)b_k(i) \qquad (20)$$

and

$$\nu_{ki} = -b_k^T(i)\mathcal{L}^{-1}[\mathbf{H}](i)b_k(i). \tag{21}$$

One sees that (19) is a system of rN scalars equations with rN scalar unknowns. Based on the fact that for each integer $\alpha \in \{1, 2, ..., rN\}$ there exists a unique pair of natural numbers $(k, i) \in \{1, 2, ..., r\} \times \{1, 2, ..., N\}$ such that $\alpha = (k-1)N + i$ we may write (19) as an equation of the form (13) on the space $\mathbb{R}^{rN}$. To this end we set $\zeta = (\zeta_1, \zeta_2, ..., \zeta_{rN})^T$, $\Psi = (\Psi_1, \Psi_2, ..., \Psi_{rN})^T$, $D = (d_{\alpha\beta})_{1 \leq \alpha, \beta \leq rN}$,

$$\zeta_\beta = b_l^T(j)X(j)b_l(j) \; if \; (l-1)N + j = \beta, \tag{22}$$

$$\Psi_\alpha = \nu_{ki} = -b_k^T(i)\mathcal{L}^{-1}[\mathbf{H}](i)b_k(i) \; if \; (k-1)N + i = \alpha \tag{23}$$

$$d_{\alpha\beta} = m_{ki,lj} = -\mu_l^2 b_k^T(i)\mathcal{L}^{-1}[\Xi_{lj}](i)b_k(i) \tag{24}$$

if $(k-1)N + i = \alpha$ and $(l-1)N + j = \beta$. With these notations (19) may be written in a compact form:

$$(I - D)\zeta = \Psi. \tag{25}$$

Since the matrix D defined by (24) depend upon the unknown parameters μ_l, for each perturbed system of type (1) one may associate a matrix $D = D(\mu)$ as before. Now we are in a position to state and proof the following result.

Theorem 4.1 *Assume $b_k(i) \neq 0$, $1 \leq k \leq r$, $1 \leq i \leq N$. Under this condition the following are equivalent:*

(i) *The perturbed system (1) corresponding to the set of parameters $\mu = (\mu_1, ..., \mu_r)$ is ESMS;*
(ii) *The nominal system (3) is ESMS and the matrix $D(\mu)$ associated to the perturbed system (1) satisfies $\rho(D(\mu)) < 1$.*

Proof. $(i) \Rightarrow (ii)$ Let $\mathbf{H} \in IntS_{n+}^N$ be arbitrary but fixed. If the perturbed system (1) is ESMS, then the nominal system (3) is also ESMS. Hence, the Eqs. (14)–(15) has a solution $\mathbf{X} = (X(1), ..., X(N))$ with $X(i) > 0$, $1 \leq i \leq N$. Consider $\zeta, \Psi \in \mathbb{R}^{rN}$ defined via (22) and (23). Since $b_k(i) \neq 0$, for all $(k, i) \in \{1, 2, ..., r\} \times \{1, 2, ..., N\}$ one deduces that $\zeta_\alpha > 0$ and $\Psi_\alpha > 0$, $1 \leq \alpha \leq rN$. Further, one associates the matrix D whose elements are computed via (24). One may check that $d_{\alpha\beta} \geq 0$, for all $1 \leq \alpha, \beta \leq rN$. So, it follows that the Eq. (25) associated to the perturbed system (1) satisfies the conditions from Proposition 3.3 (ii). Hence, $\rho(D) < 1$, this shows that the assertion (ii) from the statement holds if (i) is satisfied.

Now the implication $(ii) \Rightarrow (i)$ will be proved. Let $\mathbf{H} \in IntS_{n+}^N$ be arbitrary. It will be shown that the corresponding Eq. (4) has a solution $\mathbf{X} \in IntS_{n+}^N$. One notices that if the nominal system (3) is ESMS, then $\Psi \in Int\mathbb{R}^{rN}$ and the matrix $D \in \mathbb{R}^{rN \times rN}$ are well defined via (23) and (24), respectively. If $\rho(D) < 1$ then, based on the implication $(i) \Rightarrow (ii)$ from Proposition 3.3 it follows (25) has a unique solution $\zeta \in Int\mathbb{R}_+^{rN}$. Based on the components of the vectors Ψ, ζ and of the matrix D one may define

$$\tilde{\Psi}_{ki} = \Psi_\alpha, \tilde{\zeta}_{lj} = \zeta_\beta, \tilde{d}_{ki,lj} = d_{\alpha\beta} \tag{26}$$

if $(k,i),(l,j) \in \{1,2,...,r\} \times \{1,2,...,N\}$, $(k-1)N+i = \alpha, (l-1)N+j = \beta$. It is easy to see that $\tilde{\Psi}_{ki} = \Psi_{ki}$, $\tilde{d}_{ki,lj} = m_{ki,lj}$. With these notations one obtains the following version of the Eq. (25)

$$\tilde{\zeta}_{ki} = -\sum_{l=1}^{r}\sum_{j=1}^{N} \mu_l^2 b_k^T(i)\mathcal{L}^{-1}[\Xi_{lj}](i)b_k(i)\tilde{\zeta}_{lj} - b_k^T(i)\mathcal{L}^{-1}[\mathbf{H}](i)b_k(i) \tag{27}$$

$\forall (k,i) \in \{1,2,...,r\} \times \{1,2,...,N\}$. Defining

$$X(i) = -\sum_{l=1}^{r}\sum_{j=1}^{N} \mu_l^2 \tilde{\zeta}_{lj}\mathcal{L}^{-1}[\Xi_{lj}](i) - \mathcal{L}^{-1}[\mathbf{H}](i), 1 \le i \le N. \tag{28}$$

and setting $\mathbf{X} = (X(1),...,X(N))$, it results that $\mathbf{X} \in Int\mathcal{S}_{n+}^{N}$ since $-\mathcal{L}^{-1}[\Xi_{lj}](i) \ge 0$, $-\mathcal{L}^{-1}[\mathbf{H}](i) > 0$ and $\tilde{\zeta}_{lj} > 0$ for all $(l,j,i) \in \{1,2,...,r\} \times \{1,2,...,N\} \times \{1,2,...,N\}$. From (28) it follows

$$\mathcal{L}[\mathbf{X}] + \sum_{l=1}^{r}\sum_{j=1}^{N} \mu_l^2 \tilde{\zeta}_{lj}\Xi_{lj} + \mathbf{H} = 0. \tag{29}$$

Based on (17) it results that $\Xi_{lj}(i) = 0$ if $i \ne j$, obtaining thus the following componentwise version of (29)

$$\mathcal{L}[\mathbf{X}](i) + \sum_{l=1}^{r} \mu_l^2 \tilde{\zeta}_{li}c_l(i)c_l^T(i) + H(i) = 0, 1 \le i \le N. \tag{30}$$

From (28) with (27) it results that $\tilde{\zeta}_{ki} = b_k^T(i)X(i)b_k(i)$ for all $1 \le k \le r$, $1 \le i \le N$. Therefore one may rewrite (30) in the form

$$\mathcal{L}[\mathbf{X}](i) + \sum_{l=1}^{r} \mu_l^2 c_l(i)b_l^T(i)X(i)b_l(i)c_l^T(i) + H(i) = 0 \tag{31}$$

which is just (4). Thus the proof is complete.

Remark 4.1. The result proved in Theorem 4.1 shows that, in order to decide if the perturbed system (1) corresponding to a vector of unknown parameters $\mu_l \in [-|\tilde{\mu}_l|, |\tilde{\mu}_l|]$, $1 \le l \le r$ is ESMS, we have to check if the spectral radius of the matrix $D = D(\tilde{\mu})$ (associated via (24) to the parameters $\tilde{\mu}_l$) is less then 1. One notices that (24) may be rewritten in the form

$$d_{\alpha\beta} = \tilde{\mu}_l^2 b_k^T(i)Z_{lj}(i)b_k(i) \tag{32}$$

where $(k, i), (l, j) \in \{1, 2, ..., r\} \times \{1, 2, ..., N\}$ are such that $(k - 1)N + i = \alpha$ and $(l - 1)N + j = \beta$, $\mathbf{Z}_{lj} = (Z_{lj}(1), ..., Z_{lj}(N))$ being the unique solution of the equation

$$A^T(i)Z_{lj}(i) + Z_{lj}(i)A(i) + \sum_{\iota=1}^{N} q_{i\iota}Z_{lj}(\iota) + \Xi_{lj}(i) = 0 \qquad (33)$$

$1 \leq i \leq N$, with Ξ_{lj} defined in (17).

According with [10] one introduces the following definition.

Definition 4.1. The vector of noise intensities $\mu^0 = (\mu_1^0, \mu_2^0, ..., \mu_r^0)$ is called critical for the system (1) if the system (1) corresponding to the noise intensities $\varepsilon\mu^0$ is ESMS if $0 < \varepsilon < 1$ and it is not ESMS if $\varepsilon \geq 1$.

Remark 4.2. Denote $D(\mu)$ the matrix D corresponding to the vector $\mu = (\mu_1, ..., \mu_r)$ of the noise intensities. Based on (24) it follows that $D(\varepsilon\mu^0) = \varepsilon^2 D(\mu^0)$. Since the spectral radius of a positive matrix is an eigenvalue of that matrix one may infer that

$$\rho[D(\varepsilon\mu^0)] = \varepsilon^2\rho[D(\mu^0)].$$

If μ^0 is a critical vector of noise intensities then one obtains from Theorem 4.1 that $\rho[D(\varepsilon\mu^0)] < 1$ for all $0 < \varepsilon < 1$ obtaining thus that $\rho[D(\mu^0)] < \frac{1}{\varepsilon^2}$, $0 < \varepsilon < 1$. So, we deduce that $\rho[D(\mu^0)] \leq 1$. Since the perturbed system (1) corresponding to the vector μ^0 is not ESMS, one concludes via Theorem 4.1, that $\rho[D(\mu^0)] = 1$. Hence, the vector $\mu^0 = (\mu_1^0, ..., \mu_r^0)$ is a solution of the equation $\det(I_{rN} - D(\mu)) = 0$.

In the space $\mathbb{R}^r$ of the vector $\mu = (\mu_1, ..., \mu_r)$ the critical vectors of noise intensities μ^0 are included in the boundary of the stability region.

5   Several Special Cases

The first special case analyzed here is $r = 1$ and $N \geq 2$. In this case the system (1) becomes

$$dx(t) = A(\eta_t)x(t)dt + \mu b(\eta_t)c^T(\eta_t)x(t)dw_1(t). \qquad (34)$$

The matrix D associated to the system (34) is $D = \mu^2 D_1$ where

$$D_1 = \begin{pmatrix} b^T(1)Z_1(1)b(1) & b^T(1)Z_2(1)b(1) & ... & b^T(1)Z_N(1)b(1) \\ b^T(2)Z_1(2)b(2) & b^T(2)Z_2(2)b(2) & ... & b^T(2)Z_N(2)b(2) \\ ... & ... & ... & ... \\ b^T(N)Z_1(N)b(N) & b^T(N)Z_2(N)b(N) & ... & b^T(N)Z_N(N)b(N) \end{pmatrix} \qquad (35)$$

for each $1 \leq j \leq N$, $(Z_j(1), ..., Z_j(N))$ is the unique solution of the following equation on $\mathcal{S}_n^N$

$$A^T(i)Z_j(i) + Z_j(i)A(i) + \sum_{\iota=1}^{N} q_{i\iota}Z_j(\iota) + \Xi_j(i) = 0 \qquad (36)$$

$\Xi_j(i) = 0$ if $i \neq j$ and $\Xi_j(i) = c(j)c^T(j)$ if $i = j$.

In the special case of the system (34), the Theorem 4.1 yields to the following result.

Corollary 5.1. *If $b(i) \neq 0$, $\forall 1 \leq i \leq N$ the following are equivalent:*

(i) *The perturbed system (34) is ESMS;*
(ii) *The nominal system (3) is ESMS and the parameter μ satisfies the condition* $\mu^2 < \frac{1}{\rho[D_1]}$.

Remark 5.1. The previous Corollary shows that the exponential stability in mean square of the nominal system (3) is preserved for the perturbed system (34) if and only if the unknown parameter μ lies in the interval $(-\rho^{\frac{-1}{2}}[D_1], \rho^{\frac{-1}{2}}[D_1])$, which is the stability region in the case of perturbed system (34).

The second special case discussed here is $r \geq 1$, $N = 1$. Now, the system (1) becomes

$$dx(t) = Ax(t)dt + \sum_{l=1}^{r} \mu_l b_l c_l^T x(t)dw_l(t). \tag{37}$$

The matrix D associated to the system (37) via (24) is

$$D = \hat{D}diag(\mu_1^2, ..., \mu_r^2), \tag{38}$$

where

$$\hat{D} = \begin{pmatrix} b_1^T Z_1 b_1 & b_1^T Z_2 b_1 & ... & b_1^T Z_r b_1 \\ b_2^T Z_1 b_2 & b_2^T Z_2 b_2 & ... & b_2^T Z_r b_2 \\ ... & ... & ... & ... \\ b_r^T Z_1 b_r & b_r^T Z_2 b_r & ... & b_r^T Z_r b_r \end{pmatrix} \tag{39}$$

for each $1 \leq l \leq r$, Z_l is the unique solution of the Lyapunov equation

$$A^T Z_l + Z_l A + c_l c_l^T = 0. \tag{40}$$

The result proved in Theorem 4.1 yields the next result.

Corollary 5.2. *The following are equivalent*

(i) *The perturbed system (37) is ESMS for any value of the unknown parameters $\mu_l \in (-|\tilde{\mu}_l|; |\tilde{\mu}_l|)$, $1 \leq l \leq r$.*
(ii) *A is a Hurwitz matrix and $\rho[D] < 1$, D being defined in (38)–(39) with μ_l replaced by $\tilde{\mu}_l$, $1 \leq l \leq r$.*

Remark 5.2.(a) The result stated in Corollary 5.2 is just the main result derived in [10]. Its discrete-time version may be found in [11].

(b) Condition of the form $b_l \neq 0, 1 \leq l \leq r$ (as it is imposed in the general case in Theorem 4.1) is redundant in the case of system (37) because, if $b_{l_0} = 0$ for some l_0, it follows that the noise $w_{l_0}(t)$ does not affect the perturbed system.

In order to illustrate the above theoretical results one considers the dynamics of Hopfield neural network of form

$$\dot{v}_i(t) = a_i v_i(t) + \sum_{j=1}^{n} b_{ij} g_j (v_j(t)) + c_i, i = 1, ..., N$$

$a_i < 0$ and the *activation functions* $g_i(\cdot)$ are strictly increasing. Then its approximation around an equilibrium point v^0 is

$$\dot{x}(t) = Ax(t) + Bf(x(t))$$

where $x(t) = v(t) - v^0$, $f(x) = g(x + v^0) - g(v^0)$ and where $A = diag(a_1, ..., a_n)$ the elements of $f(\cdot)$ being *sector-type* nonlinearities satisfying $f_k(x_k) (f_k(x_k) - \mu_k x_k) \leq 0$, $k = 1, ..., n$. Then for the above system associate the linear approximation

$$
\begin{aligned}
dx(t) &= Ax(t)dt + \mu \sum_{l=1}^{r} b_l c_l^T x(t) dw_l(t), \ n = 3, \ r = 2 \\
A &= diag(-0.5, -0.5, -0.5) \\
b_1 &= [0.5, 1, 1]^T, \ c_1 = [1, 2, 1]^T, \\
b_2 &= [1, 0.25, -1]^T, \ c_2 = [0.25, -1, 1]^T
\end{aligned}
$$

Applying Corollary 5.2 it results that the above system is ESMS for all $\mu \in [-0.2857; 0.2857]$.

Acknowledgement. This paper has been partially supported by MEN-UEFISCDI, Program Partnerships, Projects PN-II-PT-PCCA- 2013-4-1349 and PN-II-PT-PCCA-2011-3.1-1560.

References

1. Chung, K.L.: Markov Chains with Stationary Transition Probabilities. Springer, Berlin (1967)
2. Costa, L.V., Fragoso, M.D., Todorov, M.G.: Continuous-Time Markov Jump Linear Systems. Springer, London (2013)
3. Doob, J.L.: Stochastic Processes. Wiley, New York (1967)
4. Huang, Y., Zhang, W., Feng, G.: Infinite horizon H_2/H_∞ control for stochastic systems with Markovian jumps. Automatica **44**, 857–863 (2008)
5. Mariton, M.: Jump Linear Systems in Automatic Control. Marcel Dekker, New York (1990)
6. Li, X., Zhou, X.Y., Rami, M.A.: Indefinite stochastic linear quadratic control with Markovian jumps in indefinite time horizon. J. Glob. Optim. **27**, 149–175 (2003)
7. Dragan, V., Morozan, T., Stoica, A.-M.: Mathematical Methods in Robust Control of Discrete-Time Linear Stochastic Systems. Springer, New York (2010)
8. Dragan, V., Morozan, T., Stoica, A.-M.: Mathematical Methods in Robust Control of Linear Stochastic Systems, 2nd edn. Springer, New York (2013)
9. Friedman, A.: Stochastic Differential Equations and Applications, vol. I. Academic, New York (1975)

10. Levit, M.V., Yakubovich, V.A.: Algebraic criterion for stochastic stability of linear systems with parametric action of white noise type. Appl. Math. Mech. **36**(1), 142–148 (1972)
11. Morozan, T.: Necessary and sufficient conditions of stability of stochastic discrete systems. Rev. Roum. Math. Pures et Appl. **18**(2), 255–267 (1973)
12. Oksendal, B.: Stochastic Differential Equations. Springer, Berlin (1998)

Optimal Control of Doubly Nonlinear Evolution Equations Governed by Subdifferentials Without Uniqueness of Solutions

M. Hassan Farshbaf-Shaker[1,2] and Noriaki Yamazaki[1,2(⊠)]

[1] Weierstrass Institute for Applied Analysis and Stochastics, Mohrenstrasse 39,
10117 Berlin, Germany
`Hassan.Farshbaf-Shaker@wias-berlin.de`
[2] Department of Mathematics, Kanagawa University, 3-27-1 Rokkakubashi,
Kanagawa-ku, Yokohama 221-8686, Japan
`noriaki@kanagawa-u.ac.jp`

Abstract. In this paper we study an optimal control problem for a doubly nonlinear evolution equation governed by time-dependent subdifferentials. We prove the existence of solutions to our equation. Also, we consider an optimal control problem without uniqueness of solutions to the state system. Then, we prove the existence of an optimal control which minimizes the nonlinear cost functional. Moreover, we apply our general result to some model problem.

Keywords: Optimal control · Doubly nonlinear evolution equations · Subdifferentials · Without uniqueness

1 Introduction

The present paper is concerned with an optimal control problem without uniqueness of solutions to a doubly nonlinear evolution equation governed by time-dependent subdifferentials in a real Hilbert space H.

In our optimal control problem, for each control f, the state system $(P; f)$ is as follows:

State system (P;f):

$$(P; f) \begin{cases} \partial \psi^t(u'(t)) + \partial \varphi(u(t)) + g(u(t)) \ni f(t) & \text{in } H \text{ for a.e. } t \in (0, T), \\ u(0) = u_0 & \text{in } H, \end{cases} \quad (1.1)$$

where $0 < T < \infty$, $u' = du/dt$ in H, $\psi^t : H \to \mathbb{R} \cup \{\infty\}$ is a time-dependent proper, l.s.c. (lower semi-continuous), convex function for each $t \in [0, T]$, $\varphi : H \to \mathbb{R} \cup \{\infty\}$ is a time-independent proper, l.s.c., convex function, $\partial \psi^t$ and $\partial \varphi$

N. Yamazaki—This work was supported by Grant-in-Aid for Scientific Research (C) No. 26400164 and No. 26400179, JSPS.

are their subdifferential in H, $g(\cdot)$ is a single-valued Lipschitz operator in H, f is a given H-valued control function and $u_0 \in H$ is a given initial data.

In this present paper, we consider the optimal control problem without uniqueness of solutions to the state system $(\mathrm{P}; f)$. To this end, let V be a real Hilbert space such that the embedding $V \hookrightarrow H$ is dense and compact. Then, we study the following optimal control problem without uniqueness of solutions to $(\mathrm{P}; f)$:

Problem (OP): Find the optimal control $f^* \in \mathcal{F}$ such that

$$J(f^*) = \inf_{f \in \mathcal{F}} J(f).$$

Here $\mathcal{F} := W^{1,2}(0,T;H) \cap L^2(0,T;V)$ is the control space and $J(f)$ is the cost functional defined by

$$J(f) := \inf_{u \in S(f)} \pi_f(u), \tag{1.2}$$

where $f \in \mathcal{F}$ is the control, $S(f)$ is the set of all solutions to $(\mathrm{P}; f)$ with the control function f. Also, u is a solution to the state system $(\mathrm{P}; f)$ and $\pi_f(u)$ is its functional defined by

$$\pi_f(u) := \frac{1}{2} \int_0^T |u(t) - u_{ad}|_H^2 dt + \frac{1}{2} \int_0^T |f(t)|_V^2 dt + \frac{1}{2} \int_0^T |f_t(t)|_H^2 dt, \tag{1.3}$$

where $u_{ad} \in L^2(0,T;H)$ is a given target profile and $|\cdot|_H$ (resp. $|\cdot|_V$) is the norm of H (resp. V).

There is vast literature on optimal control problems to (parabolic or elliptic) variational inequalities. For instance, we refer to [5, 10, 11, 17–19, 23]. In particular, Lions [18] and Neittaanmäki et al. [19, Sect. 3.1.3.1] discussed the singular control problems, which is the class of control problems characterized by not well-posed state systems.

The theory of nonlinear evolution equations are useful in the systematic study of variational inequalities. For instance, many mathematicians studied the nonlinear evolution equation of the form:

$$u'(t) + \partial\varphi^t(u(t)) \ni f(t) \quad \text{in } H \quad \text{for a.e. } t \in (0,T), \tag{1.4}$$

where $\varphi^t(\cdot) : H \to \mathbb{R} \cup \{\infty\}$ is a proper, l.s.c. and convex function. For various aspects of (1.4), we refer to [11, 14, 20, 22]. In particular, Hu–Papageorgiou [11] studied the optimal control problems to (1.4).

Also, doubly nonlinear evolution equations were studied. For instance, Kenmochi–Pawlow [15] studied the following type of doubly nonlinear evolution equations:

$$\frac{d}{dt}\partial\psi(u(t)) + \partial\varphi^t(u(t)) \ni f(t) \quad \text{in } H \quad \text{for a.e. } t \in (0,T), \tag{1.5}$$

where $\psi(\cdot) : H \to \mathbb{R} \cup \{\infty\}$ is a proper, l.s.c. and convex function. The abstract results of doubly nonlinear evolution Eq. (1.5) can be applied to elliptic-parabolic equations. Therefore, from the view point of (1.5), Hoffmann et al. [10] studied optimal control problems for quasi-linear elliptic-parabolic variational inequalities with time-dependent constraints. Also, Kadoya–Kenmochi [12] studied the optimal sharp design of elliptic-parabolic equations.

On the other hand, Akagi [1], Arai [2], Aso et al. [3,4], Colli [8], Colli–Visintin [9], Senba [21] investigated the following type of doubly nonlinear evolution Eq. (cf. (1.1)):

$$\partial\psi^t(u'(t)) + \partial\varphi(u(t)) \ni f(t) \quad \text{in } H \quad \text{for a.e. } t \in (0, T). \tag{1.6}$$

However, there was no result of optimal control for (1.1) and (1.6) since (1.1) and (1.6) are not well-posed state systems, in general. Therefore, by arguments similar to Kadoya et al. [13], more precisely, using the cost functional defined by (1.2) and (1.3), we establish the abstract theory of the optimal control problem (OP) without uniqueness of solutions to the state system (1.1).

The plan of this paper is as follows. In the next Sect. 2, we state the main result in this paper. In Sect. 3, we first give the sketch of the proof of solvability for (1.1). Also, we prove the convergence result (Proposition 3) of solutions to (P; f). Moreover, we prove the main result (Theorem 1) on the existence of the optimal control to (OP). In the final Sect. 4, we apply our abstract result to a parabolic PDE with Neumann boundary condition.

Notations

Throughout this paper, let H be a real Hilbert space with the inner product $(\cdot, \cdot)$ and norm $|\cdot|_H$, respectively. Also, let V be a real Hilbert space with the norm $|\cdot|_V$ such that the embedding $V \hookrightarrow H$ is dense and compact.

Let us here prepare some notations and definitions of subdifferential of convex functions. To this end, let E be a real Hilbert space with the inner product $(\cdot, \cdot)_E$. Then, for a proper (i.e., not identically equal to infinity), l.s.c. and convex function $\phi : E \to \mathbb{R} \cup \{\infty\}$, the effective domain $D(\phi)$ is defined by

$$D(\phi) := \{z \in E; \; \phi(z) < \infty\}.$$

The subdifferential of ϕ is a possibly multi-valued operator in E and is defined by $z^* \in \partial\phi(z)$ if and only if

$$z \in D(\phi) \quad \text{and} \quad (z^*, y - z)_E \le \phi(y) - \phi(z) \quad \text{for all} \; y \in E.$$

The next proposition is concerned with the closedness of maximal monotone operator $\partial\phi$ in E.

Proposition 1 (cf. [7, Lemma 1.2]). *Let E be a real Hilbert space with the inner product $(\cdot, \cdot)_E$. Let $\phi : E \to \mathbb{R} \cup \{\infty\}$ be a proper, l.s.c. and convex function. Also, let $[z_n, z_n^*] \in \partial\phi$ and $[z, z^*] \in E \times E$ be such that*

$$z_n \to z \; \text{weakly in} \; E \quad \text{and} \quad z_n^* \to z^* \; \text{weakly in} \; E \quad \text{as } n \to \infty.$$

Suppose that

$$\limsup_{n \to \infty} (z_n, z_n^*)_E \le (z, z^*)_E.$$

Then, it follows that $[z, z^*] \in \partial\phi$ *and* $\lim_{n \to \infty} (z_n, z_n^*)_E = (z, z^*)_E$.

For various properties and related notions of the proper, l.s.c., convex function ϕ and its subdifferential $\partial\phi$, we refer to a monograph by Brézis [6].

2   Main Theorem

We begin by defining the notion of a solution to $(P; f)$.

Definition 1. *Given* $f \in L^2(0, T; H)$ *and* $u_0 \in H$, *the function* $u : [0, T] \to H$ *is called a solution to* $(P; f)$ *on* $[0, T]$, *if the following conditions are satisfied:*

(i) $u \in W^{1,2}(0, T; H)$.
(ii) *There exist functions* $\xi \in L^2(0, T; H)$ *and* $\zeta \in L^2(0, T; H)$ *such that*

$$\xi(t) \in \partial\psi^t(u'(t)) \quad \text{in } H \quad \text{for a.e. } t \in (0, T),$$
$$\zeta(t) \in \partial\varphi(u(t)) \quad \text{in } H \quad \text{for a.e. } t \in (0, T)$$

and

$$\xi(t) + \zeta(t) + g(u(t)) = f(t) \quad \text{in } H \quad \text{for a.e. } t \in (0, T).$$

(iii) $u(0) = u_0$ *in* H.

Now, we give the assumptions on ψ^t, φ and g.

(A1) For each $t \in [0, T]$, $\psi^t(\cdot) : H \to \mathbb{R} \cup \{\infty\}$ is a proper, l.s.c. and convex function. Also, $\varphi(\cdot) : H \to \mathbb{R} \cup \{\infty\}$ is a proper, l.s.c. and convex function.
(A2) There exists a positive constant $C_1 > 0$ such that

$$\psi^t(z) \ge C_1 |z|_H^2, \quad \forall t \in [0, T], \ \forall z \in D(\psi^t).$$

(A3) There exists a positive constant $C_2 > 0$ such that

$$|z^*|_H^2 \le C_2(\psi^t(z) + 1), \quad \forall [z, z^*] \in \partial\psi^t, \ \forall t \in [0, T].$$

(A4) There are functions $\alpha \in W^{1,2}(0, T)$ and $\beta \in W^{1,1}(0, T)$ satisfying the following property: for any $s, t \in [0, T]$ with $s \le t$ and $z \in D(\psi^s)$, there exists $\tilde{z} \in D(\psi^t)$ such that

$$|\tilde{z} - z|_H \le |\alpha(t) - \alpha(s)| \left(1 + \psi^s(z)^{\frac{1}{2}}\right),$$
$$\psi^t(\tilde{z}) - \psi^s(z) \le |\beta(t) - \beta(s)| (1 + \psi^s(z)).$$

(A5) There exists a positive constant $C_3 > 0$ such that

$$\varphi(z) \geq C_3 |z|_H^2, \quad \forall z \in D(\varphi).$$

(A6) For each $r > 0$, the level set $\{z \in H; \varphi(z) \leq r\}$ is compact in H.
(A7) $g : H \to H$ is a single-valued Lipschitz operator. Namely, there is a positive
constant $L_g > 0$ such that

$$|g(z_1) - g(z_2)|_H \leq L_g |z_1 - z_2|_H, \quad \forall z_i \in H \ (i = 1, 2).$$

Remark 1. The assumption (A4) is the standard time-dependence condition of
convex functions (cf. [14, 20, 22]).

By a slight modification of [1,3], we can prove the following existence result
for problem (P;f). We give a sketch of its proof in Sect. 3.

Proposition 2 (cf. [1, Theorem 3.2], [3, Theorem 2.1]). *Assume (A1)–
(A7). Then, for each $u_0 \in D(\varphi)$ and $f \in L^2(0, T; H)$, there exists at least one
solution u to (P;f) on $[0, T]$. Moreover, there exists a positive constant $N_0 > 0$,
independent of u_0, such that*

$$\int_0^T \psi^t(u'(t))dt + \sup_{t \in [0,T]} \varphi(u(t)) \leq N_0 \left(\varphi(u_0) + |f|_{L^2(0,T;H)}^2 + 1 \right). \quad (2.1)$$

Remark 2. Colli [8, Theorem 5] and Colli–Visintin [9, Remark 2.5] showed several criteria for the uniqueness of solutions to the following type of doubly nonlinear evolution equations:

$$\partial\psi(u'(t)) + \partial\varphi(u(t)) \ni f(t) \quad \text{in } H \quad \text{for a.e. } t \in (0, T). \quad (2.2)$$

For instance, if $\partial\varphi$ is linear, positive, self-adjoint in H and $\partial\psi$ is strictly
monotone in H, we can show the uniqueness of solutions to (2.2). However, $\partial\psi^t$
and $\partial\varphi$ in (1.1) are nonlinear and not self-adjoint, and hence, the uniqueness
question to (1.1) is still open.

Now, we state the main result of this paper, which is directed to the existence
of an optimal control to (OP) without uniqueness of solutions to (P;f).

Theorem 1. *Assume (A1)–(A7) and $u_0 \in D(\varphi)$. Let u_{ad} be an element in
$L^2(0, T; H)$. Then, (OP) has at least one optimal control $f^* \in \mathcal{F}$ such that*

$$J(f^*) = \inf_{f \in \mathcal{F}} J(f).$$

3   Proof of Main Theorem 1

In this section, we give the sketch of the proof of Proposition 2 by arguments similar to Akagi [1] and Aso et al. [3]. Moreover, we prove Theorem 1.

Throughout this section, we suppose that all the assumptions of Theorem 1 hold.

Sketch of the proof of Proposition 2.

By arguments similar to Akagi [1] and Aso et al. [3], we can prove Proposition 2. In fact, for each $\lambda \in (0,1]$, we consider the following approximate problem for (P;f), denoted by (P;f)$_\lambda$:

$$(\text{P};f)_\lambda \begin{cases} \lambda u'_\lambda(t) + \partial \psi^t(u'_\lambda(t)) + \partial \varphi_\lambda(u_\lambda(t)) + g(J^\varphi_\lambda u_\lambda(t)) \ni f(t) \text{ in } H \\ \qquad\qquad\qquad\qquad\qquad\qquad\qquad\qquad \text{for a.e. } t \in (0,T), \\ u_\lambda(0) = u_0 \text{ in } H, \end{cases}$$

where $\partial \varphi_\lambda$ and $J^\varphi_\lambda := (I + \lambda \partial \varphi)^{-1}$ denote the Yosida approximation and the resolvent of $\partial \varphi$, respectively.

By Cauchy–Lipschitz–Picard's existence theorem and the fixed point argument for compact operators (e.g. the Schauder's fixed point theorem), we can prove the existence of solutions u_λ to (P;f)$_\lambda$ on $[0,T]$.

From the standard calculation, we can establish a priori estimate (cf. (2.1)) of solutions u_λ to (P;f)$_\lambda$ with respect to $\lambda \in (0,1]$. Therefore, by the limiting procedure of solutions u_λ to (P;f)$_\lambda$ as $\lambda \to 0$, we can construct the solution to (P;f) on $[0,T]$ satisfying the boundedness estimate (2.1). For a detailed argument, see [1, Sects. 4 and 5] or [3, Sects. 3 and 4], for instance.   □

Here, let us mention the result of the convergence of solutions to (P; f), which is a key proposition to proving Theorem 1.

Proposition 3. *Assume* (A1)–(A7). *Let* $\{f_n\} \subset L^2(0,T;H)$, $\{u_{0,n}\} \subset D(\varphi)$, $f \in L^2(0,T;H)$ *and* $u_0 \in D(\varphi)$. *Assume that*

$$f_n \to f \text{ strongly in } L^2(0,T;H), \qquad\qquad (3.1)$$

$$u_{0,n} \to u_0 \text{ in } H \quad \text{and} \quad \varphi(u_{0,n}) \to \varphi(u_0) \qquad\qquad (3.2)$$

as $n \to \infty$. *Let* u_n *be a solution to* (P;f_n) *on* $[0,T]$ *with initial data* $u_{0,n}$. *Then, there exist a subsequence* $\{n_k\} \subset \{n\}$ *and a function* $u \in W^{1,2}(0,T;H)$ *such that* u *is a solution to* (P;f) *on* $[0,T]$ *with initial data* u_0 *and*

$$u_{n_k} \to u \text{ in } C([0,T];H) \text{ as } k \to \infty. \qquad\qquad (3.3)$$

Proof. From the bounded estimate (2.1), (A2), (A5) and the level set compactness of φ (cf. (A6)), we derive that there are a subsequence $\{n_k\}$ of $\{n\}$ and a function $u \in W^{1,2}(0,T;H)$ such that $n_k \to \infty$,

$$\left. \begin{array}{c} u_{n_k} \to u \text{ weakly in } W^{1,2}(0,T;H), \\ \text{in } C([0,T];H), \\ \text{weakly-}* \text{ in } L^\infty(0,T;H) \end{array} \right\} \qquad\qquad (3.4)$$

as $k \to \infty$. Hence, we observe from (3.2) and (3.4) that $u(0) = u_0$ in H.

Now, let us show that u is a solution of $(P; f)$ on $[0, T]$ with initial data u_0. Since u_{n_k} is a solution of $(P; f_{n_k})$ on $[0, T]$ with initial data u_{0,n_k}, there exist functions $\xi_{n_k} \in L^2(0, T; H)$ and $\zeta_{n_k} \in L^2(0, T; H)$ such that

$$\xi_{n_k}(t) \in \partial \psi^t(u'_{n_k}(t)) \quad \text{in } H \quad \text{for a.e. } t \in (0, T), \tag{3.5}$$

$$\zeta_{n_k}(t) \in \partial \varphi(u_{n_k}(t)) \quad \text{in } H \quad \text{for a.e. } t \in (0, T), \tag{3.6}$$

$$\xi_{n_k}(t) + \zeta_{n_k}(t) + g(u_{n_k}(t)) = f_{n_k}(t) \quad \text{in } H \quad \text{for a.e. } t \in (0, T). \tag{3.7}$$

Then, it follows from (2.1) and (A3) that

$$\{\xi_{n_k}\} \text{ is bounded in } L^2(0, T; H). \tag{3.8}$$

Therefore, taking a subsequence if necessary (still denote it by $\{n_k\}$), we observe that:

$$\xi_{n_k} \to \xi \text{ weakly in } L^2(0, T; H) \text{ for some } \xi \in L^2(0, T; H) \text{ as } k \to \infty. \tag{3.9}$$

Also, it follows from (A7) and (3.4) and that

$$g(u_{n_k}) \to g(u) \text{ in } C([0, T]; H) \text{ as } k \to \infty. \tag{3.10}$$

Therefore, we infer from (3.1), (3.7), (3.8) and (3.10) that

$$\{\zeta_{n_k}\} \text{ is bounded in } L^2(0, T; H).$$

Hence, taking a subsequence if necessary (still denote it by $\{n_k\}$), we observe that:

$$\zeta_{n_k} \to \zeta \text{ weakly in } L^2(0, T; H) \text{ for some } \zeta \in L^2(0, T; H) \text{ as } k \to \infty. \tag{3.11}$$

Thus, we infer from (3.1), (3.7), (3.9), (3.10) and (3.11) that:

$$\xi + \zeta + g(u) = f \text{ in } L^2(0, T; H). \tag{3.12}$$

Also, from (3.4), (3.6), (3.11) and the demi-closedness of maximal monotone operator $\partial \varphi$ (cf. Proposition 1), we infer that

$$\zeta \in \partial \varphi(u) \text{ in } L^2(0, T; H), \tag{3.13}$$

which implies that $\zeta(t) \in \partial \varphi(u(t))$ in H for a.e. $t \in (0, T)$.

Now, we show that

$$\xi(t) \in \partial \psi^t(u'(t)) \quad \text{in } H \quad \text{for a.e. } t \in (0, T). \tag{3.14}$$

From (3.1), (3.2) and (3.4)–(3.13) we observe that

$$\limsup_{k \to \infty} \int_0^T (\xi_{n_k}(t), u'_{n_k}(t))dt$$

$$= \limsup_{k \to \infty} \int_0^T (f_{n_k}(t) - \zeta_{n_k}(t) - g(u_{n_k}(t)), u'_{n_k}(t))dt$$

$$= \limsup_{k \to \infty} \left[\int_0^T (f_{n_k}(t) - g(u_{n_k}(t)), u'_{n_k}(t))dt - \int_0^T \frac{d}{ds}\varphi(u_{n_k}(s))ds \right]$$

$$\leq \int_0^T (f(t) - g(u(t)), u'(t))dt + \limsup_{k \to \infty} (-\varphi(u_{n_k}(T)) + \varphi(u_{0,n_k}))$$

$$\leq \int_0^T (f(t) - g(u(t)), u'(t))dt - \varphi(u(T)) + \varphi(u_0)$$

$$= \int_0^T (f(t) - g(u(t)) - \zeta(t), u'(t))dt$$

$$= \int_0^T (\xi(t), u'(t))dt,$$

thus, we observe from Proposition 1, namely, the closedness of maximal monotone operator $\partial\psi^t$ that

$$\xi \in \partial\psi^t(u') \quad \text{in } L^2(0,T;H),$$

which implies that (3.14) holds. Therefore, we observe that u is a solution of (P; f) on $[0,T]$ with initial data u_0. Thus, the proof of this proposition has been completed. □

Now, let us prove the main Theorem 1 in our paper, which is the existence of an optimal control to (OP).

Proof of Theorem 1.
Note that we show the existence of an optimal control to (OP) without uniqueness of solutions to state problem (P; f)

Also note from (1.2) and (1.3) that $J(f) \geq 0$ for all $f \in \mathcal{F}$. Let $\{f_n\} \subset \mathcal{F}$ be a minimizing sequence such that

$$d^* := \inf_{f \in \mathcal{F}} J(f) = \lim_{n \to \infty} J(f_n).$$

Then, we observe that $\{J(f_n)\}$ is bounded. Therefore, by the definition (1.2) of $J(f_n)$, for each n there is a solution $u_n \in \mathcal{S}(f_n)$ such that

$$\pi_{f_n}(u_n) < J(f_n) + \frac{1}{n}.$$

Hence, we observe that $\{\pi_{f_n}(u_n)\}$ is bounded. Thus, by the definition of $\pi_{f_n}(u_n)$ (cf. (1.3)) and by the Aubin's compactness theorem (cf. [16, Chapter 1, Sect. 5]), there are a subsequence $\{n_k\} \subset \{n\}$ and a function $f^* \in \mathcal{F}$ such that

$$\left. \begin{array}{r} f_{n_k} \rightarrow f^* \quad \text{weakly in } W^{1,2}(0,T;H), \\ \text{weakly in } L^2(0,T;V), \\ \text{in } L^2(0,T;H) \end{array} \right\} \tag{3.15}$$

as $k \rightarrow \infty$,

Now, taking a subsequence if necessary, we infer from Proposition 3 that there is a solution u^* to (P;f^*) on $[0,T]$ with initial data u_0 satisfying

$$u_{n_k} \rightarrow u^* \quad \text{in } C([0,T];H) \quad \text{as } k \rightarrow \infty. \tag{3.16}$$

Therefore, it follows from (3.15)–(3.16), $u^* \in \mathcal{S}(f^*)$ and the weak lower semi-continuity of L^2–norm that

$$d^* = \inf_{f \in \mathcal{F}} J(f) \leq J(f^*) = \inf_{u \in \mathcal{S}(f^*)} \pi_{f^*}(u)$$

$$\leq \pi_{f^*}(u^*) = \frac{1}{2} \int_0^T |u^*(t) - u_{ad}|_H^2 dt + \frac{1}{2} \int_0^T |f^*(t)|_V^2 dt + \frac{1}{2} \int_0^T |f_t^*(t)|_H^2 dt$$

$$\leq \liminf_{k \to \infty} \pi_{f_{n_k}}(u_{n_k})$$

$$\leq \liminf_{k \to \infty} \left\{ J(f_{n_k}) + \frac{1}{n_k} \right\}$$

$$= \lim_{k \to \infty} J(f_{n_k}) = d^*.$$

Hence, we have $d^* = \inf_{f \in \mathcal{F}} J(f) = J(f^*)$, which implies that $f^* \in \mathcal{F}$ is an optimal control to (OP). Thus, the proof of Theorem 1 has been completed. □

4   Application

In this section, we apply Theorem 1 to the simple model problem as follows:

$$\text{(SMP)}_p \begin{cases} A(t, u_t) - \operatorname{div}\left(|\nabla u|^{p-2}\nabla u\right) + g(u) \ni f(t) & \text{in } Q := (0,T) \times \Omega, \\ \dfrac{\partial u}{\partial \nu} = 0 & \text{on } \Sigma := (0,T) \times \Gamma, \\ u(0) = u_0 & \text{a.e. in } \Omega, \end{cases}$$

where $0 < T < \infty$, Ω is a bounded domain in $\mathbb{R}^N$ ($1 \leq N < \infty$), the boundary $\Gamma := \partial\Omega$ of Ω is smooth if $N > 1$, g is Lipschitz on $\mathbb{R}$, p is a positive number with $p \geq 2$, ν is an outward normal vector on Γ and u_0 is a given initial data. Also, $A(t, \cdot)$ is the given time-dependent function defined by

$$A(t,z) := \begin{cases} z - c(t), & \text{if } z - c(t) \geq 1, \\ 1, & \text{if } 0 < z - c(t) < 1, \\ [-1,1], & \text{if } z = c(t), \\ -1, & \text{if } -1 < z - c(t) < 0, \\ z - c(t), & \text{if } z - c(t) \leq -1, \end{cases}$$

where $c(\cdot)$ is a given smooth function on $[0,T]$.

To apply the abstract result to (P;f), we put $H := L^2(\Omega)$ and $V := H^1(\Omega)$ with usual real Hilbert space structures. Define a function φ on H by

$$\varphi(z) := \begin{cases} \dfrac{1}{p}\displaystyle\int_\Omega |\nabla z(x)|^p dx + C_4, & \text{if } z \in W^{1,p}(\Omega), \\ \infty, & \text{otherwise}, \end{cases}$$

Also, for each $t \in [0,T]$, define a function ψ^t on H by

$$\psi^t(z) := \int_\Omega \hat{A}(t, z(x))dx \quad \text{for all } z \in H := L^2(\Omega),$$

where $\hat{A}(t,\cdot)$ is the primitive of $A(t,\cdot)$ such that $\hat{A}(t,\cdot) \ge 0$ for all $t \in [0,T]$.

It is not difficult to show that the assumptions (A1)–(A7) are satisfied. For instance, put $\tilde{z} = z - c(s) + c(t)$, $\alpha(t) := \int_0^t |c'(\tau)|d\tau$ and $\beta(t) \equiv 0$ for (A4) (cf. [14, Chap. 3]). Therefore, by applying Theorem 1, we can consider the control problem (OP) without uniqueness of solutins to (SMP)$_p$.

References

1. Akagi, G.: Doubly nonlinear evolution equations with non-monotone perturbations in reflexive Banach spaces. J. Evol. Eqn. **11**, 1–41 (2011)
2. Arai, T.: On the existence of the solution for $\partial\varphi(u'(t)) + \partial\psi(u(t)) \ni f(t)$. J. Fac. Sci. Univ. Tokyo Sec. IA Math. **26**, 75–96 (1979)
3. Aso, M., Frémond, M., Kenmochi, N.: Phase change problems with temperature dependent constraints for the volume fraction velocities. Nonlinear Anal. **60**, 1003–1023 (2005)
4. Aso, M., Kenmochi, N.: Quasivariational evolution inequalities for a class of reaction-diffusion systems. Nonlinear Anal. **63**, e1207–e1217 (2005)
5. Barbu, V.: Optimal Control of Variational Inequalities. Research Notes in Mathematics, vol. 100. Pitman, London (1984)
6. Brézis, H.: Opérateurs Maximaux Monotones et Semi-Groupes de Contractions dans les Espaces de Hilbert. Elsevier, Amsterdam (1973)
7. Brézis, H., Crandall, M.G., Pazy, A.: Perturbations of nonlinear maximal monotone sets in Banach space. Commun. Pure Appl. Math. **23**, 123–144 (1970)
8. Colli, P.: On some doubly nonlinear evolution equations in Banach spaces. Jpn J. Ind. Appl. Math. **9**, 181–203 (1992)
9. Colli, P., Visintin, A.: On a class of doubly nonlinear evolution equations. Commun. Partial Differ. Eqn. **15**, 737–756 (1990)
10. Hoffmann, K.-H., Kubo, M., Yamazaki, N.: Optimal control problems for elliptic-parabolic variational inequalities with time-dependent constraints. Numer. Funct. Anal. Optim. **27**, 329–356 (2006)
11. Hu, S., Papageorgiou, N.S.: Time-dependent subdifferential evolution inclusions and optimal control. Mem. Amer. Math. Soc. 632, 81 p. (1998)
12. Kadoya, A., Kenmochi, N.: Optimal shape design in elliptic-parabolic equations. Bull. Fac. Educ. Chiba Univ. **41**, 1–20 (1993)

13. Kadoya, A., Murase, Y., Kenmochi, N.: A class of nonlinear parabolic systems with environmental constraints. Adv. Math. Sci. Appl. **20**, 281–313 (2010)
14. Kenmochi, N.: Solvability of nonlinear evolution equations with time-dependent constraints and applications. Bull. Fac. Educ. Chiba Univ. **30**, 1–87 (1981)
15. Kenmochi, N., Pawlow, I.: A class of nonlinear elliptic-parabolic equations with time-dependent constraints. Nonlinear Anal. **10**, 1181–1202 (1986)
16. Lions, J.-L.: Quelques Méthodes de Résolution des Problèmes aux Limites Non Linéaires. Dunod, Gouthiers-Villars (1969)
17. Lions, J.-L.: Optimal control of systems governed by partial differential equations. Springer, New York (1971)
18. Lions, J.-L.: Contrôle des systèmes distribués singuliers. Méthodes Mathématiques de l'Informatique, vol. 13. Gauthier-Villars, Montrouge (1983)
19. Neittaanmäki, P., Sprekels, J., Tiba, D.: Optimization of Elliptic Systems: Theory and Applications. Springer Monographs in Mathematics. Springer, New York (2006)
20. Ôtani, M.: Nonlinear evolution equations with time-dependent constraints. Adv. Math. Sci. Appl. **3**, 383–399 (1994). (1993/94), Special Issue
21. Senba, T.: On some nonlinear evolution equation. Funkcial Ekvac. **29**, 243–257 (1986)
22. Yamada, Y.: On evolution equations generated by subdifferential operators. J. Fac. Sci. Univ. Tokyo Sect. IA Math. **23**, 491–515 (1976)
23. Yamazaki, N.: Convergence and optimal control problems of nonlinear evolution equations governed by time-dependent operator. Nonlinear Anal. **70**, 4316–4331 (2009)

Derivation of a Macroscopic LWR Model from a Microscopic *follow-the-leader* Model by Homogenization

Nicolas Forcadel$^{(\boxtimes)}$ and Mamdouh Zaydan

Normandie Université INSA Rouen,
Laboratoire de Mathématiques de l'INSA de Rouen,
Avenue de l'Université, 76800 Saint Etienne du Rouvray, France
nicolas.forcadel@insa-rouen.fr

Abstract. The goal of this paper is to derive a traffic flow macroscopic model from a microscopic model with a transition function. At the microscopic scale, we consider a first order model of the form "follow the leader" i.e. the velocity of each vehicle depends on the distance to the vehicle in front of it. We consider two different velocities and a transition zone. The transition zone represents a local perturbation operated by a Lipschitz function. After rescaling, we prove that the "cumulative distribution function" of the vehicles converges towards the solution of a macroscopic homogenized Hamilton-Jacobi equation with a flux limiting condition at junction which can be seen as a LWR model.

1 Introduction

The goal of this paper is to present a rigorous derivation of a traffic flow macroscopic model by homogenization of a *follow-the-leader* model, see [8,10]. The idea is to rescale the microscopic model, which describes the dynamics of each vehicle individually, in order to get a macroscopic model which describes the dynamics of density of vehicles. Several studies have been done about the connection between microscopic and macroscopic traffic flow model. This type of connection is important since it allows us to deduce macroscopic models rigorously and without using strong assumptions. We refer for example to [1–3] where the authors rescaled the empirical measure and obtained a scalar conservation law (LWR (Lighthill-Whitham-Richards) model). More recently, another kind of macroscopic models appears. These models rely on the Moskowitz function and make appear an Hamilton-Jacobi equation. This is the setting of our work which is a generalization of [6]. Indeed, authors in [6] considered a single road and one velocity throughout this road with a local perturbation at the origin while we consider two different velocities and a transition zone which can be seen as a local perturbation thats slows down the vehicles. At the macroscopic scale, we get an Hamilton-Jacobi equation with a junction condition at zero and an effective flux limiter. In order to have our homogenization result, we will construct

L. Bociu et al. (Eds.): CSMO 2015, IFIP AICT 494, pp. 272–281, 2016.
DOI: 10.1007/978-3-319-55795-3_25

the correctors. The main new technical difficulties comes from the construction of correctors and in particular the gradient estimates are more complicated from that in [6] because the gradient on the left and on the right may differ.

2   The Microscopic Model

In this paper, we consider a "follow the leader" model of the following form

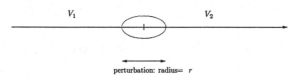

$$\dot{U}_j(t) = V_1(U_{j+1}(t) - U_j(t))\varphi\left(U_j(t)\right) + V_2(U_{j+1}(t) - U_j(t))\left(1 - \varphi\left(U_j(t)\right)\right),$$

where U_j denotes the position of the j-th vehicle and $\dot{U}_j$ its velocity. The function φ simulates the presence of a local perturbation around the origin which allows us to pass from the optimal velocity function V_1 (on the left of the origin) to V_2 (on the right). We make the following assumptions on V_1, V_2 and φ.

Assumption (A).

- (A1) $V_1, V_2 : \mathbb{R} \to \mathbb{R}^+$ are Lipschitz continuous, non-negative and non-decreasing.
- (A2) For $i = 1, 2$, there exists a $h_0^i \in (0, +\infty)$ such that

$$V_i(h) = 0 \text{ for all } h \leq h_0^i.$$

- (A3) For $i = 1, 2$, there exists a $h_{max}^i \in (0, +\infty)$ such that

$$V_i(h) = V_{imax} \text{ for all } h \geq h_{max}^i.$$

- (A4) For $i = 1, 2$, there exists a real $p_0^i \in [-1/h_0^i, 0)$ such that the function $p \mapsto pV_i(-1/p)$ is decreasing on $[-1/h_0^i, p_0^i)$ and increasing on $[p_0^i, 0)$.
- (A5) The function $\varphi : \mathbb{R} \to [0, 1]$ is Lipschitz continuous and

$$\varphi(x) = \begin{cases} 1 & \text{if } x \leq -r \\ 0 & \text{if } x > r. \end{cases}$$

3   The Homogenization Result

We introduce the "cumulative distribution function" of the vehicles:

$$\rho(t, y) = -\left(\sum_{i \geq 0} H\left(y - U_i(t)\right) + \sum_{i < 0}\left(-1 + H\left(y - U_i(t)\right)\right)\right)$$

and we make the following rescaling

$$\rho^{\varepsilon}(t,y) = \varepsilon\rho\left(t/\varepsilon, y/\varepsilon\right).$$

ρ^{ε} is a discontinuous solution of the following equation: for $(t,x) \in (0,+\infty) \times \mathbb{R}$,

$$\begin{cases} u_t^{\varepsilon} + \left(M_1^{\varepsilon}\left[\dfrac{u^{\varepsilon}(t,\cdot)}{\varepsilon}\right](x)\varphi\left(\dfrac{x}{\varepsilon}\right) + M_2^{\varepsilon}\left[\dfrac{u^{\varepsilon}(t,\cdot)}{\varepsilon}\right](x)\left(1 - \varphi\left(\dfrac{x}{\varepsilon}\right)\right)\right) \cdot |u_x^{\varepsilon}| = 0 \\ u^{\varepsilon}(0,x) = u_0(x) \end{cases} \tag{3.1}$$

where the non-local operators M_i^{ε} and M_2^{ε} are defined by

$$M_i^{\varepsilon}[U](x) = \int_{-\infty}^{+\infty} J_i(z)E\left(U(x+\varepsilon z) - U(x)\right) dz - \frac{3}{2}V_{imax} \tag{3.2}$$

with

$$E(z) = \begin{cases} 0 & \text{if } z \geq 0, \\ 1/2 & \text{if } -1 \leq z < 0 , \\ 3/2 & \text{if } z < -1. \end{cases} \quad J_1 = V_1' \text{ and } J_2 = V_2' \text{ on } \mathbb{R}. \tag{3.3}$$

We also assume that the initial condition satisfies the following assumption.

(A0) (Gradient Bound). Let $k_0 = \max\left(k_0^1, k_0^2\right)$ with $k_0^i = 1/h_0^i$. The function u_0 is Lipschitz continuous and satisfies

$$-k_0 \leq (u_0)_x \leq 0.$$

We have the following theorem (see [6]).

Theorem 1. *Assume (A0) and (A). Then, there exists a unique viscosity solution u^{ε} of (3.1). Moreover, the function u^{ε} is continuous and there exists a constant K such that*

$$u_0(x) \leq u^{\varepsilon}(t,x) \leq u_0(x) + Kt.$$

We will introduce now the macroscopic model which is a Hamilton-Jacobi equation on a junction. The Hamiltonians $\overline{H}_1$ and $\overline{H}_2$ are called effective Hamiltonians (see Proposition 2.9 in [6]) and are defined as follows: for $i = 1,2$

$$\overline{H}_i(p) = \begin{cases} -p - k_0^i & \text{for } p < -k_0^i, \\ -V_i\left(\dfrac{-1}{p}\right) \cdot |p| & \text{for } -k_0^i \leq p \leq 0, s \\ p & \text{for } p > 0, \end{cases} \tag{3.4}$$

with

$$H_0^i = \min_{p\in\mathbb{R}} \overline{H}_i(p) \quad \text{and } H_0 = \max\left(H_0^1, H_0^2\right). \tag{3.5}$$

Now we can define the limit problem. We refer to [9] for more details about existence and uniqueness of solution for this type of equation.

$$\begin{cases} u_t^0 + \overline{H}_1(u_x^0) = 0 & \text{for} \quad (t,x) \in (0,+\infty) \times (-\infty, 0) \\ u_t^0 + \overline{H}_2(u_x^0) = 0 & \text{for} \quad (t,x) \in (0,+\infty) \times (0, +\infty) \\ u_t^0 + F_{\overline{A}}(u_x^0(t,0^-), u_x^0(t,0^+)) = 0 & \text{for} \quad (t,x) \in (0,+\infty) \times \{0\} \\ u^0(0,x) = u_0(x) & \text{for} \quad x \in \mathbb{R}. \end{cases} \quad (3.6)$$

where $\overline{A}$ has to be determined and $F_{\overline{A}}$ is defined by

$$F_A(p_-, p_+) = \max\left(\overline{A}, \overline{H}_1^+(p_-), \overline{H}_2^-(p_+)\right);$$

$\overline{H}_1^+$ and $\overline{H}_2^-$ represent respectively the increasing and the decreasing part of $\overline{H}_1$ and $\overline{H}_2$. The following theorem is our main result in this paper.

Theorem 2. *There exists $\overline{A} \in \left[H_0^1, 0\right]$ such that the function u^ε defined by Theorem 1 converge locally uniformly towards the unique solution u^0 of (3.6).*

Remark 1. Formally, if we derive (3.6), we will obtain a scalar conservation law with discontinuous flux whose literature is very rich, see for example [4]. However, the passage from microscopic to macroscopic models are more difficult in this setting and in particular on networks. On the contrary, the approach proposed in this paper can be extended to models posed on networks (see [5]).

4 Correctors for the Junction

The key ingredient to prove the convergence result is to construct correctors for the junction. Given $\overline{A} \in \mathbb{R}$, we introduce two real numbers $\overline{p}_1, \overline{p}_2 \in \mathbb{R}$, such that

$$\overline{H}_2(\overline{p}_2) = \overline{H}_2^+(\overline{p}_2) = \overline{H}_1(\overline{p}_1) = \overline{H}_1^-(\overline{p}_1) = \overline{A}. \quad (4.1)$$

If $\overline{A} \leq H_0$, we then define $\overline{p}_1, \overline{p}_2 \in \mathbb{R}$ as the two real numbers satisfying

$$\overline{H}_2(\overline{p}_2) = \overline{H}_2^+(\overline{p}_2) = \overline{H}_1(\overline{p}_1) = \overline{H}_1^-(\overline{p}_1) = H_0. \quad (4.2)$$

Due to the form of $\overline{H}_1$ and $\overline{H}_2$ this two real numbers exist and are unique. We consider now the following problem: find $\lambda \in \mathbb{R}$ such that there exists a solution w of the following global-in-time Hamilton-Jacobi equation

$$(M_1[w](x) \cdot \varphi(x) + M_2[w](x) \cdot (1 - \varphi(x))) \cdot |w_x| = \lambda \quad \text{for } x \in \mathbb{R} \quad (4.3)$$

with

$$M_i[U](x) = \int_{-\infty}^{+\infty} J_i(z) E\left(U(x+z) - U(x)\right) dz - \frac{3}{2} V_{imax} \quad (4.4)$$

Theorem 3 (Existence of a global corrector for the junction). *Assume (A).*

(i) *(General properties) There exists a constant $\bar{A} \in [H_0^1, 0]$ such that there exists a solution w of (4.3) with $\lambda = \bar{A}$ and such that there exists a constant $C > 0$ and a globally Lipschitz continuous function m such that for all $x \in \mathbb{R}$,*

$$|w(x) - m(x)| \le C. \tag{4.5}$$

(ii) *(Bound from below at infinity) If $\bar{A} > H_0^1$, then there exists γ_0 such that for every $\gamma \in (0, \gamma_0)$, we have*

$$\begin{cases} w(x - h) - w(x) \ge (-\bar{p}_1 - \gamma)h - C \text{ for } x \le -r \text{ and } h \ge 0, \\ w(x + h) - w(x) \ge (\bar{p}_2 - \gamma)h - C \quad \text{ for } x \ge r \text{ and } h \ge 0. \end{cases} \tag{4.6}$$

(iii) *(Rescaling w) For $\varepsilon > 0$, we set*

$$w^\varepsilon(x) = \varepsilon w \left(\frac{x}{\varepsilon} \right),$$

then (along a subsequence $\varepsilon_n \to 0$) we have that w^ε converges locally uniformly towards a function $W = W(x)$ which satisfies

$$\begin{cases} |W(x) - W(y)| \le C|x - y| \text{ for all } x, y \in \mathbb{R}, \\ \overline{H}_1(W_x) = \bar{A} \qquad\qquad \text{ for all } x < 0, \\ \overline{H}_2(W_x) = \bar{A} \qquad\qquad \text{ for all } x > 0. \end{cases} \tag{4.7}$$

In particular, we have (with $W(0) = 0$)

$$W(x) = \bar{p}_1 x 1_{\{x < 0\}} + \bar{p}_2 x 1_{\{x > 0\}}. \tag{4.8}$$

5   Proof of Theorem 3

This section contains the proof of Theorem 3. To do this, we will construct correctors on truncated domains and then pass to the limit as the size of the domain goes to infinity. For $l \in (r, +\infty)$, $r << l$ and $r \le R << l$, we want to find $\lambda_{l,R}$, such that there exists a solution $w^{l,R}$ of

$$\begin{cases} Q_R \left(x, [w^{l,R}], w_x^{l,R} \right) = \lambda_{l,R} \text{ if } x \in (-l, l) \\ \overline{H}_1^- (w_x^{l,R}) = \lambda_{l,R} \qquad\qquad \text{ if } x \in \{-l\} \\ \overline{H}_2^+ (w_x^{l,R}) = \lambda_{l,R} \qquad\qquad \text{ if } x \in \{l\}, \end{cases} \tag{5.1}$$

with

$$Q_R(x, [U], q) = \psi_R(x) \cdot M_2[U](x) \cdot (1 - \varphi(x)) \cdot |q| + (1 - \psi_R(x)) \cdot \overline{H}_2(q) \tag{5.2}$$

$$+ \Phi_R(x) \cdot M_1[U](x) \cdot \varphi(x) \cdot |q| + (1 - \Phi_R(x)) \cdot \overline{H}_1(q) \tag{5.3}$$

and $\psi_R, \Phi_R \in C^\infty$, $\psi_R, \Phi_R : \mathbb{R} \to [0, 1]$, with

$$\psi_R \equiv \begin{cases} 1 & x \le R \\ 0 & x > R + 1 \end{cases} \quad \text{and} \quad \Phi_R \equiv \begin{cases} 1 & x \ge -R \\ 0 & x < -R - 1. \end{cases} \tag{5.4}$$

Proposition 1 (Existence of correctors on truncated domains). *There exists a unique $\lambda_{l,R} \in \mathbb{R}$ such that there exists a solutions $w^{l,R}$ of (5.1). Moreover, there exists a constant C (depending only on k_0), and a Lipschitz continuous function $m^{l,R}$, such that*

$$\begin{cases} H_0^1 \leq \lambda_{l,R} \leq 0, \\ |m^{l,R}(x) - m^{l,R}(y)| \leq C|x-y| \text{ for } x,y \in [-l,l], \\ |w^{l,R}(x) - m^{l,R}(x)| \leq C \qquad \text{ for } x \in \mathbb{R} \times [-l,l]. \end{cases} \tag{5.5}$$

Proof. We only give the main steps of the proof. Classically, we will consider the approximated problem depending on the parameter δ and then take δ to 0.

$$\begin{cases} \delta v^\delta + Q_R(x, [v^\delta], v_x^\delta) = 0 \text{ for } x \in (-l,l) \\ \delta v^\delta + \overline{H_1^-}(v_x^\delta) = 0 \qquad \text{ for } x \in \{-l\} \\ \delta v^\delta + \overline{H_2^+}(v_x^\delta) = 0 \qquad \text{ for } x \in \{l\} \end{cases} \tag{5.6}$$

Step 1: construction of barriers. Using Perron's method and 0 and $\delta^{-1}|H_0^1|$ as barriers, we deduce that there exists a continuous viscosity solution v^δ of (5.6) which satisfies

$$0 \leq v^\delta \leq \frac{|H_0^1|}{\delta}. \tag{5.7}$$

Step 2: control of the space oscillations of v^δ. The function v^δ satisfies for all $x, y \in [-l,l]$, $x \geq y$,

$$-k_0(x-y) - 1 \leq v^\delta(x) - v^\delta(y) \leq 0,$$

with $k_0 = \max(k_0^1, k_0^2)$ (see [6, Lemma 6.5]).

Step 3: construction of a Lipschitz estimate. As in [6, Lemma 6.6] we can construct a Lipschitz continuous function m^δ, such that there exists a constant C, (independent of l, R and δ) such that

$$\begin{cases} |m^\delta(x) - m^\delta(y)| \leq C|x-y| \text{ for all } x,y \in [-l,l], \\ |v^\delta(x) - m^\delta(x)| \leq C \qquad \text{ for all } x \in [-l,l]. \end{cases} \tag{5.8}$$

Step 4: passing to the limit as δ goes to 0. Classicly, taking δ to zero, we get $\lambda_{l,R}, w^{l,R}$ and $m^{l,R}$ satisfiying (5.5). The uniqueness of $\lambda_{l,R}$ is classical so we skip it. This ends the proof of Proposition 1. □

Proposition 2. *The following limits exist (up to a subsequence)*

$$\overline{A}_R = \lim_{l \to +\infty} \lambda_{l,R}, \quad \text{and} \quad \overline{A} = \lim_{R \to +\infty} \overline{A}_R.$$

Moreover, we have

$$H_0^1 \leq \overline{A}_R, \overline{A} \leq 0.$$

Proposition 3 (Control of the slopes on a truncated domain). *Assume that l and R are big enough. Let $w^{l,R}$ be the solution of (5.1) given by Proposition 1. We also assume that up to a sub-sequence $\overline{A} = \lim\limits_{R \to +\infty} \lim\limits_{l \to +\infty} \lambda_{l,R} > H_0^1$. Then there exits a $\gamma_0 > 0$ such that for all $\gamma \in (0, \gamma_0)$, there exists a constant C (independent of l and R) such that for all $x \le -r$ and $h \ge 0$*

$$w^{l,R}(x - h) - w^{l,R}(x) \ge (-\overline{p}_1 - \gamma)h - C. \tag{5.9}$$

Similarly, for all $x \ge r$ and $h \ge 0$,

$$w^{l,R}(x + h) - w^{l,R}(x) \ge (\overline{p}_2 - \gamma)h - C. \tag{5.10}$$

Proof. We only prove (5.9) since the proof for (5.10) is similar. For $\sigma > 0$ small enough, we denote by p_-^σ the real number such that

$$\overline{H}_1(p_-^\sigma) = \overline{H}_1^-(p_-^\sigma) = \lambda_{l,R} - \sigma.$$

Let us now consider the function $w^- = p_-^\sigma x$ that satisfies

$$\overline{H}_1(w_x^-) = \lambda_{l,R} - \sigma \quad \text{for } x \in \mathbb{R}.$$

We also have

$$M_1[w^-](x) = -V_1\left(\frac{-1}{p_-^\sigma}\right).$$

For all $x \in (-l, -r)$, using that $\varphi(x) = 1$ and $\psi_R(x) = 1$, we deduce that w^- satisfies

$$\begin{cases} Q_R\left(x, [w^-], w_x^-\right) = \lambda_{l,R} - \mu & \text{for } x \in (-l, -r) \\ \overline{H}_1^-(w_x^+) = \lambda_{l,R} - \mu & \text{for } x \in \{-l\}. \end{cases}$$

Using the comparaison principle, we deduce that for all $h \ge 0$, for all $x \in (-l, -r)$, we have that

$$w^{l,R}(x - h) - w^{l,R}(x) \ge -p_-^\sigma h - 2C.$$

Finally, for γ_0 and σ small enough, we can set $p_-^\sigma = \overline{p}_1 + \gamma$. $\qquad \square$

Proof of Theorem 3. The proof is performed in two steps.

Step 1: proof of (i) and (ii). The goal is to pass to the limit as $l \to +\infty$ and then as $R \to +\infty$. There exists $l_n \to +\infty$, such that

$$m^{l_n,R} - m^{l_n,R}(0) \to m^R \quad \text{as } n \to +\infty,$$

the convergence being locally uniform. We also define

$$\overline{w}^R(x) = \limsup\limits_{l_n \to +\infty}{}^* \left(w^{l_n,R} - w^{l_n,R}(0)\right),$$
$$\underline{w}^R(x) = \liminf\limits_{l_n \to +\infty}{}_* \left(w^{l_n,R} - w^{l_n,R}(0)\right).$$

Thanks to (5.5), we know that $\overline{w}^R$ and $\underline{w}^R$ are finite and satisfy

$$m^R - C \leq \underline{w}^R \leq \overline{w}^R \leq m^R + C.$$

By stability of viscosity solutions, $\overline{w}^R - 2C$ and $\underline{w}^R$ are respectively a sub and a super-solution of

$$Q_R(x, [w^R], w_x^R) = \overline{A}_R \quad \text{for } x \in \mathbb{R} \tag{5.11}$$

Therefore, using Perron's method, we can construct a solution w^R of (5.11) with $m^R, \overline{A}^R$ and w^R satisfying

$$\begin{cases} |m^R(x) - m^R(y)| \leq C|x - y| \text{ for all } x, y \in \mathbb{R}, \\ |w^R(x) - m^R(x)| \leq C \qquad \text{for } x \in \mathbb{R} \times \mathbb{R}, \\ H_0^1 \leq \overline{A}_R \leq 0. \end{cases} \tag{5.12}$$

Using Proposition 3, if $\overline{A} > H_0$, we know that there exists γ_0 and $C > 0$, such that for all $\gamma \in (0, \gamma_0)$,

$$\begin{cases} w^R(x - h) - w^R(x) \geq (-\overline{p}_1 - \gamma)h - C \text{ for all } x \leq -r, \; h \geq 0, \\ w^R(x + h) - w^R(x) \geq (\overline{p}_2 - \gamma)h - C \quad \text{for all } x \geq r, \; h \geq 0. \end{cases} \tag{5.13}$$

Passing to the limit as $R \to +\infty$ and proceeding as above, the proof is complete.

Step 2: proof of (iii). Using (4.6), we have that

$$w^\varepsilon(x) = \varepsilon m\left(\frac{x}{\varepsilon}\right) + O(\varepsilon).$$

Therefore, we can find a sequence $\varepsilon_n \to 0$, such that

$$w^{\varepsilon_n} \to W \quad \text{locally uniformly as } n \to +\infty,$$

with $W(0) = 0$. Like in [7](Appendix A.1), we have that

$$\overline{H}_1(W_x) = \overline{A} \quad \text{for } x < 0 \quad \text{and} \quad \overline{H}_2(W_x) = \overline{A} \quad \text{for } x > 0.$$

For all $\gamma \in (0, \gamma_0)$, we have that if $\overline{A} > H_0^1$ and $x > 0$,

$$W_x \geq \overline{p}_2 - \gamma,$$

where we have used (4.6). Therefore we get

$$W_x = \overline{p}_2 \quad \text{for } x > 0,$$

Similarly, we get $W_x = \overline{p}_1$ for $x < 0$. This ends the proof of Theorem 3. $\square$

6   Proof of Convergence

In this section, we will prove our homogenization result. Classicly, the proof relies on the existence of correctors. We will just prove the convergence result at the junction point since at any other point, the proof is classical using that $v = 0$ is a corrector, see [6].

Proof of Theorem 2. We introduce

$$\bar{u}(t, x) = \limsup_{\varepsilon \to 0}{}^{*} u^{\varepsilon} \quad \text{and} \quad \underline{u}(t, x) = \liminf_{\varepsilon \to 0}{}_{*} u^{\varepsilon}. \tag{6.1}$$

Let us prove that $\bar{u}$ is a sub-solution of (3.6) at the point 0, (the proof for $\underline{u}$ is similar and we skip it). The definition of viscosity solution for Hamilton-Jacobi equation is presented in Sect. 2 in [9]. We argue by contradiction and assume that there exist a test function $\Psi \in C^1(J_\infty)$ such that

$$\begin{cases} \bar{u}(\bar{t}, 0) = \Psi(\bar{t}, 0) \\ \bar{u} \le \Psi & \text{on } Q_{\bar{r}, \bar{r}}(\bar{t}, 0) \quad \text{with } \bar{r} > 0 \\ \bar{u} \le \Psi - 2\eta & \text{outside } Q_{\bar{r}, \bar{r}}(\bar{t}, 0) \text{ with } \eta > 0 \\ \Psi_t(\bar{t}, 0) + F_{\overline{A}}(\Psi_x(\bar{t}, 0^-), \Psi_x(\bar{t}, 0^+)) = \theta \text{ with } \theta > 0. \end{cases} \tag{6.2}$$

According to [9], we may assume that the test function has the following form

$$\Psi(t, x) = g(t) + \bar{p}_1 x 1_{\{x < 0\}} + \bar{p}_2 x 1_{\{x > 0\}} \quad \text{on } Q_{\bar{r}, 2\bar{r}}(\bar{t}, 0), \tag{6.3}$$

The last line in condition (6.2) becomes

$$g'(\bar{t}) + F_{\overline{A}}(\bar{p}_1, \bar{p}_2) = g'(\bar{t}) + \overline{A} = \theta. \tag{6.4}$$

Let us consider w the solution of (4.3) provided by Theorem 3, and let us denote

$$\Psi^{\varepsilon}(t, x) = \begin{cases} g(t) + w^{\varepsilon}(x) \text{ on } Q_{\bar{r}, 2\bar{r}}(\bar{t}, 0), \\ \Psi(t, x) \quad \text{outside } Q_{\bar{r}, 2\bar{r}}(\bar{t}, 0). \end{cases} \tag{6.5}$$

We claim that Ψ^{ε} is a viscosity solution on $Q_{\bar{r}, \bar{r}}(\bar{t}, 0)$ of the following problem,

$$\Psi_t^{\varepsilon} + \left(\tilde{M}_1^{\varepsilon} \left[\frac{\Psi^{\varepsilon}}{\varepsilon}(t, \cdot) \right] (x) \varphi \left(\frac{x}{\varepsilon} \right) + \tilde{M}_2^{\varepsilon} \left[\frac{\Psi^{\varepsilon}}{\varepsilon}(t, \cdot) \right] (x) \left(1 - \varphi \left(\frac{x}{\varepsilon} \right) \right) \right) \cdot |\Psi_x^{\varepsilon}| \ge \frac{\theta}{2}.$$

Indeed, let h be a test function touching φ^{ε} from below at $(t_1, x_1) \in Q_{\bar{r}, \bar{r}}(\bar{t}, 0)$, so we have that the function $\chi(y) = \frac{1}{\varepsilon}(h(t_1, \varepsilon y) - g(t_1))$ touches w from below at $\frac{x_1}{\varepsilon}$ which implies that

$$\left(\tilde{M}_1[w] \left(\frac{x_1}{\varepsilon} \right) \varphi \left(\frac{x_1}{\varepsilon} \right) + \tilde{M}_2[w] \left(\frac{x_1}{\varepsilon} \right) \left(1 - \varphi \left(\frac{x_1}{\varepsilon} \right) \right) \right) \cdot |h_x(t_1, x_1)| \ge \overline{A}. (6.6)$$

Using (6.4) and the fact that $h_t(t_1, x_1) = g'(t_1)$ and computing (6.6), we get the desired result.

Getting the Contradiction. We have that for ε small enough

$$u^\varepsilon + \eta \leq \Psi = g(t) + \overline{p}_1 x 1_{\{x<0\}} + \overline{p}_2 x 1_{\{x>0\}} \quad \text{on } \mathcal{Q}_{\overline{r},2\overline{r}}(\overline{t},0)\backslash\mathcal{Q}_{\overline{r},\overline{r}}(\overline{t},0).$$

Using the fact that $w^\varepsilon \to W$, and using (4.8), we have for ε small enough

$$u^\varepsilon + \frac{\eta}{2} \leq \Psi^\varepsilon \quad \text{on } \mathcal{Q}_{\overline{r},2\overline{r}}(\overline{t},0)\backslash\mathcal{Q}_{\overline{r},\overline{r}}(\overline{t},0).$$

Combining this with (6.5), we get that

$$u^\varepsilon + \frac{\eta}{2} \leq \Psi^\varepsilon \quad \text{outside } \mathcal{Q}_{\overline{r},\overline{r}}(\overline{t},0).$$

By the comparison principle on bounded subsets the previous inequality holds in $\mathcal{Q}_{\overline{r},\overline{r}}(\overline{t},0)$. Passing to the limit as $\varepsilon \to 0$ and evaluating the inequality at $(\overline{t},0)$, we obtain the following contradiction

$$\overline{u}(\overline{t},0) + \frac{\eta}{2} \leq \Psi(\overline{t},0) = \overline{u}(\overline{t},0).$$

$\square$

Acknowledgments. This work is co-financed by the European Union with the European regional development fund (ERDF, HN0002137) and by the Normandie Regional Council via the M2NUM project.

References

1. Aw, A., Klar, A., Rascle, M., Materne, T.: Derivation of continuum traffic flow models from microscopic follow-the-leader models. SIAM J. Appl. Math. **63**, 259–278 (2002)
2. Colombo, R.M., Rossi, E.: Rigorous estimates on balance laws in bounded domains. Acta Math. Sci. **35**, 906–944 (2015)
3. Di Francesco, M., Rosini, M.D.: Rigorous derivation of the lighthill-whitham-richards model from the follow-the-leader model as many particle limit. arXiv preprint arXiv:1404.7062 (2014)
4. Diehl, S.: On scalar conservation laws with point source and discontinuous flux function. SIAM J. Math. Anal. **26**, 1425–1451 (1995)
5. Forcadel, N., Salazar, W.: Homogenization of a discrete model for a bifurcation and application to traffic flow (2016). hal-01332787
6. Forcadel, N., Salazar, W.: A junction condition by specified homogenization of a discrete model with a local perturbation and application to traffic flow, <hal-01097085> (2015)
7. Galise, G., Imbert, C., Monneau, R.A: Junction condition by specified homogenization. arXiv preprint arXiv:1406.5283 (2014)
8. Gazis, D.C., Herman, R., Rothery, R.: Nonlinear follow-the-leader models of traffic flow. Oper. Res. **9**, 545–567 (1961)
9. Imbert, C., Monneau, R.: Flux-limited solutions for quasi-convex Hamilton-Jacobi equations on networks, <hal-00832545> (2014)
10. Prigogine, I. Herman, R.: Kinetic theory of vehicular traffic, p. 100 (1971)

Cahn–Hilliard Approach to Some Degenerate Parabolic Equations with Dynamic Boundary Conditions

Takeshi Fukao$^{(\boxtimes)}$

Department of Mathematics, Faculty of Education, Kyoto University of Education,
1 Fujinomori, Fukakusa, Fushimi-ku, Kyoto 612-8522, Japan
fukao@kyokyo-u.ac.jp

Abstract. In this paper the well-posedness of some degenerate parabolic equations with a dynamic boundary condition is considered. To characterize the target degenerate parabolic equation from the Cahn–Hilliard system, the nonlinear term coming from the convex part of the double-well potential is chosen using a suitable maximal monotone graph. The main topic of this paper is the existence problem under an assumption for this maximal monotone graph for treating a wider class. The existence of a weak solution is proved.

Keywords: Degenerate parabolic equation · Dynamic boundary condition · Weak solution · Cahn–Hilliard system

AMS (MOS) Subject Classification: 35K65 · 35K30 · 47J35

1 Introduction

The relationship between the Allen–Cahn equation [2] and the motion by mean curvature is interesting as the singular limit of the following form:

$$\frac{\partial u}{\partial t} - \Delta u + \frac{1}{\varepsilon^2}(u^3 - u) = 0 \quad \text{in } Q := (0, T) \times \Omega,$$

as $\varepsilon \searrow 0$, where $0 < T < +\infty$ and $\Omega \subset \mathbb{R}^d$ for $d = 2, 3$, which is a bounded domain with smooth boundary Γ. For example, Bronsard and Kohn presented a pioneering result in [5], and subsequently many related results have been obtained. A similar concept in this framework, the Cahn–Hilliard system [7], is connected to motion by the Mullins–Sekerka law [19] in the limit of

$$\frac{\partial u}{\partial t} - \Delta \mu = 0 \quad \text{in } Q,$$

$$\mu = -\varepsilon \Delta u + \frac{1}{\varepsilon}(u^3 - u) \quad \text{in } Q \tag{1.1}$$

as $\varepsilon \searrow 0$. For both of these, the target problems are sharp interface models in a classical sense and a powerful analysis tool seems to be the method of matched asymptotic expansions (see [1, 6, 20] and the references in these papers).

© IFIP International Federation for Information Processing 2016
Published by Springer International Publishing AG 2016. All Rights Reserved
L. Bociu et al. (Eds.): CSMO 2015, IFIP AICT 494, pp. 282–291, 2016.
DOI: 10.1007/978-3-319-55795-3_26

In this paper, we discuss this relation from a different view point. To do so, we begin with the following degenerate parabolic equation:

$$\frac{\partial u}{\partial t} - \Delta\beta(u) = g \quad \text{in } Q, \tag{1.2}$$

where $g : \Omega \to \mathbb{R}$ is a given source. This equation is characterized by the choice of $\beta : \mathbb{R} \to \mathbb{R}$. For example, if we choose β to be a piecewise linear function of the form

$$\beta(r) := \begin{cases} k_s r & r < 0, \\ 0 & 0 \le r \le L, \\ k_\ell(r - L) & r > L; \end{cases} \tag{1.3}$$

where k_s and $k_\ell > 0$ represent the heat conductivities of the solid and liquid regions, respectively, and $L > 0$ is the latent heat constant, then (1.2) is the weak formulation of the Stefan problem, or the "enthalpy formulation," where the unknown u denotes the enthalpy and $\beta(u)$ denotes the temperature. The informant of the sharp interface, in other words the Stefan condition, is hidden in the weak formulation. Another example is the weak formulation of the Hele-Shaw problem. If we choose β to be the inverse of the Heaviside function

$$\mathcal{H}(r) := \begin{cases} 0 & \text{if } r < 0, \\ [0, 1] & \text{if } r = 0, \\ 1 & \text{if } r > 0 \end{cases} \quad \text{for all } r \in \mathbb{R},$$

so that β is the multivalued function $\beta(r) := \mathcal{H}^{-1}(r) = \partial I_{[0,1]}(r)$ for all $r \in [0, 1]$, then (1.2) can be stated as

$$\xi \in \beta(u), \quad \frac{\partial u}{\partial t} - \Delta\xi = g \quad \text{in } Q,$$

where $\partial I_{[0,1]}$ is the subdifferential of the indicator function $I_{[0,1]}$ on the interval $[0, 1]$, the unknown u denotes the order parameter. Details about weak formulations may be found in Visintin [22]. Weak formulations for this kind of sharp interface model are the focus of this paper. Therefore, we use the terms "Stefan problem" and "Hele-Shaw problem" in the sense of weak formulations throughout this paper.

Recently, the author considered the approach to the following Cahn–Hilliard system for the Stefan problem in [13]:

$$\frac{\partial u}{\partial t} - \Delta\mu = 0 \quad \text{in } Q, \tag{1.4}$$

$$\mu = -\varepsilon\Delta u + \beta(u) + \varepsilon\pi(u) - f \quad \text{in } Q, \tag{1.5}$$

with a dynamic boundary condition of the form

$$\frac{\partial u}{\partial t} + \partial_\nu\mu - \Delta_\Gamma\mu = 0 \quad \text{on } \Sigma := (0, T) \times \Gamma, \tag{1.6}$$

$$\mu = \varepsilon\partial_\nu u - \varepsilon\Delta_\Gamma u + \beta(u) + \varepsilon\pi(u) - f_\Gamma \quad \text{on } \Sigma, \tag{1.7}$$

where the symbol ∂_ν denotes the normal derivative on the boundary Γ outward from Ω, the symbol Δ_Γ stands for the Laplace–Beltrami operator on Γ (see, e.g., [15, Chap. 3]), β is defined by (1.3), and $\pi : \mathbb{R} \to \mathbb{R}$ is a piecewise linear function defined by $\pi(r) := L/2$ if $r < 0$, $\pi(r) := L/2 - r$ if $0 \leq r \leq L$ and $\pi(r) := -L/2$ if $r > L$. Thanks to this choice, system (1.4)–(1.7) has the structure of a Cahn–Hilliard system. This problem originally comes from [14]. Formally, if we let $\varepsilon \searrow 0$ in (1.4)–(1.7), then we can see that the Cahn–Hilliard system (1.4)–(1.7) converges in a suitable sense to the following Stefan problem with a dynamic boundary condition:

$$\frac{\partial u}{\partial t} - \Delta\beta(u) = -\Delta f \quad \text{in } Q,$$

$$\frac{\partial u}{\partial t} + \partial_\nu \beta(u) - \Delta_\Gamma \beta(u) = \partial_\nu f_\Gamma - \Delta_\Gamma f_\Gamma \quad \text{on } \Sigma.$$

Here, we should take care of the difference between the order and position of ε in (1.1) and (1.5) even when $\beta(u) = u^3$ and $\pi(u) = -u$. In [13], β is assumed to satisfy the following condition:

β is a maximal monotone graph in $\mathbb{R} \times \mathbb{R}$, and is a subdifferential $\beta = \partial\widehat{\beta}$ of some proper, lower semicontinuous, and convex function $\widehat{\beta} : \mathbb{R} \to [0, +\infty]$ satisfying $\widehat{\beta}(0) = 0$ with some effective domain $D(\beta)$. This implies $\beta(0) = 0$. Moreover, there exist two constants $c, \tilde{c} > 0$ such that

$$\widehat{\beta}(r) \geq c|r|^2 - \tilde{c} \quad \text{for all } r \in \mathbb{R}. \tag{1.8}$$

It is easy to see that (1.2) represents a large number of problems, including the porous media equation, the nonlinear diffusion equation of Penrose–Fife type, the fast diffusion equation, and so on. However, to apply this approach from the Cahn–Hilliard system to these wider classes of the degenerate parabolic equation, the growth condition (1.8) is too strong (see, e.g. [12]). Therefore, in this paper based on the essential idea from [10], we relax the assumption in (1.8). This is the different point from the previous work [13]. See also [3,16,17] for related problems of interest.

Notation. Let $H := L^2(\Omega)$, $V := H^1(\Omega)$, $H_\Gamma := L^2(\Gamma)$ and $V_\Gamma := H^1(\Gamma)$ with the usual norms $|\cdot|_H$, $|\cdot|_V$, $|\cdot|_{H_\Gamma}$, $|\cdot|_{V_\Gamma}$ and inner products $(\cdot,\cdot)_H$, $(\cdot,\cdot)_V$, $(\cdot,\cdot)_{H_\Gamma}$, $(\cdot,\cdot)_{V_\Gamma}$, respectively, and let $\boldsymbol{H} := H \times H_\Gamma$, $\boldsymbol{V} := \{(z, z_\Gamma) \in V \times V_\Gamma : z_\Gamma = z_{|\Gamma} \text{ a.e. on } \Gamma\}$ and $\boldsymbol{W} := H^2(\Omega) \times H^2(\Gamma)$. Then $\boldsymbol{H}$, $\boldsymbol{V}$ and $\boldsymbol{W}$ are Hilbert spaces with the inner product

$$(\boldsymbol{u}, \boldsymbol{z})_{\boldsymbol{H}} := (u, z)_H + (u_\Gamma, z_\Gamma)_{H_\Gamma} \quad \text{for all } \boldsymbol{u}, \boldsymbol{z} \in \boldsymbol{H},$$

and the related norm is analogously defined as one of $\boldsymbol{V}$ or $\boldsymbol{W}$. Define $m : \boldsymbol{H} \to \mathbb{R}$ by

$$m(\boldsymbol{z}) := \frac{1}{|\Omega| + |\Gamma|} \left\{ \int_\Omega z\,dx + \int_\Gamma z_\Gamma\,d\Gamma \right\} \quad \text{for all } \boldsymbol{z} \in \boldsymbol{H},$$

where $|\Omega| := \int_\Omega 1 dx$ and $|\Gamma| := \int_\Gamma 1 d\Gamma$. The symbol V^* denotes the dual space of V, and the pair $\langle \cdot, \cdot \rangle_{V^*, V}$ denotes the duality pairing between V^* and V. Moreover, define the bilinear form $a(\cdot, \cdot) : V \times V \to \mathbb{R}$ by

$$a(u, z) := \int_\Omega \nabla u \cdot \nabla z dx + \int_\Gamma \nabla_\Gamma u_\Gamma \cdot \nabla_\Gamma z_\Gamma d\Gamma \quad \text{for all } u, z \in V,$$

where ∇_Γ denotes the surface gradient on Γ (see, e.g., [15, Chap. 3]). We introduce the subspace $H_0 := \{ z \in H : m(z) = 0 \}$ of H and $V_0 := V \cap H_0$, with their norms $|z|_{H_0} := |z|_H$ for all $z \in H_0$ and $|z|_{V_0} := a(z, z)^{1/2}$ for all $z \in V_0$. Then the duality mapping $F : V_0 \to V_0^*$ is defined by $\langle Fz, \tilde{z} \rangle_{V_0^*, V_0} := a(z, \tilde{z})$ for all $z, \tilde{z} \in V_0$ and the inner product in V_0^* is defined by $(z_1^*, z_2^*)_{V_0^*} := \langle z_1^*, F^{-1} z_2^* \rangle_{V_0^*, V_0}$ for all $z_1^*, z_2^* \in V_0^*$. Moreover, define $P : H \to H_0$ by $Pz := z - m(z)\mathbf{1}$ for all $z \in H$, where $\mathbf{1} := (1, 1)$. Thus we obtain the dense and compact embeddings $V_0 \hookrightarrow\hookrightarrow H_0 \hookrightarrow\hookrightarrow V_0^*$. See [8,9] for further details.

2   Existence of the Weak Solution

In this section, we state an existence theorem for the weak solution of a degenerate parabolic equation with a dynamic boundary condition of the following form:

$$\xi \in \beta(u), \quad \frac{\partial u}{\partial t} - \Delta \xi = g \quad \text{a.e. in } Q,$$

$$\xi_\Gamma \in \beta(u_\Gamma), \quad \xi_\Gamma = \xi_{|\Gamma}, \quad \frac{\partial u_\Gamma}{\partial t} + \partial_\nu \xi - \Delta_\Gamma \xi_\Gamma = g_\Gamma \quad \text{a.e. on } \Sigma,$$

$$u(0) = u_0 \quad \text{a.e. in } \Omega, \quad u_\Gamma(0) = u_{0\Gamma} \quad \text{a.e. on } \Gamma,$$

where β, g, g_Γ, u_0 and $u_{0\Gamma}$ satisfy the following assumptions:

(A1) β is a maximal monotone graph in $\mathbb{R} \times \mathbb{R}$, and is a subdifferential $\beta = \partial\widehat{\beta}$ of some proper, lower semicontinuous, and convex function $\widehat{\beta} : \mathbb{R} \to [0, +\infty]$ satisfying $\widehat{\beta}(0) = 0$ in some effective domain $D(\beta)$. This implies that $\beta(0) = 0$;

(A2) $g \in L^2(0, T; H_0)$;

(A3) $u_0 := (u_0, u_{0\Gamma}) \in H$ with $m_0 \in \operatorname{int} D(\beta)$, and the compatibility conditions $\widehat{\beta}(u_0) \in L^1(\Omega), \widehat{\beta}(u_{0\Gamma}) \in L^1(\Gamma)$ hold.

We remark that the growth condition of $\widehat{\beta}$ in (A1) and the regularity of u_0 in (A3) are relaxations from a previous related result [13] (cf. (1.8)).

Theorem 2.1. *Under assumptions* (A1)–(A3), *there exists at least one pair* (u, ξ) *of functions* $u \in H^1(0, T; V^*) \cap L^2(0, T; H)$ *and* $\xi \in L^2(0, T; V)$ *such that* $\xi \in \beta(u)$ *a.e. in* Q, $\xi_\Gamma \in \beta(u_\Gamma)$ *and* $\xi_\Gamma = \xi_{|\Gamma}$ *a.e. on* Σ, *and that satisfy*

$$\langle u'(t), z \rangle_{V^*, V} + \langle u'_\Gamma(t), z_\Gamma \rangle_{V^*_\Gamma, V_\Gamma} + \int_\Omega \nabla \xi(t) \cdot \nabla z dx + \int_\Gamma \nabla_\Gamma \xi_\Gamma(t) \cdot \nabla_\Gamma z_\Gamma d\Gamma$$

$$= \int_\Omega g(t) z dx + \int_\Gamma g_\Gamma(t) z_\Gamma d\Gamma \quad \text{for all } z := (z, z_\Gamma) \in V \quad (2.1)$$

for a.a. $t \in (0, T)$ *with* $u(0) = u_0$ *a.e. in* Ω *and* $u_\Gamma(0) = u_{0\Gamma}$ *a.e. on* Γ.

The continuous dependence is completely the same as in a previous result [13, Theorem 2.2]. Therefore, we devolve the uniqueness problem on [13].

3   Proof of the Main Theorem

In this section, we prove the main theorem. The strategy of the proof is similar to that of [13, Theorem 2.1]. However, to relax the assumption we use a different uniform estimate. Let us start with an approximate problem. Recall the Yosida approximation $\beta_\lambda : \mathbb{R} \to \mathbb{R}$ and the related Moreau–Yosida regularization $\widehat{\beta}_\lambda$ of $\widehat{\beta} : \mathbb{R} \to \mathbb{R}$ (see, e.g., [4]). We see that $0 \leq \widehat{\beta}_\lambda(r) \leq \widehat{\beta}(r)$ for all $r \in \mathbb{R}$. Moreover, we define the following proper, lower semicontinuous, and convex functional $\varphi : \boldsymbol{H}_0 \to [0, +\infty]$:

$$\varphi(\boldsymbol{z}) := \begin{cases} \dfrac{1}{2}\displaystyle\int_\Omega |\nabla z|^2 dx + \dfrac{1}{2}\displaystyle\int_\Gamma |\nabla_\Gamma z_\Gamma|^2 d\Gamma & \text{if } \boldsymbol{z} \in \boldsymbol{V}_0, \\ +\infty & \text{otherwise.} \end{cases}$$

The subdifferential $\partial\varphi$ on $\boldsymbol{H}_0$ is characterized by $\partial\varphi(\boldsymbol{z}) = (-\Delta z, \partial_\nu z - \Delta_\Gamma z_\Gamma)$ with $\boldsymbol{z} \in D(\partial\varphi) = \boldsymbol{W} \cap \boldsymbol{V}_0$ (see, e.g., [9, Lemma C]). By virtue of the well-known theory of evolution equations (see, e.g., [8, 9, 11, 18]), for each $\varepsilon \in (0,1]$ and $\lambda \in (0,1]$, there exist $\boldsymbol{v}_{\varepsilon,\lambda} \in H^1(0,T;\boldsymbol{H}_0) \cap L^\infty(0,T;\boldsymbol{V}_0) \cap L^2(0,T;\boldsymbol{W})$ and $\boldsymbol{\mu}_{\varepsilon,\lambda} \in L^2(0,T;\boldsymbol{V})$ such that

$$\lambda \boldsymbol{v}'_{\varepsilon,\lambda}(t) + \boldsymbol{F}^{-1}\big(\boldsymbol{v}'_{\varepsilon,\lambda}(t)\big) + \varepsilon\partial\varphi\big(\boldsymbol{v}_{\varepsilon,\lambda}(t)\big)$$
$$= \boldsymbol{P}\big(-\boldsymbol{\beta}_\lambda\big(\boldsymbol{u}_{\varepsilon,\lambda}(t)\big) - \varepsilon\boldsymbol{\pi}\big(\boldsymbol{u}_{\varepsilon,\lambda}(t)\big) + \boldsymbol{f}(t)\big) \quad \text{in } \boldsymbol{H}_0 \qquad (3.1)$$

for a.a. $t \in (0,T)$ with $\boldsymbol{v}_{\varepsilon,\lambda}(0) = \boldsymbol{v}_{0\varepsilon}$ in $\boldsymbol{H}_0$, where $\boldsymbol{v}_{0\varepsilon} \in \boldsymbol{V}_0$ solves the auxiliary problem $\boldsymbol{v}_{0\varepsilon} + \varepsilon\partial\varphi(\boldsymbol{v}_{0\varepsilon}) = \boldsymbol{v}_0 := \boldsymbol{u}_0 - m_0\boldsymbol{1}$ in $\boldsymbol{H}_0$ so that there exists a constant $C > 0$ such that

$$|\boldsymbol{v}_{0\varepsilon}|^2_{\boldsymbol{H}_0} \leq C, \quad \varepsilon|\boldsymbol{v}_{0\varepsilon}|^2_{\boldsymbol{V}_0} \leq C,$$
$$\int_\Omega \widehat{\beta}(v_{0\varepsilon} + m_0)dx \leq C, \quad \int_\Gamma \widehat{\beta}(v_{0\varepsilon} + m_0)d\Gamma \leq C. \qquad (3.2)$$

Moreover, $\boldsymbol{u}_{\varepsilon,\lambda} := \boldsymbol{v}_{\varepsilon,\lambda} + m_0\boldsymbol{1}$, $m_0 := m(\boldsymbol{u}_0)$ and $\boldsymbol{1} := (1,1)$, and $\boldsymbol{\beta}_\lambda(\boldsymbol{z}) := (\beta_\lambda(z), \beta_\lambda(z_\Gamma))$ and $\boldsymbol{\pi}(\boldsymbol{z}) := (\pi(z), \pi(z_\Gamma))$ for all $\boldsymbol{z} \in \boldsymbol{H}$, where $\pi : D(\pi) = \mathbb{R} \to \mathbb{R}$ is a Lipschitz continuous function with a Lipschitz constant L_π that breaks the monotonicity in $\beta + \varepsilon\pi$; $\boldsymbol{f} \in L^2(0,T;D(\partial\varphi))$ is the solution of $\boldsymbol{g}(t) = \partial\varphi(\boldsymbol{f}(t))$ in $\boldsymbol{H}_0$ for a.a. $t \in (0,T)$. Namely, from [9, Lemma C], we can choose $\boldsymbol{f}(t) := (f(t), f_\Gamma(t))$ to satisfy

$$\begin{cases} -\Delta f(t) = g(t) & \text{a.e. in } \Omega, \\ \partial_\nu f(t) - \Delta_\Gamma f_\Gamma(t) = g_\Gamma(t) & \text{a.e. on } \Gamma, \end{cases} \quad \text{for a.a. } t \in (0,T). \qquad (3.3)$$

3.1   Uniform Estimates for Approximate Solutions

The key strategy in the proof is to obtain uniform estimates independent of $\varepsilon > 0$ and $\lambda > 0$, after which we consider the limiting procedures $\lambda \searrow 0$ and $\varepsilon \searrow 0$. Recall (3.1) in the equivalent form

$$v'_{\varepsilon,\lambda}(s) + F(P\mu_{\varepsilon,\lambda}(s)) = 0 \quad \text{in } V_0^*, \tag{3.4}$$

$$\mu_{\varepsilon,\lambda}(s) = \lambda v'_{\varepsilon,\lambda}(s) + \varepsilon\partial\varphi(v_{\varepsilon,\lambda}(s)) + \beta_\lambda(u_{\varepsilon,\lambda}(s)) + \varepsilon\pi(u_{\varepsilon,\lambda}(s)) - f(s) \quad \text{in } H \tag{3.5}$$

for a.a. $s \in (0,T)$. Moreover, if we put $\varepsilon_0 := \min\{1, 1/(4L_\pi^2)\}$, then we have:

Lemma 3.1. *There exist positive constants M_1, M_2 independent of $\varepsilon \in (0,\varepsilon_0]$ and $\lambda \in (0,1]$ such that*

$$\lambda|v_{\varepsilon,\lambda}(t)|^2_{H_0} + |v_{\varepsilon,\lambda}(t)|^2_{V_0^*} \le M_1,$$

$$\frac{\varepsilon}{2}\int_0^t |v_{\varepsilon,\lambda}(s)|^2_{V_0}\,ds + 2\int_0^t |\widehat{\beta}_\lambda(u_{\varepsilon,\lambda}(s))|_{L^1(\Omega)}\,ds + 2\int_0^t |\widehat{\beta}_\lambda(u_{\Gamma,\varepsilon,\lambda}(s))|_{L^1(\Gamma)}\,ds \le M_2$$

for all $t \in [0,T]$.

Proof. Multiplying (3.1) by $v_{\varepsilon,\lambda}(s) \in V_0$, we have

$$\lambda(v'_{\varepsilon,\lambda}(s), v_{\varepsilon,\lambda}(s))_{H_0} + (v'_{\varepsilon,\lambda}(s), v_{\varepsilon,\lambda}(s))_{V_0^*} + \varepsilon(\partial\varphi(v_{\varepsilon,\lambda}(s)), v_{\varepsilon,\lambda}(s))_{H_0}$$
$$+ (P\beta_\lambda(v_{\varepsilon,\lambda}(s) + m_0 1), v_{\varepsilon,\lambda}(s))_{H_0} = (f(s) - \varepsilon P\pi(v_{\varepsilon,\lambda}(s) + m_0 1), v_{\varepsilon,\lambda}(s))_{H_0}$$

for a.a. $s \in (0,T)$. Using the definition of the subdifferential, we see that

$$\frac{\lambda}{2}\frac{d}{ds}|v_{\varepsilon,\lambda}(s)|^2_{H_0} + \frac{1}{2}\frac{d}{ds}|v_{\varepsilon,\lambda}(s)|^2_{V_0^*} + \frac{\varepsilon}{2}|v_{\varepsilon,\lambda}(s)|^2_{V_0}$$
$$+ |\widehat{\beta}_\lambda(u_{\varepsilon,\lambda}(s))|_{L^1(\Omega)} + |\widehat{\beta}_\lambda(u_{\Gamma,\varepsilon,\lambda}(s))|_{L^1(\Gamma)}$$
$$\le (|\Omega| + |\Gamma|)\widehat{\beta}(m_0) + \frac{1}{2}|v_{\varepsilon,\lambda}(s)|^2_{V_0^*} + |f(s)|^2_{V_0} + L_\pi^2\varepsilon^2|v_{\varepsilon,\lambda}(s)|^2_{V_0} \quad \text{for a.a. } s \in (0,T).$$

Taking $\varepsilon \in (0,\varepsilon_0]$ and using the Gronwall inequality, we obtain the existence of M_1 and M_2 independent of $\varepsilon \in (0,\varepsilon_0]$ and $\lambda \in (0,1]$ satisfying the conclusion. $\square$

Lemma 3.2. *There exists a positive constant M_3, independent of $\varepsilon \in (0,\varepsilon_0]$ and $\lambda \in (0,1]$, such that*

$$2\lambda\int_0^t |v'_{\varepsilon,\lambda}(s)|^2_{H_0}\,ds + \int_0^t |v'_{\varepsilon,\lambda}(s)|^2_{V_0^*}\,ds + \varepsilon|v_{\varepsilon,\lambda}(t)|^2_{V_0}$$
$$+ 2|\widehat{\beta}_\lambda(u_{\varepsilon,\lambda}(t))|_{L^1(\Omega)} + 2|\widehat{\beta}_\lambda(u_{\Gamma,\varepsilon,\lambda}(t))|_{L^1(\Gamma)} \le M_3,$$

$$\int_0^t |P\mu_{\varepsilon,\lambda}(s)|^2_{V_0}\,ds \le M_3 \quad \text{for all } t \in [0,T].$$

Proof. Multiplying (3.1) by $v'_{\varepsilon,\lambda}(s) \in H_0$, we have

$$\lambda|v'_{\varepsilon,\lambda}(s)|^2_{H_0} + \frac{1}{2}|v'_{\varepsilon,\lambda}(s)|^2_{V_0^*} + \varepsilon\frac{d}{ds}\varphi(v_{\varepsilon,\lambda}(s)) + \frac{d}{ds}\int_\Omega \widehat{\beta}_\lambda(u_{\varepsilon,\lambda}(s))\,dx$$
$$+ \frac{d}{ds}\int_\Gamma \widehat{\beta}_\lambda(u_{\Gamma,\varepsilon,\lambda}(s))\,d\Gamma \le L_\pi^2\varepsilon^2|v_{\varepsilon,\lambda}(s)|^2_{V_0} + |f(s)|^2_{V_0} \quad \text{for a.a. } s \in (0,T).$$

Integrating this over $(0,t)$ with respect to s, we see that there exists a positive constant M_3, independent of $\varepsilon \in (0,\varepsilon_0]$ and $\lambda \in (0,1]$, such that the first estimate holds. Next, multiplying (3.4) by $P\mu_{\varepsilon,\lambda}(s) \in V_0$ and integrating the resultant over $(0,t)$ with respect to s, we obtain the second estimate. $\square$

The previous two lemmas are essentially the same as [13, Lemmas 3.1 and 3.2]. The next uniform estimate is the point of emphasis in this paper.

Lemma 3.3. *There exists positive constant M_4, independent of $\varepsilon \in (0,1]$ and $\lambda \in (0,1]$, such that*

$$|u_{\varepsilon,\lambda}(t)|_{\boldsymbol{H}}^2 \le M_4 \left(1 + \frac{\lambda}{\varepsilon}\right), \quad |v_{\varepsilon,\lambda}(t)|_{\boldsymbol{H}_0}^2 \le M_4 \left(1 + \frac{\lambda}{\varepsilon}\right),$$

$$\lambda |v_{\varepsilon,\lambda}(t)|_{\boldsymbol{V}_0}^2 + \varepsilon \int_0^t |\partial\varphi(v_{\varepsilon,\lambda}(s))|_{\boldsymbol{H}_0}^2 ds \le M_4 \left(1 + \frac{\lambda}{\varepsilon}\right) \quad \text{for all } t \in [0,T].$$

Proof. Multiplying (3.4) by $v_{\varepsilon,\lambda}(s) \in \boldsymbol{V}_0$ and using the fact $(d/ds)m(u_{\varepsilon,\lambda}(s)) = 0$, we have

$$\left(u_{\varepsilon,\lambda}'(s), u_{\varepsilon,\lambda}(s)\right)_{\boldsymbol{H}} + a\left(\boldsymbol{\mu}_{\varepsilon,\lambda}(s), u_{\varepsilon,\lambda}(s)\right) = 0$$

for a.a. $s \in (0,T)$ (see [13, Remark 3]). On the other hand, multiplying (3.5) by $\partial\varphi(v_{\varepsilon,\lambda}(s)) \in \boldsymbol{H}_0$ and integrating by parts, we have

$$a\left(\boldsymbol{\mu}_{\varepsilon,\lambda}(s), u_{\varepsilon,\lambda}(s)\right) = \frac{\lambda}{2}\frac{d}{ds}a\left(u_{\varepsilon,\lambda}(s), u_{\varepsilon,\lambda}(s)\right) + \varepsilon|\partial\varphi(v_{\varepsilon,\lambda}(s))|_{\boldsymbol{H}_0}^2$$

$$+ \int_\Omega \beta_\lambda'(u_{\varepsilon,\lambda}(s))|\nabla u_{\varepsilon,\lambda}(s)|^2 dx + \int_\Gamma \beta_\lambda'(u_{\Gamma,\varepsilon,\lambda}(s))|\nabla_\Gamma u_{\Gamma,\varepsilon,\lambda}(s)|^2 d\Gamma$$

$$+ \varepsilon\left(\boldsymbol{\pi}(u_{\varepsilon,\lambda}(s)), \partial\varphi(u_{\varepsilon,\lambda}(s))\right)_{\boldsymbol{H}} + \left(\partial\varphi(\boldsymbol{f}(s)), u_{\varepsilon,\lambda}(s)\right)_{\boldsymbol{H}}$$

for a.a. $s \in (0,T)$. Using the Lipschitz continuity of π and (3.3), we see that there exists a positive constant C_π such that

$$\frac{d}{ds}|u_{\varepsilon,\lambda}(s)|_{\boldsymbol{H}}^2 + \lambda\frac{d}{ds}|v_{\varepsilon,\lambda}(s)|_{\boldsymbol{V}_0}^2 + \varepsilon|\partial\varphi(v_{\varepsilon,\lambda}(s))|_{\boldsymbol{H}_0}^2 \le C_\pi\left(|u_{\varepsilon,\lambda}(s)|_{\boldsymbol{H}}^2 + 1\right) + |g(s)|_{\boldsymbol{H}_0}^2$$

for a.a. $s \in (0,T)$. Then, using (3.2) and the Gronwall inequality, we deduce that

$$|u_{\varepsilon,\lambda}(t)|_{\boldsymbol{H}}^2 + \lambda|v_{\varepsilon,\lambda}(t)|_{\boldsymbol{V}_0}^2 \le \left\{|v_{0\varepsilon} + m_0 1|_{\boldsymbol{H}}^2 + \lambda|v_{0\varepsilon}|_{\boldsymbol{V}_0}^2 + C_\pi T + |g|_{L^2(0,T;\boldsymbol{H}_0)}^2\right\} e^{C_\pi T}$$

$$\le \left\{2C + 2|m_0|^2(|\Omega| + |\Gamma|) + \frac{\lambda}{\varepsilon}C + C_\pi T + |g|_{L^2(0,T;\boldsymbol{H}_0)}^2\right\} e^{C_\pi T}$$

for all $t \in [0,T]$. That is, there exists a positive constant M_4 independent of $\varepsilon \in (0,1]$ and $\lambda \in (0,1]$ such that the uniform estimates hold. $\square$

Lemma 3.4. *There exists positive constant M_5, independent of $\varepsilon \in (0,1]$ and $\lambda \in (0,1]$, such that*

$$\int_0^t |\boldsymbol{\mu}_{\varepsilon,\lambda}(s)|_{\boldsymbol{V}_0}^2 ds \le M_5 \left(1 + \frac{\lambda}{\varepsilon}\right),$$

$$\int_0^t |\beta_\lambda(u_{\varepsilon,\lambda}(s))|_{\boldsymbol{H}}^2 ds \le M_5 \left(1 + \frac{\lambda}{\varepsilon}\right) \quad \text{for all } t \in [0,T].$$

Using Lemmas 3.1 to 3.3, the proofs of these uniform estimates are completely the same as those for [9, Lemmas 4.3 and 4.4]. Therefore, we omit the proof.

3.2 Limiting Procedure

From the previous uniform estimates, we can consider the limit as $\lambda \searrow 0$. More precisely, for each $\varepsilon \in (0, \varepsilon_0]$, there exists a subsequence $\{\lambda_k\}_{k \in \mathbb{N}}$ with $\lambda_k \searrow 0$ as $k \to +\infty$ and a quadruplet $(\boldsymbol{v}_\varepsilon, \boldsymbol{v}_\varepsilon^*, \boldsymbol{\mu}_\varepsilon, \boldsymbol{\xi}_\varepsilon)$ of $\boldsymbol{v}_\varepsilon \in H^1(0, T; \boldsymbol{V}_0^*) \cap L^\infty(0, T; \boldsymbol{V}_0)$ $\cap L^2(0, T; \boldsymbol{W})$, $\boldsymbol{v}_\varepsilon^* \in L^2(0, T; \boldsymbol{H}_0)$, $\boldsymbol{\mu}_\varepsilon \in L^2(0, T; \boldsymbol{V})$, $\boldsymbol{\xi}_\varepsilon \in L^2(0, T; \boldsymbol{H})$, such that

$$\boldsymbol{v}_{\varepsilon, \lambda_k} \to \boldsymbol{v}_\varepsilon \quad \text{weakly star in } L^\infty(0, T; \boldsymbol{H}_0), \quad \boldsymbol{v}_{\varepsilon, \lambda_k} \to \boldsymbol{v}_\varepsilon \quad \text{weakly in } L^2(0, T; \boldsymbol{V}_0),$$

$$\lambda_k \boldsymbol{v}_{\varepsilon, \lambda_k}' \to \boldsymbol{0} \quad \text{in } L^2(0, T; \boldsymbol{H}_0), \quad \boldsymbol{v}_{\varepsilon, \lambda_k}' \to \boldsymbol{v}_\varepsilon' \quad \text{weakly in } L^2(0, T; \boldsymbol{V}_0^*),$$

$$\boldsymbol{u}_{\varepsilon, \lambda_k} \to \boldsymbol{u}_\varepsilon := \boldsymbol{v}_\varepsilon + m_0 \boldsymbol{1} \quad \text{weakly star in } L^\infty(0, T; \boldsymbol{V}),$$

$$\partial \varphi(\boldsymbol{v}_{\varepsilon, \lambda_k}) \to \boldsymbol{v}_\varepsilon^* \quad \text{weakly in } L^2(0, T; \boldsymbol{H}_0), \quad \boldsymbol{\mu}_{\varepsilon, \lambda_k} \to \boldsymbol{\mu}_\varepsilon \quad \text{weakly in } L^2(0, T; \boldsymbol{V}),$$

$$\beta_{\lambda_k}(\boldsymbol{u}_{\varepsilon, \lambda_k}) \to \boldsymbol{\xi}_\varepsilon \quad \text{weakly in } L^2(0, T; \boldsymbol{H}) \quad \text{as } k \to +\infty.$$

From the compactness theorem (see, e.g., [21, Sect. 8, Corollary 4]), this gives

$$\boldsymbol{v}_{\varepsilon, \lambda_k} \to \boldsymbol{v}_\varepsilon \quad \text{in } C([0, T]; \boldsymbol{H}_0), \quad \boldsymbol{u}_{\varepsilon, \lambda_k} \to \boldsymbol{u}_\varepsilon \quad \text{in } C([0, T]; \boldsymbol{H}),$$

$$\boldsymbol{\pi}(\boldsymbol{u}_{\varepsilon, \lambda_k}) \to \boldsymbol{\pi}(\boldsymbol{u}_\varepsilon) \quad \text{in } C([0, T]; \boldsymbol{H}) \quad \text{as } k \to +\infty.$$

Moreover, from the demi-closedness of $\partial \varphi$ and [4, Proposition 2.2], we see that $\boldsymbol{v}_\varepsilon^* = \partial \varphi(\boldsymbol{v}_\varepsilon)$ in $L^2(0, T; \boldsymbol{H}_0)$ and $\boldsymbol{\xi}_\varepsilon \in \beta(\boldsymbol{u}_\varepsilon)$ in $L^2(0, T; \boldsymbol{H})$. From these facts, we deduce from (3.4) and (3.5) that

$$\boldsymbol{v}_\varepsilon'(t) + \boldsymbol{F}(P\boldsymbol{\mu}_\varepsilon(t)) = \boldsymbol{0} \quad \text{in } \boldsymbol{V}_0^*, \tag{3.6}$$

$$\boldsymbol{\xi}_\varepsilon(t) \in \beta(\boldsymbol{u}_\varepsilon(t)), \quad \boldsymbol{\mu}_\varepsilon(t) = \varepsilon \partial \varphi(\boldsymbol{v}_\varepsilon(t)) + \boldsymbol{\xi}_\varepsilon(t) + \varepsilon \boldsymbol{\pi}(\boldsymbol{u}_\varepsilon(t)) - \boldsymbol{f}(t) \quad \text{in } \boldsymbol{H} \tag{3.7}$$

for a.a. $t \in (0, T)$, with $\boldsymbol{v}_\varepsilon(0) = \boldsymbol{v}_{0\varepsilon}$ in $\boldsymbol{H}$. We also have the regularity $\boldsymbol{u}_\varepsilon \in H^1(0, T; \boldsymbol{V}^*) \cap L^\infty(0, T; \boldsymbol{V}) \cap L^2(0, T; \boldsymbol{W})$. Now, taking the limit inferior as $\lambda \searrow 0$ on the uniform estimates, $\lambda / \varepsilon \searrow 0$ for all $\varepsilon \in (0, \varepsilon_0]$, and we therefore obtain the same kind of uniform estimates as in the previous lemmas independent of $\varepsilon \in (0, \varepsilon_0]$.

Proof of Theorem 2.1. By using the estimates for $\boldsymbol{v}_\varepsilon, \boldsymbol{u}_\varepsilon, \boldsymbol{\mu}_\varepsilon$ and $\boldsymbol{\xi}_\varepsilon$, there exist a subsequence $\{\varepsilon_k\}_{k \in \mathbb{N}}$ with $\varepsilon_k \searrow 0$ as $k \to +\infty$ and functions $\boldsymbol{v} \in H^1(0, T; \boldsymbol{V}_0^*) \cap L^\infty(0, T; \boldsymbol{H}_0)$, $\boldsymbol{u} \in H^1(0, T; \boldsymbol{V}^*) \cap L^\infty(0, T; \boldsymbol{H})$, $\boldsymbol{\mu} \in L^2(0, T; \boldsymbol{V})$ and $\boldsymbol{\xi} \in L^2(0, T; \boldsymbol{H})$ such that

$$\boldsymbol{v}_{\varepsilon_k} \to \boldsymbol{v} \quad \text{weakly star in } L^\infty(0, T; \boldsymbol{H}_0),$$

$$\boldsymbol{u}_{\varepsilon_k} \to \boldsymbol{u} = \boldsymbol{v} + m_0 \boldsymbol{1} \quad \text{weakly star in } L^\infty(0, T; \boldsymbol{H}), \quad \varepsilon_k \boldsymbol{v}_{\varepsilon_k} \to \boldsymbol{0} \quad \text{in } L^\infty(0, T; \boldsymbol{V}_0),$$

$$\boldsymbol{v}_{\varepsilon_k}' \to \boldsymbol{v}' \quad \text{weakly in } L^2(0, T; \boldsymbol{V}_0^*), \quad \boldsymbol{u}_{\varepsilon_k}' \to \boldsymbol{u}' \quad \text{weakly in } L^2(0, T; \boldsymbol{V}^*),$$

$$\boldsymbol{\mu}_{\varepsilon_k} \to \boldsymbol{\mu} \quad \text{weakly in } L^2(0, T; \boldsymbol{V}), \quad \boldsymbol{\xi}_{\varepsilon_k} \to \boldsymbol{\xi} \quad \text{weakly in } L^2(0, T; \boldsymbol{H}),$$

$$\varepsilon_k \boldsymbol{\pi}(\boldsymbol{u}_{\varepsilon_k}) \to \boldsymbol{0} \quad \text{in } L^\infty(0, T; \boldsymbol{H}) \quad \text{as } k \to +\infty.$$

From the Ascoli–Arzelà theorem, we also have

$$\boldsymbol{v}_{\varepsilon_k} \to \boldsymbol{v} \quad \text{in } C([0, T]; \boldsymbol{V}_0^*), \quad \boldsymbol{u}_{\varepsilon_k} \to \boldsymbol{u} \quad \text{in } C([0, T]; \boldsymbol{V}^*) \quad \text{as } k \to +\infty.$$

Now, multiplying (3.7) by $\eta \in L^2(0,T;V)$ and integrating over $(0,T)$, we obtain

$$\int_0^T \left(\mu_{\varepsilon_k}(t), \eta(t)\right)_H dt = \varepsilon_k \int_0^T a\left(v_{\varepsilon_k}(t), \eta(t)\right) dt + \int_0^T \left(\xi_{\varepsilon_k}(t), \eta(t)\right)_H dt$$

$$+ \varepsilon_k \int_0^T \left(\pi\left(u_{\varepsilon_k}(t)\right), \eta(t)\right)_H dt - \int_0^T \left(f(t), \eta(t)\right)_H dt.$$
(3.8)

Letting $k \to \infty$, we obtain

$$\int_0^T \left(\mu(t), \eta(t)\right)_H dt = \int_0^T \left(\xi(t) - f(t), \eta(t)\right)_H dt \quad \text{for all } \eta \in L^2(0,T;V),$$

namely, $\mu = \xi - f$ in $L^2(0,T;H)$. This implies the regularity of $\xi \in L^2(0,T;V)$, that is, $\xi_\Gamma = \xi|_\Gamma$ a.e. on Σ. Next, we take $\eta := u_{\varepsilon_k} \in L^2(0,T;V)$ in (3.8), so that

$$\limsup_{k\to+\infty} \int_0^T \left(\xi_{\varepsilon_k}(t), u_{\varepsilon_k}(t)\right)_H dt \le \int_0^T \langle u(t), \mu(t)\rangle_{V^*,V} dt + \int_0^T \left(f(t), u(t)\right)_H dt$$

$$= \int_0^T \left(\xi(t), u(t)\right)_H dt.$$

Thus, applying [4, Proposition 2.2] we have $\xi \in \beta(u)$ in $L^2(0,T;H)$, and so we obtain $\xi \in \beta(u)$ a.e. in Q. $\xi_\Gamma \in \beta(u_\Gamma)$ a.e. on Σ. Finally, letting $k \to +\infty$ and applying Hahn–Banach extension theorem of bounded linear functional on V to $V \times V_\Gamma$, then we see that (3.6) gives (2.1) for a.a. $t \in (0,T)$, with $u(0) = u_0$ a.e. in Ω and $u_\Gamma(0) = u_{0\Gamma}$ a.e. on Γ. $\qquad\square$

Acknowledgments. The author is indebted to professor Pierluigi Colli, who kindly gave him the opportunity for fruitful discussions. The author is supported by JSPS KAKENHI Grant-in-Aid for Scientific Research(C), Grant Number 26400164. Last but not least, the author is also grateful to the referee for the careful reading of the manuscript.

References

1. Alikakos, N.D., Bates, P.W., Chen, X.: Convergence of the Cahn-Hilliard equation to the Hele-Shaw model. Arch. Ration. Mech. Anal. **128**, 165–205 (1994)
2. Allen, S., Cahn, J.: A microscopic theory for antiphase boundary motion and its application to antiphase domain coarsing. Acta Metall. **27**, 1084–1095 (1979)
3. Andreu, F., Mazón, J.M., Toledo, J., Igbida, N.: A degenerate elliptic-parabolic problem with nonlinear dynamical boundary conditions. Interfaces Free Bound **8**, 447–479 (2006)
4. Barbu, V.: Nonlinear Differential Equations of Monotone Types in Banach Spaces. Springer, London (2010)
5. Bronsard, L., Kohn, R.V.: Motion by mean curvature as the singular limit of Ginzbrug-Landau dynamics. J. Differ. Equ. **90**, 211–237 (1991)

6. Caginalp, G., Chen, X.: Convergence of the phase field model to its sharp interface limits. Eur. J. Appl. Math. **9**, 417–445 (1998)
7. Cahn, J.W., Hilliard, J.E.: Free energy of a nonuniform system I. Interfacial free energy. J. Chem. Phys. **2**, 258–267 (1958)
8. Colli, P., Fukao, T.: Cahn-Hilliard equation with dynamic boundary conditions and mass constraint on the boundary. J. Math. Anal. Appl. **429**, 1190–1213 (2015)
9. Colli, P., Fukao, T.: Equation and dynamic boundary condition of Cahn-Hilliard type with singular potentials. Nonlinear Anal. **127**, 413–433 (2015)
10. Colli, P., Fukao, T.: Nonlinear diffusion equations as asymptotic limit of Cahn-Hilliard system. J. Differ. Equ. **260**, 6930–6959 (2016)
11. Colli, P., Visintin, A.: On a class of doubly nonlinear evolution equations. Commun. Partial Differ. Equ. **15**, 737–756 (1990)
12. Damlamian, A., Kenmochi, N.: Evolution equations generated by subdifferentials in the dual space of $(H^1(\Omega))$. Discret. Contin. Dyn. Syst. **5**, 269–278 (1999)
13. Fukao, T.: Convergence of Cahn-Hilliard systems to the Stefan problem with dynamic boundary conditions. Asymptot. Anal. **99**, 1–21 (2016)
14. Goldstein, G.R., Miranville, A., Schimperna, G.: A Cahn-Hilliard model in a domain with non-permeable walls. Phys. D **240**, 754–766 (2011)
15. Grigor'yan, A.: Heat Kernel and Analysis on Manifolds. American Mathematical Society. International Press, Boston (2009)
16. Igbida, N.: Hele-Shaw type problems with dynamical boundary conditions. J. Math. Anal. Appl. **35**, 1061–1078 (2007)
17. Igbida, N., Kirane, M.: A degenerate diffusion problem with dynamical boundary conditions. Math. Ann. **323**, 377–396 (2002)
18. Kenmochi, N., Niezgódka, M., Pawłow, I.: Subdifferential operator approach to the Cahn-Hilliard equation with constraint. J. Differ. Equ. **117**, 320–354 (1995)
19. Mullins, W., Sekerka, R.F.: Morphological stability of a particle growing by diffusion or heat flow. J. Appl. Phys. **34**, 323–329 (1963)
20. Pego, R.L.: Front migration in the nonlinear Cahn-Hilliard equation. Proc. R. Soc. Lond. A **422**, 261–278 (1989)
21. Simon, J.: Compact sets in the spaces $L^p(0, T; B)$. Ann. Math. Pura Appl. **146**(4), 65–96 (1987)
22. Visintin, A.: Models of Phase Transitions. Birkhäuser, Boston (1996)

Uniform Estimation of a Constant Issued from a Fluid-Structure Interaction Problem

Andrei Halanay[1(✉)] and Cornel Marius Murea[2]

[1] Department of Mathematics 1, University Politehnica of Bucharest,
313 Splaiul Independenței, 060042 Bucharest, Romania
halanay@mathem.pub.ro
[2] Laboratoire de Mathématiques, Informatique et Applications,
Université de Haute Alsace, 4-6, Rue des Frères Lumière,
68093 Mulhouse Cedex, France
cornel.murea@uha.fr

Abstract. We prove that the domain obtained by small perturbation of a Lipschtz domain is the union of a star-shaped domains with respect to every point of balls, such that the radius of the balls is independent of the perturbation. This result is useful in order to get uniform estimation for a fluid-structure interaction problem.

Keywords: Star-shaped domains · Fluid-structure interaction · Fictitious domain

1 Introduction

In [4,5] the existence of a solution is studied for an elastic structure immersed in an incompressible fluid.

Let $D \subset \mathbb{R}^2$ be a bounded open domain. We denote by Ω_0^S the undeformed structure domain and by $\mathbf{u} = (u_1, u_2) : \overline{\Omega}_0^S \to \mathbb{R}^2$ its displacement. A particle of the structure with initial position at the point $\mathbf{X}$ will occupy the position $\mathbf{x} = \Phi(\mathbf{X}) = \mathbf{X} + \mathbf{u}(\mathbf{X})$ in the deformed domain $\overline{\Omega}_u^S = \Phi\left(\overline{\Omega}_0^S\right)$. In [5], we have assumed that $\partial \Omega_u^S$ has the uniform cone property and the geometry of the cone is independent of $\mathbf{u}$. The fluid occupies the domain $\Omega_u^F = D \setminus \overline{\Omega}_u^S$, see Fig. 1.

It is possible to construct an uniform extension operator E from $\{\mathbf{v} \in \left(H^1\left(\Omega_u^S\right)\right)^2; \nabla \cdot \mathbf{v} = 0 \text{ in } \Omega_u^S\}$ to $\left(H_0^1(D)\right)^2$ such that

$$\nabla \cdot E(\mathbf{v}) = 0, \quad \text{in } D \tag{1}$$

$$E(\mathbf{v}) = \mathbf{v}, \quad \text{in } \Omega_u^S \tag{2}$$

$$\|E(\mathbf{v})\|_{1,D} \leq K \|\mathbf{v}\|_{1,\Omega_u^S} \tag{3}$$

where the constant $K > 0$ is independent of Ω_u^S, but it depends on the geometry of the cone.

© IFIP International Federation for Information Processing 2016
Published by Springer International Publishing AG 2016. All Rights Reserved
L. Bociu et al. (Eds.): CSMO 2015, IFIP AICT 494, pp. 292–301, 2016.
DOI: 10.1007/978-3-319-55795-3_27

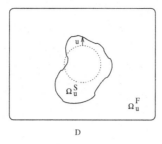

Fig. 1. Fluid and structure domains.

This construction is obtained by solving the Bogowskii problem in Ω_u^F, see for example [3], Lemma 3.1, page 121. There exists $\mathbf{w} \in \left(H_0^1\left(\Omega_u^F\right)\right)^2$ such that

$$\nabla \cdot \mathbf{w} = \mathbf{f}, \quad \text{in } \Omega_u^F \tag{4}$$

$$\mathbf{w} = 0, \quad \text{on } \partial D \tag{5}$$

$$\mathbf{w} = 0, \quad \text{on } \partial \Omega_u^S \tag{6}$$

$$|\mathbf{w}|_{1,\Omega_u^F} \le K_1 \|\mathbf{f}\|_{0,\Omega_u^F} \tag{7}$$

where $\int_{\Omega_u^F} \mathbf{f} \, dx = 0$. If Ω_u^F is star-shaped with respect to any point of a ball of radius R_u, we have the following estimation of $K_1 > 0$:

$$K_1 \le c_0 \left(\frac{diam(\Omega_u^F)}{R_u}\right)^2 \left(1 + \frac{diam(\Omega_u^F)}{R_u}\right) \tag{8}$$

where $diam(\Omega_u^F)$ is the diameter of Ω_u^F. A similar result holds if the domain Ω_u^F is union of star-shaped domains with respect to any point of a ball, see Theorem 3.1, p. 129, [3]. We have $diam(\Omega_u^F) \le diam(D)$, for all $\mathbf{u}$ such that $\overline{\Omega_u^F} \subset D$.

The aim of this paper is to prove that, under some geometrical assumptions, one can choose $R_u = R = constant$, for small $\mathbf{u}$.

2 Small Perturbation of the Boundary of a Star-Shaped Domain with Respect to Any Point of a Ball

We denote by $B_R(\mathbf{x}_0)$ the open ball of radius R centered at $\mathbf{x}_0$, i.e. $B_R(\mathbf{x}_0) = \{\mathbf{x} \in \mathbb{R}^2; \ |\mathbf{x}_0 - \mathbf{x}| < R\}$, where $|\cdot|$ is the euclidean norm. A domain Ω is star-shaped with respect to every point of a ball $B_R(x_0)$ such that $\overline{B}_R(x_0) \subset \Omega$, if and only if, for every $\mathbf{x} \in B_R(x_0)$ and $\mathbf{y} \in \Omega$, the segment with ends $\mathbf{x}$, $\mathbf{y}$ is included in Ω. A characterization of such a domain is that every ray starting from a point $\mathbf{x} \in B_R(x_0)$ intersects the boundary of the domain at only one point.

Proposition 1. *Let $a, r > 0$ be two constants such that*

$$a \ge 4r \tag{9}$$

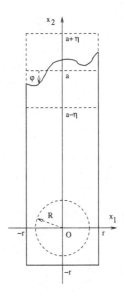

Fig. 2. A star-shaped domain with respect to every point of a ball.

and $\varphi \in W^{1,\infty}(-r,r)$ be such that

$$\exists \eta \in (0,r), \quad \sup_{x\in(-r,r)} |\varphi(x)| \leq \eta \tag{10}$$

$$\exists L > 0, \quad \sup_{x\in(-r,r)} |\varphi'(x)| \leq L. \tag{11}$$

We define the domain, see Fig. 2

$$\Omega_\varphi = \left\{(x_1,x_2) \in \mathbb{R}^2;\ -r < x_1 < r,\ -r < x_2 < a+\varphi(x_1)\right\}. \tag{12}$$

Let $R \in (0,r)$. If $L < \frac{a-2r}{2r}$, then the domain Ω_φ is star-shaped with respect to every point of the ball $B_R(0)$.

Proof. Since $R < r$, then $\overline{B}_R(0) \subset \Omega_\varphi$. Let $\mathbf{x} \in B_R(0)$. We suppose that a ray starting from this point cuts $\partial\Omega_\varphi$ in two points $\mathbf{y}$ and $\mathbf{z}$ and $\mathbf{y} \in (\mathbf{x},\mathbf{z})$, i.e. $\mathbf{y}$ is on the segment with ends $\mathbf{x}$ and $\mathbf{z}$.

Let us denote the top boundary of Ω_φ by

$$\Gamma_\varphi = \left\{(x_1, a+\varphi(x_1)) \in \mathbb{R}^2;\ -r < x_1 < r\right\}.$$

If $\mathbf{y}, \mathbf{z} \in \partial\Omega_\varphi \setminus \Gamma_\varphi$ it follows that the ray cuts the boundary of the strip

$$C = \left\{(x_1,x_2) \in \mathbb{R}^2;\ -r < x_1 < r,\ -r < x_2\right\},$$

twice, which is not true because the C is a convex set and a convex set is star-shaped with respect to every interior point.

If we have only $\mathbf{y} \in \partial\Omega_\varphi \setminus \Gamma_\varphi$, then $\mathbf{z}$ belongs to the exterior of the strip C, consequently $\mathbf{z} \notin \Gamma_\varphi$.

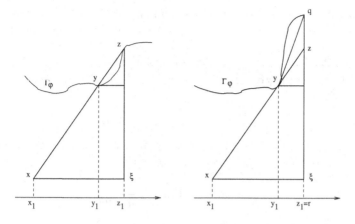

Fig. 3. Case (i) at the left and case (ii) at the right

So, we will study only two cases:

(i) $\mathbf{y}, \mathbf{z} \in \Gamma_\varphi$;
(ii) $\mathbf{y} \in \Gamma_\varphi$ and $\mathbf{z} \in \partial\Omega_\varphi \setminus \Gamma_\varphi$ (Fig. 3).

Case (i): $\mathbf{y}, \mathbf{z} \in \Gamma_\varphi$.
Let $\boldsymbol{\xi} = (\xi_1, \xi_2)$ be the point in the plane, such that the line passing throw $\boldsymbol{\xi}$ and $\mathbf{x}$ is parallel to the axis Ox_1 and the line passing throw $\boldsymbol{\xi}$ and $\mathbf{z}$ is parallel to the axis Ox_2.
We denote by α the angle $\widehat{\mathbf{xz}\boldsymbol{\xi}}$. Since $\mathbf{x} \in B_R(0)$ using (10), we get $|\mathbf{z} - \boldsymbol{\xi}| \geq a - 2r$ and $|\mathbf{x} - \boldsymbol{\xi}| \leq 2r$, then

$$\tan \alpha = \frac{|\mathbf{x} - \boldsymbol{\xi}|}{|\mathbf{z} - \boldsymbol{\xi}|} \leq \frac{2r}{a - 2r}.$$

By assumption we have $\mathbf{y}, \mathbf{z} \in \Gamma_\varphi$, then $\mathbf{y} = (y_1, a + \varphi(y_1))$ and $\mathbf{z} = (z_1, a + \varphi(z_1))$ with $-r < y_1, z_1 < r$. We have also, $\tan \alpha = \frac{|y_1 - z_1|}{|\varphi(y_1) - \varphi(z_1)|}$ and using (11) we get

$$\tan \alpha \geq \frac{1}{L}.$$

This implies that

$$\frac{2r}{a - 2r} \geq \frac{1}{L}$$

which is in contradiction with the hypothesis $L < \frac{a-2r}{2r}$.

Case (ii): $\mathbf{y} \in \Gamma_\varphi$ and $\mathbf{z} \in \partial\Omega_\varphi \setminus \Gamma_\varphi$.
As in the Case (i), we construct $\boldsymbol{\xi}$, we denote by α the angle $\widehat{\mathbf{xz}\boldsymbol{\xi}}$ and we get

$$\tan \alpha = \frac{|\mathbf{x} - \boldsymbol{\xi}|}{|\mathbf{z} - \boldsymbol{\xi}|} \leq \frac{2r}{a - 2r}.$$

Since $\mathbf{z}$ cannot belong to the bottom boundary of C, we assume that $\mathbf{z}$ belongs to the right boundary of C, so $\mathbf{z} = (r, z_2)$. We denote by $\mathbf{q}$ the point at the right top corner of Ω_φ, so $\mathbf{q} = (r, a + \varphi(r))$. We have $W^{1,\infty}(-r, r) \subset C^0([-r, r])$ (see [1], Theorem VIII.7, page 129), so $\varphi(r)$ is well defined.

Let α' be the angle $\overline{\mathbf{yq\xi}}$. Then

$$\tan \alpha' = \frac{|y_1 - r|}{|\varphi(y_1) - \varphi(r)|} \geq \frac{1}{L}.$$

If $z_2 > a + \varphi(r)$ it follows that $\mathbf{z}$ belongs to the exterior of $\overline{\Omega}_\varphi$ not to $\partial\Omega_\varphi$, so $z_2 \leq a + \varphi(r)$. In this case $\alpha \geq \alpha'$ and we get

$$\frac{1}{L} = \tan \alpha' < \tan \alpha \leq \frac{2r}{a - 2r}$$

which is in contradiction with the hypothesis $L < \frac{a-2r}{2r}$.

Proposition 2. *Let $\zeta : [-r, r] \to \mathbb{R}$ be a Lipschitz function of constant L, i.e.*

$$\forall x, y \in [-r, r], \quad |\zeta(x) - \zeta(y)| \leq L|x - y|.$$

Denote by $\Gamma_\zeta = \{(x, \zeta(x)) \in \mathbb{R}^2, \ x \in [-r, r]\}$, the graph of ζ.

Let $\mathbf{u} = (u_1, u_2) : \Gamma_\zeta \to \mathbb{R}^2$ be such that

$$\exists \eta_1 \in \left(0, \frac{r}{2}\right), \quad \sup_{\mathbf{x} \in \Gamma_\zeta} |\mathbf{u}(\mathbf{x})| \leq \eta_1 \tag{13}$$

$$\exists \eta_2 \in \left(0, \frac{1}{\sqrt{1+L^2}}\right), \ \forall \mathbf{x}, \mathbf{y} \in \Gamma_\zeta, \quad |\mathbf{u}(\mathbf{x}) - \mathbf{u}(\mathbf{y})| \leq \eta_2 |\mathbf{x} - \mathbf{y}|. \tag{14}$$

Then there exits a Lipschitz function $\varphi : \left(-\frac{r}{2}, \frac{r}{2}\right) \to \mathbb{R}$ of constant less than $\frac{L + \eta_2 \sqrt{1+L^2}}{1 - \eta_2 \sqrt{1+L^2}}$, such that its graph is included in $(\mathbf{Id} + \mathbf{u})(\Gamma_\zeta)$. If $\zeta(0) = 0$, then

$$\forall z \in \left(-\frac{r}{2}, \frac{r}{2}\right), \quad |\varphi(z)| \leq r\left(\sqrt{1+L^2} + \frac{1}{2}\right).$$

Proof. Let us introduce the function

$$\phi : [-r, r] \to \mathbb{R}, \quad \phi(x) = x + u_1(x, \zeta(x)).$$

We will prove that ϕ is strictly increasing, so it is injective. To simplify, we assume that ζ and $\mathbf{u}$ are of class C^1. We get

$$\phi'(x) = 1 + \frac{\partial u_1}{\partial x_1}(x, \zeta(x)) + \frac{\partial u_1}{\partial x_2}(x, \zeta(x))\zeta'(x).$$

Since ζ is Lipschitz function of constant L, $|\zeta'(x)| \leq L$. Similary, we have $|\nabla \mathbf{u}(\mathbf{x})| \leq \eta_2$. Using Cauchy-Schwarz we get

$$\left|\frac{\partial u_1}{\partial x_1}(x, \zeta(x)) + \frac{\partial u_1}{\partial x_2}(x, \zeta(x))\zeta'(x)\right| \leq \sqrt{\left(\frac{\partial u_1}{\partial x_1}\right)^2 + \left(\frac{\partial u_1}{\partial x_2}\right)^2}\sqrt{1+L^2}$$

$$\leq \eta_2 \sqrt{1+L^2}.$$

It follows that

$$\phi'(x) \geq 1 - \eta_2 \sqrt{1 + L^2} > 0,$$

and then ϕ is strictly increasing, so ϕ is a bijection from $[-r, r]$ to $[\phi(-r), \phi(r)]$. From (13), we obtain $\phi(-r) < -\frac{r}{2}$ and $\phi(r) > \frac{r}{2}$.

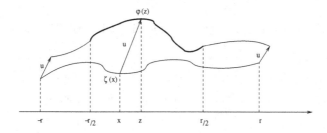

Fig. 4. The graphs of ζ and φ.

Let us introduce the function $\varphi : \left(-\frac{r}{2}, \frac{r}{2}\right) \to \mathbb{R}$ (Fig. 4),

$$\varphi(z) = \zeta(x) + u_2(x, \zeta(x)),$$

where $x = \phi^{-1}(z)$. The function is well defined because if z is in $\left(-\frac{r}{2}, \frac{r}{2}\right)$, then $x = \phi^{-1}(z)$ is in $(-r, r)$.

We will prove that φ is a Lipschitz function. Let $z, w \in \left(-\frac{r}{2}, \frac{r}{2}\right)$ and $x = \phi^{-1}(z)$, $y = \phi^{-1}(w)$.

$$\begin{aligned}
|\varphi(z) - \varphi(w)| &= |\zeta(x) - \zeta(y) + u_2(x, \zeta(x)) - u_2(y, \zeta(y))| \\
&\leq |\zeta(x) - \zeta(y)| + |u_2(x, \zeta(x)) - u_2(y, \zeta(y))| \\
&\leq L|x - y| + \eta_2 |(x, \zeta(x)) - (y, \zeta(y))| \\
&= L|x - y| + \eta_2 \sqrt{(x - y)^2 + (\zeta(x) - \zeta(y))^2} \\
&\leq \left(L + \eta_2 \sqrt{1 + L^2}\right) |x - y|.
\end{aligned}$$

We have

$$\begin{aligned}
|z - w| &= |x - y + u_1(x, \zeta(x)) - u_1(y, \zeta(y))| \\
&\geq |x - y| - |u_1(x, \zeta(x)) - u_1(y, \zeta(y))| \\
&\geq |x - y| - \eta_2 |(x, \zeta(x)) - (y, \zeta(y))| \\
&\geq |x - y| - \eta_2 \sqrt{(x - y)^2 + (\zeta(x) - \zeta(y))^2} \\
&\geq \left(1 - \eta_2 \sqrt{1 + L^2}\right) |x - y|.
\end{aligned}$$

We deduce that for all $z, w \in \left(-\frac{r}{2}, \frac{r}{2}\right)$

$$\frac{|\varphi(z) - \varphi(w)|}{|z - w|} \leq \frac{\left(L + \eta_2 \sqrt{1 + L^2}\right)}{\left(1 - \eta_2 \sqrt{1 + L^2}\right)}.$$

Let $(z, \varphi(z))$ be a point on the graph of φ, where $z \in \left(-\frac{r}{2}, \frac{r}{2}\right)$. Then there exits $x = \phi^{-1}(z) \in (-r, r)$. From the definition of ϕ and φ, we have

$$
\begin{aligned}
(z, \varphi(z)) &= (x + u_1(x, \zeta(x)), \zeta(x) + u_2(x, \zeta(x))) \\
&= (x, \zeta(x)) + (u_1(x, \zeta(x)), u_2(x, \zeta(x))) \\
&= (\mathbf{Id} + \mathbf{u})(x, \zeta(x))
\end{aligned}
$$

which proves that the graph of φ is included in $(\mathbf{Id} + \mathbf{u})(\Gamma_\zeta)$. Also

$$
\begin{aligned}
|\varphi(z)| &\leq |(z, \varphi(z))| \\
&= |(x + u_1(x, \zeta(x)), \zeta(x) + u_2(x, \zeta(x)))| \\
&\leq |(x, \zeta(x))| + |(u_1(x, \zeta(x)), u_2(x, \zeta(x)))| \\
&\leq \sqrt{x^2 + (\zeta(x))^2} + \eta_1 \leq \sqrt{r^2 + (\zeta(x))^2} + \frac{r}{2}
\end{aligned}
$$

If $\zeta(0) = 0$, since ζ is a Lipschitz function of constant L, we have

$$
|\zeta(x)| = |\zeta(x) - \zeta(0)| \leq L|x - 0| < Lr.
$$

Then $|\varphi(z)| \leq r\left(\sqrt{1 + L^2} + \frac{1}{2}\right)$.

3   Uniform Estimation of the Radius of the Ball for Small Perturbation of Lipschitz Domain

Definition 1 (see [2]). *Let r, a, L be three positive numbers and Ω an open bounded set of $\mathbb{R}^2$. We say that the boundary $\partial\Omega$ is uniform Lipschitz if, for every $\mathbf{x}_0 \in \partial\Omega$, there exits a Cartezian coordinates system $\{x_1, x_2\}$ of origin $\mathbf{x}_0$ and a Lipschitz function $\zeta : (-r, r) \to (-a, a)$ of constant L, such that $\zeta(0) = 0$,*

$$
\begin{aligned}
\partial\Omega \cap P(\mathbf{x}_0) &= \{(x_1, \zeta(x_1)), \ x_1 \in (-r, r)\}, \\
\Omega \cap P(\mathbf{x}_0) &= \{(x_1, x_2), \ x_1 \in (-r, r), \ \zeta(x_1) < x_2 < a\},
\end{aligned}
$$

where $P(\mathbf{x}_0) = (-r, r) \times (-a, a)$.

We will use the same notation as in the first section. We will prove a result similar to the Lemma 3.2, p. 40, from [3], but the radius of the balls is the same for all admissible displacements.

Theorem 1. *Let $\partial\Omega_0^S$ be an uniform Lipschitz boundary of parameters r, a, L. We set $R \in \left(0, \frac{r}{2}\right)$. There exists two constants $\eta_1 \in \left(0, \frac{r}{2}\right)$, $\eta_2 \in \left(0, \frac{1}{\sqrt{1+L^2}}\right)$ and two natural numbers $m, n \in \mathbb{N}^*$, such that for all $\mathbf{u} : \partial\Omega_0^S \to \mathbb{R}^2$, such that*

$$
\sup_{\mathbf{x} \in \partial\Omega_0^S} |\mathbf{u}(\mathbf{x})| \leq \eta_1 \tag{15}
$$

$$
\forall \mathbf{x}, \mathbf{y} \in \partial\Omega_0^S, \quad |\mathbf{u}(\mathbf{x}) - \mathbf{u}(\mathbf{y})| \leq \eta_2 |\mathbf{x} - \mathbf{y}| \tag{16}
$$

we have the decomposition

$$\Omega_u^F = \left(\bigcup_{i=1}^{m} \Omega_i\right) \bigcup \left(\bigcup_{j=m+1}^{m+n} B_{\eta_1}(\mathbf{x}_j)\right)$$

where $\Omega_i = \Omega_u^F \cap Q(\mathbf{x}_i)$ for $1 \leq i \leq m$ are star-shaped domains with respect to every point of a ball B_i of radius R, $\overline{B}_i \subset \Omega_i$, $Q(\mathbf{x}_i)$ are rectangles of center $\mathbf{x}_i \in \partial\Omega_0^S$ congruent to $\left(-\frac{r}{2}, \frac{r}{2}\right) \times (-a, a)$ and $B_{\eta_1}(\mathbf{x}_j)$ for $m+1 \leq j \leq m+n$ are balls of center $\mathbf{x}_j \in \Omega_u^F$ and radius η_1, $\overline{B}_{\eta_1}(\mathbf{x}_j) \subset \Omega_u^F$.

Proof. Let us remark that if Ω_0^S is Lipschitz of parameters r, a, L, then it is also for the parameters r', a, L, for $r' < r$. Consequently, we can choose r small enough such that $2L + \sqrt{1 + L^2} + \frac{5}{2} < \frac{a}{r}$.

Let $P(\mathbf{x})$ be the rectangle of center $\mathbf{x} \in \partial\Omega_0^S$ given by the Definition 1. We denote by $Q(\mathbf{x})$ the rectangle $\left(-\frac{r}{2}, \frac{r}{2}\right) \times (-a, a)$ using the same local coordinates. We have $\partial\Omega_0^S \subset \cup_{\mathbf{x} \in \partial\Omega_0^S} Q(\mathbf{x})$ and since $\partial\Omega_0^S$ is compact, then there exits $m \in \mathbb{N}^*$ and $\mathbf{x}_i \in \partial\Omega_0^S$, $1 \leq i \leq m$, such that

$$\partial\Omega_0^S \subset \bigcup_{i=1}^{m} Q(\mathbf{x}_i).$$

We will use the notation

$$\mathcal{O}_\eta = \{\mathbf{x} \in \mathbb{R}^2; \ d(\mathbf{x}, \partial\Omega_0^S) \leq \eta\},$$
$$\Omega_{0,\eta}^F = \{\mathbf{x} \in \Omega_0^F; \ d(\mathbf{x}, \partial\Omega_0^S) > \eta\}$$

where $d(\mathbf{x}, A) = \inf_{\mathbf{y} \in A} |\mathbf{x} - \mathbf{y}|$ is the distance function. There exists $\eta_0 > 0$, such that

$$\mathcal{O}_{\eta_0} \subset \bigcup_{i=1}^{m} Q(\mathbf{x}_i). \tag{17}$$

Set $\eta_1 = \min(\frac{\eta_0}{2}, \frac{r}{2})$ (Fig. 5).

Let $\mathbf{u} : \partial\Omega_0^S \to \mathbb{R}^2$ be such that (15) holds. Then $(\mathbf{Id}+\mathbf{u})(\partial\Omega_0^S) \subset \mathcal{O}_{\eta_1} \subset \mathcal{O}_{\eta_0}$, for all admissible $\mathbf{u}$ and it follows that

$$\Omega_u^F \subset \mathcal{O}_{\eta_0} \cup \Omega_{0,\eta_0}^F. \tag{18}$$

Since Ω_{0,η_0}^F is bounded, then there exists $n \in \mathbb{N}^*$ and $\mathbf{x}_j \in \Omega_{0,\eta_0}^F$ for $m+1 \leq j \leq m+n$, such that

$$\Omega_{0,\eta_0}^F \subset \bigcup_{j=m+1}^{m+n} B_{\eta_1}(\mathbf{x}_j). \tag{19}$$

But $\eta_1 \leq \frac{\eta_0}{2}$, so $\overline{B}_{\eta_1}(\mathbf{x}_j) \subset \Omega_u^F$. From (17), (18), (19), we get

$$\Omega_u^F = \left(\bigcup_{i=1}^{m} (\Omega_u^F \cap Q(\mathbf{x}_i))\right) \bigcup \left(\bigcup_{j=m+1}^{m+n} B_{\eta_1}(\mathbf{x}_j)\right).$$

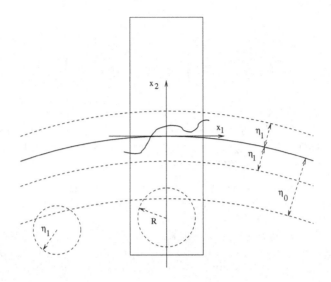

Fig. 5. The rectangle $\left(-\frac{r}{2}, \frac{r}{2}\right) \times (-a, a)$ is of center $\mathbf{x}_i \in \partial \Omega_0^S$, the ball of radius η_1 is of center $\mathbf{x}_j \in \Omega_{0,\eta_0}^F$.

Using the Proposition 2, $(\mathbf{Id} + \mathbf{u})(\partial \Omega_0^S) \cap Q(\mathbf{x}_i)$ is the graph of a Lipschitz function $\varphi : \left(-\frac{r}{2}, \frac{r}{2}\right) \to (-a, a)$ of constant

$$L' = \frac{L + \eta_2\sqrt{1 + L^2}}{1 - \eta_2\sqrt{1 + L^2}}$$

and $|\varphi| \le r\left(\sqrt{1 + L^2} + \frac{1}{2}\right)$. If $\eta_2 \in [0, \frac{1}{2\sqrt{1+L^2}})$, then $L' \in [L, 2L + 1)$.

We have $2L + 1 < \frac{a - r(\sqrt{1+L^2} + \frac{3}{2})}{r}$ by the choice of r at the begining of the proof. Using a similar argument as in the Proposition 1, we get that $\Omega_i = \Omega_u^F \cap Q(\mathbf{x}_i)$ is a star-shaped domain with respect to every point of the ball of radius R and center $(0, -a + \frac{r}{2})$.

Corollary 1. *Using the Theorem 3.1, p. 129, from [3], we get that the constant K_1 from the inequality (7) depends on $\min(\eta_1, R)$ and $diam(D)$, but it is independent on the displacement $\mathbf{u}$.*

References

1. Brezis, H.: Analyse fonctionnelle : théorie et applications. Dunod, Paris (2005)
2. Chenais, D.: On the existence of a solution in a domain identification problem. J. Math. Anal. Appl. **52**(2), 189–219 (1975)
3. Galdi, G.: An Introduction to the Mathematical Theory of the Navier-Stokes Equations: Vol. I.: Linearized Steady Problems. Springer Tracts in Natural Philosophy, vol. 38. Springer, New York (1994)

4. Halanay, A., Murea, C.M., Tiba, D.: Existence and approximation for a steady fluid-structure interaction problem using fictitious domain approach with penalization. Math. Appl. **5**(1–2), 120–147 (2013)
5. Halanay, A., Murea, C.M., Tiba, D.: Existence of a steady flow of Stokes fluid past a linear elastic structure using fictitious domain. J. Math. Fluid Mech. **18**(2), 397–413 (2016)

A Multivalued Variational Inequality with Unilateral Constraints

Piotr Kalita[1], Stanislaw Migorski[1(✉)], and Mircea Sofonea[2]

[1] Faculty of Mathematics and Computer Science, Jagiellonian University in Krakow,
ul. Łojasiewicza 6, 30348 Krakow, Poland
{kalita,migorski}@ii.uj.edu.pl
[2] Laboratoire de Mathématiques et Physique, Université de Perpignan Via Domitia,
52 Avenue Paul Alduy, 66860 Perpignan, France
sofonea@univ-perp.fr

Abstract. The present paper represents a continuation of [3]. There, we studied a new class of variational inequalities involving a pseudomonotone univalued operator and a multivalued operator, for which we obtained an existence result, among others. In the current paper we prove that this result remains valid under significantly weaker assumption on the multivalued operator. Then, we consider a new mathematical model which describes the equilibrium of an elastic body attached to a nonlinear spring on a part of its boundary. We use our abstract result to prove the weak solvability of this elastic model.

Keywords: Multivalued operator · Variational inequality · Cut-off operator · Elastic material unilateral constraint · Weak solution

1 Introduction

The theory of variational inequalities plays an important role in the study of nonlinear boundary value problems arising in Mechanics, Physics and Engineering Sciences. Based on arguments of monotonicity and convexity, it started in early sixties and has gone through substantial development since then, as illustrated in the books [1,4,9,10], for instance. There, the inequalities have been formulated in terms of univalued operators and subgradient of a convex function.

Variational inequalities with multivalued operators represent a more recent and challenging topic of nonlinear functional analysis. In particular, they represent a powerful instrument which allows to obtain new and interesting results in the study of various classes of variational-hemivariational inequalities. Such kind of inequalities involve both convex and nonconvex functions, have been introduced in [8] and have been investigated in various other works, as explained in [6,7] and the references therein.

Recently, in [3], we considered a class of stationary variational inequalities with multivalued operators, for which we proved an existence result, under a smallness assumption on the data. The proof was based on standard arguments

L. Bociu et al. (Eds.): CSMO 2015, IFIP AICT 494, pp. 302–311, 2016.
DOI: 10.1007/978-3-319-55795-3_28

of variational inequalities followed by a version of the Kakutani-Fan-Glicksberg fixed point theorem. Then, we considered a dual variational formulation of the problem, studied the link between the primal and dual formulations and provided an equivalence result. Finally, we applied this abstract formalism to the study of a frictional elastic contact model with normal compliance, for which we obtained existence and equivalence results.

The present paper represents a continuation of [3]. Our aim is twofold. The first one is to prove that the existence result of [3] remains valid under significantly weaker assumption on the corresponding multivalued operators. The second aim is to illustrate the use of this result in the study of a new mathematical model which describes the equilibrium of an elastic body attached to a nonlinear spring.

The manuscript is organized as follows. In Sect. 2 we recall the statement of the problem together with the existence result obtained in [3]. In Sect. 3, we state and prove our main abstract result, Theorem 2. Next, in Sect. 4, we introduce the elastic problem, list the assumption on the data and state its variational formulation. Finally, in Sect. 5, we apply Theorem 2 to prove its weak solvability.

2   Problem Statement

Everywhere in this paper, we assume that $(V, \| \cdot \|)$ is a reflexive Banach space, we denote by V^* its topological dual, and $\langle \cdot, \cdot \rangle$ will represent the duality pairing between V and V^*. We use 0_V for the zero element of the space V and $\| \cdot \|_*$ for the norm on V^*. Assume in addition that $(U, \| \cdot \|_U)$ is a reflexive Banach space of topological dual U^*. We denote by $\| \cdot \|_{U^*}$ the norm on U^* and by $\langle \cdot, \cdot \rangle_{U^* \times U}$ the duality pairing between U and U^*. The symbol w-U will represent the space U endowed with the weak topology while s-U^* will represent the space U^* endowed with the strong topology. For a set D in a Banach space E, we define $\|D\|_E = \sup\{ \|u\|_E \mid u \in D \}$.

Consider a set $K \subset V$, a single-valued operator $A \colon V \to V^*$, a multivalued operator $B \colon U \to 2^{U^*}$, a linear, continuous and compact operator $\iota \colon V \to U$, and a functional $f \in V^*$. We denote by $\|\iota\|$ the norm of ι in the space of linear continuous operators from V to U. With these data we state the following inequality problem.

Problem 1. Find $u \in V$ and $\xi \in U^*$ such that

$$u \in K, \quad \langle Au - f, v - u \rangle + \langle \xi, \iota(v - u) \rangle_{U^* \times U} \geq 0 \quad \text{for all} \quad v \in K,$$
$$\xi \in B(\iota u).$$

Note that, since the operator B is multivalued, we refer to Problem 1 as a multivalued variational inequality. Let $\iota^* \colon U^* \to V^*$ be the adjoint operator to ι. Then, using the definition of the subgradient of the indicator function of the set K, denoted $\partial_c I_K$, it is easy to see that Problem 1 is equivalent with the following subdifferential inclusion: find $u \in V$ such that

$$Au + \iota^* B(\iota u) + \partial_c I_K(u) \ni f \quad \text{in} \quad V^*.$$

In the study of Problem 1 we need the following assumptions.

$H(A)$: $A: V \to V^*$ is an operator such that

(1) A is coercive, i.e., $\langle Au, u \rangle \geq \alpha \|u\|^2 - \beta$ for all $u \in V$ with $\alpha, \beta > 0$.
(2) $\|Au\|_{V^*} \leq a_1 + a_2 \|u\|$ for all $u \in V$ with $a_1, a_2 > 0$.
(3) A is pseudomonotone, i.e. it is bounded and $u_n \to u$ weakly in V and
    $\limsup \langle Au_n, u_n - u \rangle \leq 0$ imply $\langle Au, u - v \rangle \leq \liminf \langle Au_n, u_n - v \rangle$ for all
    $v \in V$.
(4) A is strictly monotone on K, i.e., for all $u, v \in K$ such that $u \neq v$, we have
    $\langle Au - Av, u - v \rangle > 0$.

$H(B)$: $B: U \to 2^{U^*}$ is an operator such that

(1) B has nonempty and convex values.
(2) the graph of B is sequentially closed in $(s\text{-}U) \times (w\text{-}U^*)$ topology.
(3) $\|Bw\|_{U^*} \leq b_1 + b_2 \|w\|_U$ for all $w \in U$ with $b_1, b_2 > 0$.
(4) the smallness condition $b_2 \|\iota\|^2 < \alpha$ holds.

$H(K)$: K is a nonempty, convex and closed subset of V.

The following result, obtained in [3], provides the solvability of Problem 1.

Theorem 1. *Under hypotheses $H(A)$, $H(B)$ and $H(K)$ Problem 1 has at least one solution.*

The proof of Theorem 1 is based on arguments of elliptic variational inequalities, various estimates and a version of the Kakutani-Fan-Glicksberg fixed point theorem.

3   An Abstract Existence Result

We now consider the following assumptions on the data of the Problem 1.

$H(B)$: $B: U \to 2^{U^*}$ is an operator such that

(5) the operator $B: U \to 2^{U^*}$ is bounded, i.e., it maps bounded sets in U into
    bounded sets in U^*.
(6) there exist constants $b_3, b_4 \geq 0$ with $b_4 \|\iota\|^2 < \alpha$ such that for all $u \in U$ and
    all $\xi \in Bu$ we have

$$\langle \xi, u \rangle_{U^* \times U} \geq -b_3 - b_4 \|u\|_U^2.$$

Our main abstract existence result in the study of Problem 1 is the following.

Theorem 2. *Assume hypotheses $H(A)$, $H(B)(1)$, (2), (5) and (6), $H(K)$ and $0_V \in K$. Then Problem 1 has a solution.*

Proof. Fix $N \geq 1$ and define a cut-off operator $B^N \colon U \to 2^{U^*}$ by

$$B^N(v) = \begin{cases} Bv & \text{if } \|v\|_U \leq N \\ B\left(\frac{Nv}{\|v\|_U}\right) & \text{otherwise} \end{cases}$$

for $v \in U$. Since B satisfies assumptions $H(B)(1)$ and (2), it is straightforward to verify that the operator B^N satisfies these two assumptions as well. Moreover, since B satisfies $H(B)(5)$, it follows that B^N satisfies $H(B)(3)$ with a constant $b_1 > 0$ which depends on N and $b_2 > 0$ arbitrary small and independent of N. It is also clear that B^N satisfies $H(B)(4)$. We are now in a position to apply Theorem 1. Thus, we deduce that there exists $(u^N, \xi^N) \in V \times U^*$ such that

$$u^N \in K, \quad \langle Au^N - f, v - u^N \rangle + \langle \xi^N, \iota(v - u^N) \rangle_{U^* \times U} \geq 0 \text{ for all } v \in K, \quad (1)$$
$$\xi^N \in B^N(\iota u^N).$$

The hypothesis $0_V \in K$ allows to test (1) with $v = 0_V$. As a result, we obtain

$$\langle Au^N, u^N \rangle + \langle \xi^N, \iota u^N \rangle_{U^* \times U} \leq \|f\|_{V^*} \|u^N\|. \quad (2)$$

We estimate from below the expression $\langle \xi^N, \iota u^N \rangle_{U^* \times U}$. If $\|\iota u^N\|_U \leq N$, then $\xi^N \in Bu^N$ and

$$\langle \xi^N, \iota u^N \rangle_{U^* \times U} \geq -b_3 - b_4 \|\iota\|^2 \|u^N\|^2.$$

If, in contrast, $\|\iota u^N\|_U > N$, then $\xi_N \in B\left(\frac{N\iota u^N}{\|\iota u^N\|_U}\right)$, and assumption $H(B)(6)$ yields

$$\langle \xi^N, \iota u^N \rangle_{U^* \times U} = \left\langle \xi^N, \frac{N\iota u^N}{\|\iota u^N\|_U} \right\rangle_{U^* \times U} \frac{\|\iota u^N\|_U}{N} \geq (-b_3 - b_4 N^2)\frac{\|\iota u^N\|_U}{N}$$
$$\geq -b_3 \|\iota\| \|u^N\| - b_4 \|\iota\|^2 \|u^N\|^2.$$

In either case, we have

$$\langle \xi^N, \iota u^N \rangle_{U^* \times U} \geq -b_3 - b_3 \|\iota\| \|u^N\| - b_4 \|\iota\|^2 \|u^N\|^2. \quad (3)$$

Combining inequality (3) with $H(A)(1)$ and (2) implies

$$(\alpha - b_4 \|\iota\|^2)\|u^N\|^2 \leq b_3 + (\|f\|_{V^*} + b_3 \|\iota\|) \|u^N\|. \quad (4)$$

Next, using (4) and $H(B)(6)$, we deduce that $\|u^N\|$ is bounded by a constant independent of N. Therefore, since $\|\iota u^N\|_U \leq \|\iota\| \|u^N\|$, it follows that $\|\iota u^N\|_U$ is also bounded by a constant independent of N. We now take N large enough so that the truncation in the definition of the operator B^N is inactive. It follows that u^N also solves Problem 1, which completes the proof. $\qquad \square$

We end this section with some comments on the assumptions on the multivalued operator B. First, we note that, clearly, condition $H(B)(5)$ is significantly weaker than $H(B)(3)$. Moreover, we stress that assumption $H(B)(6)$ is weaker than $H(B)(3)$ and $H(B)(4)$. In addition, condition $H(B)(6)$ has various physical interpretations, when dealing with examples arising in mechanics. We also mention that if $\langle \xi, u \rangle_{U^* \times U} \geq 0$ for all $u \in U$ and $\xi \in Bu$, then this condition is satisfied.

4   Contact Problem in Elasticity

In this section we consider a boundary value problem which models a contact problem for elastic material and for which the abstract result of Sect. 3 can be applied.

The physical setting is the following. An elastic body occupies an open, bounded and connected set in $\Omega \subset \mathbb{R}^d$ ($d = 2, 3$) with a Lipschitz boundary $\partial\Omega = \Gamma$. The concept of measurability, used below, is considered with respect to the d-1 dimensional Hausdorff measure, denoted by m. The set Γ is partitioned into three disjoint and measurable parts Γ_1, Γ_2 and Γ_3 such that $m(\Gamma_1) > 0$. The body is clamped on Γ_1, is submitted to surface tractions on Γ_2 and is attached to a nonlinear spring on Γ_3.

We use the symbol $\mathbb{S}^d$ for the space of second order symmetric $d \times d$ matrices. The canonical inner product on $\mathbb{R}^d$ and $\mathbb{S}^d$ will be denoted by "$\cdot$" and the Euclidean norm on the space $\mathbb{S}^d$ will be denote by $\|\cdot\|_{\mathbb{S}^d}$. We also use the notation ν for the outward unit normal at Γ and, for a vector field $\boldsymbol{v}$, v_ν and $\boldsymbol{v}_\tau$ will represent the normal and tangential components of $\boldsymbol{v}$ on Γ given by $v_\nu = \boldsymbol{v} \cdot \nu$ and $\boldsymbol{v}_\tau = \boldsymbol{v} - v_\nu\nu$, respectively. The normal and tangential components of the stress field σ on the boundary are defined by $\sigma_\nu = (\sigma\nu) \cdot \nu$ and $\sigma_\tau = \sigma\nu - \sigma_\nu\nu$. The mathematical model which describe the equilibrium of the elastic body in the physical setting above is the following.

Problem 2. Find a displacement field $\boldsymbol{u}\colon \Omega \to \mathbb{R}^d$, a stress field $\sigma\colon \Omega \to \mathbb{S}^d$ and the reactive interface force $\xi\colon \Gamma_3 \to \mathbb{R}$ such that

$$\sigma = \mathcal{F}\varepsilon(\boldsymbol{u}) \quad \text{in } \Omega, \tag{5}$$

$$\text{Div}\,\sigma + \boldsymbol{f}_0 = 0 \quad \text{in } \Omega, \tag{6}$$

$$\boldsymbol{u} = 0 \quad \text{on } \Gamma_1, \tag{7}$$

$$\sigma\nu = \boldsymbol{f}_2 \quad \text{on } \Gamma_2, \tag{8}$$

$$\left.\begin{array}{l} -g_1 \leq u_\nu \leq g_2, \quad \xi \in h(u_\nu) \\ u_\nu = -g_1 \implies -\sigma_\nu \leq \xi, \\ u_\nu = g_2 \implies -\sigma_\nu \geq \xi, \\ g_1 < u_\nu < g_2 \implies -\sigma_\nu = \xi \end{array}\right\} \quad \text{on } \Gamma_3, \tag{9}$$

$$\sigma_\tau = 0 \quad \text{on } \Gamma_3. \tag{10}$$

Equation (5) is the constitutive law for elastic materials in which $\mathcal{F}$ represents the elasticity operator and $\varepsilon(\boldsymbol{u})$ denotes the linearized strain tensor. Equation (6) is the equilibrium equation in which $\boldsymbol{f}_0$ represents the density of body forces. Conditions (7) and (8) are the displacement and traction conditions, respectively, in which $\boldsymbol{f}_2$ denotes the density of surface tractions. Condition (9) represents the interface law in which h is a multivalued function which will be described below and, finally, condition (10) shows that the shear on the surface Γ_3 vanishes.

We now provide additional comments on the conditions (9) which represent the novelty of our elastic model. First, this condition shows that spring prevents the normal displacement of the body in such a way that $-g_1 \leq u_\nu \leq g_2$. When

$-g_1 < u_\nu < g_2$ the spring is active and exerts a normal reaction on the body, denoted ξ, which depends on the normal displacement, i.e., $-\sigma_\nu = \xi \in h(u_\nu)$. Note that, for physical reason, h must be negative for a positive argument (since, in this case the spring is in compression and, therefore, its reaction is towards the body) and positive for a negative argument (since, in this case the spring is in extension and, therefore, it pulls the body). A typical example is the function $h(r) = kr$ which models the behavior of a linear spring of stiffness $k > 0$. Nevertheless, from mathematical point of view, we do not need this restriction, as shown in assumption $H(h)$ below. When $u_\nu = g_2$ the spring is completely compressed, thus, besides its reaction, an additional force oriented towards the body becomes active and blocks the normal displacements u_ν. Therefore, in this case we have $-\sigma_\nu \geq \xi$. When $u_\nu = -g_1$ the spring is completely extended and, besides its reaction, an extra force pulling body becomes active, which implies that $-\sigma_\nu \leq \xi$.

In the study of Problem 2, we use standard notation for Lebesgue and Sobolev spaces. For the displacement we use the space

$$V = \{\, \boldsymbol{u} \in H^1(\Omega; \mathbb{R}^d) \mid \boldsymbol{u} = 0 \text{ on } \Gamma_1 \,\}.$$

It is well known that the trace operator $\gamma \colon V \to L^2(\partial\Omega; \mathbb{R}^d)$ is compact. Moreover, we put $U = L^2(\Gamma_3)$ and define the operator $\iota \colon V \to U$ by $\iota(\boldsymbol{u}) = u_\nu|_{\Gamma_3}$ for all $\boldsymbol{u} \in V$. For the spaces V and U, we use the notation already used for the corresponding abstract spaces in Sects. 2 and 3. In particular, $\|\cdot\|$ and $\|\cdot\|_*$ will represent the norm on V and V^*, respectively, $\langle\cdot,\cdot\rangle$ denotes the duality pairing between V and V^*, $\|\cdot\|_U$ and $\|\cdot\|_{U^*}$ are the norms on U and U^*.

We also need the following assumptions on the problem data.

$H(g):$ $g_1, g_2 \geq 0$ are two real constants.

$H(\mathcal{F}):$ $\mathcal{F} \colon \Omega \times \mathbb{S}^d \to \mathbb{S}^d$ is such that

(1) there exists $L_{\mathcal{F}} > 0$ such that $\|\mathcal{F}(\boldsymbol{x}, \varepsilon_1) - \mathcal{F}(\boldsymbol{x}, \varepsilon_2)\|_{\mathbb{S}^d} \leq L_{\mathcal{F}} \|\varepsilon_1 - \varepsilon_2\|_{\mathbb{S}^d}$ for all $\varepsilon_1, \varepsilon_2 \in \mathbb{S}^d$, a.e. $\boldsymbol{x} \in \Omega$.
(2) there exists $m_{\mathcal{F}} > 0$ such that $(\mathcal{F}(\boldsymbol{x}, \varepsilon_1) - \mathcal{F}(\boldsymbol{x}, \varepsilon_2)) \cdot (\varepsilon_1 - \varepsilon_2) \geq m_{\mathcal{F}} \|\varepsilon_1 - \varepsilon_2\|_{\mathbb{S}^d}^2$ for all $\varepsilon_1, \varepsilon_2 \in \mathbb{S}^d$, a.e. $\boldsymbol{x} \in \Omega$.
(3) $\mathcal{F}(\boldsymbol{x}, \varepsilon)$ is measurable on Ω for all $\varepsilon \in \mathbb{S}^d$.
(4) $\mathcal{F}(\boldsymbol{x}, 0_{\mathbb{S}^d}) = 0_{\mathbb{S}^d}$ a.e. $\boldsymbol{x} \in \Omega$.

$H(h):$ $h \colon \Gamma_3 \times [-g_1, g_2] \to 2^{\mathbb{R}}$ is a multifunction such that

(1) the sets $h(\boldsymbol{x}, r)$ are nonempty and convex for all $r \in [-g_1, g_2]$, a.e. $\boldsymbol{x} \in \Gamma_3$.
(2) $h(\cdot, r)$ has a measurable selection for all $r \in [-g_1, g_2]$.
(3) the graph of multifunction $h(\boldsymbol{x}, \cdot)$ is closed in $\mathbb{R}^2$ for a.e. $\boldsymbol{x} \in \Gamma_3$.
(4) $|h(\boldsymbol{x}, r)| \leq \bar{h}$ for all $r \in [-g_1, g_2]$ and a.e. $\boldsymbol{x} \in \Gamma_3$ with $\bar{h} \geq 0$.

$H(f):$ the densities of body forces and surface tractions are such that

(1) if $d = 2$, then $\boldsymbol{f}_0 \in L^{p'}(\Omega; \mathbb{R}^2)$, $\boldsymbol{f}_2 \in L^{p'}(\Gamma_2; \mathbb{R}^2)$ for some $p' \in (1, \infty)$.
(2) if $d = 3$, then $\boldsymbol{f}_0 \in L^{\frac{6}{5}}(\Omega; \mathbb{R}^3)$, $\boldsymbol{f}_2 \in L^{\frac{4}{3}}(\Gamma_2; \mathbb{R}^3)$.

We now turn to the variational formulations of Problem 2. To this end, we introduce the set of admissible displacements fields

$$K = \{\, \boldsymbol{u} \in V \mid u_\nu \in [-g_1, g_2] \ \text{ a.e. on } \Gamma_3 \,\}. \tag{11}$$

Also, we define the operator $A\colon V \to V^*$ by the formula

$$\langle A\boldsymbol{u}, \boldsymbol{v} \rangle = \int_\Omega \mathcal{F}\varepsilon(\boldsymbol{u}) \cdot \varepsilon(\boldsymbol{v}) \, dx \quad \text{for all } \ \boldsymbol{u}, \boldsymbol{v} \in V. \tag{12}$$

Note that, since $H(\mathcal{F})$ holds, the operator A is well defined. Next, we note that the hypothesis $H(f)$ implies that the mapping

$$V \ni \boldsymbol{v} \mapsto \int_\Omega \boldsymbol{f}_0(\boldsymbol{x}) \cdot \boldsymbol{v}(\boldsymbol{x}) \, dx + \int_{\Gamma_2} \boldsymbol{f}_2(\boldsymbol{x}) \cdot \boldsymbol{v}(\boldsymbol{x}) \, d\Gamma \in \mathbb{R}$$

is linear and continuous. Indeed, if $d = 2$, then the embedding $V \subset L^r(\Omega; \mathbb{R}^2)$ and the restriction of the trace operator $\gamma_{\Gamma_2}\colon V \to L^r(\Gamma_2; \mathbb{R}^2)$ is linear and continuous, for any $r \in (1, \infty)$. On the other hand, if $d = 3$, then the embedding $V \subset L^6(\Omega; \mathbb{R}^3)$ and the restriction of the trace operator $\gamma_{\Gamma_2}\colon V \to L^4(\Gamma_2; \mathbb{R}^3)$ is linear and continuous. Hence, we can define $\boldsymbol{f} \in V^*$ by the formula

$$\langle \boldsymbol{f}, \boldsymbol{v} \rangle = \int_\Omega \boldsymbol{f}_0(\boldsymbol{x}) \cdot \boldsymbol{v}(\boldsymbol{x}) \, dx + \int_{\Gamma_2} \boldsymbol{f}_2(\boldsymbol{x}) \cdot \boldsymbol{v}(\boldsymbol{x}) \, d\Gamma.$$

Next, with a slight abuse of notation, we extend the multifunction h to $\Gamma_3 \times \mathbb{R}$ by setting $h(\boldsymbol{x}, s) = h(\boldsymbol{x}, -g_1)$ and $h(\boldsymbol{x}, s) = h(\boldsymbol{x}, g_2)$ for $s < -g_1$ and $s > g_2$, respectively. We use the same symbol h to denote this extended multifunction and we introduce the multifunction $B\colon U \to 2^{U^*}$ by the formula

$$B(u) = \{\, \xi \in U^* \mid \xi(\boldsymbol{x}) \in h(\boldsymbol{x}, u(\boldsymbol{x})) \ \text{ a.e. on } \Gamma_3 \,\} \quad \text{for all } u \in U. \tag{13}$$

The variational formulation for Problem 2, obtained by using standard arguments, reads as follows.

Problem 3. Find a displacement field $\boldsymbol{u} \in V$ and a contact interface force $\xi \in U$ such that

$$\boldsymbol{u} \in K, \quad \langle A\boldsymbol{u}, \boldsymbol{v} - \boldsymbol{u} \rangle + \langle \xi, \iota(\boldsymbol{v} - \boldsymbol{u}) \rangle_{U^* \times U} \geq \langle \boldsymbol{f}, \boldsymbol{v} - \boldsymbol{u} \rangle \quad \text{for all } \boldsymbol{v} \in K,$$
$$\xi \in B(\iota\boldsymbol{u}).$$

Note that Problem 3 represents a multivaled variational inequality. Its solvability will be proved in the next section, based on the abstract existence result provided by Theorem 2.

5   Existence of the Solution

The main result in this section is the following.

Theorem 3. *Under hypotheses $H(g)$, $H(\mathcal{F})$, $H(h)$ and $H(f)$, Problem 3 has at least one solution.*

The proof of the theorem is carried out in several steps, based on two lemmas that we state and prove in what follows. The first lemma, already proved in [3], is given here for the convenience of the reader.

Lemma 1. *Assume $H(\mathcal{F})$. Then, the operator $A: V \to V^*$ defined by (12) satisfies conditions $H(A)(1)$–(4) with $\alpha = m_{\mathcal{F}}$ in $H(A)(1)$.*

Proof. By conditions $H(\mathcal{F})(1)$ and (3), we have

$$\left| \int_\Omega \mathcal{F}\varepsilon(u) \cdot \varepsilon(v) \, dx \right| \leq L_{\mathcal{F}} \|u\| \|v\| \quad \text{for all} \ \ u, v \in V.$$

This implies that $\|Au\|_* \leq L_{\mathcal{F}} \|u\|$ for all $u \in V$, which proves $H(A)(2)$. In addition, assumption $H(\mathcal{F})(2)$ yields $\langle Au - Av, u - v \rangle \geq m_{\mathcal{F}} \|u - v\|^2$ for all u, $v \in V$. This shows that condition $H(A)(4)$ is satisfied. Furthermore, for u, v, $w \in V$, by $H(\mathcal{F})(1)$, we have $\langle Au - Av, w \rangle \leq L_{\mathcal{F}} \|u - v\|_V \|w\|$ for all $u, v \in V$. This proves that $\|Au - Av\|_* \leq L_{\mathcal{F}} \|u - v\|_V$ for all $u, v \in V$, which implies that A is Lipschitz continuous and hence hemicontinuous. Since we already know that A is bounded and monotone, by Proposition 27.6 in [11], it follows that A is pseudomonotone and $H(A)(3)$ holds. By $H(\mathcal{F})(4)$, we have $A0_V = 0_V$. Thus, from (5), we get $\langle Au, u \rangle \geq m_{\mathcal{F}} \|u\|^2$ for all $u \in V$. Therefore, $H(A)(1)$ holds, which completes the proof. □

Next, we proceed with the following result.

Lemma 2. *Assume $H(g)$ and $H(h)$. Then, the multivalued operator B defined by (13) satisfies $H(B)(1)$, (2), (5) and (6).*

Proof. First we prove $H(B)(1)$. Convexity of values of B is a simple consequence of the convexity in $H(h)(1)$. To prove nonemptiness, let $u \in U$ and $(u_n)_{n=1}^\infty$ be a sequence of simple (i.e. piecewise constant) functions converging to u for a.e. $x \in \Gamma_3$. The hypothesis $H(h)(2)$ implies that the multifunction $\Gamma_3 \ni x \to h(x, u_n(x)) \subset \mathbb{R}$ has a measurable selection ξ_n for all $n \in \mathbb{N}$. From $H(h)(4)$, it follows that $\|\xi_n\|_{U^*}^2 = \int_{\Gamma_3} \xi_n(x)^2 \, d\Gamma \leq \bar{h}^2 m(\Gamma_3)$, so passing to a subsequence, if necessary, we may assume that $\xi_n \to \xi$ weakly in U^* with $\xi \in U^*$. As $\xi_n(x) \in [-\bar{h}, \bar{h}]$ for a.e. $x \in \Gamma_3$, we are in a position to apply Proposition 3.16 of [5] to obtain $\xi(x) \in \overline{\text{conv}} \limsup_{n \to \infty} (\xi_n(x))$ for a.e. $x \in \Gamma_3$, where $\limsup_{n \to \infty}$ is the Kuratowski-Painlevé upper limit of sets defined by

$$\limsup_{n \to \infty} A_n = \{ s \in \mathbb{R} \mid s_{n_k} \to s, \ s_{n_k} \in A_{n_k} \}$$

for a sequence of sets $A_n \subset \mathbb{R}$. Hence $\xi(x) \in \overline{\text{conv}} \limsup_{n \to \infty} h(x, u_n(x))$ for a.e. $x \in \Gamma_3$. From $H(h)(1)$, (3) and the pointwise convergence $u_n(x) \to u(x)$ a.e. on Γ_3, it follows that

$$\xi(x) \in \overline{\text{conv}} \limsup_{n \to \infty} h(x, u_n(x)) \subset \overline{\text{conv}} \, h(x, u(x)) = h(x, u(x)),$$

for a.e. $x \in \Gamma_3$ and ξ is the sought measurable selection.

We pass to the proof of $H(B)(2)$. Let $u_n \to u$ strongly in U and $\xi_n \to \xi$ weakly in U^* be such that $\xi_n \in B(u_n)$. We show that $\xi \in B(u)$. At least for a subsequence, not relabeled, we have $u_n(x) \to u(x)$ for a.e. $x \in \Gamma_3$. The proof that $\xi(x) \in h(x, u(x))$ for a.e. $x \in \Gamma_3$ follows the line of the corresponding proof of condition $H(B)(1)$.

By a straightforward calculation and $H(h)(4)$, we deduce that hypothesis $H(B)(5)$ holds. To prove $H(B)(6)$, let $u \in U$ and $\xi \in Bu$. From the following inequality

$$\langle \xi, u \rangle_{U^* \times U} = \int_{\Gamma_3} \xi(x)u(x)\, d\Gamma \geq -\bar{h}\int_{\Gamma_3} |u(x)|\, d\Gamma \geq -\bar{h}\, m(\Gamma_3)^{\frac{1}{2}}\|u\|_U$$
$$\geq -\frac{\alpha}{2\|\iota\|^2}\|u\|_U^2 - \frac{\|\iota\|^2\bar{h}^2 m(\Gamma_3)}{2\alpha},$$

we deduce that $H(B)(6)$. The proof is complete. □

We are now in a position to provide the proof of Theorem 3.

Proof. Note that Problem 3 is a particular case of Problem 1. Indeed, both V and U are reflexive Banach spaces, and the normal trace operator ι is linear, continuous and compact. Moreover, the set K defined by (11) is nonempty, convex, closed in V, and $0_V \in K$. In addition, from Lemma 1, it follows that $H(A)(1)$–(4) hold, and Lemma 2 shows that the operator B satisfies $H(B)(1)$, (2), (5) and (6). Theorem 3 is now a direct consequence of Theorem 2. □

Theorem 3 provides the weak solvability of the contact Problem 2, since once the displacement field is obtained by solving Problem 3, then the stress field σ is determined by using the constitutive law (5). The question of the uniqueness of the solution is left open.

Remark 1. Note that because of $H(h)(4)$, the multivalued map given by (13) satisfies hypothesis $H(B)(4)$, and Theorem 1 is sufficient to obtain the existence result in Theorem 3. However, if in the place of (9), we consider the law $-\sigma_\nu = ku_\nu$ with $k > 0$, and set $K = V$, then to apply Theorem 1, we need a smallness assumption on the constant k. On the other hand, the above law satisfies $H(B)(5)$–(6) without any limitations on the value of k.

Acknowledgement. Research supported by the Marie Curie International Research Staff Exchange Scheme Fellowship within the 7th European Community Framework Programme under Grant Agreement No. 295118, the National Science Center of Poland under the Maestro Advanced Project No. DEC-2012/06/A/ST1/00262, and the project Polonium "Mathematical and Numerical Analysis for Contact Problems with Friction" 2014/15 between the Jagiellonian University in Krakow and Université de Perpignan Via Domitia. The first and second authors are also partially supported by the International Project co-financed by the Ministry of Science and Higher Education of Republic of Poland under Grant No. W111/7.PR/2012.

References

1. Baiocchi, C., Capelo, A.: Variational and Quasivariational Inequalities: Applications to Free-Boundary Problems. Wiley, Chichester (1984)
2. Hlaváček, I., Haslinger, J., Nečas, J., Lovíšek, J.: Solution of Variational Inequalities in Mechanics. Springer, New York (1988)
3. Kalita, P., Migórski, S., Sofonea, M.: A class of subdifferential inclusions for elastic unilateral contact problems. Set-Valued Var. Anal. **24**, 355–379 (2016). doi:10.1007/s11228-015-0346-3
4. Kinderlehrer, D., Stampacchia, G.: An Introduction to Variational Inequalities and their Applications. Classics in Applied Mathematics, vol. 31. SIAM, Philadelphia (2000)
5. Migórski, S., Ochal, A., Sofonea, M.: Nonlinear Inclusions and Hemivariational Inequalities. Models and Analysis of Contact Problems. Advances in Mechanics and Mathematics. Springer, New York (2013)
6. Migórski, S., Ochal, A., Sofonea, M.: A class of variational-hemivariational inequalities in reflexive Banach spaces. J. Elast. (2016, in press). doi:10.1007/s10659-016-9600-7
7. Naniewicz, Z., Panagiotopoulos, P.D.: Mathematical Theory of Hemivariational Inequalities and Applications. Marcel Dekker, New York (1995)
8. Panagiotopoulos, P.D.: Nonconvex problems of semipermeable media and related topics. ZAMM Z. Angew. Math. Mech. **65**, 29–36 (1985)
9. Panagiotopoulos, P.D.: Inequality Problems in Mechanics and Applications. Birkhäuser, Boston (1985)
10. Sofonea, M., Matei, A.: Mathematical Models in Contact Mechanics. London Mathematical Society Lecture Note Series. Cambridge University Press, Cambridge (2012)
11. Zeidler, E.: Nonlinear Functional Analysis and Applications II B. Springer, New York (1990)

On the Solvability of a Nonlinear Tracking Problem Under Boundary Control for the Elastic Oscillations Described by Fredholm Integro-Differential Equations

Akylbek Kerimbekov[1(✉)] and Elmira Abdyldaeva[2]

[1] Faculty of Natural and Technical,
Department of Applied Mathematics and Informatics,
Kyrgyz-Russian Slavic University, Kievstr. 44, 720000 Bishkek, Kyrgyzstan
akl7@rambler.ru
http://www.krsu.edu.kg
[2] Faculty of Natural Sciences, Department of Mathematics,
Kyrgyz-Turkish Manas University, Mira Avenue 56, 720044 Bishkek, Kyrgyzstan
efa_69@mail.ru
http://www.manas.edu.kg

Abstract. In the present paper we investigate nonlinear tracking problem under boundary control for the oscillation processes described by Fredholm integro-differential equations. When we investigate this problem we use notion of a weak generalized solution of the boundary value problem. Based on the maximum principle for distributed systems we obtain optimality conditions from which follow the nonlinear integral equation of optimal control and the differential inequality. We have developed an algorithm to construct the optimization problem solution. This solving method of a nonlinear tracking problem is constructive and can be used in applications.

Keywords: Weak generalized solution · Boundary control · Functional · Maximum principle · Nonlinear integral equation · Optimization

1 Introduction

With the emergence the theory control for the systems with distributed parameters, a lot of applied problems described by integral-partial differential equations, integral equations, differential and integral-functional equations ([1], Chap. 5, pp. 193–197), [2], ([3], Chap. 16, pp. 410–414), ([4], Chap. 1, pp. 30–76), became investigate by methods of optimal control theory ([4], Chap. 6, pp. 356–383), ([5], Chap. 2, pp. 45–78), ([6], Chap. 4, pp. 281–309),[7–9]. However, the control problems described by the integral-differential equations are little learned. In this paper we investigate the boundary tracking control problem for the elastic oscillations described by the partial Fredholm integral-differential equations in partial derivatives. This problem has a number of specific properties: according

© IFIP International Federation for Information Processing 2016
Published by Springer International Publishing AG 2016. All Rights Reserved
L. Bociu et al. (Eds.): CSMO 2015, IFIP AICT 494, pp. 312–321, 2016.
DOI: 10.1007/978-3-319-55795-3_29

to the method of [10] the generalized solution of the problem is built by the solving of countable number of integral equations; the optimal control simultaneously satisfies the two relations in the form of equality and inequality, where the relation in the form of equality leads to a nonlinear integral equation, and the relation in the form of inequality is a differential with regards to the function of the external source.

The sufficient conditions for the unique solvability of specific problems were found, and algorithm was indicated for constructing solutions of nonlinear optimization problems with arbitrary precision in the form of the triple $\left(u^0(t), V^0(t,x), J[u^0(t)]\right)$, where $u^0(t)$ is the optimal control, $V^0(t,x)$ is the optimal process, $J[u^0(t)]$ is the functional's minimum value.

2   Formulation of the Optimal Control Problem and Optimality Conditions

We consider the optimization problem where it is required to minimize the integral functional

$$J[u(t)] = \int_0^T \int_Q [V(t,x) - \xi(t,x)]^2 \, dxdt + 2\beta \int_0^T M[t,u(t)]dt, \ \beta > 0 \qquad (1)$$

on the set of solutions of the boundary value problem

$$V_{tt} - AV = \lambda \int_0^T K(t,\tau)V(\tau,x)d\tau + g(t,x), \ x \in Q \subset R^n, \ 0 < t \le T, \qquad (2)$$

$$V(0,x) = \psi_1(x), \ V_t(0,x) = \psi_2(x), \ x \in Q, \qquad (3)$$

$$\Gamma V(t,x) = \sum_{i,j=1}^n a_{i,j}(x)V_{x_j}(t,x)cos(\delta, x_i) + a(x)V(t,x) \qquad (4)$$

$$= b(t,x)f[t,u(t)], \ x \in \gamma, \ 0 < t < T.$$

Here A is the elliptic operator defined by the formula

$$AV(t,x) = \sum_{i,j=1}^n \left(a_{i,j}(x)V_{x_j}(t,x)\right)_{x_i} - c(x)V(t,x), \ a_{i,j}(x) = a_{j,i}(x),$$

$$\sum_{i,j=1}^n a_{i,j}(x)\alpha_i\alpha_j \ge c_0 \sum_{i=1}^n \alpha_i^2, \ c_0 > 0$$

δ is a normal vector, emanating from the point $x \in \gamma$; $K(t,\tau)$ is a given function defined in the region $D = \{0 \le t \le T, \ 0 \le \tau \le T\}$ and satisfying the condition $\int_0^T \int_0^T K^2(t,\tau)dtd\tau < K_0 < \infty$, i.e. $K(t,\tau) \in H(D)$;

$$\psi_1(x) \in H_1(Q), \ \psi_2(x) \in H(Q), \ f_u[t,u(t)] \ne 0, \forall t \in (0,T),$$
$$\xi(t,x) \in H(Q_T), \ M[t,u(t)] \in H(0,T), \ Q_T = Q \times (0,T), \qquad (5)$$

are given functions, $a(x) \geq 0, c(x) \geq 0$ are known measurable functions; $H(X)$ is Hilbert space of functions defined on the set of X; $H_1(X)$ is the first order Sobolev space; $f_u[t, u(t)]$ is the function of boundary source which nonlinearly depends on the control function $u(t) \in H(0, T)$ and it is an element of $H(0, T)$; λ is a parameter, T is a fixed moment of time; $M_u[t, u(t)] \neq 0$ and satisfies the Lipschitz condition with respect to functional argument $u(t) \in H(0, T)$.

This problem is to find a control $u^0(t) \in H(0, T)$, for which the appropriate solution $V^0(t, x)$ of the boundary value problem (2)–(4) deviates little from the given trajectory $\xi(t, x) \in H(Q_T)$ during the entire time $t \in [0, T]$ of the control.

At the same time $u^0(t)$ is called optimal control and $V^0(t, x)$ is the optimal process.

We are looking for a solution of the boundary value problem (2)–(4) in the form of the series

$$V(t, x) = \sum_{n=1}^{\infty} V_n(t) z_n(x), \tag{6}$$

where $z_n(x)$ are generalized eigenfunctions of the boundary value problem [10]

$$D_n(\Phi, z_n) = \int_Q \left(\sum_{i,j=1}^{n} a_{i,j}(x) \Phi_{x_j} z_{nx_i} + c(x) z_n(x) \Phi(t, x) \right) dx + \int_\gamma a(x) z_n(x) \Phi(t, x) dx$$

$$= \lambda_n^2 \int_Q z_n(x) \Phi(t, x) dx;$$

$$\Gamma z_n(x) = 0, \ x \in \gamma, 0 < t < T, n = 1, 2, \ldots,$$

and they form complete orthonormal system in the Hilbert space $H(Q)$, and the corresponding eigenvalues λ_n satisfy the following conditions

$$\lambda_n \leq \lambda_{n+1}, \ \forall n = 1, 2, 3, \ldots, \ \lim_{n \to \infty} \lambda_n = \infty.$$

The Fourier coefficients $V_n(t)$ for each fixed $n = 1, 2, 3, \ldots$, satisfy the linear nonhomogeneous Fredholm integral equation of the second type

$$V_n(t) = \lambda \int_0^T K_n(t, s) V_n(s) ds + a_n(t), \tag{7}$$

where

$$K_n(t, s) = \frac{1}{\lambda_n} \int_0^t \sin\lambda_n(t - \tau) K(\tau, s) d\tau,$$

$$a_n(t) = \psi_{1n} \cos\lambda_n t + \frac{\psi_{2n}}{\lambda_n} \sin\lambda_n t + \frac{1}{\lambda_n} \int_0^t \sin\lambda_n(t - \tau)[q_n(\tau) + b_n(\tau) f[\tau, u(\tau)]] d\tau. \tag{8}$$

The solution of equation (7) we find ([11], chap. 2, pp. 98–110) by the following formula

$$V_n(t) = \lambda \int_0^T R_n(t, s, \lambda) a_n(s) ds + a_n(t), \tag{9}$$

where the resolvent $R_n(t, s, \lambda)$ of the kernel $K_n(t, s)$ is given by

$$R_n(t, s, \lambda) = \sum_{i=1}^{\infty} \lambda^{i-1} K_{n,i}(t, s), \quad K_{n,1}(t, s) = K_n(t, s)$$

and the iterated kernels $K_{n,i}(t, s)$ for each $n = 1, 2, \ldots$ are defined by the formulas

$$K_{n,i+1}(t, s) = \int_0^T K_n(t, \eta) K_{n,i}(\eta, s) d\eta, \quad i = 1, 2, 3, \ldots, \quad K_{n,1}(t, s) = K_n(t, s).$$

Resolvent $R_n(t, s, \lambda)$ is a continuous function when $|\lambda| < \frac{\lambda_1}{T\sqrt{K_0}}$ and satisfy the following estimate

$$\int_0^T R_n^2(t, s, \lambda) ds \leq \frac{TK_0}{\left(\lambda_n - |\lambda| T \sqrt{K_0}\right)^2}. \tag{10}$$

Further, taking into account (8) and (9) solution of the boundary value problem (2)–(5) can be written as

$$V(t, x) = \sum_{n=1}^{\infty} (\psi_n(t, \lambda) + \frac{1}{\lambda_n} \int_0^T \varepsilon_n(t, \eta, \lambda) b_n(\eta) f(\eta, u(\eta)) d\eta) z_n(x), \tag{11}$$

where

$$\psi_n(t, \lambda) = \psi_{1n}(\cos \lambda_n t + \lambda \int_0^T R_n(t, s, \lambda) \cos \lambda_n s ds)$$

$$+ \frac{\psi_{2n}}{\lambda_n}(\sin \lambda_n t + \lambda \int_0^T R_n(t, s, \lambda) \sin \lambda_n s ds) + \frac{1}{\lambda_n} \int_0^T \varepsilon_n(t, \eta, \lambda) g_n(\eta) d\eta, \tag{12}$$

$$\varepsilon_n(t, \eta, \lambda) = \begin{cases} \sin \lambda_n(t - \eta) + \lambda \int_\eta^T R_n(t, s, \lambda) \sin\lambda_n(s - \eta) ds, & 0 \leq \eta \leq t, \\ \\ \lambda \int_\eta^T R_n(t, s, \lambda) \sin \lambda_n(s - \eta) ds, & t \leq \eta \leq T. \end{cases} \tag{13}$$

The function (11) is an element of Gilbert space $H(Q_T)$ and weak generalized solution of boundary problem (2)–(5).

According to condition (5) each control $u(t)$ uniquely defines the controlled process $V(t, x)$. Therefore for the solution $V(t, x) + \Delta V(t, x)$ of boundary value problem (2)–(4) corresponds the control $u(t) + \Delta u(t)$, where $\Delta V(t, x)$ is the increment corresponding to the increment $\Delta u(t)$. By the method of to the maximum principle ([4], Chap. 6, pp. 356–383), ([5], Chap. 2, pp. 45–78) the increment of functional (1) can be written as

$$\Delta J[u] = J[u + \Delta u] - J[u] = -\int_0^T \Delta \Pi[t, V(t, x), \omega(t, x), u(t)] dt + \int_0^T \int_Q \Delta V^2(t, x) dx dt;$$

where

$$\Pi[t, V(t,x), \omega(t,x), u(t)] = \int_{\gamma} \omega(t,x)b(t,x)dx f[t,u(t)] - 2\beta M[t,u(t)],$$

and the function $\omega(t,x)$ is the solution of the adjoint boundary value problem

$$\omega_{tt} - A\omega = \lambda \int_0^T K(\tau,t)\omega(\tau,x)d\tau - 2[V(t,x) - \xi(t,x)], \ x \in Q, 0 < t < T,$$
$$\omega(T,x) = 0, \omega_t(T,x) = 0, \ x \in Q, \tag{14}$$
$$\Gamma\omega(t,x) = 0, \ x \in \gamma, \ 0 < t < T.$$

According to the maximum principle for systems with distributed parameters ([4], Chap. 6, pp. 356–383), ([5], Chap. 2, pp. 45–78), the optimal control is determined by the relations

$$2\beta \frac{M_u[t,u(t)]}{f_u[t,u(t)]} = \sum_{n=1}^{\infty} b_n(t)\omega_n(t), \tag{15}$$

$$f_u[t,u(t)] \left(\frac{M_u[t,u(t)]}{f_u[t,u(t)]} \right)_u > 0, \tag{16}$$

which are called *optimality conditions*.

3   Solution of the Adjoint Boundary-Value Problem

We are looking for a solution of the adjoint boundary value problem (14) in the form of the series

$$\omega(t,x) = \sum_{n=1}^{\infty} \omega_n(t)z_n(x). \tag{17}$$

The Fourier coefficients $\omega_n(t)$ for each fixed $n = 1, 2, 3, \ldots$, satisfy the linear nonhomogeneous Fredholm integral equation of the second type

$$\omega_n(t) = \lambda \int_0^T B_n(s,t)\omega_n(s)ds - \frac{2}{\lambda_n} \int_t^T \sin\lambda_n(\tau - t)[V_n(\tau) - \xi_n(\tau)]d\tau, \tag{18}$$

where

$$B_n(s,t) = \frac{1}{\lambda_n} \int_t^T \sin\lambda_n(\tau - t)K(s,\tau)d\tau.$$

The solution of equation (18) we find ([11],chap. 2, pp. 98–110) by the following formula

$$\omega_n(t) = \lambda \int_0^T P_n(s,t,\lambda) \left(-\frac{2}{\lambda_n} \int_s^T \sin\lambda_n(\tau - s)[V_n(\tau) - \xi_n(\tau)]d\tau \right) ds$$

$$-\frac{2}{\lambda_n}\int_t^T sin\lambda_n(\tau-t)[V_n(\tau)-\xi_n(\tau)]d\tau, \tag{19}$$

where the resolvent $P_n(s,t,\lambda)$ of the kernel $B_n(s,t)$ is given by

$$P_n(s,t,\lambda)=\sum_{i=1}^{\infty}\lambda^{i-1}B_{n,i}(s,t), \quad B_{n,1}(s,t)=B_n(s,t),$$

$$B_{n,i+1}(s,t)=\int_0^T B_n(\eta,t)B_{n,i}(s,\eta)d\eta, \quad i=1,2,3,...$$

and it is continuous function when $|\lambda|<\frac{\lambda_1}{T\sqrt{K_0}}$ and satisfy the following estimate

$$\int_0^T P_n^2(s,t,\lambda)d\tau\leq\frac{TK_0}{\left(\lambda_n-|\lambda|T\sqrt{K_0}\right)^2}. \tag{20}$$

Further, taking into account (17) and (19) solution of the adjoint boundary value problem can be written as

$$\omega(t,x)=-2\{-h(t,x)+E(t,x)\}, \tag{21}$$

where

$$h(t,x)=\sum_{n=1}^{\infty}\frac{1}{\lambda_n}\int_0^T b_n(t)E_n(t,\tau,\lambda)l_n(\tau,\lambda)d\tau z_n(x),$$

$$E(t,x)=\sum_{n=1}^{\infty}\frac{1}{\lambda_n}\int_0^T\left(\int_0^T b_n(t)E_n(t,\tau,\lambda)\varepsilon_n(\tau,\eta,\lambda)b_n(\eta)d\tau\right)f(\eta,u(\eta))d\eta z_n(x),$$

$$E_n(t,\tau,\lambda)=\begin{cases}\lambda\int_0^{\tau}\frac{1}{\lambda_n}P_n(s,t,\lambda)sin\lambda_n(\tau-s)ds, \quad 0\leq\tau\leq t,\\ \frac{1}{\lambda_n}sin\lambda_n(\tau-t)+\lambda\int_0^{\tau}\frac{1}{\lambda_n}P_n(s,t,\lambda)sin\lambda_n(\tau-s)ds, \quad t\leq\tau\leq T,\end{cases}$$

$$l_n(t,\lambda)=\xi_n(t)-\psi_n(t,\lambda).$$

By means of the direct calculations we have proved the following lemmas:

Lemma 1. *The function $h(t,x)$ is an element of the space $H(Q_T)$.*

Lemma 2. *Function $E(t,x)$ is an element of $H(Q_T)$.*

Based on the Lemmas 1 and 2 from (21) it follows that solution of adjoint boundary value problem (14) $\omega(t,x)$ is an element of the Hilbert space $H(Q_T)$.

4 Nonlinear Integral Equation of Optimal Control

We find the optimal control according to optimality conditions (15) and (16). We substitute in (15) the solution of adjoint boundary-value problem (14) defined by (21).

We rewrite the equality (15) in the form of

$$\beta \frac{M_u(t,u)}{f_u(t,u)} + \sum_{n=1}^{\infty} \frac{1}{\lambda_n} b_n(t) \int_0^T \int_0^T E_n(t,\tau,\lambda) \varepsilon_n(\tau,\eta,\lambda) d\tau b_n(\eta) f(\eta, u(\eta)) d\eta$$

$$= \sum_{n=1}^{\infty} \frac{1}{\lambda_n} b_n(t) \int_0^T E_n(t,\tau,\lambda) l_n(\tau,\lambda) d\tau. \tag{22}$$

Thus, the optimal control is defined as the solution of a nonlinear integral equation (22) and at the same time (15) and (16) should be carried out. Condition (5) restricts the class of functions $f[t, u(t)]$ of external influences. Therefore, we assume that the function $f[t, u(t)]$ satisfies (16) for any control $u(t) \in H(0,T)$, i.e. the optimization problem is considered in class $\{f(t, u(t))\}$ of functions satisfying (16). Nonlinear integral equation (22) is solved according to the procedure of work [7,9]. We set

$$\beta \frac{M_u(t,u)}{f_u(t,u)} = p(t). \tag{23}$$

According to condition (16) control function $u(t)$ is uniquely determined from equality (23), i.e. there is a such function φ that ([12], Chap. 8, pp. 467–480)

$$u(t) = \varphi(t, p(t), \beta). \tag{24}$$

By (23) and (24) we rewrite the equation (22) in the operator form

$$p(t) + G[p, \lambda] = h(t, \lambda) \tag{25}$$

where

$$G[p, \lambda] = \sum_{n=1}^{\infty} \frac{1}{\lambda_n} b_n(t) \int_0^T \left(\int_0^T E_n(t,\tau,\lambda) \varepsilon(\tau,\eta,\lambda) d\tau \right) \times b_n(\eta) f[\eta, \varphi(\eta, p(\eta), \beta)] d\eta,$$

$$h(t, \lambda) = \sum_{n=1}^{\infty} \frac{1}{\lambda_n} b_n(t) \int_0^T E_n(t,\tau,\lambda) l_n(\tau,\lambda) d\tau.$$

Now we investigate the questions of unique solvability of the operator equation (25).

Lemma 3. *The function $p(t)$ is an element of the space $H(0,T)$.*

Proof. By (23) we have the estimate

$$\sup \left| \frac{M_u(t,u)}{f_u(t,u)} \right| \leq N, \ \forall t \in [0,T].$$

Since $u(t) \in H(0,T)$, the statement of the lemma follows by the inequality

$$\int_0^T |p(t)|^2 dt \leq \beta^2 \int_0^T \left| \frac{M_u(t,u)}{f_u(t,u)} \right|^2 dt \leq \beta^2 N^2 T^2 < \infty.$$

Lemma 4. *The operator $G[p(t)]$ which defined by the formula (25) maps the space $H(0,T)$ into itself, i.e. it is an element of the space $H(0,T)$.*

Proof. Taking into account the following estimations

$$\int_0^T \epsilon_n^2(t,\eta,\lambda) d\eta \leq 2T \left(1 + \frac{\lambda^2 T^2 K_0}{(\lambda_1 - |\lambda| \sqrt{T^2 K_0})^2} \right),$$

$$\int_0^T E_n(t,\tau,\lambda) d\tau \leq \frac{2T}{\lambda_n^2} \left(1 + \frac{\lambda^2 T^2 K_0}{(\lambda_1 - |\lambda| \sqrt{T^2 K_0})^2} \right),$$

we obtain the assertion of lemma from following inequality

$$\int_0^T G^2[p,\lambda] = \int_0^T \left\{ \sum_{n=1}^\infty \frac{1}{\lambda_n} b_n(t) \int_0^T \left(\int_0^T E_n(t,\tau,\lambda) \varepsilon_n(\tau,\eta,\lambda) d\tau \right) \right.$$

$$\left. \times b_n(\eta) f[\eta, \varphi(\eta, p(\eta), \beta)] d\eta \right\}^2 dt \leq \int_0^T \sum_{n=1}^\infty \frac{1}{\lambda_n^2} b_n^2(t)$$

$$\times \sum_{n=1}^\infty \left\{ \int_0^T \int_0^T E_n(t,\tau,\lambda) \varepsilon_n(\tau,\eta,\lambda) d\tau b_n(\eta) f[\eta, \varphi(\eta, p(\eta), \beta)] d\eta \right\}^2 dt$$

$$\leq \int_0^T \sum_{n=1}^\infty \frac{1}{\lambda_n^2} b_n^2(t) \sum_{n=1}^\infty \int_0^T \int_0^T E_n^2(t,\tau,\lambda) d\tau \int_0^T \varepsilon_n^2(\tau,\eta,\lambda) d\tau b_n^2(\eta) d\eta$$

$$\times \int_0^T f^2[\eta, \varphi(\eta, p(\eta), \beta)] d\eta dt \leq \int_0^T \sum_{n=1}^\infty \frac{1}{\lambda_n^2} b_n^2(t)$$

$$\times \sum_{n=1}^\infty \int_0^T \frac{2T}{\lambda_n^2} \left\{ 1 + \frac{\lambda^2 T^2 K_0}{(\lambda_1 - |\lambda| \sqrt{T^2 K_0})^2} \right\} 2T \left\{ 1 + \frac{\lambda^2 T^2 K_0}{(\lambda_n - |\lambda| \sqrt{T^2 K_0})^2} \right\} b_n^2(\eta) d\eta$$

$$\times \int_0^T f^2[\eta, \varphi(\eta, p(\eta), \beta)] d\eta dt \leq \frac{1}{\lambda_1^2} \left(\int_0^T \sum_{n=1}^\infty b_n^2(t) dt \right)^2 \frac{(2T)^2}{\lambda_1^2}$$

$$\times \left\{ 1 + \frac{\lambda^2 T^2 K_0}{(\lambda_1 - |\lambda| \sqrt{T^2 K_0})^2} \right\}^2 \|f[\eta, \varphi(\eta, p(\eta), \beta)]\|_{H(0,T)}^2 < \infty.$$

Lemma 5. *Suppose that the conditions*

$$\|f[t, u(t)] - f[t, \bar{u}(t)]\|_{H(0,T)} \leq f_0 \|u(t) - \bar{u}(t)\|_{H(0,T)}, \quad f_0 > 0,$$

$$\|\varphi[t, p(t), \beta] - \varphi[t, \bar{p}(t), \beta]\|_{H(0,T)} \leq \varphi_0(\beta)\|p(t) - \bar{p}(t)\|_{H(0,T)}, \quad \varphi_0(\beta) > 0$$

are satisfied. Then if the condition

$$\gamma = C_0 f_0 \varphi_0(\beta) < 1,$$

is met, the operator $G[p, \lambda]$ is contractive. Here

$$C_0 = \frac{2T}{\lambda_1^2}\left(1 + \frac{\lambda^2 T^2 K_0}{\left(\lambda_1 - |\lambda|\sqrt{T^2 K_0}\right)^2}\right)\|b(t, x)\|_{H(\gamma_T)}^2.$$

Proof. The proof of this theorem follows from Lemma 4 by the following inequality, i.e. the following inequality is fulfilled

$$\|G[p, \lambda] - G[p, \tilde{p}, \lambda]\|_{H(0,T)}^2 \leq C_0^2 \|f[t, u(t)] - f[t, \tilde{u}(t)]\|_{H(0,T)}^2$$

$$\leq C_0^2 f_0^2 \|\varphi[t, p(t), \beta] - \varphi[t, \tilde{p}(t), \beta]\|_{H(0,T)}^2 \leq C_0^2 f_0^2 \varphi_0^2(\beta)\|p(t) - \tilde{p}(t)\|_{H(0,T)}^2.$$

Theorem 1. *Suppose that conditions (5), Lemma 5 and $|\lambda| < \frac{\lambda_1}{T\sqrt{K_0}}$ are satisfied. Then operator equation (25) has a unique solution in the space $H(0, T)$.*

Proof. According to Lemmas 3 and 4, operator equation (25) can be considered in the space $H(0, T)$. According to Lemma 5 operator $G(p)$ is contractive. Since the Hilbert space $H(0, T)$ is a complete metric space, by the theorem on principle of contracting mappings ([12], Chap. 1, pp. 43–53) the operator $G(p)$ has a unique fixed point, i.e. operator equation (25) has unique solution.

The solution of operator equation (25) can be found by the method of successive approximations, i.e. n th approximation of the solution is found by the formula

$$p_n = h - G[p_{n-1}], \quad n = 1, 2, 3, \ldots,$$

where $p_0(t)$ is an arbitrary element of the space $H(0, T)$. For the exact solution $\bar{p}(t) = \lim_{n \to \infty} p_n(t)$ we have the following estimate

$$\|\bar{p}(t) - p_n(t)\| \leq \frac{\gamma^n}{1 - \gamma}\|h - G[p_0(t)] - p_0(t)\|_{H(0,T)}$$

or when $h = p_0(t)$

$$\|\bar{p}(t) - p_n(t)\|_{H(0,T)} \leq \frac{\gamma^n}{1 - \gamma}\|G[p_0(t)]\|_{H(0,T)},$$

where $0 < \gamma < 1$ is the contraction constant. The exact solution can be found as the limit of the approximate solutions, i.e. substituting this solution in (24) we find the optimal control

$$u^0(t) = \varphi[t, \bar{p}(t), \beta].$$

We find the optimal process $V^0(t,x)$, i.e. the solution of boundary value problem (2)–(4), corresponding to the optimal control $u^0(t,x)$, according to (6) from the formula

$$V^0(t,x) = \sum_{n=1}^{\infty} \left(\lambda \int_0^T R_n(t,s,\lambda) a_n^0(s) ds - a_n^0(t) \right) z_n(x),$$

$$a_n^0(t) = \psi_{1n} \cos\lambda_n t + \frac{\psi_{2n}}{\lambda_n} \sin\lambda_n t + \frac{1}{\lambda_n} \int_0^t \sin\lambda_n(t-\tau)[q_n(\tau) + b_n(\tau) f[\tau, u^0(\tau)]] d\tau.$$

The minimum value of the functional (1) is calculated by the formula

$$J[u^0(t)] = \int_0^T \int_Q \left[V^0(t,x) - \xi(t,x) \right]^2 dx dt + 2\beta \int_0^T M[t, u^0(t)] dt.$$

The found triple $(u^0(t), V^0(t,x), J[u^0(t)])$ is a solution of the nonlinear optimization problem.

References

1. Volterra, V.: Theory of Functionals and of Integral and Integro-Differential Equations. Courier Corporation, New York (2005)
2. Vladimirov, V.S.: Mathematical problems of speed transport theory of particles. J. Trudy MIAN **61**, 3–158 (1961). (in Russian)
3. Richtmyer, R.D.: Principles of Advanced Mathematical Physics, vol. 1. Springer, Heidelberg (1978)
4. Egorov, A.I.: Optimal Control of Thermal and Diffusion Processes, 500 p. Nauka, Moscow (1978). (in Russian)
5. Komkov, V.: Optimal Control Theory for the Damping of Vibrations of Simple Elastic Systems. Springer, Heidelberg (1972)
6. Lions, J.L.: Contrle optimal de systmes gouverns par des equations aux derives partielles. Dunod Gauthier-Villars, Paris (1968)
7. Kerimbekov, A.K.: On solvability of the nonlinear optimal control problem for processes described by the semi-linear parabolic equations. In: Proceedings World Congress on Engineering 2011, London, UK, 6–8 July 2011, vol. 1, pp. 270–275 (2011)
8. Khurshudyan, A.Z.: On optimal boundary and distributed control of partial integrodifferential equations. Arch. Control Sci. **24(LX)**(1), 5–25 (2014)
9. Kerimbekov, A.: On the solvability of a nonlinear optimal control problem for the thermal processes described by fredholm integro-differential equations. In: Mityushev, V.V., Ruzhansky, M.V. (eds.) Current Trends in Analysis and Its Applications. A Series of Trends in Mathematics, pp. 803–811. Springer International Publishing, Cham (2015)
10. Plotnikov, V.I.: The energy inequality and the over-determination property of systems of eigenfunction. J. Math. USSR math. ser. **32**(4), 743–755 (1968)
11. Krasnov, M.V.: Integral Equations, 303p. Nauka, Moscow (1975). (in Russian) – 224 p. (2003)
12. Lusternik, L.A., Sobolev, V.I.: Elements of Functional Analysis, 520 p. Nauka, Moscow (1965)

Observability of a Ring Shaped Membrane via Fourier Series

Vilmos Komornik[1], Paola Loreti[2], and Michel Mehrenberger[1(✉)]

[1] Université de Strasbourg, 7 rue René Descartes, F-67084 Strasbourg, France
{komornik,mehrenbe}@math.unistra.fr
[2] Dipartimento di Scienze di Base e Applicate per l'Ingegneria,
Sapienza Università di Roma, Via A. Scarpa 16, I-00161 Roma, Italy
paola.loreti@sbai.uniroma1.it

Abstract. We study the inverse Ingham type inequality for a wave equation in a ring. This leads to a conjecture on the zeros of Bessel cross product functions. We motivate the validity of the conjecture through numerical results. We do a complete analysis in the particular case of radial initial data, where an improved time of observability is available.

1 Introduction

We study the solution $u = u(t, r, \theta)$ of the wave equation in an annulus Ω of small radius a and big radius b:

$$
\begin{cases}
u'' = \Delta u, & 0 < t < T, \quad x \in \Omega \\
u = 0, & 0 < t < T, \quad x \in \Gamma = \partial\Omega \\
u(0, x) = u_0(x), & \partial_t u(0, x) = u_1(x), \quad x \in \Omega.
\end{cases}
\tag{1}
$$

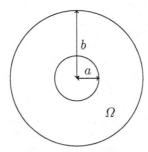

We have the following classical observability estimate (see [7] for example):

Proposition 1. *There exists $T_0 > 0$ such that for $T > T_0$ the system (1) is observable: there exists a constant $c > 0$ such that we have*

$$
\|u_0\|^2_{H^1_0(\Omega)} + \|u_1\|^2_{L^2(\Omega)} \le c \int_0^T \int_\Gamma |\partial_\nu u(t, x)|^2 \, d\Gamma dt,
\tag{2}
$$

for all $(u_0, u_1) \in H^1_0(\Omega) \times L^2(\Omega)$.

L. Bociu et al. (Eds.): CSMO 2015, IFIP AICT 494, pp. 322–330, 2016.
DOI: 10.1007/978-3-319-55795-3_30

This proposition is a special case of a general result, which can be proved by micro-local analysis [2] with the critical time $T_0 = 2\sqrt{b^2 - a^2}$, according to the geometrical ray condition.

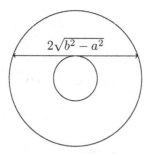

Our aim here is to study this problem using a Fourier series approach [5,6] Note that such an approach has been tackled for the whole disk (see [7]); the case of an annulus leads to new difficulties. In particular, we are not able to treat the general case, as it relies on very precise estimates of the location of the zeros of Bessel cross product functions, that we did not find in the literature. Instead, we state a conjecture on these zeros and give some numerical results to support the conjecture. In the case of radial functions or more generally functions with a limited number of modes in the angle direction, we are able to get the observability estimates with this method, even for smaller times $T > 2(b - a)$, using an asymptotic result of MacMahon (see [1], p. 374).

In Sect. 2, we give the expression of the solution. In Sect. 3, we formulate a theorem for the special case of radial functions. Section 4 is devoted to the statement of the conjecture and its numerical illustration. Finally, we prove in Sect. 5 the theorem of Sect. 3.

2   Expression of the Solution

Let J_ν (resp. Y_ν) be the Bessel functions of first (resp. second) kind of order ν. We recall the following proposition (see [7,8]).

Proposition 2. *(i) Let $\nu \in \mathbb{R}^*$ and $0 < \alpha < 1$. The nonnegative zeros $\gamma_{\nu,k}$, $k \in \mathbb{N}^*$ of the Bessel cross product function*

$$H_{\nu,\alpha}(x) = Y_\nu(x)J_\nu(\alpha x) - J_\nu(x)Y_\nu(\alpha x),$$

form a strictly increasing sequence:

$$0 < \gamma_{\nu,\alpha,1} < \gamma_{\nu,\alpha,2} < \dots \gamma_{\nu,\alpha,k} < \dots, \quad k \in \mathbb{N}^*.$$

(ii) The eigenfunctions of the Laplacian corresponding to (1) are $R_{k,m}(r)e^{im\theta}$ for $k \in \mathbb{N}^*$ and $m \in \mathbb{N}$, where

$$R_{k,m}(r) = Y_m(\gamma_{m,\alpha,k})J_m(\frac{r}{b}\gamma_{m,\alpha,k}) - J_m(\gamma_{m,\alpha,k})Y_m(\frac{r}{b}\gamma_{m,\alpha,k}). \tag{3}$$

(iii) For a dense set of initial data the solution of (1) *is given by the formula*

$$u(t,r,\theta) = \sum_{m=0}^{\infty} \sum_{k=1}^{\infty} R_{k,m}(r) \left(c_{k,m}^{+} e^{im\theta} e^{i\frac{\gamma_{m,\alpha,k}}{b}t} + c_{k,m}^{-} e^{-im\theta} e^{-i\frac{\gamma_{m,\alpha,k}}{b}t} \right) \quad (4)$$

with complex coefficients $c_{k,m}^{\pm}$, *all but finitely of which vanish.*

3   Observability Estimates for Radial Solutions

We recall from [1] the following estimate of MacMahon (1894):

$$\gamma_{\nu,\alpha,k} = \frac{\pi k}{1-\alpha} + \frac{4\nu^2 - 1}{8\pi\alpha}\frac{1-\alpha}{k} + O\left(\frac{1-\alpha}{k}\right)^3. \quad (5)$$

Thanks to this estimation, we can obtain the following theorem.

Theorem 1. *Let* $T > 2(b-a)$. *For each positive integer* M *there exists a constant* $c_M > 0$ *such that*

$$\|u_0\|_{H_0^1(\Omega)}^2 + \|u_1\|_{L^2(\Omega)}^2 \le c_M \int_0^T \int_\Gamma |\partial_\nu u(t,x)|^2 \, d\Gamma dt, \quad (6)$$

for all solutions of (1) *of the form* (4) *with* $c_{k,m}^{\pm} = 0$ *whenever* $m \ge M$.

Remark 1. – This theorem covers the case of radial initial data, corresponding to the case $M = 1$.
- If $2\sqrt{b^2 - a^2} > T > 2(b-a)$, then the constant c_M tends to zero as $M \to \infty$ because $T_0 = 2\sqrt{b^2 - a^2}$ is the critical observability time for general initial data.
- The theorem remains true if the integral in (6) is taken only over the outer boundary (the circle of radius b: see the estimate (16)) below.
- We may obtain similar results by changing the boundary conditions. For example, we may take homogeneous Neumann condition on the inner boundary, and observe the solution only on the outer boundary (see [4] for the corresponding asymptotic gap estimate that is needed).

4   Conjecture and Numerical Illustration

We then state the following conjecture.

Conjecture 1. *Let* $0 < \alpha < 1$ *and* $\nu > 1/2$.

- There exists a positive integer $k_\nu(\alpha)$ such that $\gamma_{\nu,\alpha,k+1} - \gamma_{\nu,\alpha,k}$ is decreasing for $k \le k_\nu(\alpha)$ and increasing for $k \ge k_\nu(\alpha)$.
- We have $\gamma_{\nu,\alpha,k_\nu(\alpha)+1} - \gamma_{\nu,\alpha,k_\nu(\alpha)} \ge \pi \left(1 - \alpha^2\right)^{-1/2}$.

Using a Bessel Zeros Computer [3] we can evaluate the zeros $\gamma_{\nu,\alpha,k}$ for several parameters ν, α and k.

We plot on Fig. 1 (top and bottom left) the values $k_\nu(\alpha)$ versus ν for different values of α. We observe that $k_\nu(\alpha)$ increases with ν for a fixed α. The dependence seems to be almost linear, except for small values of α where the dependence seems to be quadratic (see Fig. 1, bottom left). On Fig. 1 (bottom right), we see the relative difference with the gap $\frac{\pi}{\sqrt{1-\alpha^2}}$ (corresponding to the optimal value of Proposition 1). We observe that $\gamma_{\nu,\alpha,k_\nu(\alpha)+1} - \gamma_{\nu,\alpha,k_\nu(\alpha)}$ decreases and approaches to this gap as ν increases.

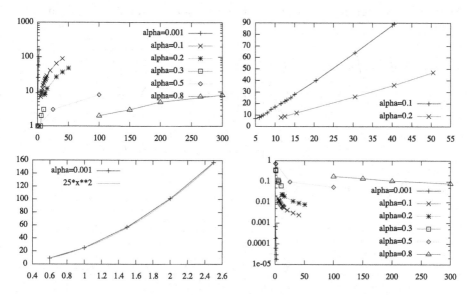

Fig. 1. Index $k_\nu(\alpha)$ where minimal gap occurs versus $\nu \geq 1/2$ (top left, logarithmic plot for $\alpha \in \{0.001, 0.1, 0.2, 0.3, 0.5, 0.8\}$; top right, standard plot for $\alpha = 0.1$ and $\alpha = 0.2$; bottom left, $\alpha = 0.001$ and comparison with function $0.25\,\nu^2$). Difference $\frac{\gamma_{\nu,\alpha,k_\nu(\alpha)+1} - \gamma_{\nu,\alpha,k_\nu(\alpha)}}{\pi/\sqrt{1-\alpha^2}} - 1$ versus ν in logarithmic scale (bottom right).

5   Proof of Theorem 1

We first express the norm in terms of the Fourier coefficients. We have

$$u(t,r,\theta) = \sum_{m=0}^{\infty} \sum_{k=1}^{\infty} R_{k,m}(r) \left(c_{k,m}^+ e^{im\theta} e^{i\frac{\gamma_{m,\alpha,k}}{b}t} + c_{k,m}^- e^{-im\theta} e^{-i\frac{\gamma_{m,\alpha,k}}{b}t} \right)$$

with

$$R_{k,m}(r) = Y_m(\gamma_{m,\alpha,k}) J_m\left(\frac{r}{b}\gamma_{m,\alpha,k}\right) - J_m(\gamma_{m,\alpha,k}) Y_m\left(\frac{r}{b}\gamma_{m,\alpha,k}\right).$$

This leads to the equalities

$$u_0(r,\theta) = u(0,r,\theta) = \sum_{m=0}^{\infty} \sum_{k=1}^{\infty} R_{k,m}(r) \left(c_{k,m}^+ e^{im\theta} + c_{k,m}^- e^{-im\theta} \right).$$

and

$$u_1(r,\theta) = \partial_t u(0,r,\theta) = \sum_{m=0}^{\infty} \sum_{k=1}^{\infty} \frac{i\gamma_{m,\alpha,k}}{b} R_{k,m}(r) e^{im\theta} \left(c_{k,m}^+ e^{im\theta} - c_{k,m}^- e^{-im\theta} \right).$$

Now, using the orthogonality of the eigenvectors of the Laplacian operator we obtain

$$\begin{aligned}
\|u_1\|_{L^2(\Omega)}^2 &= \int_{r=a}^{b} \int_{\theta=0}^{2\pi} |u_1(r,\theta)|^2 \, r \, dr \, d\theta \\
&= 2\pi \sum_{m=1}^{\infty} \sum_{k=1}^{\infty} \left| \frac{\gamma_{m,\alpha,k}}{b} \right|^2 \left(\left| c_{k,m}^+ \right|^2 + \left| c_{k,m}^- \right|^2 \right) \int_a^b |R_{k,m}(r)|^2 \, r \, dr \\
&\quad + 2\pi \sum_{k=1}^{\infty} \left| \frac{\gamma_{0,\alpha,k}}{b} \right|^2 \left| c_{k,0}^+ - c_{k,0}^- \right|^2 \int_a^b |R_{k,0}(r)|^2 \, r \, dr,
\end{aligned}$$

and a similar computation gives

$$\begin{aligned}
\int_\Omega |u_0|^2 \, d\Omega &= 2\pi \sum_{m=1}^{\infty} \sum_{k=1}^{\infty} \left(\left| c_{k,m}^+ \right|^2 + \left| c_{k,m}^- \right|^2 \right) \int_a^b |R_{k,m}(r)|^2 \, r \, dr \\
&\quad + 2\pi \sum_{k=1}^{\infty} \left(\left| c_{k,0}^+ + c_{k,0}^- \right|^2 \right) \int_a^b |R_{k,0}(r)|^2 \, r \, dr.
\end{aligned}$$

Since

$$\|u_0\|_{H_0^1(\Omega)}^2 = \int_\Omega |\nabla u_0|^2 \, d\Omega + \int_\Omega |u_0|^2 \, d\Omega,$$

we have to compute also the first integral on the right side. We have

$$\int_\Omega |\nabla u_0|^2 \, d\Omega = \int_{r=a}^{b} \int_{\theta=0}^{2\pi} |\partial_r u_0|^2 \, r \, dr \, d\theta + \int_{r=a}^{b} \int_{\theta=0}^{2\pi} \frac{1}{r} |\partial_\theta u_0|^2 \, dr \, d\theta.$$

Since

$$\partial_r u_0(r,\theta) = \partial_r u(0,r,\theta) = \sum_{m=0}^{\infty} \sum_{k=1}^{\infty} R_{k,m}'(r) \left(c_{k,m}^+ e^{im\theta} + c_{k,m}^- e^{-im\theta} \right)$$

and

$$\partial_\theta u_0(r,\theta) = \partial_\theta u(0,r,\theta) = \sum_{m=0}^{\infty} \sum_{k=1}^{\infty} im R_{k,m}(r) \left(c_{k,m}^+ e^{im\theta} - c_{k,m}^- e^{-im\theta} \right),$$

using the orthogonality of the eigenfunctions, we get

$$
\int_{r=a}^{b}\int_{\theta=0}^{2\pi}|\partial_r u_0|^2\, r\, dr\, d\theta = 2\pi \sum_{m=1}^{\infty}\sum_{k=1}^{\infty}\left(\left|c_{k,m}^{+}\right|^2+\left|c_{k,m}^{-}\right|^2\right)\int_{a}^{b}\left|R'_{k,m}(r)\right|^2 r\, dr,
$$

$$
+2\pi\sum_{k=1}^{\infty}\left(\left|c_{k,0}^{+}+c_{k,0}^{-}\right|^2\right)\int_{a}^{b}\left|R'_{k,0}(r)\right|^2 r\, dr
$$

and

$$
\int_{r=a}^{b}\int_{\theta=0}^{2\pi}\frac{1}{r}|\partial_\theta u_0|^2\, dr\, d\theta = 2\pi\sum_{m=1}^{\infty}\sum_{k=1}^{\infty}\left(\left|c_{k,m}^{+}\right|^2+\left|c_{k,m}^{-}\right|^2\right)m^2\int_{a}^{b}\frac{1}{r}|R_{k,m}(r)|^2\, dr.
$$

Using all these results we obtain the equalities

$$
E_0 := \|u_0\|_{H_0^1(\Omega)}^2 + \|u_1\|_{L^2(\Omega)}^2 = 2\pi\sum_{m=1}^{\infty}\sum_{k=1}^{\infty}\left(\left|c_{k,m}^{+}\right|^2+\left|c_{k,m}^{-}\right|^2\right)
$$

$$
\left(\int_{a}^{b}\left(1+\left|\frac{\gamma_{m,\alpha,k}}{b}\right|^2+\frac{m^2}{r^2}\right)|R_{k,m}(r)|^2 r\, dr+\int_{a}^{b}|R'_{k,m}(r)|^2 r\, dr\right)
$$

$$
+2\pi\sum_{k=1}^{\infty}(\int_{a}^{b}\left(\left|c_{k,0}^{+}+c_{k,0}^{-}\right|^2+\left|c_{k,0}^{+}-c_{k,0}^{-}\right|^2\left|\frac{\gamma_{0,\alpha,k}}{b}\right|^2\right)|R_{k,0}(r)|^2 r\, dr
$$

$$
+\left|c_{k,0}^{+}+c_{k,0}^{-}\right|^2\int_{a}^{b}\left|R'_{k,0}(r)\right|^2 r\, dr) \tag{7}
$$

and

$$
\int_{\Gamma}|\partial_\nu u(t,x)|^2\, d\Gamma = \sum_{\ell\in\{a,b\}}\int_{0}^{2\pi}|\partial_r u(t,\ell,\theta)|^2\, d\theta
$$

$$
= 2\pi\sum_{\ell\in\{a,b\}}\sum_{m=1}^{\infty}\left(\left|\sum_{k=1}^{\infty}c_{k,m}^{+}R'_{k,m}(\ell)e^{i\frac{\gamma_{m,\alpha,k}}{b}t}\right|^2+\left|\sum_{k=1}^{\infty}c_{k,m}^{-}R'_{k,m}(\ell)e^{-i\frac{\gamma_{m,\alpha,k}}{b}t}\right|^2\right)
$$

$$
+2\pi\sum_{\ell\in\{a,b\}}\left(\left|\sum_{k=1}^{\infty}c_{k,0}^{+}R'_{k,0}(\ell)e^{i\frac{\gamma_{0,\alpha,k}}{b}t}+\sum_{k=1}^{\infty}c_{k,0}^{-}R'_{k,0}(\ell)e^{-i\frac{\gamma_{0,\alpha,k}}{b}t}\right|^2\right).
$$

Since asymptotic gap is $\frac{1}{b}\frac{\pi}{1-\alpha}=\frac{\pi}{b-a}$ by the formula (5), we may apply Ingham's theorem (more precisely its version due to Haraux, see, e.g., [7]) to deduce from the last equality the existence of $C_{1,M}>0$ such that

$$
\int_{\Gamma}|\partial_\nu u(t,x)|^2\, d\Gamma \geq C_{1,M}\sum_{m=0}^{\infty}\sum_{k=1}^{\infty}\left(\left|c_{k,m}^{+}\right|^2+\left|c_{k,m}^{-}\right|^2\right)\sum_{\ell\in\{a,b\}}|R'_{k,m}(\ell)|^2,
$$

for all complex sequences $(c_{k,m}^{\pm})$ with $c_{k,m}^{\pm}=0$ whenever $m\geq M$.

Now it remains to prove that there exists another constant $C_{2,M} > 0$ such that

$$\sum_{\ell \in \{a,b\}} |R'_{k,m}(\ell)|^2$$

$$\geq C_{2,M} \left(\int_a^b \left(1 + \left| \frac{\gamma_{m,\alpha,k}}{b} \right|^2 + \frac{m^2}{r^2} \right) |R_{k,m}(r)|^2 \, r dr + \int_a^b |R'_{k,m}(r)|^2 \, r dr \right), \quad (8)$$

for $m \geq 1$, and a similar inequality for $m = 0$.
We adapt an argument used in [7], p. 107. Let y satisfy the Bessel equation

$$x^2 y'' + x y' + (x^2 - m^2) y = 0. \quad (9)$$

Let $c > 0$. Multiplying the equation by $2y'$ and integrating over (ca, cb) (instead of $(0, c)$ as in [7]), we get

$$\int_{ca}^{cb} (2x^2 y' y'' + 2x(y')^2 + 2(x^2 - m^2) yy') dx = 0.$$

Integrating by parts the first and third terms, we obtain

$$\int_{ca}^{cb} (-2x(y')^2 + 2x(y')^2 + 2xy^2) dx = [x^2 y'^2 + (x^2 - m^2) y^2]_{ca}^{cb},$$

so that

$$2 \int_{ca}^{cb} xy^2 \, dx = [x^2 y'^2 + (x^2 - m^2) y^2]_{ca}^{cb}.$$

The change of variable $x = cr$ transforms this into

$$2c^2 \int_a^b r y^2(cr) dr = [x^2 y'^2 + (x^2 - m^2) y^2]_{ca}^{cb}. \quad (10)$$

Recall that

$$R_{k,m}(r) = Y_m(\gamma_{m,\alpha,k}) J_m\left(\frac{r}{b} \gamma_{m,\alpha,k}\right) - J_m(\gamma_{m,\alpha,k}) Y_m\left(\frac{r}{b} \gamma_{m,\alpha,k}\right).$$

Now, we define

$$y_{k,m}(x) = Y_m(\gamma_{m,\alpha,k}) J_m(x) - J_m(\gamma_{m,\alpha,k}) Y_m(x)$$

which satisfies (9), and thus we have (10) with $y = y_{k,m}$ and $c = \frac{\gamma_{m,\alpha,k}}{b}$.
We have

$$y(cr) = R_{k,m}(r), \; y(ca) = R_{k,m}(a) = 0, \; y(cb) = R_{k,m}(b) = 0,$$
$$cy'(cr) = R'_{k,m}(r)$$

and thus

$$2c^2 \int_a^b r R_{k,m}(r)^2 dr = b^2 R'_{k,m}(b)^2 - a^2 R'_{k,m}(a)^2. \quad (11)$$

From (3) we have the relation

$$r^2 R_{k,m}''(r) + r R_{k,m}'(r) - m^2 R_{k,m}(r) = -c^2 r^2 R_{k,m}(r). \tag{12}$$

Integrating by parts and using the relations $R_{k,m}(b) = R_{k,m}(a) = 0$ we obtain that

$$\int_a^b R_{k,m}'(r)^2 r \, dr = - \int_a^b r R_{k,m}(r) \left(R_{k,m}''(r) + \frac{1}{r} R_{k,m}'(r) \right) dr.$$

Using the relation (12) hence we conclude that

$$\int_a^b R_{k,m}'(r)^2 r \, dr = - \int_a^b r R_{k,m}(r)^2 \left(-c^2 + \frac{m^2}{r^2} \right) dr. \tag{13}$$

Setting

$$A = \int_a^b \left(1 + c^2 + \frac{m^2}{r^2} \right) |R_{k,m}(r)|^2 \, r \, dr + \int_a^b |R_{k,m}'(r)|^2 \, r \, dr$$

we deduce from the above relations that

$$A = (1 + 2c^2) \int_a^b |R_{k,m}(r)|^2 \, r \, dr$$

$$= \frac{1 + 2c^2}{2c^2} \left(b^2 R_{k,m}'(b)^2 - a^2 R_{k,m}'(a)^2 \right). \tag{14}$$

Since

$$\inf_{k, |m| \le M} |\gamma_{k,\alpha,m}| > 0 \tag{15}$$

by (5) and the inequalities $\gamma_{k,\alpha,m} > 0$, we conclude that

$$A \le C_{3,M} R_{k,m}'(b)^2 \le C_{3,M} \left(R_{k,m}'(b)^2 + R_{k,m}'(a)^2 \right) \tag{16}$$

for a suitable constant $C_{3,M} > 0$. This proves (8) for $m \ge 1$.
For $m = 0$, using (13) we deduce from (7) that

$$2\pi \sum_{k=1}^\infty \left(\int_a^b \left(\left| c_{k,0}^+ + c_{k,0}^- \right|^2 + \left| c_{k,0}^+ - c_{k,0}^- \right|^2 \left| \frac{\gamma_{0,\alpha,k}}{b} \right|^2 \right) |R_{k,0}(r)|^2 \, r \, dr \right.$$

$$\left. + \left| c_{k,0}^+ + c_{k,0}^- \right|^2 \int_a^b |R_{k,0}'(r)|^2 \, r \, dr \right)$$

$$= 2\pi \sum_{k=1}^\infty \left(\int_a^b \left(\left| c_{k,0}^+ + c_{k,0}^- \right|^2 + \left| c_{k,0}^+ - c_{k,0}^- \right|^2 \left| \frac{\gamma_{0,\alpha,k}}{b} \right|^2 \right) |R_{k,0}(r)|^2 \, r \, dr \right.$$

$$\left. + \left| c_{k,0}^+ + c_{k,0}^- \right|^2 \left| \frac{\gamma_{0,\alpha,k}}{b} \right|^2 \int_a^b |R_{k,0}(r)|^2 \, r \, dr \right)$$

$$\asymp \sum_{k=1}^\infty \left(\left| c_{k,0}^+ \right|^2 + \left| c_{k,0}^- \right|^2 \right) \left| \frac{\gamma_{0,\alpha,k}}{b} \right|^2 \int_a^b |R_{k,0}(r)|^2 \, r \, dr,$$

and we conclude as before.

References

1. Abramowitz, M., Stegun, I.A. (eds.): Handbook of Mathematical Functions. Dover, New York (1972)
2. Bardos, C., Lebeau, G., Rauch, J.: Sharp suffcient conditions for the observation, control and stabilization of waves from the boundary. SIAM J. Control Optim. **30**, 1024–1065 (1992)
3. http://cose.math.bas.bg/webMathematica/webComputing/BesselZeros.jsp
4. Gottlieb, H.P.W.: Eigenvalues of the Laplacian with Neumann boundary conditions. J. Austr. Math. Soc. Ser. B **24**, 435–438 (1983)
5. Haraux, A.: Séries lacunaires et contrôle semi-interne des vibrations d'une plaque rectangulaire. J. Math. Pures Appl. **68**, 457–465 (1989)
6. Ingham, A.E.: Some trigonometrical inequalities with applications in the theory of series. Math. Z. **41**, 367–379 (1936)
7. Komornik, V., Loreti, P.: Fourier Series in Control Theory. Springer, Heidelberg (2005)
8. Lebedev, N.N., Skalskaya, I.P., Uflyand, Y.S.: Worked Problems in Applied Mathematics, Problem 121. Dover, New York (1979)

A Mixed Approach to Adjoint Computation with Algorithmic Differentiation

Kshitij Kulshreshtha[1]([✉]), Sri Hari Krishna Narayanan[2], and Tim Albring[3]

[1] Universität Paderborn, Paderborn, Germany
kshitij@math.upb.de
[2] Argonne National Laboratory, Argonne, IL, USA
[3] Technische Universität Kaiserslautern, Kaiserslautern, Germany

Abstract. Various algorithmic differentiation tools have been developed and applied to large-scale simulation software for physical phenomena. Until now, two strictly disconnected approaches have been used to implement algorithmic differentiation (AD), namely, source transformation and operator overloading. This separation was motivated by different features of the programming languages such as Fortran and C++. In this work we have for the first time combined the two approaches to implement AD for C++ codes. Source transformation is used for core routines that are repetitive, where the transformed source can be optimized much better by modern compilers, and operator overloading is used to interconnect at the upper level, where source transformation is not possible because of complex language constructs of C++. We have also devised a method to apply the mixed approach in the same application semi-automatically. We demonstrate the benefit of this approach using some real-world applications.

Keywords: Algorithmic differentiation · Adjoint computation

1 Introduction

Solution techniques for optimal control and optimal design problems rely on the correct and efficient computation of the adjoint state. Various analytical and numerical techniques have been devised to compute these derivatives in the past. One of the emerging techniques for the computation of derivatives on modern computers is algorithmic differentiation (AD) [6]. Despite differentiation being a badly conditioned operation in general, research has shown [3] that the process of algorithmic differentiation is well behaved and the derivatives obtained are accurate to within round-off errors. This situation is in contrast to numerical derivatives computed by using finite-differencing techniques, where the difference step size is of critical importance.

Algorithmic differentiation assumes that functions are evaluated by using a finitely terminating evaluation procedure consisting of simple arithmetic operations $\{+, -, /, *\}$ and elementary function evaluations $\{\sqrt{\ }, \sin, \cos, \exp, \log, \ldots\}$.

L. Bociu et al. (Eds.): CSMO 2015, IFIP AICT 494, pp. 331–340, 2016.
DOI: 10.1007/978-3-319-55795-3_31

Since the analytical derivatives of such arithmetic operations and elementary functions are well known, these can be introduced in the evaluation procedure almost mechanically, and the chain rule of differentiation can be applied to propagate the derivatives from one variable to another in the evaluation. In [6] various modes of propagation of derivatives as well as methods to implement tools are discussed in great detail. Here we present two techniques of AD, namely, source transformation and operator overloading.

Source transformation: Source transformation AD tools generate a new source code that computes the derivatives of an input source code. The output code must be compiled and executed in order to compute the derivatives. Tools such as ADIFOR [2], Tapenade [7], and OpenAD [11] can be used to generate derivative code for functions written in Fortran. Tapenade and ADIC [9] are examples of source transformation AD tools for C. In this work, we use ADIC to differentiate input source code portions written in C. In the output code, active variables are declared as objects of DERIV_TYPE, and runtime functions are used to propagate derivatives between them. When the output code is compiled with an appropriate driver and runtime library provided by ADIC, the Jacobian matrix can be computed.

Because such tools perform source code analysis, they can identify algorithmically active and passive variables and portions of the code. Furthermore, compilers can optimize the output code, resulting in high performance. However, no tool can generate derivative code for complete C++ input. C++ contains features such as polymorphism, inheritance, and templates that cannot be resolved statically, precluding the generation of correct derivative code.

Operator overloading: In an object-oriented language such as C++ the concept of operator overloading is well known. Several tools have been developed in recent years for AD using C++ operator overloading. ADOL-C [13] is a well-known open source AD tool with many features and high flexibility and has been successfully used to compute derivatives in a large number of simulation codes. The most important manual change required for using ADOL-C in any simulation is to convert the datatype of the real values to the special datatype adouble defined in the ADOL-C library. All operations executed after a call to trace_on() and before a call to trace_off() are recorded in an internal representation called the *trace*. Before the actual computation takes place, the independent variables are marked by assigning them values using the special <<= operator. Similarly the final dependent variables are marked by extracting their values using the special >>= operator. The trace can then be used in any mode of AD (i.e., forward or reverse) in order to compute first or higher derivatives. Several easy-to-use drivers for computing the derivative information from the trace are available. The most-used drivers are gradient(), jacobian(), and hessian(). For further usage details see [13].

The creation of the trace is the most crucial part of the whole program; and depending on the complexity of the functions being traced, the trace can become large and thus has the most impact on the memory consumption of the program.

Where the trace does not fit into a prescribed amount of memory (RAM), it spills over automatically to the disk as trace files, thereby reducing the performance of the implementation severely. Past attempts at reducing the memory requirement for certain applications include using checkpointing strategies [4,5]. For a number of problems, however, checkpointing is not applicable.

We propose a mixed approach that uses both operator overloading and source transformation to differentiate an input code that is largely written in C++ but whose computationally intensive portions are written in a C-like manner. Our approach employs operator overloading for most of the application and source transformation for the C-like portions. Because the computationally intensive portions contribute most to the trace, using source transformation instead for these portions leads to a smaller trace and better performance. We have made changes to both ADIC and ADOL-C and written a preprocessor that enables the approach to be semi-automated. The rest of the paper is organized as follows. Section 2 presents the details of the mixed approach. Section 3 presents experimental results on two applications and, Sect. 4 discusses future work.

2   Mixed Approach

The process of converting an ADOL-C instrumented application to use ADIC in certain parts is the following: (1) The user identifies a computationally intensive and C-like function from the input based on performance analysis or experience. (2) This function must be treated as an externally differentiated function (EDF) by ADOL-C. For this purpose, annotations are added to the input to support extraction of the EDF and its callees for differentiation by ADIC. Additional annotations are used to generate wrappers functions and files to copy data between ADOL-C data structures and the EDF. (3) ADIC is used to differentiate the EDF and provide forward- and reverse-mode differentiated code for it. (4) The EDF input, output, wrapper files, and original ADOL-C code are then built together. The rest of this section elaborates on the concepts of the EDF and the changes we made to ADOL-C and ADIC to support the mixed approach.

Externally differentiated functions in ADOL-C: The individual arithmetic operations and mathematical function evaluations of an EDF are not recorded on the ADOL-C trace. Instead the actual implementation of the differentiated EDF is provided via user defined function pointers that implement a certain predefined signature. As one can see in Fig. 1 the EDF replaces a large part of the trace by repeated calls to itself, which reduces the size of the trace. When ADOL-C processes the trace and arrives at a call to the EDF, ADOL-C calls the corresponding user-provided forward mode- or reverse mode- derivative code to obtain the derivatives.

ADOL-C previously maintained the EDF interface using a special structure `struct ext_diff_fct` that is registered to the ADOL-C core on a per function basis. Implementations for the forward- and reverse-mode first-order derivative computations are set up in this structure as function pointers that have a

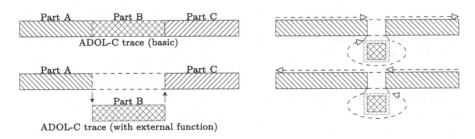

Fig. 1. ADOL-C trace of a simple and externally differentiated function (left). Repeated evaluation of an external function in forward and reverse mode (right)

particular signature as defined in the header file <adolc/externfcts.h>. The limitation of this interface is that it expects all inputs as well as all outputs to the EDF to be passed as two contiguously allocated arrays.

The design of the adouble type in ADOL-C creates an internal representation of the executed code at runtime. In order to do so most efficiently, adouble objects are allocated in a memory pool whereever there is unused space. Unless the pool is exhausted, new memory is not allocated. This design makes the allocation of large contiguous arrays an expensive operation, because of the need for finding a suitable chunk of unused space in the memory pool. Several smaller contiguous arrays, on the other hand, can be allocated more easily. Therefore we designed a second version of the EDF interface structure struct ext_diff_fct_v2 that supports providing several input arrays and several output arrays, each not necessarily of the same size. We also added extra integer-valued input parameters and an opaque object-valued input/output parameter that do not have an effect on the differentiation process outside the EDF. These changes required adjusting the signatures of the forward- and reverse-mode implementations for the EDF. The signatures now contain the number of input and output vectors, the sizes of each of these, their values, the corresponding tangents or adjoints, extra integer-valued input arrays, and an opaque context object if needed (see Fig. 2(a)). However, the process of registration and setup of the function pointers stays the same as in the original EDF interface and can even be encapsulated in a separate routine (see Fig. 2(b)), which is called once before the function is required to be evaluated. The ADOL-C evaluation of the complete structure would then look something like the code in Fig. 2(c).The user-provided functions edf->fov_forward() or edf->fov_reverse() are called during the evaluation of the jacobian() at the appropriate point.

Runtime support for ADIC generated code: To support the mixed approach's use of forward-and-reverse mode AD in a single execution instance, we recoded ADIC's runtime library in C++ and used namespaces to separate forward- and reverse-mode derivative manipulation routines. The namespace usage is inserted into the ADIC-generated code by using simple postprocessing scripts. The DERIV_TYPE structure was rewritten to be a class that supports both dynamic and static allocation of the grad array within DERIV_TYPE. Because dynamic

```
// primal function signature
int myfunc_v2 (int iArrLen, int *iArr, int nin, int nout, int *insz, double **x, int *outsz,
    double **y, void* ctx);
// first order forward implemetation signature
int myfunc_forward_v2(int iArrLen, int* iArr, int nin, int nout, int *insz, double **x, int
    ndir, double ***Xp, int *outsz, double **y, double ***Yp, void* ctx);
// first order reverse implementation signature
int myfunc_reverse_v2(int iArrLen, int* iArr, int nout, int nin, int *outsz, int dir, double
    ***Up, int *insz, double ***Zp, double **x, double **y, void* ctx);
```

(a)

```
ext_diff_fct_v2 * reg_ext_fct_myfunc(){
    ext_diff_fct_v2 *edf
        = reg_ext_fct(myfunc_v2);
    edf->zos_forward = myfunc_v2;
    edf->fov_forward = myfunc_forward_v2;
    edf->fov_reverse = myfunc_reverse_v2;
    // similar for scalar modes
    ...
    return edf;
}
```

```
trace_on(tag);
... // evaluations
if (firsttime) edf = reg_ext_fct_myfunc();
call_ext_fct(edf,...);
... // further evaluations
trace_off();
...
jacobian(tag,...); // when required
```

(b)   (c)

Fig. 2. (a) Signatures of the forward- and reverse-mode wrapper routines; (b) per-routine registration of EDF; (c) calling an EDF in ADOL-C instrumented code

allocation for every DERIV_TYPE object can be expensive, we created a memory manager that allocates a large amount of memory from the heap and then allocates the grad array of an object from this pool. We matched ADIC's layout of grad array to ADOL-C's layout of tangents and adjoints for the input and output vectors. Therefore only pointers are copied, and ADIC reuses memory already allocated in ADOL-C.

User annotations and preprocessing: User annotations have two purposes: First, they identify an EDF and its callees for extraction and subsequent differentiation by ADIC. The annotations surround the EDF and its callees, as shown in Fig. 3(a). The extraction of code is necessary because ADIC requires the C code to be isolated from the C++ code that it does not differentiate. Additional user editing may be required to obtain code that is appropriate for differentiation by ADIC. Second, annotations are used to generate the interface code for arguments of the EDF. The annotation identifies inputs, outputs, and their respective sizes or extra integers required for the computation, as well as the position of each formal parameter in the argument list of the EDF. This information helps generate wrapper code to transfer data between ADOL-C data structures and ADIC-generated code. These annotations are written directly as Python tuples, as seen in Fig. 3(b). Each tuple contains the name of the formal argument, followed by its size and the position in the formal argument list. The size itself is a list of length 0, 1, or 2, depending on whether the argument represents a scalar, a vector, or a matrix. Integer arguments are always scalars. The size may also contain references to values stored in the integers list. Several interface definitions, and thus multiple EDF structures, may also be used in any application.

```
/*@ declare doubletype=adouble @*/
 // since ADIC doesn't know adouble

/*@ begin adic_extract global @*/
... // global defines, variables etc.
    // required by ADIC routines
/*@ end adic_extract @*/

/*@ begin adic_extract @*/
... // lower level computational routines
    // differentiated by ADIC
/*@ end adic_extract @*/

/*@ begin adic_extract replace=rk_iter
    type=void @*/
... // top level interface routine
    // differentiated by ADIC
/*@ end adic_extract @*/
```

```
void rk_iter(double h, adouble *y, adouble
    **k, adouble *rhs, int n, adouble *u,
    int m, adouble *yt, adouble *ynew)
{
/*@ begin adic_export interface
name = 'rk_iter'
iarr = [ ('n',5), ('m',7) ]
input = [ ('h', [], 1),
          ('y', ['2*nDe+5'], 2),
          ('u', ['5'],6) ]
output = [ ('k', ['stage','2*nDe+5'], 3),
           ('rhs', ['2*nDe+5'], 4),
           ('yt', ['2*nDe+5'], 8),
           ('ynew',['2*nDe+5'], 9)]
@*/
... // original ADOL-C computation code
/*@ end adic_export @*/
}
```

(a)   (b)

Fig. 3. (a) Annotations for extracting code for ADIC processing; (b) annotations describing the interface routine to generate wrapper code

3   Applications

We have tested the mixed approach on two applications. The following describes each application briefly and provides the results obtained by using the mixed approach.

Periodic adsorption process: The periodic adsorption process was studied from an optimization point of view in [8,12]. A system of PDAEs in time and space with periodic boundary conditions models the cyclic steady state of a process, where a fluid is preferentially absorbed on the surface of a sorbent bed. This leads to dense Jacobians that dominate the computation time (see [8]). Therefore, previous works have used inexact Jacobians (for example, [12]). Using AD, however, we compute the equality and inequality constraint Jacobians as well as the objective gradient exactly up to machine precision.

The PDAE system is discretized in space by using a finite-volume approach, and the resulting system of ODEs is then integrated in time by using a Runge-Kutta method. This Runge-Kutta iteration in the implementation was determined to be a suitable EDF for differentiation by ADIC, particularly because this routine is repeatedly called at each time step of the simulation and has a C-like implementation. The annotations for declaring this interface routine are shown in Fig. 3(b). Two other lower-level routines for computing the right-hand side of the ODE system are also processed by ADIC. The overall problem size depends on the spatial and temporal discretization (N_{space} and N_{time}). In Table 1 we show the memory required by the trace files created on disk in a purely ADOL-C implementation for various problem sizes, which are absent in the mixed approach. Both approaches preallocate memory of size 2.3 GB in all cases. The absence of trace files on disk in the mixed approach shows that the trace was small enough in all cases to fit into the preallocated memory.

Table 1. Sizes of trace files created on disk in a purely ADOL-C implementation of periodic adsorbtion process that are not present in a mixed approach

N_{space}	N_{time}		
	2000	3000	5000
20	576 MB	863 MB	2511 MB
30	856 MB	2240 MB	3734 MB
50	2472 MB	3707 MB	7146 MB

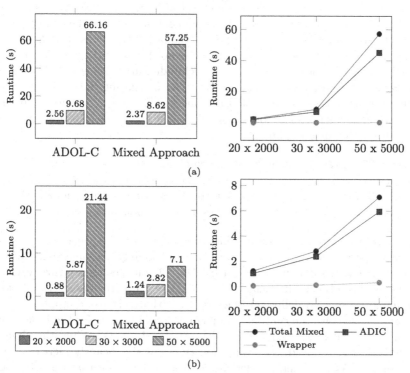

Fig. 4. Time required (in seconds) to compute (a) Jacobian of equality constraints in forward mode; (b) Jacobian of inequality constraints and gradient of objective in reverse mode

Additionally, the runtimes of the mixed approach show improvement over a pure ADOL-C implementation. In Fig. 4(a) the runtimes required in the computation of a equality constraint Jacobian with forward mode are plotted in the left figure for certain problem sizes. In the right side is the time required in the mixed approach is divided into the time required in the wrapper code of the EDF and the ADIC-processed part of the EDF. The same runtimes for the computation of the inequality constraint Jacobian and the objective gradient are shown in Fig. 4(b).

Fluid dynamics – airfoil simulation: Recently, AD was successfully applied to the open source multiphysics suite SU2 [10], which uses a highly modular C++ code structure, to design an efficient adjoint solver [1] for optimization. The implementation is based on the fixed-point formulation of the underlying solver and requires only the recording of one iteration using the converged flow solution. Therefore, at least for steady-state problems, the necessity for checkpointing is eliminated. Still, because of the nature of operator overloading, the memory requirements increase by approximately a factor of 10 compared with the direct flow solver.

SU2 is based on a finite-volume method and offers several well-established combinations of spatial and temporal methods for discretizing the flow equations. Either the steady Euler or the Navier-Stokes equation can be used as the physical model. For this work we have used a second-order central discretization plus an artificial dissipation term (Jameson-Schmidt-Turkel scheme, JST) for the convective terms and a least-squares method for evaluating the gradients needed for the viscous terms. The explicit Euler method is used to advance in pseudo-time until convergence. The following two routines were identified as promising use cases for the mixed approach:

1. `CCentJST_Flow::ComputeResidual(su2double*val_residual)`:
   per edge convective residual, projects convective flux on the cell-face normal.
2. `CEulerSolver::SetPrimitive_Gradient_LS(CGeometry *geometry)`:
   per node gradient of non-conservative variables using least-squares (only Navier-Stokes).

Both routines contain mainly C-like code, which can be processed by ADIC. A potential drawback, however, is that they use mainly class member variables as input. Another difficulty is posed by calls of routines that return variables from other class objects. In such cases we manually copy the data back and forth into simple arrays and define interface routines that take extra inputs.

Figure 5(a) shows the runtime and memory requirements for the Euler solver with a 2D airfoil in transonic flow with 10,216 elements. While the time for tracing is clearly reduced, the evaluation time has significantly increased. This indicates that ADIC-generated derivative code is slower for that case compared with ADOL-C. However there is a decrease of disk usage, solely due to trace files, and an increase in the used RAM. For a 3D airfoil with 582,752 elements the workload for each element is much higher. In that case the total runtime for tracing plus evaluation decreases in the mixed approach as shown in Fig. 5(b). Still, a large fraction of the evaluation time is in the ADIC-generated code. Disk usage reduces by approximately 10% while the used RAM increases insignificantly. For the Navier-Stokes solver with a 2D airfoil with 13937 elements, as shown in Fig. 5(c), the time for tracing reduces by 36%, and the evaluation time increases slightly, resulting in a total reduction by 16%. Furthermore, the total memory usage decreases by 15%.

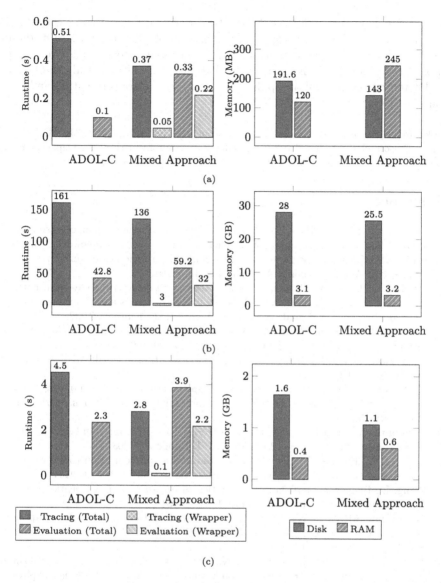

Fig. 5. Runtime and memory requirements for a 2D Euler case (a), 3D Euler case (b) and 2D Navier-Stokes case (c).

4   Conclusions and Future Work

We have implemented a mixed approach to AD that uses the operator overloading approach to differentiate most of an application and source transformation to differentiate just the computationally intensive portions. The user identifies these portions to be processed by ADIC and uses annotations and a preprocessor to generate code that interfaces ADOL-C's internal data structures with ADIC generated code. The mixed approach has been applied successfully to medium-sized and large-sized applications, resulting in lower memory usage. We plan to apply the mixed approach to more applications. We will also study the benefit of differentiating most of an application using source transformation and only some C++ portions using ADOL-C.

Acknowledgments. This work was funded in part by a grant from DAAD Project Based Personnel Exchange Programme and by a grant from the U.S. Department of Energy, Office of Science, under contract DE-AC02-06CH11357.

References

1. Albring, T.A., Sagebaum, M., Gauger, N.R.: Development of a consistent discrete adjoint solver in an evolving aerodynamic design framework. In: 16th AIAA/ISSMO Multidisciplinary Analysis and Optimization Conference. American Institute of Aeronautics and Astronautics, June 2015
2. Bischof, C., Carle, A., Corliss, G., et al.: Generating derivative codes from Fortran programs. Technical report. MCS-P263-0991, Mathematics and Computer Science Division, Argonne National Laboratory, Argonne, IL, February 1992
3. Griewank, A., Kulshreshtha, K., Walther, A.: On the numerical stability of algorithmic differentiation. Computing **94**(2–4), 125–149 (2012)
4. Griewank, A., Walther, A.: Optimal program execution reversal. Aust. Math. Soc. ANZIAM **42**, C627–C652 (2000)
5. Griewank, A., Walther, A.: Revolve: an implementation of checkpointing for the reverse or adjoint mode of computational differentiation. ACM Trans. Math. Softw. **26**(1), 19–45 (2000)
6. Griewank, A., Walther, A.: Evaluating Derivatives. Principles and Techniques of Algorithmic Differentiation, 2nd edn. SIAM, Bangkok (2008)
7. Hascoet, L., Pascual, V.: The Tapenade automatic differentiation tool: principles, model, and specification. ACM Trans. Math. Softw. **39**(3), 1–43 (2013)
8. Jiang, L., Biegler, L., Fox, G.: Optimization of pressure swing adsorption systems for air separation. AIChE J. **49**, 1140–1157 (2003)
9. Narayanan, S.H.K., Norris, B., Winnicka, B.: ADIC2: Development of a component source transformation system for differentiating C and C++. Procedia Comput. Sci. **1**(1), 1845–1853 (2010). ICCS 2010
10. Palacios, F., Alonso, J., Duraisamy, K., et al.: Stanford University Unstructured (SU2): an open-source integrated computational environment for multi-physics simulation and design. In: 51st AIAA Aerospace Sciences Meeting including the New Horizons Forum and Aerospace Exposition. American Institute of Aeronautics and Astronautics (2013)
11. Utke, J., Naumann, U., Fagan, M., et al.: OpenAD/F: a modular open-source tool for automatic differentiation of Fortran codes. ACM Trans. Math. Softw. **34**(4), 18:1–18:36 (2008)
12. Vetukuri, S.R., Biegler, L.T., Walther, A.: An inexact trust-region algorithm for the optimization of periodic adsorption processes. I EC Res. **49**, 12004–12013 (2010)
13. Walther, A., Griewank, A.: Getting started with ADOL-C. In: Naumann, U., Schenk, O. (eds.) Combinatorial Scientific Computing. Chapman-Hall, London (2012)

Sparsity Constrained Image Restoration: An Approach Using the Newton Projection Method

Germana Landi[(✉)]

Department of Mathematics, University of Bologna, Bologna, Italy
germana.landi@unibo.it

Abstract. Image restoration under sparsity constraints has received increased attention in recent years. This problem can be formulated as a nondifferentiable convex optimization problem whose solution is challenging. In this work, the non-differentiability of the objective is addressed by reformulating the image restoration problem as a nonnegatively constrained quadratic program which is then solved by a specialized Newton projection method where the search direction computation only requires matrix-vector operations. A comparative study with state-of-the-art methods is performed in order to illustrate the efficiency and effectiveness of the proposed approach.

Keywords: ℓ_1-norm based regularization · Newton projection method · Image restoration · Inverse problems

1 Introduction

This work is concerned with the general problem of image restoration under sparsity constraints formulated as

$$\min_{\mathbf{x}} \ \phi(\mathbf{x}) = \frac{1}{2}\|\mathbf{A}\mathbf{x} - \mathbf{b}\|_2^2 + \lambda\|\mathbf{x}\|_1 \tag{1}$$

where $\mathbf{A}$ is a $m \times n$ real matrix (usually $m \le n$), $\mathbf{x} \in \mathbb{R}^n$, $\mathbf{b} \in \mathbb{R}^m$ and λ is a positive parameter. (Throughout the paper, $\|\cdot\|$ will denote the Euclidean norm). In some image restoration applications, $\mathbf{A} = \mathbf{K}\mathbf{W}$ where $\mathbf{K} \in \mathbb{R}^{m\times n}$ is a discretized linear operator and $\mathbf{W} \in \mathbb{R}^{n\times n}$ is a transformation matrix from a domain where the image is *a priori* known to have a sparse representation. The variable $\mathbf{x}$ contains the coefficients of the unknown image and the data $\mathbf{b}$ is the measurements vector which is assumed to be affected by Gaussian white noise intrinsic to the detection process. The formulation (1) is usually referred to as *synthesis formulation* since it is based on the synthesis equation of the unknown image from its coefficients $\mathbf{x}$.

The penalization of the ℓ_1-norm of the coefficients vector $\mathbf{x}$ in (1) simultaneously favors sparsity and avoids overfitting. For this reason, sparsity constrained

© IFIP International Federation for Information Processing 2016
Published by Springer International Publishing AG 2016. All Rights Reserved
L. Bociu et al. (Eds.): CSMO 2015, IFIP AICT 494, pp. 341–350, 2016.
DOI: 10.1007/978-3-319-55795-3_32

image restoration has received considerable attention in the recent literature and has been successfully used in various areas. The efficient solution of problem (1) is a critical issue since the nondifferentiability of the ℓ_1-norm makes standard unconstrained optimization methods unusable. Among the current state-of-the-art methods there are gradient descent-type methods as TwIST [5], SparSA [14], FISTA [2] and NESTA [3]. GPSR [6] is a gradient-projection algorithm for the equivalent convex quadratic program obtained by splitting the variable $\mathbf{x}$ in its positive and negative parts. Fixed-point continuation methods [9], as well as methods based on Bregman iterations [7] and variable splitting, as SALSA [1], have also been recently proposed. In [12], the classic Newton projection method is used to solve the bound-constrained quadratic program formulation of (1) obtained by splitting $\mathbf{x}$. A Modified Newton projection (MNP) method has been recently proposed in [11] for the *analysis formulation* of the ℓ_1-regularized least squares problem where $\mathbf{W}$ is the identity matrix and $\mathbf{x}$ represents the image itself. The MNP method uses a fair regularized approximation to the Hessian matrix so that products of its inverse and vectors can be computed at low computational cost. As a result, the only operations required for the search direction computation are matrix-vector products.

The main contribution of this work is to extend the MNP method of [11], developed for the case $\mathbf{W} = \mathbf{I}_n$, to the synthesis formulation of problem (1) where $\mathbf{W} \neq \mathbf{I}_n$. In the proposed approach, problem (1) is firstly formulated as a nonnegatively constrained quadratic programming problem by splitting the variable $\mathbf{x}$ into the positive and negative parts. Then, the quadratic program is solved by a special purpose MNP method where a fair regularized approximation to the Hessian matrix is proposed so that products of its inverse and vectors can be computed at low computational cost. As a result, the search direction can be efficiently obtained. The convergence of the proposed MNP method is analyzed. Even if the size of the problem is doubled, the low computational cost per iteration and less iterative steps make MNP quite efficient. The performance of MNP is evaluated on several image restoration problems and is compared with that of some state-of-the-art methods. The results of the comparative study show that MNP is competitive and in some cases is outperforming some state-of-the-art methods in terms of computational complexity and achieved accuracy.

The rest of the paper can be outlined as follows. In Sect. 2, the quadratic program formulation of (1) is derived. The MNP method is presented and its convergence is analyzed in Sect. 3. In this section, the efficient computation of the search direction is also discussed. In Sect. 4, the numerical results are presented. Conclusions are given in Sect. 5.

2   Nonnegatively Constrained Quadratic Program Formulation

The proposed approach firstly needs to reformulate (1) as a nonnegatively constrained quadratic program (NCQP). The NCQP formulation is obtained by splitting the variable $\mathbf{x}$ into its positive and negative parts [6], i.e.

$$\mathbf{x} = \mathbf{u} - \mathbf{v}, \quad \mathbf{u} = \max(\mathbf{x}, 0), \quad \mathbf{v} = \max(-\mathbf{x}, 0).$$

Problem (1) can be written as the following NCQP:

$$\min_{(\mathbf{u},\mathbf{v})} \mathcal{F}(\mathbf{u}, \mathbf{v}) = \frac{1}{2}\|\mathbf{A}(\mathbf{u} - \mathbf{v}) - \mathbf{b}\|^2 + \lambda \mathbf{1}^H \mathbf{u} + \lambda \mathbf{1}^H \mathbf{v}$$

$$\text{s.t.} \quad \mathbf{u} \geq 0, \ \mathbf{v} \geq 0 \tag{2}$$

where $\mathbf{1}$ denotes the n-dimensional column vector of ones. The gradient $\mathbf{g}$ and Hessian $\mathbf{H}$ of $\mathcal{F}(\mathbf{u}, \mathbf{v})$ are respectively defined by

$$\mathbf{g} = \begin{bmatrix} \mathbf{A}^H \mathbf{A}(\mathbf{u} - \mathbf{v}) - \mathbf{A}^H \mathbf{b} + \lambda \mathbf{1} \\ -\mathbf{A}^H \mathbf{A}(\mathbf{u} - \mathbf{v}) + \mathbf{A}^H \mathbf{b} + \lambda \mathbf{1} \end{bmatrix}, \quad \mathbf{H} = \begin{bmatrix} \mathbf{A}^H \mathbf{A} & -\mathbf{A}^H \mathbf{A} \\ -\mathbf{A}^H \mathbf{A} & \mathbf{A}^H \mathbf{A} \end{bmatrix}. \tag{3}$$

We remark that the computation of the objective function and its gradient values requires only one multiplication by $\mathbf{A}$ and one by $\mathbf{A}^H$, nevertheless the double of the problem size. Since $\mathbf{H}$ is positive semidefinite, we propose to approximate it with the positive definite matrix $\mathbf{H}_\tau$:

$$\mathbf{H}_\tau = \begin{bmatrix} \mathbf{A}^H \mathbf{A} + \tau \mathbf{I}_n & -\mathbf{A}^H \mathbf{A} \\ -\mathbf{A}^H \mathbf{A} & \mathbf{A}^H \mathbf{A} + \tau \mathbf{I}_n \end{bmatrix} \tag{4}$$

where τ is a positive parameter and $\mathbf{I}$ is the identity matrix of size n.

Proposition 21. *Let* $\sigma_1, \sigma_2, \ldots, \sigma_n$ *be the nonnegative eigenvalues of* $\mathbf{A}$ *in nonincreasing order:*

$$\sigma_1 \geq \sigma_2 \geq \ldots \geq \sigma_m \geq \sigma_{m+1} = \ldots = \sigma_n = 0. \tag{5}$$

Then, $\mathbf{H}_\tau$ *is a positive definite matrix whose eigenvalues are*

$$2\sigma_1 + \tau, \ 2\sigma_2 + \tau, \ \ldots, \ 2\sigma_n + \tau, \ \tau, \ \ldots, \ \tau. \tag{6}$$

The proof is immediate since the spectrum of $\mathbf{H}_\tau$ is the union of the spectra of $\mathbf{A}^H \mathbf{A} + \tau \mathbf{I} + \mathbf{A}^H \mathbf{A}$ and $\mathbf{A}^H \mathbf{A} + \tau \mathbf{I} - \mathbf{A}^H \mathbf{A}$.

The following proposition shows that an explicit formula for the inverse of $\mathbf{H}_\tau$ can be derived.

Proposition 22. *The inverse of the matrix* $\mathbf{H}_\tau$ *is the matrix* $\mathbf{M}_\tau$ *defined as*

$$\mathbf{M}_\tau = \frac{1}{\tau} \mathbf{M}_1 \mathbf{M}_2 \tag{7}$$

where

$$\mathbf{M}_1 = \begin{bmatrix} \mathbf{A}^H \mathbf{A} + \tau \mathbf{I} & \mathbf{A}^H \mathbf{A} \\ \mathbf{A}^H \mathbf{A} & \mathbf{A}^H \mathbf{A} + \tau \mathbf{I} \end{bmatrix}, \quad \mathbf{M}_2 = \begin{bmatrix} (2\mathbf{A}^H \mathbf{A} + \tau \mathbf{I})^{-1} & 0 \\ 0 & (2\mathbf{A}^H \mathbf{A} + \tau \mathbf{I})^{-1} \end{bmatrix}.$$

Proof. We have

$$\frac{1}{\tau}\mathbf{H}_\tau\mathbf{M}_1 = \frac{1}{\tau}\begin{bmatrix} (\mathbf{A}^H\mathbf{A}+\tau\mathbf{I})^2 - (\mathbf{A}^H\mathbf{A})^2 & 0 \\ 0 & (\mathbf{A}^H\mathbf{A}+\tau\mathbf{I})^2 - (\mathbf{A}^H\mathbf{A})^2 \end{bmatrix}$$

$$= \frac{1}{\tau}\begin{bmatrix} 2\tau\mathbf{A}^H\mathbf{A}+\tau^2\mathbf{I} & 0 \\ 0 & 2\tau\mathbf{A}^H\mathbf{A}+\tau^2\mathbf{I} \end{bmatrix} = \mathbf{M}_2^{-1}. \tag{8}$$

Similarly, we can prove that

$$\frac{1}{\tau}\mathbf{M}_1\mathbf{H}_\tau = \mathbf{M}_2^{-1}. \tag{9}$$

We now show that $\mathbf{M}_1\mathbf{M}_2 = \mathbf{M}_2\mathbf{M}_1$. We have

$$\mathbf{M}_1\mathbf{M}_2 = \begin{bmatrix} (\mathbf{A}^H\mathbf{A})(2\mathbf{A}^H\mathbf{A}+\tau\mathbf{I})^{-1} & (\mathbf{A}^H\mathbf{A})(2\mathbf{A}^H\mathbf{A}+\tau\mathbf{I})^{-1} \\ (\mathbf{A}^H\mathbf{A})(2\mathbf{A}^H\mathbf{A}+\tau\mathbf{I})^{-1} & (\mathbf{A}^H\mathbf{A})(2\mathbf{A}^H\mathbf{A}+\tau\mathbf{I})^{-1} \end{bmatrix}$$

$$+ \begin{bmatrix} \tau(2\mathbf{A}^H\mathbf{A}+\tau\mathbf{I})^{-1} & 0 \\ 0 & \tau(2\mathbf{A}^H\mathbf{A}+\tau\mathbf{I})^{-1} \end{bmatrix}$$

$$\mathbf{M}_2\mathbf{M}_1 = \begin{bmatrix} (2\mathbf{A}^H\mathbf{A}+\tau\mathbf{I})^{-1}(\mathbf{A}^H\mathbf{A}) & (2\mathbf{A}^H\mathbf{A}+\tau\mathbf{I})^{-1}(\mathbf{A}^H\mathbf{A}) \\ (2\mathbf{A}^H\mathbf{A}+\tau\mathbf{I})^{-1}(\mathbf{A}^H\mathbf{A}) & (2\mathbf{A}^H\mathbf{A}+\tau\mathbf{I})^{-1}(\mathbf{A}^H\mathbf{A}) \end{bmatrix}$$

$$+ \begin{bmatrix} \tau(2\mathbf{A}^H\mathbf{A}+\tau\mathbf{I})^{-1} & 0 \\ 0 & \tau(2\mathbf{A}^H\mathbf{A}+\tau\mathbf{I})^{-1} \end{bmatrix}.$$

After some simple algebra, it can be proved that

$$(\mathbf{A}^H\mathbf{A})(2\mathbf{A}^H\mathbf{A}+\tau\mathbf{I})^{-1} = (2\mathbf{A}^H\mathbf{A}+\tau\mathbf{I})^{-1}(\mathbf{A}^H\mathbf{A}).$$

and thus

$$\mathbf{M}_1\mathbf{M}_2 = \mathbf{M}_2\mathbf{M}_1. \tag{10}$$

From (8), (9) and (10), it follows

$$\mathbf{H}_\tau\mathbf{M}_\tau = \frac{1}{\tau}\mathbf{H}_\tau\mathbf{M}_1\mathbf{M}_2 = \mathbf{M}_2^{-1}\mathbf{M}_2 = \mathbf{I}_{2n}$$

$$\mathbf{M}_\tau\mathbf{H}_\tau = \frac{1}{\tau}\mathbf{M}_1\mathbf{M}_2\mathbf{H}_\tau = \frac{1}{\tau}\mathbf{M}_2\mathbf{M}_1\mathbf{H}_\tau = \mathbf{M}_2^{-1}\mathbf{M}_2^{-1} = \mathbf{I}_{2n}.$$

3   The Modified Newton Projection Method

The Algorithm. The Newton projection method [4] for problem (2) can be written as

$$\begin{bmatrix} \mathbf{u}^{(k+1)} \\ \mathbf{v}^{(k+1)} \end{bmatrix} = \left[\begin{bmatrix} \mathbf{u}^{(k)} \\ \mathbf{v}^{(k)} \end{bmatrix} - \alpha^{(k)}\mathbf{p}^{(k)}\right]^+, \quad \mathbf{p}^{(k)} = \mathbf{S}^{(k)}\mathbf{g}^{(k)}, \quad \mathbf{g}^{(k)} = \begin{bmatrix} \mathbf{g}_u^{(k)} \\ \mathbf{g}_v^{(k)} \end{bmatrix} \tag{11}$$

where $[\cdot]^+$ denotes the projection on the positive orthant, $\mathbf{g}_u^{(k)}$ and $\mathbf{g}_v^{(k)}$ respectively indicate the partial derivatives of $\mathcal{F}$ with respect to $\mathbf{u}$ and $\mathbf{v}$ at the current iterate. The scaling matrix $\mathbf{S}^{(k)}$ is a partially diagonal matrix with respect to the index set $\mathcal{A}^{(k)}$ defined as

$$\mathcal{A}^{(k)} = \left\{ i \mid 0 \le y_i^{(k)} \le \varepsilon^{(k)} \text{ and } g_i^{(k)} > 0 \right\}$$

$$\mathbf{y}^{(k)} = \begin{bmatrix} \mathbf{u}^{(k)} \\ \mathbf{v}^{(k)} \end{bmatrix}, \; \varepsilon^{(k)} = \min\{\varepsilon, w^{(k)}\}, \; w^{(k)} = \|\mathbf{y}^{(k)} - [\mathbf{y}^{(k)} - \mathbf{g}^{(k)}]^+\|$$

and ε is a small positive parameter.

The step-length $\alpha^{(k)}$ is computed with the Armijo rule along the projection arc [4]. Let $\mathbf{E}^{(k)}$ and $\mathbf{F}^{(k)}$ be the diagonal matrices [13] such that

$$\{\mathbf{E}^{(k)}\}_{ii} = \begin{cases} 1, & i \notin \mathcal{A}^{(k)}; \\ 0, & i \in \mathcal{A}^{(k)}; \end{cases}, \quad \mathbf{F}^{(k)} = \mathbf{I}_{2n} - \mathbf{E}^{(k)}.$$

In MNP, we propose to define the scaling matrix $\mathbf{S}^{(k)}$ as

$$\mathbf{S}^{(k)} = \mathbf{E}^{(k)} \mathbf{M}_\tau \mathbf{E}^{(k)} + \mathbf{F}^{(k)}. \tag{12}$$

Therefore, the complexity of computation of the search direction $\mathbf{p}^{(k)} = \mathbf{S}^{(k)} \mathbf{g}^{(k)}$ is mainly due to one multiplication of the inverse Hessian approximation $\mathbf{M}_\tau$ with a vector since matrix-vector products involving $\mathbf{E}^{(k)}$ and $\mathbf{F}^{(k)}$ only extracts some components of the vector and do not need not to be explicitly performed. We remark that, in the Newton projection method proposed in [4], the scaling matrix is $\mathbf{S}^{(k)} = \left(\mathbf{E}^{(k)} \mathbf{H}_\tau \mathbf{E}^{(k)} + \mathbf{F}^{(k)}\right)^{-1}$ which requires to extract a submatrix of $\mathbf{H}_\tau$ and then to invert it.

Convergence Analysis. As proved in [4], the convergence of Newton-type projection methods can be proved under the general following assumptions which basically require the scaling matrices $\mathbf{S}^{(k)}$ to be positive definite matrices with uniformly bounded eigenvalues.

A1 The gradient $\mathbf{g}$ is Lipschitz continuous on each bounded set of $\mathbb{R}^{2n}$.
A2 There exist positive scalars c_1 and c_2 such that

$$c_1 \|\mathbf{y}\|^2 \le \mathbf{y}^H \mathbf{S}^{(k)} \mathbf{y} \le c_2 \|\mathbf{y}\|^2, \; \forall \mathbf{y} \in \mathbb{R}^{2n}, \; k = 0, 1, \dots$$

The key convergence result is provided in Proposition 2 of [4] which is restated here for the shake of completeness.

Proposition 31. *[4, Proposition 2] Let $\{[\mathbf{u}^{(k)}, \mathbf{v}^{(k)}]\}$ be a sequence generated by iteration (11) where $\mathbf{S}^{(k)}$ is a positive definite symmetric matrix which is diagonal with respect to $\mathcal{A}^{(k)}$ and α^k is computed by the Armijo rule along the projection arc. Under assumptions A1 and A2 above, every limit point of a sequence $\{[\mathbf{u}^{(k)}, \mathbf{v}^{(k)}]\}$ is a critical point with respect to problem (2).*

Since the objective $\mathcal{F}$ of (2) is twice continuously differentiable, it satisfies assumption A1. From Propositions 21 and 22, it follows that $\mathbf{M}_\tau$ is a symmetric positive definite matrix and hence, the scaling matrix $\mathbf{S}^{(k)}$ defined by (12) is a positive definite symmetric matrix which is diagonal with respect to $\mathcal{A}^{(k)}$. The global convergence of the MNP method is therefore guaranteed provided $\mathbf{S}^{(k)}$ verifies assumption A2.

Proposition 32. *Let $\mathbf{S}^{(k)}$ be the scaling matrix defined as $\mathbf{S}^{(k)} = \mathbf{E}^{(k)}\mathbf{H}_\tau\mathbf{E}^{(k)} + \mathbf{F}^{(k)}$. Then, there exist two positive scalars c_1 and c_2 such that*

$$c_1\|\mathbf{y}\|^2 \le \mathbf{y}^H\mathbf{S}^{(k)}\mathbf{y} \le c_2\|\mathbf{y}\|^2, \quad \forall \mathbf{y} \in \mathbb{R}^{2n}, \quad k = 0, 1, \ldots$$

Proof. Proposition 21 implies that the largest and smallest eigenvalue of $\mathbf{M}_\tau$ are respectively $1/\tau$ and $1/(2\sigma_1 + \tau)$, therefore

$$\frac{1}{2\sigma_1 + \tau}\|\mathbf{y}\|^2 \le \mathbf{y}^H\mathbf{M}_\tau\mathbf{y} \le \frac{1}{\tau}\|\mathbf{y}\|^2, \quad \forall \mathbf{y} \in \mathbb{R}^{2n}. \tag{13}$$

We have $\mathbf{y}^H\mathbf{S}^{(k)}\mathbf{y} = \left(\mathbf{E}^{(k)}\mathbf{y}\right)^H\mathbf{M}_\tau\left(\mathbf{E}^{(k)}\mathbf{y}\right) + \mathbf{y}^H\mathbf{F}^{(k)}\mathbf{y}$. From (13) it follows that

$$\frac{\|\mathbf{E}^{(k)}\mathbf{y}\|^2}{2\sigma_1 + \tau} + \mathbf{y}^H\mathbf{F}^{(k)}\mathbf{y} \le \left(\mathbf{E}^{(k)}\mathbf{y}\right)^H\mathbf{M}_\tau\left(\mathbf{E}^{(k)}\mathbf{y}\right) + \mathbf{y}^H\mathbf{F}^{(k)}\mathbf{y} \le \frac{\|\mathbf{E}^{(k)}\mathbf{y}\|^2}{\tau} + \mathbf{y}^H\mathbf{F}^{(k)}\mathbf{y}.$$

Moreover we have $\mathbf{y}^H\mathbf{F}^{(k)}\mathbf{y} = \sum_{i \in \mathcal{A}^{(k)}} y_i^2$, $\|\mathbf{E}^{(k)}\mathbf{y}\|^2 = \sum_{i \notin \mathcal{A}^{(k)}} y_i^2$; hence:

$$\frac{1}{2\sigma_1 + \tau}\sum_{i \notin \mathcal{A}^{(k)}} y_i^2 + \sum_{i \in \mathcal{A}^{(k)}} y_i^2 \le \mathbf{y}^H\mathbf{S}^{(k)}\mathbf{y} \le \frac{1}{\tau}\sum_{i \notin \mathcal{A}^{(k)}} y_i^2 + \sum_{i \in \mathcal{A}^{(k)}} y_i^2$$

and

$$\min\{\frac{1}{2\sigma_1 + \tau}, 1\}\|\mathbf{y}\|^2 \le \mathbf{y}^H\mathbf{S}^{(k)}\mathbf{y} \le \max\{\frac{1}{\tau}, 1\}\|\mathbf{y}\|^2.$$

The thesis follows by setting $c_1 = \min\{\frac{1}{2\sigma_1 + \tau}, 1\}$ and $c_2 = \max\{\frac{1}{\tau}, 1\}$.

Computing the Search Direction. We suppose that $\mathbf{K}$ is the matrix representation of a spatially invariant convolution operator with periodic boundary conditions so that $\mathbf{K}$ is a block circulant with circulant blocks (BCCB) matrix and matrix-vector products can be efficiently performed via the FFT. Moreover, we assume that the columns of $\mathbf{W}$ form an orthogonal basis for which fast sparsifying algorithms exist, such as a wavelet basis, for example. Under these assumptions, $\mathbf{A}$ is a full and dense matrix but the computational cost of matrix-vector operations with $\mathbf{A}$ and $\mathbf{A}^H$ is relatively cheap. As shown by (12), the computation of the search direction $\mathbf{p}^{(k)} = \mathbf{S}^{(k)}\mathbf{g}^{(k)}$ requires the multiplication of a vector by $\mathbf{M}_\tau$. Let $[\mathbf{z}, \mathbf{w}] \in \mathbb{R}^{2n}$ be a given vector, then it immediately follows that

$$\mathbf{M}_\tau\begin{bmatrix}\mathbf{z}\\\mathbf{w}\end{bmatrix} = \frac{1}{\tau}\begin{bmatrix}\mathbf{A}^H\mathbf{A}\left(2\mathbf{A}^H\mathbf{A} + \tau\mathbf{I}\right)^{-1}(\mathbf{z} + \mathbf{w}) + \tau\left(2\mathbf{A}^H\mathbf{A} + \tau\mathbf{I}\right)^{-1}\mathbf{z}\\\mathbf{A}^H\mathbf{A}\left(2\mathbf{A}^H\mathbf{A} + \tau\mathbf{I}\right)^{-1}(\mathbf{z} + \mathbf{w}) + \tau\left(2\mathbf{A}^H\mathbf{A} + \tau\mathbf{I}\right)^{-1}\mathbf{w}\end{bmatrix}. \tag{14}$$

Formula (14) needs the inversion of $2\mathbf{A}^H\mathbf{A} + \tau\mathbf{I}$. Our experimental results indicate that the search direction can be efficiently and effectively computed as follows. Using the Sherman-Morrison-Woodbury formula, we obtain

$$(2\mathbf{A}^H\mathbf{A} + \tau\mathbf{I})^{-1} = \frac{1}{\tau}\left(\mathbf{I} - \mathbf{W}^H\mathbf{K}^H(\mathbf{K}\mathbf{K}^H + \frac{\tau}{2})^{-1}\mathbf{K}\mathbf{W}\right) \quad (15)$$

$$\mathbf{A}^H\mathbf{A}(2\mathbf{A}^H\mathbf{A} + \tau\mathbf{I})^{-1} = \frac{1}{\tau}\left(\mathbf{W}^H(\mathbf{K}^H\mathbf{K} - \mathbf{K}^H\mathbf{K}\mathbf{K}^H(\mathbf{K}\mathbf{K}^H + \frac{\tau}{2})^{-1}\mathbf{K})\mathbf{W}\right). \quad (16)$$

Substituting (15) and (16) in (14), we obtain

$$\mathbf{A}^H\mathbf{A}(2\mathbf{A}^H\mathbf{A} + \tau\mathbf{I})^{-1}(\mathbf{z} + \mathbf{w}) + \tau(2\mathbf{A}^H\mathbf{A} + \tau\mathbf{I})^{-1}\mathbf{z} =$$
$$\frac{1}{\tau}\mathbf{W}^H(\mathbf{K}^H\mathbf{K} - \mathbf{K}^H\mathbf{K}\mathbf{K}^H(\mathbf{K}\mathbf{K}^H + \frac{\tau}{2})^{-1}\mathbf{K})(\mathbf{W}\mathbf{z} + \mathbf{W}\mathbf{w})$$
$$- \mathbf{W}^H\mathbf{K}^H(\mathbf{K}\mathbf{K}^H + \frac{\tau}{2})^{-1}\mathbf{K})\mathbf{W}\mathbf{z} + \mathbf{z} \quad (17)$$

$$\mathbf{A}^H\mathbf{A}(2\mathbf{A}^H\mathbf{A} + \tau\mathbf{I})^{-1}(\mathbf{z} + \mathbf{w}) + \tau(2\mathbf{A}^H\mathbf{A} + \tau\mathbf{I})^{-1}\mathbf{w} =$$
$$\frac{1}{\tau}\mathbf{W}^H(\mathbf{K}^H\mathbf{K} - \mathbf{K}^H\mathbf{K}\mathbf{K}^H(\mathbf{K}\mathbf{K}^H + \frac{\tau}{2})^{-1}\mathbf{K})(\mathbf{W}\mathbf{z} + \mathbf{W}\mathbf{w})$$
$$- \mathbf{W}^H\mathbf{K}^H(\mathbf{K}\mathbf{K}^H + \frac{\tau}{2})^{-1}\mathbf{K})\mathbf{W}\mathbf{w} + \mathbf{w}. \quad (18)$$

Since $\mathbf{K}$ is BCCB, it is diagonalized by the Discrete Fourier Transform (DFT), i.e. $\mathbf{K} = \mathbf{U}^H\mathbf{D}\mathbf{U}$ where $\mathbf{U}$ denotes the unitary matrix representing the DFT and $\mathbf{D}$ is a diagonal matrix. Thus, we have:

$$\mathbf{K}^H(\mathbf{K}\mathbf{K}^H + \frac{\tau}{2})^{-1}\mathbf{K} = \mathbf{U}^H\left(\frac{|\mathbf{D}|^2}{|\mathbf{D}|^2 + \frac{\tau}{2}}\right)\mathbf{U} \quad (19)$$

$$\mathbf{K}^H\mathbf{K} - \mathbf{K}^H\mathbf{K}\mathbf{K}^H(\mathbf{K}\mathbf{K}^H + \frac{\tau}{2})^{-1}\mathbf{K} = \mathbf{U}^H\left(|\mathbf{D}|^2 - \frac{|\mathbf{D}|^4}{|\mathbf{D}|^2 + \frac{\tau}{2}}\right)\mathbf{U}. \quad (20)$$

Substituting (19) and (20) in (17) and (18), we obtain

$$\mathbf{A}^H\mathbf{A}(2\mathbf{A}^H\mathbf{A} + \tau\mathbf{I})^{-1}(\mathbf{z} + \mathbf{w}) + \tau(2\mathbf{A}^H\mathbf{A} + \tau\mathbf{I})^{-1}\mathbf{z} =$$
$$\frac{1}{\tau}\mathbf{W}^H\mathbf{U}^H\left((|\mathbf{D}|^2 - \frac{|\mathbf{D}|^4}{|\mathbf{D}|^2 + \frac{\tau}{2}})(\mathbf{U}\mathbf{W}\mathbf{z} + \mathbf{U}\mathbf{W}\mathbf{w}) - (\frac{|\mathbf{D}|^2}{|\mathbf{D}|^2 + \frac{\tau}{2}})\mathbf{U}\mathbf{W}\mathbf{z}\right) + \mathbf{z} \quad (21)$$

$$\mathbf{A}^H\mathbf{A}(2\mathbf{A}^H\mathbf{A} + \tau\mathbf{I})^{-1}(\mathbf{z} + \mathbf{w}) + \tau(2\mathbf{A}^H\mathbf{A} + \tau\mathbf{I})^{-1}\mathbf{w} =$$
$$\frac{1}{\tau}\mathbf{W}^H\mathbf{F}^H\left((|\mathbf{D}|^2 - \frac{|\mathbf{D}|^4}{|\mathbf{D}|^2 + \frac{\tau}{2}})(\mathbf{U}\mathbf{W}\mathbf{z} + \mathbf{U}\mathbf{W}\mathbf{w}) - (\frac{|\mathbf{D}|^2}{|\mathbf{D}|^2 + \frac{\tau}{2}})\mathbf{U}\mathbf{W}\mathbf{w}\right) + \mathbf{w}.$$
$$(22)$$

Equations (14), (21) and (22) show that, at each iteration, the computation of the search direction $\mathbf{p}^{(k)}$ requires two products by $\mathbf{W}$, two products by $\mathbf{W}^H$, two products by $\mathbf{U}$ and two products by $\mathbf{U}^H$. The last products can be performed efficiently by using the FFT algorithm.

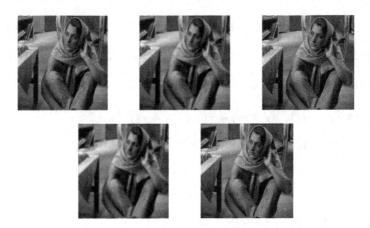

Fig. 1. Top line: exact image (left), Gaussian blurred image (middle) and MNP reconstruction (right). Bottom line: out-of-focus blurred blurred image (left) and MNP reconstruction (right). The noise level is NL $= 7.5 \cdot 10^{-3}$.

4   Numerical Results

In this section, we present the numerical results of some image restoration test problems. The numerical experiments aim at illustrating the performance of MNP compared with some state-of-the-art methods as SALSA [1], CGIST [8], the nonmonotonic version of GPSR [6], and the Split Bregman method [7]. Even if SALSA has been shown to outperform GPSR [1], we consider GPSR in our comparative study since it solves, as MNP, the quadratic program (2). The Matlab source code of the considered methods, made publicly available by the authors, has been used in the numerical experiments. The numerical experiments have been executed in Matlab R2012a on a personal computer with an Intel Core i7-2600, 3.40 GHz processor.

The numerical experiments are based on the well-known *Barbara* image (Fig. 1), whose size is 512 × 512 and whose pixels have been scaled into the range between 0 and 1. In our experiments, the matrix **W** represents an orthogonal Haar wavelet transform with four levels. For all the considered methods, the initial iterate $\mathbf{x}^{(0)}$ has been chosen as $\mathbf{x}^{(0)} = \mathbf{W}^H \mathbf{b}$; the regularization parameter λ has been heuristically chosen. In MNP, the parameter τ of the Hessian approximation has been fixed at $\tau = 100\lambda$. This value has been fixed after a wide experimentation and has been used in all the presented numerical experiments.

The methods iteration is terminated when the relative distance between two successive objective values becomes less than a tolerance tol_ϕ. A maximum number of 100 iterations has been allowed for each method.

In the first experiment, the *Barbara* image has been convolved with a Gaussian PSF with variance equal to 2, obtained with the code psfGauss from [10], and then, the blurred image has been corrupted by Gaussian noise with noise level equal to $7.5 \cdot 10^{-3}$ and $1.5 \cdot 10^{-2}$. (The noise level NL is defined as

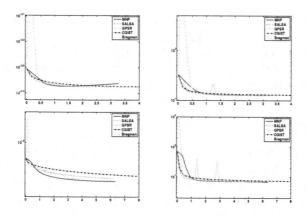

Fig. 2. MSE (left column) and objective function (right column) histories of MNP (*blue solid line*), SALSA (*magenta dashed line*), GPSR (*red dotted line*), CGIST (*black dashdotted line*) and Split Bregman (*cyan dashdotted line*) methods. Top line: Guassian blur; bottom line: out-of-focus blur. The noise level is $NL = 1.5 \cdot 10^{-2}$. (Color figure online)

Table 1. Restoring the noised and blurred *Barbara* image: numerical results.

tol_ϕ	Method	Gaussian blur						Out-of-focus blur					
		NL = 0.0075			NL = 0.015			NL = 0.0025			NL = 0.0075		
		MSE	Obj	Time	MSE	Obj	Time	MSE	Obj	Time	MSE	Obj	Time
10^{-1}	MNP	$4.18 \cdot 10^{-3}$	7.33	1.15	$4.25 \cdot 10^{-3}$	13.26	1.17	$2.99 \cdot 10^{-3}$	3.04	2.68	$3.79 \cdot 10^{-3}$	7.73	1.42
	SALSA	$4.34 \cdot 10^{-3}$	7.86	0.47	$4.45 \cdot 10^{-3}$	15.05	0.36	$4.35 \cdot 10^{-3}$	3.96	0.47	$4.85 \cdot 10^{-3}$	9.78	0.47
	GPSR	$4.31 \cdot 10^{-3}$	7.49	0.89	$4.35 \cdot 10^{-3}$	13.76	0.89	$4.83 \cdot 10^{-3}$	5.41	0.66	$4.86 \cdot 10^{-3}$	9.90	0.64
	CGIST	$4.37 \cdot 10^{-3}$	8.09	0.72	$4.46 \cdot 10^{-3}$	14.92	0.56	$4.90 \cdot 10^{-3}$	5.33	0.90	$5.00 \cdot 10^{-3}$	10.36	0.72
	Breg.	$4.34 \cdot 10^{-3}$	7.50	0.72	$4.43 \cdot 10^{-3}$	14.32	0.75	$3.61 \cdot 10^{-3}$	3.03	0.48	$4.58 \cdot 10^{-3}$	8.81	0.72
10^{-2}	MNP	$4.17 \cdot 10^{-3}$	7.03	1.68	$4.25 \cdot 10^{-3}$	13.00	1.70	$2.47 \cdot 10^{-3}$	2.54	5.66	$3.45 \cdot 10^{-3}$	7.29	2.48
	SALSA	$4.21 \cdot 10^{-3}$	7.04	1.19	$4.31 \cdot 10^{-3}$	13.55	0.92	$3.25 \cdot 10^{-3}$	2.78	2.07	$4.20 \cdot 10^{-3}$	7.98	1.62
	GPSR	$4.22 \cdot 10^{-3}$	6.84	1.62	$4.34 \cdot 10^{-3}$	13.67	1.00	$3.94 \cdot 10^{-3}$	3.16	2.39	$4.03 \cdot 10^{-3}$	7.74	1.59
	CGIST	$4.24 \cdot 10^{-3}$	7.19	1.84	$4.32 \cdot 10^{-3}$	13.61	1.51	$3.87 \cdot 10^{-3}$	3.28	4.82	$4.34 \cdot 10^{-3}$	8.22	2.82
	Breg.	$4.23 \cdot 10^{-3}$	6.98	1.31	$4.33 \cdot 10^{-3}$	13.49	1.22	$2.85 \cdot 10^{-3}$	2.58	1.33	$4.15 \cdot 10^{-3}$	7.85	1.56
10^{-3}	MNP	$4.21 \cdot 10^{-3}$	6.54	3.42	$4.35 \cdot 10^{-3}$	12.77	3.26	$2.34 \cdot 10^{-3}$	2.49	8.92	$3.16 \cdot 10^{-3}$	6.88	6.38
	SALSA	$4.14 \cdot 10^{-3}$	6.60	4.09	$4.22 \cdot 10^{-3}$	12.79	3.46	$2.45 \cdot 10^{-3}$	2.46	6.74	$3.46 \cdot 10^{-3}$	7.05	6.27
	GPSR	$4.20 \cdot 10^{-3}$	6.84	1.90	$4.22 \cdot 10^{-3}$	124	2.04	$3.81 \cdot 10^{-3}$	2.57	5.19	$3.40 \cdot 10^{-3}$	6.94	4.24
	CGIST	$4.14 \cdot 10^{-3}$	6.65	7.13	$4.22 \cdot 10^{-3}$	12.82	5.71	$2.86 \cdot 10^{-3}$	2.59	18.81	$3.52 \cdot 10^{-3}$	7.11	11.65
	Breg.	$4.16 \cdot 10^{-3}$	6.57	4.38	$4.23 \cdot 10^{-3}$	12.79	3.76	$2.21 \cdot 10^{-3}$	2.41	4.15	$3.44 \cdot 10^{-3}$	7.02	6.01

$NL := \|\boldsymbol{\eta}\| / \|\mathbf{A}\mathbf{x}_{original}\|$ where $\mathbf{x}_{original}$ is the original image and $\boldsymbol{\eta}$ is the noise vector.) In the second experiment, the *Barbara* image has been corrupted by out-of-focus blur, obtained with the code `psfDefocus` from [10] and by Gaussian noise with noise level equal to $2.5 \cdot 10^{-3}$ and $7.5 \cdot 10^{-3}$. The degraded images and the MNP restorations are shown in Fig. 1 for $NL = 7.5 \cdot 10^{-3}$. Table 1 reports the Mean Squared Error (MSE) values, the objective function values and the CPU times in seconds obtained by using the stopping tolerance values

$tol_\phi = 10^{-1}, 10^{-2}, 10^{-3}$. In Fig. 2, the MSE behavior and the decreasing of the objective function versus time (in seconds) are illustrated.

The reported numerical results indicate that MNP is competitive with the considered state-of-the-art and, in terms of MSE reduction, MNP reaches the minimum MSE value very early.

5   Conclusions

In this work, the MNP method has been proposed for sparsity constrained image restoration. In order to gain low computational complexity, the MNP method uses a fair approximation of the Hessian matrix so that the search direction can be computed efficiently by only using FFTs and fast sparsifying algorithms. The results of numerical experiments show that MNP may be competitive with some state-of-the-art methods both in terms of computational efficiency and accuracy.

References

1. Afonso, M.V., Bioucas-Dias, J.M., Figueiredo, M.A.T.: Fast image recovery using variable splitting and constrained optimization. IEEE Trans. Image Process. **19**, 2345–2356 (2010)
2. Beck, A., Teboulle, M.: A fast iterative shrinkage-thresholding algorithm for linear inverse problems. SIAM J. Imaging Sci. **2**(1), 183–202 (2009)
3. Becker, S., Bobin, J., Candès, E.J.: NESTA: a fast and accurate first-order method for sparse recovery. SIAM J. Imaging Sci. **4**(1), 1–39 (2011)
4. Bertsekas, D.: Projected Newton methods for optimization problem with simple constraints. SIAM J. Control Optim. **20**(2), 221–245 (1982)
5. Bioucas-Dias, J., Figueiredo, M.: A new twist: two-step iterative shrinkage/thresholding algorithms for image restoration. IEEE Trans. Image Process. **16**(12), 299173004 (2007)
6. Figueiredo, M.A.T., Nowak, R.D., Wright, S.J.: Gradient projection for sparse reconstruction: application to compressed sensing and other inverse problems. IEEE J. Sel. Top. Signal Process. **1**, 586–597 (2007)
7. Goldstein, T., Osher, S.: The split Bregman method for L1 regularized problems. Siam J. Imaging Sci. **2**(2), 323–343 (2009)
8. Goldstein, T., Setzer, S.: High-order methods for basis pursuit. Technical report CAM report 10–41, UCLA (2010)
9. Hale, E.T., Yin, W., Zhang, Y.: Fixed-point continuation for L1-minimization: methodology and convergence. SIAM J. Optim. **19**, 1107–1130 (2008)
10. Hansen, P.C., Nagy, J., O'Leary, D.P.: Deblurring Images. Matrices, Spectra and Filtering. SIAM, Philadelphia (2006)
11. Landi, G.: A modified Newton projection method for ℓ_1-regularized least squares image deblurring. J. Math. Imaging Vis. **51**(1), 195–208 (2015)
12. Schmidt, M., Kim, D., Sra, S.: Projected Newton-type methods in machine learning. In: Optimization for Machine Learning. MIT Press (2011)
13. Vogel, C.R.: Computational Methods for Inverse Problems. SIAM, Philadelphia (2002)
14. Wright, S., Nowak, R., Figueiredo, M.: Sparse reconstruction by separable approximation. IEEE Trans. Signal Process. **57**(7), 2479172493 (2009)

The Hilbert Uniqueness Method for a Class of Integral Operators

Paola Loreti and Daniela Sforza$^{(\boxtimes)}$

Dipartimento di Scienze di Base e Applicate per l'Ingegneria,
Sapienza Università di Roma, Via Antonio Scarpa 16, 00161 Roma, Italy
{paola.loreti,daniela.sforza}@sbai.uniroma1.it

Abstract. The Hilbert Uniqueness Method introduced by J.-L. Lions in 1988 has great interest among scientists in the control theory, because it is a basic tool to get controllability results for evolutive systems. Our aim is to outline the Hilbert Uniqueness Method for first order coupled systems in the presence of memory terms in general Hilbert spaces. At the end of the paper we give some applications of our general results.

Keywords: Coupled systems · Convolution kernels · Reachability

1 Introduction

It is well known that heat equations with memory of the following type

$$y_t = \alpha \triangle y + \int_0^t K(t-s)\triangle y(s)ds, \tag{1}$$

with $\alpha > 0$, cannot be controlled to rest for large classes of memory kernels and controls, see e.g. [3,4]. The motivation for that kind of results is due to the smoothing effect of the solutions, because (1) is a parabolic equation when the constant α before the Laplacian is positive.

On the other hand the class of the partial integro-differential equations changes completely if in the Eq. (1) one takes $\alpha = 0$. The physical model relies on the Cattaneo's paper [1]. Indeed, in [1] to overcome the fact that the solutions of the heat equation propagate with infinite speed, Cattaneo proposed the following equation

$$y_t = \int_0^t K(t-s)\triangle y(s)ds, \tag{2}$$

with $K(t) = e^{-\gamma t}$, γ being a positive constant. The interest for equations of the type (2) is in the property of the solutions to have finite propagation speed, the same property of the solutions of wave equations.

L. Bociu et al. (Eds.): CSMO 2015, IFIP AICT 494, pp. 351–359, 2016.
DOI: 10.1007/978-3-319-55795-3_33

From a mathematical point of view, a natural question is to study integro-differential equations of the type

$$u_t + \int_0^t M(t-s)\triangle^2 u(s)ds = 0\,,$$

where $M(t)$ is a suitable kernel, locally integrable on $(0,+\infty)$, and $\triangle^2$ denotes the biharmonic operator, that is in the N-dimensional case

$$\triangle^2 u = \sum_{i=1}^N \sum_{j=1}^N \partial_{ii}^2 \partial_{jj}^2 u\,.$$

The Hilbert Uniqueness Method has been introduced by Lions, see [7,8], to study control problems for partial differential systems. That method has been largely used in the literature, see e.g. [5].

Inspired by those problems, the goal of the present paper is to describe the Hilbert Uniqueness Method, for coupled hyperbolic equations of the first order with memory in a general Hilbert space, when the integral kernels involved are general functions $k_1, k_2 \in L^1(0,T)$ and integral terms also occur in the coupling:

$$\begin{cases} u_{1t} + \displaystyle\int_0^t k_1(t-s)\mathcal{A}u_1(s)ds + \mathcal{L}_1(1*u_2) = 0 \\[3mm] u_{2t} + \displaystyle\int_0^t k_2(t-s)\mathcal{A}^2 u_2(s)ds + \mathcal{L}_2(1*u_1) = 0 \end{cases} \quad \text{in } (0,T)\,,$$

In another context, in [2] the authors study the exact controllability of the equation

$$y_t = \int_0^t K(t-s)\triangle y(s)ds + u\chi_\omega \qquad \text{in} \quad (0,T)\times\Omega, \tag{3}$$

where ω is a given nonempty open subset of Ω. The hyperbolic nature of (3) allows to show its exact controllability under suitable conditions on the waiting time T and the controller ω, thanks to observability inequalities for the solutions of the dual system obtained by means of Carleman estimates.

For a different approach leading to solve control problems for hyperbolic systems, we refer to [6,11].

2   The Hilbert Uniqueness Method

Let H be a real Hilbert space with scalar product $\langle\,\cdot\,,\,\cdot\,\rangle$ and norm $\|\cdot\|$.

We consider a linear operator $\mathcal{A} : D(\mathcal{A}) \subset H \to H$ with domain $D(\mathcal{A})$, $k_1, k_2 \in L^1(0,T)$ and $\mathcal{L}_i$ $(i=1,2)$ linear operators on H with domain $D(\mathcal{L}_i) \supset D(\mathcal{A})$. We assume that $\mathcal{L}_2$ is self-adjoint and $\mathcal{L}_1$ is self-adjoint on a subset of its domain that will be precised later.

Moreover, let H_1 be another real Hilbert space with scalar product $\langle\,\cdot\,,\,\cdot\,\rangle_{H_1}$ and norm $\|\cdot\|_{H_1}$ and $\mathcal{B} \in L(H_0; H_1)$, where H_0 is a space such that $D(\mathcal{A}) \subset H_0 \subset H$. In the applications $\mathcal{B}$ could be, for example, a trace operator.

We take into consideration the following first order coupled system with memory

$$\begin{cases} u_{1t} + \displaystyle\int_0^t k_1(t-s)\mathcal{A}u_1(s)ds + \mathcal{L}_1(1*u_2) = 0 \\[4mm] u_{2t} + \displaystyle\int_0^t k_2(t-s)\mathcal{A}^2 u_2(s)ds + \mathcal{L}_2(1*u_1) = 0 \end{cases} \quad \text{in } (0,T), \quad (4)$$

with null initial conditions

$$u_1(0) = u_2(0) = 0, \tag{5}$$

and satisfying

$$\mathcal{B}u_1(t) = g_1(t), \quad \mathcal{B}u_2(t) = 0, \quad \mathcal{B}\mathcal{A}u_2(t) = g_2(t), \quad t \in (0,T). \tag{6}$$

For a reachability problem we mean the following.

Definition 1. *Given $T > 0$ and $u_{10}, u_{20} \in H$, a reachability problem consists in finding $g_i \in L^2(0,T; H_1)$, $i = 1, 2$ such that the weak solution u of problem (4)–(6) verifies the final conditions*

$$u_1(T) = u_{10}, \quad u_2(T) = u_{20}. \tag{7}$$

One can solve such reachability problems by means of the Hilbert Uniqueness Method. To show that, we proceed as follows.

To begin with, we assume the following conditions.

Assumptions (H1)

1. There exists a self-adjoint positive linear operator A on H with dense domain $D(A)$ satisfying

$$D(A) \subset D(\mathcal{A}), \quad \mathcal{A}x = Ax \;\; \forall x \in D(A), \quad D(\sqrt{A}) = \mathrm{Ker}(\mathcal{B}).$$

2. $\mathcal{L}_2$ is self-adjoint and $\mathcal{L}_1$ is self-adjoint on $D(A) \cap \mathrm{Ker}(\mathcal{B})$, that is

$$\langle \mathcal{L}_1\varphi, \xi \rangle = \langle \varphi, \mathcal{L}_1\xi \rangle, \quad \forall \varphi, \xi \in D(A) \cap \mathrm{Ker}(\mathcal{B}). \tag{8}$$

3. There exists $D_\nu \in L(H_0; H_1)$ such that the following identity holds
$$\langle \mathcal{A}\varphi, \xi \rangle = \langle \varphi, A\xi \rangle - \langle \mathcal{B}\varphi, D_\nu\xi \rangle_{H_1}, \quad \forall \varphi \in D(\mathcal{A}), \xi \in D(A). \tag{9}$$

Now, we consider the *adjoint* system of (4), that is, the following coupled system

$$\begin{cases} z_{1t} - \displaystyle\int_t^T k_1(s-t)Az_1(s)ds - \int_t^T L_2 z_2(s)ds = 0 \\[4mm] z_{2t} - \displaystyle\int_t^T k_2(s-t)A^2 z_2(s)ds - \int_t^T L_1 z_1(s)ds = 0 \end{cases} \quad \text{in } (0,T), \quad (10)$$

with given final data

$$z_1(T) = z_{1T}, \quad z_2(T) = z_{2T}. \tag{11}$$

We assume that for final data sufficiently regular an existence and regularity result for the solution of (10)–(11) holds. Precisely:

Theorem 1. *For any $z_{1T} \in D(A)$ and $z_{2T} \in D(A^2)$ there exists a unique solution (z_1, z_2) of (10)–(11)   such that $z_1 \in C^1([0,T], H) \cap C([0,T], D(A))$ and $z_2 \in C^1([0,T], H) \cap C([0,T], D(A^2))$.*

That type of result will be true in the applications, taking into account that backward problems are equivalent to forward problems by means of a change of the variable t into $t - T$.

If Theorem 1 holds true, then the regularity of the solution (z_1, z_2) of (10)–(11) and assumption (H1)-3 allow to obtain the following properties: the functions $D_\nu z_i$, $i = 1, 2$, belong to $C(0, T; H_1)$, because $D(A) \subset D(\mathcal{A}) \subset H_0$. So, we can consider the nonhomogeneous problem

$$
\begin{cases}
\phi_1'(t) + \displaystyle\int_0^t k_1(t-s)\mathcal{A}\phi_1(s)ds + \mathcal{L}_1(1 * \phi_2) = 0 \\[2mm]
\phi_2'(t) + \displaystyle\int_0^t k_2(t-s)\mathcal{A}^2\phi_2(s)ds + \mathcal{L}_2(1 * \phi_1) = 0 \\[2mm]
\hspace{6cm} \text{in } (0,T) \\[2mm]
\mathcal{B}\phi_1(t) = \displaystyle\int_t^T k_1(s-t)D_\nu z_1(s)ds, \\[2mm]
\mathcal{B}\phi_2(t) = 0, \quad \mathcal{B}\mathcal{A}\phi_2(t) = \displaystyle\int_t^T k_2(s-t)D_\nu z_2(s)ds \\[2mm]
\phi_1(0) = \phi_2(0) = 0.
\end{cases}
\tag{12}
$$

If (ϕ_1, ϕ_2) denotes the solution of problem (12), then we can introduce the following linear operator on $H \times H$:

$$
\Psi(z_{1T}, z_{2T}) = (\phi_1(T), \phi_2(T)), \qquad (z_{1T}, z_{2T}) \in D(A) \times D(A^2).
$$

We will prove the next result.

Theorem 2. *If (ξ_1, ξ_2) is the solution of the system*

$$
\begin{cases}
\xi_1'(t) - \displaystyle\int_t^T k_1(s-t)A\xi_1(s)ds - \displaystyle\int_t^T \mathcal{L}_2\xi_2(s)ds = 0, \\[2mm]
\hspace{6cm} \text{in } (0,T) \\[2mm]
\xi_2'(t) - \displaystyle\int_t^T k_2(s-t)A^2\xi_2(s)ds - \displaystyle\int_t^T \mathcal{L}_1\xi_1(s)ds = 0, \\[2mm]
\xi_1(T) = \xi_{1T}, \quad \xi_2(T) = \xi_{2T},
\end{cases}
$$

where $(\xi_{1T}, \xi_{2T}) \in D(A) \times D(A^2)$, then the identity

$$
\langle \Psi(z_{1T}, z_{2T}), (\xi_{1T}, \xi_{2T}) \rangle
$$

$$
= \int_0^T \langle \mathcal{B}\phi_1(t), \int_t^T k_1(s-t)D_\nu\xi_1(s)\, ds \rangle_{H_1} dt
$$

$$
+ \int_0^T \langle \mathcal{B}\mathcal{A}\phi_2(t), \int_t^T k_2(s-t)D_\nu\xi_2(s)\, ds \rangle_{H_1} dt,
\tag{13}
$$

holds true.

Proof. We multiply the first equation in (12) by $\xi_1(t)$ and integrate on $[0, T]$, so we have

$$\int_0^T \langle \phi_1', \xi_1 \rangle \, dt + \int_0^T \langle \int_0^t k_1(t-s)\mathcal{A}\phi_1(s) \, ds, \xi_1 \rangle \, dt$$

$$+ \int_0^T \langle \mathcal{L}_1(1 * \phi_2), \xi_1 \rangle \, dt = 0. \tag{14}$$

In the second term of the above identity we change the order of integration and, since $\xi_1(t) \in D(A)$, we can use (9) to get

$$\int_0^T \langle \int_0^t k_1(t-s)\mathcal{A}\phi_1(s) \, ds, \xi_1(t) \rangle dt = \int_0^T \int_s^T k_1(t-s)\langle \mathcal{A}\phi_1(s), \xi_1(t) \rangle \, dt \, ds$$

$$= \int_0^T \langle \phi_1(s), \int_s^T k_1(t-s) A\xi_1(t) \, dt \rangle \, ds$$

$$- \int_0^T \langle \mathcal{B}\phi_1(s), \int_s^T k_1(t-s)D_\nu\xi_1(t) \, dt \rangle_{H_1} \, ds.$$

Note that, in virtue of assumption (H1)-1, one has $D(A) \subset D(\mathcal{A}) \cap \mathrm{Ker}(\mathcal{B})$; so, changing again the order of integration and applying (8), we obtain

$$\int_0^T \langle \mathcal{L}_1(1 * \phi_2), \xi_1 \rangle \, dt = \int_0^T \langle \phi_2(s), \int_s^T \mathcal{L}_1\xi_1(t) \, dt \rangle \, ds.$$

If we integrate by parts the first term in (14) and take into account the previous two identities, then, in view also of $\phi_1(0) = 0$, we get

$$\langle \phi_1(T), \xi_1(T) \rangle - \int_0^T \langle \phi_1(t), \xi_1'(t) \rangle \, dt + \int_0^T \langle \phi_1(t), \int_t^T k_1(s-t)A\xi_1(s) \, ds \rangle \, dt$$

$$- \int_0^T \langle \mathcal{B}\phi_1(t), \int_t^T k_1(s-t)D_\nu\xi_1(s) \, ds \rangle_{H_1} \, dt$$

$$+ \int_0^T \langle \phi_2(t), \int_t^T \mathcal{L}_1\xi_1(s) \, ds \rangle \, dt = 0.$$

As a consequence of the former equation and

$$\xi_1'(t) - \int_t^T k_1(s-t)A\xi_1(s)ds = \int_t^T \mathcal{L}_2\xi_2(s)ds,$$

we obtain

$$\langle \phi_1(T), \xi_{1T} \rangle - \int_0^T \langle \mathcal{B}\phi_1(t), \int_t^T k_1(s-t)D_\nu\xi_1(s) \, ds \rangle_{H_1} \, dt$$

$$+ \int_0^T \langle \phi_2(t), \int_t^T \mathcal{L}_1\xi_1(s) \, ds \rangle \, dt - \int_0^T \langle \phi_1(t), \int_t^T \mathcal{L}_2\xi_2(s)ds \rangle \, dt = 0. \tag{15}$$

In a similar way, we multiply the second equation in (12) by $\xi_2(t)$ and integrate on $[0, T]$: if we integrate by parts the first term, take into account that

$\phi_2(0) = 0$ and change the order of integration in the other two terms, then we have

$$
\langle \phi_2(T), \xi_{2T} \rangle - \int_0^T \langle \phi_2(t), \xi_2'(t) \rangle \, dt
$$

$$
+ \int_0^T \int_s^T k_2(t - s) \langle \mathcal{A}^2 \phi_2(s), \xi_2(t) \rangle \, dt \, ds \tag{16}
$$

$$
+ \int_0^T \int_s^T \langle \mathcal{L}_2 \phi_1(s), \xi_2(t) \rangle \, dt \, ds = 0.
$$

Now, we observe that from (9) it follows for any $\varphi \in D(\mathcal{A}^2)$ and $\xi \in D(\mathcal{A}^2)$

$$
\langle \mathcal{A}^2 \varphi, \xi \rangle = \langle \varphi, \mathcal{A}^2 \xi \rangle - \langle \mathcal{B}\varphi, D_\nu \mathcal{A}\xi \rangle_{H_1} - \langle \mathcal{B}\mathcal{A}\varphi, D_\nu \xi \rangle_{H_1}.
$$

Putting the above equation into (16) and taking into account that the operator $\mathcal{L}_2$ is self-adjoint yield

$$
\langle \phi_2(T), \xi_{2T} \rangle - \int_0^T \langle \phi_2(t), \xi_2'(t) \rangle \, dt
$$

$$
+ \int_0^T \langle \phi_2(s), \int_s^T k_2(t - s) \mathcal{A}^2 \xi_2(t) \, dt \rangle \, ds
$$

$$
- \int_0^T \langle \mathcal{B}\mathcal{A}\phi_2(s), \int_s^T k_2(t - s) D_\nu \xi_2(t) \, dt \rangle_{H_1} \, ds
$$

$$
+ \int_0^T \langle \phi_1(s), \int_s^T \mathcal{L}_2 \xi_2(t) \, dt \rangle \, ds = 0.
$$

In virtue of

$$
\xi_2'(t) - \int_t^T k_2(s - t) \mathcal{A}^2 \xi_2(s) ds = \int_t^T \mathcal{L}_1 \xi_1(s) ds,
$$

we get

$$
\langle \phi_2(T), \xi_{2T} \rangle - \int_0^T \langle \mathcal{B}\mathcal{A}\phi_2(s), \int_s^T k_2(t - s) D_\nu \xi_2(t) \, dt \rangle_{H_1} \, ds
$$

$$
+ \int_0^T \langle \phi_1(t), \int_t^T \mathcal{L}_2 \xi_2(s) ds \rangle \, dt - \int_0^T \langle \phi_2(t), \int_t^T \mathcal{L}_1 \xi_1(s) ds \rangle \, dt = 0. \tag{17}
$$

If we sum Eqs. (15) and (17), then we have

$$
\langle \Psi(z_{1T}, z_{2T}), (\xi_{1T}, \xi_{2T}) \rangle = \langle \phi_1(T), \xi_{1T} \rangle + \langle \phi_2(T), \xi_{2T} \rangle
$$

$$
= \int_0^T \langle \mathcal{B}\phi_1(t), \int_t^T k_1(s - t) D_\nu \xi_1(s) \, ds \rangle_{H_1} \, dt \tag{18}
$$

$$
+ \int_0^T \langle \mathcal{B}\mathcal{A}\phi_2(t), \int_t^T k_2(s - t) D_\nu \xi_2(s) \, ds \rangle_{H_1} \, dt,
$$

that is, (13) holds true. $\qquad\qquad\qquad\qquad\qquad\qquad\qquad\qquad\qquad\square$

If we take $(\xi_{1T}, \xi_{2T}) = (z_{1T}, z_{2T})$ in (13), then we have

$$\langle \Psi(z_{1T}, z_{2T}), (z_{1T}, z_{2T}) \rangle$$
$$= \int_0^T \left(\left| \int_t^T k_1(s-t)D_\nu z_1(s) \, ds \right|_{H_1}^2 + \left| \int_t^T k_2(s-t)D_\nu z_2(s) \, ds \right|_{H_1}^2 \right) dt.$$

Consequently, we can introduce a semi-norm on the space $D(A) \times D(A^2)$. Precisely, if we consider, for any $(z_{1T}, z_{2T}) \in D(A) \times D(A^2)$, the solution (z_1, z_2) of the system (10)–(11), then we define

$$\|(z_{1T}, z_{2T})\|_F^2 :=$$
$$\int_0^T \left(\left| \int_t^T k_1(s-t)D_\nu z_1(s) \, ds \right|_{H_1}^2 + \left| \int_t^T k_2(s-t)D_\nu z_2(s) \, ds \right|_{H_1}^2 \right) dt. \tag{19}$$

We observe that $\| \cdot \|_F$ is a norm if and only if the following uniqueness theorem holds.

Theorem 3. *If (z_1, z_2) is the solution of problem (10)–(11) such that*

$$\int_t^T k_1(s-t)D_\nu z_1(s) \, ds = \int_t^T k_1(s-t)D_\nu z_2(s) \, ds = 0, \quad \text{on } [0,T],$$

then

$$z_1 = z_2 = 0 \quad \text{in } [0,T].$$

The validity of Theorem 3 is the starting point for the application of the Hilbert Uniqueness Method. Indeed, if we assume that Theorem 3 holds true, then we can define the Hilbert space F as the completion of $D(A) \times D(A^2)$ for the norm $\| \cdot \|_F$. Thanks to (13) and (19) we have

$$\langle \Psi(z_{1T}, z_{2T}), (\xi_{1T}, \xi_{2T}) \rangle = \langle (z_{1T}, z_{2T}), (\xi_{1T}, \xi_{2T}) \rangle_F$$
$$\forall (z_{1T}, z_{2T}), (\xi_{1T}, \xi_{2T}) \in D(A) \times D(A^2), \tag{20}$$

where $\langle \cdot, \cdot \rangle_F$ denotes the scalar product associated with the norm $\| \cdot \|_F$.
Consequently,

$$\left| \langle \Psi(z_{1T}, z_{2T}), (\xi_{1T}, \xi_{2T}) \rangle \right| \leq \|(z_{1T}, z_{2T})\|_F \, \|(\xi_{1T}, \xi_{2T})\|_F$$
$$\forall (z_{1T}, z_{2T}), (\xi_{1T}, \xi_{2T}) \in D(A) \times D(A^2).$$

Thanks to the above inequality, the operator Ψ can be extended uniquely to a linear continuous operator, denoted again by Ψ, from F into its dual space F'. By (20) it follows that

$$\langle \Psi(z_{1T}, z_{2T}), (\xi_{1T}, \xi_{2T}) \rangle = \langle (z_{1T}, z_{2T}), (\xi_{1T}, \xi_{2T}) \rangle_F$$
$$\forall (z_{1T}, z_{2T}), (\xi_{1T}, \xi_{2T}) \in F,$$

and, as a consequence, we have that the operator $\Psi : F \to F'$ is an isomorphism.

Moreover, the key point to characterize the space F is to establish observability estimates of the following type

$$\int_0^T \left(\left| \int_t^T k_1(s-t)D_\nu z_1(s)\, ds \right|_{H_1}^2 + \left| \int_t^T k_2(s-t)D_\nu z_2(s)\, ds \right|_{H_1}^2 \right) dt \tag{21}$$

$$\asymp \|z_{1T}\|_{F_1}^2 + \|z_{2T}\|_{F_2}^2$$

for suitable spaces F_1, F_2. In that case, the uniqueness result stated by Theorem 3 holds true, so the operator $\Psi : F \to F'$ is an isomorphism, and in virtue of (19) and (21) we get

$$F = F_1 \times F_2$$

with the equivalence of the respective norms. Finally, we are able to solve the reachability problem (4)–(7) for $(u_{10}, u_{20}) \in F_1' \times F_2'$.

3   Applications

Example 1. Let $H = L^2(0, \pi)$ be endowed with the usual scalar product and norm. In [9] we take $\mathcal{A} = \frac{d^2}{dx^2}$ with null Dirichlet boundary conditions, $k_1(t) = \frac{\beta}{\eta}e^{-\eta t} + 1 - \frac{\beta}{\eta}$, $k_2 \equiv 1$. We examine the case in which $\mathcal{L}_i = a_i I$, with $a_i \in \mathbb{R}$, $i = 1, 2$ and I the identity operator on H.

By writing the solutions as Fourier series, we are able to prove Theorems 1 and 3, thanks also to some properties of the solutions of integral equations. In particular, by showing suitable Ingham type estimates, we prove observability estimates of the type (21) where $F = H_0^1(0, \pi) \times H_0^1(0, \pi)$. Therefore, we can deduce reachability results by means of the Hilbert Uniqueness Method.

Example 2. We consider $H = L^2(0, \pi)$ endowed with the usual scalar product and norm. In [9] we take $\mathcal{A} = \frac{d^2}{dx^2}$ with null Dirichlet boundary conditions, $k_1(t) = \frac{\beta}{\eta}e^{-\eta t} + 1 - \frac{\beta}{\eta}$, $k_2 \equiv 1$, $\mathcal{L}_1 = a_1 \frac{d^2}{dx^2}$ and $\mathcal{L}_2 = a_2 I$ with $a_i \in \mathbb{R}$, $i = 1, 2$.

Example 3. Let $H = L^2(\Omega)$ be endowed with the usual scalar product and norm. In [10] we take $\mathcal{A} = \Delta$ with null Dirichlet boundary conditions, $k_1(t) = \frac{\beta}{\eta}e^{-\eta t} + 1 - \frac{\beta}{\eta}$, $k_2 \equiv 1$, $\mathcal{L}_1 = a_1 \Delta$ and $\mathcal{L}_2 = a_2 I$ with $a_i \in \mathbb{R}$, $i = 1, 2$.

References

1. Cattaneo, C.: Sulla conduzione del calore. Atti Sem. Mat. Fis. Univ. Modena **3**, 83–101 (1949)
2. Fu, X., Yong, J., Zhang, X.: Controllability and observability of a heat equation with hyperbolic memory kernel. J. Differ. Equ. **247**, 2395–2439 (2009)
3. Guerrero, S., Imanuvilov, O.Y.: Remarks on non controllability of the heat equation with memory. ESAIM Control Optim. Calc. Var. **19**, 288–300 (2013)

4. Ivanov, S., Pandolfi, L.: Heat equation with memory: lack of controllability to rest. J. Math. Anal. Appl. **355**, 1–11 (2009)
5. Komornik, V.: Exact Controllability and Stabilization: The Multiplier Method. RAM: Research in Applied Mathematics. Masson/Wiley, Paris/Chichester (1994)
6. Lasiecka, I., Triggiani, R.: Exact controllability of the wave equation with Neumann boundary control. Appl. Math. Optim. **19**, 243–290 (1989)
7. Lions, J.-L.: Contrôlabilité Exacte, Perturbations et Stabilisation de Systèmes Distribués. Tome 1. Contrôlabilité Exacte, with appendices by E. Zuazua, C. Bardos, G. Lebeau and J. Rauch. Rech. Math. Appl., vol. 8. Masson, Paris (1988)
8. Lions, J.-L.: Contrôlabilité Exacte, Perturbations et Stabilisation de Systèmes Distribués. Tome 2. Perturbations. Rech. Math. Appl., vol. 9. Masson, Paris (1988)
9. Loreti, P., Sforza, D.: Control problems for weakly coupled systems with memory. J. Differ. Equ. **257**, 1879–1938 (2014)
10. Loreti, P., Sforza, D.: Observability of N-dimensional integro-differential systems. Discret. Contin. Dyn. Syst. Ser. S **9**, 745–757 (2016)
11. Triggiani, R.: Exact boundary controllability on $L_2(\Omega) \times H^{-1}(\Omega)$ of the wave equation with Dirichlet boundary control acting on a portion of the boundary $\partial \Omega$, and related problems. Appl. Math. Optim. **18**, 241–277 (1988)

Game Control Problem for Systems of Distributed Equations

Vyacheslav Maksimov[1,2(✉)]

[1] Moscow State University, Moscow, Russia
maksimov@imm.uran.ru
[2] Institute of Mathematics and Mechanics of UB RAS, Yekaterinburg, Russia

Abstract. We consider a game problem of guaranteed positional control for a distributed system described by the phase field equations under incomplete information on system's phase states. This problem is investigated from the viewpoint of the first player (the partner). For this player, a procedure for forming feedback controls is specified. This procedure is stable with respect to informational noises and computational errors and is based on the method of extremal shift and the method of stable sets from the theory of guaranteed positional control. It uses the idea of stable dynamical inversion of controlled systems.

Keywords: Guaranteed control · Phase field equations

1 Introduction

The control theory for distributed systems has been intensively developed in recent time as a part of mathematical control theory. At present, there exists a number of monographs devoted to control problems for distributed systems [1–3]. As a rule, the emphasis is on program control problems in the case when all system's parameters are precisely specified. Along with this, the investigation of control problems for systems with uncontrollable disturbances (the problems of game control) is also reasonable. Similar problems have been poorly investigated. In the early 70es, N.N. Krasovskii suggested an effective approach to solving game (guaranteed) control problems, which is based on the formalism of positional strategies. The detailed description of this approach for dynamical systems described by ordinary differential equations is given in [4]. The goal of the present work is to illustrate possibilities of this approach for investigating a game problem for systems described by the phase field equations.

We consider a system modeling the solidification process and governed by the phase field equations (introduced in [5])

$$
\begin{aligned}
\frac{\partial}{\partial t}\psi + l\frac{\partial}{\partial t}\varphi &= k\Delta_L\psi + Bu - Cv \quad \text{in} \quad Q = \Omega \times (t_0, \vartheta], \\
\tau\frac{\partial}{\partial t}\varphi &= \xi^2\Delta_L\varphi + g(\varphi) + \psi, \quad \vartheta = \text{const} < +\infty,
\end{aligned}
\tag{1}
$$

V. Maksimov—This work was supported by the Russian Science Foundation (project 14-01-00539).

Published by Springer International Publishing AG 2016. All Rights Reserved
L. Bociu et al. (Eds.): CSMO 2015, IFIP AICT 494, pp. 360–369, 2016.
DOI: 10.1007/978-3-319-55795-3_34

with the boundary condition $\frac{\partial}{\partial n}\psi = \frac{\partial}{\partial n}\varphi = 0$ on $\partial\Omega \times (t_0, \vartheta]$ and the initial condition $\psi(t_0) = \psi_0$, $\varphi(t_0) = \varphi_0$ in Ω. Here, $\Omega \subset \mathbb{R}^n$ is a bounded domain with the sufficiently smooth boundary $\partial\Omega$, Δ_L is the Laplace operator, $\partial/\partial n$ is the outward normal derivative, $(U, |\cdot|_U)$ and $(V, |\cdot|_V)$ are Banach spaces, $B \in \mathcal{L}(U; H)$ and $C \in \mathcal{L}(V; H)$ are linear continuous operators, $H = L_2(\Omega)$, and the function $g(z)$ is the derivative of a so-called potential $G(z)$. We assume that $g(z) = az + bz^2 - cz^3$.

Systems of form (1) have been investigated by many authors. In what follows, for the sake of simplicity, we assume that $k = \xi = \tau = c = 1$. Further, we assume that the following conditions are fulfilled: (A1) the domain $\Omega \subset \mathbb{R}^n$, $n = 2, 3$, has the boundary of C^2-class; (A2) the coefficients a and b are elements of the space $L_\infty(T \times \Omega)$, $T = [t_0, \vartheta]$, and vrai sup $c(t, \eta) > 0$ for $(t, \eta) \in [t_0, \vartheta] \times \Omega$; (A3) the initial functions ψ_0 and φ_0 are such that $\{\psi_0, \varphi_o\} \in \mathcal{R} = \{\psi, \varphi \in W_\infty^2(\Omega) : \frac{\partial}{\partial n}\psi = \frac{\partial}{\partial n}\varphi = 0$ on $\partial\Omega\}$.

Introduce the notation: $W_p^{2,1}(Q) = \left\{u \mid u, \frac{\partial u}{\partial \eta_i}, \frac{\partial^2 u}{\partial \eta_i \partial \eta_j}, \frac{\partial u}{\partial t} \in L^p(Q)\right\}$ for $p \in [1, \infty)$ is the standard Sobolev space with the norm $\|u\|_{W_p^{2,1}(Q)}$; $(\cdot, \cdot)_H$ and $|\cdot|_H$ are the scalar product and the norm in H, respectively. Let some initial state $x_0 = \{\psi_0, \varphi_0\}$ and functions $u(\cdot) \in L_\infty(T; U)$ and $v(\cdot) \in L_\infty(T; V)$ be fixed. A solution of system (1), $x(\cdot; t_0, x_0, u(\cdot), v(\cdot)) = \{\psi(\cdot; t_0, \psi_0, u(\cdot), v(\cdot)), \varphi(\cdot; t_0, \varphi_0, u(\cdot), v(\cdot))\}$, is a unique function $x(\cdot) = x(\cdot; t_0, x_0, u(\cdot), v(\cdot)) \in V_T^{(1)} = V_1 \times V_1$, $V_1 = W_2^{2,1}(Q)$ satisfying relations (1). By virtue of the corresponding embedding theorem, without loss of generality, one can assume that the space $V_T^{(1)}$ is embedded into the space $C(T; H \times H)$. Therefore, the element $x(t) = \{\psi(t), \varphi(t)\} \in H \times H$ is the phase state of system (1) at the time t. The following theorem takes place

Theorem 1. [6, p. 25, Assertion 5] *Let conditions (A1)–(A3) be fulfilled. Then for any $u(\cdot) \in L_\infty(T; U)$ and $v(\cdot) \in L_\infty(T; V)$ there exists a unique solution of system (1).*

The paper is devoted to the investigation of the game control problem. Let us give the informal formulation of this problem. Let a uniform net $\Delta = \{\tau_i\}_{i=0}^m$, $\tau_i = \tau_{i-1} + \delta$, $\tau_0 = t_0$, $\tau_m = \vartheta$ with a diameter $\delta = \tau_i - \tau_{i-1}$ be fixed on a given time interval T. Let a solution of system (1) be unknown. At the times $\tau_i \in \Delta$, a part of the phase states $x(\tau_i)$ (namely $\phi(\tau_i)$) is inaccurately measured. The results of measurements $\xi_i^h \in H$, $i \in [1 : m-1]$, satisfy the inequalities

$$|\xi_i^h - \phi(\tau_i)|_H \leq h. \tag{2}$$

Here, $h \in (0, 1)$ is a level of informational noise. Let the following quality criterion be given: $I(x(\cdot), u(\cdot)) = \sigma(x(\vartheta)) + \int_{t_0}^{\vartheta} \chi(t, x(t), u(t))\, dt$, where $\sigma : H \times H \to \mathbb{R}$ and $\chi : T \times H \times H \times U \to \mathbb{R}$ are given functions satisfying the local Lipschitz conditions. Let also a prescribed value of the criterion, number I_*, be fixed. The control problem under consideration consists in the following. There are two players-antagonists controlling system (1) by means of u and v, respectively.

One of them is called a partner; another, an opponent. Let $P \subset U$ and $E \subset V$ be given convex bounded and closed sets. The problem undertaken by the partner is as follows. It is necessary to construct a law (a strategy) for forming the control u (with values from P) by the feedback principle (on the base of measurements of $\varphi(\tau_i)$) in order to provide the prescribed value of the quality criterion for any (unknown) realization $v = v(\cdot)$.

To form the control u providing the solution of the problem, along with the information on the "part" of coordinates of the solution of system (1) (namely, on the values ξ_i^h satisfying inequalities (2)), it is necessary to obtain some additional information on the coordinate $\psi(\cdot)$, which is missing. To get such a piece of information during the control process, it is reasonable, following the approach developed in [7–9], to introduce an auxiliary controlled system. This system is described by a parabolic equation (the form is specified below). The equation has an output $w_*(t)$, $t \in T$, and an input $p^h(t)$, $t \in T$. The input $p^h(\cdot)$ is some new auxiliary control; it should be formed by the feedback principle in such a way that $p^h(\cdot)$ "approximates" the unknown coordinate $\psi(\cdot)$ in the mean square metric. Thus, along with the block of forming the control in the real system (it is called an controller), we need to incorporate into the control contour one more block (it is called an identifier) allowing to reconstruct the missing coordinate $\psi(\cdot)$ in real time. Note that, in essence, the identifier block solves a dynamical inverse problem, namely, the problem of (approximate) reconstruction of the unknown coordinate $\psi(\cdot)$. In he recent time, the theory of inverse problems for distributed systems has been intensively developed. Among the latest investigations, it is possible to mark out the research [10].

2   Problem statement

Before passing to the problem formulation, we give some definitions. Furthermore, we denote by $u_{a,b}(\cdot)$ the function $u(t)$, $t \in [a, b]$, considered as a whole. The symbol $P_{a,b}(\cdot)$ stands for the restriction of the set $P_T(\cdot)$ onto the segment $[a, b] \subset T$. Any strongly measurable functions $u(\cdot) : T \to P$ and $v(\cdot) : T \to E$ are called program controls of the partner and opponent, respectively. The sets of all program controls of the partner and opponent are denoted by the symbols $P_T(\cdot)$ and $E_T(\cdot) : P_T(\cdot) = \{u(\cdot) \in L_2(T;U) : u(t) \in P \text{ a.e. } t \in T\}$, $E_T(\cdot) = \{v(\cdot) \in L_2(T;V) : v(t) \in E \text{ a.e. } t \in T\}$. Elements of the product $T \times \mathcal{H}$ are called positions, $\mathcal{H} = H \times H \times \mathbb{R} \times H \times H \times \mathbb{R}$. Any function (perhaps, multifunction) $\mathcal{U} : T \times \mathcal{H} \to P$ is said to be a positional strategy of the partner. The positional strategy corrects the controls at discrete times given by some partition of the interval T. Any function $\mathcal{V} : T \times H \times H \to H$ is said to be a strategy of reconstruction. The strategy $\mathcal{V}$ is formed in order to reconstruct the unknown component $\psi(\cdot)$.

Consider the following ordinary differential equation

$$\dot{q}(t) = \chi(t, x(t), u(t)), \quad q(t_0) = 0. \tag{3}$$

Introducing this new variable q, we reduce the control problem of Bolza type to a control problem with a terminal quality criterion of the form $I = \sigma(x(\vartheta)) + q(\vartheta)$.

In this case, the controlled system consists of phase field Eq. (1) and ordinary differential Eq. (3).

The scheme of an algorithm for solving the problem undertaken by the partner is as follows. In the beginning, auxiliary systems M_1 and M_2 (models) are introduced. The system M_1 has an input $u^*(\cdot)$ and an output $w(\cdot)$; the system M_2, an input $p^h(\cdot)$ and an output $w_*(\cdot)$, respectively. The model M_2 with its control law $\mathcal{V}$ forms the identifier, whereas the model M_1 and system (1) (with their control laws) form the controller. The process of synchronous feedback control of systems (1), (3), M_1, and M_2 is organized on the interval T. This process is decomposed into $m - 1$ identical steps. At the ith step carried out during the time interval $[\tau_i, \tau_{i+1})$, the following actions are fulfilled. First, at the time τ_i, according to some chosen rules $\mathcal{V}$ and $\mathcal{U}$, the elements $p_i^h \in \mathcal{V}(\tau_i, \xi_i^h, w_*(\tau_i))$, $u_i^h \in \mathcal{U}(\tau_i, \xi_i^h, p_i^h, \psi_i^h, w(\tau_i))$ are calculated. Here, ψ_i^h is the result of measuring $q(\tau_i)$. Then (till the moment τ_{i+1}), the control $p^h(t) = p_i^h$, $\tau_i \le t < \tau_{i+1}$, is fed onto the input of the system M_2; the control $u^h(t) = u_i^h$, $\tau_i \le t < \tau_{i+1}$, onto the input of system (1), (3). Under the action of these controls, as well as of the given control $u^*(t)$, $\tau_i \le t < \tau_{i+1}$, and the unknown control of the opponent $v(t)$, $\tau_i \le t < \tau_{i+1}$, the states $x(\tau_{i+1})$, $q(\tau_i)$, $w(\tau_{i+1})$, and $w_*(\tau_{i+1})$ are realized at the time τ_{i+1}. The procedure stops at the time ϑ.

Let models M_1 and M_2 with phase trajectories $w(\cdot)$ and $w_*(\cdot)$ be fixed. A solution $x(\cdot)$ of system (1) starting from an initial state (t_*, x_*) and corresponding to piecewise constant controls $u^h(\cdot)$ and $p^h(\cdot)$ (formed by the feedback principle) and to a control $v_{t_*,\vartheta}(\cdot) \in E_{t_*,\vartheta}(\cdot)$ is called an (h, Δ)-motion $x_{\Delta,w}^h(\cdot) = x_{\Delta,w}^h(\cdot; t_*, x_*, \mathcal{U}, \mathcal{V}, v_{t_*,\vartheta}(\cdot))$. This motion is generated by the positional strategies $\mathcal{U}$ and $\mathcal{V}$. Thus, the motions $x_{\Delta,w}^h(\cdot)$, $q_\Delta^h(\cdot), w(\cdot)$, and $w_*(\cdot)$ are formed simultaneously. So, for $t \in [\tau_i, \tau_{i+1})$, we define $x_{\Delta,w}^h(t) = x(t; \tau_i, x_{\Delta,w}^h(\tau_i), u_{\tau_i,\tau_{i+1}}^h(\cdot), v_{\tau_i,\tau_{i+1}}(\cdot))$, $q_\Delta^h(t) = q(t; \tau_i, q_\Delta^h(\tau_i), u_{\tau_i,\tau_{i+1}}^h(\cdot))$, $w(t) = w(t; \tau_i, w(\tau_i), u_{\tau_i,\tau_{i+1}}^*(\cdot)), w_*(t) = w_*(t; \tau_i, w_*(\tau_i), p_{\tau_i,\tau_{i+1}}^h(\cdot))$, where

$$u^h(t) = u_i^h \in \mathcal{U}(\tau_i, \xi_i^h, p_i^h, q^h(\tau_i), w(\tau_i)), \quad p^h(t) = p_i^h \in \mathcal{V}(\tau_i, \xi_i^h, w_*(\tau_i)) \qquad (4)$$

$$\text{for } t \in [\tau_i, \tau_{i+1}),$$

$$i \in [i(t_*) : m - 1], \quad |\xi_i^h - \phi_{\Delta,w}^h(\tau_i)|_H \le h, \quad |\psi_i^h - g_\Delta^h(\tau_i)| \le h, \qquad (5)$$

$$i(t_*) = \min\{i : \tau_i > t_*\}, \quad u^h(t) = u_*^h \in P, \quad p^h(t) = p_*^h \in H \text{ for } t \in [t_*, \tau_{i(t_*)}).$$

The set of all (h, Δ)-motions is denoted by $X_h(t_*, x_*, \mathcal{U}, \mathcal{V}, \Delta, w)$.

Problem. It is necessary to construct a positional strategy $\mathcal{U} : T \times \mathcal{H} \to P$ of the partner and a positional strategy $\mathcal{V} : T \times H \times H \to H$ of reconstruction with the following properties: whatever a value $\varepsilon > 0$ and a disturbance $v_T(\cdot) \in E_T(\cdot)$, one can find (explicitly) numbers $h_* > 0$ and $\delta_* > 0$ such that the inequalities

$$|I(x_{\Delta,w}^h(\cdot), u_T^h(\cdot)) - I_*| \le \varepsilon \quad \forall x_{\Delta,w}^h(\cdot) \in X_h(t_0, x_0, \mathcal{U}, \mathcal{V}, \Delta, w) \qquad (6)$$

are fulfilled uniformly with respect to all measurements ξ_i^h and ψ_i^h with properties (5), if $h \le h_*$ and $\delta = \delta(\Delta) \le \delta_*$.

3   Algorithm for Solving the Problem

To solve the Problem, we use ideas from [4], namely, the method of a priori stable sets. In our case, this method consists in the following. Let a trajectory of model M_1, $w(\cdot)$, possessing the property $\sigma(w_1(\vartheta)) + w_2(\vartheta) = I_*$ be known. Then, a feedback strategy $\mathcal{U}$ providing tracking the prescribed trajectory of M_1 by the trajectory of real system (1) is constructed. This means that the (h, Δ)-motion $x_\Delta^h(\cdot)$ formed by the feedback principle (see (4)) remains at a "small" neighborhood of the trajectory $w(\cdot)$ during the whole interval T. This property of the (h, Δ)-motion allows us to conclude that the chosen strategy solves the considered control problem.

Let us pass to the realization of this scheme. Define $\Phi(t, x, u, v) = \{Bu - Cv, \chi(t, x, u)\} \in H \times \mathbb{R}$, $\Phi_u(t, x, v) = \bigcup_{u \in P} \Phi(t, x, u, v)$, $H_*(t; x) = \bigcap_{v \in E} \Phi_u(t, x, v)$, $H_*(\cdot; x) = \{u(\cdot) \in L_2(T; H \times \mathbb{R}) : u(t) \in H_*(t; x)$ for a. a. $t \in T\}$. As a model M_1, we take the system including two subsystems, i.e., the phase field equation

$$\frac{\partial}{\partial t} w^{(1)} + l \frac{\partial}{\partial t} w^{(2)} = \Delta_L w^{(1)} + u_1 \quad \text{in} \quad \Omega \times (t_0, \vartheta], \tag{7}$$

$$\frac{\partial}{\partial t} w^{(2)} = \Delta_L w^{(2)} + g(w^{(2)}) + w^{(1)}$$

with the boundary condition $\frac{\partial}{\partial n} w^{(1)} = \frac{\partial}{\partial n} w^{(2)} = 0$ on $\partial\Omega \times (t_0, \vartheta]$ and the initial condition $w^{(1)}(t_0) = \psi_0$, $w^{(2)}(t_0) = \varphi_0$ in Ω, as well as the ordinary differential equation

$$\dot{w}^{(3)}(t) = u_2(t), \quad w^{(3)} \in \mathbb{R}, \quad w^{(3)}(0) = 0. \tag{8}$$

By the symbol $w(\cdot)$, we denote the solution of system (7), (8). Then, the model M_1 has the control $u(\cdot) = \{u_1(\cdot), u_2(\cdot)\}$. As a model M_2, we use the equation

$$\frac{\partial}{w_*} \partial t = \Delta_L w_* + p^h + g(w_*) \quad \text{in} \quad \Omega \times (t_0, \vartheta] \tag{9}$$

with the boundary condition $\frac{\partial w_*}{\partial n} = 0$ on $\partial\Omega \times (t_0, \vartheta]$ and the initial condition $w_*(t_0) = \varphi_0$ in Ω.

Condition 1. There exists a control $u(\cdot) = u^*(\cdot) = \{u_1^*(\cdot), u_2^*(\cdot)\} \in H(t; w^{(1)}(t), w^{(2)}(t))$ for a.a. $t \in T$ such that $I_* = \sigma(w^{(1)}(\vartheta)) + w^{(2)}(\vartheta)$.

The strategies $\mathcal{U}$ and $\mathcal{V}$ (see (4)) are defined in such a way:

$$\mathcal{U}(t, \xi, p, \psi, w) = \arg\min\{L(u, y) + (\psi - w^{(3)})\chi(t, \xi, p, u) : u \in P\}, \tag{10}$$

$$\mathcal{V}(t, \xi, w_*) = \arg\min\{l(t, \alpha, u, s) : u \in U_d\}, \tag{11}$$

where $w = \{w^{(1)}, w^{(2)}, w^{(3)}\}$, $L(u, y) = (-y, Bu)_H$, $y = w^{(1)} - p + l(w^{(2)} - \xi)$, $l(t, \alpha, u, s) = \exp(-2b_* t)(s, u)_H + \alpha |u|_H^2$, $s = w_* - \xi$, $b_* = |a + 1/3b|_{L_\infty(Q)}$, $\alpha = \alpha(h)$, $U_d = \{p(\cdot) \in L_2(T; H) : |p(t)|_H \leq d$ for a.a. $t \in T\}$, $d \geq \sup_{t \in T}\{|\psi(t; t_0, x_0, u(\cdot), v(\cdot))|_H : u(\cdot) \in P_T(\cdot), v(\cdot) \in E_T(\cdot)\}$.

Condition 2. Let $h, \delta(h)$, and $\alpha(h)$ satisfy the conditions: $\alpha(h) \to 0, \delta(h) \to 0, (h + \delta(h))\alpha^{-1}(h) \to 0$ as $h \to 0$.

Let us pass to the description of the algorithm for solving the Problem. Namely, we describe the procedure of forming the (h, Δ)-motion $x^h_{\Delta,w}(\cdot) = \{\psi^h_{\Delta,w}(\cdot), \varphi^h_{\Delta,w}(\cdot)\}$ and trajectory $g^h_\Delta(\cdot)$ corresponding to some fixed partition Δ and the strategies $\mathcal{U}$ and $\mathcal{V}$, see (10) and (11). Before the algorithm starts, we fix a value $h \in (0,1)$, a function $\alpha = \alpha(h)\colon (0,1) \to (0,1)$, and a partition $\Delta_h = \{\tau_{h,i}\}_{i=0}^{m_h}$ with diameter $\delta = \delta(h) = \tau_{i+1} - \tau_i, \tau_i = \tau_{h,i}, \tau_{h,0} = t_0, \tau_{h,m_h} = \vartheta$. The work of the algorithm is decomposed into $m_h - 1$ identical steps. We assume that

$$u^h(t) = u^h_0 \in \mathcal{U}(t_0, \xi^h_0, p^h_0, 0, w(t_0)), \quad p^h(t) = p^h_0 \in \mathcal{V}(t_0, \xi^h_0, \varphi_0)$$

$(|\xi^h_0 - \varphi_0|_H \leq h)$ on the interval $[t_0, \tau_1)$. Under the action of these piecewise-constant controls as well as of an unknown disturbance $v_{t_0,\tau_1}(\cdot)$, the (h, Δ)-motion $\{x^h_{\Delta,w}(\cdot)\}_{t_0,\tau_1} = \{\psi^h_{\Delta,w}(\cdot; t_0, \psi_0, u^h_{t_0,\tau_1}(\cdot), v_{t_0,\tau_1}(\cdot)), \varphi^h_{\Delta,w}(\cdot, t_0, \varphi_0, u^h_{t_0,\tau_1}(\cdot), v_{t_0,\tau_t}(\cdot))\}_{t_0,\tau_1}$ of system (1), the trajectory $\{q^h_\Delta(\cdot)\}_{t_0,\tau} = \{q(\cdot; t_0, q(t_0)), u^h_{t_0,\tau_1}(\cdot))\}_{t_0,\tau_1}$ of Eq. (3), the trajectory $\{w_*(\cdot)\}_{t_0,\tau_1} = \{w_*(\cdot; t_0, \phi_0, p^h_{t_0,\tau_1}(\cdot))\}_{t_0,\tau_1}$ of the model M_2, and the trajectory $\{w(\cdot)\}_{t_0,\tau_1} = \{w(\cdot; t_0, x_0, u^*_{t_0,\tau_1}(\cdot))\}_{t_0,\tau_1}$ of the model M_1 are realized. At the time $t = \tau_1$, we determine u^h_1 and p^h_1 from the conditions

$$u^h_1 \in \mathcal{U}(\tau_1, \xi^h_1, p^h_1, \psi^h_1, w(\tau_1)), \quad p^h_1 \in \mathcal{V}(\tau_1, \xi^h_1, w_*(\tau_1))$$

$(|\xi^h_1 - \varphi^h_{\Delta,w}(\tau_1)|_H \leq h, |\psi^h_1 - g^h_\Delta(\tau_1)| \leq h)$, i.e., we assume that $u^h(t) = u^h_1$ and $p^h(t) = p^h_1$ for $t \in [\tau_1, \tau_2)$. Then, we calculate the realization of the (h, Δ)-motion

$$\{x^h_{\Delta,w}(\cdot)\}_{\tau_1,\tau_2} = \{\psi^h_{\Delta,w}(\cdot; \tau_1, \psi^h_{\Delta,w}(\tau_1), u^h_{\tau_1,\tau_2}(\cdot), v_{\tau_1,\tau_2}(\cdot)), \\ \varphi^h_{\Delta,w}(\cdot; \tau_1, \varphi^h_{\Delta,w}(\tau_1), u^h_{\tau_1,\tau_2}(\cdot), v_{\tau_1,\tau_2}(\cdot))\}_{\tau_1,\tau_2},$$

the trajectory $\{q^h_\Delta(\cdot)\}_{\tau_1,\tau_2} = \{q(\cdot; \tau_1, q^h_\Delta(\tau_1), p^h_{\tau_1,\tau_2}(\cdot))\}_{\tau_1,\tau_2}$ of Eq. (3), the trajectory $\{w_*(\cdot)\}_{\tau_1,\tau_2} = \{w_*(\cdot; \tau_1, w_*(\tau_1), p^h_{\tau_1,\tau_2}(\cdot))\}_{\tau_1,\tau_2}$ of the model M_2, and the trajectory $\{w(\cdot)\}_{\tau_1,\tau_2} = \{w(\cdot; \tau_1, w(\tau_1), u^*_{\tau_1,\tau_2}(\cdot))\}_{\tau_1,\tau_2}$ of the model M_1.

Let the (h, Δ)-motion $x^h_{\Delta,w}(\cdot)$, the trajectory $q^h_\Delta(\cdot)$ of Eq. (3), the trajectory $w_*(\cdot)$ of the model M_2, and the trajectory $w(\cdot)$ of the model M_1 be defined on the interval $[t_0, \tau_i]$. At the time $t = \tau_i$, we assume that

$$u^h_i \in \mathcal{U}(\tau_i, \xi^h_i, p^h_i, \psi^h_i, w(\tau_i)), \quad p^h_i \in \mathcal{V}(\tau_i, \xi^h_i, w_*(\tau_i)) \qquad (12)$$

$(|\xi^h_i - \varphi^h_{\Delta,w}(\tau_i)|_H \leq h, |\psi^h_i - g^h_\Delta(\tau_i)| \leq h)$, i.e., we set $u^h(t) = u^h_i$ and $p^h(t) = p^h_i$ for $t \in [\tau_i, \tau_{i+1})$. As a result of the action of these controls and an unknown disturbance $v_{\tau_i,\tau_{i+1}}(\cdot)$, the (h, Δ)-motion

$$\{x^h_{\Delta,w}(\cdot)\}_{\tau_i,\tau_{i+1}} = \{\psi^h_{\Delta,w}(\cdot; \tau_i, \psi^h_{\Delta,w}(\tau_i), u^h_{\tau_i,\tau_{i+1}}(\cdot), v_{\tau_i,\tau_{i+1}}(\cdot)), \\ \varphi(\cdot; \tau_i, \varphi^h_{\Delta,w}(\tau_i), u^h_{\tau_i,\tau_{i+1}}(\cdot), v_{\tau_i,\tau_{i+1}}(\cdot))\}_{\tau_i,\tau_{i+1}},$$

the trajectory $\{q_\Delta^h(\cdot)\}_{\tau_i,\tau_{i+1}} = \{q(\cdot;\tau_i, q_\Delta^h(\tau_i), p_{\tau_i,\tau_{i+1}}^h(\cdot)\}_{\tau_i,\tau_{i+1}}$ of Eq. (3), the trajectory $\{w_*(\cdot)\}_{\tau_i,\tau_{i+1}} = \{w_*(\cdot;\tau_i, w_*(\tau_i), p_{\tau_i,\tau_{i+1}}^h(\cdot))\}_{\tau_i,\tau_{i+1}}$ of the model M_2, and the trajectory $\{w(\cdot)\}_{\tau_i,\tau_{i+1}} = \{w(\cdot;\tau_i, w(\tau_i), u_{\tau_i,\tau_{i+1}}^*(\cdot))\}_{\tau_i,\tau_{i+1}}$ of the model M_1 are realized on the interval $[\tau_i, \tau_{i+1}]$. This procedure stops at the time ϑ.

Theorem 2. *Let Conditions 1 and 2 be fulfilled. Let also the models M_1 and M_2 be specified by relations (7), (8), and (9), respectively. Then, the strategies $\mathcal{U}$ and $\mathcal{V}$ of form (10) and (11) solve the Problem.*

Proof. To prove the theorem, we estimate the variation of the functional

$$\Lambda(t, x_{\Delta,w}^h(\cdot), q_\Delta^h(\cdot), w(\cdot)) = \Lambda^0(x_{\Delta,w}^h(t), q_\Delta^h(t), w(t))$$

$$+ 0.5 \int_0^t \left\{ \int_\Omega |\nabla \pi^h(\varrho, \eta)|^2 \, d\eta + l^2 \int_\Omega |\nabla \nu^h(\varrho, \eta)|^2 \, d\eta \right\} d\varrho,$$

where $x_{\Delta,w}^h(\cdot) = \{\psi_{\Delta,w}^h(\cdot), \phi_{\Delta,w}^h(\cdot)\}$, $w(\cdot) = \{w^{(1)}(\cdot), w^{(2)}(\cdot), w^{(3)}(\cdot)\}$, $\pi^h(t) = w^{(1)}(t) - \psi_{\Delta,w}^h(t)$, $\nu^h(t) = w^{(2)}(t) - \varphi_{\Delta,w}^h(t)$, $g^h(t) = \pi^h(t) + l\nu^h(t)$, $\Lambda_1(x_{\Delta,w}^h(t), w(t)) = 0.5|g^h(t)|_H^2 + 0.5l^2|\nu^h(t)|_H^2$, $\lambda(q_\Delta^h(t), w^{(3)}(t)) = 0.5|q_\Delta^h(t) - w^{(3)}(t)|^2$, $\Lambda^0(x_{\Delta,w}^h(t), q_\Delta^h(t), w(t)) = \Lambda_1(x_{\Delta,w}^h(t), w(t)) + \lambda(q_\Delta^h(t), w^{(3)}(t))$. It is easily seen that the functions $\pi^h(\cdot)$ and $\nu^h(\cdot)$ are solutions of the system

$$\frac{\partial \pi^h(t,\eta)}{\partial t} + l\frac{\partial \nu^h(t,\eta)}{\partial t} = \Delta_L \pi^h(t,\eta) + u_1^*(t,\eta) - (Bu^h)(t,\eta) + (Cv)(t,\eta), \quad (13)$$

$$\frac{\partial \nu^h(t,\eta)}{\partial t} = \Delta_L \nu^h(t,\eta) + R^h(t,\eta)\nu^h(t,\eta) + \pi^h(t,\eta) \quad \text{in } \Omega \times (t_0, \vartheta],$$

with the initial condition $\pi^h(t_0) = \nu^h(t_0) = 0$ in Ω and with the boundary condition $\frac{\partial \pi^h}{\partial n} = \frac{\partial \nu^h}{\partial n} = 0$ on $\partial\Omega \times (t_0, \vartheta]$. Here, $R^h(t,\eta) = a(t,\eta) + b(t,\eta)(w^{(1)}(t,\eta) + \varphi_{\Delta,w}^h(t,\eta)) - ((w^{(1)}(t,\eta))^2 + w^{(1)}(t,\eta)\varphi_{\Delta,w}^h(t,\eta) + (\varphi_{\Delta,w}^h)^2(t,\eta))$. Multiplying scalarly the first equation of (13) by $g^h(t)$, and the second one, by $\nu^h(t)$, we obtain

$$(g^h(t), g_t^h(t))_H + \int_\Omega \{|\nabla \pi^h(t,\eta)|^2 + l\nabla \pi^h(t,\eta)\nabla \nu^h(t,\eta)\} \, d\eta$$

$$= (g^h(t), u_1^*(t) - Bu^h(t) + Cv(t))_H, \quad (14)$$

$$(\nu^h(t), \nu_t^h(t))_H + \int_\Omega |\nabla \nu^h(t,\eta)|^2 \, d\eta \leq (\pi^h(t), \nu^h(t))_H + b|\nu^h(t)|_H^2 \quad \text{for a.a. } t \in T.$$

Here, we use the inequality $\underset{(t,\eta)\in T\times\Omega}{\text{vrai max}} \{a(t,\eta) + b(t,\eta)(v_1 + v_2) - (v_1^2 + v_1 v_2 + v_2^2)\} \leq b$, which is valid for any $v_1, v_2 \in \mathbb{R}$. It is evident that the inequality

$$\int_\Omega l(\nabla \pi^h(t,\eta), \nabla \nu^h(t,\eta)) \, d\eta \geq -0.5 \int_\Omega \{|\nabla \pi^h(t,\eta)|^2 + l^2|\nabla \nu^h(t,\eta)|^2\} \, d\eta \quad (15)$$

is fulfilled for a.a. $t \in T$. Let us multiply the first inequality of (14) by l^2 and add to the second one. Taking into account (15), we have for a.a. $t \in T$

$$(g^h(t), g^h_t(t))_H + l^2(\nu^h(t), \nu^h_t(t))_H + 0.5 \int_\Omega \{|\nabla \pi^h(t, \eta)|^2 + l^2 |\nabla \nu^h(t, \eta)|^2\} \, d\eta \quad (16)$$

$$\leq (g^h(t), u_1^*(t) - Bu^h(t) + Cv(t))_H + l^2(\pi^h(t), \nu^h(t))_H + bl^2 |\nu^h(t)|^2_H.$$

Note that $\pi^h(t) = g^h(t) - l\nu^h(t)$. In this case, for a.a. $t \in T$

$$(\pi^h(t), \nu^h(t))_H + b|\nu^h(t)|^2_H = (g^h(t) - l\nu^h(t), \nu^h(t))_H + b|\nu^h(t)|^2_H \quad (17)$$

$$= (g^h(t), \nu^h(t))_H + (b - l)|\nu^h(t)|^2_H \leq 0.5(|g^h(t)|^2_H + (0.5 + |b - l|)|\nu^h(t)|^2_H.$$

Combining (16) and (17), we obtain

$$\frac{d}{dt} \Lambda^0(x^h_{\Delta,w}(t), q^h_\Delta(t), w(t)) + 0.5 \int_\Omega \{|\nabla \pi^h(t, \eta)|^2 + l^2 |\nabla \nu^h(t, \eta)|^2\} \, d\eta \quad (18)$$

$$\leq 2l^2 \lambda^2 \Lambda^0(x^h_{\Delta,w}(t), q^h_\Delta(t), w(t)) + \gamma^{(1)}_t + \gamma^{(2)}_t,$$

where $\gamma^{(1)}_t = (q^h_\Delta(t) - w^{(3)}(t))(\chi(t, x^h_{\Delta,w}(t), u^h(t)) - u_2^*(t)), \gamma^{(2)}_t = (g^h(t), u_1^*(t) - Bu^h(t) + Cv(t))_H$ for a.a. $t \in T$. For a.a. $t \in \delta_i = [\tau_i, \tau_{i+1}]$, it is easily seen that

$$\gamma^{(1)}_t \leq (q^h_\Delta(\tau_i) - w^{(3)}(\tau_i))(\chi(t, x^h_{\Delta,w}(t), u^h(t) - u_2^*(t)) + k_0(t - \tau_i)^{1/2}, \quad (19)$$

$$|\chi(t, x^h_{\Delta,w}(t), u^h(t)) - \chi(\tau_i, \xi^h_i, p^h_i, u^h(t))| \leq k_1\{h + (t - \tau_i)^{1/2} + |p^h_i - \psi^h_{\Delta,w}(t)|^2_H\}. \quad (20)$$

From (19) and (20) we have for a.a. $t \in \delta_i$

$$\gamma^{(1)}_t \leq (q^h_\Delta(\tau_i) - w^{(3)}(\tau_i))(\chi(\tau_i, \xi^h_i, \psi^h_i, u^h(t)) - u_2^*(t)) \quad (21)$$

$$+ k_2 |q^h_\Delta(\tau_i) - w^{(3)}(\tau_i)|\{h + (t - \tau_i)^{1/2} + |p^h_i - \psi^h_{\Delta,w}(t)|_H\}.$$

Estimate the last term in the right-hand side of inequality (18). For a.a. $t \in [\tau_i, \tau_{i+1}]$

$$|g^h(t) - y^h_i|_H = |\pi^h(t) + l\nu^h(t) - y^h_i|_H \leq \lambda_{1,i}(t) + \lambda_{2,i}(t), \quad (22)$$

where $y^h_i = w^{(1)}(\tau_i) - p^h_i - l(w^{(2)}(\tau_i) - \xi^h_i), \lambda_{1,i}(t) = |w^{(1)}(t) - \psi^h_{\Delta,w}(t) - w^{(1)}(\tau_i) + p^h_i|_H, \lambda_{2,i}(t) = l|w^{(2)}(t) - \varphi^h_{\Delta,w}(t) - w^{(2)}(\tau_i) + \xi^h_i|_H$. For a.a. $t \in \delta_i = [\tau_i, \tau_{i+1}]$, it is easily seen that $\lambda_{1,i}(t) \leq |p^h_i - \psi^h_{\Delta,w}(t)|_H + \int_{\tau_i}^t |\dot{w}^{(1)}(\tau)|_H \, d\tau, \lambda_{2,i}(t) \leq lh +$

$l \int_{\tau_i}^t \{|\dot{\varphi}^h_{\Delta,w}(\tau)|_H + |\dot{w}^{(2)}(\tau)|_H\} \, d\tau$. From this equation and (22), for a.a. $t \in \delta_i$, it follows that

$$|g^h(t) - y^h_i|_H \leq lh + \int_{\tau_i}^t \{l|\dot{\varphi}^h_{\Delta,w}(\tau)|_H + |\dot{w}^{(1)}(\tau)|_H + |\dot{w}^{(2)}(\tau)|_H\} \, d\tau + |p^h_i - \psi^h_{\Delta,w}(t)|_H.$$

$$(23)$$

By virtue of (11), taking into account results of [8,9], from estimate (23) we derive

$$\sum_{i=0}^{m-1} \int_{\tau_i}^{\tau_{i+1}} M(t;\tau_i)\, dt \le k_3(h+\delta) + k_4 \int_{t_0}^{\vartheta} |p^h(\tau) - \psi_{\Delta,w}^h(\tau)|_H\, d\tau \le k_5 \nu^{1/2}(h), \quad (24)$$

where $M(t;\tau_i) = |g^h(t) - y_i^h|_H \{|Bu_i^h|_H + |Cv(t)|_H + |u_1^*(t)|_H\}$ for a.a. $t \in \delta_i$, $\nu(h) = (h + \delta(h) + \alpha(h))^{1/2} + (h + \delta(h))\alpha^{-1}(h)$. Note that, for a.a. $t \in \delta_i$,

$$\gamma_t^{(2)} \le (y_i^h, u_1^*(t) - Bu_i^h + Cv(t))_H + M(t;\tau_i). \quad (25)$$

Let the symbol $(\cdot,\cdot)_{H\times\mathbb{R}}$ denote the scalar product in the space $H \times \mathbb{R}$. Let us define elements v_i^e from the conditions

$$\inf_{u\in P}(s_i, \Phi(\tau_i, p_i^h, \xi_i^h, u, v_i^e))_{H\times\mathbb{R}} \ge \sup_{v\in E}\inf_{u\in P}(s_i, \Phi(\tau_i, \xi_i^h, u, v))_{H\times\mathbb{R}} - h, \quad (26)$$

where $\Phi(\tau_i, p_i^h, \xi_i^h, u_i^h, v(t)) = \{Bu_i^h - Cv(t), \chi(\tau_i, \xi_i^h, p_i^h, u_i^h)\}$, $s_i = \{-y_i^h, \psi_i^h - w^{(3)}(\tau_i)\}$. It is obvious (see Condition 1) that $u_*(t) \in H(t, w^{(1)}(t), w^{(2)}(t)) \subset \bigcup_{u\in P} \Phi(t, w^{(1)}(t), w^{(2)}(t), u, v_i^e)$ for a. a. $t \in [\tau_i, \tau_{i+1})$. Then, for a.a. $t \in \delta_i$, there exists a control $u^{(1)}(t) \in P$ such that

$$\Phi(t, w^{(1)}(t), w^{(2)}(t), u^{(1)}(t), v_i^e) = u^*(t) \quad \text{for a. a.} \quad t \in [\tau_i, \tau_{i+1}]. \quad (27)$$

Using the rule of definition of the strategy $\mathcal{U}$, we deduce that

$$(s_i, \Phi(\tau_i, p_i^h, \xi_i^h, u_i^h, v(t)))_{H\times\mathbb{R}} \le \inf_{u\in P}\sup_{v\in E}(s_i, \Phi(\tau_i, p_i^h, \xi_i^h, u, v))_{H\times\mathbb{R}} + h. \quad (28)$$

Here (see (12)), $u_i^h \in \mathcal{U}(\tau_i, \xi_i^h, p_i^h, \psi_i^h, w(\tau_i))$; $v_{\tau_i,\tau_{i+1}}(\cdot)$ is an unknown realization of the control of the opponent. In turn, from (26) we have

$$\sup_{v\in E}\inf_{u\in P}(s_i, \Phi(\tau_i, p_i^h, \xi_i^h, u, v))_{H\times\mathbb{R}} \le \inf_{u\in P}(s_i, \Phi(\tau_i, p_i^h, \xi_i^h, u, v_i^e))_{H\times\mathbb{R}} + h. \quad (29)$$

Moreover, it is evident that the equality

$$\inf_{u\in P}\sup_{v\in E}(s_i, \Phi(\tau_i, p_i^h, \xi_i^h, u, v))_{H\times\mathbb{R}} = \sup_{v\in E}\inf_{u\in P}(s_i, \Phi(\tau_i, p_i^h, \xi_i^h, u, v))_{H\times\mathbb{R}} \quad (30)$$

is valid. From (28)–(30) we have

$$(s_i, \Phi(\tau_i, p_i^h, \xi_i^h, u_i^h, v(t)))_{H\times\mathbb{R}} \le \inf_{u\in P}(s_i, \Phi(\tau_i, p_i^h, \xi_i^h, u, v_i^e))_{H\times\mathbb{R}} + 2h \quad (31)$$

$$\le (s_i, \Phi(t, p_i^h, \xi_i^h, u^{(1)}(t), v_i^e))_{H\times\mathbb{R}} + 2h + L|\psi_i^h - w^{(3)}(\tau_i)|(t - \tau_i).$$

Here, L is the Lipschitz constant of the function $\chi(\cdot)$ in the corresponding domain. In this case, for a.a. $t \in \delta_i$, it follows from (27), (31) that

$$(s_i, \Phi(\tau_i, p_i^h, \xi_i^h, u_i^h, v(t)) - u^*(t))_{H\times\mathbb{R}} \quad (32)$$

$$\leq 2h + L|\psi_i^h - w^{(3)}(\tau_i)|\{t - \tau_i + |\xi_i^h - w^{(1)}(t)|_H + |p_i^h - w^{(2)}(t)|_H\}.$$

By virtue of (32) and (21), for the interval $[\tau_i, \tau_{i+1}]$, we derive

$$\gamma_t^{(1)} + \gamma_t^{(2)} \tag{33}$$

$$\leq \pi_i^*(t) + k_6\{h^2 + t - \tau_i + \Lambda^0(x_{\Delta,w}^h(t), q_\Delta^h(t), w(t)) + |p_i^h - \psi_{\Delta,w}^h(t)|_H^2\},$$

where

$$\pi_i^*(t) = (s_i, \Phi(\tau_i, p_i^h, \xi_i^h, u_i^h, v(t)) - u^*(t))_{H \times \mathbb{R}}.$$

We deduce from inequalities (18) and (33) that

$$\frac{d\Lambda(t, x_{\Delta,w}^h(\cdot), q_\Delta^h(\cdot), w(\cdot))}{dt} \leq M(t; \tau_i) \tag{34}$$

$$+k_6\Lambda(t, x_{\Delta,w}^h(\cdot), q_\Delta^h(\cdot), w(\cdot)) + k_7\{h^2 + t - \tau_i + |p_i^h - \psi_{\Delta,w}^h(t)|_H^2\} \quad \text{for a.a. } t \in \delta_i.$$

Using (34) and (24), by virtue of the Gronwall lemma, we obtain for $t \in T$

$$\Lambda(t, x_{\Delta,w}^h(\cdot), q_{\Delta,w}^h(\cdot), w(\cdot)) \leq k_8 \sum_{i=0}^{m-1} \left\{ \int_{\tau_i}^{\tau_{i+1}} M(t; \tau_i)\, dt + \delta(h^2 + \delta) \right\} \leq k_9 \nu^{1/2}(h).$$

The statement of the theorem follows from the last inequality.

References

1. Fattorini, H.O.: Infinite Dimensional Optimization and Control Theory. Cambridge University Press, Cambridge (1999)
2. Lasiecka, I., Triggiani, R.: Control Theory for Partial Differential Equations: Continuous and Approximation Theories. Part I: Abstract Parabolic Systems. Part II: Abstract Hyperbolic-like Systems Over a Finite Time Horizon. Cambridge University Press, Cambridge (2000)
3. Tröltzsch, F.: Optimal Control of Partial Differential Equations. Theory, Methods and Applications. AMS, Providence (2010)
4. Krasovskii, N., Subbotin, A.: Game-Theoretical Control Problems. Springer, Berlin (1988)
5. Caginalp, G.: An analysis of a phase field model of a free boundary. Arch. Rat. Mech. Analysis. **92**, 205–245 (1986)
6. Hoffman, K.-H., Jiang, L.S.: Optimal control problem of a phase field model for solidification. Numer. Funct. Anal. and Optimiz. **13**(1–2), 11–27 (1992)
7. Maksimov, V.I.: Dynamical Inverse Problems of Distributed Systems. VSP, Boston (2002)
8. Maksimov, V., Tröltzsch, F.: On an inverse problem for phase field equations. Dokl. Akad. Nauk. **396**(3), 309–312 (2004)
9. Maksimov, V., Tröltzsch, F.: Dynamical state and control reconstruction for a phase field model. Dyn. Continuous Discrete Impulsive Syst. A: Math. Anal. **13**(3–4), 419–444 (2006)
10. Liu, S., Triggiani, R.: Inverse problem for a linearized Jordan-Moore-Gibson-Thompson equation. In: Favini, A., Fragnelli, G., Mininni, R.M. (eds.) New Prospects in Direct, Inverse and Control Problems for Evolution Equations, vol. 10, pp. 305–352. Springer, Heidelberg (2014)

On the Weak Solvability and the Optimal Control of a Frictional Contact Problem with Normal Compliance

Andaluzia Matei[(✉)]

Department of Mathematics, University of Craiova, A.I. Cuza street, 13, 200585
Craiova, Romania
andaluziamatei@inf.ucv.ro
http://math.ucv.ro/~amatei/

Abstract. In the present work we consider a frictional contact model with normal compliance. Firstly, we discuss the weak solvability of the model by means of two variational approaches. In a first approach the weak solution is a solution of a quasivariational inequality. In a second approach the weak solution is a solution of a mixed variational problem with solution-dependent set of Lagrange multipliers. Nextly, the paper focuses on the boundary optimal control of the model. Existence results, an optimality condition and some convergence results are presented.

Keywords: Contact model · Friction · Normal compliance · Weak solutions · Optimal control · Optimality condition · Convergences

1 Introduction

The present paper focuses on the weak solvability and the boundary optimal control of the following contact model.

Problem 1. Find a displacement field $\boldsymbol{u} : \Omega \to \mathbb{R}^3$ and a stress field $\boldsymbol{\sigma} : \Omega \to \mathbb{S}^3$ such that

$$\text{Div}\,\boldsymbol{\sigma} + \boldsymbol{f}_0 = 0 \quad \text{in } \Omega,$$

$$\boldsymbol{\sigma} = \mathcal{F}\varepsilon(\boldsymbol{u}) \quad \text{in } \Omega,$$

$$\boldsymbol{u} = 0 \text{ on } \Gamma_1,$$

$$\boldsymbol{\sigma}\nu = \boldsymbol{f}_2 \quad \text{on } \Gamma_2,$$

$$-\sigma_\nu = p_\nu(u_\nu - g_a) \quad \text{on } \Gamma_3,$$

$$\left. \begin{array}{l} \|\boldsymbol{\sigma}_\tau\| \le p_\tau(u_\nu - g_a), \\ \boldsymbol{\sigma}_\tau = -p_\tau(u_\nu - g_a) \frac{\boldsymbol{u}_\tau}{\|\boldsymbol{u}_\tau\|} \quad \text{if} \quad \boldsymbol{u}_\tau \neq 0 \end{array} \right\} \quad \text{on } \Gamma_3.$$

A. Matei—This work was supported by a grant of the Romanian National Authority for Scientific Research, CNCS-UEFISCDI, project number PN-II-ID-PCE-2011-3-0257, and by LEA Math Mode CNRS-IMAR.

© IFIP International Federation for Information Processing 2016
Published by Springer International Publishing AG 2016. All Rights Reserved
L. Bociu et al. (Eds.): CSMO 2015, IFIP AICT 494, pp. 370–379, 2016.
DOI: 10.1007/978-3-319-55795-3_35

Notice that Ω is a bounded domain of $\mathbb{R}^3$ with smooth enough boundary, partitioned in three measurable part Γ_1, Γ_2 and Γ_3; $u_\nu = \boldsymbol{u} \cdot \nu$, $\boldsymbol{u}_\tau = \boldsymbol{u} - u_\nu \nu$, $\sigma_\nu = (\boldsymbol{\sigma}\nu) \cdot \nu$, $\boldsymbol{\sigma}_\tau = \boldsymbol{\sigma}\nu - \sigma_\nu\nu$, "." denotes the inner product of two vectors, $\|\cdot\|$ denotes the Euclidean norm, ν is the unit outward normal vector.

Problem 1 is a contact problem with the normal compliance condition, associated to the Coulomb's law of dry friction. A normal compliance condition was firstly proposed in [17]. Then, such a contact condition was used in many models, see e.g. the papers [2,7–9,20].

In the normal compliance contact condition

$$-\sigma_\nu = p_\nu(u_\nu - g_a) \quad \text{on} \quad \Gamma_3,$$

p_ν is a nonnegative prescribed function which vanishes for negative argument and $g_a > 0$ denotes the gap (the distance between the body and the obstacle on the normal direction). When $u_\nu < g_a$ there is no contact and the normal pressure vanishes. When there is contact then $u_\nu - g_a$ is positive and represents a measure of the interpenetration of the asperities. Then, the normal compliance condition shows that the foundation exerts a pressure on the body which depends on the penetration. For details on the physical significance of the model we refer to [22].

The rest of the paper has the following structure. Section 2 is devoted to the weak solvability of the model by means of two variational approaches. In Sect. 3 we discuss an optimal control problem which consists of leading the stress tensor as close as possible to a given target, by acting with a control on a part of the boundary.

There are several works concerning the optimal control of variational inequalities, see for instance [3,4,6,10,15,16,18,23]. Nevertheless, only few works are devoted to the optimal control of contact problems, see [1,5,14]. The present paper adds a new contribution.

2   On the Weak Solvability of the Model

In this section we shall indicate two variational approaches in the study of Problem 1. Let us make the following assumptions.

Assumption 1. $\mathcal{F} : \Omega \times \mathbb{S}^3 \to \mathbb{S}^3$, $\mathcal{F}(\boldsymbol{x}, \varepsilon) = (\mathcal{F}_{ijkl}(\boldsymbol{x})\varepsilon_{jk})$ for all $\varepsilon = (\varepsilon_{ij}) \in \mathbb{S}^3$, a.e. $\boldsymbol{x} \in \Omega$. $\mathcal{F}_{ijkl} = \mathcal{F}_{jikl} = \mathcal{F}_{klij} \in L^\infty(\Omega)$, $1 \le i,j,k,l \le 3$. There exists $m_\mathcal{F} > 0$ such that $\mathcal{F}(\boldsymbol{x}, \boldsymbol{\tau}) : \boldsymbol{\tau} \ge m_\mathcal{F}\|\boldsymbol{\tau}\|^2$ for all $\boldsymbol{\tau} \in \mathbb{S}^3$, a.e. $\boldsymbol{x}$ in Ω.

Assumption 2. $\boldsymbol{f}_0 \in L^2(\Omega)^3$, $\boldsymbol{f}_2 \in L^2(\Gamma_2)^3$.

Assumption 3. $p_e : \Gamma_3 \times \mathbb{R} \to \mathbb{R}_+$ ($e \in \{\nu, \tau\}$). There exists $L_e > 0$ such that $|p_e(\boldsymbol{x}, r_1) - p_e(\boldsymbol{x}, r_2)| \le L_e |r_1 - r_2|$ for all $r_1, r_2 \in \mathbb{R}$ a.e. $\boldsymbol{x} \in \Gamma_3$. The mapping $\boldsymbol{x} \mapsto p_e(\boldsymbol{x}, r)$ is measurable on Γ_3, for any $r \in \mathbb{R}$ and $p_e(\boldsymbol{x}, r) = 0$ for all $r \le 0$, a.e. $\boldsymbol{x} \in \Gamma_3$.

Assumption 4. $g_a \in L^2(\Gamma_3)$, $g_a(\boldsymbol{x}) \ge 0$, a.e. $\boldsymbol{x} \in \Gamma_3$.

Assumption 5. $m_{\mathcal{F}} > c_0^2(L_\nu + L_\tau)$.

Assumption 5 is a smallness assumption which was introduced mainly for mathematical reasons. However, for some materials and frictional contact conditions we have appropriate constants $m_{\mathcal{F}}$, L_ν and L_τ which fulfill Assumption 5. Notice that " : " denotes the inner product of two tensors and $c_0 = c_0(\Omega, \Gamma_1, \Gamma_3) > 0$ is a "trace constant" such that:

$$\|v\|_{L^2(\Gamma_3)^3} \le c_0 \|v\|_V \quad \text{for all } v \in V, \tag{1}$$

where

$$V = \{ v \in H^1(\Omega)^3 \mid v = 0 \text{ a.e. on } \Gamma_1 \}.$$

In a first approach the weak solution is a solution of a quasivariational inequality having as unknown the displacement field.

Problem 2. Find a displacement field $u \in V$ such that

$$(Au, v - u)_V + j(u, v) - j(u, u) \ge (f, v - u)_V \quad \text{for all } v \in V. \tag{2}$$

Herein,

$$A : V \to V \quad (Au, v)_V = (\mathcal{F}\varepsilon(u), \varepsilon(v))_{L^2(\Omega)_s^{3\times3}},$$

$$j : V \times V \to \mathbb{R} \quad j(u, v) = \int_{\Gamma_3} p_\nu(u_\nu - g_a)|v_\nu|\, d\Gamma + \int_{\Gamma_3} p_\tau(u_\nu - g_a)\|v_\tau\|\, d\Gamma,$$

$$(f, v)_V = \int_\Omega f_0 \cdot v\, dx + \int_{\Gamma_2} f_2 \cdot v\, d\Gamma. \tag{3}$$

We have the following existence and uniqueness result.

Theorem 1. *Under Assumptions 1–5, Problem 2 has a unique weak solution.*

For the proof we refer to Theorem 5.30 in [22].

The second approach is a mixed variational approach. The mixed variational formulations are related to modern numerical techniques in order to approximate the weak solutions of contact models. Referring to numerical techniques for approximating weak solutions of contact problems via saddle point technique, we send the reader to, e.g., [19,24,25]. The functional frame is the following one.

$$V = \{v \in H^1(\Omega)^3 \mid v = 0 \text{ a.e. on } \Gamma_1\};$$
$$S = \{w|_{\Gamma_3} \mid w \in V\};$$
$$D = S'.$$

Notice that $w|_{\Gamma_3}$ denotes the restriction of the trace of the element $w \in V$ to Γ_3. Thus, $S \subset H^{1/2}(\Gamma_3; \mathbb{R}^3)$ where $H^{1/2}(\Gamma_3; \mathbb{R}^3)$ is the space of the restrictions on Γ_3 of traces on Γ of functions of $H^1(\Omega)^3$. We use the Sobolev-Slobodeckii norm

$$\|\zeta\|_S = \left(\int_{\Gamma_3} \int_{\Gamma_3} \frac{\|\zeta(x) - \zeta(y)\|^2}{\|x - y\|^3}\, ds_x\, ds_y \right)^{1/2}.$$

For each $\zeta \in S$, $\zeta_\nu = \zeta \cdot \nu$ and $\zeta_\tau = \zeta - \zeta_\nu \nu$ a.e. on Γ_3.

Let us consider $\boldsymbol{f} \in V$, see (3), and let us define two bilinear forms $a(\cdot,\cdot)$ and $b(\cdot,\cdot)$ as follows:

$$a(\cdot,\cdot): V \times V \to \mathbb{R}, \quad a(\boldsymbol{u},\boldsymbol{v}) = \int_\Omega \mathcal{F}\varepsilon(\boldsymbol{u}(\boldsymbol{x})) : \varepsilon(\boldsymbol{v}(\boldsymbol{x}))\, dx;$$
$$b(\cdot,\cdot): V \times D \to \mathbb{R} \quad b(\boldsymbol{v},\mu) = \langle \mu, \boldsymbol{v}|_{\Gamma_3}\rangle.$$

Also, we define a variable set $\Lambda = \Lambda(\varphi)$,

$$\Lambda(\varphi) = \{\mu \in D \,|\, \langle \mu, \boldsymbol{v}|_{\Gamma_3}\rangle$$
$$\leq \int_{\Gamma_3} (p_\nu(\boldsymbol{x}, \varphi_\nu(\boldsymbol{x}) - g_a)|v_\nu(\boldsymbol{x})| + p_\tau(\boldsymbol{x}, \varphi_\nu(\boldsymbol{x}) - g_a)\|\boldsymbol{v}_\tau(\boldsymbol{x})\|)d\Gamma \quad \boldsymbol{v} \in V\}.$$

Notice that $\langle \cdot,\cdot \rangle$ denotes the duality pairing between D and S.

The second variational formulation of Problem 1 is the following one.

Problem 3. Find $\boldsymbol{u} \in V$ and $\lambda \in \Lambda(\boldsymbol{u}) \subset D$ such that

$$a(\boldsymbol{u},\boldsymbol{v}) + b(\boldsymbol{v},\lambda) = (\boldsymbol{f},\boldsymbol{v})_V \quad \text{for all } \boldsymbol{v} \in V,$$
$$b(\boldsymbol{u},\mu - \lambda) \quad\quad \leq 0 \quad\quad \text{for all } \mu \in \Lambda(\boldsymbol{u}).$$

In this second approach, a weak solution is a pair consisting of the displacement field and a Lagrange multiplier related to the friction force.

Theorem 2. *Under Assumptions 1–4, Problem 3 has at least one solution.*

The proof of Theorem 2, based on the abstract results we have got in [11], can be found in the very recent paper [12].

Remark 1. Treating the model in the first approach we can prove the existence and the uniqueness of the weak solution. But, the approximation of the weak solution is based on a regularization/penalization technique. Treating the model in the second approach we are led to a generalized saddle point problem. Recall that, for weak formulations in Contact Mechanics via saddle point problems, efficient algorithms can be written in order to approximate the weak solution (see primal-dual active set strategies). But, there are a few open questions here:

- the study of the uniqueness of the weak solution of the mixed variational formulation Problem 3;
- a priori error estimates; algorithms.

3   Boundary Optimal Control

Let us discuss in this section a boundary optimal control problem related to our contact problem.

For a fixed function $\boldsymbol{f}_0 \in L^2(\Omega)^3$, we consider the following *state problem.*

(PS1) Let $f_2 \in L^2(\Gamma_2)^3$ (called control) be given. Find $u \in V$ such that

$$(Au, v - u)_V + j(u, v) - j(u, u) \geq \int_\Omega f_0(x) \cdot (v(x) - u(x)) \, dx \qquad (4)$$

$$+ \int_{\Gamma_2} f_2(x) \cdot (v(x) - u(x)) \, d\Gamma \quad \text{for all } v \in V.$$

According to Theorem 1, for every control $f_2 \in L^2(\Gamma_2)^3$, the *state problem* (PS1) has a unique solution $u \in V$, $u = u(f_2)$. In addition, the following estimation takes place:

$$\|u\|_V \leq \frac{1}{m_{\mathcal{F}}}(\|f_0\|_{L^2(\Omega)^3} + c_0\|f_2\|_{L^2(\Gamma_2)^3}),$$

where $m_{\mathcal{F}}$ is the constant in Assumption 1 and c_0 appears in (1).

Now, we would like to act a control on Γ_2 such that the resulting stress σ be as close as possible to a given target

$$\sigma_d = \mathcal{F}\varepsilon(u_d)$$

where u_d is a given function.

Let Q_∞ be the real Banach space

$$Q_\infty = \{\mathcal{F} = \mathcal{F}_{ijkl} \mid \mathcal{F}_{ijkl} = \mathcal{F}_{jikl} = \mathcal{F}_{klij} \in L^\infty(\Omega), \ 1 \leq i,j,k,l \leq 3\}$$

endowed with the norm $\|\mathcal{F}\|_\infty = \max_{1 \leq i,j,k,l \leq 3} \|\mathcal{F}_{ijkl}\|_{L^\infty(\Omega)}$. According to [22], page 97,

$$\|\mathcal{F}\tau\|_{L^2(\Omega)^{3\times 3}_s} \leq 3\|\mathcal{F}\|_\infty \|\tau\|_{L^2(\Omega)^{3\times 3}_s} \quad \text{for all } \tau \in L^2(\Omega)^{3\times 3}_s.$$

Therefore,

$$\|\sigma - \sigma_d\|_{L^2(\Omega)^{3\times 3}_s} \leq 3 \max_{1 \leq i,j,k,l \leq 3} \|\mathcal{F}_{ijkl}\|_{L^\infty(\Omega)} \|u - u_d\|_V.$$

Thus, σ and σ_d will be close to one another if the difference between the functions u and u_d is small in the sense of V−norm.

To give an example of a target of interest, u_d, we can consider $u_d = 0$. In this situation, by acting a control on Γ_2, the tension σ is small in the sense of L^2- norm, even if the volume forces f_0 does not vanish in Ω.

Let $\alpha, \beta > 0$ be two positive constants and let us define the following functional

$$L : L^2(\Gamma_2)^3 \times V \to \mathbb{R}, \quad L(f_2, u) = \frac{\alpha}{2}\|u - u_d\|_V^2 + \frac{\beta}{2}\|f_2\|_{L^2(\Gamma_2)^3}^2.$$

Furthermore, we denote

$$\mathcal{V}_{ad} = \{[u, f_2] \mid [u, f_2] \in V \times L^2(\Gamma_2), \text{ such that (4) is verified}\}.$$

(POC1) Find $[u^*, f_2^*] \in \mathcal{V}_{ad}$ such that $L(f_2^*, u^*) = \min_{[u, f_2] \in \mathcal{V}_{ad}} \left\{L(f_2, u)\right\}$.

A solution of (POC1) is called *an optimal pair*. The second component of the optimal pair is called *an optimal control*.

Theorem 3. *Problem (POC1) has at least one solution* $(\boldsymbol{u}^*, \boldsymbol{f}_2^*)$.

Let us fix $\rho > 0$ and $\boldsymbol{f}_0 \in L^2(\Omega)^3$.

We introduce the following regularized state problem.

(PS2) *Let* $\boldsymbol{f}_2 \in L^2(\Gamma_2)^3$ *(called regularized control) be given. Find* $\boldsymbol{u} \in V$ *such that*

$$(A\boldsymbol{u}, \boldsymbol{v} - \boldsymbol{u})_V + j_\rho(\boldsymbol{u}, \boldsymbol{v}) - j_\rho(\boldsymbol{u}, \boldsymbol{u}) \geq (\boldsymbol{f}_0, \boldsymbol{v} - \boldsymbol{u})_{L^2(\Omega)^3} \tag{5}$$
$$+ (\boldsymbol{f}_2, \boldsymbol{v} - \boldsymbol{u})_{L^2(\Gamma_2)^3} \qquad \text{for all } \boldsymbol{v} \in V.$$

Herein, $j_\rho : V \times V \to \mathbb{R}$ is defined as follows,

$$j_\rho(\boldsymbol{u}, \boldsymbol{v}) = \int_{\Gamma_3} p_\nu^\rho(\boldsymbol{x}, u_\nu(\boldsymbol{x}) - g_a(\boldsymbol{x}))(\sqrt{(v_\nu(\boldsymbol{x}))^2 + \rho^2} - \rho)d\Gamma$$

$$+ \int_{\Gamma_3} p_\tau^\rho(\boldsymbol{x}, u_\nu(\boldsymbol{x}) - g_a(\boldsymbol{x}))(\sqrt{\|\boldsymbol{v}_\tau(\boldsymbol{x})\|^2 + \rho^2} - \rho)d\Gamma \qquad \text{for all } \boldsymbol{u}, \boldsymbol{v} \in V,$$

where p_e^ρ, $e \in \{\nu, \tau\}$, satisfies the following assumptions.

Assumption 6 $p_e^\rho : \Gamma_3 \times \mathbb{R} \to \mathbb{R}_+$. *The mapping* $\boldsymbol{x} \mapsto p_e^\rho(\boldsymbol{x}, r)$ *is measurable on* Γ_3 *for any* $r \in \mathbb{R}$, *and* $p_e^\rho(\boldsymbol{x}, r) = 0$ *for all* $r \leq 0$, *a.e.* $\boldsymbol{x} \in \Gamma_3$.

Assumption 7 $p_e^\rho(\boldsymbol{x}, \cdot) \in C^1(\mathbb{R})$ *a.e. on* $\boldsymbol{x} \in \Gamma_3$. *There exists* $M_e > 0$ *such that* $|p_e^\rho(\boldsymbol{x}, r)| \leq M_e$ *for all* $r \in \mathbb{R}$, *a.e.* $\boldsymbol{x} \in \Gamma_3$. *In addition* $|\partial_2 p_e^\rho(\boldsymbol{x}, r)| \leq L_e$ *for all* $r \in \mathbb{R}$, *a.e.* $\boldsymbol{x} \in \Gamma_3$.

Assumption 8. *There exists* $G_e : \mathbb{R}_+ \to \mathbb{R}_+$ $(e \in \{\nu, \tau\})$ *such that* $|p_e^\rho(\boldsymbol{x}, r) - p_e(\boldsymbol{x}, r)| \leq G_e(\rho)$ *for all* $r \in \mathbb{R}$, *a.e.* $\boldsymbol{x} \in \Gamma_3$ *and* $\lim_{\rho \to 0} G_e(\rho) = 0$.

Notice that the functional $j_\rho(\cdot, \cdot)$ has the following properties:

- for all $\boldsymbol{u}, \boldsymbol{v} \in V$, $j_\rho(\boldsymbol{u}, \boldsymbol{v}) \geq 0$; $j_\rho(\boldsymbol{u}, 0_V) = 0$;
- for all $\boldsymbol{u} \in V$, $j_\rho(\boldsymbol{u}, \cdot) : V \to \mathbb{R}$ is a convex and Gâteaux differentiable functional;
- $j_\rho(\eta_1, \boldsymbol{v}_2) - j_\rho(\eta_1, \boldsymbol{v}_1) + j_\rho(\eta_2, \boldsymbol{v}_1) - j_\rho(\eta_2, \boldsymbol{v}_2) \leq c_0^2(L_\nu + L_\tau)\|\eta_1 - \eta_2\|_V \|\boldsymbol{v}_1 - \boldsymbol{v}_2\|_V$ for all $\eta_1, \eta_2, \boldsymbol{v}_1, \boldsymbol{v}_2 \in V$.
- for all $\boldsymbol{u}, \boldsymbol{v} \in V$, there exists $\nabla_2 j_\rho(\boldsymbol{u}, \boldsymbol{v}) \in V$ such that

$$\lim_{h \to 0} \frac{j_\rho(\boldsymbol{u}, \boldsymbol{v} + h\boldsymbol{w}) - j_\rho(\boldsymbol{u}, \boldsymbol{v})}{h} = (\nabla_2 j_\rho(\boldsymbol{u}, \boldsymbol{v}), \boldsymbol{w})_V \qquad \text{for all } \boldsymbol{w} \in V.$$

$$(\nabla_2 j_\rho(\boldsymbol{u}, \boldsymbol{v}), \boldsymbol{w})_V = \int_{\Gamma_3} p_\nu^\rho(\boldsymbol{x}, u_\nu(\boldsymbol{x}) - g_a(\boldsymbol{x})) \frac{v_\nu(\boldsymbol{x})w_\nu(\boldsymbol{x})}{\sqrt{(v_\nu(\boldsymbol{x}))^2 + \rho^2}} d\Gamma$$

$$+ \int_{\Gamma_3} p_\tau^\rho(\boldsymbol{x}, u_\nu(\boldsymbol{x}) - g_a(\boldsymbol{x})) \frac{\boldsymbol{v}_\tau(\boldsymbol{x}) \cdot \boldsymbol{w}_\tau(\boldsymbol{x})}{\sqrt{\|\boldsymbol{v}_\tau(\boldsymbol{x})\|^2 + \rho^2}} d\Gamma.$$

The regularized state problem has a unique solution $u^\rho \in V$ that depends Lipschitz continuously on $\boldsymbol{f}$. This is a straightforward consequence of an abstract result in the theory of the quasivariational inequalities, see e.g. Theorem 3.7, in [21].

For every $\boldsymbol{f}_2 \in L^2(\Gamma_2)^3$, the problem (PS2) has a unique solution $\boldsymbol{u} \in V$, $\boldsymbol{u} = \boldsymbol{u}(\boldsymbol{f}_2)$. In addition,

$$\|\boldsymbol{u}\|_V \le \frac{1}{m_{\mathcal{F}}}(\|\boldsymbol{f}_0\|_{L^2(\Omega)^3} + c_0\|\boldsymbol{f}_2\|_{L^2(\Gamma_2)^3}).$$

There exists an unique $\boldsymbol{z} \in V$ such that

$$(\boldsymbol{z}, \boldsymbol{v})_V = \int_\Omega \boldsymbol{f}_0 \cdot \boldsymbol{v} \, dx \quad \text{for all } \boldsymbol{v} \in V.$$

Furthermore, there exists an unique $\boldsymbol{y}(\boldsymbol{f}_2) \in V$ such that

$$(\boldsymbol{y}(\boldsymbol{f}_2), \boldsymbol{v})_V = \int_{\Gamma_2} \boldsymbol{f}_2 \cdot \boldsymbol{v} \, d\Gamma \quad \text{for all } \boldsymbol{v} \in V.$$

Let $\boldsymbol{u} \in V$ be the unique solution of (PS2).
Let us define

$$\partial_2 j_\rho(\boldsymbol{u}, \boldsymbol{u}) = \{\boldsymbol{\zeta} \in V \mid j_\rho(\boldsymbol{u}, \boldsymbol{v}) - j_\rho(\boldsymbol{u}, \boldsymbol{u}) \ge (\boldsymbol{\zeta}, \boldsymbol{v} - \boldsymbol{u})_V \quad \text{for all } \boldsymbol{v} \in V\}.$$

Therefore,
$$\boldsymbol{z} + \boldsymbol{y}(\boldsymbol{f}_2) - A\boldsymbol{u} \in \partial_2 j_\rho(\boldsymbol{u}, \boldsymbol{u}).$$

Since $j_\rho(\cdot, \cdot)$ is convex and Gâteaux differentiable in the second argument, we can write
$$\partial_2 j_\rho(\boldsymbol{u}, \boldsymbol{u}) = \{\nabla_2 j_\rho(\boldsymbol{u}, \boldsymbol{u})\}.$$

Thus, we are led to the following operatorial equation

$$A\boldsymbol{u} + \nabla_2 j_\rho(\boldsymbol{u}, \boldsymbol{u}) = \boldsymbol{z} + \boldsymbol{y}(\boldsymbol{f}_2).$$

Let us define the admissible set,

$$\mathcal{V}_{ad}^\rho = \{[\boldsymbol{u}, \boldsymbol{f}_2] \mid [\boldsymbol{u}, \boldsymbol{f}_2] \in V \times L^2(\Gamma_2)^3, \text{ such that (5) is verified}\}.$$

Using the functional L, we introduce the *regularized optimal control problem*,
(POC2) *Find* $[\bar{\boldsymbol{u}}, \bar{\boldsymbol{f}}_2] \in \mathcal{V}_{ad}^\rho$ *such that* $L(\bar{\boldsymbol{f}}_2, \bar{\boldsymbol{u}}) = \min_{[\boldsymbol{u}, \boldsymbol{f}_2] \in \mathcal{V}_{ad}^\rho} \{L(\boldsymbol{f}_2, \boldsymbol{u})\}.$

Theorem 4. *The problem (POC2) has at least one solution* $(\bar{\boldsymbol{u}}, \bar{\boldsymbol{f}}_2)$.

A solution of (POC2) is called *a regularized optimal pair* and the second component $\bar{\boldsymbol{f}}_2$ is called *a regularized optimal control*.
The following result hold true.

Theorem 5 *(An optimality condition). Any regularized optimal control $\bar{f}_2$ verifies*

$$\bar{f}_2 = -\frac{1}{\beta}\gamma(p(\bar{f}_2)),$$

where γ is the trace operator and $p(\bar{f}_2)$ is the unique solution of the variational equation

$$a(u(\bar{f}_2) - u_d, w)_V = (p(\bar{f}_2), Aw + D_2^2 j_\rho(u(\bar{f}_2), u(\bar{f}_2))w)_V \quad for\ all\ w \in V,$$

$u(\bar{f}_2)$ *being the solution of (PS2) with $f_2 = \bar{f}_2$.*

Herein, for all $v \in V$, writing u instead of $u(\bar{f}_2)$,

$$(D_2^2 j_\rho(u, u)v, w)_V = \int_{\Gamma_3} \partial_2 p_\nu^\rho(x, u_\nu(x) - g_a(x)) \frac{u_\nu(x)v_\nu(x)w_\nu(x)}{\sqrt{u_\nu(x)^2 + \rho^2}}\, d\Gamma$$

$$+ \int_{\Gamma_3} \partial_2 p_\tau^\rho(x, u_\nu(x) - g_a(x)) \frac{u_\tau(x) \cdot w_\tau(x)v_\nu(x)}{\sqrt{\|u_\tau(x)\|^2 + \rho^2}}\, d\Gamma$$

$$+ \int_{\Gamma_3} + p_\nu^\rho(x, u_\nu(x) - g_a(x)) \frac{v_\nu(x)w_\nu(x)\rho^2}{(u_\nu(x)^2 + \rho^2)^{3/2}}\, d\Gamma$$

$$+ \int_{\Gamma_3} p_\tau^\rho(x, u_\nu(x) - g_a(x)) \frac{v_\tau(x) \cdot w_\tau(x)(\|u_\tau\|^2 + \rho^2) - (u_\tau \cdot w_\tau)(u_\tau \cdot v_\tau)}{(\|u_\tau(x)\|^2 + \rho^2)^{3/2}}\, d\Gamma.$$

The main tool in the proof of Theorem 5 is a Lions's Theorem, which we recall here for the convenience of the reader.

Theorem 6. *Let $\mathcal{B}$ be a Banach space, X and Y two reflexive Banach spaces. Let also be given two C^1 functions $F : \mathcal{B} \times X \to Y$, $L : \mathcal{B} \times X \to \mathbb{R}$. We suppose that, for all $\beta \in \mathcal{B}$,*
  (i) There exists a unique $\tilde{u}(\beta)$ such that $F(\beta, \tilde{u}(\beta)) = 0$,
  (ii) $\partial_2 F(\beta, \tilde{u}(\beta))$ is an isomorphism from X onto Y.
Then, $J(\beta) = L(\beta, \tilde{u}(\beta))$ is differentiable and, for every $\zeta \in \mathcal{B}$,

$$\frac{dJ}{d\beta}(\beta)\zeta = \partial_1 L(\beta, \tilde{u}(\beta))\zeta - \langle p(\beta), \partial_1 F(\beta, \tilde{u}(\beta))\zeta \rangle_{Y', Y},$$

where $p(\beta) \in Y'$ is the adjoint state, unique solution of

$$\left[\partial_2 F(\beta, \tilde{u}(\beta))\right]^* p(\beta) = \partial_2 L(\beta, \tilde{u}(\beta))\ in\ X'.$$

For the proof of Theorem 6 we refer to, e.g., [1].
Let us indicate in the last part of this section two convergence results. The first one involves the unique solution of the regularized state problem (PS2) and the unique solution of the state problem (PS1).

Theorem 7. *Let* $\rho > 0$, $\boldsymbol{f}_0 \in L^2(\Omega)^3$ *and* $\boldsymbol{f}_2 \in L^2(\Gamma_2)^3$ *be given. If* $\boldsymbol{u}^\rho$, $\boldsymbol{u} \in V$ *are the solutions of problems* $(PS2)$ *and* $(PS1)$, *respectively, then,*

$$\boldsymbol{u}^\rho \to \boldsymbol{u} \; in \; V \; as \; \rho \to 0.$$

Next, we have a convergence result involving the solutions of the problems (POC2) and (POC1).

Theorem 8. *Let* $[\bar{\boldsymbol{u}}^\rho, \bar{\boldsymbol{f}}_2^{\;\rho}]$ *be a solution of the problem (POC2). Then, there exists a solution of the problem (POC1),* $[\boldsymbol{u}^*, \boldsymbol{f}_2^*]$, *such that*

$$\bar{\boldsymbol{u}}^\rho \to \boldsymbol{u}^* \; in \; V \; as \; \rho \to 0,$$
$$\bar{\boldsymbol{f}}_2^{\;\rho} \rightharpoonup \boldsymbol{f}_2^* \; in \; L^2(\Gamma_2)^3 \; as \; \rho \to 0.$$

Theorems 3–8 are new results; their proofs will be published in [13].

Let us mention here some open questions:
- $\bar{\boldsymbol{f}}_2^{\;\rho} \to \boldsymbol{f}_2^*$ in $L^2(\Gamma_2)^3$ as $\rho \to 0$;
- an optimality condition for (PS1);
- to study the boundary optimal control of the model by means of the mixed variational formulation.

References

1. Amassad, A., Chenais, D., Fabre, C.: Optimal control of an elastic contact problem involving Tresca friction law. Nonlinear Anal. **48**, 1107–1135 (2002)
2. Andersson, L.-E.: A quasistatic frictional problem with normal compliance. Nonlinear Anal. TMA **16**, 347–370 (1991)
3. Barbu, V.: Optimal Control of Variational Inequalities. Pitman Advanced Publishing, Boston (1984)
4. Bonnans, J.F., Tiba, D.: Pontryagin's principle in the control of semiliniar elliptic variational inequalities. Appl. Math. Optim. **23**(1), 299–312 (1991)
5. Capatina, A., Timofte, C.: Boundary optimal control for quasistatic bilateral frictional contact problems. Nonlinear Anal. Theory Methods Appl. **94**, 84–99 (2014)
6. Friedman, A.: Optimal control for variational inequalities. SIAM J. Control Optim. **24**(3), 439–451 (1986)
7. Kikuchi, N., Oden, J.T.: Contact Problems in Elasticity: A Study of Variational Inequalities and Finite Element Methods. SIAM, Philadelphia (1988)
8. Klarbring, A., Mikelič, A., Shillor, M.: Frictional contact problems with normal compliance. Int. J. Eng. Sci. **26**, 811–832 (1988)
9. Klarbring, A., Mikelič, A., Shillor, M.: A global existence result for the quasistatic frictional contact problem with normal compliance. In: del Piero, G., Maceri, F. (eds.) Unilateral Problems in Structural Analysis, vol. 4, pp. 85–111. Birkhäuser, Boston (1991)
10. Lions, J.-L.: Contrôle optimale des systèmes gouvernés par des équations aux dérivées partielles. Dunod, Paris (1968)
11. Matei, A.: On the solvability of mixed variational problems with solution-dependent sets of Lagrange multipliers. Proc. Royal Soc. Edinburgh Sect. Math. **143**(05), 1047–1059 (2013)

12. Matei, A.: Weak solutions via lagrange multipliers for contact models with normal compliance. Konuralp J. Math. **3**(2) (2015)
13. Matei, A., Micu, S.: Boundary optimal control for a frictional contact problem with normal compliance, submitted
14. Matei, A., Micu, S.: Boundary optimal control for nonlinear antiplane problems. Nonlinear Anal. Theory Methods Appli. **74**(5), 1641–1652 (2011). doi:10.1016/j. na.2010.10.034. ISSN 0362-546X
15. Mignot, R.: Contrôle dans les inéquations variationnelles elliptiques. J. Func. Anal. **22**, 130–185 (1976)
16. Mignot, R., Puel, J.-P.: Optimal control in some variational inequalities. SIAM J. Control Optim. **22**, 466–476 (1984)
17. Martins, J.A.C., Oden, J.T.: Existence and uniqueness results for dynamic contact problems with nonlinear normal and friction interface laws. Nonlinear Anal. TMA **11**, 407–428 (1987)
18. Neittaanmaki, P., Sprekels, J., Tiba, D.: Optimization of Elliptic Systems: Theory and Applications. Springer Monographs in Mathematics. Springer, New York (2006)
19. Hild, P., Renard, Y.: A stabilized lagrange multiplier method for the finite element approximation of contact problems in elastostatics. Numer. Math. **115**, 101–129 (2010)
20. Rochdi, M., Shillor, M., Sofonea, M.: Quasistatic viscoelastic contact with normal compliance and friction. J. Elast. **51**, 105–126 (1998)
21. Sofonea, M., Matei, A.: Variational Inequalities with Applications. A Study of Antiplane Frictional Contact Problems. Advances in Mechanics and Mathematics. Springer, New York (2009)
22. Sofonea, M., Matei, A.: Mathematical Models in Contact Mechanics, London Mathematical Society. Lecture Note Series, vol. 398. Cambridge University Press, Cambridge (2012)
23. Sokolowski, J., Zolesio, J.P.: Introduction to Shape Optimization. Shape Sensitivity Analysis. Springer, Berlin (1991)
24. Wohlmuth, B.: A mortar finite element method using dual spaces for the lagrange multiplier. SIAM J. Numeri. Anal. **38**, 989–1012 (2000)
25. Wohlmuth, B.: Discretization Methods and Iterative Solvers Based on Domain Decomposition. Lecture Notes in Computational Science and Engineering, vol. 17. Springer, Heidelberg (2001)

Multimaterial Topology Optimization of Variational Inequalities

Andrzej Myśliński[✉]

Systems Research Institute, ul. Newelska 6, 01-447 Warsaw, Poland
andrzej.myslinski@ibspan.waw.pl

Abstract. The paper is concerned with the analysis and the numerical solution of the multimaterial topology optimization problems for bodies in unilateral contact. The contact phenomenon with Tresca friction is governed by the elliptic variational inequality. The structural optimization problem consists in finding such topology of the domain occupied by the body that the normal contact stress along the boundary of the body is minimized. The original cost functional is regularized using the multiphase volume constrained Ginzburg-Landau energy functional. The first order necessary optimality condition is formulated. The optimal topology is obtained as the steady state of the phase transition governed by the generalized Allen-Cahn equation. The optimization problem is solved numerically using the operator splitting approach combined with the projection gradient method. Numerical examples are provided and discussed.

Keywords: Topology optimization · Unilateral contact problems · Phase field regularization · Operator splitting method

1 Introduction

Multimaterial topology optimization aims to find the optimal distribution of several elastic materials in a given design domain to minimize a criterion describing the mechanical or the thermal properties of the structure or its cost under constraints imposed on the volume or the mass of the structure [1]. In recent years multiple phases topology optimization problems have become subject of the growing interest [1,5,15]. The use of multiple number of phases during design of engineering structures opens a new opportunities in the design of smart and advanced structures in material science and/or industry. In contrast to single material design the use of multiple number of materials extends the design space and may lead to better design solutions.

Analytical and numerical aspects of the multimaterial structural optimization are subject of intensive research (see references in [1,15]). Many methods including the homogenization method [2], the Solid Isotropic Material Penalization (SIMP) method [3] or different methods based on the level set approach [8,11,12], successful in single material optimization, have been extended to deal

ⓒ IFIP International Federation for Information Processing 2016
Published by Springer International Publishing AG 2016. All Rights Reserved
L. Bociu et al. (Eds.): CSMO 2015, IFIP AICT 494, pp. 380–389, 2016.
DOI: 10.1007/978-3-319-55795-3_36

with the multimaterial optimization. The extension of these methods faces several challenges. A crucial issue in the solution of the multimaterial optimization problems is the lack of physically based parametrization of the phases mixture [1,15]. Although in the literature are proposed different material interpolation schemes, in general, they may influence the optimization path in terms of the computational efficiency and the final design. The level set methods can eliminate the need of the material interpolation schemes provided that interphase interfaces are actually tracked explicitly [1]. Among others, in [1] a multimaterial topology optimization problem for the elliptic equation has been solved using the level set method. The elasticity tensor has been smeared out using the signed distance function. In [15] similar optimization problem has been solved numerically using a generalized Allen–Cahn equation.

The paper is concerned with the structural topology optimization of systems governed by the variational inequalities. The class of such systems includes among others unilateral contact phenomenon [9] between the surfaces of the elastic bodies. This optimization problem consists in finding such topology of the domain occupied by the body that the normal contact stress along the boundary of the body is minimized. In literature [11] this problem usually is considered as two-phase material optimization problem with voids treated as one of the materials. In the paper the domain occupied by the body is assumed to consist from several elastic materials rather than two materials. Material fraction function is a variable subject to optimization. The regularization of the objective functional by the multiphase volume constrained Ginzburg-Landau energy functional is used. The derivative formula of the cost functional with respect to the material fraction function is calculated and is employed to formulate a necessary optimality condition for the topology optimization problem. The cost functional derivative is also used to formulate a gradient flow equation for this functional in the form of the generalized Allen–Cahn equation governing the evolution of the material phases. Two step operator splitting approach [15] is used to solve this gradient flow equation. The optimal topology is obtained as a steady state solution to this equation. Finite difference and finite element methods are used as the approximation methods. Numerical examples are reported and discussed.

2   Problem Formulation

Consider deformations of an elastic body occupying two–dimensional bounded domain Ω with the smooth boundary Γ (see Fig. 1). The body is subject to body forces $f(x) = (f_1(x), f_2(x))$, $x \in \Omega$. Moreover, the surface tractions $p(x) = (p_1(x), p_2(x))$, $x \in \Gamma$, are applied to a portion Γ_1 of the boundary Γ. The body is clamped along the portion Γ_0 of the boundary Γ and the contact conditions are prescribed on the portion Γ_2. Parts Γ_0, Γ_1, Γ_2 of the boundary Γ satisfy: $\Gamma_i \cap \Gamma_j = \emptyset$, $i \neq j$, $i, j = 0, 1, 2$, $\Gamma = \bar{\Gamma}_0 \cup \bar{\Gamma}_1 \cup \bar{\Gamma}_2$.

The domain Ω is assumed to be occupied by $s \geq 2$ distinct isotropic elastic materials. Each material is characterized by Young modulus. The voids are considered as one of the phases, i.e., as a weak material characterized by low

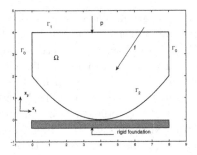

Fig. 1. Elastic body occupying domain Ω in unilateral contact with the foundation.

value of Young modulus [2]. The materials distribution is described by a phase field vector $\rho = \{\rho_m\}_{m=1}^{s}$ where the local fraction field $\rho_m = \rho_m(x) : \Omega \to R$, $m = 1, ..., s$, corresponds to the contributing phase. The phase field approach allows for a certain mixing between different materials. This mixing is restricted only to a small interfacial region. In order to ensure that the phase field vector describes the fractions the following pointwise bound constraints called in material science the Gibbs simplex [5,15] are imposed on every ρ_m

$$\alpha_m \leq \rho_m \leq \beta_m, \text{ for } m = 1, ..., s, \text{ and } \sum_{m=1}^{s} \rho_m = 1, \tag{1}$$

where the constants $0 \leq \alpha_m \leq \beta_m \leq 1$ are given and the summation operator is understood componentwise. The second condition in (1) ensures that no overlap and gap of fractions are allowed in the expected optimal domain. Moreover the total spatial amount of material fractions satisfies

$$\int_{\Omega} \rho_m(x)dx = w_m \mid \Omega \mid, \; 0 \leq w_m \leq 1, \text{ for } m = 1, ..., s, \text{ and } \sum_{m=1}^{s} w_m = 1. \tag{2}$$

The parameters w_m are user defined and $\mid \Omega \mid$ denotes the volume of the domain Ω. From the equality (1) it results that $\rho_s = 1 - \sum_{m=1}^{s-1} \rho_m$ and the fraction ρ_s may be removed from the set of the design functions. Therefore from now on the unknown phase field vector ρ is redefined as $\rho = \{\rho_m\}_{m=1}^{s-1}$. Due to the simplicity and robustness the SIMP material interpolation model [3,4] is used. Following this model the elastic tensor $\mathcal{A}(\rho) = \{a_{ijkl}(\rho)\}_{i,j,l=1}^{2}$ of the material body is assumed to be a function depending on the fraction function ρ:

$$\mathcal{A}(\rho) = \sum_{m=1}^{s} g(\rho_m)\mathcal{A}_m = \sum_{m=1}^{s-1} g(\rho_m)\mathcal{A}_m + g(1 - \sum_{m=1}^{s-1} \rho_m)\mathcal{A}_s, \tag{3}$$

with $g(\rho_m) = \rho_m^3$. The constant stiffness tensor $\mathcal{A}_m = \{\tilde{a}_{ijkl}^{m}\}_{i,j,k,l=1}^{2}$ characterizes the m-th elastic material of the body. For detailed discussion of the interpolation of the material elasticity tensor see [1,4,7]. It is assumed, that elements

a_{ijkl} and $\tilde{a}^m_{ijkl}(x)$, $i, j, k, l = 1, 2$, $m = 1, ..., s$, of the elasticity tensors $\mathcal{A}$ and $\mathcal{A}_m$, respectively, satisfy [9] usual symmetry, boundedness and ellipticity conditions. Denote by $u = (u_1, u_2)$, $u = u(x)$, $x \in \Omega$, the displacement of the body and by $\sigma(x) = \{\sigma_{ij}(u(x))\}$, $i, j = 1, 2$, the stress field in the body. Consider elastic bodies obeying Hooke's law, i.e., for $x \in \Omega$ and $i, j, k, l = 1, 2$,

$$\sigma_{ij}(u(x)) = a_{ijkl}(\rho)e_{kl}(u(x)), \quad e_{kl}(u(x)) \stackrel{def}{=} \frac{1}{2}(u_{k,l}(x) + u_{l,k}(x)), \quad (4)$$

where $u_{k,l}(x) = \frac{\partial u_k(x)}{\partial x_l}$. We use here and throughout the paper the summation convention over repeated indices [9]. The stress field σ satisfies the system of equations in the domain Ω [9]

$$-\sigma_{ij}(x)_{,j} = f_i(x) \quad \sigma_{ij}(x)_{,j} = \frac{\partial \sigma_{ij}(x)}{\partial x_j}, \quad x \in \Omega, \ i, j = 1, 2. \quad (5)$$

The following boundary conditions are imposed on the boundary $\partial\Omega$

$$u_i(x) = 0 \quad \text{on} \quad \Gamma_0, \quad \sigma_{ij}(x)n_j = p_i \quad \text{on} \quad \Gamma_1, \ i, j = 1, 2, \quad (6)$$

$$(u_N + v) \leq 0, \quad \sigma_N \leq 0, \quad (u_N + v)\sigma_N = 0 \quad \text{on} \ \Gamma_2, \quad (7)$$

$$|\sigma_T| \leq 1, \quad u_T\sigma_T + |u_T| = 0 \quad \text{on} \ \Gamma_2, \quad (8)$$

where $n = (n_1, n_2)$ is the unit outward versor to the boundary Γ. Here $u_N = u_i n_i$ and $\sigma_N = \sigma_{ij}n_i n_j$, $i, j = 1, 2$, represent [9] the normal components of displacement u and stress σ, respectively. The tangential components of displacement u and stress σ are given [9] by $(u_T)_i = u_i - u_N n_i$ and $(\sigma_T)_i = \sigma_{ij}n_j - \sigma_N n_i$, $i, j = 1, 2$, respectively. $|u_T|$ denotes the Euclidean norm in R^2 of the tangent vector u_T. A gap between the bodies is described by a given function v.

Let us formulate contact problem (5)–(8) in the variational form. Denote by $V_{sp} = \{z \in H^1(\Omega; R^2) : z_i = 0 \text{ on } \Gamma_0, \ i = 1, 2\}$ and $K = \{z \in V_{sp} : z_N + v \leq 0 \text{ on } \Gamma_2\}$ the space and the set of kinematically admissible displacements and by $\Lambda = \{\zeta \in L^2(\Gamma_2; R^2) : |\zeta| \leq 1\}$ the set of tangential tractions on Γ_2. Variational formulation of problem (5)–(8) has the form: for a given $(f, p, \rho) \in L^2(\Omega; R^2) \times L^2(\Gamma_2; R^2) \times L^\infty(\Omega; R^{s-1}) \cap H^1(\Omega; R^{s-1})$ find a pair $(u, \lambda) \in K \times \Lambda$ satisfying for $i, j, k, l = 1, 2$

$$\int_\Omega a_{ijkl}(\rho)e_{ij}(u)e_{kl}(\varphi_i - u_i)dx - \int_\Omega f_i(\varphi_i - u_i)dx -$$

$$\int_{\Gamma_1} p_i(\varphi_i - u_i)ds + \int_{\Gamma_2} \lambda_i(\varphi_{Ti} - u_{Ti})ds \geq 0 \quad \forall\varphi_i \in K, \quad (9)$$

$$\int_{\Gamma_2} (\zeta_i - \lambda_i)u_{Ti}ds \leq 0 \quad \forall\zeta_i \in \Lambda. \quad (10)$$

The function λ is interpreted as a Lagrange multiplier corresponding to the term $|u_T|$ in equality constraint in (8) [9]. This function is equal to tangent stress along the boundary Γ_2, i.e., $\lambda = -\sigma_{T|\Gamma_2}$ and belongs to space $H^{-1/2}(\Gamma_2; R^2)$. Here following [9] function λ is assumed to be more regular, i.e., $\lambda \in L^2(\Gamma_2; R^2)$. Recall from [9,14] under assumptions imposed on the elasticity tensors by standard arguments for a given $(f, p, \rho) \in L^2(\Omega; R^2) \times L^2(\Gamma_2; R^2) \times L^\infty(\Omega; R^{s-1})$ there exists a unique solution $(u, \lambda) \in K \times \Lambda$ to the system (9)–(10).

3   Phase Field Based Topology Optimization Problem

Before formulating a structural optimization problem for the system (9)–(10) let us introduce a set U_{ad}^ρ of the admissible fraction functions:

$$U_{ad}^\rho = \{\rho \in L^\infty(\Omega; R^{s-1}) \cap H^1(\Omega; R^{s-1}) : 1 - \beta_s \le \sum_{m=1}^{s-1} \rho_m \le 1 - \alpha_s,$$

$$\alpha_m \le \rho_m \le \beta_m, \int_\Omega \rho_m dx = w_m \mid \Omega \mid \text{ for } m = 1, ..., s - 1\}. \quad (11)$$

The set U_{ad}^ρ is assumed to be nonempty. The aim of the structural optimization of the bodies in contact is to reduce the normal contact stress responsible for wear, vibrations of the contacting surfaces or generated noise. The structural optimization problem with normal contact stress functional is difficult to analyze it and to solve it numerically. Therefore following [11] we shall use the cost functional $J_\eta : H^1(\Omega) \to R$ approximating the normal contact stress on the contact boundary Γ_2

$$J_\eta(u(\rho)) = \int_{\Gamma_2} \sigma_N(u(\rho))\eta_N(x)ds, \quad (12)$$

depending on a given auxiliary bounded function $\eta(x) \in M^{st}$. The set $M^{st} = \{\eta = (\eta_1, \eta_2) \in H^1(\Omega; R^2) : \eta_i \le 0 \text{ on } \Omega, i = 1, 2, \parallel \eta \parallel_{H^1(\Omega; R^2)} \le 1\}$. Functions σ_N and η_N are the normal components of the stress field σ and the function η, respectively. The optimization problem consisting in finding such $\rho \in U_{ad}^\rho$ to minimize the functional $J_\eta(u(\rho))$ in general has no solutions [2,6,7,9, 14,15]. In order to ensure the existence of optimal solutions let us regularize the cost functional (12) by adding to it a regularizing term $E(\rho) : U_{ad}^\rho \to R$ rather than the standard perimeter term [2,7,14]

$$J(\rho, u(\rho)) = J_\eta(u(\rho)) + E(\rho). \quad (13)$$

The Ginzburg-Landau free energy functional $E(\rho)$ is expressed as [7,15]

$$E(\rho) = \sum_{m=1}^{s-1} \int_\Omega \psi(\rho_m)d\Omega, \quad \psi(\rho_m) = \frac{\gamma\epsilon}{2} \mid \nabla\rho_m \mid^2 + \frac{\gamma}{\epsilon}\psi_B(\rho_m), \quad (14)$$

where $\epsilon > 0$ is a real constant governing the width of the intrefaces, $\gamma > 0$ is a real parameter related to the interfacial energy density. Moreover $\nabla\rho_m \cdot n = 0$ on Γ for each m. The function $\psi_B(\rho_m) = \rho_m^2(1 - \rho_m)^2$ is a double-well potential [7] which characterizes the concentration of the material phases [15]. The structural optimization problem for the system (9)–(10) takes the form: *find $\rho^\star \in U_{ad}^\rho$ such that*

$$J(\rho^\star, u^\star) = \min_{\rho \in U_{ad}^\rho} J(\rho, u(\rho)), \tag{15}$$

where $u^\star = u(\rho^\star)$ denotes a solution to the state system (9)–(10) depending on $\rho^\star \in L^\infty(\Omega; R^{s-1}) \cap H^1(\Omega; R^{s-1})$ and the set U_{ad}^ρ is given by (11). The existence of an optimal solution $\rho^\star \in U_{ad}^\rho$ to the problem (15) follows by classical arguments (see [5, 6, 14]).

4   Necessary Optimality Condition

Let us apply the Lagrangian approach to compute the derivative of the cost functional (13) with respect to the function ρ. The Lagrangian $L(\rho) = L(\rho, u, \lambda, p^a, q^a) : L^\infty(\Omega; R^{s-1}) \cap H^1(\Omega; R^{s-1}) \cap U_{ad}^\rho \times H^1(\Omega; R^2) \times L^2(\Gamma_2; R^2) \times K_1 \times \Lambda_1$ associated to the problem (15) is expressed for $i, j, k, l = 1, 2$ as

$$L(\rho, u, \lambda, p^a, q^a) = J_\eta(u(\rho)) + E(\rho) + \int_\Omega a_{ijkl}(\rho) e_{ij}(u) e_{kl}(p^a) dx -$$

$$\int_\Omega f_i p_i^a dx - \int_{\Gamma_1} p_i p_i^a ds + \int_{\Gamma_2} \lambda_i p_{Ti}^a ds + \int_{\Gamma_2} q_i^a u_{Ti} ds. \tag{16}$$

Using the same approach as in proof of [14, Theorem 4.35, p. 219] an adjoint state $(p^a, q^a) \in K_1 \times \Lambda_1$ for $i, j, k, l = 1, 2$ is defined as the solution to the system

$$\int_\Omega a_{ijkl}(\rho) e_{ij}(\eta + p^a) e_{kl}(\varphi) dx + \int_{\Gamma_2} q_i^a \varphi_{Ti} ds = 0 \quad \forall \varphi_i \in K_1, \tag{17}$$

$$\int_{\Gamma_2} \zeta_i (p_{Ti}^a + \eta_{Ti}) ds = 0 \quad \forall \zeta_i \in \Lambda_1. \tag{18}$$

The sets K_1 and Λ_1 are given by $K_1 = \{\xi \in V_{sp} : \xi_N = 0 \text{ on } A^{st}\}$ as well as $\Lambda_1 = \{\zeta \in \Lambda : \zeta(x) = 0 \text{ on } B_1 \cup B_2 \cup B_1^+ \cup B_2^+\}$, while the coincidence set $A^{st} = \{x \in \Gamma_2 : u_N + v = 0\}$. Moreover the other sets are determined as $B_1 = \{x \in \Gamma_2 : \lambda(x) = -1\}$, $B_2 = \{x \in \Gamma_2 : \lambda(x) = +1\}$, $\tilde{B}_i = \{x \in B_i : u_N(x) + v = 0\}$, $i = 1, 2$, $B_i^+ = B_i \setminus \tilde{B}_i$, $i = 1, 2$. From [12], [14, Theorems 4.16, 4.27] it follows the mapping $\rho \to u(\rho)$ is Gâteaux differentiable. Using (9)–(10) and (16)–(18) the derivative of the Lagrangian L with respect to ρ is determined for all $\zeta \in H^1(\Omega; R^{s-1})$ and $i, j, k, l = 1, 2$ as

$$\int_\Omega \frac{\partial J}{\partial \rho}(\rho, u) \zeta dx = \int_\Omega \frac{\partial L}{\partial \rho}(\rho, u, \lambda, p^a, q^a) \zeta dx = \sum_{m=1}^{s-1} \int_\Omega [\gamma \epsilon \nabla \rho_m \cdot \nabla \zeta_m +$$

$$\frac{\gamma}{\epsilon} \psi_B'(\rho_m) \zeta_m] dx + \int_\Omega [a'_{ijkl}(\rho_m) e_{ij}(u) e_{kl}(p^a + \eta) - f_i(p_i^a + \eta_i)] \zeta_m dx. \tag{19}$$

From (1) and (3) it results the derivatives of the function $\psi_B(\rho_m)$ and the tensor element $a_{ijkl}(\rho_m)$ with respect to ρ_m are equal to $\psi_B'(\rho_m) = 4\rho_m^3 - 6\rho_m^2 + 2\rho_m$ and

$a'_{ijkl}(\rho) = 3\rho_m^2 \tilde{a}_{ijkl}^m - 3\rho_s^2 \tilde{a}_{ijkl}^s$, respectively. Using (19) the necessary optimality condition to the optimization problem (15) takes the form [2], [13, Lemma 2.21]:

Theorem 1. *Let U_{ad}^ρ be a nonempty closed convex subset of $H^1(\Omega; R^{s-1})$ and $\rho^\star \in U_{ad}^\rho$ be an optimal solution to the structural optimization problem (15). Then*

$$\int_\Omega \frac{\partial J}{\partial \rho}(\rho^\star, u^\star)(\rho - \rho^\star)dx \geq 0 \quad \forall \rho \in U_{ad}^\rho. \tag{20}$$

The functions $(u^\star, \lambda^\star) \in K \times \Lambda$ and $(p^{a\star}, q^{a\star}) \in K_1 \times \Lambda_1$ in the derivative formula (19) denote the solutions to the systems (9)–(10) and (17)–(18) for $\rho = \rho^\star$.

Using the orthogonal projection operator $P_{U_{ad}^\rho} : L^2(\Omega; R^{s-1}) \rightarrow U_{ad}^\rho$ from $L^2(\Omega; R^{s-1})$ on the set U_{ad}^ρ condition (20) can be written [15] in the form: if $\rho^\star \in U_{ad}^\rho$ is an optimal solution to the structural optimization problem (15), then for $\mu \in R$ and $\mu > 0$

$$P_{U_{ad}^\rho}[\rho^\star - \mu \frac{\partial J(\rho^\star, u^\star)}{\partial \rho}] - \rho^\star = 0. \tag{21}$$

Recall [7] the structural optimization problem (15) can be considered as a phase transition setting problem consisting in such evolution of the phases to minimize the cost functional (13) with respect to the initial configuration. In order to describe the evolution of phases in time let us assume that the phase field vector ρ depends not only on $x \in \Omega$ but also on time variable $t \in [0, T]$, $T > 0$ is a given constant, i.e., $\rho = \rho(x, t) = \{\rho_m(x, t)\}_{m=1}^{s-1}$. The variable t may be interpreted as an artificial time or iteration number in the computational algorithm [7,15]. Using the right hand side of (21) let us formulate the constrained gradient flow equation of Allen-Cahn type [5,7,15] for the cost functional (13): *find function $\rho \in U_{ad}^\rho$ satisfying the initial boundary value problem:*

$$\frac{\partial \rho}{\partial t} = -P_{U_{ad}^\rho}[\rho - \mu \frac{\partial J(\rho, u)}{\partial \rho}] + \rho \quad \text{in } \Omega, \forall t \in [0, T), \tag{22}$$

$$\nabla \rho \cdot n = 0 \qquad \text{on } \partial\Omega, \forall t \in [0, T), \tag{23}$$

$$\rho(0, x) = \rho_0(x) \qquad \text{in } \Omega, t = 0, \tag{24}$$

with $\rho_0(x) = \{\rho_{0m}(x)\}_{m=1}^{s-1}$ denoting a given $H^1(\Omega; R^{s-1})$ regular function. For $\rho_0 \in H^1(\Omega; R^{s-1})$ the system (22)–(24) possesses a solution $\rho \in L^\infty(0, T; H^1(\Omega)) \cap H^1(0, T; L^2(\Omega; R^{s-1}))$ (see [10]). The stationary solutions of (22)–(24) fulfill the first order necessary optimality conditions (20) or (21) for the problem (15) [5,7]. For $\frac{\partial \rho}{\partial t} = 0$ the right hand side of the Eq. (22) vanishes and $\rho(x, t) = \rho^\star(x, t)$ is an optimal solution to the problem (15).

For the sake of numerical calculations we reformulate the initial boundary value problem (22)–(24) using the operator splitting approach [5,15]. Remark, the cost functional (13) may be represented as a sum of two functionals, i.e., $J(\rho, u) = J_1(\rho, u) + J_2(\rho)$ given by $J_1(\rho, u) = J_\eta(u(\rho)) + \sum_{m=1}^{s-1} \int_\Omega \frac{\gamma}{\epsilon} \psi_B(\rho_m) d\Omega$ and $J_2(\rho) = \sum_{m=1}^{s-1} \int_\Omega \frac{\gamma\epsilon}{2} |\nabla\rho_m|^2 d\Omega$. The derivatives of these functionals result from formula (19). Assume the time interval $[0, T]$ is divided into N subintervals with stepsize $\Delta t = t_{k+1} - t_k$, $k = 1, ..., N$ and $\rho_k = \rho(t_k)$ is known. The design variable ρ_{k+1} at the next time step t_{k+1} is calculated in two substeps. First the trial value $\tilde{\rho}$ is calculated from the gradient flow Eq. (22) for the functional J_1 only. Next this solution is updated to ensure its $H^1(\Omega)$ regularity [5] by solving the gradient flow Eq. (22) for the functional J_2 only with the boundary condition (23), i.e.,

$$\frac{\partial \tilde{\rho}}{dt} = -P_{U_{ad}^\rho}[\tilde{\rho} - \frac{\partial J_1(\tilde{\rho}, u)}{\partial \rho}] + \tilde{\rho}, \quad \tilde{\rho}(t_k) = \rho_k, \quad t_k < t \leq t_{k+1}. \tag{25}$$

$$\frac{\partial \rho}{dt} = -\frac{\partial J_2(\rho)}{\partial \rho}, \quad \rho(t_k) = \tilde{\rho}_{k+1}, \quad t_k < t \leq t_{k+1}. \tag{26}$$

5   Numerical Results

The topology optimization problem (15) has been discretized and solved numerically. Time derivatives are approximated by the forward finite difference. Piecewise constant and piecewise linear finite element method is used as discretization method in space variables. The derivative of the double well potential is linearized with respect to ρ_m. Primal-dual active set method has been used to solve the state and adjoint systems (5)–(8) and (17)–(18). The initial boundary value problem (22)–(24) has been solved in two steps according to scheme (25)–(26). The algorithms are programmed in Matlab environment. As an example a body occupying 2D domain

$$\Omega = \{(x_1, x_2) \in R^2 : 0 \leq x_1 \leq 8 \land 0 < v(x_1) \leq x_2 \leq 4\}, \tag{27}$$

is considered. The boundary Γ of the domain Ω is divided into three disjoint pieces $\Gamma_0 = \{(x_1, x_2) \in R^2 : x_1 = 0, 8 \land 0 < v(x_1) \leq x_2 \leq 4\}$, $\Gamma_1 = \{(x_1, x_2) \in R^2 : 0 \leq x_1 \leq 8 \land x_2 = 4\}$, $\Gamma_2 = \{(x_1, x_2) \in R^2 : 0 \leq x_1 \leq 8 \land v(x_1) = x_2\}$. The domain Ω and the boundary Γ_2 depend on the function $v(x_1) = 0.125 \cdot (x_1 - 4)^2$. Domain Ω is filled with $s = 3$ elastic materials. The Poisson's ratio of each material is $\nu = .3$. The Young's moduli of materials are: $E_1 = 6 \cdot E_0$, $E_2 = 3 \cdot E_0$ and $E_3 = E_0$, $E_0 = 2.1 \cdot 10^{11}$ Pa. The parameters w_1, w_2, w_3 are equal to .25, .5 and .25 respectively. As an initial design ρ_0 a feasible design with the uniform material distribution has been taken. The body is loaded by the boundary traction $p_1 = 0$, $p_2 = -5.6 \cdot 10^6$ N along the boundary Γ_1, the body forces $f_i = 0$, $i = 1, 2$. The auxiliary function η is selected as a piecewise linear on Ω and is approximated by a piecewise linear function. The domain Ω is divided into 80×40 grid. The parameters ε and γ are equal to the mesh size and to 0.5, respectively. The total number of iterations k_{max} in the optimization algorithm has been set to 90. It is approximately equivalent to final time $T = 125$ s.

Figure 2 presents the optimal topology domain obtained by solving structural optimization problem (15) using the necessary optimality condition (22)–(24). The areas with the weak phases appear in the central part of the body and near the fixed edges. The areas with the strong phases appear close to the contact

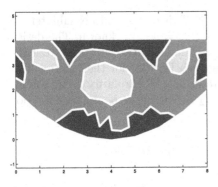

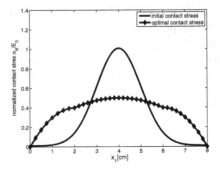

Fig. 2. Optimal material distribution in domain Ω^*.

Fig. 3. Initial and optimal normal contact stress.

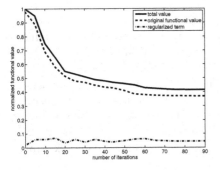

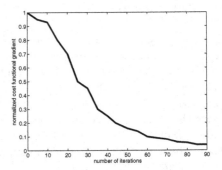

Fig. 4. Convergence of the cost functional value.

Fig. 5. Convergence of the cost functional gradient.

zone and along the edges. The rest of the domain is covered with the intermediate phase. The obtained normal contact stress for the optimal topology is almost constant along the contact boundary and has been significantly reduced comparing to the initial one (see Fig. 3). The convergence of the cost functional value and its gradient as a function of the number of iterations are shown on Figs. 4 and 5, respectively. The cost functional value decreases almost monotonically when the number of iterations increases. At the beginning this decrease is significant and finally the cost functional value is almost steady. The regularized functional value at first is increasing and next rapidly decreasing (see Fig. 4). Similarly, after a few initial iterations the gradient of the cost functional also almost monotonically decreases to reach the steady state (Fig. 5).

6   Conclusions

The topology optimization problem for elastic contact problem with Tresca friction has been solved numerically in the paper. Obtained numerical results indicate that the optimal topologies are qualitatively comparable to the results reported in other phase-field topology optimization methods. Since the optimization problem is non-convex it has possibly many local solutions dependent on initial estimate. Gradient flow method employed in H^1 space is more regular and efficient than standard Allen-Cahn approach.

References

1. Allaire, G., Dapogny, C., Delgado, G., Michailidis, G.: Multi-phase structural optimization via a level set method. ESAIM - Control Optimisation Calc. Var. **20**, 576–611 (2014)
2. Allaire, G.: Shape optimization by the homogenization method. Springer, New York (2001)
3. Bendsoe, M.P., Sigmund, O.: Topology Optimization: Theory, Methods, and Applications. Springer, Berlin (2004)
4. Blank, L., Butz, M., Garcke, H., Sarbu, L., Styles, V.: Allen-Cahn and Cahn-Hiliard variational inequalities solved with optimization techniques. In: Leugering, G., Engell, S., Griewank, A., Hinze, M., Rannacher, R., Schulz, V., Ulbrich, M., Ulbrich, S. (eds.) Constrained Optimization and Optimal Control for Partial Differential Equations, International Series of Numerical Mathematics, vol. 160, pp. 21–35. Birkhäuser, Basel (2012)
5. Blank, L., Farshbaf-Shaker, M.H., M., Garcke, H., Rupprecht, C., Styles, V.: Multimaterial phase field approach to structural topology optimization. In: Leugering, G., Benner, P., Engell, S., Griewank, A., Harbrecht, H., Hinze, M., Rannacher, R., Ulbrich, S.: Trends in PDE Constrained Optimization. International Series of Numerical Mathematics 165, pp. 231–246. Birkhäuser, Basel (2014)
6. Burger, M., Stainko, R.: Phase-field relaxation of topology optimization with local stress constraints. SIAM J. Control. Optim. **45**, 1447–1466 (2006)
7. Dede, L., Boroden, M.J., Hughes, T.J.R.: Isogeometric analysis for topology optimization with a phase field model. Arch. Comput. Methods Eng. **19**(3), 427–465 (2012)
8. van Dijk, N.P., Maute, K., Langlaar, M., van Keulen, F.: Level-set methods for structural topology optimization: a review. Struct. Multi. Optim. **48**, 437–472 (2013)
9. Haslinger, J., Mäkinen, R.: Introduction to Shape Optimization. Theory Approximation, and Computation. SIAM Publications, Philadelphia (2003)
10. Karali, G., Ricciardi, T.: On the convergence of a fourth order evolution equation to the Allen Cahn equation. Nonlinear Anal. **72**, 4271–4281 (2010)
11. Myśliński, A.: Piecewise constant level set method for topology optimization of unilateral contact problems. Adv. Eng. Softw. **80**, 25–32 (2015)
12. Myśliński, A.: Level set method for optimization of contact problems. Eng. Anal. Bound. Elem. **32**, 986–994 (2008)
13. Tröltzsch, F.: Optimal Control of Partial Differential Equations: Theory, Methods and Applications. Graduate Studies in Mathematics, vol. 112. American Mathematical Society, Providence (2010)
14. Sokołowski, J., Zolesio, J.P.: Introduction to Shape Optimization. Shape Sensitivity Analysis. Springer, Berlin (1992)
15. Tavakoli, R.: Multimaterial topology optimization by volume constrained Allen-Cahn system and regularized projected steepest descent method. Comput. Meth. Appl. Mech. Eng. **276**, 534–565 (2014)

Implicit Parametrizations and Applications

Mihaela Roxana Nicolai[1(✉)] and Dan Tiba[1,2]

[1] Institute of Mathematics, Romanian Academy,
P.O.BOX 1-764, 014700 Bucharest, Romania
roxana.nicolai@gmail.com, dan.tiba@imar.ro
[2] Academy of Romanian Scientists,
Splaiul Independenței 54, 050094 Bucharest, Romania

Abstract. We discuss recent constructive parametrizations approaches for implicit systems, via systems of ordinary differential equations. We also present the notion of generalized solution, in the critical case and indicate some numerical examples in dimension two and three, using MatLab. In shape optimizations problems, using this method, we introduce general optimal control formulations in the boundary observation case. This extends previous work of the authors on optimal design problems with distributed cost functional.

Keywords: Implicit systems · Local parametrization · Shape optimization · Boundary observation

1 Introduction

The implicit function theorem or the inverse function theorem ensure as well the local existence of implicit parametrizations for the solution of implicit systems, under classical assumptions, [2,3]. Recently, in authors' papers, [8,14], explicit constructions (via iterated ordinary differential equations) of implicit parametrizations, in dimensions two and three, have been discussed. A possible extension of such constructions to arbitrary dimension is investigated in the preprint [15]. Moreover, this new approach allows the introduction of the notion of generalized solution, solving implicit systems under C^1 hypotheses, in the critical case.

Such considerations have impact in shape optimization problems (fixed domain methods) where implicit representations of domains play an essential role. The aim is to obtain a general method to solve optimal design problems via optimal control theory. In the case of Dirichlet boundary conditions, a theoretical analysis together with numerical experiments are reported in [5,7]. For a general background, we quote [6]. Notice that this method is essentially different from the level set method (for instance, no Hamilton-Jacobi equation is necessary in [5,7]).

This paper is organized as follows. In Sect. 2, we recall briefly the implicit parametrization approach. Some numerical examples in the critical case, in dimension two and three are outlined in Sect. 3. The last section discusses possible applications in shape optimization.

© IFIP International Federation for Information Processing 2016
Published by Springer International Publishing AG 2016. All Rights Reserved
L. Bociu et al. (Eds.): CSMO 2015, IFIP AICT 494, pp. 390–398, 2016.
DOI: 10.1007/978-3-319-55795-3_37

2 Implicit Parametrizations and Generalized Solution

We limit the presentation to the case of one implicit equation in dimension three:

$$f(x, y, z) = 0, \quad in \ \Omega \subset \mathbb{R}^3, \tag{1}$$

Ω an open connected subset. We assume that $f \in \mathcal{C}^1(\Omega)$ and there is $(x^0, y^0, z^0) \in \Omega$ such that (1) is satisfied. We also impose, for the moment, that (x^0, y^0, z^0) is noncritical for f, i.e. $\nabla f(x^0, y^0, z^0) \neq 0$. To fix ideas, we assume:

$$f_x(x^0, y^0, z^0) \neq 0. \tag{2}$$

Later, we shall remove this hypothesis and discuss the critical case. For the general situation, we quote [15].

We associate to (1) the following systems of first order partial differential equations (with independent variables t and s) of iterated type:

$$
\begin{aligned}
x'(t) &= -f_y(x(t), y(t), z(t)), & t &\in I_1, \\
y'(t) &= f_x(x(t), y(t), z(t)), & t &\in I_1, \\
z'(t) &= 0, & t &\in I_1,
\end{aligned}
\tag{3}
$$

$$x(0) = x^0, \ y(0) = y^0, \ z(0) = z^0; \tag{4}$$

$$
\begin{aligned}
\dot{\varphi}(s, t) &= -f_z(\varphi(s, t), \psi(s, t), \xi(s, t)), & s &\in I_2(t), \\
\dot{\psi}(s, t) &= 0, & s &\in I_2(t), \\
\dot{\xi}(s, t) &= f_x(\varphi(s, t), \psi(s, t), \xi(s, t)), & s &\in I_2(t), \\
\varphi(0, t) &= x(t), \ \psi(0, t) = y(t), \ \xi(0, t) = z(t), & t &\in I_1.
\end{aligned}
\tag{5}
$$

$$ \tag{6}$$

The iterated character of the PDE system (3)–(6) consists in the fact that the coupling between (3)–(6) is made just via the initial conditions (6). This very weak coupling, together with the presence of just one derivative in each equation create the possibility to solve (3)–(6) as ODE systems. Namely, the system (3)–(4) is indeed of ODE type. The system (5)–(6) has the t independent variable as a parameter, entering via the initial conditions and may be interpreted as an ODE system with parameters. The existence may be obtained via the Peano theorem since $f \in \mathcal{C}^1(\Omega)$. Moreover, one may infer via this theorem and some simple calculations that the local existence interval $I_2(t)$ may be chosen independently of $t \in I_1$, i.e. $I_2(t) = I_2$. Under slightly stronger regularity assumptions, for instance ∇f locally Lipschitzian, we also obtain uniqueness for (3)–(6). For very weak assumptions in this sense, see [18] or [1], since the system (3)–(6) has divergence free right-hand side.

The fact that (3)–(6) provides a parametrization of the solution of (1), around (x^0, y^0, z^0) is stated in the following theorem, proved in [8]:

Theorem 1. *Assume that $f \in \mathcal{C}^2(\Omega)$ and I_1, I_2 are sufficiently small. Then, $(\varphi, \psi, \xi) : I_1 \times I_2 \to \Omega$ is a regular transformation on its image.*

Remark 1. The systems (3)–(4), respectively (5)–(6) are of Hamiltonian type, in fact. In dimension two, for the implicit equation $g(x, y) = 0$, just one Hamiltonian system may be used, to obtain the parametrization (see [13, 14]):

$$a'(t) = -g_y(a(t), b(t)), \ t \in I_1,$$
$$b'(t) = g_x(a(t), b(t)), \ t \in I_1, \tag{7}$$
$$a(0) = a^0, \ b(0) = b^0,$$

In arbitrary dimension, the solution is more involved [15].

Remark 2. The advantage of Theorem 1 or other implicit parametrization results, is exactly their explicit character. Moreover, the fact that the solution is obtained via systems of ordinary differential equations makes it possible to use maximal solutions. Theorem 1 has a local character, but in applications, the maximal existence intervals I_1, I_2 may be very large. In many cases, one may obtain even global solutions [8]. In comparison with implicit function theorems, removing the restrictive requirement that the solution is in function form, allows to obtain a more complete description of the manifold corresponding to (1).

The above construction provides the basis for the introduction of the generalized solution of (1), in the critical case and we recall it briefly here, for reader's convenience. We remove hypothesis (2), that is the point (x^0, y^0, z^0) may be critical. We notice that the following weaker property is valid in general: there is $(x^n, y^n, z^n) \in \Omega$ such that

$$(x^n, y^n, z^n) \rightarrow (x^0, y^0, z^0) \text{ such that } \nabla f(x^n, y^n, z^n) \neq 0. \tag{8}$$

If (8) is not valid, then $\nabla f(x^0, y^0, z^0) = 0$ in a neighborhood V of (x^0, y^0, z^0) and, consequently, $f(x, y, z) = 0$ in V. This is a trivial situation of no interest. For general implicit systems, a similar property may be stated, expressing the fact that the equations of the system have to be functionally independent [15].

Due to (8), one can construct the solution of (3)–(6) with initial condition in (x^n, y^n, z^n). We denote by $T^n \subset \Omega$, the set described by $(\varphi^n, \psi^n, \xi^n)$ obtained in (5), in this case. Since f is in $C^1(\Omega)$, T^n may be assumed compact. By truncation (or imposing Ω to be bounded), we get $\{T^n\}$ bounded with respect to n. On a subsequence denoted by α, we have:

$$T^n \rightarrow T_\alpha$$

in the Hausdorff-Pompeiu metric, where $T_\alpha \subset \Omega$ is some compact subset, [6].

Definition 1. *In the general (critical) case, we call $T = \cup T_\alpha$ to be the generalized solution of (1), where the union is taken after all the sequences and subsequences as above.*

It can be shown that $(x^0, y^0, z^0) \in T$, any point in T satisfies (1) and that Definition 1 provides the usual solution in the classical nonsingular case, [15]. That is, Definition 1 is an extension of the classical notion of solution. If (x^0, y^0, z^0)

is an isolated critical point, then again T coincides (locally) with the solution of (1). Otherwise, T may be just the boundary of the solution set of (1), according to [15]. A complete description of the level sets (even of positive measure) of a function, may be obtained in this way.

An algorithm for the approximation of generalized solutions is detailed in [9]. In the next section we indicate some computational examples in this sense.

3   Numerical Examples

All the examples in this section were performed with MatLab. We consider just the critical case.

Example 1. Let $g(x,y) = x^3 - siny$, with the critical point $(x_0, y_0) = \left(0, \dfrac{\pi}{2}\right)$.

We are in the critical case: $g(x_0, y_0) = 0$, $\nabla g(x_0, y_0) = (0,0)$. According to Definition 1, we have to solve (7) with approximating initial conditions $\left(0, \dfrac{\pi}{2}\right) + \left(\pm\dfrac{1}{100}, \mp\dfrac{1}{100}\right)$, $\left(0, \dfrac{\pi}{2}\right) + \left(\pm\dfrac{1}{100}, \pm\dfrac{1}{100}\right)$. The result is shown in Fig. 1. In Fig. 2 we perform a zoom around the critical point. The four curves corresponding to these initial conditions cannot be distinguished visually, from each other.

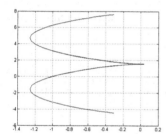

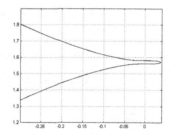

Fig. 1. $g(x,y)=x^3-\sin y, (x_0, y_0)=(0, \dfrac{\pi}{2})$

Fig. 2. Zoom around $(x_0, y_0) = (0, \dfrac{\pi}{2})$

Example 2. Let $f(x, y, z) = xyz$, with the initial point $(x_0, y_0, z_0) = (0,0,0)$.

We are in the critical case: $f(x_0, y_0, z_0) = 0$ and $\nabla f(x_0, y_0, z_0) = 0$. Consider the approximating initial conditions $\left(\dfrac{1}{50}, \dfrac{1}{50}, \dfrac{1}{50}\right)$, $\left(-\dfrac{1}{50}, -\dfrac{1}{50}, \dfrac{1}{50}\right)$, $\left(-\dfrac{1}{50}, \dfrac{1}{50}, \dfrac{1}{50}\right)$, $\left(\dfrac{1}{50}, -\dfrac{1}{50}, \dfrac{1}{50}\right)$.

In Fig. 3 we show the results for the system (3)–(4), the four curves corresponding to the four initial conditions. In Fig. 4 we superpose the results, corresponding to each of the previous four curves as initial conditions, obtained for the system (5)–(6) together with the following alternative (see [9]) system:

$$\dot{\varphi} = 0, \qquad\qquad\qquad s \in I_2$$
$$\dot{\psi} = -f_z(\varphi, \psi, \xi), \qquad\qquad s \in I_2, \qquad (9)$$
$$\dot{\xi} = f_y(\varphi, \psi, \xi), \qquad\qquad s \in I_2,$$
$$\varphi(0) = x(t), \ \psi(0) \ = \ y(t), \ \xi(0) \ = \ z(t),$$

The dark colours correspond to the points where both systems produce solutions.

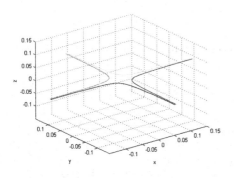

Fig. 3. Ex. 2: the first system   **Fig. 4.** Ex. 2: the second systems

Using the symmetry, the result can be extended for $z < 0$.

Example 3. Let $f(x, y, z) = (10x^2 - y^2 - z)(x^2 - 10y^2 - z)$, with the initial point $(x_0, y_0, z_0) = (0, 0, 0)$.

We are again in the critical case: $f(x_0, y_0, z_0) = 0$ and $\nabla f(x_0, y_0, z_0) = 0$. Consider the approximating initial conditions $\left(\dfrac{1}{10}, \dfrac{1}{10}, \dfrac{9}{100}\right)$, $\left(-\dfrac{1}{10}, \dfrac{1}{10}, \dfrac{9}{100}\right)$, $\left(\dfrac{1}{10}, \dfrac{1}{10}, -\dfrac{9}{100}\right)$, $\left(-\dfrac{1}{10}, \dfrac{1}{10}, -\dfrac{9}{100}\right)$, $\left(\dfrac{1}{10}, -\dfrac{1}{10}, -\dfrac{9}{100}\right)$, $\left(-\dfrac{1}{10}, -\dfrac{1}{10}, -\dfrac{9}{100}\right)$.

In Fig. 5 we represent the solutions of the systems (3)–(4), respectively (5)–(6) for the initial condition $\left(\dfrac{1}{10}, \dfrac{1}{10}, \dfrac{9}{100}\right)$. In Fig. 6 we have put together the solutions of the second system corresponding to all the six approximating initial conditions. In fact, for the initial conditions with negative third coordinate we have used the variant (9) of the system (5)–(6) that avoids $y = ct$, specific to (5)–(6), and improves the graphical representation. We have intendedly represented, for clarity, just a small number of integral curves.

4   Applications in Shape Optimization

A typical shape optimization problem has the form:

$$\min_{\Omega} \int_{\Lambda} j(y_\Omega(x), x)dx \qquad (10)$$

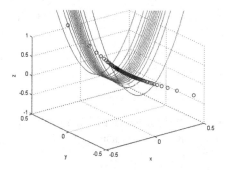

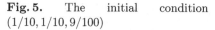

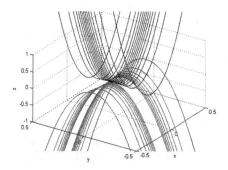

Fig. 5. The initial condition $(1/10, 1/10, 9/100)$

Fig. 6. The solution corresponding to the six initial conditions

subject to

$$- \Delta y_\Omega = f \quad in \ \Omega \tag{11}$$

$$y_\Omega = 0 \quad on \ \partial\Omega \tag{12}$$

Here $E \subset \Omega \subset D \subset \mathbb{R}^d$, $d \in \mathbb{N}$, are bounded Lipschitzian domains, with E and D given and Ω unknown, $f \in L^2(D)$, Λ is either E or Ω and $j : \mathbb{R} \times D \to \mathbb{R}$ satisfies appropriate measurability and other hypotheses [6]. More general state equation, other boundary conditions may be considered as well. Traditional solution methods are boundary variations, speed method, topological asymptotics, the mapping method, the level set method etc., [11,12]. In the papers [5,7,10] functional variations for domains are introduced and studied. The idea is to assume that the admissible domains for (10)–(12) are defined implicitly by

$$\Omega = \Omega_g = int\{x \in D; g \in \mathcal{C}(\bar{D}); g(x) \geq 0\} \tag{13}$$

Here g is in some subset of admissible functions G_{ad}, corresponding to the family of admissible domains $\Omega = \Omega_g$. For instance, since we impose that $E \subset \Omega = \Omega_g$, then we have to require

$$g(x) \geq 0, \quad \forall x \in E. \tag{14}$$

Other constraints may be added according to the envisaged applications.

Due to the representation (13), one may use the functional variations of Ω, in the direction of $h \in \mathcal{C}(\bar{D})$:

$$\Omega_{g+\lambda h} = int\{x \in D; g + \lambda h \geq 0\}, \tag{15}$$

where $\lambda > 0$ is a scalar and h satisfies similar conditions as in (13), (14). Notice that the variations (15) of $\Omega = \Omega_g$ may be very complex, combining boundary and topological variations. The function g is a shape (or level) function, but the approach (13)–(15) is not to be confused with the level set method (we have no artificial time in the definition of Ω_g or in g, we need no Hamilton-Jacobi equation, etc.).

An essential ingredient in using general functional variations in the problem (10)–(12) is the following approximation of (11), (12):

$$- \Delta y_\varepsilon + \frac{1}{\varepsilon} \left(1 - H^\varepsilon(g)\right) y_\varepsilon = f, \qquad in \ D, \qquad (16)$$

$$y_\varepsilon = 0, \qquad on \ \partial D, \qquad (17)$$

where H is the Heaviside function and H^ε is some regularization. Such penalization/regularization approaches were used already in [4]. The basic approximation result in this case, according to [5], is the following:

Theorem 2. *If* $\Omega = \Omega_g$ *is of class* $\mathcal{C}$, *then* $y_\varepsilon/\Omega_g \to y_{\Omega_g}$ *weakly in* $H^1(\Omega_g)$, *on a subsequence.*

Domains of class $\mathcal{C}$, roughly speaking, have just continuous boundary. A thorough presentation with applications can be found in [16].

Based on Theorem 2, one can approximate the problem (10)–(12) by the problem (10), (16), (17). If $\Lambda = E$, this formulation is already a self-contained optimal control problem with unknown g. If $\Lambda = \Omega$, then we approximate as well (10) by

$$\int_D H^\varepsilon(g(x)) j(y_\varepsilon(x), x) dx. \qquad (18)$$

Notice that the formulation (16)–(18) excludes the explicit presence of the unknown geometry of Ω and is in fact an optimal control problem with the control g acting in the coefficients of the lower order terms of the state equation (16). Solving for the optimal g_ε immediately yields the optimal geometry Ω_{g_ε} as a level set. In the formulation (16)–(18) one can easily use functional variations as in (15). See [5,7,10] for numerical examples as well.

The case of boundary observation (cost) needs a special treatment based on the developments from Sect. 2 and we shall briefly comment on this, just in dimension 2.

We also fix, without loss of generality, some point $x_0 = (x_0^1, x_0^2) \in D \backslash E \subset \mathbb{R}^2$ such that $g(x_0) = 0$ for any admissible $g \in \mathcal{C}^1(D)$. We assume as well that

$$g(x) = 0 \ \Rightarrow \nabla g(x) \neq 0, \qquad (19)$$

$$g(x) \neq 0, \qquad on \ \partial D, \qquad (20)$$

to avoid the presence of critical points on $\partial \Omega_g$ and the corresponding admissible domains will not "touch" ∂D. We shall denote by Λ_g, the connected component of $\partial \Omega_g$, containing x_0 (the presence of multiply connected domains Ω_g is allowed in our setting).

The above setting together with the condition (14) defines the set G_{ad} of admissible controls in (16)–(18). It is possible to work without (20), but this would complicate the exposition. Finally, we impose that $G_{ad} \subset W^{3,\infty}(D)$. Notice that the obtained class of admissible geometries remains very rich.

Then, by Sect. 2, relation (7), we know that Λ_g can be parametrized by the solution of the Hamiltonian system:

$$x_1'(t) = -\frac{\partial g}{\partial x_2}(x_1(t), x_2(t)), \ t \in I_g, \tag{21}$$

$$x_2'(t) = \frac{\partial g}{\partial x_1}(x_1(t), x_2(t)), \ t \in I_g,$$

$$(x_1(0), x_2(0)) = x_0. \tag{22}$$

By Proposition 3 in [14], due to the boundedness of D and (19), (20), we may assume that the solution of (21), (22) is periodic and Λ_g is a closed curve, for any $g \in G_{ad}$. The interval I_g may be assumed to start in 0 and have the period length (that can be easily determined numerically in applications).

We comment now the following example:

$$J(g) = \int_{\Lambda_g} \left(\frac{\partial y_\varepsilon}{\partial n}\right)^2 d\sigma = \int_{I_g} \left[\frac{\partial y_\varepsilon}{\partial n}(x_1(t), x_2(t))\right]^2 \sqrt{x_1'(t)^2 + x_2'(t)^2} dt, \tag{23}$$

which is a typical case in boundary observation problems and y_ε solves (16), (17). By (21), (22) and simple computations, we have:

$$J(g) = \int_{I_g} [\nabla y_\varepsilon(x_1(t), x_2(t)) \cdot \nabla g(x_1(t), x_2(t))]^2 |\nabla g(x_1(t), x_2(t))|^{-1} dt. \tag{24}$$

Notice that the cost functional in (23)–(24) makes sense since $G_{ad} \subset W^{3,\infty}(D)$ which ensures the regularity of Λ_g. Together with (19), (20) and (16), (17), we have that $y_\varepsilon \in H^2(D)$. The formulation (16), (17), (24) again removes any direct reference to the geometric unknowns. Under regularity assumptions, one can consider functional variations (15) and compute directional derivatives (by the chain rule). In the case of thickness optimization problems for shells, a numerical approach based on directional derivatives is used in [17]. A detailed study of shape optimization problems with boundary observation, including the adjoint equation method and computational examples will be performed in a forthcoming paper based on the above approach. The computation of the adjoint equation, in the simpler case of distributed observation in E, is performed in [5].

Remark 3. The advantage of our approach is given by the generality of the admissible variations (combining boundary and topological perturbations without "prescribing" their topological type) and the fact that optimal control theory, in the fixed domain D, may be applied.

Acknowledgments. The work of both authors was supported by Grant 145/2011 of CNCS Romania.

References

1. DiPerna, R.J., Lions, P.L.: Ordinary differential equations, transport theory and Sobolev spaces. Invent. Math. **98**, 511–547 (1989)

2. Dontchev, A.L., Rockafellar, R.T.: Implicit Functions and Solution Mappings. Springer, New York (2009)
3. Krantz, S.G., Parks, H.R.: The Implicit Functions Theorem. Birkhäuser, Boston (2002)
4. Mäkinen, R., Neittaanmäki, P., Tiba, D.: On a fixed domain approach for a shape optimization problem. In: Ames W.F., van Houwen, P.J. (eds.) Computational and Applied Mathematics II: Differential Equations, North Holland, Amsterdam, pp. 317–326 (1992)
5. Neittaanmäki, P., Pennanen, A., Tiba, D.: Fixed domain approaches in shape optimization problems with Dirichlet boundary conditions. Inverse Prob. **25**(5), 1–18 (2009)
6. Neittaanmäki, P., Sprekels, J., Tiba, D.: Optimization of Elliptic Systems. Theory and Applications. Springer Monographs in Mathematics. Springer, New York (2006)
7. Neittaanmäki, P., Tiba, D.: Fixed domain approaches in shape optimization problems. Inverse Prob. **28**(9), 1–35 (2012)
8. Nicolai, M.R., Tiba, D.: Implicit functions and parametrizations in dimension three: generalized solution. DCDA-A **35**(6), 2701–2710 (2015). doi:10.3934/dcds.2015.35.2701.
9. Nicolai, M.R.: An algorithm for the computation of the generalized solution for implicit systems. Ann. Acad. Rom. Sci. Ser. Math. Appl. **7**(2), 310–322 (2015)
10. Philip, P., Tiba, D.: A penalization and regularization technique in shape optimization. SIAM J. Control Optim. **51**(6), 4295–4317 (2013)
11. Sethian, J.A.: Level Set Methods. Cambridge Univ. Press, Cambridge (1996)
12. Sokolowsky, J., Zolesio, J.-P.: Introduction to Shape Optimization. Shape Sensitivity Analysis. Springer, Berlin (1992)
13. Thorpe, J.A.: Elementary Topics in Differential Geometry. Springer, New York (1979)
14. Tiba, D.: The implicit functions theorem and implicit parametrizations. Ann. Acad. Rom. Sci. Ser. Math. Appl. **5**(1–2), 193–208 (2013). http://www.mathematics-and-its-applications.com/preview/june2013/data/art%2010.pdf
15. Tiba, D.: A Hamiltonian approach to implicit systems, generalized solutions and applications in optimization (2016). http://arxiv.org/abs/1408.6726v4.pdf
16. Tiba, D.: Domains of class C: properties and applications. Ann. Univ. Buchar. Math. Ser. **4(LXII)**(1), 89–102 (2013)
17. Ziemann, P.: Optimal thickness of a cylindrical shell. Ann. Acad. Rom. Sci. Ser. Math. Appl. **6**(2), 214–234 (2014). http://www.mathematics-and-its-applications.com/preview/november_2014/data/articol6.pdf
18. Zuazua, E.: Log-Lipschitz regularity and uniqueness of the flow for a field in $[W_{loc}^{n/p+1}(R^n)]^n$. CRAS Paris, Ser I **335**, 17–22 (2002)

Approximate Riesz Representatives of Shape Gradients

Alberto Paganini$^{(\boxtimes)}$ and Ralf Hiptmair

Seminar for Applied Mathematics, ETH Zurich, Zurich, Switzerland
alberto.paganini@sam.math.ethz.ch

Abstract. We study finite element approximations of Riesz representatives of shape gradients. First, we provide a general perspective on its error analysis. Then, we focus on shape functionals constrained by elliptic boundary value problems and H^1-representatives of shape gradients. We prove linear convergence in the energy norm for linear Lagrangian finite element approximations. This theoretical result is confirmed by several numerical experiments.

Keywords: Shape gradients · Finite element approximations

1 Introduction

A *shape functional* is a real map defined on a set of admissible shapes. The goal of *shape optimization* is to modify an initial shape so that a shape functional attains an extremal value. A common approach is to employ steepest descent algorithms [8, Chap. 3.4]. Shapes may be parameterized by C^1-mappings acting on reference configurations. Then the shape gradient is a linear continuous operator on the non-reflexive Banach space C^1, and the concept of steepest descent may not be well-defined; see [7, P. 103]. A compromise is to replace "steepest descents" with Riesz representatives of shape gradients with respect to a Hilbert space X. Henceforth, we refer to these representatives as *X-representatives*.

After recalling basic definitions of shape calculus, we provide a general perspective on error analysis in the energy norm for finite element approximations of X-representatives of shape gradients. Then, we zero in on shape functionals constrained to elliptic boundary value problems. For this case, insight into shape Hessians [12,14] suggests to select representatives of shape gradients with respect to $X = H_0^1(D)$, where D is a hold-all domain that encloses the initial guess Ω; see Fig. 1. For the choice $X = H_0^1(D)$, it is natural to consider discretization by means of linear Lagrangian finite elements [2,14,15]. We show that linear Lagrangian finite element approximations of H^1-representatives of shape gradients converge linearly with respect to the mesh width. Additionally, this convergence rate does not deteriorate when state and adjoint variables are replaced by linear Lagrangian finite elements solutions. This is an improvement on the result presented in [2], which involves approximations of state and adjoint

L. Bociu et al. (Eds.): CSMO 2015, IFIP AICT 494, pp. 399–409, 2016.
DOI: 10.1007/978-3-319-55795-3_38

variables with quadratic finite elements. Finally, we provide numerical evidence of the linear convergence rate predicted.

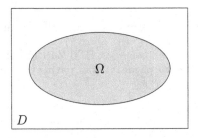

Fig. 1. The hold-all domain D encloses the domain Ω.

2 Shape Functionals and Shape Gradients

Let $\Omega \subset \mathbb{R}^d$, $d = 2, 3$, be an open bounded domain with piecewise smooth boundary $\partial\Omega$, and let $\mathcal{J}(\Omega) \in \mathbb{R}$ be a real-valued quantity of interest associated to it. One is often interested in shape sensitivity, which quantifies the impact of small perturbations of $\partial\Omega$ on the value $\mathcal{J}(\Omega)$.

We model perturbations of the domain Ω through maps of the form

$$T_\mathcal{V}(\mathbf{x}) := \mathbf{x} + \mathcal{V}(\mathbf{x}), \quad \mathbf{x} \in \mathbb{R}^d, \tag{1}$$

where $\mathcal{V}$ is a vector field in $C^1(\mathbb{R}^d; \mathbb{R}^d)$. It can easily be proved that the map (1) is a diffeomorphism for $\|\mathcal{V}\|_{C^1} < 1$ [8, Lemma 6.13].

The value $\mathcal{J}(\Omega)$ is interpreted as the realization of a *shape functional*, a real map

$$\mathcal{J} : \mathcal{V} \mapsto \mathcal{J}(T_\mathcal{V}(\Omega))$$

defined on the ball $\{\mathcal{V} \in C^1(\mathbb{R}^d; \mathbb{R}^d); \|\mathcal{V}\|_{C^1} < 1\}$. Clearly, $\mathcal{J}(\Omega) = \mathcal{J}(T_0(\Omega))$.

The sensitivity of $\mathcal{J}(\Omega)$ with respect to the perturbation direction $\mathcal{V}$ is given by the *Eulerian derivative* of the shape functional $\mathcal{J}$ in the direction $\mathcal{V}$, that is,

$$d\mathcal{J}(\Omega; \mathcal{V}) := \lim_{s \searrow 0} \frac{\mathcal{J}(T_{s \cdot \mathcal{V}}(\Omega)) - \mathcal{J}(\Omega)}{s}. \tag{2}$$

We say that the shape functional is *shape differentiable* if Formula (2) defines a linear and bounded operator $\mathcal{V} \mapsto d\mathcal{J}(\Omega; \mathcal{V})$. In literature, this operator is called *shape gradient* [9, Chap. 9, Sect. 3.4]. As mentioned in the introduction, X-representatives of shape gradients can be employed to solve shape optimization problems, that is, to find

$$\Omega^* \in \operatorname*{argmin}_{\Omega \in U_{\mathrm{ad}}} \mathcal{J}(\Omega),$$

where U_{ad} denotes a set of admissible shapes.

Often, the quantity of interest takes the form

$$J(\Omega) = \int_B \alpha \nabla(u-g) \cdot \nabla(u-g) + \beta(u-g)^2 \, d\boldsymbol{x}, \tag{3}$$

where the *state function* u is the solution of a boundary value problem stated on Ω, $B \subset \Omega$, α and β are two real constants and g is a sufficiently smooth target function. In this work, $u \in H_0^1(\Omega)$ is the (weak) solution of the elliptic boundary value problem with homogeneous Dirichlet boundary conditions

$$-\Delta u + u = f \quad \text{in } \Omega, \qquad u = 0 \quad \text{on } \partial\Omega, \tag{4}$$

that is,

$$\int_\Omega \nabla u \cdot \nabla v + uv \, d\boldsymbol{x} = \int_\Omega fv \, d\boldsymbol{x} \quad \text{for all } v \in H_0^1(\Omega), \tag{5}$$

where $f \in H^1(\Omega)$. For the sake of brevity, we set $g = 0$. Then, the shape gradient of the shape functional associated to (3) and constrained to (5) reads [4, Formula (2.9)]

$$dJ(\Omega; \mathcal{V}) = \int_\Omega \Big((\nabla f \cdot \mathcal{V})p + \nabla u \cdot (\mathbf{D}\mathcal{V} + \mathbf{D}\mathcal{V}^T)\nabla p$$

$$+ \operatorname{div} \mathcal{V}(fp + \chi_B(\alpha \nabla u \cdot \nabla u + \beta u^2) - \nabla u \cdot \nabla p - up) \Big) \, d\boldsymbol{x}, \tag{6}$$

where the *adjoint function* $p \in H_0^1(\Omega)$ is the solution of

$$\int_\Omega \nabla p \cdot \nabla v + pv \, d\boldsymbol{x} = \int_B \alpha \nabla u \cdot \nabla v + \beta uv \, d\boldsymbol{x} \quad \text{for all } v \in H_0^1(\Omega). \tag{7}$$

Formula (3) is a prototypical PDE-constrained shape functional. In this work, Formula (6) is used as test case for proving convergence estimates and performing numerical experiments.

Remark 1. Formula (6) holds even if homogeneous Dirichlet boundary conditions in (4) are replaced by homogeneous Neumann boundary conditions, in which case the test and the trial spaces in (5) and (7) are replaced with $H^1(\Omega)$.

Remark 2. For the sake of simplicity, we restrict our considerations to homogeneous boundary conditions. However, we expect that the results of this work hold true for (sufficiently regular) inhomogeneous boundary conditions, too. Note that Formula (6) should be adjusted accordingly; see [4, Sect. 2].

3   Error Analysis for Finite Element Representatives

3.1   The General Case

Let $(\cdot, \cdot)_X$ denote the inner product of a Hilbert space X, and let us assume that the shape gradient dJ is well-defined on X. The X-representative $\mathcal{V}^X$ of dJ can be computed by solving

$$(\mathcal{V}^X, \mathcal{W})_X = dJ(\Omega; \mathcal{W}) \quad \text{for all } \mathcal{W} \in X.$$

Next, for an index set $\mathcal{N}$, we introduce a family $\{X_n\}_{n\in\mathcal{N}}$ of finite-dimensional subspaces of X. Let $\{\mathcal{V}^{X_n}\}_{n\in\mathcal{N}}$ be a sequence of approximate X-representatives of $d\mathcal{J}$ defined by

$$(\mathcal{V}^{X_n}, \mathcal{W}_n)_X = d\mathcal{J}(\Omega; \mathcal{W}_n) \quad \text{for all } \mathcal{W}_n \in X_n.$$

By Cea's Lemma [11, Theorem 2.4.1], there exists a constant $C > 0$ independent of n such that

$$\|\mathcal{V}^X - \mathcal{V}^{X_n}\|_X \le C \inf_{\mathcal{W}_n \in X_n} \|\mathcal{V}^X - \mathcal{W}_n\|_X. \tag{8}$$

By and large, the shape gradient of a PDE-constrained shape functional depends also on the state and the adjoint variables u and p. These functions are solutions of boundary value problem. Usually, only numerical approximations u_h and p_h are available. In that case, the approximate X-representative $\mathcal{V}^{X_n}$ has to be replaced with the solution $\mathcal{V}_h^{X_n}$ of

$$(\mathcal{V}_h^{X_n}, \mathcal{W}_n)_X = d\mathcal{J}_h(\Omega; \mathcal{W}_n) \quad \text{for all } \mathcal{W}_n \in X_n, \tag{9}$$

where $d\mathcal{J}_h$ is an approximation of the operator $d\mathcal{J}$ obtained by replacing the functions u and p with their numerical approximations u_h and p_h.

By Strang Lemma [11, Theorem 4.1.1], the estimate (8) should be corrected by adding a consistency term, that is,

$$\|\mathcal{V}^X - \mathcal{V}_h^{X_n}\|_X \le C\left(\inf_{\mathcal{W}_n \in X_n} \|\mathcal{V}^X - \mathcal{W}_n\|_X \right.$$
$$\left. + \sup_{\mathcal{W}_n \in X_n} \frac{|d\mathcal{J}(\Omega; \mathcal{W}_n) - d\mathcal{J}_h(\Omega; \mathcal{W}_n)|}{\|\mathcal{W}_n\|_X} \right) \tag{10}$$

for a constant $C > 0$ independent of n and h.

3.2 H^1-Representatives and Linear Lagrangian Finite Elements

A popular approach in shape optimization consists of replacing the initial domain Ω with a polygon/polyhedron equipped with a finite element mesh Ω_h. This mesh is used to compute linear Lagrangian finite element approximations of the functions u and p. Then, the coordinates of the mesh nodes are (iteratively) updated according to the shape gradient [8, Chap. 6.5]. This is equivalent to extending Ω_h to a mesh D_h that covers a hold-all domain D and choosing linear Lagrangian finite elements to construct the finite-dimensional subspace X_n. Formula (10), standard finite element estimates, and Proposition 1 readily imply that, for this discretization, the approximate H^1-representative of (6) satisfies

$$\|\mathcal{V}^X - \mathcal{V}_h^{X_n}\|_{H^1(D)} = \mathcal{O}(h), \tag{11}$$

which is the main result of this work.

Proposition 1. *Let $\Omega \subset \mathbb{R}^d$ be a polyhedral domain, let $f \in W^{1,4}(\Omega)$ in (5), and let us assume that the solution u of (5) satisfies*

$$\|u\|_{W^{2,4}(\Omega)} \leq C\|f\|_{L^4(\Omega)}. \tag{12}$$

Let $(V_h)_{h \in (0,1]}$ be a family of $H_0^1(\Omega)$-conforming piecewise linear Lagrangian finite element spaces built on a quasi-uniform family of simplicial meshes $(\mathcal{T}^h)_{h \in (0,1]}$, that is, a family of meshes such that

$$\max\{\operatorname{diam}(T) : T \in \mathcal{T}^h\} \leq h\operatorname{diam}(\Omega)$$

and

$$\min\{\operatorname{diam}(B_T) : T \in \mathcal{T}^h\} \geq \rho h\operatorname{diam}(\Omega) \quad \text{for all } h \in (0,1],$$

for a $\rho > 0$, where B_T is the largest ball contained in the simplex T [10, Definition 4.4.13]. Let $u_h, p_h \in V_h$ be solutions of

$$\int_\Omega \nabla u_h \cdot \nabla v_h + u_h v_h \, d\boldsymbol{x} = \int_\Omega f v_h \, d\boldsymbol{x} \qquad \text{for all } v_h \in V_h, \tag{13}$$

$$\int_\Omega \nabla p_h \cdot \nabla v_h + p_h v_h \, d\boldsymbol{x} = \int_B \alpha \nabla u_h \cdot \nabla v_h + \beta u_h v_h \, d\boldsymbol{x} \quad \text{for all } v_h \in V_h, \tag{14}$$

where $\alpha, \beta \in \mathbb{R}$, $B \subset \Omega$, and $\alpha = 0$ or $B = \Omega$, if $d = 3$. Let $d\mathcal{J}_h(\Omega; W_n)$ denote the operator defined by Formula (6) with u and p replaced by u_h and p_h, respectively. Then,

$$\sup_{W_n \in X_n} \frac{|d\mathcal{J}(\Omega; W_n) - d\mathcal{J}_h(\Omega; W_n)|}{\|W_n\|_{H^1(D)}} \leq C(\Omega, f, u, p)h \tag{15}$$

for a constant $C(\Omega, f, u, p) > 0$ independent of n and h.

Proof. First of all, note that

$$d\mathcal{J}(\Omega; W_n) - d\mathcal{J}_h(\Omega; W_n) = \int_\Omega (\nabla f \cdot W_n + f \operatorname{div} W_n)(p - p_h) \, d\boldsymbol{x}$$

$$+ \int_\Omega \nabla u \cdot (\mathbf{D}W_n + \mathbf{D}W_n^T)\nabla p - \nabla u_h \cdot (\mathbf{D}W_n + \mathbf{D}W_n^T)\nabla p_h \, d\boldsymbol{x}$$

$$+ \int_\Omega \operatorname{div} W_n (\nabla u_h \cdot \nabla p_h + u_h p_h - \nabla u \cdot \nabla p - up) \, d\boldsymbol{x}$$

$$+ \int_B \operatorname{div} W_n \left(\alpha(\nabla u \cdot \nabla u - \nabla u_h \cdot \nabla u_h) + \beta(u^2 - u_h^2)\right) \, d\boldsymbol{x}. \tag{16}$$

We recall that, for generic functions $q_0 \in L^2(\Omega)$ and $q_1, q_2 \in L^4(\Omega)$, the Cauchy-Schwarz inequality implies

$$\|q_0 q_1 q_2\|_{L^1(\Omega)} \leq \|q_0\|_{L^2(\Omega)} \|q_1 q_2\|_{L^2(\Omega)} \leq \|q_0\|_{L^2(\Omega)} \|q_1\|_{L^4(\Omega)} \|q_2\|_{L^4(\Omega)}. \tag{17}$$

Thus, the first integral in (16) may be estimated as follows[1]

$$\left| \int_\Omega (\nabla f \cdot \mathcal{W}_n + f \operatorname{div} \mathcal{W}_n)(p - p_h) \, d\boldsymbol{x} \right| \leq C \|\mathcal{W}_n\|_{H^1(\Omega)} \|f\|_{W^{1,4}(\Omega)} \|p - p_h\|_{L^4(\Omega)}.$$

The second integral in (16) may be estimated as follows

$$\left| \int_\Omega \nabla u \cdot (\mathbf{D}\mathcal{W}_n + \mathbf{D}\mathcal{W}_n^T) \nabla p - \nabla u_h \cdot (\mathbf{D}\mathcal{W}_n + \mathbf{D}\mathcal{W}_n^T) \nabla p_h \, d\boldsymbol{x} \right|$$

$$= \left| \int_\Omega \nabla (u - u_h) \cdot (\mathbf{D}\mathcal{W}_n + \mathbf{D}\mathcal{W}_n^T) \nabla p + \nabla u_h \cdot (\mathbf{D}\mathcal{W}_n + \mathbf{D}\mathcal{W}_n^T) \nabla (p - p_h) \, d\boldsymbol{x} \right|$$

$$\leq C \|\mathcal{W}_n\|_{H^1(\Omega)} \left(\|u - u_h\|_{W^{1,4}(\Omega)} \|p\|_{W^{1,4}(\Omega)} + \|u_h\|_{W^{1,4}(\Omega)} \|p - p_h\|_{W^{1,4}(\Omega)} \right).$$

The third and the fourth integral in (16) may be estimated similarly.
Stability of the Ritz projection with respect to $W^{1,4}(\Omega)$ [3][2]

$$\|u_h\|_{W^{1,4}(\Omega)} \leq C \|u\|_{W^{1,4}(\Omega)} \tag{18}$$

implies $\|u - u_h\|_{W^{1,4}(\Omega)} = \mathcal{O}(h)$. To show

$$\|p_h\|_{W^{1,4}(\Omega)} \leq C \|p\|_{W^{1,4}(\Omega)}, \tag{19}$$

which in turn implies $\|p - p_h\|_{W^{1,4}(\Omega)} = \mathcal{O}(h)$, it is necessary to repeat the proof of (18) given in [3] tracking the consistency term

$$\int_\Omega \nabla(p - p_h) \cdot \nabla g_h^z + (p - p_h) g_h^z \, d\boldsymbol{x} = \int_B \alpha \nabla(u - u_h) \cdot \nabla g_h^z + \beta(u - u_h) g_h^z \, d\boldsymbol{x}. \tag{20}$$

The discrete Green's function $g_h^z \in V_h$ is given in [3] and satisfies $\|g_h^z\|_{H^1(\Omega)} = \mathcal{O}(h^{-d/2})$. By the Cauchy-Schwarz inequality and standard finite element estimates,

$$\left| \int_B \alpha \nabla(u - u_h) \cdot \nabla g_h^z + \beta(u - u_h) g_h^z \, d\boldsymbol{x} \right| = \mathcal{O}\left((|\alpha|h + |\beta|h^2) h^{-d/2} \right). \tag{21}$$

The stability result (19) holds if (21) is bounded independently of h. For this reason, we need to set $\alpha = 0$ when $d = 3$, unless $B = \Omega$. In this latter case, by Galerkin orthogonality, (20) is bounded by $\|(\beta - \alpha)(u - u_h) g_h^z\|_{L^1(\Omega)}$.   □

Remark 3. In Proposition 1, we assume $W^{2,4}$-regularity of the solution u of (5). This assumption is made to achieve linear convergence with respect to h in the estimate (15). However, a three-dimensional polyhedral domain must satisfy tight geometric conditions for u to be in $W^{2,4}$ [6, Theorem 7.1]. Nevertheless, in [5] the authors show $W^{1,\infty}$-stability of the Ritz projection for general convex polyhedral domains. Therefore, we expect that (in the latter case) the right-hand side of (15) can be replaced with a term of order $\mathcal{O}(h^\alpha)$, where the rate α depends on the regularity of u and satisfies $0 < \alpha \leq 1$.

---

[1] Henceforth, C denotes a positive generic constant independent of n and h.
[2] The assumption $\Omega \subset \mathbb{R}^2$ made in [3] can be replaced by $\Omega \subset \mathbb{R}^3$; cf. [10, Chap. 8].

Remark 4. In [4,13], the authors show that one can expect superconvergence in the approximation of the shape gradient dJ. In particular, they show that

$$|dJ(\Omega; \mathcal{W}) - dJ_h(\Omega; \mathcal{W})| \le C\|\mathcal{W}\|_{W^{2,4}(\Omega)} h^2. \tag{22}$$

However, in the right-hand side of (22) appears the $W^{2,4}(\Omega)$-norm of $\mathcal{W}$. Note that to prove convergence in the approximation of a H^1-representative of dJ, the upper bound of

$$|dJ(\Omega; \mathcal{W}) - dJ_h(\Omega; \mathcal{W})|$$

cannot involve a norm stronger than the H^1-norm; see Eq. (10).

Remark 5. By the *Hadamard structure theorem* [9, Chap. 9, Theorem 3.6], most shape gradients admit representatives $\mathfrak{g}(\Omega)$ in the space of distributions $\mathcal{D}^k(\partial\Omega)$, that is,

$$dJ(\Omega; \mathcal{V}) = \langle \mathfrak{g}(\Omega), \gamma_{\partial\Omega} \mathcal{V} \cdot \mathbf{n} \rangle_{\mathcal{D}^k(\partial\Omega)}, \tag{23}$$

where $\gamma_{\partial\Omega} \mathcal{V} \cdot \mathbf{n}$ is the normal component of $\mathcal{V}$ on the boundary $\partial\Omega$. For instance, if $u, p \in H^2(\Omega)$, Formula (6) is equivalent to [4, Formula (2.10)]

$$dJ(\Omega; \mathcal{V}) = \int_{\partial\Omega} (\mathcal{V} \cdot \mathbf{n}) \left(\alpha \nabla u \cdot \nabla u + \beta u^2 + \frac{\partial p}{\partial \mathbf{n}} \frac{\partial u}{\partial \mathbf{n}} \right) dS. \tag{24}$$

We advise against the use of $\mathfrak{g}(\Omega)$ (which corresponds to the $L^2(\partial\Omega)$-representative of dJ) to define descent directions because L^2-representatives might bristle with undesirable oscillations [1].

4 Numerical Experiments

We provide numerical evidence of the estimate (11). We employ linear Lagrangian finite elements on quasi-uniform triangular meshes. The experiments are performed in MATLAB and are partly based on the library LehrFEM developed at ETHZ. Mesh generation and uniform refinement are performed with the functions `initmesh` and `refinemesh` of the MATLAB PDE Toolbox [16]. The boundary of computational domains is approximated by a polygon, which is generally believed not to affect the convergence of linear finite elements [10, Sect. 10.2]. For domains with curved boundaries, the refined mesh is always adjusted to fit the boundary. Integrals in the domain are computed with a 3-point quadrature rule of order 3 in each triangle and line integrals with a 6-point Gauss quadrature on each segment.

We consider three different geometries for the domain Ω (see Fig. 2):

1. A disc of radius $\sqrt{6/5}$ centered in $(0.01, 0.02)$.
2. A triangle with corners located at

$$(-\sqrt{6/5}, -\sqrt{6/5}), (\sqrt{6/5}, -\sqrt{6/5}), (-\sqrt{6/5}, \sqrt{6/5}).$$

3. A circular sector of radius $\sqrt{6/5}$ centered in $(0.01, 0.02)$ of angle $0.9 \cdot 2\pi$.

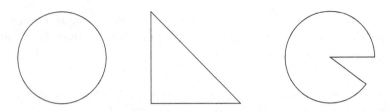

Fig. 2. The domain Ω is chosen to be either a disc or a triangle or a sector.

The source function in (5) is

$$f(x,y) = \cos(x + \pi/4).$$

The hold-all domain D is a square with edges of length 3 centered in the origin. The region of interest B is the whole domain Ω. We set $\alpha = 0$ and $\beta = 1$ in (3).

The reference value $\mathcal{V}^X$ is approximated by computing $\mathcal{V}_h^{X_n}$ on a mesh with an extra level of refinement. In light or Remark 5, we employ both Formula (6) and Formula (24) to evaluate the right-hand side $d\mathcal{J}_h$ in (9). To avoid biased results, we display convergence history of $\|\mathcal{V}^X - \mathcal{V}_h^{X_n}\|_{H^1(D)}$ both with self- and cross-comparison.

In Fig. 3, we plot the convergence history when the domain Ω is either a disc (first row) or a triangle (second row). As predicted by (11), we observe linear convergence when the right-hand side in (9) is evaluated according to (6). Interestingly, using Formula (24) seems not to affect the convergence rate. The same behavior is observed when homogeneous Dirichlet boundary conditions are replaced by homogeneous Neumann boundary conditions. Note that, in this latter case, the boundary-integral counterpart of Formula (6) reads [4]

$$d\mathcal{J}(\Omega; \mathcal{V}) = \int_{\partial\Omega} \mathcal{V} \cdot \mathbf{n} \left(\nabla u \cdot \nabla(\alpha u - p) + u(\beta u - p) + fp \right) dS. \qquad (25)$$

For the sake of brevity, we omit these plots.

In Fig. 4 (first row), we plot the convergence history when the domain Ω is a sector. This domain does not guarantee that u and p are in $H^2(\Omega)$ because it has a re-entrant corner. We observe that the convergence rates decrease to fractional values. This is a consequence of the lower regularity of the functions u and p. Additionally, the convergence rates depend on the formula used to evaluate $d\mathcal{J}_h$. In particular, in the cross-comparison, the convergence line saturates when Formula (6) is used. This may be due to a poor accuracy of the reference solution. However, we point out that Formulas (6) and (24) may not be equivalent due to the lack of regularity of the functions u and p; cf. Remark 5. Curiously, for homogeneous Neumann boundary conditions, the presence of the re-entrant corner seems to have a milder impact on convergence rates; see Fig. 4 (second row). However, note that the approximate algebraic convergence rates of

$$\|u - u_h\|_{H^1(\Omega)} \quad \text{and} \quad \|p - p_h\|_{H^1(\Omega)}$$

with respect to h drop to 0.67 and 0.62, respectively.

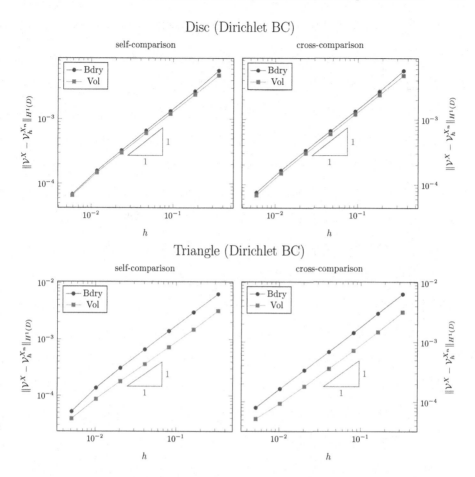

Fig. 3. Convergence history when Ω is a disc (*first row*) and a triangle (*second row*). Line ─■─ refers to evaluation of $d\mathcal{J}_h$ according to Formula (6); line ─●─ to Formula (24). We observe the linear convergence rate predicted by (11).

By the Hadamard structure theorem (see Remark 5), vector fields $\mathcal{W}_n$ associated to interior nodes of the mesh Ω_h lie in the kernel of $d\mathcal{J}$. However, these vector fields are not in the kernel of $d\mathcal{J}_h$ because u and p are replaced by finite element approximations. Schulz et al. [14] report that this numerical error might largely affect the computation of the Riesz representative. Although we have not experienced this issue, we have repeated the numerical experiments by setting to zero the values of $d\mathcal{J}_h(\Omega; \mathcal{W}_n)$ for all $\mathcal{W}_n$ associated to interior nodes of Ω_h. We have not observed any significative difference in the results. Thus, we acknowledge that computational resources might be saved by dropping the evaluation of $d\mathcal{J}_h(\Omega; \mathcal{W}_n)$ for vector fields associated to interior nodes of Ω_h.

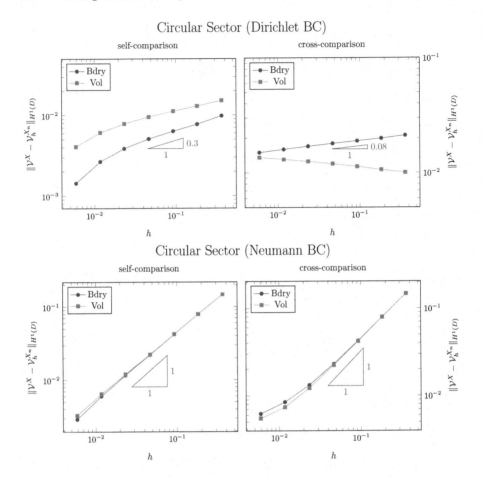

Fig. 4. Convergence history when Ω is a sector. Line $\blacksquare$ refers to evaluation of $d\mathcal{J}_h$ according to Formula (6); line $\bullet$ to Formula (24)(in the *first row*) and to Formula (25) (in the *second row*). For Dirichlet boundary conditions, convergence rates decay to fractional values.

5   Conclusion

Most shape optimization algorithms rely on Riesz representatives of shape gradients with respect to a chosen Hilbert space. Numerical discretization is inevitable when the shape functional is constrained to a boundary value problem. Formula (10) indicates how to estimate the discretization error when the Riesz representative is computed on a finite-dimensional trial space and the shape gradient can be evaluated only approximately.

For linear Lagrangian approximations of H^1-representatives, Proposition 1 implies that the discretization error decays linearly with respect to the mesh width h. This convergence behavior is observed in several numerical experiments.

As a consequence of the Hadamard structure theorem, most shape gradients can be equivalently formulated as boundary or volume integrals. Although

Proposition 1 relies on the volume formulation of the shape gradient, we have observed linear convergence independently of the formula employed to evaluate $d\mathcal{J}$. However, we advise to rely on the volume-based formula because it imposes lower regularity assumptions on the state and the adjoint variables [4,9,15].

Acknowledgments. The work of A. Paganini was partly supported by ETH Grant CH1-02 11-1.

References

1. Hiptmair, R., Paganini, A.: Shape optimization by pursuing diffeomorphisms. Comput. Methods Appl. Math. **15**(3), 291–305 (2015)
2. Murai, D., Azegami, H.: Error analysis of H1 gradient method for shape-optimization problems of continua. JSIAM Lett. **5**, 29–32 (2013)
3. Rannacher, R., Scott, R.: Some optimal error estimates for piecewise linear finite element approximations. Math. Comput. **38**, 437–445 (1982)
4. Hiptmair, R., Paganini, A., Sargheini, S.: Comparison of approximate shape gradients. BIT Numer. Math. **55**(2), 459–485 (2015)
5. Guzmán, J., Leykekhman, D., Rossmann, J., Schatz, A.H.: Hölder estimates for Green's functions on convex polyhedral domains and their applications to finite element methods. Numer. Math. **112**(2), 221–243 (2009)
6. Maz'ya, V.G., Romann, J.: Weighted Lp estimates of solutions to boundary value problems for second order elliptic systems in polyhedral domains. Z. Angew. Math. Mech. **83**(7), 435467 (2003)
7. Hinze, M., Pinnau, R., Ulbrich, M., Ulbrich, S.: Optimization with PDE Constraints. Springer, New York (2009)
8. Allaire, G.: Conception Optimale de Structures. Springer, Berlin (2007)
9. Delfour, M.C., Zolésio, J.P.: Shapes and geometries. Metrics, analysis, differential calculus, and optimization. Society for Industrial and Applied Mathematics (SIAM), Philadelphia (2011)
10. Brenner, S.C., Ridgway Scott, L.R.: The Mathematical Theory of Finite Element Methods. Springer, New York (2008)
11. Ciarlet, P.G.: The Finite Element Method for Elliptic Problems. Society for Industrial and Applied Mathematics (SIAM), Philadelphia (2002)
12. Eppler, K., Harbrecht, H.: Shape optimization for free boundary problems-analysis and numerics. In: Leugering, G., Engell, S., Griewank, A., Hinze, M., Rannacher, R., Schulz, V., Ulbrich, M., Ulbrich, S. (eds.) Constrained Optimization and Optimal Control for Partial Differential Equations. International Series of Numerical Mathematics, vol. 160, pp. 277–288. Birkhäuser/Springer, Basel (2012)
13. Paganini, A.: Approximate shape gradients for interface problems. In: Pratelli, A., Leugering, G. (eds.) New Trends in Shape Optimization. International Series of Numerical Mathematics, vol. 166, pp. 217–227. Springer International Publishing, Heidelberg (2015)
14. Schulz, V.H., Siebenborn, M., Welker, K.: Efficient PDE constrained shape optimization based on Steklov–Poincaré-type metrics. SIAM J. Optim. **26**(4), 2800–2819 (2016). doi:10.1137/15M1029369
15. Laurain, A., Sturm, K.: Domain expression of the shape derivative and application to electrical impedance tomography. WIAS Preprint No. 1863 (2013)
16. MATLAB and Partial Differential Equation Toolbox (R2015a), The MathWorks Inc., Natick, Massachussets, United States (2015)

Exploitation of the Value Function in a Bilevel Optimal Control Problem

Konstantin Palagachev and Matthias Gerdts[✉]

Institut für Mathematik und Rechneranwendung (LRT),Universität der Bundeswehr,
Werner-Heisenberg-Weg 39, 85577 Neubiberg, Germany
{konstantin.palagachev,matthias.gerdts}@unibw.de,
http://www.unibw.de/lrt1/gerdts

Abstract. The paper discusses a class of bilevel optimal control problems with optimal control problems at both levels. The problem will be transformed to an equivalent single level problem using the value function of the lower level optimal control problem. Although the computation of the value function is difficult in general, we present a pursuit-evasion Stackelberg game for which the value function of the lower level problem can be derived even analytically. A direct discretization method is then used to solve the transformed single level optimal control problem together with some smoothing of the value function.

Keywords: Bilevel optimal control · Value function · Pursuit-evasion Stackelberg game

1 Introduction

Bilevel optimization problems occur in various applications, e.g. in locomotion and biomechanics, see [1,2,15,20], in optimal control under safety constraints, see [12,18,19], or in Stackelberg dynamic games, compare [10,24]. An abstract bilevel optimization problem (BOP) reads as follows:

Minimize $F(x,y)$ with respect to $(x,y) \in X \times Y$ subject to the constraints

$$G(x,y) \in K, \qquad H(x,y) = 0, \qquad y \in M(x),$$

where $M(x)$ is the set of minimizers of the lower level optimization problem

$$Minimize \quad f(x,y) \quad w.r.t. \quad y \in Y \quad s.t. \quad g(x,y) \in C, \quad h(x,y) = 0.$$

Herein, X, Y are (finite or infinite) Banach spaces, $F, f : X \times Y \to \mathbb{R}$, $H : X \times Y \to V^u$, $h : X \times Y \to V^\ell$, $G : X \times Y \to W^u$, $g : X \times Y \to W^\ell$ are

---

The work is supported by Munich Aerospace e.V.

L. Bociu et al. (Eds.): CSMO 2015, IFIP AICT 494, pp. 410–419, 2016.
DOI: 10.1007/978-3-319-55795-3_39

sufficiently smooth functions into Banach spaces V^u, V^ℓ, W^u, W^ℓ, and $K \subset W^u$, $C \subset W^\ell$ are convex and closed cones.

Bilevel optimization problems turn out to be very challenging with regard to both, the investigation of theoretical properties and numerical methods, compare [8]. Necessary conditions have been investigated, e.g., in [9,25]. Typical solution approaches aim at reducing the bilevel structure into a single stage optimization problem. In the MPCC approach a single level optimization problem subject to complementarity constraints (MPCC) is obtained by replacing the lower level problem by its first order necessary conditions, compare [1]. However, if the lower level problem is non-convex, the MPCC is not equivalent in general to the original bilevel problem since non-optimal stationary points or non-global solutions may satisfy the necessary conditions as well. Still, the approach is often used owing to a well-established theory and the availability of numerical methods for MPCCs, especially for finite dimensional problems.

In this paper we focus on an equivalent transformation of the bilevel problem to a single level problem (see [7] for an alternative way). The equivalence can be guaranteed by exploitation of the value function $V : X \to \mathbb{R}$ of the lower level problem, which is defined as

$$V(x) := \inf_{y \in Y} \{f(x,y) \mid g(x,y) \in C, h(x,y) = 0\}.$$

An equivalent reformulation of the bilevel optimization problem is then given by the following single level problem, compare [22,25,26]:

Minimize $F(x,y)$ w.r.t. $(x,y) \in X \times Y$ subject to the constraints

$$G(x,y) \in K, \ H(x,y) = 0, \ g(x,y) \in C, \ h(x,y) = 0, \ f(x,y) \leq V(x).$$

The advantage of the value function approach is its equivalence with the bilevel problem. On the downside one has to be able to compute the value function, which in general might be intractable. Moreover, the value function is non-smooth in general (often Lipschitz continuous) and hence suitable methods from non-smooth optimization are required to solve the resulting single level problem. In Sect. 2 we discuss a class of bilevel optimal control problems that fit into the problem class BOP. In Sect. 3 we we are able to derive an analytical expression for the value function for an example and present numerical results. The new contribution of this paper is the discussion of a particular example, which combines the analytical expression of the value function of the lower level problem and a direct discretization method for the reformulated single level problem. This problem may serve as a test problem for theoretical and numerical investigations. The problem exhibits already most features of more challenging problems such as non-convexity, pure state constraints on the upper level problem as well as control constraints on both levels.

2   A Class of Bilevel Optimal Control Problems

Let $T > 0$, be the fixed final time, $X := W^{1,\infty}([0,T], \mathbb{R}^{n_x}) \times L^\infty([0,T], \mathbb{R}^{n_u}) \times \mathbb{R}^{n_p}$, $n_x, n_u, n_p \in \mathbb{N}_0$, $Y := W^{1,\infty}([0,T], \mathbb{R}^{n_y}) \times L^\infty([0,T], \mathbb{R}^{n_v}) \times \mathbb{R}^{n_q}$,

$n_y, n_v, n_q \in \mathbb{N}_0$, where $L^\infty([0,T], \mathbb{R}^n)$ denotes the Banach space of essentially bounded vector-valued functions from $[0,T]$ into $\mathbb{R}^n$ and $W^{1,\infty}([0,T], \mathbb{R}^n)$ is the Banach space of absolutely continuous vector-valued functions from $[0,T]$ into $\mathbb{R}^n$ with essentially bounded first derivatives. Moreover, let the Banach spaces $V^u := L^\infty([0,T], \mathbb{R}^{n_x}) \times \mathbb{R}^{n_H}$, $V^\ell := L^\infty([0,T], \mathbb{R}^{n_v}) \times \mathbb{R}^{n_h}$, $n_H, n_h \in \mathbb{N}_0$, and the closed convex cones $W^u := \{k \in L^\infty([0,T], \mathbb{R}^{n_G}) \mid k(t) \leq 0 \text{ a.e. in } [0,T]\}$, $W^\ell := \{k \in L^\infty([0,T], \mathbb{R}^{n_g}) \mid k(t) \leq 0 \text{ a.e. in } [0,T]\}$, $n_G, n_g \in \mathbb{N}_0$, be given. Let

$$J, j : \mathbb{R}^{n_x} \times \mathbb{R}^{n_y} \times \mathbb{R}^{n_p} \times \mathbb{R}^{n_q} \to \mathbb{R},$$
$$F : \mathbb{R}^{n_x} \times \mathbb{R}^{n_y} \times \mathbb{R}^{n_u} \times \mathbb{R}^{n_v} \times \mathbb{R}^{n_p} \times \mathbb{R}^{n_q} \to \mathbb{R}^{n_x},$$
$$f : \mathbb{R}^{n_y} \times \mathbb{R}^{n_v} \times \mathbb{R}^{n_p} \times \mathbb{R}^{n_q} \to \mathbb{R}^{n_y},$$
$$\Psi : \mathbb{R}^{n_x} \times \mathbb{R}^{n_y} \times \mathbb{R}^{n_x} \times \mathbb{R}^{n_y} \times \mathbb{R}^{n_p} \times \mathbb{R}^{n_q} \to \mathbb{R}^{n_H},$$
$$\psi : \mathbb{R}^{n_x} \times \mathbb{R}^{n_y} \times \mathbb{R}^{n_x} \times \mathbb{R}^{n_y} \times \mathbb{R}^{n_p} \times \mathbb{R}^{n_q} \to \mathbb{R}^{n_h},$$
$$S : \mathbb{R}^{n_x} \times \mathbb{R}^{n_y} \times \mathbb{R}^{n_u} \times \mathbb{R}^{n_v} \times \mathbb{R}^{n_p} \times \mathbb{R}^{n_q} \to \mathbb{R}^{n_G},$$
$$s : \mathbb{R}^{n_y} \times \mathbb{R}^{n_v} \times \mathbb{R}^{n_p} \times \mathbb{R}^{n_q} \to \mathbb{R}^{n_g}.$$

be sufficiently smooth mappings. With these definitions the following class of bilevel optimal control problems (BOCP) subject to control-state constraints and boundary conditions fits into the general bilevel optimization problem BOP.

Minimize $J(x(T), y(T), p, q)$ w.r.t. $(x, u, p, y, v, q) \in X \times Y$ subject to the constraints

$$x'(t) = F(x(t), y(t), u(t), v(t), p, q), \tag{1}$$
$$0 = \Psi(x(0), y(0), x(T), y(T), p, q), \tag{2}$$
$$0 \geq S(x(t), y(t), u(t), v(t), p, q), \tag{3}$$
$$(y, v, q) \in M(x(0), x(T), p)$$

where $M(x(0), x(T), p)$ is the set of minimizers of the lower level problem $OCP_L(x(0), x(T), p)$:

Minimize $j(x(T), y(T), p, q)$ w.r.t. $(y, v, q) \in Y$ subject to the constraints

$$y'(t) = f(y(t), v(t), p, q), \tag{4}$$
$$0 = \psi(x(0), y(0), x(T), y(T), p, q), \tag{5}$$
$$0 \geq s(y(t), v(t), p, q). \tag{6}$$

Herein, $(x, u, p) \in X$ are the state, the control, and the parameter vector of the upper level problem and $(y, v, q) \in Y$ are the state, the control, and the parameter vector of the lower level problem. Please note that the lower level problem only depends on the initial and terminal states $x(0), x(T)$ and the parameter vector p of the upper level problem. The value function V is then a mapping from $\mathbb{R}^{n_x} \times \mathbb{R}^{n_x} \times \mathbb{R}^{n_p}$ into $\mathbb{R}$ defined by

$$V(x_0, x_T, p) := \inf_{(y,v,q) \in Y} \left\{ j(x_T, y(T), p, q) \; \middle| \; \begin{array}{l} y'(t) = f(y(t), v(t), p, q), \\ 0 = \psi(x_0, y(0), x_T, y(T), p, q), \\ 0 \geq s(y(t), v(t), p, q) \end{array} \right\}.$$

Remark 1. In a formal way the problem class can be easily extended in such a way that the lower level dynamics f and the lower level control-state constraints s depend on x, u as well. However, in the latter case the value function of the lower level problem would then be a functional $V : X \to \mathbb{R}$, i.e. a functional defined on the Banach space X rather than a functional defined on the finite dimensional space $\mathbb{R}^{n_x} \times \mathbb{R}^{n_x} \times \mathbb{R}^{n_p}$. Computing the mapping $V : X \to \mathbb{R}$ numerically would be computationally intractable in most cases.

Using the value function V we arrive at the following equivalent single level optimal control problem subject to control-state constraints, smooth boundary conditions, and an in general non-smooth boundary condition with the value function.

Minimize $J(x(T), y(T), p, q)$ w.r.t. $(x, u, p, y, v, q) \in X \times Y$ subject to the constraints (1)-(3), (4)-(6), and

$$j(x(T), y(T), p, q) \leq V(x(0), x(T), p). \qquad (7)$$

It remains to compute the value function V and to solve the potentially non-smooth single level optimal control problem. Both are challenging tasks owing to non-smoothness and non-convexity. The value function sometimes can be derived analytically as we shall demonstrate in Sect. 3. Otherwise, if Bellman's optimality principle applies, the value function satisfies a Hamilton-Jacobi-Bellman (HJB) equation, see [3]. Various methods exist for its numerical solution, compare [4,11, 14,17,21]. The HJB approach is feasible if the state dimension n_y does not exceed 5 or 6. If no analytical formula is available and if the HJB approach is not feasible, then a pointwise evaluation of V at $(x(0), x(T), p)$ can be realized by using suitable optimal control software, e.g. [13]. However, if the lower level problem is non-convex, then it is usually not possible to guarantee global optimality by such an approach. The single level problem can be approached by the non-smooth necessary conditions in [5,6]. Alternatively, direct discretization methods may be applied. The non-smoothness in V in (7) has to be taken into account by, e.g., using bundle type methods, see [23], or by smoothing the value function and applying standard software. Finally, the HJB approach could also be applied to the single level problem again.

3   A Follow-the-leader Application

We consider a pursuit-evasion dynamic Stackelberg game of two vehicles moving in the plane. Throughout we assume that the evader knows the optimal strategy of the pursuer and can optimize its own's strategy accordingly. This gives rise to a bilevel optimal control problem. The lower level player (=pursuer P) aims to capture the upper level player (=evader E) in minimum time T. The evader aims to minimize a linear combination of the negative capture time $-T$ and its control effort. The players have individual dynamics and constraints. The coupling occurs through capture conditions at the final time.

3.1   The Bilevel Optimal Control Problem

The evader E aims to solve the following optimal control problem, called the upper level problem (OCP$_U$):

Minimize

$$-T + \int_0^T \frac{\alpha_1}{2} w(t)^2 + \frac{\alpha_2}{2} a(t)^2 dt \tag{8}$$

subject to the constraints

$$x_E'(t) = v_E(t) \cos \psi(t), \quad x_E(0) = x_{E,0}, \quad x_E(T) = x_P(T), \tag{9}$$

$$y_E'(t) = v_E(t) \sin \psi(t), \quad y_E(0) = y_{E,0}, \quad y_E(T) = y_P(T), \tag{10}$$

$$\psi'(t) = \frac{v_E(t)}{\ell} \tan \delta(t), \quad \psi(0) = \psi_0, \tag{11}$$

$$\delta'(t) = w(t), \quad \delta(0) = \delta_0, \tag{12}$$

$$v_E'(t) = a(t), \quad v_E(0) = v_{E,0}, \tag{13}$$

$$v_E(t) \in [0, v_{E,max}], \quad w(t) \in [-w_{max}, w_{max}], \quad a(t) \in [a_{min}, a_{max}], \tag{14}$$

$$(x_P, y_P, T) \in M(x_E(T), y_E(T)),$$

where $M(x_E(T), y_E(T))$ denotes the set of minimizers of the lower level problem $OCP_L(x_E(T), y_E(T))$ below.

The equations of motion of E describe a simplified car model of length $\ell > 0$ moving in the plane. The controls are the steering angle velocity w and the acceleration a with given bounds $\pm w_{max}$, a_{min}, and a_{max}, respectively. The velocity v_E is bounded by the state constraint $v_E(t) \in [0, v_{E,max}]$ with a given bound $v_{E,max} > 0$. The position of the car's rear axle is given by $z_E = (x_E, y_E)^\top$ and its velocity by v_E. ψ denotes the yaw angle and $\alpha_1, \alpha_2 \geq 0$ are weights in the objective function. The initial state is fixed by the values $x_{E,0}, y_{E,0}, \psi_0, \delta_0, v_{E,0}$. The final time T is determined by the lower level player P, who aims to solve the following optimal control problem, called the lower level problem OCP$_L(x_{E,T}, y_{E,T})$ with its set of minimizers denoted by $M(x_{E,T}, y_{E,T})$:

Minimize $T = \int_0^T 1 dt$ subject to the constraints

$$z_P'(t) = v_P(t), \quad z_P(0) = z_{P,0}, \quad z_P(T) = (x_{E,T}, y_{E,T})^\top, \tag{15}$$

$$v_P'(t) = u_P(t), \quad v_P(0) = v_P(T) = 0, \tag{16}$$

$$u_{P,i}(t) \in [-u_{max}, u_{max}], \quad i = 1, 2. \tag{17}$$

Herein, $z_P = (x_P, y_P)^\top$, $v_P = (v_{P,1}, v_{P,2})^\top$, and $u_P = (u_{P,1}, u_{P,2})^\top$ denote the position vector, the velocity vector, and the acceleration vector, respectively, of P in the two-dimensional plane. $z_{P,0} = (x_{P,0}, y_{P,0})^\top \in \mathbb{R}^2$ is a given initial position. $u_{max} > 0$ is a given control bound for the acceleration. The dynamics of the pursuer allow to move in x and y direction independently, which models, e.g., a robot with omnidirectional wheels.

3.2   The Lower-Level Problem and Its Value Function

The lower level problem admits an analytical solution. To this end, the Hamilton function (regular case only) reads as

$$\mathcal{H}(z_P, v_P, u_P, \lambda_z, \lambda_v) = 1 + \lambda_z^\top v_P + \lambda_v^\top u_P.$$

The first order necessary optimality conditions for a minimum $(\hat{z}_P, \hat{v}_P, \hat{u}_P, \hat{T})$ are given by the minimum principle, compare [16]. There exist adjoint multipliers λ_z, λ_v with

$$\lambda_z'(t) = -\nabla_{z_P}\mathcal{H}[t] = 0, \qquad \lambda_v'(t) = -\nabla_{v_P}\mathcal{H}[t] = -\lambda_z(t),$$

and

$$\mathcal{H}(\hat{z}_P(t), \hat{v}_P(t), \hat{u}_P(t), \lambda_z(t), \lambda_v(t)) \leq \mathcal{H}(\hat{z}_P(t), \hat{v}_P(t), u_P, \lambda_z(t), \lambda_v(t))$$

for all $u_P \in [-u_{max}, u_{max}]^2$ for almost every $t \in [0, \hat{T}]$. The latter implies

$$\hat{u}_{P,i}(t) = \begin{cases} u_{max}, & \text{if } \lambda_{v,i}(t) < 0 \\ -u_{max}, & \text{if } \lambda_{v,i}(t) > 0 \\ \text{singular}, & \text{if } \lambda_{v,i}(t) = 0 \text{ on some interval,} \end{cases} \qquad i = 1, 2.$$

The adjoint equations yield $\lambda_z(t) = c_z$ and $\lambda_v(t) = -c_z t + c_v$ with constants $c_z, c_v \in \mathbb{R}^2$. A singular control component $\hat{u}_{P,i}$ with $i \in \{1,2\}$ can only occur if $c_{z,i} = c_{v,i} = 0$. In this case, the minimum principle provides no information on the singular control except feasibility. Notice furthermore that not all control components can be singular since this would lead to trivial multipliers in contradiction to the minimum principle. Hence, there is at least one index i for which the control component $\hat{u}_{P,i}$ is non-singular. In the non-singular case there can be at most one switch of each component $\hat{u}_{P,i}$, $i \in \{1,2\}$, in the time interval $[0, \hat{T}]$, since $\lambda_{v,i}$ is linear in time. The switching time $\hat{t}_{s,i}$ for the i-th control component computes to $\hat{t}_{s,i} = c_{v,i}/c_{z,i}$ if $c_{z,i} \neq 0$. We discuss several cases for non-singular controls.

Case 1: No switching occurs in $\hat{u}_{P,i}$, i.e. $\hat{u}_{P,i}(t) \equiv \pm u_{max}$ for $i \in \{1,2\}$. By integration we obtain $\hat{v}_{P,i}(t) = \pm u_{max}t$ and thus $\hat{v}_{P,i}(\hat{T}) \neq 0$ in contradiction to the boundary conditions. Consequently, each non-singular control component switches exactly once in $[0, \hat{T}]$.

Case 2: The switching structure for control component $i \in \{1,2\}$ is

$$\hat{u}_{P,i}(t) = \begin{cases} u_{max}, & \text{if } 0 \leq t < \hat{t}_{s,i}, \\ -u_{max}, & \text{otherwise.} \end{cases}$$

By integration and the boundary conditions we find

$$\hat{v}_{P,i}(t) = \begin{cases} u_{max}t, & \text{if } 0 \leq t < \hat{t}_{s,i} \\ u_{max}(2\hat{t}_{s,i} - t), & \text{otherwise} \end{cases}$$

$$\hat{z}_{P,i}(t) = \begin{cases} \hat{z}_{P,i}(0) + \frac{1}{2}u_{max}t^2, & \text{if } 0 \leq t < \hat{t}_{s,i} \\ \hat{z}_{P,i}(0) + u_{max}\left(\hat{t}_{s,i}^2 - \frac{1}{2}(2\hat{t}_{s,i} - t)^2\right), & \text{otherwise.} \end{cases}$$

The boundary conditions for $\hat{v}_{P,i}(\hat{T})$ and $\hat{z}_{P,i}(\hat{T})$ yield

$$\hat{T}_i = 2\hat{t}_{s,i} \quad \text{and} \quad \hat{t}_{s,i} = \sqrt{\frac{\hat{z}_{P,i}(\hat{T}) - \hat{z}_{P,i}(0)}{u_{max}}} \text{ if } \hat{z}_{P,i}(\hat{T}) - \hat{z}_{P,i}(0) \geq 0.$$

Case 3: The switching structure for control component $i \in \{1,2\}$ is

$$\hat{u}_{P,i}(t) = \begin{cases} -u_{max}, & \text{if } 0 \leq t < \hat{t}_{s,i}, \\ u_{max}, & \text{otherwise.} \end{cases}$$

This case can be handled analogously to Case 2 and we obtain

$$\hat{T}_i = 2\hat{t}_{s,i} \quad \text{and} \quad \hat{t}_{s,i} = \sqrt{\frac{\hat{z}_{P,i}(0) - \hat{z}_{P,i}(\hat{T})}{u_{max}}} \text{ if } \hat{z}_{P,i}(0) - \hat{z}_{P,i}(\hat{T}) \geq 0.$$

The above analysis reveals the shortest times $\hat{T}_i$, $i \in \{1,2\}$, in which the i-th state can reach its terminal boundary condition. The minimum time $\hat{T}$ for a given terminal position is thus given by the value function V of $\text{OCP}_L(x_{E,T}, y_{E,T})$ (=minimum time function) with

$$V(x_{E,T}, y_{E,T}) = \max\{\hat{T}_1, \hat{T}_2\} = 2 \max\left\{ \sqrt{\frac{|x_{P,0} - x_{E,T}|}{u_{max}}}, \sqrt{\frac{|y_{P,0} - y_{E,T}|}{u_{max}}} \right\}. \tag{18}$$

That is, the final time is defined by the component i with the largest distance $|\hat{z}_{P,i}(\hat{T}) - \hat{z}_{P,i}(0)|$. For this component, the control is of bang-bang type with one switch at the midpoint of the time interval. The remaining control can be singular and it is not uniquely defined. The value function is locally Lipschitz continuous except at the point $(x_{E,T}, y_{E,T}) = (x_{P,0}, y_{P,0})$, compare Fig. 1. This point, however, is of minor interest because interception takes place immediately.

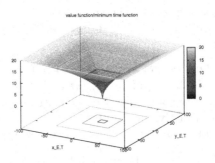

value function/minimum time function

Fig. 1. Value function of lower level problem with data $x_{P,0} = y_{P,0} = 0$, $u_{max} = 1$.

The equivalent single level problem (SL-OCP) reads as follows:

Minimize (8) subject to the constraints (9)-(14), (15)-(17) with $(x_{E,T}, y_{E,T})^\top = (x_E(T), y_E(T))^\top$ and the non-smooth constraint

$$T \leq V(x_E(T), y_E(T)). \tag{19}$$

3.3   Numerical Results

For the numerical solution of the single level problem SL-OCP we applied the direct shooting method OCPID-DAE1, [13]. The non-smooth constraint $T \leq V(x_E(T), y_E(T))$ with V from (18) was replaced by a continuously differentiable constraint which was obtained by smoothing the maximum function and the absolute value function in (18). Figure 2 shows a numerical solution of the pursuit-evasion Stackelberg bilevel optimal control problem for the data $v_{E,0} = 10$, $\psi_E(0) = \pi/4$, $\alpha_1 = 10$, $\alpha_2 = 0$, $w_{max} = 0.5$, $v_{E,max} = 20$, $a_{min} = -5$, $a_{max} = 1$, $u_{max} = 5$, $N = 50$, $T \approx 18.01$. Figure 3 shows several trajectories for

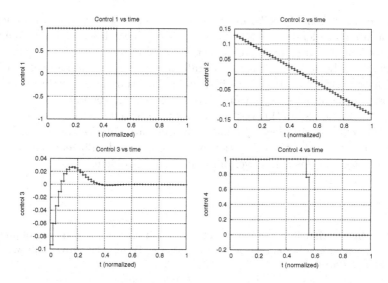

Fig. 2. Numerical results for the bilevel optimal control problem: Trajectories of pursuer (lines with '+') and evader (lines with boxes, top), controls of the pursuer (middle), controls of the evader (bottom).

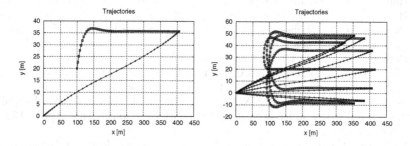

Fig. 3. Left picture: Trajectories of pursuer (lines with '+') and evader (lines with boxes) for $\psi_E(0) = \pi/4$. Right picture: Trajectories of the pursuer (lines with '+') and the evader (lines with boxes) for different initial yaw angles of the evader.

the pursuer and the evader for different initial yaw angles covering the interval $[0, 2\pi)$.

Remark 2. The constraint (19) may become infeasible under discretization. Instead, the value function V_h of the discretized lower level optimal control problem should be used. However, since V_h is hardly available for all kinds of discretizations, we use instead the relaxed constraint $T \leq V(x_E(T), y_E(T)) + \varepsilon$ with some $\varepsilon > 0$.

4 Conclusions and Outlook

The paper discusses a specific bilevel optimal control problem and its reformulation as an equivalent single level problem using the value function of the lower level problem. For a sample problem it is possible to compute the value function analytically and to solve the overall bilevel problem numerically using a direct discretization method. This first numerical study leaves many issues open that have to be investigated in future research for the general problem setting. Amongst them are smoothness properties of the value function, representation of subdifferentials, the development of appropriate solution methods for nonsmooth problems, and the derivation of necessary (and sufficient) conditions of optimality for the class of bilevel optimal control problems.

References

1. Albrecht, S.: Modeling and numerical solution of inverse optimal control problems for the analysis of human motions. Ph.D. thesis, Technische Universität München, München (2013)
2. Albrecht, S., Leibold, M., Ulbrich, M.: A bilevel optimization approach to obtain optimal cost functions for human arm movements. Numer. Algebra Control Optim. **2**(1), 105–127 (2012)
3. Bardi, M., Capuzzo-Dolcetta, I.: Optimal control and viscosity solutions of Hamilton-Jacobi-Bellman equations. Reprint of the 1997 original. Birkhäuser, Basel (2008)
4. Bokanowski, O., Desilles, A., Zidani, H.: ROC-HJ: Reachability analysis and optimal control problems - Hamilton-Jacobi equations. Technical report, Universite Paris Diderot, ENSTA ParisTech, Paris (2013)
5. Clarke, F.: Functional Analysis Calculus of Variations and Optimal Control. Graduate Texts in Mathematics, vol. 264. Springer, Heidelberg (2013)
6. de Pinho, M., Vinter, R.B.: Necessary conditions for optimal control problems involving nonlinear differential algebraic equations. J. Math. Anal. Appl. **212**, 493–516 (1997)
7. Dempe, S., Gadhi, N.: A new equivalent single-level problem for bilevel problems. Optimization **63**(5), 789–798 (2014)
8. Dempe, S.: Foundations of Bilevel Programming. Kluwer Academic Publishers, Dordrecht (2002)
9. Dempe, S., Zemkoho, A.B.: KKT reformulation and necessary conditions for optimality in nonsmooth bilevel optimization. SIAM J. Optim. **24**(4), 1639–1669 (2014)

10. Ehtamo, H., Raivio, T.: On applied nonlinear and bilevel programming for pursuit-evasion games. J. Optim. Theory Appl. **108**(1), 65–96 (2001)
11. Falcone, M., Ferretti, R.: Convergence analysis for a class of high-order semi-Lagrangian advection schemes. SIAM J. Numer. Anal. **35**(3), 909–940 (1998)
12. Fisch, F.: Development of a framework for the solution of high-fidelity trajectory optimization problems and bilevel optimal control problems. Ph.D. thesis, Technische Universität München, München (2011)
13. Gerdts, M.: OCPID-DAE1 - optimal control and parameter identification with differential-algebraic equations of index 1. Technical report, User's Guide, Engineering Mathematics, Department of Aerospace Engineering, University of the Federal Armed Forces at Munich (2013). http://www.optimal-control.de
14. Grüne, L.: An adaptive grid scheme for the discrete hamilton-jacobi-bellman equation. Numer. Math. **75**(3), 319–337 (1997)
15. Hatz, K.: Efficient numerical methods for hierarchical dynamic optimization with application to cerebral palsy gait modeling. Dissertation, Univ. Heidelberg, Heidelberg, Naturwissenschaftlich-Mathematische Gesamtfakultät (2014)
16. Ioffe, A.D., Tihomirov, V.M.: Theory of Extremal Problems. Studies in Mathematics and its Applications, vol. 6. North-Holland Publishing Company, Amsterdam (1979)
17. Jiang, G.-S., Peng, D.: Weighted ENO schemes for Hamilton-Jacobi equations. SIAM J. Sci. Comput. **21**(6), 2126–2143 (2000)
18. Knauer, M.: Bilevel-Optimalsteuerung mittels hybrider Lösungsmethoden am Beispiel eines deckengeführten Regalbediengerätes in einem Hochregallager. Ph.D. thesis, University of Bremen, Bremen (2009)
19. Knauer, M.: Fast and save container cranes as bilevel optimal control problems. Math. Comput. Model. Dyn. Syst. **18**(4), 465–486 (2012)
20. Mombaur, K.D.: Stability optimization of open-loop controlled walking robots. University Heidelberg, Naturwissenschaftlich-Mathematische Gesamtfakultät, Heidelberg (2001)
21. Osher, S., Shu, C.W.: High-order essentially nonoscillatory schemes for Hamilton-Jacobi equations. SIAM J. Numer. Anal. **28**(4), 907–922 (1991)
22. Outrata, J.V.: On the numerical solution of a class of Stackelberg problems. Z. Oper. Res. **34**(4), 255–277 (1990)
23. Schramm, H., Zowe, J.: A version of the bundle idea for minimizing a nonsmooth function: conceptual idea, convergence analysis, numerical results. SIAM J. Optim. **2**(1), 121–152 (1992)
24. Stackelberg, H.: The Theory of Market Economy. Oxford University Press, Oxford (1952)
25. Ye, J.J.: Necessary conditions for bilevel dynamic optimization problems. SIAM J. Control Optim. **33**(4), 1208–1223 (1995)
26. Ye, J.J.: Optimal strategies for bilevel dynamic problems. SIAM J. Control Optim. **35**(2), 512–531 (1997)

Multibody Dynamics with Unilateral Constraints: Computational Modelling of Soft Contact and Dry Friction

Laetitia Paoli[✉]

Institut Camille Jordan, UMR CNRS 5208, University of Lyon, University Jean Monnet, 23 Rue Michelon, 42023 Saint-Etienne Cedex 2, France
laetitia.paoli@univ-st-etienne.fr

Abstract. We consider a system of rigid bodies subjected to unilateral constraints with soft contact and dry friction. When the constraints are saturated, velocity jumps may occur and the dynamics is described in generalized coordinates by a second-order measure differential inclusion for the unknown configurations. Observing that the right velocity obeys a minimization principle, a time-stepping algorithm is proposed. It allows to construct a sequence of approximate solutions satisfying at each time-step a discrete contact law which mimics the behaviour of the system in case of collision. In case of tangential contact, dry friction may lead to indeterminacies such as the famous Painlevé's paradoxes. By a precise study of the asymptotic properties of the scheme, it is shown that the limit of the approximate trajectories exhibits the same kind of indeterminacies.

Keywords: Unilateral constraints · Coulomb's law · Measure differential inclusion · Time-stepping scheme

1 Description of the Problem

We consider a discrete mechanical system with d degrees of freedom. We denote by $q \in \mathbb{R}^d$ its representative point in generalized coordinates and by $M(q)$ its inertia operator. We assume that the system is subjected to unilateral constraints characterized by the geometrical inequality

$$g(q) \leq 0 \quad \text{(non penetration condition)}$$

with a smooth (at least C^1) function g such that ∇g does not vanish in a neighborhood of $\{q \in \mathbb{R}^d;\ g(q) = 0\}$. Let us denote by $\langle \cdot, \cdot \rangle$ the Euclidean inner product in $\mathbb{R}^d$. If for some instant $t > 0$ the constraints are saturated, i.e. $g(q(t)) = 0$, then

$$\langle \dot{q}^-(t), \nabla g(q(t)) \rangle \geq 0, \quad \langle \dot{q}^+(t), \nabla g(q(t)) \rangle \leq 0$$

L. Bociu et al. (Eds.): CSMO 2015, IFIP AICT 494, pp. 420–429, 2016.
DOI: 10.1007/978-3-319-55795-3_40

and the velocity may be discontinuous. It follows that $u = \dot{q}$ is a function of Bounded Variations and the dynamics is described by the Measure Differential Equation

$$M(q)du = f(t, q, u)dt + r \tag{1}$$

where du is the Stieljes measure associated to u and r is the reaction due to the constraints. Of course a reaction is applied to the system only when a contact occurs and we have a complementarity condition

$$g\big(q(t)\big) < 0 \Longrightarrow r = 0.$$

Furthermore, we assume that the contact is non-adhesive which yields

$$\langle r, \nabla g\big(q(t)\big)\rangle \leq 0 \quad \text{if } g\big(q(t)\big) = 0.$$

In the frictionless case we get

$$r \in -N\big(q(t)\big) \quad \text{if } g\big(q(t)\big) = 0$$

where $N\big(q(t)\big)$ is the normal cone to the set of admissible configurations at $q(t)$ and in the frictional case

$$r \in C\big(q(t)\big) \quad \text{if } g\big(q(t)\big) = 0$$

where $C\big(q(t)\big)$ is the so-called friction cone at $q(t)$. Hence we may rewrite (1) as a Measure Differential Inclusion ([10,17])

$$M(q)du - f(t, q, u)dt \in \mathcal{R}(q) \tag{2}$$

where

$$\mathcal{R}(q) = \begin{cases} \{0\} & \text{if } g(q) < 0, \\ \mathbb{R}^+\big(n(q) + D_1(q)\big) & \text{if } g(q) \geq 0, \end{cases}$$

and $n(q) = -\dfrac{\nabla g(q)}{\|\nabla g(q)\|}$ and $D_1(q)$ is the disc of center 0 and radius μ in $\big(\mathbb{R}n(q)\big)^{\perp}$ with $\mu = 0$ (frictionless constraints) or $\mu > 0$ (Coulomb's friction).

Let us assume also soft contact i.e.

$$u^+(t) \in T\big(q(t)\big) = \big(\mathbb{R}n(q(t))\big)^{\perp} \quad \text{if } g\big(q(t)\big) = 0. \tag{3}$$

We infer that

$$u^+(t) \in \big(u^-(t) + M^{-1}\big(q(t)\big)\mathcal{R}\big(q(t)\big)\big) \cap T\big(q(t)\big) \quad \text{if } g\big(q(t)\big) = 0. \tag{4}$$

In the frictionless case we may decompose $u^{\pm}(t)$ as

$$u^{\pm}(t) = u_N^{\pm}(t) + u_T^{\pm}(t), \quad u_N^{\pm}(t) \in \mathbb{R}M^{-1}\big(q(t)\big)n\big(q(t)\big), \quad u_T^{\pm}(t) \in T\big(q(t)\big),$$

with

$$u_N^\pm(t) = \frac{\langle u^\pm(t), n(q(t)) \rangle}{\langle n(q(t)), M^{-1}(q(t)) n(q(t)) \rangle}.$$

With (3)–(4) we get

$$u_N^+(t) = 0, \quad u_T^+(t) = u_T^-(t)$$

which is equivalent to

$$u^+(t) = \text{Proj}_{q(t)}\big(0, \big(u^-(t) + M^{-1}(q(t)) \mathcal{R}(q(t))\big) \cap T(q(t))\big) \text{ if } g(q(t)) = 0$$

where $\text{Proj}_{q(t)}$ denotes the projection relatively to the kinetic metric at $q(t)$.

In the frictional case, when $M(q) \equiv m\text{Id}_{\mathbb{R}^d}$, $m > 0$, Coulomb's law ([3,11,12]) yields

$$-u^+(t) \in \partial \psi_{r_N D_1(q(t))}(r_T) \quad \text{if } g(q(t)) = 0 \tag{5}$$

where r_N and r_T are respectively the projection relatively to the Euclidean metric of r on $\mathbb{R}n(q(t))$ and $T(q(t))$ and $\partial \psi_{r_N D_1(q(t))}$ is the indicatrix function of the disc of radius μr_N and center 0 in $T(q(t))$. Reminding that $\mathcal{R}(q(t)) = \mathbb{R}^+\big(n(q(t)) + D_1(q(t))\big)$, (5) is equivalent to

$$-u^+(t) \in \text{Proj}\big(T(q(t)), \partial \psi_{\mathcal{R}(q(t))}(r)\big)$$

which can be rewritten as

$$u^+(t) = \text{Proj}\big(0, \big(u^-(t) + M^{-1}(q(t)) \mathcal{R}(q(t))\big) \cap T(q(t))\big) \text{ if } g(q(t)) = 0. \tag{6}$$

Let us emphasize that (4) and (6) imply that $u^+(t) \neq u^-(t)$ only if $g(q(t)) = 0$ and $\langle u^-(t), n(q(t)) \rangle < 0$, i.e. only in case of collision. Moreover $u^+(t)$ is defined as the Argmin of the kinematically admissible right velocities. The same property holds when $M(q) \neq m\text{Id}_{\mathbb{R}^d}$ and $\mu > 0$ and we still have ([6,12])

$$u^+(t) = \text{Proj}_{q(t)}\big(0, \big(u^-(t) + M^{-1}(q(t)) \mathcal{R}(q(t))\big) \cap T(q(t))\big) \tag{7}$$

if $g(q(t)) = 0$ and $\langle u^-(t), n(q(t)) \rangle < 0$. On the contrary, when $g(q(t)) = 0$ and $\langle u^-(t), n(q(t)) \rangle = 0$, (4) yields

$$u^+(t) \in u^-(t) + \big(M^{-1}(q(t)) \mathcal{R}(q(t)) \cap T(q(t))\big)$$

and velocity jumps without collision may occur if $M^{-1}(q(t)) \mathcal{R}(q(t)) \cap T(q(t)) \neq \{0\}$. Such phenomena can easily be observed when we consider the model problem of a slender rod in contact at one edge with an horizontal obstacle, leading to the famous Painlevé's paradoxes ([13,14]): there exists a subset $\mathcal{A}(q(t), u^-(t))$, containing $u^-(t)$ but not reduced to this single point, such that any value of $u^+(t) \in \mathcal{A}(q(t), u^-(t))$ solves the problem (see for instance [1,4,8] or more recently [2,7,12]). Hence, in case of tangential contact with dry friction and non-trivial inertia operator, the dynamics exhibits indeterminacies.

2 Computational Modelling: The *Contact Dynamics* Approach

In order to solve numerically the problem, the *Contact Dynamics* approach has been introduced by Moreau in the mid 80's ([10–12]). The core idea is to avoid any regularization of the unilateral constraints and to build a time-stepping scheme by combining an Euler discretization of the measure differential inclusion (2) on each interval $[t_i, t_{i+1}]$ with an impulsional form of the contact law at t_{i+1}. More precisley the approximate position is updated as

$$q_{i+1} = q_i + hu_i, \quad h = t_{i+1} - t_i$$

and a "free" left velocity at t_{i+1} is defined by

$$v_{i+1} = u_i + hM^{-1}(q_{i+1})f(t_{i+1}, q_{i+1}, u_i).$$

Then u_{i+1} is the right velocity at t_{i+1} given by

$$M(q_{i+1})(u_{i+1} - u_i) - hf(t_{i+1}, q_{i+1}, u_i) \in \mathcal{R}(q_{i+1}), \quad u_{i+1} \in T(q_{i+1})$$

and

$$u_{i+1} = \mathcal{S}(q_{i+1}, v_{i+1})$$

where $\mathcal{S}$ is a discrete analogous of the contact law. Starting from the definition of $\mathcal{R}(q_{i+1})$, we get immediately

$$\mathcal{S}(q_{i+1}, v_{i+1}) = v_{i+1} \quad \text{if } g(q_{i+1}) < 0.$$

If $g(q_{i+1}) \geq 0$ and $\langle v_{i+1}, n(q_{i+1}) \rangle > 0$, the left velocity v_{i+1} points inward and

$$\mathcal{S}(q_{i+1}, v_{i+1}) = v_{i+1}.$$

If $g(q_{i+1}) \geq 0$ and $\langle v_{i+1}, n(q_{i+1}) \rangle < 0$, we may interpret t_{i+1} as a collision instant and with (6) we get

$$\mathcal{S}(q_{i+1}, v_{i+1}) = \text{Proj}_{q_{i+1}}\left(0, \left(v_{i+1} + M^{-1}(q_{i+1})\mathcal{R}(q_{i+1})\right) \cap T(q_{i+1})\right).$$

Finally, if $g(q_{i+1}) \geq 0$ and $\langle v_{i+1}, n(q_{i+1}) \rangle = 0$, we get

$$u_{i+1} \in v_{i+1} + \left(M^{-1}(q_{i+1})\mathcal{R}(q_{i+1}) \cap T(q_{i+1})\right).$$

Hence u_{i+1} is uniquely defined if $M^{-1}(q_{i+1})\mathcal{R}(q_{i+1}) \cap T(q_{i+1}) = \{0\}$ and we have $u_{i+1} = \mathcal{S}(q_{i+1}, v_{i+1}) = v_{i+1}$. On the contrary, if $M^{-1}(q_{i+1})\mathcal{R}(q_{i+1}) \cap T(q_{i+1}) \neq \{0\}$, t_{i+1} may be interpreted as a discrete tangential contact with possible indeterminacies and there is not any natural choice for u_{i+1}.

In [10–12] Moreau proposed $u_{i+1} = \mathcal{S}(q_{i+1}, v_{i+1}) = v_{i+1}$ if $g(q_{i+1}) \leq 0$, $\langle v_{i+1}, n(q_{i+1}) \rangle = 0$ and $M^{-1}(q_{i+1})\mathcal{R}(q_{i+1}) \cap T(q_{i+1}) \neq \{0\}$. It follows that $\mathcal{S}$ is defined as

$$\mathcal{S}(q, u^-) = \begin{cases} v & \text{if } v \in V(q), \\ \text{Proj}_q\left(0, \left(v + M^{-1}(q)\mathcal{R}(q)\right) \cap T(q)\right) & \text{otherwise,} \end{cases}$$

with

$$V(q) = \begin{cases} \mathbb{R}^d & \text{if } g(q) < 0, \\ \{v \in \mathbb{R}^d; \ \langle v, n(q) \rangle \geq 0\} & \text{if } g(q) \geq 0. \end{cases}$$

For this scheme the following stability property holds.

Proposition 1. *For all $i \geq 0$, let $r_{i+1} = M(q_{i+1})(u_{i+1} - u_i) - hf(t_{i+1}, q_{i+1}, u_i)$ and $\| \cdot \|_{q_{i+1}}$ be the kinetric norm at t_{i+1} defined by $\|v\|_{q_{i+1}} = \langle v, M(q_{i+1})v \rangle^{1/2}$ for all $v \in \mathbb{R}^d$. Then, $r_{i+1} \in \mathcal{R}(q_{i+1})$, $\langle u_{i+1}, r_{i+1} \rangle \leq 0$ and*

$$\|u_{i+1}\|_{q_{i+1}} \leq \|u_i\|_{q_{i+1}} + h\|M^{-1/2}(q_{i+1})\|\|f(t_{i+1}, q_{i+1}, u_i)\|.$$

Proof. If $r_{i+1} = 0$ the result is obvious.
Otherwise, $v_{i+1} \notin V(q_{i+1})$ and $\langle v_{i+1}, n(q_{i+1}) \rangle < 0$. By definition of $\mathcal{S}$ we get

$$u_{i+1} = \text{Proj}_{q_{i+1}}\left(0, \left(v_{i+1} + M^{-1}(q_{i+1})\mathcal{R}(q_{i+1})\right) \cap T(q_{i+1})\right).$$

Hence

$$\langle -u_{i+1}, M(q_{i+1})(v - u_{i+1}) \rangle \leq 0 \quad \forall v \in \left(v_{i+1} + M^{-1}(q_{i+1})\mathcal{R}(q_{i+1})\right) \cap T(q_{i+1}).$$

By choosing $v = v_{i+1} + \lambda M^{-1}(q_{i+1})n(q_{i+1})$ with

$$\lambda = -\frac{\langle v_{i+1}, n(q_{i+1}) \rangle}{\langle n(q_{i+1}), M^{-1}(q_{i+1})n(q_{i+1}) \rangle}$$

we obtain

$$\langle -u_{i+1}, M(q_{i+1})(v_{i+1} - u_{i+1}) \rangle = \langle u_{i+1}, r_{i+1} \rangle \leq 0.$$

Furthermore

$$\|u_{i+1}\|_{q_{i+1}} \leq \|v\|_{q_{i+1}} \leq \|v_{i+1}\|_{q_{i+1}} = \|u_i + hM^{-1}(q_{i+1})f(t_{i+1}, q_{i+1}, u_i)\|_{q_{i+1}}$$
$$\leq \|u_{i+1}\|_{q_{i+1}} + h\|M^{-1}(q_{i+1})f(t_{i+1}, q_{i+1}, u_i)\|_{q_{i+1}}$$
$$\leq \|u_{i+1}\|_{q_{i+1}} + h\|M^{-1/2}(q_{i+1})\|\|f(t_{i+1}, q_{i+1}, u_i)\|.$$

Observing that the inequality $\langle u_{i+1}, r_{i+1} \rangle \leq 0$ corresponds to a dissipativity property, we infer that the scheme reproduces at the discrete level the main features of the dynamics except the indeterminacies of Coulomb's law. Indeed, $\mathcal{S}(q, v) = v$ if $g(q) = 0$, $\langle v, n(q) \rangle = 0$ and the discrete contact law does not lead to any velocity jump in case of tangential contact.

Starting from Proposition 1, the convergence of the scheme has been established by Monteiro-Marques in [9] when $M(q) \equiv m\text{Id}_{\mathbb{R}^d}$ ($m > 0$) and $\mu \geq 0$ and by Dzonou and Monteiro-Marques in [5] when $M(q) \not\equiv m\text{Id}_{\mathbb{R}^d}$ and $\mu = 0$. In both cases we have $M^{-1}(q)\mathcal{R}(q) \cap T(q) = \{0\}$ for all $q \in \mathbb{R}^d$ such that $g(q) = 0$ and the difficulty due to indeterminacies is avoided. Nevertheless the convergence has also been proved recently when $M(q) \not\equiv m\text{Id}_{\mathbb{R}^d}$ and $\mu > 0$ ([15]), so a natural question arises: is it possible to recover with such a scheme velocity jumps without collisions at the limit when h tends to zero? A first answer has been given by Moreau in [12]: numerical simulations show that the approximated trajectories exhibit the plurality of solutions given by Coulomb's law.

3   Asymptotic Properties of the Discrete Contact Law

Let us assume from now on that $M(q) \neq m\mathrm{Id}_{\mathbb{R}^d}$ $(m > 0)$ and $\mu > 0$. Then the limit trajectory will satisfy in case of convergence the following property

$$u^+(t) \in \lim_{\varepsilon \to 0} \{ \mathcal{S}(q_\varepsilon, u_\varepsilon^-); \ q_\varepsilon \in B(q, \varepsilon), \ u_\varepsilon^- \in B(u^-(t), \varepsilon) \}$$

We may recover the indeterminacies of Coulomb's law if, for all (q, u^-) such that $g(q) = 0$, $\langle u^-, n(q) \rangle = 0$ and $M^{-1}(q)\mathcal{R}(q) \cap T(q) \neq \{0\}$, we have

$$\lim_{\varepsilon \to 0} d_H \big(\mathcal{A}_\varepsilon(q, u^-), \mathcal{A}(q, u^-) \big) = 0 \tag{8}$$

with

$$\mathcal{A}_\varepsilon(q, u^-) = \{ \mathcal{S}(q_\varepsilon, u_\varepsilon^-); \ q_\varepsilon \in B(q, \varepsilon), \ u_\varepsilon^- \in B(u^-, \varepsilon) \}.$$

Let us recall that the Hausdorff distance between two subsets A and B of $\mathbb{R}^d$ is defined as

$$d_H(A, B) = \max \big(e(A, B), e(B, A) \big)$$

where

$$
\begin{aligned}
e(A, B) &= \sup_{a \in A} \mathrm{dist}(a, B) = \sup_{a \in A} \inf_{b \in B} \|a - b\| \quad \text{(the excess of } A \text{ from } B), \\
e(B, A) &= \sup_{b \in B} \mathrm{dist}(b, A) = \sup_{b \in B} \inf_{a \in A} \|b - a\| \quad \text{(the excess of } B \text{ from } A).
\end{aligned}
$$

Hence (8) can be decomposed as

$$\lim_{\varepsilon \to 0} \sup \{ \mathrm{dist} \big(\mathcal{S}(q_\varepsilon, u_\varepsilon^-), \mathcal{A}(q, u^-) \big); \ q_\varepsilon \in B(q, \varepsilon), \ u_\varepsilon^- \in B(u^-, \varepsilon) \} = 0 \tag{9}$$

which can be interpreted as an *asymptotic consistency* property of the discrete contact law $\mathcal{S}$ and

$$\lim_{\varepsilon \to 0} \sup \{ \mathrm{dist} \big(v, \mathcal{A}_\varepsilon(q, u^-) \big); \ v \in \mathcal{A}(q, u^-) \} = 0 \tag{10}$$

which can be interpreted as an *asymptotic indeterminacy* of the scheme.

In the one-dimensional friction case, i.e. when $\mathrm{Dim} \big(\mathrm{Span}(D_1(q)) \big) = 1$, then

$$
\mathcal{A}(q, u^-) =
\begin{cases}
\{u^-, \tilde{u}\} \text{ if } \min_{w \in D_1(q)} \langle n(q), M^{-1}(q)(n(q) + w) \rangle < 0, \\
[u^-, \tilde{u}] \text{ if } \min_{w \in D_1(q)} \langle n(q), M^{-1}(q)(n(q) + w) \rangle = 0
\end{cases}
$$

with

$$\tilde{u} = \mathrm{Proj}_q \big(0, \big(u^- + M^{-1}(q)\mathcal{R}(q) \big) \cap T(q) \big)$$

(see [12]). Then we can prove that (9) is satisfied while (10) is not always true and depends on the evolution of the mappings $q_\varepsilon \mapsto \mathcal{R}(q_\varepsilon)$ and $q_\varepsilon \mapsto n(q_\varepsilon)$ in a neighborhood of the contact point. More precisely let us assume that

(H1) the mapping M is of class C^1 from $\mathbb{R}^d$ to the set of symmetric positive definite $d \times d$ matrices;

(H2) the function g belongs to $C^1(\mathbb{R}^d)$, ∇g is locally Lipschitz continuous and does not vanish in a neighbourhood of $\{q \in \mathbb{R}^d; g(q) = 0\}$;

(H3) for all $q \in \mathbb{R}^d$, $D_1(q)$ is a closed, bounded, convex subset of $\mathbb{R}^d$ such that $0 \in D_1(q)$ and the multivalued mapping $q \mapsto D_1(q)$ is Hausdorff continuous. Furthermore, $\nabla g(q) \notin \mathrm{Span}(D_1(q))$ for all $q \in \mathbb{R}^d$ such that $\nabla g(q) \neq 0$.

We denote by K the set of admissible configurations i.e.

$$K = \{q \in \mathbb{R}^d; \ g(q) \leq 0\}.$$

Now let $(q, u^-) \in \mathbb{R}^d \times \mathbb{R}^d$ such that $g(q) \geq 0$, $\langle u^-, n(q) \rangle = 0$ and $M^{-1}(q)\mathcal{R}(q) \cap T(q) \neq \{0\}$. With assumption (H2) there exists $r_q > 0$ such that the mapping

$$n : \begin{cases} \overline{B}(q, r_q) \to \mathbb{R}^d \\ q' \mapsto n(q') = -\dfrac{\nabla g(q')}{\|\nabla g(q')\|} \end{cases}$$

is well defined and Lipschitz continuous. Let us assume moreover that, possibly reducing r_q, we have

(H4) $\dim(\mathrm{Span}(D_1(q'))) = 1$ for all $q' \in \overline{B}(q, r_q)$.

We may observe that $M^{-1}(q)\mathcal{R}(q) \cap T(q) \neq \{0\}$ if and only if there exists $w \in D_1(q)$ such that $\langle n(q), M^{-1}(q)(n(q) + w) \rangle = 0$. Hence we introduce the mapping $\gamma : \overline{B}(q, r_q) \to \mathbb{R}$ defined by

$$\gamma(q') = \min_{w' \in D_1(q')} \langle n(q'), M^{-1}(q')(n(q') + w') \rangle \quad \forall q' \in \overline{B}(q, r_q).$$

With the previous assumptions we obtain that γ is continuous at q and $\gamma(q) \leq 0$. Then we have the following result:

Theorem 1. [16] *If $\gamma(q) < 0$ or $\gamma(q) = 0$ and for all $\varepsilon \in (0, r_q)$ there exists $q_\varepsilon \in B(q, \varepsilon) \setminus (\mathrm{Int}(K) \cup \{q\})$ such that $\gamma(q_\varepsilon) > 0$, we have*

$$\lim_{\varepsilon \to 0} d_H(\mathcal{A}(q, u^-), \mathcal{A}_\varepsilon(q, u^-)) = 0.$$

Otherwise, if $\gamma(q) = 0$ and there exists $\varepsilon_q \in (0, r_q)$ such that $\gamma(q') \leq 0$ for all $q' \in B(q, \varepsilon_q) \setminus \mathrm{Int}(K)$, then $d_H(\mathcal{A}(q, u^-), \mathcal{A}_\varepsilon(q, u^-))$ does not tend to zero as ε tends to zero if $\tilde{u} \neq u^-$ and we only have

$$\lim_{\varepsilon \to 0} d_H(\{u^-, \tilde{u}\}, \mathcal{A}_\varepsilon(q, u^-)) = 0.$$

Idea of the proof: For all $\varepsilon \in (0, r_q)$ and $(q_\varepsilon, u_\varepsilon^-) \in B(q, \varepsilon) \times B(u^-, \varepsilon)$ such that $u_\varepsilon^+ = \mathcal{S}(q_\varepsilon, u_\varepsilon^-) \neq u_\varepsilon^-$ we have

$$g(q_\varepsilon) \geq 0, \quad \langle u_\varepsilon^-, n(q_\varepsilon) \rangle < 0$$

and $u_\varepsilon^+ = \tilde{u}_\varepsilon$ with

$$\tilde{u}_\varepsilon = \text{proj}_{q_\varepsilon}\big(0, \big(u_\varepsilon^- + M^{-1}(q_\varepsilon)\mathcal{R}(q_\varepsilon)\big) \cap T(q_\varepsilon)\big).$$

By using the same kind of arguments as in Proposition 1, we obtain that $(\tilde{u}_\varepsilon)_{r_q > \varepsilon > 0}$ is bounded. Moreover we can decompose $\tilde{u}_\varepsilon$ as follows

$$\tilde{u}_\varepsilon = u_\varepsilon^- + \lambda_\varepsilon M^{-1}(q_\varepsilon)\big(n(q_\varepsilon) + w_\varepsilon\big),$$

with $\lambda_\varepsilon > 0$ and $w_\varepsilon \in D_1(q_\varepsilon)$ such that

$$\big\langle n(q_\varepsilon), M^{-1}(q_\varepsilon)\big(n(q_\varepsilon) + w_\varepsilon\big)\big\rangle = 0$$

for all $\varepsilon \in (0, r_q)$. Using assumption (H4) (or assumption (H'4) see below) we infer that there exists a unique vector $\tilde{w} \in D_1(q)$ such that $\big\langle n(q), M^{-1}(q)\big(n(q) + \tilde{w})\big\rangle = 0$ and with assumptions (H1)–(H3) we obtain

$$\lim_{\varepsilon \to 0} w_\varepsilon = \tilde{w}.$$

Moreover, the boundedness of $(\tilde{u}_\varepsilon)_{r_q > \varepsilon > 0}$ implies that $(\lambda_\varepsilon)_{r_q > \varepsilon > 0}$ is also bounded and we infer that the adherence values of $(\tilde{u}_\varepsilon)_{r_q > \varepsilon > 0}$ belong to $u^- + \big(M^{-1}(q)\mathcal{R}(q) \cap T(q)\big)$.

If furthermore there exists $\varepsilon_q \in (0, r_q)$ such that $\gamma(q') \leq 0$ for all $q' \in B(q, \varepsilon_q)\backslash\text{Int}(K)$, then we can prove that $(\tilde{u}_\varepsilon)_{r_q > \varepsilon > 0}$ admits a unique adherence value given by $\tilde{u}$. It follows that

$$\lim_{\varepsilon \to 0} d_H\big(\{u^-, \tilde{u}\}, \mathcal{A}_\varepsilon(q, u^-)\big) = 0.$$

On the contrary, if for any $\varepsilon \in (0, r_q)$ there exists $q_\varepsilon \in B(q, \varepsilon)\backslash\big(\text{Int}(K)\cup\{q\}\big)$ such that $\gamma(q_\varepsilon) > 0$, then, for any $\bar{v} \in [u^-, \tilde{u}]\backslash\{u^-\}$, $\text{dist}\big(\bar{v}, \mathcal{A}_\varepsilon(q, u^-)\big)$ tends to zero as ε tends to zero. Indeed, we may construct a sequence $(q_{\varepsilon_n}, u_{\varepsilon_n}^-)_{n \geq 1}$ with $(\varepsilon_n)_{n \geq 1}$ decreasing to zero such that $(q_{\varepsilon_n}, u_{\varepsilon_n}^-) \in B(q, \varepsilon_n) \times B(u^-, \varepsilon_n)$, and $\gamma(q_{\varepsilon_n}) > 0$ for all $n \geq 1$ and $(\mathcal{S}(q_{\varepsilon_n}, u_{\varepsilon_n}^-))_{n \geq 1}$ converges to $\bar{v}$. It follows that

$$\lim_{\varepsilon \to 0} d_H\big([u^-, \tilde{u}], \mathcal{A}_\varepsilon(q, u^-)\big) = 0.$$

Then we conclude by using the continuity of γ at q and the definition of $\mathcal{A}(q, u^-)$.

We infer that the discrete contact law always satisfies the asymptotic consistency property while the asymptotic indeterminacy of the scheme holds only if $\gamma(q) < 0$ or $\gamma(q) = 0$ and for all $\varepsilon \in (0, r_q)$ there exists $q_\varepsilon \in B(q, \varepsilon)\backslash\big(\text{Int}(K)\cup\{q\}\big)$ such that $\gamma(q_\varepsilon) > 0$ if $\tilde{u} \neq u^-$. In the latter case, any $\bar{v} \in [u^-, \tilde{u}]\backslash\{u^-, \tilde{u}\}$ is the limit of a sequence of post-collision velocities $(\mathcal{S}(q_{\varepsilon_n}, u_{\varepsilon_n}^-) = \tilde{u}_{\varepsilon_n})_{n \geq 1}$ with $(\varepsilon_n)_{n \geq 1}$ decreasing to zero and $(q_{\varepsilon_n}, u_{\varepsilon_n}^-) \in B(q, \varepsilon_n) \times B(u^-, \varepsilon_n)$ for all $n \geq 1$. Hence, for all $n \geq 1$, $\mathcal{S}(q_{\varepsilon_n}, u_{\varepsilon_n}^-) = \tilde{u}_{\varepsilon_n}$ is defined as the Argmin of the kinetic norm of the admissible right velocities at $(q_{\varepsilon_n}, u_{\varepsilon_n}^-)$ but $\bar{v}$ is not the Argmin of the kinetic norm of the admissible right velocities at (q, u^-). It means that the minimization property (7) defining post-collision velocities is not continuous at (q, u^-) and it appears that this mathematical propery is deeply related to the indeterminacies of Coulomb's law.

Finally let us emphasize that (H3) allows to take into account both isotropic and anisotropic friction. Moreover the conclusions of Theorem 1 are still valid when (H4) is replaced by

(H'4) $D_1(q')$ is strictly convex for any $q' \in \overline{B}(q, r_q)$ i.e. for any w_1 and w_2 belonging to $D_1(q')$ such that $w_1 \neq w_2$, and for any $\gamma \in (0, 1)$, $\gamma w_1 + (1 - \gamma)w_2$ belongs to the relative interior of $D_1(q')$, i.e. there exists a open subset $\mathcal{O}$ of $\mathbb{R}^d$ such that

$$\gamma w_1 + (1 - \gamma)w_2 \in \mathcal{O} \cap \mathrm{Span}\big(D_1(q')\big) \subset D_1(q')$$

which is always true for the classical isotropic Coulomb's friction characterized by a friction coefficient $\mu > 0$.

References

1. Beghin, H.: Sur certains problèmes de frottement. Nouv. Ann. de Math. **2**, 305–312 (1923–24)
2. Brogliato, B.: Nonsmooth Mechanics. Communications and Control Engineering, 2nd edn. Springer, Heidelberg (1999)
3. Coulomb, C.A.: Théorie des machines simples. Bachelier, Paris (1821)
4. Delassus, E.: Considérations sur le frottement de glissement. Nouv. Ann. de Math. **20**, 485–496 (1920). (4ème série)
5. Dzonou, R., Monteiro-Marques, M.: Sweeping process for inelastic impact problem with a general inertia operator. Eur. J. Mech. A. Solids **26**, 474–490 (2007)
6. Erdmann, M.: On a representation of friction in configuration space. Int. J. Robot. Res. **13**, 240–271 (1994)
7. Genot, F., Brogliato, B.: New results on Painlevé paradoxes. Eur. J. Mech. A. Solids **18**(4), 653–677 (1999)
8. Lecornu, L.: Sur la loi de Coulomb. C.R. Acad. Sci. Paris **140**, 847–848 (1905)
9. Monteiro-Marques, M.: Differential Inclusions in Non-Smooth Mechanical Problems Shocks and Dry Friction. Birkhauser, Boston (1993)
10. Moreau, J.J.: Standard inelastic shocks and the dynamics of unilateral constraints. In: Del Piero, G., Maceri, F. (eds.) Unilateral Problems in Structural Analysis. ICMS, vol. 288, pp. 173–221. Springer, Vienna (1985). doi:10.1007/978-3-7091-2632-5_9
11. Moreau, J.J.: Dynamique de systèmes à liaisons unilatérales avec frottement sec éventuel, essais numériques. Preprint 85–1 2nd edn. LMGC Montpellier (1986)
12. Moreau, J.J.: Unilateral contact and dry friction in finite freedom dynamics. In: Moreau, J.J., Panagiotopoulos, P.D. (eds.) Nonsmooth Mechanics and Applications. ICMS, vol. 302, pp. 1–82. Springer, Vienna (1988). doi:10.1007/978-3-7091-2624-0_1
13. Painlevé, P.: Sur les lois du frottement de glissement. C.R. Acad. Sci. Paris **121**, 112–115 (1895)

14. Painlevé, P.: Sur les lois du frottement de glissement. C.R. Acad. Sci. Paris **141**, 401–405 and 546–552 (1905)
15. Paoli, L.: Vibro-impact problems with dry friction - Part I: existence result. SIAM J. Math. Anal. **47**(5), 3285–3313 (2015)
16. Paoli, L.: Multibody dynamics with unilateral constraints and dry friction: how the contact dynamics approach may handle Coulomb's law indeterminacies? J. Convex Anal. **23**(3), 849–876 (2016)
17. Schatzman, M.: A class of nonlinear differential equations of second order in time. Nonlinear Anal. Theory Methods Appl. **2**, 355–373 (1978)

A Complex Mathematical Model with Competition in Leukemia with Immune Response - An Optimal Control Approach

I.R. Rădulescu[✉], D. Cândea, and A. Halanay

Department of Mathematics and Informatics,
POLITEHNICA University of Bucharest,
Splaiul Independentei 313, 060042 Bucharest, Romania
nicola_rodica@yahoo.com

Abstract. This paper investigates an optimal control problem associated with a complex nonlinear system of multiple delay differential equations modeling the development of healthy and leukemic cell populations incorporating the immune system. The model takes into account space competition between normal cells and leukemic cells at two phases of the development of hematopoietic cells. The control problem consists in optimizing the treatment effect while minimizing the side effects. The Pontryagin minimum principle is applied and important conclusions about the character of the optimal therapy strategy are drawn.

Keywords: Leukemia · Asymmetric division · Competition · Optimal control · Treatment

1 Introduction

Leukemia is a cancer of the blood and bone marrow, characterized by large and uncontrolled growth of white blood cells. The most studied type of leukemia, Chronic myelogenous leukemia (CML), involves granular leukocyte precursors, namely the myelocyte line. The trigger of CML is a chromosomal abnormality, called the Philadelphia chromosome, that occurs in all cell lineages in about 90% of cases. The product of this chromosome is the formation of the BCR-ABl fusion protein which is thought to be responsible for the dysfunctional regulation of myelocyte proliferation and other features of CML. The standard treatment of CML in recent years is Imatinib, a molecular targeted drug [15], that has the effect that almost all patients attain hematological remission [13] and 75% attain cytogenetic remission.

Nowadays, it is well known that both innate and adaptive immunity are implicated in the defense mechanisms against cancer and recent progress in cancer immunology suggest that the immune system plays a fundamental role in tumor progression [23]. In CML, the biological literature reveals that T cells may play

© IFIP International Federation for Information Processing 2016
Published by Springer International Publishing AG 2016. All Rights Reserved
L. Bociu et al. (Eds.): CSMO 2015, IFIP AICT 494, pp. 430–441, 2016.
DOI: 10.1007/978-3-319-55795-3_41

an important role in stemming the expansion of leukemic cells. This happens because leukemic cells express antigens that are immunogenic and can be recognized by cytotoxic T cells (CD8+ T cells or CTLs). In the paper [22], the authors found that leukemia-specific effectors CTLs were able to eliminate leukemic stem cells (LSCs) in vitro and in vivo in a setting with minimal leukemia load. The role of CD4+ T cells in leukemia is less clear, although in [3] the authors ascertained that some CML patients under imatinib-induced remission develop an anti-leukemia immune response involving both CD4+ and CD8+ T cells. Hence, our goal in this paper is to capture in a mathematical model the underlying dynamics of this disease by considering the evolution of healthy and leukemic cell populations along with one of the most important component of the cellular immune response to CML, namely T cell response.

Even if a variety of mathematical papers have applied a range of modeling approaches to study tumor-immune interactions in general (see, for example the recent review [7]), only a few described the specific leukemia-immune interaction. Leukemia-immune models have been formulated using mostly ordinary differential equations (ODE) [16,17] or delay differential equations (DDE) [2,5,12,19]. Some models that specifically study the immune response to CML are [2,12,16,18]. However, none of the above papers have considered competition between healthy and leukemic cell populations, which is an important factor in CML dynamics.

2   Assumptions on the Model

In this paper, we use a five-dimensional system of DDEs. The first four equations describing the healthy and leukemic cell populations are based on the Mackey and collaborators models of hematopoiesis [14]. In the present model, we consider two types of hematopoietic cell populations: stem-like cell populations consisting of stem cells and progenitors with self-renew ability and mature cell populations formed by differentiated cells without self-renew ability. We include two types of cell populations, healthy and leukemic, each of them with two subpopulations of cells: stem-like and mature. In this way, we introduce space competition between the normal cell population and the CML one. We underline that in the case of leukemic cell populations mature cells are mostly unable to perform their functions.

The main difference from the Mackey model is considering the competition between healthy and CML cell populations and the fact that three types of division of a stem-like cell are considered: asymmetric division, symmetric self renewal and differentiation. In this paper we use the notation $\alpha = h, l$ with h for healthy and l for leukemia. Consequently, we assume that a fraction $\eta_{1\alpha}$, $\alpha = h, l$, of stem-like cell population is susceptible to asymmetric division: one daughter cell proceeds to differentiate and the other re-enters the stem cell compartment. A fraction $\eta_{2\alpha}$, $\alpha = h, l$, is susceptible to differentiate symmetrically with both cells that result following a phase of maturation and the fraction $1 - \eta_{1\alpha} - \eta_{2\alpha}$, $\alpha = h, l$, is susceptible to self-renewal so both cells that results after mitosis are stem-like cells. This four-dimensional competition model was introduced by Radulescu et al. in [20].

The fifth equation of the system models the anti-leukemia T cell immune response. The T cell population considered in this paper consists only of activated anti-leukemia T cells (involving both CD4+ and CD8+ T cells), which actively interact with CML cells. We do not consider other parts of the immune response, as the population of antigen presenting cells or the levels of citokines production. We assume that after encountering a leukemic cell, a T cell has two possibilities: either it inhibits the leukemic cell and activates a feedback function to stimulate the production of new T cells, or it is inhibited itself by the leukemic cell.

3   Description of the Model

The state variables of the model are the healthy cell populations x_1 - stem-like and x_2 - mature, the CML cell populations x_3 - stem-like and x_4 - more differentiated and x_5 - the population of anti-leukemia T cells. The delays for healthy and leukemia stem-like cells are τ_1 and τ_3 for the duration of the cell cycle, independent of the type of division, and τ_2 and τ_4 for the time necessary for differentiation into mature leukocytes for healthy and, respectively, leukemia cells. τ is the duration of the cell cycle for T-cells and $\tau_5 = n\tau$ with n the number of antigen depending divisions. We denote $X_{\tau_i} = X(t - \tau_i)$, where $X = (x_1, x_2, x_3, x_4, x_5)$.

The optimal control model is

$$
\begin{aligned}
\dot{x}_1 &= f_1(x_1, x_2, x_3, x_4, x_{1\tau_1}, x_{2\tau_1}, x_{3\tau_1}, x_{4\tau_1}) \\
\dot{x}_2 &= f_2(x_2, x_{1\tau_2}, x_{2\tau_2}, x_{4\tau_2}) \\
\dot{x}_3 &= f_3(x_1, x_2, x_3, x_4, x_5, x_{1\tau_3}, x_{2\tau_3}, x_{3\tau_3}, x_{4\tau_3}, u_1, k_{1l}, u_{1\tau_3}) \\
\dot{x}_4 &= f_4(x_3, x_4, x_5, x_{2\tau_4}, x_{3\tau_4}, x_{4\tau_4}, k_{2l}, u_{1\tau_4}) \\
\dot{x}_5 &= f_5(x_4, x_5, x_{4\tau_5}, x_{5\tau_5}, u_2)
\end{aligned}
\tag{1}
$$

where

$$
\begin{aligned}
f_1 =\ & -\gamma_{1h} x_1 - (\eta_{1h} + \eta_{2h}) k_h (x_2 + x_4) x_1 - (1 - \eta_{1h} - \eta_{2h}) \beta_h (x_1 + x_3) x_1 + \\
& + 2e^{-\gamma_{1h}\tau_1}(1 - \eta_{1h} - \eta_{2h})\beta_h(x_{1\tau_1} + x_{3\tau_1})x_{1\tau_1} + \eta_{1h}e^{-\gamma_{1h}\tau_1}k_h(x_{2\tau_1} + x_{4\tau_1})x_{1\tau_1} \\
f_2 =\ & -\gamma_{2h} x_2 + A_h(2\eta_{2h} + \eta_{1h})k_h(x_{2\tau_2} + x_{4\tau_2})x_{1\tau_2} \\
f_3 =\ & -(\gamma_{1l} + \mathbf{f}_{1a})x_3 - [(\eta_{1l} + \eta_{2l})k_l((x_2 + x_4)\,\mathbf{f}_{u_1}) + (1 - \eta_{1l} - \eta_{2l})\beta_l((x_1 + x_3)\,\mathbf{f}_{u_1})]x_3 + \\
& + [2e^{-(\gamma_{1l}+\tilde{\mathbf{f}}_{1a})\tau_3}(1 - \eta_{1l} - \eta_{2l})\beta_l((x_{1\tau_3} + x_{3\tau_3})\,\mathbf{f}_{u_{1\tau_3}}) + \\
& + \eta_{1l}e^{-(\gamma_{1l}+\tilde{\mathbf{f}}_{1a})\tau_3}k_l((x_{2\tau_3} + x_{4\tau_3})\,\mathbf{f}_{u_{1\tau_3}})]x_{3\tau_3} - b_1 x_3 x_5 l_1(x_3 + x_4) \\
f_4 =\ & -(\gamma_{2l} + \mathbf{f}_{2a})x_4 + A_l(2\eta_{2l} + \eta_{1l})k_l((x_{2\tau_4} + x_{4\tau_4})\,\mathbf{f}_{u_{\tau_4}})x_{3\tau_4} - b_2 x_4 x_5 l_1(x_3 + x_4) \\
f_5 =\ & a_1 - a_2 x_5 - a_3 \mathbf{f}_{u_2} x_5 l_2(x_4) + 2^{n_1} a_4 x_{5\tau_5} l_2(x_{4\tau_5})
\end{aligned}
$$

subject to minimization of the cost functional

$$
\min J(u),
\tag{2}
$$

where

$$J(u) = g(x(T)) + L(t, u(t), x_5(t))$$

with $g(x(T)) = A_1 x_3(T) + A_2 x_4(T) + E_1 x_3(T)/x_1(T) + E_2 x_4(T)/x_2(T)$ – being the weighted sum of the final tumor population and the ratio the ratio of leukemia cells and the healthy ones and

$$L(t, u(t), x_5(t)) = \int_0^T [B_1 u_1(t) + B_2 u_2(t) + C_1 k_1(t) + C_2 k_2(t) - D x_5(t)] \, dt \; -$$

the cumulative drug toxicity and T cell amount.

The history of the state variables is given by $X(\theta) = \varphi(\theta), \theta \in [-\tau_{\max}, 0], \tau_{\max} = \max(\tau_1, \tau_2, \tau_3, \tau_4, \tau_5)$.

The healthy and leukemic blood cell populations are seen in competition for resources and this is reflected in the fact that both feedback laws for self-renewal and differentiation depend on the sum of healthy and leukemia cells. Consequently, the rate of self-renewal is $\beta_\alpha(x_1 + x_3) = \beta_{0\alpha} \dfrac{\theta_{1\alpha}^{m_\alpha}}{\theta_{1\alpha}^{m_\alpha} + (x_1 + x_3)^{m_\alpha}}$, with $\beta_{0\alpha}$ the maximal rate of self-renewal and $\theta_{1\alpha}$ half of the maximal value and the rate of differentiation is $k_\alpha(x_2 + x_4) = k_{0\alpha} \dfrac{\theta_{2\alpha}^{n_\alpha}}{\theta_{2\alpha}^{n_\alpha} + (x_2 + x_4)^{n_\alpha}}$, with $k_{0\alpha}$ the maximal rate of differentiation and $\theta_{2\alpha}$ is half of the maximal value. The rest of the parameters for healthy and CML cell populations are: $\gamma_{1\alpha}$ - the natural apoptosis, A_α - an amplification factor and m_α, n_α parameters that control the sensitivity of β_α respectively k_α to changes in the size of stem-like and respectively mature populations. Table 1 contains a complete description of parameters of the model.

To model the influence of T cells on CML cells, we consider the feedback function $l_1(y) = \dfrac{1}{b_5 + y}$. Consequently, the last terms of the third and fourth equations represent the inhibition of CML cells by anti-leukemia T cells. We assumed that the inhibition of CML cell population by T cells increases with the number of leukemic cells up to a certain level and then reaches a maximal value of inhibition. A further increase in CML population will not modify this value.

As concerns the fifth equation, the first term a_1 is the natural supply of naive T cells, while the second term $-a_2 x_5$ indicates that T cells exit the population through death at the rate a_2. Leukemia cells suppress anti-leukemia immune response. The precise mechanism is unknown, but it is assumed that the level of down regulation depends on the current leukemia population, so we consider that the immune system is regulated by the feedback function $l_2(y) = \dfrac{y}{b_5 + y^2}$. This function ensures that T cells are stimulated by CML cells only if leukemia cell population has values in a certain range, called "the optimal load zone" (see [12]). We take the rate of antigen stimulation as a feedback function depending on the level of the mature leukemic population, $l_2(x_4)$ and the third and the fourth terms, $-a_3 u_2 x_5 l_2(x_4)$ and $2^{n_1} a_4 x_{5\tau_6} l_2(x_{4\tau_5})$ gives the rate at which naive T cells leave and re-enter the effector state after finishing the minimal developmental program of n_1 cell divisions (due to antigen stimulation). The time delay $\tau_5 = n_1\tau$ is the duration of this program. These terms represent the loss and respectively the production of T cells due to the competition with leukemic cells.

The treatment targeting the BCR-ABL gene is supposed to affect the apoptosis and the proliferation rates of leukemia cells [11]. In view of this fact, we consider the treatment functions $f_{u_1} = \dfrac{1}{1 - u_1}$, $f_{1a} = (\gamma_{1h} - \gamma_{1l})\,k_{1l}$ and $f_{2a} = c\gamma_{2h}k_{2l}$, with $u_1, k_{1l}, k_{2l} : [0, T] \rightarrow [0, 1]$, where $u_1(t), k_{1l}(t), k_{2l}(t)$ are the treatment effects.

The action of treatment on the proliferation rate will be considered through f_{u_1} in the function of self-renew β_l and in the function of differentiation or asymmetric division k_l. Note that, in this way, both β_l and k_l became decreasing functions of u_1. For more details, see [21]. The treatment acts on the apoptosis through the function f_{1a} on the stem cells and through the function f_{2a} on the mature ones. Also, from the law of the mass, we have $\tilde{f}_{1a} = \int\limits_{t-\tau_1}^{t} k_{1l}(s)ds$.

Moreover, it seems that, in vivo, Imatinib is the trigger of complex mechanisms, some of them able to promote T cell expansion (see [22]). Imatinib's effect on T cell population is introduced in the form of a treatment function $f_{u_2} = 1 - u_2$, with the stimulatory effect $u_2 : [0, T] \rightarrow [0, 1]$. If no drug is given, then $f_{u_2} = 1$ and a maximal effect takes place for $u_2(t) = 1$, when T cell population is no longer inhibited by CML cell population.

The **existence of an optimal control** follows since one can transform the given problem into an optimal control problem for a system of ODEs whose solutions will be bounded together with their derivatives on compact intervals (see [1]).

4   Discretization of the Optimal Problem

In this section, we apply the numerical procedure from Gollmann and Maurer [10], in order to solve the delay optimal control problem (1) + (2) (see also [8,9]). For that matter, we write the cost functional in the Mayer form

$$J(u, x) = h(x(T)), \quad x = (x_1, x_2, x_3, x_4, x_5) \in R^5.$$

We introduce the additional state variable z through the equation

$$\dot{z}(t) = B_1 u_1(t) + B_2 u_2(t) + C_1 k_1(t) + C_2 k_2(t) - D x_5(t), \quad z(0) = 0.$$

Then, the cost functional (2) is rewritten as

$$J(u, x, z) = g(x(T)) + z(T).$$

In the following, let $\tau > 0$ be such that $\tau_1 = j_1\tau$, $\tau_2 = j_2\tau$, $\tau_3 = j_3\tau$, $\tau_4 = j_4\tau$, $\tau_5 = j_5\tau$, $j_i \in N^*$, $i = \overline{1,5}$, $T = N\tau$ and use the Euler integration method with a uniform step size $\tau > 0$. Of course, τ can be refined in order to obtain an appropriate smaller step-size. Using the grid points $t_i = i\tau$, $i = \overline{0, N}$ and the approximations $x_1(t_i) \simeq x_{1i} \in R$, $x_2(t_i) \simeq x_{2i} \in R$, $x_3(t_i) \simeq x_{3i} \in R$, $x_4(t_i) \simeq x_{4i} \in R$, $x_5(t_i) \simeq x_{5i} \in R$, $u_1(t_i) \simeq u_{1i}$, $u_2(t_i) \simeq u_{2i}$ and $k_1(t_i) \simeq k_{1i}$, $k_2(t_i) \simeq k_{2i}$ the treatment function f_{1a} becomes $\sum_{j=1}^{j_3} k_{1l_{i-j}}\tau$ and the delay control problem $(1) + (2)$ is transformed into the nonlinear programming problem (NLP)

$$Minimize \; J = g(x_N) + z_N \tag{3}$$

subject to

$$
\begin{cases}
x_{1i} - x_{1i+1} + \tau f_1(x_{1i}, x_{2i}, x_{3i}, x_{4i}, x_{1i-j_1}, x_{2i-j_1}, x_{3i-j_1}, x_{4i-k_1}) = 0 \\
x_{2i} - x_{2i+1} + \tau f_2(x_{2i}, x_{1i-j_2}, x_{2i-j_2}, x_{4i-j_2}) = 0 \\
x_{3i} - x_{3i+1} + \tau f_3(x_{1i}, x_{2i}, x_{3i}, x_{4i}, x_{5i}, x_{1i-j_3}, x_{2i-j_3}, x_{3i-j_3}, x_{4i-j_3}, u_{1i}, \\
\quad k_{1i}, \sum_{j=1}^{j_3} k_{1i-j}\tau, u_{1i-j_3}) = 0 \\
x_{4i} - x_{4i+1} + \tau f_4(x_{4i}, x_{5i}, x_{2i-j_4}, x_{3i-j_4}, x_{4i-j_4}, k_{2li}, u_{1i-j_4}) = 0 \\
x_{5i} - x_{5i+1} + \tau f_5(x_{4i}, x_{5i}, x_{4i-j_5}, x_{5i-j_5}, u_{2i}) = 0 \\
z_i - z_{i+1} + \tau(B_1 u_{1i} + B_2 u_{2i} + C_1 k_{1i} + C_2 k_{2i} - D x_{5i}) = 0
\end{cases}
\tag{4}
$$

$$
- u_{1i} \le 0, u_{1i} - 1 \le 0, -u_{2i} \le 0, u_{2i} - 1 \le 0,
$$
$$
-k_{1i} \le 0, k_{1i} - 1 \le 0, -k_{2i} \le 0, k_{2i} - 1 \le 0, i = \overline{0, N-1}.. \tag{5}
$$

Herein, the initial value profiles $\varphi_1, \varphi_2, \varphi_3, \varphi_4$ and φ_5 give the values

$$x_{1_{-i}} := \varphi_1(-i\tau), \; i = \overline{0, l_1}, x_{2_{-i}} := \varphi_2(-i\tau), \; i = \overline{0, l_2}, x_{3_{-j}} := \varphi_3(-i\tau), \; i = \overline{0, l_3}$$

$$x_{4_{-j}} := \varphi_4(-i\tau), \; i = \overline{0, l_4}, x_{5_{-j}} := \varphi_5(-i\tau), \; i = \overline{0, l_5}.$$

The variable to be optimized is represented by the vector $w = (u_{1_0}, u_{2_0}, k_{1_0}, k_{2_0}, x_{1_1}, ..., x_{5_1}, z_1, .., u_{1N-1}, u_{2N-1}, k_{1N-1}, k_{2N-1}, x_{1N}, .., x_{5N}, z_N) \in R^{10N}$.

5   Numerical Results

In the following figures, we plotted the trajectories of the healthy, respectively CML cell populations for the competition system, showing a comparison between the dynamics of a system without treatment and the dynamics of a system subject to optimal treatment. In the following simulations two aspects are combined, resulting four distinct manifestations of the disease:

- starting treatment in two different stages of the disease: a less severe stage when the population of leukemic cells and the healthy cells coexist (the leukemia cell population is still small) (S1) and a stage where healthy cell population disappeared and the number of leukemic cells is already very high (S2);
- two configurations of parameters describing two different forms of the disease for patients, configuration 1 and configuration 2 (see Table 1). The configuration 2 corresponds to a more serious disease.

Considering multiple effects of treatment, on the apoptosis of leukemic stem cells, leukemic mature cells, proliferation rate and immune system, simulations show the impact of the disease on various optimal control solutions for four hypothetical patients (see Figs. 1, 2, 3 and 4 for comparison between the dynamics of a system without treatment and with optimal reatment).

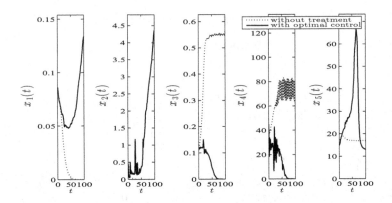

Fig. 1. Simulations start from S1 for configuration 1 of parameters. Dashed line represents the dynamics of a system without treatment and continuous line represents the dynamics of a system with optimal treatment.

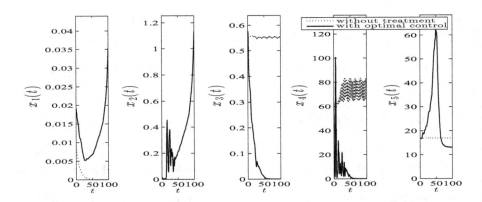

Fig. 2. Simulations start from S2 for configuration 1 of parameters. Dashed line represents the dynamics of a system without treatment and continuous line represents the dynamics of a system with optimal treatment.

In the Figs. 5, 6, 7 and 8 the controls k_{1l}, k_{2l}, u_1, u_2 represent the influence of drug on the apoptosis of leukemic stem cells, leukemic mature cells, proliferation rate and immune system. The value of cost functional was improved in all situations (see figures). To solve the problem of optimal control the *Matlab* solver for NLP problems *fmincon* was used, selecting the '*interior-point*' solver.

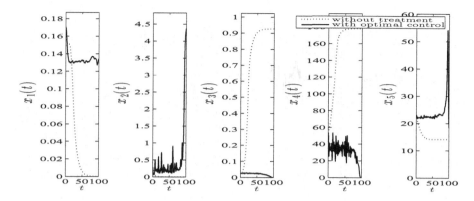

Fig. 3. Simulations start from S1 for configuration 2 of parameters. Dashed line represents the dynamics of a system without treatment and continuous line represents the dynamics of a system with optimal treatment.

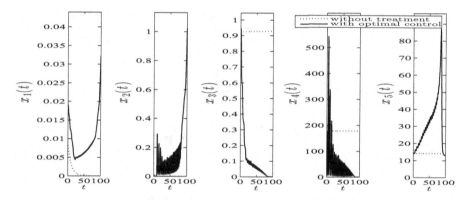

Fig. 4. Simulations start from S2 for configuration 2 of parameters. Dashed line represents the dynamics of a system without treatment and continuous line represents the dynamics of a system with optimal treatment.

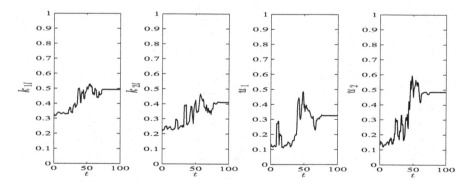

Fig. 5. Controls for configuration 1 of parameters, simulations start from S1. The cost function was improved from 4000 to 2700.

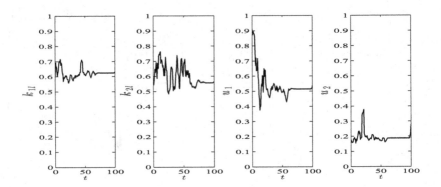

Fig. 6. Controls for configuration 1 of parameters, simulations start from S2. The cost function was improved from 5100 to 4100.

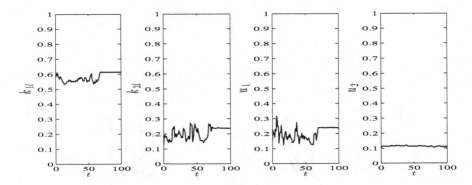

Fig. 7. Controls for configuration 2 of parameters, simulations start from S1. The cost function was improved from 3400 to1600.

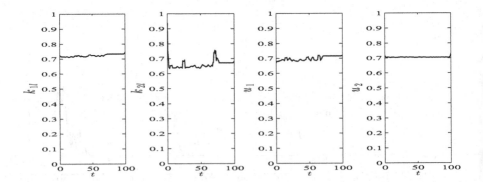

Fig. 8. Controls for configuration 2 of parameters, simulations start from S2. The cost function was improved from 2700 to 2100.

6 Conclusions

In this paper, an optimal control model for CML with the influence of the immune system and treatment was investigated. Based on clinical evidences and assumptions, the effects of Imatinib, the current first line treatment in CML, was considered. These effects include the decrease of leukemic proliferation and

Table 1. Description of parameters

Par.	Description	Conf. 1	Conf. 2
β_{0h}	Maximal value of the β_h feedback function (days^{-1})	1.77	1.77
β_{0l}	Maximal value of the β_l feedback function $(days^{-1})$	2	2.27
k_{0h}	Maximal value of the k_h feedback function $(days^{-1})$	0.1	0.1
k_{0l}	Maximal value of the k_l feedback function $(days^{-1})$	0.4	0.8
m_h	Hill coefficient of the β_h feedback function	4	4
m_l	Hill coefficient of the β_l feedback function	4	4
n_h	Hill coefficient of the k_h feedback function	3	3
n_l	Hill coefficient of the k_l feedback function	3	3
θ_{1h}	Parameter for the β_h feedback function ($10^6 cells/kg$)	1.6	1.6
θ_{2h}	Parameter for the k_h feedback function ($10^6 cells/kg$)	12	12
θ_{1l}	Parameter for the β_l feedback function ($10^6 cells/kg$)	0.5	0.5
θ_{2l}	Parameter for the k_l feedback function ($10^6 cells/kg$)	36	36
γ_{1h}	Loss of stem cells due to mortality for healthy cells $(days^{-1})$	0.1	0.1
γ_{1l}	Loss of stem cells due to mortality for leukemic cells $(days^{-1})$	0.04	0.01
η_{1h}	Rate of asymmetric division for healthy cells	0.7	0.7
η_{1l}	Rate of asymmetric division for leukemic cells	0.1	0.1
η_{2h}	Rate of symmetric division for healthy cells	0.1	0.1
η_{2l}	Rate of symmetric division for leukemic cells	0.7	0.7
γ_{2h}	Instant mortality of mature normal leukocytes $(days^{-1})$	2.4	2.4
γ_{2l}	Instant mortality of mature leukemic leukocytes $(days^{-1})$	1.5	0.15
A_h	Amplification factor for normal leukocytes	829	829
A_l	Amplification factor for leukemic leukocytes	1843	3686
b_1	Loss of leukemic stem cells due to cytotoxic T cells	0.3	0.3
b_2	Loss of mature leukemic leukocytes due to cytotoxic T cells	0.6	0.6
b_3	Standard half-saturation in a Michaelis-Menten law	36	36
b_4	Standard half-saturation in a Michaelis-Menten law	36	36
a_1	Anti-leukemia T-cell supply rate	3	3
a_2	Anti-leukemia T-cell death rate	0.23	0.23
a_3	Coefficient of influence due to leukemic cells	0.3	0.3
a_4	Probab. that T cell survives the encounter with a leukemia cell	0.9	0.9
n_1	The number of antigen depending divisions	2	2
τ_1	Duration of cell cycle for normal stem cells (days)	2.8	2.8
τ_2	Duration of cell cycle for normal leukocytes (days)	3.5	3.5
τ_3	Duration of cell cycle for leukemic stem cells (days)	2.1	2.1
τ_4	Duration of cell cycle for leukemic leukocytes (days)	2.8	2.8
τ_5	Duration of one T cell division (days)	1.4	1.4

differentiation, the increase of leukemic apoptosis and some influences on the anti-leukemia immune response.

From Figs. 1, 2, 3 and 4 one can see one can see the decline of leukemic cells (i.e. molecular remission) and the increase of the number of healthy cells after approximately three months of treatment. Depending on the level of cell populations at diagnosis (S1 or S2) and on the leukemia severity (i.e. configuration 1 or configuration 2) the evolution of healthy cell population to a normal amount is more or less rapid.

The plots of optimal controls (Figs. 5, 6, 7 and 8) exhibit an optimal control effect different for four hypothetical patients. One can observe that the drug influence is slightly different for various manifestations of the disease. Consequently, for an optimal effect of treatment, the prescribed dose should be adapted considering the parameter's disease of a certain patient and the leukemic burden at diagnosis. Although the identification of most of the values parameters of the disease is a daunting task, there are some which can be computed by means of current methods and they might provide an important indication concerning dose adjustment and therapy management.

Acknowledgments. This work was supported by CNCS-ROMANIA Grant ID-PCE-2011-3-0198.

References

1. Benosman, C.: Control of the dynamics of chronic myeloid leukemia by Imatinib. Ph.D. thesis (2010)
2. Berezansky, L., Bunimovich-Mendrazitsky, S., Domoshnitsky, A.: A mathematical model with time-varying delays in the combined treatment of chronic myeloid leukemia. Adv. Differ. Equ. **217**, 257–266 (2012)
3. Chen, C.I., Maecker, H.T., Lee, P.P.: Development and dynamics of robust T-cell responses to CML under imatinib treatment. Blood **111**, 5342–5349 (2008)
4. Cortes, J., Talpaz, M., O'Brien, S., Jones, D., Luthra, R., et al.: Molecular responses in patients with chronic myelogenous leukemia in chronic phase treated with imatinib mesylate. Clin. Cancer Res. **11**(9), 3425–3432 (2005)
5. DeConde, R., Kim, P.S., Levy, D., Lee, P.P.: Post-transplantation dynamics of the immune response to chronic myelogenous leukemia. J. Theor. Biol. **236**, 39–59 (2005)
6. Delitala, M., Lorenzi, T., Melensi, M.: A structured population model of competition between cancer cells and T cells under immunotherapy. In: Eladdadi, A., Kim, P., Mallet, D. (eds.) Mathematical Models of Tumor-Immune System Dynamics. Springer Proceedings in Mathematics & Statistics, vol. 107, pp. 47–58. Springer, New York (2014). doi:10.1007/978-1-4939-1793-8_3
7. Eftimie, R., Bramson, J.L., Earn, D.J.: Interactions between the immune system and cancer: a brief review of non-spatial mathematical models. Bull. Math. Biol. **73**, 2–32 (2011)
8. Gollmann, L., Kern, D., Maurer, H.: Optimal control problems with control and state delays and applications to growth processes. In: IIASA Symposium on Applications of Dynamic Systems to Economic Growth with Environment, Luxemburg, 7–8 November 2008

9. Gollmann, L., Kern, D., Maurer, H.: Optimal control problems with delays in state and control variables subject to mixed control state constraints. Optim. Control Appl. Meth. **30**, 341–365 (2009)
10. Gollmann, L., Maurer, H.: Theory and applications of optimal control problems with multiple time-delays. J. Ind. Manag. Optim. **10**(2), 413–441 (2014)
11. Gottschalk, S., Anderson, N., Hainz, C., et al.: Imatinib (STI571)-mediated changes in glucose metabolism in human leukemia BCR-ABL-positive cells. Clin. Cancer Res. **10**, 6661–6668 (2004)
12. Kim, P.S., Lee, P.P., Levy, D.: Dynamics and potential impact of the immune response to chronic myelogenous leukemia. PLoS Comput. Biol. **4**, e1000 (2008)
13. Lee, S.J.: Chronic myelogenous leukemia. Br. J. Haematol. **111**(4), 993–1009 (2000)
14. Mackey, M.C., Ou, C., Pujo-Menjouet, L., Wu, J.: Periodic oscillations of blood cell population in chronic myelogenous leukemia. SIAM J. Math. Anal. **38**, 166–187 (2006)
15. Marley, S., Gordon, M.: Chronic myeloid leukemia: stem cell derived but progenitor cell driven. Clin. Sci. **109**, 13–25 (2006)
16. Moore, H., Li, N.K.: A mathematical model for chronic myelogenous leukemia (CML) and T cell interaction. J. Theor. Biol. **225**, 513–523 (2004)
17. Nanda, S., dePillis, L.G., Radunskaya, A.E.: B cell chronic lymphocytic leukemia – a model with immune response. Discret. Contin. Dyn. Syst. Ser. B **18**, 1053–1076 (2013)
18. Neiman, B.: A mathematical model of chronic myelogenous leukaemia. Master's thesis University College, Oxford University (2002)
19. Peet, M.M., Kim, P.S., Niculescu, S.I., Levy, D.: New computational tools for modeling chronic myelogenous leukemia. Math. Model. Nat. Phenom. **4**(2), 119–139 (2009)
20. Radulescu, I.R., Candea, D., Halanay, A.: A study on stability and medical implications for a complex delay model for CML with cell competition and treatment. J. Theor. Biol. **363**, 30–40 (2014)
21. Rădulescu, I.R., Cândea, D., Halanay, A.: A control delay differential equations model of evolution of normal and leukemic cell populations under treatment. In: Pötzsche, C., Heuberger, C., Kaltenbacher, B., Rendl, F. (eds.) CSMO 2013. IAICT, vol. 443, pp. 257–266. Springer, Heidelberg (2014). doi:10.1007/978-3-662-45504-3_25
22. Riether, C., Schurch, C.M., Ochsenbein, A.F.: Regulation of hematopoietic and leukemic stem cells by the immune system. Cell Death Differ. **22**, 187–198 (2015)
23. Topalian, S.L., Weiner, G.J., Pardoll, D.M.: Cancer immunotherapy comes of age. J. Clin. Oncol. **29**, 4828–4836 (2011)

A Contact Model for Piezoelectric Beams

Á. Rodríguez-Arós[1(✉)], M.T. Cao-Rial[2,3], and M. Sofonea[4]

[1] Departamento de Métodos Matemáticos e Representación,
Universidade da Coruña, Paseo de Ronda 51, 15011 A Coruña, Spain
angel.aros@udc.es
[2] Technological Institute for Industrial Mathematics (ITMATI),
Campus Vida, 15782 Santiago de Compostela, Spain
[3] Departamento de Matemáticas, Universidade da Coruña, A Coruña, Spain
[4] Laboratoire de Mathématiques et Physique, Université de Perpignan Via Domitia,
52 Avenue de Paul Alduy, 66860 Perpignan, France

Abstract. We consider a mathematical model which describes the equilibrium of an electro-elastic beam in contact with an electrically conductive foundation. The model is constructed by coupling the beam equation with the one dimensional piezoelectricity system obtained in [13]. We state the unique weak solvability of the model as well as the continuous dependence of the weak solution with respect to the data. We also introduce a discrete scheme for which we perform the numerical analysis, including convergence and error estimates results. Finally, we present numerical simulations in the study of a test problem.

Keywords: Beam · Elasticity · Piezoelectricity · Contact · Normal compliance · Finite element method

1 Introduction

Piezoelectric materials belong to the family of "smart materials" and are characterized by a coupling of mechanical and electrical properties. Thus, electric charges can be observed on a piezoelectric body subjected to the action of external forces and, conversely, an electric potential applied on a piezoelectric body gives rise to stresses and strains. These properties of piezoelectric materials make them suitable to be used as sensors and actuators in various industrial settings and real-world applications. For this reason, the interest in the analysis of mathematical models with piezoelectric materials is currently increasing.

The construction of appropriate models to describe the behaviour of thin deformable bodies like plates, shells and beams, represents an important topic in

The work of Á. Rodríguez-Arós was supported by the project "Modelización y simulación numérica de sólidos y fluidos en dominios con pequeñas dimensiones. Aplicaciones en estructuras, biomecánica y aguas someras", MTM2012-36452-C02-01 financed by the Spanish Ministry of Economía y Competitividad with the participation of FEDER.

L. Bociu et al. (Eds.): CSMO 2015, IFIP AICT 494, pp. 442–451, 2016.
DOI: 10.1007/978-3-319-55795-3_42

Solid Mechanics. By using asymptotic analysis, several classical reduced models have been mathematically justified over the years. Pioneering work for modelling of thin linearly isotropic piezoelectric beams was performed in [1,5,8,13–15,17]. Models for elastic beams in contact with a foundation have been justified in [7,9,16], based on the ideas in [12].

The present paper represents a continuation of our previous works. Here we analyse, both mathematically and numerically, a model for an elastic piezoelectric beam in contact with a deformable conductive foundation. The manuscript is structured as follows. In Sect. 2 we describe the physical setting together with the corresponding mathematical model. Then, we list the assumptions on the data and state our main results in the analysis of the model. In Sect. 3 we formulate the discrete problem by using Finite Elements and provide existence, uniqueness, convergence and error estimates results. We also provide a brief description of the corresponding numerical algorithm. Finally, in Sect. 4 we present numerical simulations which highlight the performances of the algorithm and describe the effects of the different parameters on the solution.

2   Problem Statement

We consider an elastic piezoelectric beam of length $L > 0$, cross-section area A, Young Modulus E and Inertia Moment I. The beam is clamped at both ends and subjected to axial and vertical external forces $\bar{f}$ and $f^{\perp}$, respectively. As a result, it may enter in contact with a conductive deformable foundation. Based on our previous works [9,13,16], we associate to this physical setting the following mathematical model.

Problem 1. Find a bending field $\xi : [0, L] \to \mathbb{R}$, a stretching field $u : [0, L] \to \mathbb{R}$ and an electric potential $q : [0, L] \to \mathbb{R}$ such that

$$(C^{\perp}\xi'')'' = f^{\perp} - p(\xi - s) - \mu_1 R(q)\tilde{R}(\xi - s), \tag{2.1}$$

$$- (Pu')' - (\varepsilon q')' = \mu_2(\tilde{R}(\xi - s))^2, \tag{2.2}$$

$$- (\bar{C}u')' + (Pq')' = \bar{f}, \tag{2.3}$$

$$\xi(0) = \xi'(0) = \xi(L) = \xi'(L) = 0, \tag{2.4}$$

$$q(0) = q_0, \ q(L) = q_L, \tag{2.5}$$

$$u(0) = u(L) = 0. \tag{2.6}$$

Here, P and $\varepsilon > 0$ denote the piezoelectric coefficient and the electric permittivity coefficient, respectively, and $C^{\perp} = EI > 0$, $\bar{C} = EA > 0$. Moreover, $p(\cdot) : \mathbb{R} \to \mathbb{R}_+$ denotes the normal compliance function, which vanishes for negative arguments, and $\mu_1 > 0$, $\mu_2 > 0$ represent coefficients of the system. The dependence of the various functions on $x \in [0, L]$ is not indicated explicitly and the symbol $'$ stands for the derivative with respect to this spatial variable.

We now briefly describe the equations and conditions (2.1)–(2.6). First, equation (2.1) is the beam equation in normal compliance contact with a foundation, with an initial gap s. It also contains an additional term on the right hand side, which describes the electric charges from the obstacle to the beam, when the contact arises. Here we use the notation $\tilde{R}(z) = (R(z))_+$, for all $z \in \mathbb{R}$, where $r_+ = max\{r, 0\}$ denotes the nonnegative part of r and R is the truncation operator given by

$$R(z) = \begin{cases} -M & \text{if } z < -M, \\ z & \text{if } -M \le z \le M, \\ M & \text{if } z > M, \end{cases}$$

$M > 0$ being a positive constant that depends on the characteristic length of the system. Equations (2.2)–(2.3) are the piezoelectric equations for elastic beams presented in [13], and (2.4)–(2.6) represent the boundary conditions, in which q_0 and q_L denote the electric potentials applied on both ends of the beam.

For the analysis of Problem 1 we use the standard notation for Sobolev and Lebesgue spaces. In particular, we denote by $\|\cdot\|_0$, $\|\cdot\|_1$, $\|\cdot\|_2$ and $\|\cdot\|_\infty$ the norms on the spaces $L^2(0, L)$, $H^1(0, L)$, $H^2(0, L)$ and $L^\infty(0, L)$, respectively. Let $\varphi = q - \hat{q}$, where $\hat{q}$ is a lifting function of q_0 and q_L in $H^1(0, L)$ (see [13]). Note that

$$\hat{q} \in \mathcal{C}^0([0, L]) \quad \text{and} \quad \max_{x \in [0,L]}\{\hat{q}(x)\} = \max\{q_0, q_L\}.$$

Moreover, without loss of generality, we assume that q_0, $q_L > 0$. As a consequence, it follows that $\hat{q} > 0$ as well. We also assume that

$$s \in \mathcal{C}^1([0, L]), \; s(0) = s'(0) = s(L) = s'(L) = 0, \tag{2.7}$$

and we denote $\eta = \xi - s$. Then, using a standard procedure, it is easy to obtain the following variational formulation of problem (2.1)–(2.6).

Problem 2. Find $\eta \in H_0^2(0, L)$, $u \in H_0^1(0, L)$, $\varphi \in H_0^1(0, L)$, such that

$$\int_0^L C^\perp \eta'' \zeta'' dx + \int_0^L \mu_1 R(\varphi + \hat{q}) \tilde{R}(\eta) \zeta dx + \int_0^L p(\eta) \zeta dx$$

$$= \int_0^L f^\perp \zeta dx - \int_0^L C^\perp s'' \zeta'' dx, \tag{2.8}$$

$$\int_0^L Pu'\psi' dx + \int_0^L \varepsilon\varphi'\psi' dx - \int_0^L \mu_2(\tilde{R}(\eta))^2 \psi dx = -\int_0^L \varepsilon\hat{q}'\psi' dx, \tag{2.9}$$

$$\int_0^L \bar{C}u'v' dx - \int_0^L P\varphi'v' dx = \int_0^L \bar{f}v dx + \int_0^L P\hat{q}'v' dx, \tag{2.10}$$

for all $\zeta \in H_0^2(0, L)$, v, $\psi \in H_0^1(0, L)$.

In the study of Problem 2 we assume the following hypotheses.

$$\bar{f}, f^\perp \in L^2(0, L), \tag{2.11}$$

$$C^\perp, P, \varepsilon, \bar{C}, \mu_1, \mu_2 \in L^\infty(0, L), \tag{2.12}$$

$$C^\perp \ge C_0^\perp > 0, \; \varepsilon \ge \varepsilon_0 > 0, \; \bar{C} \ge \bar{C}_0 > 0, \; \mu_1 > 0, \; \mu_2 > 0 \text{ a.e. in } (0, L), \tag{2.13}$$

Moreover, we assume that the normal compliance function p satisfies

$$
\begin{cases}
\text{(a) } p : [0, L] \times \mathbb{R} \to \mathbb{R}_+, \\[4pt]
\text{(b) There exists } c_p > 0 \text{ such that} \\
\quad |p(x, r_1) - p(x, r_2)| \leq c_p |r_1 - r_2|, \\
\quad \forall r_1, r_2 \in \mathbb{R}, \text{ a.e. } x \in (0, L), \\[4pt]
\text{(c) There exists } m_p \geq 0 \text{ such that} \\
\quad (p(x, r_1) - p(x, r_2))(r_1 - r_2) \geq m_p |r_1 - r_2|^2, \\
\quad \forall r_1, r_2 \in \mathbb{R}, \text{ a.e. } x \in (0, L), \\[4pt]
\text{(d) The mapping } p(\cdot, r) : x \mapsto p(x, r) \text{ is measurable on } [0, L], \\
\quad \text{for all } r \in \mathbb{R}, \\[4pt]
\text{(e) The mapping } p(\cdot, r) : x \mapsto p(x, r) \text{ vanishes for all } r \leq 0,
\end{cases}
\tag{2.14}
$$

and, in addition,

$$
\frac{C_0^\perp}{c_2^2 M} > \frac{3}{2} \|\mu_1\|_\infty + \|\mu_2\|_\infty, \qquad \frac{\varepsilon_0}{c_1^2 M} > \frac{1}{2} \|\mu_1\|_\infty + \|\mu_2\|_\infty. \tag{2.15}
$$

Next, we consider the space $X(0, L) = H_0^2(0, L) \times H_0^1(0, L) \times H_0^1(0, L)$, which is a real Hilbert space with the canonical inner product, denoted by $(\cdot, \cdot)_{X(0,L)}$. We then define the operator $A : X(0, L) \to X(0, L)$ and the element $\boldsymbol{F} \in X(0, L)$ by equalities

$$
(A\boldsymbol{x}, \boldsymbol{y})_{X(0,L)} = \int_0^L C^\perp \eta'' \zeta'' dx + \int_0^L \mu_1 R(\varphi + \hat{q}) \tilde{R}(\eta) \zeta \, dx + \int_0^L p(\eta) \zeta \, dx
$$

$$
+ \int_0^L P u' \psi' dx + \int_0^L \varepsilon \varphi' \psi' dx - \int_0^L \mu_2 (\tilde{R}(\eta))^2 \psi \, dx + \int_0^L \bar{C} u' v' dx
$$

$$
- \int_0^L P \varphi' v' dx \qquad \forall \boldsymbol{x} = (\eta, \varphi, u), \; \boldsymbol{y} = (\zeta, \psi, v) \in X(0, L), \tag{2.16}
$$

$$
(\boldsymbol{F}, \boldsymbol{y})_{X(0,L)} = \int_0^L f^\perp \zeta \, dx + \int_0^L \bar{f} v \, dx - \int_0^L C^\perp s'' \zeta'' dx - \int_0^L \varepsilon \hat{q}' \psi' dx
$$

$$
+ \int_0^L P \hat{q}' v' dx \qquad \forall \boldsymbol{y} = (\zeta, \psi, v) \in X(0, L). \tag{2.17}
$$

Then, an equivalent formulation of Problem 2 is as follows.

Problem 3. Find $\boldsymbol{x} = (\eta, \varphi, u) \in X(0, L)$ such that

$$
(A\boldsymbol{x}, \boldsymbol{y}) = (\boldsymbol{F}, \boldsymbol{y})_{X(0,L)} \qquad \forall \boldsymbol{y} \in X(0, L). \tag{2.18}
$$

Our main result in the study of Problem 3 is the following.

Theorem 1. *Under the assumptions* (2.7), (2.11)–(2.15), *there exists a unique solution* $\boldsymbol{x} = (\eta, \varphi, u) \in X(0, L)$ *to Problem 3. Moreover, if* $\boldsymbol{x}_i = (\eta_i, u_i, \varphi_i)$

represents the solution of Problem 3 for the data $\{q_{0i}, q_{Li}\}$, $\{s_i\}$ *and* $\{f_i^{\perp}, \bar{f}_i\}$ *verifying the assumptions (2.7) and (2.11), $i = 1, 2$, then there exists $C > 0$ such that*

$$\|\boldsymbol{x}_1 - \boldsymbol{x}_2\|_{X(0,L)} \leq C(\|f_1^{\perp} - f_2^{\perp}\|_0 + \|s_1 - s_2\|_2 + \|\bar{f}_1 - \bar{f}_2\|_0 + \|\hat{q}_1 - \hat{q}_2\|_1). \quad (2.19)$$

The proof of this theorem is based on arguments of monotonicity and will be included in our forthcoming paper [10]. Besides the unique weak solvability of Problem 1, it provides the continuous dependence of the solution with respect to the boundary data, the initial gap and the external forces.

3   Numerical Analysis

We now turn to the numerical analysis of the problem. To this end let $0 < h < L$, $N(h) \in \mathbb{N}$ and let $0 = x_0^h < x_1^h < \ldots < x_i^h < x_{i+1}^h < \ldots < x_{N(h)}^h = L$ be a partition of the interval $[0, L]$ in $N(h)$ intervals with maximum length h. We denote by Θ^h the set of all elements and $K_i^h = [x_i^h, x_{i+1}^h] \in \Theta^h$, $0 \leq i < N(h)$. We consider the finite element spaces

$$V_1^h(0, L) = \{\xi^h \in C^0([0, L]), \xi_{|K_i^h}^h \in P_1(K_i^h), \ 0 \leq i < N(h), \ \xi^h(0) = \xi^h(L) = 0\},$$

$$V_3^h(0, L) = \{\xi^h \in C^1([0, L]), \xi_{|K_i^h}^h \in P_3(K_i^h), \ 0 \leq i < N(h),$$

$$\xi^h(0) = \xi^h(L) = (\xi^h)'(0) = (\xi^h)'(L) = 0\},$$

where $P_k(K^h)$ represents the space of polynomials of degree less or equal than k restricted to K^h. It is straightforward to see that $V_3^h(0, L) \subset H_0^2(0, L)$ and $V_1^h(0, L) \subset H_0^1(0, L)$. Let $X^h(0, L) = V_3^h(0, L) \times V_1^h(0, L) \times V_1^h(0, L) \subset X(0, L)$, and let $P_{X^h} : X(0, L) \to X^h(0, L)$ denote the projection operator. Then, the discrete version of Problem 3 can be formulated as follows:

Problem 4. Find $\boldsymbol{x}^h = (\eta^h, \varphi^h, u^h) \in X^h(0, L)$ such that

$$(A\boldsymbol{x}^h, \boldsymbol{y}^h)_{X(0,L)} = (\boldsymbol{F}, \boldsymbol{y}^h)_{X(0,L)} \qquad \forall \boldsymbol{y}^h = (\zeta^h, \psi^h, v^h) \in X^h(0, L).$$

Using arguments similar to those used in the proof of Theorem 1 it follows that Problem 4 has a unique solution $\boldsymbol{x}^h = (\eta^h, \varphi^h, u^h) \in X^h(0, L)$. In addition, the following *a priori* error estimation holds:

$$\|\boldsymbol{x} - \boldsymbol{x}^h\|_{X(0,L)} \leq C\|\boldsymbol{x} - \boldsymbol{y}^h\|_{X(0,L)} \qquad \forall \boldsymbol{y}^h \in X^h(0, L), \quad (3.1)$$

where, recall, $\boldsymbol{x} = (\eta, \varphi, u) \in X(0, L)$ is the solution of Problem 3.

For a given $K_i^h = [x_i^h, x_{i+1}^h] \in \Theta^h$, we use a local nodal notation, so $K_i^h = [x_0^{i,h}, x_1^{i,h}]$. Denote by $\Pi_1^h : C([0, L]) \to V_1^h(0, L)$ the Lagrange global interpolation operator, i.e.

$$\Pi_1^h v_{|K_i^h} = \Pi_1^h{}_{|K_i^h} v \qquad \forall K_i^h \in \Theta^h, \ v \in C([0, L]),$$

where $\Pi^h_{1|K^h_i} : C(K^h_i) \to P_1(K^h_i)$ represents the local Lagrange interpolation operator. Similarly, denote by $\Pi^h_3 : C^1([0,L]) \to V^h_3(0,L)$ the Hermite global interpolation operator, i.e.

$$\Pi^h_3 v_{|K^h_i} = \Pi^h_{3|K^h_i} v \qquad \forall K^h_i \in \Theta^h \; v \in C^1([0,L]),$$

where $\Pi^h_{3|K^h_i} : C^1(K^h_i) \to P_3(K^h_i)$ is the local Hermite interpolation operator. Then, the following interpolation error estimations holds:

$$\|v - \Pi^h_k v\|_{m,K} \le C\, h^l_K |v|_{r,K} \qquad \forall K \in \Theta^h, \tag{3.2}$$

where $v \in H^r(K)$, $\Pi^h_k v$ denotes its corresponding interpolant in $V^h_k(0,L)$, $l = \min\{k+1-m, r-m\}$ and $C > 0$ is a constant which does not depend on v, k and h. For the proof of (3.2) see, for instance, [4,6].

Finally, let $\Pi^h : C^1([0,L]) \times C^0([0,L]) \times C^0([0,L]) \to X^h(0,L)$ denote the global interpolation operator given by $\Pi^h(\boldsymbol{y}) = (\Pi^h_3(\zeta), \Pi^h_1(\psi), \Pi^h_1(v))$ for all $\boldsymbol{y} = (\zeta, \psi, v) \in C^1([0,L]) \times C^0([0,L]) \times C^0([0,L])$. Therefore, given $\{r_1, r_2, r_3\} \subset \mathbb{N}$, from (3.2) we find that

$$\|\boldsymbol{y} - \Pi^h \boldsymbol{y}\|_{X(0,L)} \le C\, (h^{r_1} \|\zeta\|_2 + h^{r_2} \|\psi\|_1 + h^{r_3} \|v\|_1) \le C\, h^l \|\boldsymbol{y}\|_{X(0,L)},$$

for all $\boldsymbol{y} = (\zeta, \psi, v) \in H^{2+r_1}(0,L) \times H^{1+r_2}(0,L) \times H^{1+r_3}(0,L)$, where $l = \min\{r_1, r_2, r_3\}$. Furthermore, by using the previous error estimation and density arguments in (3.1), we conclude that

$$\lim_{h \to 0} \|\boldsymbol{x} - \boldsymbol{x}^h\|_{X(0,L)} = 0,$$

which represents a convergence result for our discrete scheme.

Algorithm Implementation. The algorithm for the numerical solution of Problem 4 is based on a fixed point strategy to compute the bending, the stretching and the electric potential, iteratively. The novelty lies in the method we use to solve the contact problem on each step. The description of this method represents our main aim in the rest of this section. In order to simplify it, note that the numerical discretization of (2.8) fits in the following general framework:

Find $\xi^h \in E^h$ such that $L \in A\xi^h + BHB^*\xi^h$, where $H : E^h \to 2^{E^h}$ is a maximal monotone operator, $B \in \mathcal{L}(E^h, V^{h'}_3)$, $A \in \mathcal{L}(V^h_3, V^{h'}_3)$ and $L \in V^{h'}_3$.

Here and below the symbol E' denotes the dual of E. To solve this problem we use a penalization algorithm of the Uzawa family whose performance is improved by combining it with the Newton method. We restrict ourselves to describe the main steps of the algorithm, and refer the reader to [2,11] for further details. Let H^ω_μ denote the Yosida approximation of the operator $H^\omega = H - \omega I$, with $\omega\mu < 1$. Then the algorithm introduced in [2] is the following:

Let $q^{h,0}$, $\xi^{h,0}$ be arbitrary. Given $q^{h,n}$ and $\xi^{h,n-1}$, compute $q^{h,n+1}$ and $\xi^{h,n}$ such that

$$\begin{cases} A\xi^{h,n} + \omega BB^*\xi^{h,n} = L - Bq^{h,n}, \\ q^{h,n+1/2} = H^\omega_\mu \left[B^*\xi^{h,n} + \mu q^{h,n}\right], \\ q^{h,n+1} = \rho q^{h,n+1/2} + (1-\rho)q^{h,n}. \end{cases} \tag{3.3}$$

The efficiency of this algorithm is well known, although its convergence is quite slow. For this reason, following [3], we combine it with the Newton method which accelerates its convergence. First, note that if $\omega = 0$, then $\omega\mu < 1$ and

$$H_\mu^0(x) = H_\mu(x) = \frac{k}{1+k\mu}(I - \Pi_{E^h})(x).$$

Therefore, we need to solve the system

$$A\xi^{h,n+1} = L - Bq^{h,n+1}, \tag{3.4}$$

$$q^{h,n+1} = H_\mu(B\xi^{h,n+1} + \mu q^{h,n+1}). \tag{3.5}$$

Besides, given $\phi \in E^h$, we have $H_\mu(\phi)_{|C} = G(\phi_{|C})$ where

$$G(p) = \begin{cases} 0 & \text{if } p \le s \\ \frac{k}{1+k\mu}(p-s)(x) & \text{if } p > s \end{cases}.$$

We approximate $G(p_0)$ by $G(p_1) + V(p_0 - p_1) + O(|p_0 - p_1|^2)$, where

$$V \in \partial G(p_0) = \begin{cases} 0 & \text{if } p_0 < s \\ \left[0, \frac{k}{1+k\mu}\right] & \text{if } p_0 = s \\ \frac{k}{1+k\mu} & \text{if } p_0 > s \end{cases}. \tag{3.6}$$

Thus, taking $p_0 = B\xi^{h,n+1} + \mu q^{h,n+1}$ and $p_1 = B\xi^{h,n} + \mu q^{h,n}$, we obtain that

$$q^{h,n+1} = H_\mu(p_0) = G(B\xi^{h,n} + \mu q^{h,n}) + V(B\xi^{h,n+1} + \mu q^{h,n+1} - B\xi^{h,n} + \mu q^{h,n}).$$

Next, we use the subsets of the mesh Θ^h given by

$$\Omega_n^+ = \{K_i^h \in \Theta^h; (B\xi^{h,n} + \mu q^{h,n})_{|K_i^h} > s\}, \quad \Omega_n^- = \Theta^h \setminus \Omega_n^+, \tag{3.7}$$

and we take into account the values of V in (3.6) to find that

$$q^{h,n+1} = \begin{cases} 0 & \text{on } \Omega_n^- \\ \frac{k}{1+k\mu}(B\xi^{h,n+1} + \mu q^{h,n+1}) - \frac{k}{1+k\mu}s & \text{on } \Omega_n^+ \end{cases},$$

which implies that $q^{h,n+1} = 0$ on Ω_n^- and $q^{h,n+1} = k(B\xi^{h,n+1} - s)$ on Ω_n^+. Recall that our aim is to solve the system (3.4)–(3.5). To this end, substituting $q^{h,n+1}$ in (3.4) by this value obtained by applying the Newton approximation to (3.5), we obtain the following algorithm:

Let $q^{h,0}$, $\xi^{h,0}$ be arbitrary. Given $q^{h,n}$, $\xi^{h,n}$ and Ω_n^+, compute $q^{h,n+1}$, $\xi^{h,n+1}$ and Ω_{n+1}^+ such that

$$\begin{cases} \int_\Omega A\xi^{h,n+1}\zeta dx + \int_{\Omega_n^+} kB\xi^{h,n+1}\zeta dx = \int_\Omega L\zeta dx + \int_{\Omega_n^+} ks\zeta dx, \\ q^{h,n+1} = \begin{cases} 0 & \text{on } \Omega_n^- \\ k(B\xi^{h,n+1} - s) & \text{on } \Omega_n^+ \end{cases}, \\ \Omega_{n+1}^+ = \{K_i^h \in \Theta^h; (B\xi^{h,n+1} + \mu q^{h,n+1})_{|K_i^h} > s\}. \end{cases} \tag{3.8}$$

We finally recall that even though the algorithm above has been presented in a general framework, in our work it is highly simplified since the operator B is, in fact, the identity operator.

4   Numerical Simulations

In this section we will show the numerical results obtained for several simulations designed to highlight the performance of the algorithm. We consider a beam of length $L = 1\,m$, and a uniformly spaced mesh with element size $h = 0.01$. The choice of values for the various parameters is the following:

$$E = 1 \times 10^5 \, \frac{N}{m^2}, \; I = 0.05 \; m^4, \; A = 1 \; m^2, \; P = 100 \, \frac{N \cdot m}{V},$$

$$\mu_1 = \mu_2 = 1 \, \frac{N}{V \cdot m^2}, \; \varepsilon = 1 \, \frac{N \cdot m^2}{V}, \; \bar{f} = 1 \times 10^6 \, \frac{N}{n}, \; f^{\perp} = 1 \times 10^6 \, \frac{N}{m}.$$

Our results are presented in Figs. 1, 2, 3 and 4 where the bending, the stretching, the deformation and the electric potential of the beam are plotted.

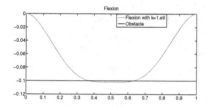

Fig. 1. Bending.

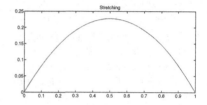

Fig. 2. Stretching.

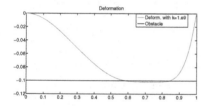

Fig. 3. Deformation.

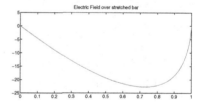

Fig. 4. Electric potential.

Next, in Fig. 5, we illustrate the influence of the stiffness coefficient on the bending of the beam. We start with the value $k = 1$ for the stiffness coefficient and increase it up to the value $k = 1 \times 10^{16}$. Our results show that more the obstacle is stiff, less the penetration is. We also note that the influence of the obstacle arises only for values of the stiffness coefficient larger than the Young modulus of the beam, E. We also plot the solutions obtained by using both the method in (3.3) and its Newton improvement formulated in (3.8), for $k = 1 \times 10^{16}$. As expected, the two solutions are practically the same.

Now, in order to show the improvement of the Newton method versus the original algorithm, we represent in Fig. 6 the number of iterations needed to achieve convergence for both algorithms, with various tolerances. As we can

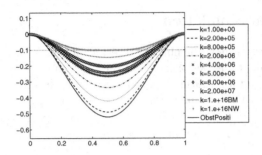

Fig. 5. Bending for several coefficients of stiffness

see, the convergence of the Newton method is always achieved in about ten iterations, while the Bermúdez-Moreno algorithm (3.3) needs more than one thousand iterations, if the tolerance is smaller than 1×10^{-3}.

In Figs. 7, 8 and 9 we show the convergence of the solutions for bending, stretching and electric potential, respectively, as the mesh size decreases. We take the solution obtained for $h = 1 \times 10^{-4}$ as "exact" solution. We note that the convergence for the electric potential and the stretching field is linear while the convergence for the bending field is slower.

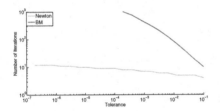

Fig. 6. Iterations.

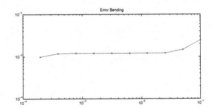

Fig. 7. Bending error.

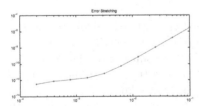

Fig. 8. Stretching error

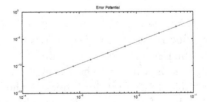

Fig. 9. Electric potential error

References

1. Bellis, S., Imperiale, S.: Dynamical 1D models of passive piezoelectric sensors. Math. Mech. Solids **19**(5), 451–476 (2014)
2. Bermúdez, A., Moreno, C.: Duality methods for solving variational inequalities. Comput. Math. Appl. **7**(1), 43–58 (1981)
3. Cao, M.T., Moreno, C., Quintela, P.: Simulation of Rayleigh waves in cracked plates. Math. Methods Appl. Sci. **30**, 15–42 (2007)
4. Ciarlet, P.G.: The Finite Element Method for Elliptic Problems. Studies in Mathematics and its Applications, 4th edn. North-Holland Publishing Co., Amsterdam (1978)
5. Figueiredo, I., Leal, C.: A generalized piezoelectric Bernoulli-Navier anisotropic rod model. J. Elast. **31**, 85–106 (2006)
6. Hughes, T.J.R.: The Finite Element Method. Prentice Hall Inc., Englewood Cliffs (1987)
7. Irago, H., Viaño, J.M., Rodríguez-Arós, Á.: Asymptotic derivation of frictionless contact models for elastic rods on a foundation with normal compliance. Nonlinear Anal. Real World Appl. **14**, 852–866 (2013)
8. Ribeiro, C.: Asymptotic derivation of models for anisotropic piezoelectric beams and shallow arches. Ph.D. thesis, Departament of Mathematics and Applications, Minho University (2009)
9. Rodríguez-Arós, Á., Viaño, J.M.: A bending-stretching model in adhesive contact for elastic rods obtained by using asymptotic methods. Nonlinear Anal.-RWA **22**, 632–644 (2015)
10. Rodríguez-Arós, Á., Cao-Rial, M.T., Sofonea, M.: Mathematical analysis of a piezoelectric elastic beam in contact with a deformable foundation. Preprint
11. Viaño, J.M.: Análisis numérico de un método numérico con elementos finitos para problemas de contacto unilateral sin rozamiento en lasticidad: Aproximación y resolución de los problemas discretos. Rev. Internac. Métod. Numér. Cálc. Diseñ. Ingr. **2**, 63–86 (1986)
12. Viaño, J.M.: The one-dimensional obstacle problem as approximation of the three-dimensional Signorini problem. Bull. Math. Soc. Sci. Math. Roumanie (N.S.) **48**, 243–258 (2005)
13. Viaño, J., Figueiredo, J., Ribeiro, C., Rodríguez-Arós, Á.: A model for bending and stretching of piezoelectric rods obtained by asymptotic analysis. Zeitschrift fr angewandte Mathematik und Physik **66**(3), 1207–1232 (2015)
14. Viaño, J.M., Ribeiro, C., Figueiredo, J.: Asymptotic modelling of a piezoelectric beam. In: Proceedings of Congreso de Métodos Numéricos en Ingeniería, Granada, Spain, vol. A310, pp. 1–17 (2005)
15. Viaño, J.M., Ribeiro, C., Figueiredo, J.: Asymptotic modelling of a piezoelectric beam. In: Proceedings of II ECCOMAS Thematic Conference of Smart Materials and Structures, pp. 1–12. Lisbon, Portugal, EO26MOD (2005)
16. Viaño, J.M., Rodríguez-Arós, Á., Sofonea, M.: Asymptotic derivation of quasistatic frictional contact models with wear for elastic rods. J. Math. Anal. Appl. **401**, 641–653 (2013)
17. Weller, T., Licht, C.: Asymptotic modeling of linearly piezoelectric slender rods. C. R. Mecanique **336**, 572–577 (2008)

Optimization of Engineering Processes Including Heating in Time-Dependent Domains

Alfred Schmidt[1]([✉]), Eberhard Bänsch[2], Mischa Jahn[1], Andreas Luttmann[1], Carsten Niebuhr[1], and Jost Vehmeyer[1]

[1] Zentrum für Technomathematik, Universität Bremen, Bremen, Germany
{schmidt,mischa,andreasl,niebuhr,vehmeyer}@math.uni-bremen.de
[2] Department Mathematik, Friedrich-Alexander-Universität Erlangen-Nürnberg,
Erlangen, Germany
baensch@math.fau.de
http://www.math.uni-bremen.de/zetem
http://www.mso.math.fau.de/applied-mathematics-3

Abstract. We present two models for engineering processes, where thermal effects and time-dependent domains play an important role. Typically, the parabolic heat equation is coupled with other equations. Challenges for the optimization of such systems are presented.

The first model describes a milling process, where material is removed and heat is produced by the cutting, leading to thermomechanical distortion. Goal is the minimization of these distortions.

The second model describes the melting and solidification of metal, where the geometry is a result of free-surface flow of the liquid and the microstructure of the re-solidified material is important for the quality of the produced preform.

Keywords: Optimization with PDEs · Time-dependent domain · Heat equation · Thermoelasticity · Free surface flow

1 Introduction

The optimization of industrial engineering processes often lead to the treatment of coupled, nonlinear systems of PDEs. Here we want to investigate applications where an important part of the nonlinearities is created by time-dependent domains, which are not prescribed but who are part of the solution itself. The treatment of such models in the context of PDE constraints in optimal control problems typically generates additional challenges in both, the solution of the forward problems and the treatment and storage of adjoint solutions.

In the following, we present two models for engineering processes, where heating of metal workpieces and time-dependent domains play an important role. Thus, a parabolic heat equation is coupled with other equations. Challenges for the optimization of such coupled systems are presented.

The first model describes a milling process, where material is removed and heat is produced by the cutting. This leads to a thermomechanical distortion of

L. Bociu et al. (Eds.): CSMO 2015, IFIP AICT 494, pp. 452–461, 2016.
DOI: 10.1007/978-3-319-55795-3_43

the workpiece during the cutting process and leads to an incorrect removal of material. An optimization of the cutting path and speed, varying chip thickness and thus heat production, etc., should give reduced distortion during the process and lead to a correct workpiece shape.

The second model describes the melting and solidification of metal heated by a laser beam. Due to free-surface flow, the shape of the liquid part depends on capillary boundary conditions, and heat transport on the flow field. The microstructure of the re-solidified material, which is important for subsequent process steps, depends on the temperature gradients near the moving liquid-solid interface and its velocity. Accelerating the process for mass production on the one hand and improving the microstructure on the other hand compete for an optimized process.

Both optimization problems can be formulated in an abstract setting as

$$\min_{u \in U_{ad}} J(u,y) \quad \text{under the constraint} \quad y = S(u) \tag{1}$$

where u denotes the control, U_{ad} the set of admissible controls, y the state, J the error functional to be minimized, and S is the control-to-state operator, given by a nonlinear PDE. J is typically given by a deviation of y (or something derived from it) from a desired function y_d plus some regularization by a norm of u, like

$$J(u,y) = d(y,y_d) + \alpha \|u\|_p^p. \tag{2}$$

For both applications, we will first state and describe the primal problem giving the solution operator S, and later cover some details of the associated optimization problem.

2   Thermomechanics of Milling Processes

2.1   Application

During a milling process, heat is produced by the cutting tool and transfered into the workpiece, and mechanical load is generated by cutting forces. Due to the resulting thermomechanical deformation of the workpiece, the final shape of the processed workpiece deviates from the desired shape, making a postprocessing finishing necessary. Deformations are relatively large especially when producing fine structures like thin walls for lightweight constructions. In order to reduce the shape deviation, an optimization of the tool path and other process parameters is desirable, taking into account the thermomechanical deformations.

2.2   Model

The mathematical model for the thermomechanics of the process includes thermoelasticity of the workpiece, energy and forces introduced by the process, and most importantly the cutting process itself, which leads to a time-dependent domain, whose shape influences the process and vice versa. As typical under the assumption of small deformations, the model is formulated in a reference configuration.

Let $\Omega(t) \subset \mathbb{R}^3$ denote the time-dependent domain in the reference configuration, $Q_T := \{(x,t) : x \in \Omega(t), t \in (0,T)\}$ the space-time-domain, $\theta : Q_T \to \mathbb{R}$ the temperature, and $\mathbf{v} : Q_T \to \mathbb{R}^3$ the deformation.

The change of geometry by material removal as well as energy and forces introduced by the cutting process are provided by a process model [3,5], taking into account the tool path and velocity, chip thickness, temperature and deformation, and other global and local parameters and properties. Here, we rely on a macroscopic model where microscopic processes like chip formation are not directly considered, but their effects considered via the process model. Let us denote by $\Gamma(t) \subset \partial\Omega(t)$ the contact zone of the cutting tool at time t, $\Gamma_T := \{(x,t) : x \in \Gamma(t), t \in (0,T)\}$, $q_\Gamma : \Gamma_T \to \mathbb{R}$ the normal heat flux, and $g_\Gamma : \Gamma_T \to \mathbb{R}^3$ the forces introduced at the cutting surface.

In the notion of optimal control problems, the state y consists of the domain, temperature, and deformation, $y = (\Omega, \theta, \mathbf{v})$, while the control u consists of the process parameters like tool path, feed rate, rotational velocity, etc. The material removal and thus the domain $\Omega(t)$ depend on the cutting process (control u) and the deformation $\mathbf{v}$.

The coupled model includes the parabolic heat equation and quasistatic, elliptic thermoelasticity

$$\dot{\theta} - \nabla \cdot (\kappa\nabla\theta) = 0, \tag{3}$$
$$-\nabla \cdot \sigma = f_v(\theta) \tag{4}$$

on Ω_T with stress tensor $\sigma = 2\mu D\mathbf{v} + \lambda tr(D\mathbf{v})\mathbb{I}$ and $D\mathbf{v} = \frac{1}{2}(\nabla\mathbf{v} + \nabla\mathbf{v}^T)$ the symmetric gradient or strain tensor. On the contact zone Γ_T, we have boundary conditions for heat flux and mechanical forces given by the process model,

$$\kappa\nabla\theta \cdot \mathbf{n} = q_\Gamma(u, \theta, \mathbf{v}), \tag{5}$$
$$\sigma\mathbf{n} = \mathbf{g}_\Gamma(u, \theta, \mathbf{v}). \tag{6}$$

The workpiece is clamped, which is reflected by Dirichlet conditions on a subset $\Gamma_D \subset \partial\Omega \setminus \Gamma_T$,

$$\mathbf{v} = 0 \quad \text{on } \Gamma_D, \tag{7}$$

while cooling conditions and free deformation apply on the rest of the boundary,

$$\kappa\nabla\theta \cdot \mathbf{n} = r(\theta_{ext} - \theta), \tag{8}$$
$$\sigma\mathbf{n} = 0. \tag{9}$$

Initially, the temperature is typically constant at room temperature, thus $\theta = \theta_0$ on $\Omega(0)$.

2.3   Numerical Discretization of the Forward Problem

The system (3–9) of PDEs is discretized using a finite element method on an adaptively locally refined tetrahedral mesh [10, 13], using piecewise polynomial functions for the temperature and the components of deformation. Time-discretzation is based on a semi-implicit time stepping scheme.

The time-dependent domain $\Omega(t)$ is approximated by a subset $\Omega_h(t)$ of the triangulation, where the completely cut off elements are ignored. The cutting process is simulated by a dexel method [11], which is able to compute the interaction of the tool with the (deformed) workpiece very efficiently, giving $\tilde{\Omega}_h(t)$ and $\tilde{\Gamma}_h(t)$. At the same time, chip thickness and other cutting parameters are computed and the process model returns approximations of the heat flux $q_{\Gamma,h}$ and forces $\mathbf{g}_{\Gamma,h}$ at the cutting surface $\tilde{\Gamma}_h(t)$. Based on that, the finite element method computes $\Omega_h(t)$ and $\Gamma_h(t)$ and projects the boundary data onto $\Gamma_h(t)$. Finite element approximations of temperature and deformation are computed on $\Omega_h(t)$. This is done in every time step of the finite element method. The overall method is described in [4, 5].

Figure 1 shows the mesh, temperature, and deformation from the simulation of a milling process. The mesh is adaptively refined in order to approximate the geometry $\Omega(t)$ well by $\Omega_h(t)$. The process removes layers of material to mill a pocket into a rectangular bar of metal. Especially the final thin backward wall is prone to deformations larger than the given tolerance, making it hard to produce the desired shape.

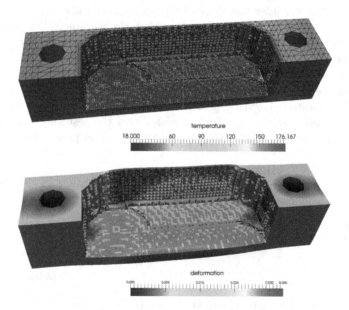

Fig. 1. Simulation of a milling process: mesh and temperature (top) and deformation (bottom, amplified by a factor 100). The tool is at the moment cutting near the backward left lower corner.

2.4   Optimization

The goal of a process optimization is a compensation and reduction of geometry errors while not slowing down the process too much. Control is given by the variation of process parameters like tool path and velocity, speed of tool rotation, etc. Such variations result in changes of the tool entry situation and thus, reflected by the process model, changes in heat source and cutting forces, and finally a change in the thermomechanic deformation. Admissible controls in U_{ad} are defined by restrictions on the machining process.

Given a prescribed final geometry Ω_d of the workpiece, the optimization functional should include the deviation of the process geometry from the desired one, as well as the process duration. Considering the geometry error, different criteria are possible, especially comparing geometries during the whole cutting process or only in the end. For the latter, this would nevertheless include geometry error terms over time, as material which was removed before cannot be added later on again.

Another approach to an error functional includes geometry deviations near the cutting zone Γ_T during the whole process:

$$J(u, \Omega, \theta, \mathbf{v}) = \int_0^T \|\Omega(t) - \Omega_d(t)\|_{\Gamma(t)}^2 + \lambda\|u\|^2. \tag{10}$$

We show the effect of such an error functional in a model situation, where a L-shaped geometry is produced from a rectangular plate. This can be seen as a slice through the original workpiece, see the left of Fig. 2. The non-optimized control does not consider the thermal extension and leads to increased material removal resulting in a recessed surface after the workpiece has cooled down. Figure 2 shows on the right the geometry error in the contact zone over the process time. The general optimal control problem requires to find the spatial tool positions and the cutting parameters which minimize (10). In a first investigation of the model problem, simple raster milling is performed and control is given by traditional setting parameters, i.e. cutting depth, radial and tangential feed and cutting velocity. Figure 3 shows the resulting surface for the non-optimized process and for the process with optimal parameters.

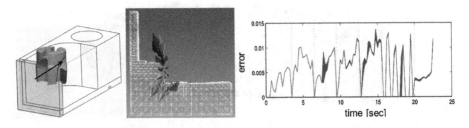

Fig. 2. Model geometry (left), deformation $\mathbf{v}$ in the cutting zone at $t = 10.5\,s$ (middle), and maximal deformation in the cutting zone $\Gamma(t)$ over time (right).

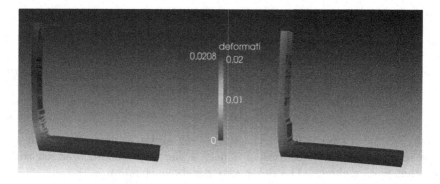

Fig. 3. L-shaped workpiece near the end of the milling process, with initial control (left) and optimized parameters (right). Colors depict the modulus of deformation. (Color figure online)

Adjoint Problem. The optimization shown above was done with the standard MATLAB optimization toolbox which just calls the finite element package to solve the forward problem. For a more involved algorithm, the computation and storage of the adjoint solution would be used. The corresponding system of adjoint problems consists of the adjoint (backward) heat equation on the time-dependent domain (now given from the forward problem), coupled to the quasistatic adjoint elasticity equation. Due to the rather long process time, the three-dimensional time-dependent domain, and adaptive meshes for approximation of domain and solution, the handling of such adjoint solutions in the optimization procedure is a challenge.

3   Material Accumulation by Laser Heating

3.1   Application

For the production of micro components (like micro-valves, etc., with diameters smaller than 1 mm) by cold forming, a necessary pre-forming step is to accumulate enough material for a subsequent cold forming step. This can be done by partial melting and solidification of a half-finished product like a thin wire [14]. Due to the small scale, the dominant surface tension of the melted material leads to a nearly spherical form, leading to an accumulated solid sphere attached to the wire after solidification which is called preform. Due to the industrial background, very high process speeds are requested. However, for the subsequent forming step, the microstructure of the material is important. Thus, besides the speed of the process and an accurate size of the preform, its microstructure is part of the optimization goal. Formation of dendritic structures and their spacings, or other phases, during the solidification are strongly influenced by the liquid-solid interface velocity and local temperature gradients [9]. Thus, temperature, phase transitions, and the geometry are important aspects of the corresponding optimization problem.

3.2 Model

We consider the time-dependent domain $\Omega(t)$ consisting of the solid and partially melted parts of the material (metal). Melting and solidification are typically modeled by the Stefan problem, including temperature θ and energy density e as variables in the space-time-domain $Q_T := \{(x,t) : x \in \Omega(t), t \in (0,T)\}$:

$$\dot{e} + \mathbf{v} \cdot \nabla e - \nabla \cdot (\kappa \nabla \theta) = 0, \qquad \theta = \beta(e), \tag{11}$$

where $\beta(s) := c_1 \min(s,0) + c_2 \max(0, s - L)$ and L denotes the latent heat of solid-liquid phase transitions. β is only a piecewise-smooth function with a constant part, making (11) a degenerate parabolic equation.

The liquid subdomain $\Omega_l(t)$ is given by all points where the temperature is above the melting temperature (which is assumed to be 0 after some scaling),

$$\Omega_l(t) := \{x \in \Omega(t) : \theta(x,t) > 0\}. \tag{12}$$

The shape of the melted (and later on re-solidified) material accumulation is mainly influenced by the surface tension of the liquid, together with gravitational forces etc., which means free-surface flow. Parabolic Navier-Stokes equations with capillary boundary condition is the main model component for this, with solenoidal velocity field $\mathbf{v}$ and pressure p in $\Omega_l(t)$,

$$\dot{\mathbf{v}} + \mathbf{v} \cdot \nabla \mathbf{v} - \nabla \cdot \sigma = f_v(\theta), \qquad \nabla \cdot \mathbf{v} = 0, \tag{13}$$

with stress tensor $\sigma = \frac{1}{Re} D\mathbf{v} - p\mathbb{I}$. Here, f_v denotes the forces introduced by gravity due to a temperature-dependent density and the Boussinesq approximation. The shape of the liquid subdomain is given through the capillary boundary condition on the free surface $\Gamma(t)$ of the melted subdomain, where the surface tension (proportional to the mean curvature of the surface) balances the normal stress. In a differential geometric PDE formulation, the mean curvature vector H_Γ is given by the Laplace-Beltrami-operator $-\Delta_\Gamma$ applied to the coordinates of the surface (represented by the embedding $id : \Gamma(t) \to \mathbb{R}^3$), giving another (nonlinear) elliptic equation in the coupled system. Additionally, the normal component of the fluid velocity should be equal to the normal velocity V_Γ of the capillary surface. Both lead to the following equations on $\Gamma(t)$:

$$\sigma \mathbf{n} = \frac{1}{We} H_\Gamma = -\frac{1}{We} \Delta_\Gamma id, \qquad \mathbf{v} \cdot \mathbf{n} = V_\Gamma. \tag{14}$$

On the solid-liquid interface and in the solid subdomain, the velocity vanishes,

$$\mathbf{v} = 0 \qquad \text{in } \Omega(t) \setminus \Omega_l(t). \tag{15}$$

The heating is done through a laser pointing at a spot on the boundary, modeled by a time- and space-dependent energy density q_L, and cooling conditions apply on the whole boundary,

$$\kappa \nabla \theta \cdot \mathbf{n} = q_L + r(\theta_{ext} - \theta) \qquad \text{on } \partial\Omega(t). \tag{16}$$

A more detailed description of the model can be found in [8].

3.3   Numerical Discretization of the Forward Problem

The forward problem with given boundary conditions is discretized by a finite element method which combines a Stefan problem solver with a free-surface Navier-Stokes solver. The latter is based on the Navier code [1], the combined approach is described in more detail in [2, 7]. Locally refined (triangular or tetrahedral) meshes are needed in order to approximate the large variations in temperature near the heating zone and the surrounding of the solid-liquid interface sufficiently well, while keeping the overall numerical costs acceptable. Due to the big changes in geometry, starting from a thin wire and ending in a relatively large spherical accumulation, several remeshings are necessary during the simulation in order to avoid a degeneration of mesh elements.

Figure 4 shows a typical mesh, temperature field, and velocity field during the melting, with the laser pointing to the center bottom of the material. Due to rotational symmetry, a 2D FEM with triangular meshes could be used. As the wire is melted from below, the growing sphere is moving upwards and thus the velocity vectors are pointing upwards, too. In Fig. 5, we show several stages during the solidification after the heating is switched off (from a simulation with shorter heating period). In contrast to our results for energy dissipation during the heating process [6], the cooling after switching off the laser is mainly done by conducting heat upwards into the wire, which results in a downward movement of the interface. The additional cooling by the boundary conditions leads later on to an additional solidification at the boundary of the liquid region. During solidification, the shape does not change much anymore, so the velocities are typically quite small and not shown here.

Fig. 4. Melting the end of a thin wire: liquid material accumulation with mesh, solid-liquid interface, and liquid flow field.

3.4   Optimization

The goals for an optimization of the process are on the one hand a high speed in order to be able to produce thousands of micro preform parts in a short time, while generating material accumulations of the desired size, and on the other

Fig. 5. Material accumulation at the end of a thin wire: Solid-liquid interface and isothermal lines at 5 different times during solidification.

hand generating a microstructure which is well suited for the subsequent cold forming step and usage properties of the work piece. Models for microstructure evolution during solidification show a dependency of the microstructure quality on the speed V_{Γ_I} of the liquid-solid interface $\Gamma_I(t)$ and the temperature gradients near the interface [9]. A higher speed and larger gradient typically result in a microstructure which gives better forming characteristics and useful properties. Thus, the error functional has to include parts for geometry approximation, for speed, and for the microstructure generation during solidification,

$$J(u, \Omega, \theta, \mathbf{v}) = \int_0^T \|\Omega(t) - \Omega_d(t)\|^2 \tag{17}$$
$$+\lambda_1 \int_0^T \left(\|V_{\Gamma_I} - V_d\|_{\Gamma_I(t)}^2 + \|\nabla\theta - G_d\|_{\Gamma_I(t)}^2 \right) + \lambda_2 \|u\|^2.$$

The control parameters u are given by the time-dependent intensity and location of the laser spot, thus they enter the system of equations via the heating energy density $q_L(u)$. An additional control variable might be the intensity r of outer cooling, for example by adjusting the flow velocity of a cooling gas. U_{ad} is given by control restrictions which follow from limits of the process, like an upper bound of the laser power for preventing an evaporation of the material.

Adjoint Problem. For an efficient implementation of the optimization procedure, the solution of the adjoint problem will be used. Here, the system of adjoint equations includes the adjoint (linearized) Navier-Stokes system on the prescribed time dependent domain (from the forward solution), together with the (linear) Laplace-Beltrami equation on the prescribed capillary boundary. For the formulation of the adjoint Stefan problem, a regularization could be used which was used in [12] for the derivation of an aposteriori error estimate.

As in our first application, also here the efficient handling of the adjoint problem including adaptively refined meshes with remeshings for the time dependent domain and the corresponding system of solutions in the domain and on the capillary surface poses a challenge for the overall numerical optimization method.

Acknowledgments. The authors gratefully acknowledge the financial support by the DFG (German Research Foundation) for the subproject A3 within the Collaborative Research Center SFB 747 "Micro cold forming" and the project MA1657/21-3 within the Priority Program 1480 "Modeling, Simulation and Compensation of thermal effects for complex machining processes".

Furthermore, we thank the Bremen Institute for Applied Beam Technology (BIAS) and the Institute of Production Engineering and Machine Tools Hanover (IFW) for cooperation.

References

1. Bänsch, E.: Finite element discretization of the Navier-Stokes equations with a free capillary surface. Numer. Math. **88**, 203–235 (2001)
2. Bänsch, E., Paul, J., Schmidt, A.: An ALE finite element method for a coupled Stefan problem and Navier-Stokes equations with free capillary surface. Int. J. Numer. Methods Fluids **71**, 1282–1296 (2013)
3. Denkena, B., Maaß, P., Niederwestberg, D., Vehmeyer, J.: Identification of the specific cutting force for complex cutting tools under varying cutting conditions. Int. J. Mach. Tools Manuf. **3**, 82–83 (2014)
4. Denkena, B., Schmidt, A., Henjes, J., Niederwestberg, D., Niebuhr, C.: Modeling a thermomechanical NC-simulation. Procedia CIRP **8**, 69–74 (2013)
5. Denkena, B., Schmidt, A., Maaß, P., Niederwestberg, D., Niebuhr, C., Vehmeyer, J.: Prediction of temperature induced shape deviations in dry milling. Procedia CIRP **31**, 340–345 (2015)
6. Jahn, M., Brüning, H., Schmidt, A., Vollertsen, F.: Energy dissipation in laser-based free form heading: a numerical approach. Prod. Eng. Res. Dev. **8**, 51–61 (2014)
7. Jahn, M., Luttmann, A., Schmidt, A., Paul, J.: Finite element methods for problems with solid-liquid-solid phase transitions and free melt surface. PAMM **12**, 403–404 (2012)
8. Jahn, M., Schmidt, A.: Finite element simulation of a material accumulation process including phase transitions and a capillary surface. Technical report 12–03, ZeTeM, Bremen (2012)
9. Kurz, W., Fisher, J.D.: Fundamentals of Solidification. Trans Tech Publications, Switzerland-Germany-UK-USA (1986)
10. Niebuhr, C., Niederwestberg, D., Schmidt, A.: Finite element simulation of macroscopic machining processes - implementation of time-dependent domain and boundary conditions, 14 p., Berichte aus der Technomathematik 14–01, University of Bremen (2014)
11. Niederwestberg, D., Denkena, B., Böß, V., Ammermann, C.: Contact zone analysis based on multidexel workpiece model and detailed tool geometry. Procedia CIRP **4**, 41–45 (2012)
12. Nochetto, R.H., Schmidt, A., Verdi, C.: A posteriori error estimation and adaptivity for degenerate parabolic problems. Math. Comput. **69**, 1–24 (2000)
13. Schmidt, A., Siebert, K.G.: Design of Adaptive Finite Element Software: The Finite Element Toolbox ALBERTA. LNCSE, vol. 42. Springer, Heidelberg (2005)
14. Vollertsen, F., Walther, R.: Energy balance in laser based free form heading. CIRP Ann. **57**, 291–294 (2008)

Generalized Solutions of Hamilton – Jacobi Equation to a Molecular Genetic Model

Nina Subbotina[1,2]([✉]) and Lyubov Shagalova[1]

[1] Krasovskii Institute of Mathematics and Mechanics, Ural Branch of RAS,
S. Kovalevskaya str., 16, 620990 Ekaterinburg, Russia
subb@uran.ru, shag@imm.uran.ru
[2] Ural Federal University named after the first President of Russia B.N. Yeltsin,
Ekaterinburg, Russia

Abstract. A boundary value problem with state constraints is under consideration for a nonlinear noncoercive Hamilton-Jacobi equation. The problem arises in molecular biology for the Crow – Kimura model of genetic evolution. A new notion of continuous generalized solution to the problem is suggested. Connections with viscosity and minimax generalized solutions are discussed. In this paper the problem is studied for the case of additional requirements to structure of solutions. Constructions of the solutions with prescribed properties are provided and justified via dynamic programming and calculus of variations. Results of simulations are exposed.

Keywords: Hamilton – Jacobi equation · Method of characteristics · Generalized solutions · Viscosity solutions · State constraints

1 Introduction

In [1] a new way to study molecular evolution has been proposed. According to this way dynamics of the Crow – Kimura model for molecular evolution can be analyzed via the following HJE

$$\partial u/\partial t + H(x, \partial u/\partial x) = 0, \tag{1}$$

where the Hamiltonian $H(\cdot)$ has the form

$$H(x,p) = -f(x) + 1 - \frac{1+x}{2}e^{2p} - \frac{1-x}{2}e^{-2p}. \tag{2}$$

The function $f(\cdot)$ in (2) is given and called fitness. Equation (1) is considered for $t \geq 0$, $-1 \leq x \leq 1$. It is also assumed that an initial function $u_0 : \mathbb{R} \to \mathbb{R}$ is given such that

$$u(0,x) = u_0(x), \quad x \in [-1;1]. \tag{3}$$

In [1] problem (1)–(3) was studied for input data $u_0(x) = -a(x - x_0)^2$, $a > 0$, $f(x) = x^2$ and physical interpretations were used.

© IFIP International Federation for Information Processing 2016
Published by Springer International Publishing AG 2016. All Rights Reserved
L. Bociu et al. (Eds.): CSMO 2015, IFIP AICT 494, pp. 462–471, 2016.
DOI: 10.1007/978-3-319-55795-3_44

The classical method for solving PDE of the first order in Cauchy problem is the method of characteristics (see, e.g. [2]). This method reduces integration of PDEs to integration of the characteristic system of ODEs.

The characteristic system for problem (1)–(3) has the form

$$\dot{x} = H_p(x,p) = -(1+x)e^{2p} + (1-x)e^{-2p},$$
$$\dot{p} = -H_x(x,p) = f'(x) + (e^{2p} - e^{-2p})/2, \tag{4}$$
$$\dot{z} = pH_p(x,p) - H(x,p),$$

with initial conditions

$$x(0,y) = y, \quad p(0,y) = u_0'(y), \quad z(0,y) = u_0(y), \qquad y \in [-1;1]. \tag{5}$$

Here $H_x(x,p) = \partial H(x,p)/\partial x$, $H_p(x,p) = \partial H(x,p)/\partial p$, $f'(x) = \partial f(x)/\partial x$. Solutions of the system (4)–(5) are called characteristics. Components $x(\cdot,y)$, $p(\cdot,y)$ and $z(\cdot,y)$ of the solution are called state, conjugate, and value characteristics, respectively.

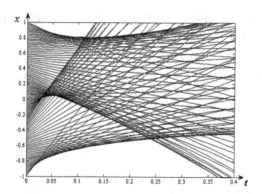

Fig. 1. State characteristics for the case $f(x) = -0.25x^2$, $u_0(x) = 0.25(x-0.5)^2 - 0.1\cos 2\pi x$.

The method of characteristics can be applied to constructions of solutions for problem (1)–(3) in such a neighborhood of the initial manifold (5) where state characteristics don't cross. As a rule, characteristics for problem (1)–(3) are nonextendable to the whole time axis and can cross each other. Moreover, there are points in strip $t \geq 0$, $-1 \leq x \leq 1$. where solution of (1)–(3) should be found, and where the state characteristics do not pass. An example of such a behavior of state characteristics is presented on Fig. 1.

So, one can see that solutions of the problem (1)–(3) should be understood in a generalized sense.

In [3], we introduced a concept of continuous generalized solutions (see Definition 1 below) and proved it's existence in problem (1)–(3) using tools of Nonsmooth Analysis and results of the Optimal Control Theory. It was also shown that the generalized solution is not unique.

In this paper, we consider problem (1)–(3) with additional requirements to the structure of solutions, see [4,5]. Namely, we need to construct a continuous solution in the strip $t \geq 0$, $-1 \leq x \leq 1$ in such a way that it coincides with a solution obtained by the method of characteristics in a domain part where the characteristics defined by (4) and (5) pass.

The paper is organized as follows. In Sect. 2, the definition of a continuous generalized solution is introduced, and the results on its existence are presented. In Sect. 3 we state the problem of constructing the generalized solution with prescribed properties, give sufficient conditions under which the problem can be solved, and formulate auxiliary results on which solving is based. A scheme for constructing the generalized solution and results of a simulation are presented in Sects. 4 and 5 respectively. And, in Sect. 6, we compare our generalized solution with viscosity solutions.

2 Continuous Generalized Solutions to the Problem with State Constraints

2.1 The Problem with State Constraints and Definition of a Generalized Solution

Let $T > 0$ be such an instant that characteristics (4), (5) are extendable up to T, and $x(\cdot, y)$, $p(\cdot, y)$, $z(\cdot, y)$ are continuous on $[0, T]$ for all $y \in [-1; 1]$. Exact estimates for intervals of extendibility are obtained in [4,6].

We consider problem (1)–(3) on the restricted closed domain

$$\overline{\Pi}_T = [0; T] \times [-1; 1],$$

and also use the notations

$$\Pi_T = (0; T) \times (-1; 1), \quad \Gamma_T = \{(t, x) | \ t \in (0, T), \ x = \pm 1\}.$$

In the HJEs' theory various concepts of generalized solutions have been introduced (see, e.g. [7–9]). Note that definitions of generalized solutions to HJEs in open areas were applied to problems with state constraints as additional requirements to solutions on the border were imposed. These requirements play a role of boundary conditions. Unfortunately, results of the theories of generalized solutions are inapplicable to the problem (1)–(3). In particular, one of the key conditions under which the known theorems on existense of a generalized viscosity solution [8,10] has been proved is the coercivity of the Hamiltonian (see (22) below). And the theory of minimax solutions [9] is not developed for problems with state constraints. So, below a new definition of a generalized solution is introduced [3]. This definition is based on the minimax and viscosity approahes and uses the following tools of nonsmooth analysis [10,11].

Let W be a set in $\mathbb{R}^2$. Denote by $\overline{W}$ the closure of this set, by $C(W)$—the class of functions continuous on the set W.

Let $u(\cdot) \in C(\overline{W})$ and $(t, x) \in \overline{W}$. The subdifferential of the function $u(\cdot)$ at (t, x) is the set

$$D^-u(t,x) = \left\{ (a,s) \in \mathbb{R} \times \mathbb{R} \;\middle|\; \liminf_{\substack{(\tau,y)\to(t,x) \\ (\tau,y)\in\overline{W}}} \frac{u(\tau,y)-u(t,x)-a(\tau-t)-s(y-x)}{|\tau-t|+|y-x|} \geq 0 \right\}.$$

The superdifferential of the function $u(\cdot)$ at (t,x) is the set

$$D^+u(t,x) = \left\{ (a,s) \in \mathbb{R} \times \mathbb{R} \;\middle|\; \limsup_{\substack{(\tau,y)\to(t,x) \\ (\tau,y)\in\overline{W}}} \frac{u(\tau,y)-u(t,x)-a(\tau-t)-s(y-x)}{|\tau-t|+|y-x|} \leq 0 \right\}.$$

Let $\text{Dif}(u)$ be the set of points where the function $u(\cdot) \in C(\overline{W})$ is differentiable. For a given set $M \subset \mathbb{R}^2$, the symbol $\text{co}M$ means its convex hull [13]. Let us define the set

$$\partial u(t,x) = \text{co}\left\{ (a,s) \;\middle|\; a = \lim_{i\to\infty} \frac{\partial u(t_i,x_i)}{\partial t}, \; s = \lim_{i\to\infty} \frac{\partial u(t_i,x_i)}{\partial x}; \right.$$

$$\left. (t_i,x_i) \to (t,x) \text{ as } i \to \infty, \quad (t_i,x_i) \in \text{Dif}(u) \right\}.$$

Definition 1. *A continuous function $u(\cdot) : \overline{\Pi}_T \to \mathbb{R}^2$ is called a generalized solution to problem (1)–(3) iff it satisfies the initial condition (3) and the following relations are true*

$$a + H(x,s) \leq 0, \quad \forall (a,s) \in D^+u(t,x), \forall (t,x) \in \Pi_T, \tag{6}$$

$$a + H(x,s) \geq 0, \quad \forall (a,s) \in D^-u(t,x), \forall (t,x) \in \Pi_T, \tag{7}$$

$$a + H(x,s) \geq 0, \quad \forall (a,s) \in D^-u(t,x) \cap \partial u(t,x), \forall (t,x) \in \Gamma_T. \tag{8}$$

2.2 Existence of Generalized Solutions

The following statement was proved in [4] by using tools of Mathematical Theory of Optimal Control [14] and the method of generalized characteristics [15,16].

Theorem 1. *Let input data $u_0(\cdot) : [-1,1] \to \mathbb{R}$ and $f(\cdot) : [-1,1] \to \mathbb{R}$ be continuously differentiable functions. Let a function $\varphi(t,x) : \mathbb{R}^2 \to \mathbb{R}$ be also continuously differentiable and satisfy the relations*

$$\varphi(0,x) = u_0(x) \;\; \forall\, x \in [-1,1];$$

$$\frac{\partial\varphi(t,\pm 1)}{\partial t} + H\left(\pm 1, \frac{\partial\varphi(t,\pm 1)}{\partial x} \right) = 0 \;\; \forall\, t \geq 0. \tag{9}$$

Then there exists a solution $u(t,x)$ of problem (1)–(3) in sense of Definition 1. The solution has the form

$$u(t,x) = \max_{x(t,y^\sharp)=x} \left[\varphi(t^\sharp,y^\sharp) + \right.$$

$$\left. + \int_{t^\sharp}^t p(\tau,y^\sharp)H_p(x(\tau,y^\sharp),p(\tau,y^\sharp)) - H(x(\tau,y^\sharp),p(\tau,y^\sharp))d\tau \right], \tag{10}$$

for all $(t, x) \in \overline{\Pi}_T$, where $t^\sharp \in [0, T]$. If $t^\sharp = 0$ then $y^\sharp = y \in [-1, 1]$; if $t^\sharp > 0$, then $y^\sharp = \pm 1$. The functions $(x(\cdot, y^\sharp), p(\cdot, y^\sharp)) : [t^\sharp, t] \to \mathbb{R}^2$ are solutions for the system composed of the first two equations of characteristic system (4) with initial conditions

$$x(t^\sharp, y^\sharp) = y^\sharp, \quad p(t^\sharp, y^\sharp) = \frac{\partial \varphi(t^\sharp, y^\sharp)}{\partial y} = p_0(t^\sharp, y^\sharp).$$

To obtain $u(t, x)$ in accordance with (10), one should consider the set of all state characteristics $x(\cdot, y^\sharp)$ passing through the point (t, x), namely, $x(t, y^\sharp) = x$. Note that the generalized solution to problem (1)–(3) is not unique because of wide choice of functions $\varphi(\cdot)$ in Theorem 1.

3   Solutions with Prescribed Properties

Here, we consider a problem to construct the generalized solution of some particular structure.

Let $x^-(t) = x(t, -1)$ and $x^+(t) = x(t, +1)$, $t \in [0, T]$ be the state characteristics started at $t = 0$ from the points $x = -1$ and $x = 1$, respectively. Below, we assume that the following condition is satisfied.

A. For the state characteristics $x(\cdot, y)$ with initial conditions (5) at $t = 0$ the inequalities are valid

$$-1 \leq x^-(t) \leq x(t, y) \leq x^+(t) \leq 1, \quad \forall y \in [-1, 1], \forall t \in [0, T].$$

Define the subdomains

$$\begin{aligned} G_+ &= \{(t, x) \mid t \in [0, T], \ x^+(t) \leq x \leq 1\}, \\ G_0 &= \{(t, x) \mid t \in [0, T], \ x^-(t) \leq x \leq x^+(t)\}. \\ G_- &= \{(t, x) \mid t \in [0, T], \ -1 \leq x \leq x^-(t)\}. \end{aligned} \tag{11}$$

So, under the assumption **A**, we get

$$\overline{\Pi}_T = G_+ \cup G_0 \cup G_-.$$

The goal of the work is to construct the generalized solution to problem (1)–(3) such that it has the following form in G_0:

$$u(t, x) = \max_{x(t, y) = x} [u_0(y) + \int_0^t p(\tau) H_p(x(\tau), p(\tau)) - H(x(\tau), p(\tau)) d\tau], \tag{12}$$

where $x(t) = x(t, y)$, $p(t) = p(t, y)$, $t \geq 0$, are state and conjugate characteristics, respectively, which satisfy at $t = 0$ the initial conditions

$$x(0, y) = y, \quad p(0, y) = u_0'(y), \quad y \in [-1, 1].$$

3.1 Sufficient Conditions

To solve the problem (1)–(3) with the requirement (12) we introduce the following additional assumptions on input data.

B1. The derivative $u_0'(\cdot) : [-1, 1] \to \mathbb{R}$ is continuous and satisfies the inequalities

$$u_0'(1) < 0, \quad u_0'(-1) > 0.$$

B2. The derivative $f'(\cdot) : [-1, 1] \to \mathbb{R}$ is continuous and monotone nondecreasing. It satisfies the inequalities

$$2f'(1) + e^{2u_0'(1)} < e^{-2u_0'(1)}, \quad -2f'(-1) + e^{-2u_0'(-1)} < e^{2u_0'(-1)}.$$

3.2 Auxiliary Problems of Calculus of Variations

Consider the following two problems of Calculus of Variations over the set of all continuously differentiable functions $x(\cdot) : [0, T] \to \mathbb{R}$

$$I(x(\cdot)) = \int_0^{\bar{t}} H^*(x(\tau), \dot{x}(\tau)) d\tau \mapsto \max \tag{13}$$

$$x_1(0) = 1, \quad x_1(\bar{t}) = \bar{x}, \quad (\bar{t}, \bar{x}) \in G_+; \tag{14}$$

$$x_2(0) = -1, \quad x_2(\bar{t}) = \bar{x}, \quad (\bar{t}, \bar{x}) \in G_- \tag{15}$$

where

$$H^*(x(t), \dot{x}(t)) = \inf_{p \in R} [p\dot{x}(t) - H(x(t), p)]. \tag{16}$$

The following assertions are proven in [4,5], where conditions **B1-B2** are essential.

Theorem 2. *For any interior point $(\bar{t}, \bar{x}) \in G_+$ there exists a unique extremal $x = x(t)$ of the problem (13), (14), (16). The extremal coinsides with a state characteristic $x(\cdot; 0, 1, p_0)$ satisfying the initial conditions $x(0) = 1, p(0) = p_0 \in (-\infty, u_0'(1))$ where initial value p_0 can be defined uniquely from the condition $x(\bar{t}) = \bar{x}$. The functional (13) attains its strong maximum at this extremal.*

Theorem 3. *For any interior point $(\bar{t}, \bar{x}) \in G_-$ there exists a unique extremal $x = x(t)$ of the problem (13), (15), (16). The extremal coinsides with a state characteristic $x(\cdot; 0, -1, p_0)$ satisfying the initial conditions $x(0) = -1, \quad p(0) = p_0 \in (u_0'(-1), \infty)$ where initial value p_0 can be defined uniquely from the condition $x(\bar{t}) = \bar{x}$. The functional (13) attains its strong maximum at this extremal.*

Theorem 4. *For any boundary point $(\bar{t}, 1), 0 < \bar{t} \le T$ in G_+, the maximum of functional (13), (16) is attained at a state characteristic $x(\cdot; 0, 1, p_0)$, such that the characteristics $x(\cdot; 0, 1, p_0), p(\cdot; 0, 1, p_0)$ are nonextendable from the interval $[0, t^*(p_0) = \bar{t})$ to the right. The maximum of functional (13), (16) is equal to $I(x(\cdot; 0, 1, p_0)) = (f(1) - 1)\bar{t}$.*

Theorem 5. *For any boundary point* $(\bar{t}, -1), 0 < \bar{t} \leq T$ *in* G_-, *the maximum of the functional (13)–(16) is attained at a state characteristic* $x(\cdot; 0, -1, p_0)$, *such that characteristics* $x(\cdot; 0, -1, p_0), p(\cdot; 0, -1, p_0)$ *are nonextendable from the interval* $[0, t^*(p_0) = \bar{t})$ *to the right. The maximum of functional (13), (16) is equal to* $I(x(\cdot; 0, -1, p_0)) = (f(-1) - 1)\bar{t}$.

4   Construction of the Generalized Solution

The generalized solution of the problem (1)–(3) with the prescribed property (12) has the following form in G_0:

$$u(t, x) = \max_{x(t,y)=x} [\int_0^t p(\tau) H_p(x(\tau), p(\tau)) - H(x(\tau), p(\tau))d\tau + u_0(y)], \qquad (17)$$

where $x(t) = x(t, y), \; p(t) = p(t, y), \; t \geq 0$, are state and conjugate characteristics satisfied initial conditions

$$x(0, y) = y, \quad p(0, y) = \partial u_0(y)/\partial x, \quad y \in [-1, 1].$$

Let $(t_*, x_*) \in G_+$ and $x_* < 1$. We assign

$$u(t_*, x_*) = u_0(1) + \int_0^{t_*} \left[p(\tau) H_p(x(\tau), p(\tau)) - H(x(\tau), p(\tau)) \right] d\tau, \qquad (18)$$

where $x(t) = x^+(t, p_0(t_*, x_*)), \; p(t) = p^+(t, p_0(t_*, x_*))$ is the solution of the problem of Calculus Variations (13), (14), (16).

For $x_* = 1, \; 0 \leq t_* \leq T$, we set

$$u(t_*, 1) = u_0(1) + (f(1) - 1)t_*. \qquad (19)$$

Let $(t_*, x_*) \in G_-$ and $x_* > -1$. We assign

$$u(t_*, x_*) = u_0(-1) + \int_0^{t_*} \left[p(\tau) H_p(x(\tau), p(\tau)) - H(x(\tau), p(\tau)) \right] d\tau, \qquad (20)$$

where $x(t) = x^-(t, p_0(t_*, x_*)), \; p(t) = p^-(t, p_0(t_*, x_*))$ is the solution of the problem of Calculus Variations (13), (15), (16).

For $x_* = -1, \; 0 \leq t_* \leq T$, we set

$$u(t_*, -1) = u_0(-1) + (f(-1) - 1)t_*. \qquad (21)$$

So, we have defined function $u(\cdot)$ for all points from $\overline{\Pi}_T$ by relations (17)–(21). Following the Cauchy method of characteristics [2], one can show that $u(\cdot)$ is continuously differentiable at any interior point $(t, x) \in G_+ \cup G_-$, and the gradient of $u(\cdot)$ is equal to $(-H(x, p(t)), p(t))$. Theorems 2–5 imply that $u(\cdot)$ is continuous in $\overline{\Pi}_T$, and inequalities (6)–(7) are valid. Below, in Sect. 6.2, we will show that $D^- u(t, x) \cap \partial u(t, x) = \varnothing$, $(t, x) \in \Gamma_T$. So, $u(\cdot)$ is a generalized solution of problem (1)–(3) in sense of Definition 1. It follows from (17) that $u(\cdot)$ satisfies the prescribed property (12).

5   Numerical Example

Results of simulation for the input data $u_0(x) = -0.02x^2 + 0.001 \cos 2\pi x$, $f(\cdot) = -0.5x^2$ are presented in Fig. 2. One can easily check that these input data satisfy the conditions **B1, B2**.

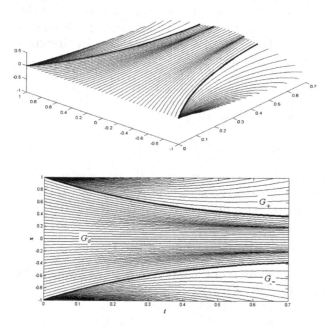

Fig. 2. The graph of the generalized solution and its projection on the (t, x)-plane for the input data $u_0(x) = -0.02x^2 + 0.001 \cos 2\pi x$, $f(x) = -0.5x^2$.

6   Comparison with Viscosity Solution

One can see that Definition 1 coincides with the definition of viscosity solution in the interior points of the region $\overline{\Pi}_T$. The difference between these definitions is evident at boundary points, namely, on the set Γ_T. In condition (8), the inequality holds for such points (a, s) of the subdifferential $D^- u(t, x)$ which at the same time belong to the set $\partial u(t, x)$. In contrast to Definition 1, the notion of viscosity solution [10] for Eq. (1) on the set $\overline{\Pi}_T$ requires that this solution satisfies inequality (8) at the boundary points $(t, x) \in \Gamma_T$ for all $(a, s) \in D^- u(t, x)$.

6.1   On Coercivity Condition

Conditions ensuring existence and uniqueness for viscosity solutions on the compact domain were obtained in [10]. One of the key conditions for the existence of viscosity solutions (see [10, 12]) is the coercivity of the Hamiltonian:

$$H(x, p) \to +\infty \quad \text{as} \quad |p| \to \infty \tag{22}$$

It can be easily checked that the Hamiltonian (2) does not satisfy condition (22), for example, at $x = 1$ and $x = -1$. Therefore, the existence theorems for viscosity solutions in the sense of paper [10] cannot be used in the problem (1)–(3).

Moreover, the notion of generalized viscosity solution is inapplicable to the initial value problem (1)–(3) on the compact set $\overline{\Pi}_T$: If a viscosity solution would satisfy the condition $D^- u(t_*, x_*) \neq \varnothing$ at some point $(t_*, x_*) \in \Gamma_T$, then inequality (8) would not hold in this case. Indeed, let $0 \leq t_* \leq T$ and, for definiteness, $x_* = 1$. Let $(a, s) \in D^- u(t_*, x_*)$. One can use the above definition of subdifferential to get the following inclusion:

$$(a, s + k) \in D^- u(t_*, x_*) \quad \forall k \geq 0.$$

So, if condition (8) would hold then

$$a + H(1, s + k) = a - f(x) + 1 - e^{2(s+k)} \geq 0, \qquad \forall k \geq 0,$$

which is obviously unfair. Therefore, we use the intersection of the subdifferential with the set $\partial u(t, x)$ (see condition (8)) in Definition 1 of a continuous generalized solution to the initial value problem (1)–(3) on the compact set $\overline{\Pi}_T$.

6.2   Structure of Subdifferentials on the Border

Let's consider the structure of the sets $D^- u(t, x)$ and $\partial u(t, x)$ for the function $u(\cdot)$ defined by (17)–(20) if $(t, x) \in \Gamma_T$.

In the case $0 < t < T$, $x = 1$, we have
$D^- u(t, x) = D^- u(t, 1) = \{(f(1) - 1, s) | s \in R, s \geq 0\}$,
$\partial u(t, x) = \partial u(t, 1) = \{(-H(1, -\infty), -\infty)\} = \{(f(1) - 1, -\infty)\}$.

In the case $0 < t < T$, $x = -1$, we have
$D^- u(t, x) = D^- u(t, -1) = \{(f(-1) - 1, s) | s \in R, s \leq 0\}$,
$\partial u(t, x) = \partial u(t, -1) = \{(-H(-1, \infty), \infty)\} = \{(f(-1) - 1, \infty)\}$.

Thus, for generalized solution (17)–(20)

$$D^- u(t, x) \cap \partial u(t, x) = \varnothing, \quad (t, x) \in \Gamma_T.$$

Acknowledgments. This work was partially supported by the Russian Foundation for Basic Research (project no. 14-01-00168) and the Complex Program of Ural Branch of RAS (project no. 15-16-1-11).

References

1. Saakian, D.B., Rozanova, O., Akmetzhanov, A.: Dynamics of the Eigen and the Crow-Kimura models for molecular evolution. Phys. Rev. E - Stat. Nonlinear Soft Matter Phys. **78**(4), 041908 (2008)
2. Courant, R., Hilbert, D.: Methods of Mathematical Physics, vol. 2. Interscience, New York (1962)

3. Subbotina, N.N., Shagalova, L.G.: On a solution to the Cauchy problem for the Hamilton-Jacobi equation with state constraints. Trudy Inst. Mat. i Mekh. UrO RAN **17**(2), 191–208 (2011). (in Russian)
4. Subbotina, N.N., Shagalova, L.G.: Construction of a continuous minimax/viscosity solution of the Hamilton Jacobi Bellman equation with nonextendable characteristics. Trudy Inst. Mat. i Mekh. UrO RAN **20**(4), 247–257 (2014). (in Russian)
5. Subbotina, N.N., Shagalova, L.G.: On the continuous extension of a generalized solution of the Hamilton-Jacobi equation by characteristics that form a central field of extremals. Trudy Inst. Mat. i Mekh. UrO RAN **21**(2), 220–235 (2015). (in Russian)
6. Shagalova, L.: Applications of dynamic programming to generalized solutions for Hamilton - Jacobi equations with state constraints. SOP Trans. Appl. Math. **1**(2), 70–83 (2014)
7. Kruzhkov, S.N.: Generalized solutions of nonlinear equations of the first order with several variables. I. Matematicheskii Sbornik (N.S.) **70(112)**(3), 394–415 (1966). (in Russian)
8. Crandall, M.G., Lions, P.-L.: Viscosity solutions of Hamilton-Jacobi equations. Trans. Am. Math. Soc. **277**(1), 1–42 (1983)
9. Subbotin, A.I.: Generalized Solutions of First Order PDEs: The Dynamical Optimization Perspective. Birkhauser, Boston (1995)
10. Capuzzo-Dolcetta, I., Lions, P.-L.: Hamilton-Jacobi equations with state constraints. Trans. AMS **318**(2), 643–683 (1990)
11. Clarke, F.: Optimization and Nonsmooth Analysis. Wiley, New York (1983)
12. Crandall, M.G., Newcomb, R.: Viscosity solutions of Hamilton-Jacobi equations at the boundary. Proc. AMS **94**(2), 283–290 (1985)
13. Rockafellar, R.T.: Convex Analysis. Princeton University Press, Princeton (1970)
14. Pontryagin, L.S., Boltyanskii, V.G., Gamkrelidze, R.V., Mishchenko, E.F.: The Mathematical Theory of Optimal Processes. Pergamon Press, Oxford (1964)
15. Subbotina, N.N.: The method of characteristics for Hamilton-Jacobi equation and its applications in dynamical optimization. Modern Math. Appl. **20**, 2955–3091 (2004)
16. Subbotina, N.N., Kolpakova, E.A., Tokmantsev, T.B., Shagalova, L.G.: The Method of Characteristics for Hamilton-Jacobi-Bellman Equation. RIO UrO RAN, Ekaterinburg (2013). (in Russian)

Dynamic Programming to Reconstruction Problems for a Macroeconomic Model

Nina N. Subbotina$^{(\boxtimes)}$, Timofey B. Tokmantsev, and Eugenii A. Krupennikov

Krasovskii Institute of Mathematics and Mechanics,
Ural Branch of the Russian Academy of Sciences, Ekaterinburg, Russia
subb@uran.ru

Abstract. Perturbed inverse reconstruction problems for controlled dynamic systems are under consideration. A sample history of the actual trajectory is known. This trajectory is generated by a control, which isn't known. Moreover, the deviation of the samples from the actual trajectory satisfies the known estimate of the sample error. The inverse problem with perturbed (inaccurate) sample of trajectory consists of reconstructing trajectories which are close to the actual trajectory in C. Controls generating the trajectories should be close in L_2 to the normal control generating the actual trajectory and have the least norm in L_2. A numerical method for solving this problem is suggested. The application of the suggested method is illustrated by the graphics.

Keywords: Inverse problem · Positive and negative discrepancy · Optimal control problem · Dynamic programming

1 Introduction

A model of macroeconomics is considered due to works by Al'brekht [1]. The model has the form of two nonlinear ordinary differential equations. The right-hand sides of the equations depend on control parameters. The rate of taxation, the refunding rate and the currency exchange course are included in control parameters because they determine economical conditions for production activity. A sample history of the actual trajectory of the model is known. A numerical method is suggested and verified to reconstruct the actual trajectory and the control generating it. It is based on the method of the dynamic programming. Results of numerical calculations of the solutions of the inverse problem are represented for statistic data obtained from a reports of companies sent to local statistic offices in Russia.

2 Macroeconomic Model

Consider a model of a macroeconomic system, where the symbol x_1 denotes the gross product, x_2 denotes production costs, G denotes profits.

© IFIP International Federation for Information Processing 2016
Published by Springer International Publishing AG 2016. All Rights Reserved
L. Bociu et al. (Eds.): CSMO 2015, IFIP AICT 494, pp. 472–481, 2016.
DOI: 10.1007/978-3-319-55795-3_45

Let dynamics of $x_1(t)$, $x_2(t)$ be of the form

$$\frac{dx_1}{dt} = u_1(t)\frac{\partial G(x_1,x_2)}{\partial x_1},$$
$$\frac{dx_2}{dt} = u_2(t)\frac{\partial G(x_1,x_2)}{\partial x_2} \tag{1}$$

on a time interval $t \in [0,T]$. Here $u_1(t)$, $u_2(t)$ are control parameters, satisfying the geometric restrictions

$$|u_1| \le U_1, \qquad |u_2| \le U_2, \tag{2}$$

where $U_1 > 0$, $U_2 > 0$ are constants.

3   Known Data

We have got the following statistic data in the form of a table of parameters x_1^*, x_2^*, G^* measured at given instants t_i, $t_i = 0, 1, \ldots, N$, $t_0 = 0$, $t_N = T$,

$$
\begin{array}{lll}
x_1^*(t_0), & x_1^*(t_1),\ldots, & x_1^*(t_N), \\
x_2^*(t_0), & x_2^*(t_1),\ldots, & x_2^*(t_N), \\
G^*(t_0), & G^*(t_1),\ldots, & G^*(t_N),
\end{array}
$$

where $x_1^*(t_i)$, $x_2^*(t_i)$ are measurements of the actual trajectory $x_{1*}(\cdot)$, $x_{2*}(\cdot)$ of the system (1) on the interval $[0,T]$.

4   Hypothesis

Following the Albrekht's works, we assume that the mathematical model of the measured dynamics meets the following assertions

- the structure of the function $G(x_1,x_2)$ has the form of the polynomial

$$G(x_1,x_2) = x_1 x_2 (a_0 + a_1 x_1 + a_2 x_2). \tag{3}$$

- the given statistic data are measurements of the actual trajectory $x_*(\cdot) = (x_{1*}(\cdot), x_{2*}(\cdot))$ and profit function $G(x_*(\cdot))$ with errors, while estimate δ on admissible errors is known.

$$|x_{1*}(t_i) - x_1^*(t_i)| \le \delta, \qquad |x_{2*}(t_i) - x_2^*(t_i)| \le \delta,$$
$$|G(x_{1*}(t_i), x_{2*}(t_i)) - G^*(t_i)| \le \delta, \quad i = 0, 1, \ldots, N. \tag{4}$$

- such smooth continuous interpolations $y(\cdot) = (y_1(\cdot), y_2(\cdot))$ of the data $x^*(t_i) = (x_1^*(t_i), x_2^*(t_i))$, $i = 0, 1, \ldots, N$ are defined, that

$$\left|\frac{d^2 y_i(t)}{dt^2}\right| \le K, \quad K > 0, \ t \in [0,T], \ i = 1, 2, . \tag{5}$$

$$\|y(\cdot) - x_*(\cdot)\|_c \to 0, \quad \text{as } \delta \to 0. \tag{6}$$

5   Reconstruction Problems

The inverse problems are identification problem and reconstruction problem for the model, which supposes reconstructing such trajectories $x^\delta(\cdot)$ of system (1) generated by measurable controls $u^\delta(\cdot)$, satisfying (2), that

$$\|x^\delta(\cdot) - x_*(\cdot)\|_C = \max_{t \in [0,T]} \|x^\delta(t) - x_*(t)\| \to 0, \text{ as } \delta \to 0;$$

$$\|u^\delta(\cdot) - u_*(\cdot)\|_{L_2}^2 = \int_0^T \|u^\delta(t) - u_*(t)\|^2 dt \to 0, \text{ as } \delta \to 0;$$

where $x_*(\cdot) = (x_{1*}(\cdot), x_{2*}(\cdot))$ is the actual trajectory on $[0,T]$ generated by "normal" control $u_*(\cdot) = (u_{1*}(\cdot), u_{2*}(\cdot))$, which has the minimal norm in $L_2([0,T], R^2)$. The method suggested below is based on the dynamic programming [2] for auxiliary optimal control problems. It can be interpreted as a modification of Tikhonov method [3]. The other approach to solutions of the inverse problems with the help of optimal feedbacks [4] in auxiliary optimal control problems was suggested in works by Osipov and Kryazhimskii [5].

6   Identification Problem for the Function $G(x_1, x_2)$

At first we consider the identification problem for parameters a_0, a_1, a_2 of the polynomial

$$G(x_1, x_2) = x_1 x_2 (a_0 + a_1 x_1 + a_2 x_2)$$

to obtain the best correspondence with the given statistic materials.

In order to do this, we apply the least square method to the statistic data

$$\sum_{i=0}^N [G^*(t_i) - G(x_1^*(t_i), x_2^*(t_i))]^2 \longrightarrow \min_{(a_0, a_1, a_2)}.$$

7   Auxiliary Optimal Control Problems (AOCPs)

We introduce the following AOCPs to solve the reconstruction problem. Consider dynamics of the form

$$\frac{dx_1}{dt} = u_1 \frac{\partial G(x_1, x_2)}{\partial x_1},$$

$$\frac{dx_2}{dt} = u_2 \frac{\partial G(x_2, x_2)}{\partial x_2}, \qquad (7)$$

$$t \in [0,T], \qquad u = (u_1, u_2) \in P,$$

$$P = \{|u_1| \le U_1, \qquad |u_2| \le U_2, \}. \qquad (8)$$

The set of admissible controls is defined as

$$U_{[0,T]} = \{\forall u(\cdot) \colon [0,T] \to P \text{ — measurable}\}.$$

We introduce the α-regularized positive discrepancy functional

$$I^+_{0,x^0_1,x^0_2}(u(\cdot)) = \int_0^T \frac{[(y_1(t) - x_1(t))^2 + (y_2(t) - x_2(t))^2]}{2}$$

$$+ \alpha^2 \frac{(u_1^2(t) + u_2^2(t))}{2} dt, \tag{9}$$

where α is a small parameter. The functions $y_1(\cdot), y_2(\cdot)$ are interpolations of statistic data.

We also consider the α-regularized negative discrepancy functional

$$I^-_{0,x^0_1,x^0_2}(u(\cdot)) = \int_0^T -\frac{[(y_1(t) - x_1(t))^2 + (y_2(t) - x_2(t))^2]}{2}$$

$$+ \alpha^2 \frac{(u_1^2(t) + u_2^2(t))}{2} dt. \tag{10}$$

8   Optimal Results in AOCPs

Let small parameters $\alpha > 0, \delta > 0$ be fixed and interpolations $y_1(\cdot), y_2(\cdot)$ of the statistic data be known. The aim of the AOCPs at an initial state $t = 0$, $x_1(0) = x^0_1$, $x_2(0) = x^0_2$ is to minimize the cost functionals (10), (9) under the condition

$$x_1(T) = y_1(T), \quad x_2(T) = y_2(T). \tag{11}$$

The optimal results in the class $U_{[0,T]}$ are equal to

$$V^\pm(0, x^0_1, x^0_2) = \inf_{u(\cdot) \in U_{[0,T]}} I^\pm_{t_0,x^0_1,x^0_2}(u(\cdot)). \tag{12}$$

8.1   Hamiltonian

Let's consider the AOCP for the negative discrepancy functional (10). Let us denote

$$\omega_1(x) = \omega_1(x_1, x_2) = \frac{\partial G(x_1(t), x_2(t))}{\partial x_1} = a_0 x_2 + 2a_1 x_1 x_2 + a_2 x_2^2,$$

$$\omega_2(x) = \omega_2(x_1, x_2) = \frac{\partial G(x_1(t), x_2(t))}{\partial x_2} = a_0 x_1 + a_1 x_1^2 + 2a_2 x_1 x_2.$$

$$H^\alpha(t, x_1, x_2, s_1, s_2) = \min_{u \in P} \Big[s_1 u_1 \omega_1(x_1, x_2) + s_2 u_2 \omega_2(x_1, x_2)$$

$$+ \frac{\alpha^2(u_1^2 + u_2^2)}{2} - \frac{(x_1 - y_1(t))^2 + (x_2 - y_2(t))^2}{2} \Big]$$

$$= \Big[s_1 u_1^0 \omega_1(x_1, x_2) + s_2 u_2^0 \omega_2(x_1, x_2) + \frac{\alpha^2(u_1^{0^2} + u_2^{0^2})}{2}$$

$$- \frac{(x_1 - y_1(t))^2 + (x_2 - y_2(t))^2}{2} \Big]. \tag{13}$$

where for $i = 1, 2$,

$$u_i^0(x, s) = \begin{cases} -U_i, & \text{if} \quad -\frac{s_i \omega_i(x(t))}{\alpha^2} \leq -U_i, \\ -\frac{s_1 \omega_i(x(t))}{\alpha^2}, & \text{if} \quad -\frac{s_1 \omega_i(x(t))}{\alpha^2} \in [-U_i, U_i], \\ U_i, & \text{if} \quad -\frac{s_1 \omega_i(x(t))}{\alpha^2} \geq U_i. \end{cases}$$

So, for the simple case

$$u_i^0(x, s) \in [-U_i, U_i], \qquad i = 1, 2, \tag{14}$$

we get Hamiltonian of the form

$$H^\alpha(t, x_1, x_2, s_1, s_2) = -\frac{1}{2\alpha^2}(s_1^2 + s_2^2) - \frac{(x_1 - y_1(t))^2 + (x_2 - y_2(t))^2}{2}.$$

8.2   Characteristics

Necessary optimality conditions for the AOCPs has the following form [6,7]: the characteristic system

$$\frac{dx_i}{dt} = \frac{\partial H^\alpha(t, x, s)}{\partial s_i}, \quad \frac{ds_i}{dt} = -\frac{\partial H^\alpha(t, x, s)}{\partial x_i}, \quad i = 1, 2, \quad t \in [0, T], \tag{15}$$

and the boundary conditions

$$x_i(T) = y_i(T), \quad s_i(T) = \xi_i, \quad \left| \frac{\omega_i^2(x(T))\xi_i}{\alpha^2} - \dot{y}_i(T) \right| \leq \delta, \quad i = 1, 2. \tag{16}$$

8.3   Characteristics for the Simple Case

Restrictions U_1, U_2 for admissible controls are usually unknown. To simplify the explanations we assume that U_1, U_2 are large enough to let interpolations $y(t)$ provide the simple case (14) with boundary conditions (16).

The characteristic system for the simple case has the form:

$$\begin{aligned} \frac{dx_1(t)}{dt} &= -\frac{s_1(t)}{\alpha^2} \omega_1^2(x_1(t), x_2(t)), \\ \frac{dx_2(t)}{dt} &= -\frac{s_2(t)}{\alpha^2} \omega_2^2(x_1(t), x_2(t)), \\ \frac{ds_1(t)}{dt} &= x_1(t) - y_1(t) \\ &+ \frac{s_1^2(t)}{\alpha^2} F_1(x_1(t), x_2(t)) + \frac{s_2^2(t)}{\alpha^2} F_2(x_1(t), x_2(t)), \\ \frac{ds_2(t)}{dt} &= x_2(t) - y_2(t) \\ &+ \frac{s_1^2(t)}{\alpha^2} F_3(x_1(t), x_2(t)) + \frac{s_2^2(t)}{\alpha^2} F_4(x_1(t), x_2(t)), \end{aligned} \tag{17}$$

where

$$\begin{aligned} F_1(x_1, x_2) &= 2a_1 x_2 (a_0 x_2 + 2a_1 x_1 x_2 + a_2 x_2^2), \\ F_2(x_1, x_2) &= (a_0 + 2a_1 x_1 + 2a_2 x_2)(a_0 x_1 + a_1 x_1^2 + 2a_2 x_1 x_2), \\ F_3(x_1, x_2) &= (a_0 + 2a_1 x_1 + 2a_2 x_2)(a_0 x_2 + 2a_1 x_1 x_2 + a_2 x_2^2), \\ F_4(x_1, x_2) &= 2a_2 x_1 (a_0 x_1 + a_1 x_1^2 + 2a_2 x_1 x_2), \end{aligned}$$

boundary conditions

$$x_1(T) = y_1(T), \ x_2(T) = y_2(T), \tag{18}$$
$$\xi_1^- \leq s_1(T) = \xi_1 \leq \xi_1^+, \ \xi_2^- \leq s_2(T) = \xi_2 \leq \xi_2^+,$$

where

$$\xi_i^- = -\frac{\dot{y}_i(T)\alpha^2}{\omega_i(y(T))^2} - \frac{\delta\alpha^2}{\omega_i(y(T))^2},$$
$$\xi_i^+ = -\frac{\dot{y}_i(T)\alpha^2}{\omega_i(y(T))^2} + \frac{\delta\alpha^2}{\omega_i(y(T))^2}, \tag{19}$$
$$i = 1, 2.$$

9   Solutions of Inverse Problems

Let us pick such characteristics (15)–(19) $x_\delta^\alpha(\cdot)$ and the realizations of extremal feedbacks $u_\delta^\alpha[t] = u^\alpha(t, x_\delta^\alpha(t), s_\delta^\alpha(t))$, generating them, which satisfy the relations:

$$\|x(0,\xi) - y(0)\| \leq \alpha + \delta,$$
$$I_{0,x_\delta^\alpha(0)}^\pm(u_\delta^\alpha[\cdot]) = \min_{\|x(0,\xi)-y(0)\|\leq\alpha+\delta} I_{0,x(0)}^\pm(u^\alpha(\cdot)) = V^\pm(0, x_\delta^\alpha(0)),$$
$$u^\alpha(t) = u^\alpha(t, x(t,\xi), s(t,\xi)), \quad t \in [0,T]. \tag{20}$$

We have got that these characteristics $x_\delta^\alpha(\cdot, \xi)$ and controls $u_\delta^\alpha[\cdot]$, generating them, provide solutions to the inverse problems [8–10].

9.1   Assumptions

A1   Such constants $\alpha_0 > 0$, $\delta_0 > 0$ exist that state characteristics $x_1(t, \xi)$ and $x_2(t, \xi)$ of the form (15)–(19) for all $t \in [0, T]$ belong to the compact set Φ:

$$\Phi \supset \Phi(\delta, \alpha) \quad \forall \, \delta, \, \alpha: \quad 0 < \delta \leq \delta_0, \quad 0 < \alpha \leq \alpha_0, \tag{21}$$
$$\Phi(\delta, \alpha) = \Big\{ (t, x): t \in [0, T], \ x = x(t, \xi),$$
$$x(T, \xi) = y(T), \ \left| \frac{\omega_i^2(x(T))\xi_i}{\alpha^2} - \dot{y}_i(T) \right| \leq \delta, \quad i = 1, 2 \Big\}.$$

A2   For $(x_1, x_2) \in \Phi$ such constants $\underline{\omega}_i > 0$, $\bar{\omega}_i > 0$, $i = 1, 2$ exist, that

$$0 < \underline{\omega}_1^2 \leq \omega_1^2(x_1(t), x_2(t)) \leq \bar{\omega}_1^2, \, 0 < \underline{\omega}_2^2 \leq \omega_2^2(x_1(t), x_2(t)) \leq \bar{\omega}_2^2, \quad t \in [0, T].$$

9.2   Note

In the example below, one can choose such $\alpha_0 > 0$, $\delta_0 > 0$, $r > 0$, that

$$\Phi = \Phi^r = \{(t, x): t \in [0, T], \ \|x - y(t)\| \leq r\},$$
$$\min_{0 \leq t \leq T} y_i(t) > 3r > 0, \quad i = 1, 2,$$

and assumptions *A1–A2* are true.

9.3   Main Result

Let us consider AOCPs for the system (7), (8) at initial states

$$x(0) \in \{x : \|x - y(0)\| \le \delta + \alpha\}$$

where the aim is to reach the target set $\{T, x = y(T)\}$ and minimize the functional (10).

The following assertions are proven [9, 10].

Lemma 1. *Let $x_\delta^\alpha(t)$ be a solution of the AOCP (7), (8), (10). Let $u_\delta^\alpha(t)$ be a control generating $x_\delta^\alpha(t)$. If conditions **A1–A2** are true in the problem, then such constant $c > 0$ exists that the following estimate takes place:*

$$I_{0,x_\delta^\alpha(0)}(u_\delta^\alpha(\cdot)) \le I_{0,x_*(0)}(u_*(\cdot)) + \zeta(\alpha, \delta), \quad \zeta(\alpha, \delta) = c\delta(\delta^2 + \alpha^2 U_*^2),$$

where $U_* = \max\{U_1, U_2\}$.

We introduce the functions

$$\phi(\alpha, \delta, h) = TMh\left(\frac{TMh}{2} + 2\delta + \alpha + \zeta(\alpha, \delta)\right), \qquad \rho(h) = nU_*T(K + M)h,$$

where K, M are constant parameters.

Let us denote numerical approximations of the solution $x_\delta^\alpha(\cdot)$, $u_\delta^\alpha(\cdot)$ of AOCP (7), (8), (10) as $x_h(\cdot)$, $u_h(\cdot)$.

Theorem 1. *Let conditions **A1** – **A2** be true in AOCP (7), (8), (10). Then there exists such constants $M > 0$, $K > 0$ and parameters $h = h(\delta) > 0$, $\alpha = \alpha(\delta) > 0$, $\delta > 0$, satisfying the conditions $\lim_{\delta \to 0} h(\delta) = 0$, $\lim_{\delta \to 0} \alpha(\delta) = 0$,*

$$\lim_{\delta \to 0} \frac{2}{\alpha^2}\left(\phi(\alpha, \delta, h) + \rho(h) + \frac{T}{2}(Mh + \alpha + 2\delta + \zeta(\alpha, \delta))^2\right) = 0, \qquad (22)$$

that the following relations are true

$$\lim_{\delta \to 0} \|x_{h(\delta)}(\cdot) - x_*(\cdot)\|_C = 0, \quad \lim_{\delta \to 0} \|u_{h(\delta)}(\cdot) - u_*(\cdot)\|_{L_2} = 0.$$

10   Numerical Experiments

Results of application of the suggested numerical method via AOCP with the functional $I_{(0,x(0))}^-(u(\cdot))$ are exposed on the Figs. 1, 2, 3, 4, 5 and 6 below.

Note that the results obtained via AOCP with the functional $I_{(0,x(0))}^+(u(\cdot))$ are not so satisfying (see Figs. 7 and 8). This is because of the properties of characteristics in the considered AOCPs.

We used the data on the industry of the Ural Region in Russia for the period 1970–1985 (10000 Rubles = 1) due to paper [1]:

t	Year	Gross regional product x_1^*	Costs x_2^*	Profit G^*
0	1970	37.88	21.69	6.17
1	1971	40.63	23.70	6.31
2	1972	43.25	25.45	6.68
3	1973	46.00	27.30	6.98
4	1974	49.33	29.44	7.04
5	1975	53.04	32.16	7.27
6	1976	57.03	35.01	7.62
7	1977	59.85	36.92	8.00
8	1978	62.72	38.69	8.27
9	1979	63.45	38.76	8.42
10	1980	65.74	39.96	8.61
11	1981	65.90	39.75	8.21
12	1982	69.22	41.31	9.65
13	1983	64.52	37.86	9.28
14	1984	71.03	42.04	10.26
15	1985	74.69	45.05	10.76

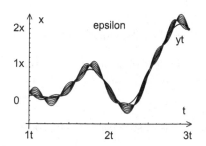

Fig. 1. Trajectory bundle obtained with $\alpha^2 = 10^{-4}$, $t \in [1, 1.5]$

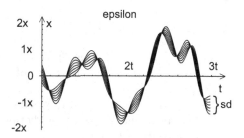

Fig. 2. Controls bundle obtained with $\alpha^2 = 10^{-4}$, $t \in [1, 1.5]$.

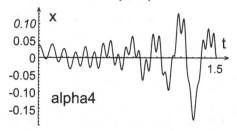

Fig. 3. Discrepancy $x_1(t) - y_1(t)$ with $\alpha^2 = 10^{-4}$, $t \in [0, 1.5]$

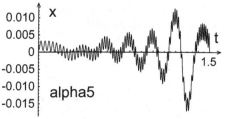

Fig. 4. Discrepancy $x_1(t) - y_1(t)$ with $\alpha^2 = 10^{-5}$, $t \in [0, 1.5]$

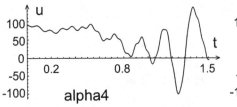

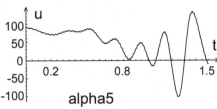

Fig. 5. Control $u_1(t)$ with $\alpha^2 = 10^{-4}$, $t \in [0, 1.5]$

Fig. 6. Control $u_1(t)$ with $\alpha^2 = 10^{-5}$, $t \in [0, 1.5]$

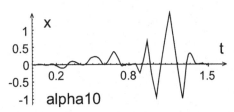

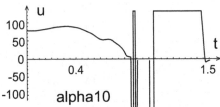

Fig. 7. Discrepancy $x_1(t) - y_1(t)$ for functional $I^+(\cdot)$ with $\alpha^2 = 10^{-10}$, $t \in [0, 1.5]$

Fig. 8. Control $u_1(t)$ for functional $I^+(\cdot)$ with $\alpha^2 = 10^{-10}$, $t \in [0, 1.5]$

11   Perspectives

The suggested numerical method can be applied in the following directions.

- Identification and reconstruction of dynamic models of production activity for single firms, various branches of industry or industry and economics of a region.
- Investigating properties of the examined object.
- A short-term and long-term prediction and analysis of scenarios of the process development in the future.
- Analysis of the production plan and construction of feedback controls realizing the plan.

Acknowledgements. The work is supported partially by RFBR (projects No. 14–01–00168, No. 14-01-00486).

References

1. Al'brekht E.G.: Methods of construction and identification of mathematical models for macroeconomic processes. Electron. J. Invest. Russ. http://zhurnal.ape.relarn. ru/articles/2002/005.pdf. (in Russian)
2. Bellman, R.: Dynamic Programming. Princeton University Press, Princeton (1957). 340 p

3. Tihonov, A.N.: Solution of incorrectly formulated problems and the regularization method. Sov. Math. **4**, 1035–1038 (1963)
4. Krasovskii, N.N., Subbotin, A.I.: Game–Theoretical Control Problems. Springer, New York (1987)
5. Osipov, Y.S., Kryazhimskii, A.V.: Inverse Problems for Ordinary Differential Equations: Dynamical Solutions. Gordon and Breach, London (1995)
6. Pontryagin, L.S., Boltyanskii, V.G., Gamkrelidze, R.V., Mishchenko, E.F.: The Mathematical Theory of Optimal Processes. Nauka, Moscow (1961). 392 p. Wiley, New York (1962)
7. Clarke, F.: Optimization and Nonsmooth Analysis. Wiley, New York (1983). 308 p
8. Subbotina, N.N., Kolpakova, E.A., Tokmantsev, T.B., Shagalova, L.G.: The Method of Characteristics for Hamilton-Jacobi-Bellman Equation. RIO UrO RAN, Ekaterinburg (2013). 144 p. (in Russian)
9. Krupennikov, E.A.: Validation of a solution method for the problem of reconstructing the dynamics of a mecroeconomic system. Proc. Inst. Math. Mech. **21**(2), 100–112 (2015). RIO UrO RAN, Ekaterinburg (in Russian)
10. Subbotina, N.N., Tokmantsev, T.B.: A study of the stability of solutions to inverse problems of dynamics of control systems under perturbations of initial data. Proc. Steklov Inst. Math. **291**(Suppl. 1), S1–S17 (2015)

Stable Sequential Pontryagin Maximum Principle as a Tool for Solving Unstable Optimal Control and Inverse Problems for Distributed Systems

Mikhail Sumin[✉]

Faculty of Mechanics and Mathematics, Nizhnii Novgorod State University,
Gagarin Ave. 23, 603950 Nizhnii Novgorod, Russia
m.sumin@mail.ru

Abstract. This article is devoted to studying dual regularization method as applied to parametric convex optimal control problem of controlled third boundary-value problem for parabolic equation with boundary control and with equality and inequality pointwise state constraints. These constraints are understood as ones in the Hilbert space L_2. A major advantage of the constraints of the original problem which are understood as ones in L_2 is that the resulting dual regularization algorithm is stable with respect to errors in the input data and leads to the construction of a minimizing approximate solution in the sense of J. Warga. Simultaneously, this dual algorithm yields the corresponding necessary and sufficient conditions for minimizing sequences, namely, the stable, with respect to perturbation of input data, sequential or, in other words, regularized Lagrange principle in nondifferential form and Pontryagin maximum principle for the original problem. Regardless of the fact that the stability or instability of the original optimal control problem, they stably generate a minimizing approximate solutions for it. For this reason, we can interpret these regularized Lagrange principle and Pontryagin maximum principle as tools for direct solving unstable optimal control problems and reducing to them unstable inverse problems.

Keywords: Optimal boundary control · Parabolic equation · Minimizing sequence · Dual regularization · Stability · Pontryagin maximum principle

1 Introduction

Pontryagin maximum principle is the central result of all optimal control theory, including optimal control for differential equations with partial derivatives. Its statement and proof assume, first of all, that the optimal control problem is considered in an ideal situation, when its input data are known exactly. However, in the vast number of important practical problems of optimal control, as well

Published by Springer International Publishing AG 2016. All Rights Reserved
L. Bociu et al. (Eds.): CSMO 2015, IFIP AICT 494, pp. 482–492, 2016.
DOI: 10.1007/978-3-319-55795-3_46

as numerous problems reducing to optimal control problems, the requirement of exact defining input data is very unnatural, and in many undoubtedly interest cases is simply impracticable. In similar problems, we can not, strictly speaking, to take as an approximation to the solution of the initial (unperturbed) problem with the exact input data, a control formally satisfying the maximum principle in the perturbed problem. The reason of such situation lies in the natural insta-bility of optimization problems with respect to perturbation of its input data. As a typical property of optimization problems in general, including constrained ones, instability fully manifests itself in optimal control problems (see., e.g., [1]). As a consequence, the mentioned above instability implies "instability" of the classical optimality conditions, including the conditions in the form of Pontrya-gin maximum principle. This instability manifests itself in selecting by them of arbitrarily distant "perturbed" optimal elements from their unperturbed coun-terparts in the case of an arbitrarily small perturbations of the input data. The above applies, in full measure, both to discussed below optimal control problem with pointwise state constraints for linear parabolic equation in divergent form, and to the classical optimality conditions in the form of the Lagrange principle and the Pontryagin maximum principle for this problem.

In this paper we discuss how to overcome the problem of instability of the classical optimality conditions in optimal control problems in the way of apply-ing dual regularization method (see., e.g., [2–4]) and simultaneous transition to the concept of minimizing sequence of admissible elements as the main concept of optimization theory. In the role of the last, acts the concept of the minimiz-ing approximate solution in the sense of Warga [5]. The main attention in the paper is given to the discussion of the so-called regularized or, in other words, stable, with respect to perturbation of input data, sequential Lagrange princi-ple in the nondifferential form and Pontryagin maximum principle. Regardless of the stability or instability of the original optimal control problem, they sta-bly generate minimizing approximate solutions for it. For this reason, we can interpret the regularized Lagrange principle and Pontryagin maximum princi-ple that are obtained in the article as tools for direct solving unstable opti-mal control problems and reducing to them unstable inverse problems [1,6,7]. Thus, they contribute to a significant expansion of the range of applicability of the theory of optimal control in which a central role belongs to classic con-structions of the Lagrange and Hamilton-Pontryagin functions. Finally, we note that discussed in this article regularized Lagrange principle in the nondiffer-ential form and Pontryagin maximum principle may have another kind, more convenient for applications [7]. Justification of these alternative forms of the regularized Lagrange principle and Pontryagin maximum principle is based on the so-called method of iterative dual regularization [2,3]. In this case, they take the form of iterative processes with the corresponding stopping rules when the error of input data is fixed and finite. Here these alternative forms are not considered.

2   Statement of Optimal Control Problem

We consider the fixed-time parametric optimal control problem

$$(P^\delta_{p,r}) \qquad g^\delta_0(\pi) \to \min, \quad \pi \equiv (u,w) \in \mathcal{D} \subset L_2(Q_T) \times L_2(S_T),$$

$$g^\delta_1(\pi)(x,t) \equiv \varphi^\delta_1(x,t) z^\delta[\pi](x,t) = h^\delta(x,t) + p(x,t) \quad \text{for a.e. } (x,t) \in Q,$$
$$g^\delta_2(\pi)(x,t) \equiv \varphi^\delta_2(x,t,z^\delta[\pi](x,t)) \le r(x,t) \quad \text{for a.e. } (x,t) \in Q$$

with equality and inequality pointwise state constraints understood as ones in the Hilbert space $\mathcal{H} \equiv L_2(Q)$; $\mathcal{D} \equiv \{u \in L_2(Q_T) : u(x,t) \in U \text{ for a.e. } (x,t) \in Q_T\} \times \{w \in L_2(S_T) : w(x,t) \in W \text{ for a.e. } (x,t) \in S_T\}$; U, $W \subset \mathbb{R}^1$ are convex compact sets. In this problem, $p \in \mathcal{H}$ and $r \in \mathcal{H}$ are parameters; $g^\delta_0 : L_2(Q_T) \times L_2(S_T)$ is a continuous convex functional, $Q \subset \overline{Q}_{\iota,T}$ is a compact set without isolated points with a nonempty interior, $\iota \in (0,T)$, $Q = \operatorname{cl} \operatorname{int} Q$; and $z^\delta[\pi] \in V^{1,0}_2(Q_T) \cap C(\overline{Q}_T)$ is a weak solution [8,9] to the third boundary-value problem[1]

$$z_t - \frac{\partial}{\partial x_i}(a_{i,j}(x,t) z_{x_j}) + a^\delta(x,t) z + u(x,t) = 0, \tag{1}$$

$$z(x,0) = v^\delta_0(x), \ x \in \Omega, \quad \frac{\partial z}{\partial N} + \sigma^\delta(x,t) z = w(x,t), \ (x,t) \in S_T,$$

corresponding to the pair $\pi \equiv (u,w)$. The superscript δ in the input data of Problem $(P^\delta_{p,r})$ indicates that these data are exact ($\delta = 0$) or perturbed ($\delta > 0$), i.e., they are specified with an error, $\delta \in [0,\delta_0]$, where $\delta_0 > 0$ is a fixed number.

For definiteness, as target functional we take terminal one

$$g^\delta_0(\pi) \equiv \int_\Omega G^\delta(x, z^\delta[\pi](x,T)) dx.$$

The input data for Problem $(P^0_{p,r})$ are assumed to meet the following conditions:

(a) It is true that $a_{i,j} \in L_\infty(Q_T)$, $i,j = 1,\ldots,n$, $a^\delta \in L_\infty(Q_T)$, $\sigma^\delta \in L_\infty(S_T)$, $v^\delta_0 \in C(\overline{\Omega})$,

$$\nu|\xi|^2 \le a_{i,j}(x,t)\xi_i\xi_j \le \mu|\xi|^2 \quad \forall(x,t) \in Q_T, \quad \nu,\mu > 0,$$

$a^\delta(x,t) \ge C_0$ for a.e. $(x,t) \in Q_T$, $\sigma^\delta(x,t) \ge C_0$ for a.e. $(x,t) \in S_T$;

(b) It is true that φ^δ_1, $h^\delta \in L_\infty(Q)$; $\varphi^\delta_2 : Q \times \mathbb{R}^1 \to \mathbb{R}^1$ is Lebesgue measurable function that is continuous and convex with respect to z for a.e. $(x,t) \in Q$, $\varphi^\delta_2(\cdot,\cdot,z(\cdot,\cdot)) \in L_\infty(Q) \ \forall z \in C(Q)$; $G^\delta : \Omega \times \mathbb{R}^1 \to \mathbb{R}^1$ is Lebesgue measurable function that is continuous and convex with respect to z for a.e. $x \in \Omega$, $G^\delta(\cdot, z(\cdot,T)) \in L_\infty(\Omega) \ \forall z(\cdot,T) \in C(Q)$;

(c) $\Omega \subset \mathbb{R}^n$ be a bounded domain with Lipschitz boundary S.

---

[1] Here and below, we use the notations for the sets Q_T, S_T, $Q_{i,T}$ and also for functional spaces and norms of their elements adopted in monograph [8].

Assume that the following estimates hold:

$$|G^\delta(x,z) - G^0(x,z)| \le C_M \delta \ \forall\, (x,z) \in \Omega \times S_M^1, \quad \|\varphi_1^\delta - \varphi_1^0\|_{\infty,Q} \le C\delta, \quad (2)$$

$$\|h^\delta - h^0\|_{\infty,Q} \le C\delta, \ |\varphi_2^\delta(x,t,z) - \varphi_2^0(x,t,z)| \le C_M \delta \ \forall\, (x,t,z) \in Q \times S_M^1,$$

$$\|a^\delta - a^0\|_{\infty,Q_T} \le C\delta, \ |v_0^\delta - v_0^0|_{\overline{\Omega}}^{(0)} \le C\delta, \ \|\sigma^\delta - \sigma^0\|_{\infty,S_T} \le C\delta,$$

where $C, C_M > 0$ are independent of δ; $S_M^n \equiv \{x \in \mathbb{R}^n : |x| < M\}$. Let's note, that the conditions on the input data of Problem $(P_{p,r}^\delta)$, and also the estimates of deviations of the perturbed input data from the exact ones can be weakened.

In this paper we use for discussing the main results, related to the stable sequential Lagrange principle and Pontryagin maximum principle in Problem $(P_{p,r}^0)$, a scheme of studying the similar optimization problems in the papers [10, 11] for a system of controlled ordinary differential equations. In these works, both spaces of admissible controls and spaces, where lie images of the operators that define the pointwise state constraints, represented as Hilbert spaces of square-integrable functions. For this reason, we put the set $\mathcal{D}$ of admissible controls π into a Hilbert space also, i.e., assume that $\mathcal{D} \subset Z \equiv L_2(Q_T) \times L_2(S_T)$, $\|\pi\| \equiv (\|u\|_{2,Q_T}^2 + \|w\|_{2,S_T}^2)^{1/2}$. At the same time, we note that the conditions on the input data of Problem $(P_{p,r}^\delta)$ allow formally to consider that the operators g_1^δ, g_2^δ, specifying the state constraints of the problem, act into space $L_p(Q)$ with any index $p \in [1,+\infty]$. However, in this paper, taking into account the above remark, we will put images of these functional operators in the Hilbert space $L_2(Q) \equiv \mathcal{H}$.

Suppose that Problem $(P_{p,r}^0)$ has a solution (which is unique if g_0^0 is strictly (strongly) convex). Its solutions are denoted by $\pi_{p,r}^0 \equiv (u_{p,r}^0, w_{p,r}^0)$, and the set of all such solutions is designated as $U_{p,r}^0$. Define the Lagrange functional, a set of its minimizers and the concave dual problem

$$L_{p,r}^\delta(\pi,\lambda,\mu) \equiv g_0^\delta(\pi) + \langle \lambda, g_1^\delta(\pi) - h^\delta - p \rangle + \langle \mu, g_2^\delta(\pi) - r \rangle, \ \pi \in \mathcal{D},$$

$$U^\delta[\lambda,\mu] \equiv \mathrm{Argmin}\, \{L_{p,r}^\delta(\pi,\lambda,\mu) : \pi \in \mathcal{D}\} \, \forall (\lambda,\mu) \in \mathcal{H} \times \mathcal{H}_+,$$

$$V_{p,r}^\delta(\lambda,\mu) \to \sup, \ (\lambda,\mu) \in \mathcal{H} \times \mathcal{H}_+, \ V_{p,r}^\delta(\lambda,\mu) \equiv \inf_{\pi \in \mathcal{D}} L_{p,r}^\delta(\pi,\lambda,\mu).$$

Since the Lagrange functional is continuous and convex for any pair $(\lambda,\mu) \in \mathcal{H} \times \mathcal{H}_+$ and the set $\mathcal{D}$ is bounded, the dual functional $V_{p,r}^\delta$, is obviously defined and finite for any $(\lambda,\mu) \in \mathcal{H} \times \mathcal{H}_+$.

The concept of a minimizing approximate solution in the sense of Warga [5] is of great importance for the design of a dual regularizing algorithm for problem $(P_{p,r}^0)$. Recall that a minimizing approximate solution is a sequence $\pi^i \equiv (u^i, w^i)$, $i = 1,2,\ldots$ such that $g_0^0(\pi^i) \le \beta(p,r) + \delta^i$, $\pi^i \in \mathcal{D}_{p,r}^{0,\epsilon^i}$ for some nonnegative number sequences δ^i and ϵ^i, $i = 1,2,\ldots$, that converge to zero. Here, $\beta(p,r)$ is the generalized infimum, i.e., an S-function:

$$\beta(p,r) \equiv \lim_{\epsilon \to +0} \beta_\epsilon(p,r), \ \beta_\epsilon(p,r) \equiv \inf_{\pi \in \mathcal{D}_{p,r}^{0,\epsilon}} g_0^0(\pi), \ \beta_\epsilon(p,r) \equiv +\infty \text{ if } \mathcal{D}_{p,r}^{0,\epsilon} = \emptyset,$$

$$\mathcal{D}_{p,r}^{\delta,\epsilon} \equiv \{\pi \in \mathcal{D} : \|g_1^\delta(\pi) - h^\delta - p\|_{2,Q} \le \epsilon, \min_{z \in \mathcal{H}_-} \|g_2^\delta(\pi) - r - z\|_{2,Q} \le \epsilon\}, \; \epsilon \ge 0 \,,$$

$$\mathcal{D}_{p,r}^{00} \equiv \mathcal{D}_{p,r}^0, \; \mathcal{H}_- \equiv \{z \in L_2(Q) : z(x,t) \le 0 \text{ for a.e. } (x,t) \in Q\}, \; \mathcal{H}_+ \equiv -\mathcal{H}_-.$$

Obviously, in the general situation, $\beta(p,r) \le \beta_0(p,r)$, where $\beta_0(p,r)$ is the classical value of the problem. However, in the case of Problem $(P_{p,r}^0)$, we have $\beta(p,r) = \beta_0(p,r)$. Simultaneously, we may asset that $\beta : L_2(Q) \times L_2(Q) \to \mathbb{R}^1 \cup \{+\infty\}$ is a convex and lower semicontinuous function. Note here that the existence of a minimizing approximate solution in Problem $(P_{p,r}^0)$ obviously implies its solvability.

From the conditions (a)–(c) and the theorem on the existence of a weak solution of the third boundary-value problem for a linear parabolic equation of the divergent type (see [8, chap. III, Sect. 5] and also [12]), it follows that the direct boundary-value problem (1) and the corresponding adjoint problem are uniquely solvable in $V_2^{1,0}(Q_T)$.

Proposition 1. *For any pair* $(u,w) \in L_2(Q_T) \times L_2(S_T)$ *and any* $T > 0$ *the direct boundary-value problem* (1) *is uniquely solvable in* $V_2^{1,0}(Q_T)$ *and the estimate*

$$|z^\delta[\pi]|_{Q_T} + \|z^\delta[\pi]\|_{2,S_T} \le C_T(\|u\|_{2,Q_T} + \|v_0^\delta\|_{2,\Omega} + \|w\|_{2,S_T}),$$

takes place, where the constant C_T *is independent of* $\delta \ge 0$ *and pair* $\pi \equiv (u,w) \in L_2(Q_T) \times L_2(S_T)$. *Also the adjoint problem*

$$-\eta_t - \frac{\partial}{\partial x_j} a_{i,j}(x,t)\eta_{x_i} + a^\delta(x,t)\eta = \chi(x,t),$$

$$\eta(x,T) = \psi(x), \; x \in \Omega, \qquad \frac{\partial \eta}{\partial N} + \sigma^\delta(x,t)\eta = \omega(x,t), \; (x,t) \in S_T$$

is uniquely solvable in $V_2^{1,0}(Q_T)$ *for any* $\chi \in L_2(Q_T)$, $\psi \in L_2(\Omega)$, $\omega \in L_2(S_T)$ *and any* $T > 0$. *Its solution is denoted as* $\eta[\chi,\psi,\omega]$. *Simultaneously, the estimate*

$$|\eta^\delta[\chi,\psi,\omega]|_{Q_T} + \|\eta^\delta[\chi,\psi,\omega]\|_{2,S_T} \le C_T^1(\|\chi\|_{2,Q_T} + \|\psi\|_{2,\Omega} + \|\omega\|_{2,S_T}),$$

is true, where the constant C_T^1 *is independent of* $\delta \ge 0$ *and a triple* (χ,ψ,ω).

Simultaneously, from conditions (a)–(c) and the theorems on the existence of a weak (generalized) solution of the third boundary-value problem for a linear parabolic equation of the divergent type (see, e.g., [9]), it follows that the direct boundary-value problem is uniquely solvable in $V_2^{1,0}(Q_T) \cap C(\overline{Q}_T)$.

Proposition 2. *Let us* $l > n + 1$. *For any pair* $(u,w) \in L_l(Q_T) \times L_l(S_T)$ *and any* $T > 0$, $\delta \in [0,\delta_0]$ *the direct boundary-value problem* (1) *is uniquely solvable in* $V_2^{1,0}(Q_T) \cap C(\overline{Q}_T)$ *and the estimate*

$$|z^\delta[\pi]|_{\overline{Q}_T}^{(0)} \le C_T(\|u\|_{l,Q_T} + |v_0^\delta|_{\Omega}^{(0)} + \|w\|_{l,S_T}),$$

takes place, where the constant C_T *is independent of pair* $\pi \equiv (u,w)$ *and* δ.

Further, the minimization problem for Lagrange functional

$$L_{p,r}^{\delta}(\pi, \lambda, \mu) \to \min, \ \pi \in \mathcal{D} \quad \text{when } (\lambda, \mu) \in L_2(Q) \times L_2^+(Q) \qquad (3)$$

plays the central role in all subsequent constructions. It is usual problem without equality and inequality constraints. It is solvable as a minimization problem for weakly semicontinuous functional on the weak compact set $\mathcal{D} \subset L_2(Q_T) \times L_2(S_T)$. Here, the weak semicontinuity is a consequence of the convexity and continuity with respect to π of the Lagrange functional. Minimizers $\pi^{\delta}[\lambda, \mu] \in U^{\delta}[\lambda, \mu]$ for this optimal control problem satisfy the Pontryagin maximum principle under supplementary assumption of the existence of Lebesgue measurable with respect to $(x,t) \in Q$ for all $z \in \mathbb{R}^1$ and continuous with respect to z for a.e. x, t gradients $\nabla_z \varphi_2^{\delta}(x,t,z)$, $\nabla_z G^{\delta}(x,z)$ with the estimates $|\nabla_z \varphi_2^{\delta}(x,t,z)| \leq C_M$, $|\nabla_z G^{\delta}(x,z)| \leq C_M \ \forall z \in S_M^1$ where $C_M > 0$ is independent of δ. Due to the estimates of the Propositions 1 and 2 and to the so called two-parameter variation [13] of the pair $\pi^{\delta}[\lambda, \mu]$ that is needle-shaped with respect to control u and classical with respect to control w the following lemma is true.

Lemma 1. *Let $H(y, \eta) \equiv -\eta y$ and the additional condition that specified above is fulfilled. Any pair $\pi^{\delta}[\lambda, \mu] = (u^{\delta}[\lambda, \mu], w^{\delta}[\lambda, \mu]) \in U^{\delta}[\lambda, \mu]$, $(\lambda, \mu) \in L_2(Q) \times L_2^+(Q)$ satisfies to (usual) Pontryagin maximum principle in the problem (3): for $\pi = \pi^{\delta}[\lambda, \mu]$ the following maximum relations*

$$H(u(x,t), \eta^{\delta}(x,t)) = \max_{u \in U} H(u, \eta^{\delta}(x,t)) \text{ for a.e. } Q_T, \qquad (4)$$

$$H(w(s,t), \eta^{\delta}(s,t)) = \max_{w \in W} H(w, \eta^{\delta}(s,t)) \text{ for a.e. } S_T$$

hold, where $\eta^{\delta}(x,t)$, $(x,t) \in Q_T$ is a solution for $\pi = \pi^{\delta}[\lambda, \mu]$ of the adjoint problem

$$-\eta_t - \frac{\partial}{\partial x_j}(a_{i,j}(x,t)\eta_{x_i}) + a^{\delta}(x,t)\eta =$$
$$\varphi_1^{\delta}(x,t)\lambda(x,t) + \nabla_z \varphi_2^{\delta}(x,t,z^{\delta}[\pi](x,t))\mu(x,t), \ (x,t) \in Q_T,$$
$$\eta(x,T) = \nabla_z G^{\delta}(x, z^{\delta}[\pi](x,T)), \ x \in \Omega, \quad \frac{\partial \eta(x,t)}{\partial \mathcal{N}} + \sigma^{\delta}(x,t)\eta = 0, \ (x,t) \in S_T.$$

Remark 1. Note that here and below, if the functions φ_1^{δ}, $\nabla_z \varphi_2^{\delta}(\cdot, \cdot, z(\cdot, \cdot))$, $\lambda, \mu \in L_2(Q)$ are considered on the entire cylinder Q_T, we set that the equalities $\varphi_1^{\delta}(x,t) = \nabla_z \varphi_2^{\delta}(x,t,z(x,t)) = \lambda(x,t) = \mu(x,t) = 0$ take place for $(x,t) \in Q_T \backslash Q$; the same notation is preserved if these functions are taken on the entire cylinder.

In the next section we construct minimizing approximate solutions for Problem $(P_{p,r}^0)$ from the elements $\pi^{\delta}[\lambda, \mu]$, $(\lambda, \mu) \in L_2(Q) \times L_2^+(Q)$. As consequence, this construction leads us to various versions of the stable sequential Lagrange principle and Pontragin maximum principle. In the case of strong convexity and subdifferentiability of the target functional g_0^0, these versions are statements about stable approximations of the solutions of Problem $(P_{p,r}^0)$ in the metric of

$Z \equiv L_2(Q_T) \times L_2(S_T)$ by the points $\pi^\delta[\lambda, \mu]$. Due to the estimates (2) and the Propositions 1 and 2 we may assert that the estimates

$$|g_0^\delta(\pi) - g_0^0(\pi)| \le C_1\delta \ \forall \pi \in \mathcal{D}, \quad \|g_1^\delta(\pi) - g_1^0(\pi)\|_{2,Q} \le C_2\delta(1 + \|\pi\|) \ \forall \pi \in Z, \quad (5)$$

$$\|h^\delta - h^0\|_{2,Q} \le C\delta, \quad \|g_2^\delta(\pi) - g_2^0(\pi)\|_{2,Q} \le C_3\delta \ \forall \pi \in \mathcal{D},$$

hold, in which the constants C_1, C_2, $C_3 > 0$ are independent of $\delta \in (0, \delta_0]$, π.

3   Stable Sequential Pontryagin Maximum Principle

In this section we discuss the so-called regularized or, in other words, stable, with respect errors of input data, sequential Pontryagin maximum principle for Problem $(P_{p,r}^0)$ as necessary and sufficient condition for elements of minimizing approximate solutions. Simultaneously, this condition we may treat as one for existence of a minimizing approximate solutions in Problem $(P_{p,r}^0)$ with perturbed input data or as condition of stable construction of a minimizing sequence in this problem. The proof of the necessity of this condition is based on the dual regularization method [2–4] that is stable algorithm of constructing a minimizing approximate solutions in Problem $(P_{p,r}^0)$. Sketches of the proofs for the theorems in this section (Theorems 1, 2 and 3) and some comments may be found in [14, 15].

3.1   Dual Regularization for Optimal Control Problem with Pointwise State Constraints

The estimates (5) give a possibility to organize for constructing a minimizing approximate solution in Problem $(P_{p,r}^0)$ the procedure of the dual regularization in accordance with a scheme of the paper [11]. In accordance with this scheme the dual regularization consists in the direct solving dual problem to Problem $(P_{p,r}^0)$ and its Tikhonov stabilization

$$R_{p,r}^{\delta,\alpha(\delta)}(\lambda, \mu) \equiv V_{p,r}^\delta(\lambda, \mu) - \alpha(\delta)\|(\lambda, \mu)\|^2 \to \max, \quad (\lambda, \mu) \in L_2(Q) \times L_2^+(Q)$$

under consistency condition $\delta/\alpha(\delta) \to 0$, $\alpha(\delta) \to 0$, $\delta \to 0$. This dual regularization leads to constructing minimizing approximate solution in Problem $(P_{p,r}^0)$ from the elements $\pi^\delta[\lambda_{p,r}^{\delta,\alpha(\delta)}, \mu_{p,r}^{\delta,\alpha(\delta)}] \in \operatorname{Argmin}\{L_{p,r}^\delta(\pi, \lambda, \mu) : \pi \in \mathcal{D}\}$, where $(\lambda_{p,r}^{\delta,\alpha}, \mu_{p,r}^{\delta,\alpha}) \equiv \operatorname{argmax}\{R_{p,r}^{\delta,\alpha}(\lambda, \mu) : (\lambda, \mu) \in L_2(Q) \times L_2^+(Q)\}$ and $\delta \to 0$.

We may assert that the following "convergence" theorem for the dual regularization method in Problem $(P_{p,r}^0)$ is valid.

Theorem 1. *Regardless of the properties of the solvability of the dual problem to Problem $(P_{p,r}^0)$ or, in other words, regardless of the properties of the subdifferential $\partial\beta(p, r)$ (it is empty or not empty), it is true that exist elements $\pi^\delta \in U^\delta[\lambda_{p,r}^{\delta,\alpha(\delta)}, \mu_{p,r}^{\delta,\alpha(\delta)}]$ such that the relations*

$$g_0^0(\pi^\delta) \to g_0^0(\pi_{p,r}^0), \ g_1^0(\pi^\delta) - h^0 - p \to 0, \ g_2^0(\pi^\delta) - r \le \kappa(\delta), \ \|\kappa(\delta)\| \to 0, \ \delta \to 0,$$
$$\langle(\lambda_{p,r}^{\delta,\alpha(\delta)}, \mu_{p,r}^{\delta,\alpha(\delta)}), (g_1^\delta(\pi^\delta) - h^\delta - p, g_2^\delta(\pi^\delta) - r)\rangle \to 0, \ \delta \to 0$$

hold, in which the inequality $g_2^\delta(\pi^\delta) - r \le \kappa(\delta)$ is understood in the sense of ordering on a cone of nonpositive functions in $L_2(Q)$. Simultaneously, the equality

$$\lim_{\delta \to +0} V_{p,r}^0(\lambda_{p,r}^{\delta,\alpha(\delta)}, \mu_{p,r}^{\delta,\alpha(\delta)}) = \sup_{(\lambda,\mu) \in \mathcal{H} \times \mathcal{H}_+} V_{p,r}^0(\lambda,\mu)$$

is valid. If the dual of Problem $(P_{p,r}^0)$ is solvable, then the limit relation $(\lambda_{p,r}^{\delta,\alpha(\delta)}, \mu_{p,r}^{\delta,\alpha(\delta)}) \to (\lambda_{p,r}^0, \mu_{p,r}^0), \delta \to 0$ is valid also, where $(\lambda_{p,r}^0, \mu_{p,r}^0)$ denotes minimum-norm solution of the dual problem.

This theorem may be proved in exact accordance with a scheme of proving the similar theorem in [11]. We note only that, as in [11], this proving uses a weak continuity of the operators g_1^δ, g_2^δ that is consequence of the conditions on the input data of Problem $(P_{p,r}^0)$ and a regularity of the bounded solutions of the boundary-value problem (1) inside of the cylinder Q_T [8, chap. III, Theorem 10.1].

3.2 Stable Sequential Lagrange Principle for Optimal Control Problem with Pointwise State Constraints

We formulate in this subsection the necessary and sufficient condition for existence of a minimizing approximate solution in Problem $(P_{p,r}^0)$. Also, it can be called by stable sequential Lagrange principle in nondifferential form for this problem. Simultaneously, as we deal only with regular Lagrange function, the formulated theorem may be called by Kuhn-Tucker theorem in nondifferential form. Note that the necessity of the conditions of formulated below theorem follows from the Theorem 1. At the same time, their sufficiency is a simple consequence of the convexity of Problem $(P_{p,r}^0)$ and the conditions on its input data. A verification of these propositions for similar situation of the convex programming problem in a Hilbert space may be found in [1,7].

Theorem 2. *Regardless of the properties of the subdifferential $\partial\beta(p,r)$ (it is empty or not empty) or, in other words, regardless of the properties of the solvability of the dual problem to Problem $(P_{p,r}^0)$, necessary and sufficient conditions for Problem $(P_{p,r}^0)$ to have a minimizing approximate solution is that there is a sequence of dual variables $(\lambda^k, \mu^k) \in \mathcal{H} \times \mathcal{H}_+$, $k = 1, 2, \ldots$, such that $\delta^k \|(\lambda^k, \mu^k)\| \to 0, k \to \infty$, and relations*

$$\pi^{\delta^k}[\lambda^k, \mu^k] \in \mathcal{D}_{p,r}^{\delta^k, \epsilon^k}, \quad \epsilon^k \to 0, \; k \to \infty, \tag{6}$$

$$\langle (\lambda^k, \mu^k), (g_1^{\delta^k}(\pi^{\delta^k}[\lambda^k, \mu^k]) - h^{\delta^k} - p, g_2^{\delta^k}(\pi^{\delta^k}[\lambda^k, \mu^k]) - r) \rangle \to 0, \; k \to \infty \tag{7}$$

hold for some elements $\pi^{\delta^k}[\lambda^k, \mu^k] \in U^{\delta^k}[\lambda^k, \mu^k]$. The sequence $\pi^{\delta^k}[\lambda^k, \mu^k]$, $k = 1, 2, \ldots$, is the desired minimizing approximate solution and each of its weak limit points is a solution of Problem $(P_{p,r}^0)$. As $(\lambda^k, \mu^k) \in \mathcal{H} \times \mathcal{H}_+$, $k = 1, 2, \ldots$, we can use the sequence of the points $(\lambda_{p,r}^{\delta^k,\alpha(\delta^k)}, \mu_{p,r}^{\delta^k,\alpha(\delta^k)})$, $k = 1, 2, \ldots$, generated by the dual regularization method of the Theorem 1. If the dual of Problem $(P_{p,r}^0)$

is solvable, the sequence $(\lambda^k, \mu^k) \in \mathcal{H} \times \mathcal{H}_+$, $k = 1, 2, \ldots$, *should be assumed to be bounded. The limit relation*

$$V^0_{p,r}(\lambda^k, \mu^k) \to \sup_{(\lambda,\mu)\in\mathcal{H}\times\mathcal{H}_+} V^0_{p,r}(\lambda, \mu) \tag{8}$$

holds as a consequence of the relations (6) *and* (7). *Furthermore, each weak limit point* (*if such points exist*) *of the sequence* $(\lambda^k, \mu^k) \in \mathcal{H} \times \mathcal{H}_+$, $k = 1, 2, \ldots$ *is a solution of the dual problem* $V^0_{p,r}(\lambda, \mu) \to \max$, $(\lambda, \mu) \in \mathcal{H} \times \mathcal{H}_+$.

Remark 2. If the functional g^0_0 is strongly convex and subdifferentiable on $\mathcal{D}$ then from the weak convergence of the unique in this case elements $\pi^{\delta^k}[\lambda^k, \mu^k]$ to unique element $\pi^0_{p,r}$ as $k \to \infty$, and numerical convergence $g^0_0(\pi^{\delta^k}[\lambda^k, \mu^k]) \to g^0_0(\pi^0_{p,r})$, $k \to \infty$ follows the strong convergence $\pi^{\delta^k}[\lambda^k, \mu^k] \to \pi^0_{p,r}$, $k \to \infty$. Problem $(P^0_{p,r})$ with the strongly convex g^0_0 for linear system of ordinary differential equations but with exact input data is studied in [10].

3.3   Stable Sequential Pontryagin Maximum Principle for Optimal Control Problem with Pointwise State Constraints

Denote by $U^\delta_{max}[\lambda, \mu]$ a set of the elements $\pi \in \mathcal{D}$ that satisfy all relations of the maximum principle (4) of the Lemma 1. Under the supplementary condition of existence of continuous with respect to z gradients $\nabla_z \varphi^\delta_2(x, t, z)$, $\nabla_z G^\delta(x, z)$ with corresponding estimates, it follows that the proposition of the Theorem 2 may be rewritten in the form of the stable sequential Pontryagin maximum principle. It is obviously that the equality $U^\delta_{max}[\lambda, \mu] = U^\delta[\lambda, \mu]$ takes place under mentioned supplementary condition.

Theorem 3. *Regardless of the properties of the subdifferential* $\partial\beta(p, r)$ (*it is empty or not empty*) *or, in other words, regardless of the properties of the solvability of the dual problem to Problem* $(P^0_{p,r})$, *necessary and sufficient conditions for Problem* $(P^0_{p,r})$ *to have a minimizing approximate solution is that there is a sequence of dual variables* $(\lambda^k, \mu^k) \in \mathcal{H} \times \mathcal{H}_+$, $k = 1, 2, \ldots$, *such that* $\delta^k \|(\lambda^k, \mu^k)\| \to 0$, $k \to \infty$, *and relations* (6) *and* (7) *hold for some elements* $\pi^{\delta^k}[\lambda^k, \mu^k] \in U^{\delta^k}_{max}[\lambda^k, \mu^k]$. *Moreover, the sequence* $\pi^{\delta^k}[\lambda^k, \mu^k]$, $k = 1, 2, \ldots$, *is the desired minimizing approximate solution and each of its weak limit points is a solution of Problem* $(P^0_{p,r})$. *As* $(\lambda^k, \mu^k) \in \mathcal{H} \times \mathcal{H}_+$, $k = 1, 2, \ldots$, *we can use the sequence of the points* $(\lambda^{\delta^k, \alpha(\delta^k)}_{p,r}, \mu^{\delta^k, \alpha(\delta^k)}_{p,r})$, $k = 1, 2, \ldots$, *generated by the dual regularization method of the Theorem 1. If the dual of Problem* $(P^0_{p,r})$ *is solvable, the sequence* $(\lambda^k, \mu^k) \in \mathcal{H} \times \mathcal{H}_+$, $k = 1, 2, \ldots$, *should be assumed to be bounded. The limit relation* (8) *holds as a consequence of the relations* (6) *and* (7).

Remark 3. When the inequality constraint in Problem $(P^0_{p,r})$ is absent, i.e., $(P^0_{p,r}) = (P^0_p)$, and $\varphi_2(x, t) = r \equiv 0$, $\varphi_1(x, t) \equiv 1$, the target functional g^0_0 is taken, for example, in the form $g^0_0(\pi) \equiv \|\pi\|^2 \equiv \|u\|^2 + \|w\|^2$ then Problem (P^0_p) acquires the typical form of unstable inverse problem. In this case the stable sequential Pontryagin maximum principle of the Theorem 3 becomes a tool for the direct solving such unstable inverse problem.

Remark 4. In important partial case of Problem $(P_{p,r}^0) = (P_r^0)$, when it has only the inequality constraint $(\varphi_1^\delta(x,t) = h^\delta(x,t) = p(x,t) = 0,\ (x,t) \in Q)$, "weak" passage to the limit in the relations of the Theorem 3 leads to usual for similar optimal control problems Pontryagin maximum principle (see, e.g., [9,16]) with nonnegative Radon measures in the input data of the adjoint equation.

Acknowledgments. This work was supported by the Russian Foundation for Basic Research (project no. 15-47-02294-r_povolzh'e_a) and by the Ministry of Education and Science of the Russian Federation within the framework of project part of state tasks in 2014–2016 (code no. 1727).

References

1. Sumin, M.I.: Stable sequential convex programming in a Hilbert space and its application for solving unstable problems. Comput. Math. Math. Phys. **54**, 22–44 (2014)
2. Sumin, M.I.: A regularized gradient dual method for the inverse problem of a final observation for a parabolic equation. Comput. Math. Math. Phys. **44**, 1903–1921 (2004)
3. Sumin, M.I.: Duality-based regularization in a linear convex mathematical programming problem. Comput. Math. Math. Phys. **46**, 579–600 (2007)
4. Sumin, M.I.: Regularized parametric Kuhn-Tucker theorem in a Hilbert space. Comput. Math. Math. Phys. **51**, 1489–1509 (2011)
5. Warga, J.: Optimal Control of Differential and Functional Equations. Academic Press, New York (1972)
6. Sumin, M.I.: Dual regularization and Pontryagin's maximum principle in a problem of optimal boundary control for a parabolic equation with nondifferentiable functionals. Proc. Steklov Inst. Math. **275**(Suppl.), S161–S177 (2011)
7. Sumin, M.I.: On the stable sequential Kuhn-Tucker theorem and its applications. Appl. Math. **3**, 1334–1350 (2012)
8. Ladyzhenskaya, O.A., Solonnikov, V.A., Ural'tseva, N.N.: Linear and Quasilinear Equations of Parabolic Type. American Mathematical Society, Providence (1968)
9. Casas, E., Raymond, J.-P., Zidani, H.: Pontryagin's principle for local solutions of control problems with mixed control-state constraints. SIAM J. Control Optim. **39**, 1182–1203 (2000)
10. Sumin, M.I.: Parametric dual regularization for an optimal control problem with pointwise state constraints. Comput. Math. Math. Phys. **49**, 1987–2005 (2009)
11. Sumin, M.I.: Stable sequential Pontryagin maximum principle in optimal control problem with state constraints. In: Proceedings of XIIth All-Russia Conference on Control Problems, pp. 796–808. Institute of Control Sciences of RAS, Moscow (2014)
12. Plotnikov, V.I.: Existence and uniqueness theorems and a priori properties of weak solutions. Dokl. Akad. Nauk SSSR **165**, 33–35 (1965)
13. Sumin, M.I.: The first variation and Pontryagin's maximum principle in optimal control for partial differential equations. Comput. Math. Math. Phys. **49**, 958–978 (2009)

14. Sumin, M.I.: Stable sequential Pontryagin maximum principle in optimal control for distributed systems. In: International Conference "Systems Dynamics and Control Processes" Dedicated to the 90th Anniversary of Academician N.N. Krasovskii (Ekaterinburg, Russia, 15–20 September 2014), pp. 301–308. Ural Federal University, Ekaterinburg (2015)
15. Sumin, M.I.: Subdifferentiability of value functions and regularization of Pontryagin maximum principle in optimal control for distributed systems. Tambov State University reports. Series: Natural and Technical Sciences, vol. 20, pp. 1461–1477 (2015)
16. Raymond, J.-P., Zidani, H.: Pontryagin's principle for state-constrained control problems governed by parabolic equations with unbounded controls. SIAM J. Control Optim. **36**, 1853–1879 (1998)

Double Convergence of a Family of Discrete Distributed Mixed Elliptic Optimal Control Problems with a Parameter

Domingo Alberto Tarzia$^{(\boxtimes)}$

CONICET - Depto. Matemática, FCE, Univ. Austral, Paraguay 1950,
S2000FZF Rosario, Argentina
DTarzia@austral.edu.ar

Abstract. The convergence of a family of continuous distributed mixed elliptic optimal control problems (P_α), governed by elliptic variational equalities, when the parameter $\alpha \to \infty$ was studied in Gariboldi - Tarzia, Appl. Math. Optim., 47 (2003), 213-230 and it has been proved that it is convergent to a distributed mixed elliptic optimal control problem (P). We consider the discrete approximations $(P_{h\alpha})$ and (P_h) of the optimal control problems (P_α) and (P) respectively, for each $h > 0$ and $\alpha > 0$. We study the convergence of the discrete distributed optimal control problems $(P_{h\alpha})$ and (P_h) when $h \to 0$, $\alpha \to \infty$ and $(h, \alpha) \to (0, +\infty)$ obtaining a complete commutative diagram, including the diagonal convergence, which relates the continuous and discrete distributed mixed elliptic optimal control problems $(P_{h\alpha})$, (P_α), (P_h) and (P) by taking the corresponding limits. The convergent corresponds to the optimal control, and the system and adjoint system states in adequate functional spaces.

Keywords: Double convergence · Distributed optimal control problems · Elliptic variational equalities · Mixed boundary conditions · Numerical analysis · Finite element method · Fixed points · Optimality conditions · Error estimations

1 Introduction

The purpose of this paper is to do the numerical analysis, by using the finite element method, of the convergence of the continuous distributed mixed optimal control problems with respect to a parameter (the heat transfer coefficient) given in [10, 11] obtaining a double convergence when the parameter of the finite element method goes to zero and the heat transfer coefficient goes to infinity.

We consider a bounded domain $\Omega \subset \mathbb{R}^n$ whose regular boundary $\Gamma = \partial\Omega = \Gamma_1 \cup \Gamma_2$ consists of the union of two disjoint portions Γ_1 and Γ_2 with meas $(\Gamma_1) > 0$. We consider the following elliptic partial differential problems with mixed boundary conditions, given by:

$$-\Delta u = g \quad \text{in } \Omega; \quad u = b \quad \text{on } \Gamma_1; -\frac{\partial u}{\partial n} = q \quad \text{on } \Gamma_2, \tag{1}$$

© IFIP International Federation for Information Processing 2016
Published by Springer International Publishing AG 2016. All Rights Reserved
L. Bociu et al. (Eds.): CSMO 2015, IFIP AICT 494, pp. 493–504, 2016.
DOI: 10.1007/978-3-319-55795-3_47

$$-\Delta u = g \quad \text{in } \Omega; \; -\frac{\partial u}{\partial n} = \alpha(u - b) \quad \text{on } \Gamma_1; \; -\frac{\partial u}{\partial n} = q \quad \text{on } \Gamma_2 \qquad (2)$$

where g is the internal energy in Ω, $b = Const. > 0$ is the temperature on Γ_1 for the system (1) and the temperature of the external neighborhood on Γ_1 for the system (2) respectively, q is the heat flux on Γ_2 and $\alpha > 0$ is the heat transfer coefficient on Γ_1. The systems (1) and (2) can represent the steady-state two-phase Stefan problem for adequate data [21, 22]. We consider the following continuous distributed optimal control problem (P) and a family of continuous distributed optimal control problems (P_α) for each parameter $\alpha > 0$, defined in [10], where the control variable is the internal energy g in Ω, that is: Find the continuous distributed optimal controls $g_{op} \in H = L^2(\Omega)$ and $g_{\alpha_{op}} \in H$ (for each $\alpha > 0$) such that:

$$\text{Problem (P): } J\big(g_{op}\big) = \min_{g \in H} J(g), \quad \text{Problem } (P_\alpha) : J_\alpha\big(g_{\alpha_{op}}\big) = \min_{g \in H} J_\alpha(g) \qquad (3)$$

where the quadratic cost functional $J, J_\alpha : H \to \mathbb{R}_0^+$ are defined by [2, 18, 26]:

$$(a)\; J(g) = \frac{1}{2}\|u_g - z_d\|_H^2 + \frac{M}{2}\|g\|_H^2, \quad (b)\; J_\alpha(g) = \frac{1}{2}\|u_{\alpha g} - z_d\|_H^2 + \frac{M}{2}\|g\|_H^2 \qquad (4)$$

with $M > 0$ and $z_d \in H$ given, $u_g \in K$ and $u_{\alpha g} \in V$ are the state of the systems defined by the mixed ellliptic differential problems (1) and (2) respectively whose elliptic variational equalities are given by [16]:

$$u_g \in K : \quad a\big(u_g, v\big) = (g, v) - \int_{\Gamma_2} qv d\gamma, \quad \forall v \in V_0 \qquad (5)$$

$$u_{\alpha g} \in V : \quad a_\alpha\big(u_{\alpha g}, v\big) = (g, v) - \int_{\Gamma_2} qv d\gamma + \alpha \int_{\Gamma_1} bv d\gamma, \quad \forall v \in V \qquad (6)$$

and their adjoint system states $p_g \in V$ and $p_{\alpha g} \in V$ are defined by the following elliptic variational equalities:

$$\begin{array}{l} (a)\; p_g \in V_o : a\big(p_g, v\big) = \big(u_g - z_d, v\big), \forall v \in V_0; \\ (b)\; p_{\alpha g} \in V : a_\alpha\big(p_{\alpha g}, v\big) = \big(u_{\alpha g} - z_d, v\big), \forall v \in V \end{array} \qquad (7)$$

with the spaces and bilinear forms defined by:

$$V = H^1(\Omega), \; V_0 = \{v \in V, v/\Gamma_1 = 0\}, \; K = b + V_0, \; H = L^2(\Omega), \; Q = L^2(\Gamma_2) \quad (8)$$

$$a(u, v) = \int_\Omega \nabla u. \nabla v dx, \quad a_\alpha(u, v) = a(u, v) + \alpha \int_{\Gamma_1} uv d\gamma, \quad (u, v) = \int_\Omega uv \, dx \quad (9)$$

where the bilinear, continuous and symmetric forms a and a_α are coercive on V_0 and V respectively, that is [16]:

$$\exists \lambda > 0 \text{ such that } \lambda \|v\|_V^2 \leq a(v,v), \quad \forall v \in V_0 \tag{10}$$

$$\exists \lambda_\alpha = \lambda_1 \min(1,\alpha) > 0 \text{ such that } \lambda_\alpha \|v\|_V^2 \leq a_\alpha(v,v), \quad \forall v \in V \tag{11}$$

and $\lambda_1 > 0$ is the coercive constant for the bilinear form a_1[16, 21].

The unique continuous distributed optimal energies g_{op} and $g_{\alpha_{op}}$ have been characterized in [10] as a fixed point on H for a suitable operators W and W_α over their optimal adjoint system states $p_{g_{op}} \in V_0$ and $p_{\alpha g_{\alpha_{op}}} \in V$ defined by:

$$W, W_\alpha : H \to H \quad \text{such that} \quad (a) \; W(g) = -\frac{1}{M} p_g, \quad (b) \; W_\alpha(g) = -\frac{1}{M} p_{\alpha g}. \tag{12}$$

The limit of the optimal control problem (P_α) when $\alpha \to \infty$ was studied in [10] and it was proven that:

$$\lim_{\alpha \to \infty} \left\| u_{\alpha g_{\alpha_{op}}} - u_{g_{op}} \right\|_V = 0, \quad \lim_{\alpha \to \infty} \left\| p_{\alpha g_{\alpha_{op}}} - p_{g_{op}} \right\|_V = 0, \quad \lim_{\alpha \to \infty} \left\| g_{\alpha_{op}} - g_{op} \right\|_H = 0 \tag{13}$$

for a large constant $M > 0$ by using the characterization of the optimal controls as fixed points through operators (12a) and (12b); this restrictive hypothesis on data was eliminated in [11] by using the variational formulations. We can summary the conditions (13) saying that the distributed optimal control problems (P_α) converges to the distributed optimal control problem (P) when $\alpha \to +\infty$.

Now, we consider the finite element method and a polygonal domain $\Omega \subset \mathbb{R}^n$ with a regular triangulation with Lagrange triangles of type 1, constituted by affine-equivalent finite element of class C^0 being h the parameter of the finite element approximation which goes to zero [3, 7]. Then, we discretize the elliptic variational equalities for the system states (6) and (5), the adjoint system states (7a) and (7b), and the cost functional (4a, b) respectively. In general, the solution of a mixed elliptic boundary problem belongs to $H^r(\Omega)$ with $1 < r \leq 3/2 - \varepsilon$ ($\varepsilon > 0$) but there exist some examples which solutions belong to $H^r(\Omega)$ with $2 \leq r$ [1, 17, 20]. Note that mixed boundary conditions play an important role in various applications, e.g. heat conduction and electric potential problems [12].

The goal of this paper is to study the numerical analysis, by using the finite element method, of the convergence results (13) corresponding to the continuous distributed elliptic optimal control problems (P_α) and (P) when $\alpha \to +\infty$. The main result of this paper can be characterized by the following result:

Theorem 1. We have the following complete commutative diagram which relates the continuous distributed mixed optimal control problems (P_α) and (P), with the discrete distributed mixed optimal control problems $(P_{h\alpha})$ and (P_h) and it is obtained by taking the limits $h \to 0, \alpha \to +\infty$ and $(h,\alpha) \to (0,+\infty)$, as in Fig. 1, where $g_{h\alpha_{op}}$, $u_{h\alpha g_{h\alpha_{op}}}$ and $p_{h\alpha g_{h\alpha_{op}}}$ are respectively the optimal control, the system and the adjoint system

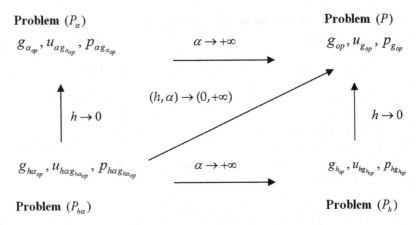

Fig. 1. Relationship among optimal control problems $(P_{h\alpha})$, (P_α), (P_h) and (P) by taking the limits $h \to 0$, $\alpha \to +\infty$ and $(h, \alpha) \to (0, +\infty)$.

states of the discrete distributed mixed optimal control problem $(P_{h\alpha})$ for each $h > 0$ and $\alpha > 0$, and the double convergence is the diagonal one.

The study of the limit $h \to 0$ of the discrete solutions of optimal control problems can be considered as a classical limit, see [4–6, 8, 9, 13–15, 19, 23, 24, 27, 28] but the limit $\alpha \to +\infty$, for each $h > 0$, and the double limit $(h, \alpha) \to (0, +\infty)$ can be considered as a new ones.

The paper is organized as follows. In Sect. 2 we define the discrete elliptic variational equalities for the state systems u_{hg} and $u_{h\alpha g}$, we define the discrete distributed cost functional J_h and $J_{h\alpha}$, we define the discrete distributed optimal control problems (P_h) and $(P_{h\alpha})$, and the discrete elliptic variational equalities for the adjoint state systems p_{hg} and $p_{h\alpha g}$ for each $h > 0$ and $\alpha > 0$, and we obtain properties for the discrete optimal control problems (P_h) and $(P_{h\alpha})$. In Sect. 3 we study the classical convergences of the discrete distributed optimal control problems (P_h) to (P), and $(P_{h\alpha})$ to (P_α) when $h \to 0$ (for each $\alpha > 0$) and the estimations for the discrete cost functional J_h and $J_{h\alpha}$. In Sect. 4 we study the new convergence of the discrete distributed optimal control problems $(P_{h\alpha})$ to (P_h) when $\alpha \to +\infty$ for each $h > 0$ and we obtain a commutative diagram which relates the continuous and discrete distributed mixed optimal control problems $(P_{h\alpha})$, (P_α), (P_h) and (P) by taking the limits $h \to 0$ and $\alpha \to +\infty$. In Sect. 5 we study the new double convergence of the discrete distributed optimal control problems $(P_{h\alpha})$ to (P) when $(h, \alpha) \to (0, +\infty)$ and we obtain the diagonal convergence in the previous commutative diagram.

2 Discretization by Finite Element Method and Properties

We consider the finite element method and a polygonal domain $\Omega \subset \mathbb{R}^n$ with a regular triangulation with Lagrange triangles of type 1, constituted by affine-equivalent finite element of class C^0 being h the parameter of the finite element approximation which goes to zero [3, 7]. We can take h equal to the longest side of the triangles $T \in \tau_h$ and we can approximate the sets V, V_0 and K by:

$$V_h = \{v_h \in C^0(\bar{\Omega})/v_h/T \in P_1(T), \forall T \in \tau_h\}, V_{0h} = \{v_h \in V_h/v_h/\Gamma_1 = 0\}; K_h$$
$$= b + V_{0h} \tag{14}$$

where P_1 is the set of the polymonials of degree less than or equal to 1. Let $\pi_h : C^0(\bar{\Omega}) \to V_h$ be the corresponding linear interpolation operator. Then there exists a constant $c_0 > 0$ (independent of the parameter h) such that [3]:

$$(a) \; \|v - \pi_h(v)\|_H \leq c_0 h^r \|v\|_r; (b) \; \|v - \pi_h(v)\|_V \leq c_0 h^{r-1} \|v\|_r; \forall v \in H^r(\Omega), 1 < r \leq 2. \tag{15}$$

We define the discrete cost functional $J_h, J_{h\alpha} : H \to \mathbb{R}_0^+$ by the following expressions:

$$(a) \; J_h(g) = \frac{1}{2} \|u_{hg} - z_d\|_H^2 + \frac{M}{2} \|g\|_H^2, \quad (b) \; J_{h\alpha}(g) = \frac{1}{2} \|u_{h\alpha g} - z_d\|_H^2 + \frac{M}{2} \|g\|_H^2 \tag{16}$$

where u_{hg} and $u_{h\alpha g}$ are the discrete system states defined as the solution of the following discrete elliptic variational equalities [16, 24]:

$$u_{hg} \in K_h : \quad a(u_{hg}, v_h) = (g, v_h) - \int_{\Gamma_2} q v_h d\gamma, \quad \forall v_h \in V_{0h}, \tag{17}$$

$$u_{h\alpha g} \in V_h : \quad a_\alpha(u_{h\alpha g}, v_h) = (g, v_h) - \int_{\Gamma_2} q v_h d\gamma + \alpha \int_{\Gamma_1} b v_h d\gamma, \quad \forall v_h \in V_h. \tag{18}$$

The corresponding discrete distributed optimal control problems consists in finding $g_{h_{op}}, g_{h\alpha_{op}} \in H$ such that:

$$(a) \; \text{Problem } (P_h) : J_h(g_{h_{op}}) = \underset{g \in H}{Min} \; J_h(g),$$
$$(b) \; \text{Problem } (P_{h\alpha}) : J_{h\alpha}(g_{h\alpha_{op}}) = \underset{g \in H}{Min} \; J_{h\alpha}(g) \tag{19}$$

and their corresponding discrete adjoint states p_{hg} and $p_{h\alpha g}$ are defined respectively as the solution of the following discrete elliptic variational equalities:

$$p_{hg} \in V_{0h} : \quad a(p_{hg}, v_h) = (u_{hg} - z_d, v_h), \quad \forall v_h \in V_{0h} \tag{20}$$

$$p_{h\alpha g} \in V_h : \quad a_\alpha(p_{h\alpha g}, v_h) = (u_{h\alpha g} - z_d, v_h), \quad \forall v_h \in V_h \tag{21}$$

Remark 1. We note that the discrete (in the n-dimensional space) distributed optimal control problem (P_h) and $(P_{h\alpha})$ are still infinite dimensional optimal control problems since the control space is not discretized.

Lemma 2.

(i) *There exist unique solutions $u_{hg} \in K_h$ and $p_{hg} \in V_{0h}$, and $u_{h\alpha g} \in V_h$ and $p_{h\alpha g} \in V_h$ of the elliptic variational equalities (17) and (20), (18), and (21) respectively $\forall g \in H, \forall q \in Q, b > 0$ on Γ_1.*

(ii) *The operators $g \in H \to u_{hg} \in V$, and $g \in H \to u_{h\alpha g} \in V$ are Lipschitzians. The operators $g \in H \to p_{hg} \in V_{0g}$, and $g \in H \to p_{h\alpha g} \in V_h$ are Lipschitzians and strictly monotone operators.*

Proof. We use the Lax-Milgram Theorem, the variational equalities (17), (18), (20) and (21), the coerciveness (10) and (11) and following [10, 18, 25]. $\square$

Theorem 3.

(i) *The discrete cost functional J_h and $J_{h\alpha}$ are H - elliptic and strictly convexe applications, that is $(\forall g_1, g_2 \in H, \forall t \in [0,1])$:*

$$(1-t)J_h(g_2) + tJ_h(g_1) - J_h(tg_1 + (1-t)g_2) \geq M\frac{t(1-t)}{2}\|g_2 - g_1\|_H^2 \quad (22)$$

$$(1-t)J_{h\alpha}(g_2) + tJ_{h\alpha}(g_1) - J_{h\alpha}(tg_1 + (1-t)g_2) \geq M\frac{t(1-t)}{2}\|g_2 - g_1\|_H^2 \quad (23)$$

(ii) *There exist a unique optimal controls $g_{h_{op}} \in H$ and $g_{h\alpha_{op}} \in H$ that satisfy the optimization problems (19a) and (19b) respectively.*

(iii) *J_h and $J_{h\alpha}$ are Gâteaux differentiable applications and their derivatives are given by the following expressions:*

$$(a)\ J_h'(g) = Mg + p_{hg}, \quad (b)\ J_{h\alpha}'(g) = Mg + p_{h\alpha g}, \quad \forall g \in H, \quad \forall h > 0 \quad (24)$$

(iv) *The optimality condition for the optimization problems (19a) and (19b) are given by:*

$$(a)\ J_h'(g_{h_{op}}) = 0 \Leftrightarrow g_{h_{op}} = -\frac{1}{M}p_{hg_{h_{op}}}; \ (b)\ J_{h\alpha}'(g_{h\alpha_{op}}) = 0 \Leftrightarrow g_{h\alpha_{op}}$$
$$= -\frac{1}{M}p_{h\alpha g_{h\alpha op}} \quad (25)$$

(v) *J_h' and $J_{h\alpha}'$ are Lipschitzians and strictly monotone operators.*

Proof. We use the definitions (16a, b), the elliptic variational equalities (17) and (18) and the coerciveness (10) and (11), following [10, 18, 25]. $\square$

We define the operators:

$$W_h, W_{h\alpha} : H \to H \text{ such that } (a)\ W_h(g) = -\frac{1}{M}P_{hg}, \quad (b)\ W_{h\alpha}(g) = -\frac{1}{M}P_{h\alpha g}. \quad (26)$$

Theorem 4. *We have that:*

(i) W_h *and* $W_{h\alpha}$ *are Lipschitzian operators, and* W_h *(* $W_{h\alpha}$*) is a contraction operator if and only if M is large, that is:*

$$(a)\ M > \frac{1}{\lambda^2}, \quad (b)\ M > \frac{1}{\lambda_\alpha^2}. \quad (27)$$

(ii) *If M verifies the inequalities (27a, b) then the discrete distributional optimal control* $g_{h_{op}} \in H$ *(* $g_{h\alpha_{op}} \in H$*) is obtained as the unique fixed point of* W_h *(*$W_{h\alpha}$*), i.e.:*

$$
\begin{aligned}
g_{h_{op}} &= -\frac{1}{M}P_{hg_{h_{op}}} \Leftrightarrow W_h\big(g_{h_{op}}\big) = g_{h_{op}}, \\
g_{h\alpha_{op}} &= -\frac{1}{M}P_{h\alpha g_{h\alpha_{op}}} \Leftrightarrow W_{h\alpha}\big(g_{h\alpha_{op}}\big) = g_{h\alpha_{op}}.
\end{aligned}
\quad (28)
$$

Proof. We use the definitions (25a, b), and the properties (25a, b) and Lemma 2. □

3 Convergence of the Discrete Distributed Optimal Control Problems (P_h) to (P) and $(P_{h\alpha})$ to (P_α) When $h \to 0$

We obtain the following error estimations between the continuous and discrete solutions:

Theorem 6. *We suppose the continuous system states and adjoint system states have the regularities* $u_g, u_{\alpha g_{\alpha_{op}}} \in H^r(\Omega)$ *and* $p_g, p_{\alpha g_{\alpha_{op}}} \in H^r(\Omega)$ *(*$1 < r \le 2$*). If M verifies the inequalities (27a, b) then we have the following error bonds:*

$$\left\|g_{h_{op}} - g_{op}\right\|_H \le ch^{r-1}, \ \left\|u_{hg_{h_{op}}} - u_{g_{op}}\right\|_V \le ch^{r-1}, \ \left\|p_{hg_{h_{op}}} - p_{g_{op}}\right\|_V \le ch^{r-1} \quad (29)$$

$$
\begin{aligned}
&\left\|g_{h\alpha_{op}} - g_{\alpha_{op}}\right\|_H \le ch^{r-1}, \quad \left\|u_{h\alpha g_{h\alpha_{op}}} - u_{\alpha g_{\alpha_{op}}}\right\|_V \le ch^{r-1}, \\
&\left\|p_{h\alpha g_{h\alpha_{op}}} - p_{\alpha g_{\alpha_{op}}}\right\|_V \le ch^{r-1}
\end{aligned}
\quad (30)
$$

where c's are constants independents of h.

Proof. It is useful to use the restriction $\alpha > 1$ by splitting a_α by [21, 24, 25].

$$a_\alpha(u, v) = a_1(u, v) + (\alpha - 1) \int_{\Gamma_1} uvd\gamma \tag{31}$$

but then it can be replaced by $\alpha \geq \alpha_0$ for any $\alpha_0 > 0$. We follow a similar method to the one developed in [25] for Neumann boundary optimal control problems by using the elliptic variational equalities (17), (18), (20) and (21), the thesis holds.   □

Remark 2. If M verifies the inequalities (27a, b) we can obtain the convergence in Theorem 6 by using the characterization of the fixed point (28a, b), and the uniqueness of the optimal controls $g_{op} \in H$ and $g_{\alpha_{op}} \in H$.

Now, we give some estimations for the discrete cost functional $J_{h\alpha}$ and J_h.

Lemma 7. *If M verifies the inequality (27a, b) and the continuous system states and adjoint system states have the regularities* $u_g, u_{\alpha g} \in H^r(\Omega)$ $p_g, p_{\alpha g} \in H^r(\Omega)(1 < r \leq 2)$ *then we have the following error bonds:*

$$\frac{M}{2}\left\|g_{h_{op}} - g_{op}\right\|_H^2 \leq J(g_{h_{op}}) - J(g_{op}) \leq Ch^{2(r-1)},$$
$$\frac{M}{2}\left\|g_{h\alpha_{op}} - g_{\alpha_{op}}\right\|_H^2 \leq J_\alpha(g_{h\alpha_{op}}) - J_\alpha(g_{\alpha_{op}}) \leq Ch^{2(r-1)} \tag{32}$$

$$\frac{M}{2}\left\|g_{h_{op}} - g_{op}\right\|_H^2 \leq J_h(g_{op}) - J_h(g_{h_{op}}) \leq Ch^{2(r-1)};$$
$$\frac{M}{2}\left\|g_{h\alpha_{op}} - g_{\alpha_{op}}\right\|_H^2 \leq J_{h\alpha}(g_{\alpha_{op}}) - J_{h\alpha}(g_{h\alpha_{op}}) \leq Ch^{2(r-1)} \tag{33}$$

$$\left|J_h(g_{op}) - J(g_{op})\right| \leq Ch^{r-1}, \quad \left|J_h(g_{h_{op}}) - J(g_{op})\right| \leq Ch^{r-1} \tag{34}$$

$$\left|J_{h\alpha}(g_{op}) - J_\alpha(g_{op})\right| \leq Ch^{r-1}, \quad \left|J_{h\alpha}(g_{h\alpha_{op}}) - J_\alpha(g_{\alpha_{op}})\right| \leq Ch^{r-1} \tag{35}$$

where C's are constants independents of h and α.

Proof. Estimations (32) and (33) follow from the estimations (29), and the equalities (similar relationship for J and J_α):

$$J_\alpha(g_{h\alpha_{op}}) - J_\alpha(g_{\alpha_{op}}) = \frac{1}{2}\left\|u_{h\alpha g_{h\alpha_{op}}} - u_{\alpha g_{op}}\right\|_H^2 + \frac{M}{2}\left\|g_{h\alpha_{op}} - g_{\alpha_{op}}\right\|_H^2 \tag{36}$$

$$J_{h\alpha}(g_{\alpha_{op}}) - J_{h\alpha}(g_{h\alpha_{op}}) = \frac{1}{2}\left\|u_{h\alpha g_{h_{op}}} - u_{h\alpha g_{h\alpha_{op}}}\right\|_H^2 + \frac{M}{2}\left\|g_{h\alpha_{op}} - g_{\alpha_{op}}\right\|_H^2 \tag{37}$$

$$\left|J_{h\alpha}(g) - J_\alpha(g)\right| \leq \left(\frac{1}{2}\left\|u_{h\alpha g} - u_{\alpha g}\right\|_H + \left\|u_{\alpha g} - z_d\right\|_H\right)\left\|u_{h\alpha g} - u_{\alpha g}\right\|_H, \quad \forall g \in H. \tag{38}$$

□

4 Convergence of the Discrete Optimal Control Problems $(P_{h\alpha})$ to (P_h) When $\alpha \to +\infty$

Theorem 9. We have the following limits:

$$\lim_{\alpha \to +\infty} \left\| u_{h\alpha g_{h\alpha op}} - u_{h g_{hop}} \right\|_V = \lim_{\alpha \to +\infty} \left\| p_{h\alpha g_{h\alpha op}} - p_{h g_{hop}} \right\|_V = \lim_{\alpha \to +\infty} \left\| g_{h\alpha op} - g_{hop} \right\|_H$$
$$= 0, \forall h > 0. \tag{39}$$

Proof. We omit this proof because we prefer to prove the next one with more details.

5 Double Convergence of the Discrete Distributed Optimal Control Problem $(P_{h\alpha})$ to (P) When $(h, \alpha) \to (0, +\infty)$

For the discrete distributed optimal control problem $(P_{h\alpha})$ we will now consider the double limit $(h, \alpha) \to (0, +\infty)$.

Theorem 10. We have the following limits:

$$\lim_{(h,\alpha) \to (0,+\infty)} \left\| u_{h\alpha g_{h\alpha op}} - u_{g_{op}} \right\|_V = \lim_{(h,\alpha) \to (0,+\infty)} \left\| p_{h\alpha g_{h\alpha op}} - p_{g_{op}} \right\|_V$$
$$= \lim_{(h,\alpha) \to (0,+\infty)} \left\| g_{h\alpha op} - g_{op} \right\|_H = 0 \tag{40}$$

Proof. From now on we consider that c's represent positive constants independents simultaneously of $h > 0$ and $\alpha > 0$ (see (31)). We show a sketch of the proof by obtaining the following estimations (for $\forall h > 0$ and $\forall \alpha > 1$):

$$\|u_{h0}\|_V \le c_1, \quad \|u_{h\alpha 0}\|_V \le c_2, \quad (\alpha - 1) \int_{\Gamma_1} (u_{h\alpha 0} - b)^2 d\gamma \le c_3 \tag{41}$$

$$\|g_{h\alpha op}\|_H \le c_4, \quad \|u_{h\alpha g_{h\alpha op}}\|_H \le c_5, \quad \|g_{hop}\|_H \le c_6 \tag{42}$$

$$\|u_{h g_{hop}}\|_V \le c_7, \quad \|u_{h\alpha g_{h\alpha op}}\|_V \le c_8, \quad (\alpha - 1) \int_{\Gamma_1} \left(u_{h\alpha g_{h\alpha op}} - b \right)^2 d\gamma \le c_9 \tag{43}$$

$$\|p_{h g_{hop}}\|_V \le c_{10}, \quad \|p_{h\alpha g_{h\alpha op}}\|_V \le c_{11}, \quad (\alpha - 1) \int_{\Gamma_1} p_{h\alpha g_{h\alpha op}}^2 d\gamma \le c_{12}. \tag{44}$$

For example, the constant c_{11} is a positive constant independent simultaneously of $h > 0$ and $\alpha > 0$, and it is given by the following expression:

$$c_{11} = \|z_d\|_H \left[\frac{1}{\lambda_1} \left(1 + \frac{1}{\sqrt{M}} \left(\frac{1}{\lambda_1} + \frac{1}{\lambda} + \frac{1}{\lambda\lambda_1} \right) \right) + \frac{1}{\lambda} \left(1 + \frac{1}{\lambda_1} \right) \left(1 + \frac{1}{\lambda\sqrt{M}} \right) \right]$$

$$+ b \left[\frac{1}{\lambda_1} \left(1 + \frac{1}{\lambda_1} \right) \left(1 + \frac{1}{\lambda\sqrt{M}} + \frac{1}{\lambda_1\sqrt{M}} \right) + \frac{1}{\lambda} \left(1 + \frac{1}{\lambda_1} \right) \left(1 + \frac{1}{\lambda\sqrt{M}} \right) \right]$$

$$+ \|q\|_Q \|\gamma_0\| \left[\frac{1}{\lambda_1} \left[\frac{1}{\lambda_1} + \frac{1}{\lambda} \left(1 + \frac{1}{\lambda_1} \right) + \frac{1}{\sqrt{M}} \left(\frac{1}{\lambda\lambda_1} + \frac{1}{\lambda^2} \left(1 + \frac{1}{\lambda_1} \right) + \frac{1}{\lambda_1^2} \left(1 + \frac{1}{\lambda} \right) \right) \right] \right.$$

$$\left. + \frac{1}{\lambda^2} \left(1 + \frac{1}{\lambda_1} \right) \left(1 + \frac{1}{\lambda\sqrt{M}} \right) \right] \tag{45}$$

Therefore, from the above estimations we have that:

$$\exists f \in H/g_{h\alpha_{op}} \longrightarrow f \text{ in } H \text{ weak as } (h, \alpha) \to (0, +\infty) \tag{46}$$

$$\exists \eta \in V/u_{h\alpha g_{h\alpha op}} \longrightarrow \eta \text{ in } V \text{ weak } (H \text{ strong}) \text{ as } (h, \alpha) \to (0, +\infty) \text{ with } \eta/\Gamma_1 = b \tag{47}$$

$$\exists \xi \in V/p_{h\alpha g_{h\alpha op}} \longrightarrow \xi \text{ in } V \text{ weak } (H \text{ strong}) \text{ as } (h, \alpha) \to (0, +\infty) \text{ with } \xi/\Gamma_1 = 0 \tag{48}$$

$$\exists f_h \in H/g_{h\alpha_{op}} \longrightarrow f_h \text{ in } H \text{ weak as } \alpha \to +\infty \tag{49}$$

$$\exists \eta_h \in V/u_{h\alpha g_{h\alpha op}} \longrightarrow \eta_h \text{ in } V \text{ weak (in } H \text{ strong}) \text{ as } \alpha \to +\infty \text{ with } \eta_h/\Gamma_1 = b \tag{50}$$

$$\exists \xi_h \in V/p_{h\alpha g_{h\alpha op}} \longrightarrow \xi_h \text{ in } V \text{ weak (in } H \text{ strong}) \text{ as } \alpha \to +\infty \text{ with } \xi_h/\Gamma_1 = 0 \tag{51}$$

$$\exists f_\alpha \in H/g_{h\alpha_{op}} \longrightarrow f_\alpha \text{ in } H \text{ weak as } h \to 0 \tag{52}$$

$$\exists \eta_\alpha \in V/u_{h\alpha g_{h\alpha op}} \longrightarrow \eta_\alpha \text{ in } V \text{ weak (in } H \text{ strong}) \text{ as } h \to 0 \text{ with } \eta_\alpha/\Gamma_1 = b \tag{53}$$

$$\exists \xi_\alpha \in V/p_{h\alpha g_{h\alpha op}} \longrightarrow \xi_\alpha \text{ in } V \text{ weak (in } H \text{ strong}) \text{ as } h \to 0 \text{ with } \xi_\alpha/\Gamma_1 = 0 \tag{54}$$

$$\exists f^* \in H/g_{h_{op}} \longrightarrow f^* \text{ in } H \text{ weak as } h \to 0 \tag{55}$$

$$\exists \eta^* \in V/u_{hg_{hop}} \longrightarrow \eta^* \text{ in } V \text{ weak } (H \text{ strong}) \text{ as } h \to 0 \text{ with } \eta^*/\Gamma_1 = b \tag{56}$$

$$\exists \xi^* \in V/p_{hg_{hop}} \longrightarrow \xi^* \text{ in } V \text{ weak } (H \text{ strong}) \text{ as } h \to 0 \text{ with } \xi^*/\Gamma_1 = 0 \tag{57}$$

Taking into account the uniqueness of the distributed optimal control problems $(P_{h\alpha})$, (P_α), (P_h) and (P), and the uniqueness of the elliptic variational equalities corresponding to their state systems we get

$$\eta_h = u_{hf_h} = u_{hg_{hop}}, \quad \xi_h = p_{hf_h} = p_{hg_{hop}}, \quad f_h = g_{hop} \tag{58}$$

$$\eta_\alpha = u_{\alpha f_\alpha} = u_{\alpha g_{\alpha op}}, \quad \xi_\alpha = p_{\alpha f_\alpha} = p_{\alpha g_{\alpha op}}, \quad f_\alpha = g_{\alpha op} \tag{59}$$

$$\eta = \eta^* = u_f = u_{g_{op}}, \quad \xi = \xi^* = p_f = p_{g_{op}}, \quad f = f^* = g_{op}. \tag{60}$$

Now, by using [11] we obtain

$$\lim_{\alpha \to +\infty} \left\| f_\alpha - g_{op} \right\|_H = 0, \quad \lim_{\alpha \to +\infty} \left\| \eta_\alpha - u_{g_{op}} \right\|_V = 0, \quad \lim_{\alpha \to +\infty} \left\| \xi_\alpha - p_{g_{op}} \right\|_V = 0 \tag{61}$$

and therefore the three double limits (40) hold when $(h, \alpha) \to (0, +\infty)$.

Proof of Theorem 1. It is a consequence of the properties (29), (30), (39), (40) and [10, 11].

Remark 3. We note that this double convergence is a novelty with respect to the recent results obtained for a family of discrete Neumann boundary optimal control problems [25].

Acknowledgements. The present work has been partially sponsored by the Projects PIP No 0534 from CONICET - Univ. Austral, Rosario, Argentina, and AFOSR-SOARD Grant FA9550-14-1-0122.

References

1. Azzam, A., Kreyszig, E.: On solutions of elliptic equations satisfying mixed boundary conditions. SIAM J. Math. Anal. **13**, 254–262 (1982)
2. Bergounioux, M.: Optimal control of an obstacle problem. Appl. Math. Optim. **36**, 147–172 (1997)
3. Brenner, S., Scott, L.R.: The Mathematical Theory of Finite Element Methods. Springer, New York (2008)
4. Casas, E., Mateos, M.: Uniform convergence of the FEM. Applications to state constrained control problems. Comput. Appl. Math. **21**, 67–100 (2002)
5. Casas, E., Mateos, M.: Dirichlet control problems in smooth and nonsmooth convex plain domains. Control Cybern. **40**, 931–955 (2011)
6. Casas, E., Raymond, J.P.: Error estimates for the numerical approximation of Dirichlet boundary control for semilinear elliptic equations. SIAM J. Control Optim. **45**, 1586–1611 (2006)
7. Ciarlet, P.G.: The Finite Element Method for Elliptic Problems. SIAM, Philadelphia (2002)
8. Deckelnick, K., Günther, A., Hinze, M.: Finite element approximation of ellliptic control problems with constraints on the gradient. Numer. Math. **111**, 335–350 (2009)
9. Deckelnick, K., Hinze, M.: Convergence of a finite element approximation to a state-constrained ellliptic control problem. SIAM J. Numer. Anal. **45**, 1937–1953 (2007)
10. Gariboldi, C.M., Tarzia, D.A.: Convergence of distributed optimal controls on the internal energy in mixed elliptic problems when the heat transfer coefficient goes to infinity. Appl. Math. Optim. **47**, 213–230 (2003)

11. Gariboldi, C.M., Tarzia, D.A.: A new proof of the convergence of the distributed optimal controls on the internal energy in mixed elliptic problems. MAT – Serie A **7**, 31–42 (2004)
12. Haller-Dintelmann, R., Meyer, C., Rehberg, J., Schiela, A.: Hölder continuity and optimal control for nonsmooth elliptic problems. Appl. Math. Optim. **60**, 397–428 (2009)
13. Hintermüller, M., Hinze, M.: Moreau-Yosida regularization in state constrained ellliptic control problems: error estimates and parameter adjustement. SIAM J. Numer. Anal. **47**, 1666–1683 (2009)
14. Hinze, M.: A variational discretization concept in control constrained optimization: the linear-quadratic case. Comput. Optim. Appl. **30**, 45–61 (2005)
15. Hinze, M., Matthes, U.: A note on variational discretization of elliptic Nuemann boundary control. Control Cybern. **38**, 577–591 (2009)
16. Kinderlehrer, D., Stampacchia, G.: An Introduction to Variational Inequalities and Their Applications. SIAM, Philadelphia (2000)
17. Lanzani, L., Capogna, L., Brown, R.M.: The mixed problem in L^P for some two-dimensional Lipschitz domain. Math. Ann. **342**, 91–124 (2008)
18. Lions, J.L.: Contrôle optimal des systèmes gouvernés par des équations aux dérivées partielles. Dunod, Paris (1968)
19. Mermri, E.B., Han, W.: Numerical approximation of a unilateral obstacle problem. J. Optim. Theory Appl. **153**, 177–194 (2012)
20. Shamir, E.: Regularization of mixed second order elliptic problems. Isr. J. Math. **6**, 150–168 (1968)
21. Tabacman, E.D., Tarzia, D.A.: Sufficient and/or necessary condition for the heat transfer coefficient on Γ_1 and the heat flux on Γ_2 to obtain a steady-state two-phase Stefan problem. J. Diff. Equ. **77**, 16–37 (1989)
22. Tarzia, D.A.: An inequality for the constant heat flux to obtain a steady-state two-phase Stefan problem. Eng. Anal. **5**, 177–181 (1988)
23. Tarzia, D.A.: Numerical analysis for the heat flux in a mixed elliptic problem to obtain a discrete steady-state two-phase Stefan problem. SIAM J. Numer. Anal. **33**, 1257–1265 (1996)
24. Tarzia, D.A.: Numerical analysis of a mixed elliptic problem with flux and convective boundary conditions to obtain a discrete solution of non-constant sign. Numer. Methods PDE **15**, 355–369 (1999)
25. Tarzia, D.A.: A commutative diagram among discrete and continuous boundary optimal control problems. Adv. Diff. Equ. Control Process. **14**, 23–54 (2014)
26. Tröltzsch, F.: Optimal Control of Partial Differential Equations. Theory, Methods and Applications. American Mathematical Society, Providence (2010)
27. Yan, M., Chang, L., Yan, N.: Finite element method for constrained optimal control problems governed by nonlinear elliptic PDEs. Math. Control Relat. Fields **2**, 183–194 (2012)
28. Ye, Y., Chan, C.K., Lee, H.W.J.: The existence results for obstacle optimal control problems. Appl. Math. Comput. **214**, 451–456 (2009)

Modeling and Control of Low-Frequency Electromagnetic Fields in Multiply Connected Conductors

Fredi Tröltzsch[1(✉)] and Alberto Valli[2]

[1] Institut für Mathematik, Technische Universität Berlin, D-10623 Berlin, Germany
fredi.troeltzsch@math.tu-berlin.de
[2] Dipartimento di Matematica, Università di Trento, 38123 Trento, Italy
valli@science.unitn.it

Abstract. We consider a particular model for electromagnetic fields in the context of optimal control. Special emphasis is laid on a non-standard H-based formulation of the equations of low-frequency electromagnetism in multiply connected conductors. By this technique, the low-frequency Maxwell equations can be solved with reduced computational complexity. We show the well-posedness of the system and derive the sensitivity analysis for different models of controls.

Keywords: Electromagnetic fields · Maxwell equations · Eddy current equations · H-based approximation · Low frequency approximation · Optimal control

1 Introduction

In this paper, we suggest an H-based eddy current formulation of the time-harmonic Maxwell equations, where a standard scalar elliptic equation is given in the insulator and a vector formulation is only needed in the conductor. This approach is theoretically slightly more complicated than the well known vector potential ansatz. However, we think that the computational savings can be considerable, if the computational domain Ω must be large. We apply this H-based formulation to the optimal control of electric and magnetic fields and discuss associated optimality conditions. Special emphasis is laid on a variety of models for controls.

Optimal control of electromagnetic fields is a quite active subject, important for various applications. We mention only the control of induction heating as in [8,9,16], the optimal control of MHD processes as in [3–7], optimal control problems for time-harmonic eddy current problems as in [10,11], inverse problems for electromagnetic fields as in [2], or the control of magnetic fields in flow measurement as in [12,13] and refer to [15] for more references.

F. Tröltzsch—The first author was supported by Centro Internazionale di Ricerca Matematica (CIRM-FBK) Trento and by the Einstein Center for Mathematics Berlin (ECMath).

L. Bociu et al. (Eds.): CSMO 2015, IFIP AICT 494, pp. 505–516, 2016.
DOI: 10.1007/978-3-319-55795-3_48

2   The Model for the Electromagnetic Fields

2.1   Time-Harmonic Maxwell and Eddy Current Equations

The main quantities in our eddy current formulation, are the magnetic field H, the electric field E, and the (total) current J that is the sum of the generated current and an impressed current J_e. By the generalized Ohm's law, we have

$$J = \sigma E + J_e, \tag{1}$$

where σ is the electrical conductivity, that is assumed to be a symmetric and (uniformly) positive definite matrix in the conducting region and to vanish in the insulating region. We assume that the entries of σ are bounded and measurable real functions on the conducting domain Ω_C.

We consider a time-harmonic model and assume that J_e is an alternating current of the form $J_e(\mathbf{x}, t) = J(\mathbf{x}) \cos(\omega t + \phi)$, where J is a real vector function that accounts for direction and strength of the current, ω is the angular frequency and ϕ is the phase angle. Expressing these quantities in a complex setting, we have

$$J_e(\mathbf{x}, t) = \mathrm{Re}\left[J(\mathbf{x})e^{i\omega t + i\phi}\right] = \mathrm{Re}\left[\mathbf{J}_e(\mathbf{x})e^{i\omega t}\right].$$

The complex vector function $\mathbf{J}_e = J e^{i\phi}$ will be our control; we assume that it is supported in the conducting region, namely, it is vanishing inside the non-conducting region. This time-periodic impressed current J_e generates associated time-periodic solutions in the form

$$E(\mathbf{x}, t) = \mathrm{Re}\left[\mathbf{E}(\mathbf{x})e^{i\omega t}\right], \qquad H(\mathbf{x}, t) = \mathrm{Re}\left[\mathbf{H}(\mathbf{x})e^{i\omega t}\right].$$

Inserting these quantities in the full Maxwell equations and assuming that the displacement current term $\partial(\varepsilon E)/\partial t$, ε being the electric permittivity, can be neglected, one arrives in a standard way at the following *time-harmonic eddy current system*

$$\begin{aligned} \mathrm{curl}\,\mathbf{H} - \sigma\mathbf{E} &= \mathbf{J}_e \\ \mathrm{curl}\,\mathbf{E} + i\omega\mu\mathbf{H} &= \mathbf{0} \end{aligned} \tag{2}$$

that holds in the whole space $\mathbb{R}^3$. Here, μ is the magnetic permeability, a uniformly positive definite matrix that is assumed to have bounded and measurable real functions as entries on the holdall domain Ω.

2.2   Eddy Current Formulation in Weak and Strong Form

Assumption 1 (Geometry). In the paper, $\Omega \subset \mathbb{R}^3$ is a bounded and simply connected Lipschitz domain with connected boundary Γ; Ω is the "holdall" computational domain containing all conductors. The subdomain $\Omega_C \subset \Omega$ that denotes the conductor is a bounded Lipschitz set. We require that Ω_C is the union of finitely many disjoint open and connected sets $(\Omega_C)_l$, $l \in \{1, \ldots, k\}$, the so-called (connected) components of Ω_C. Assume further that $\mathrm{cl}\,\Omega_C \cap \partial\Omega = \emptyset$. The set $\Omega_I := \Omega \setminus \mathrm{cl}\,\Omega_C$ stands for the non-conducting domain. For simplicity, it is assumed to be connected.

Definition 1. *Let* $g \in \mathbb{N} \cup \{0\}$ *be the number of all "handles" of* Ω_I *(precisely, the rank of the first homology group of* cl Ω_I, *or, equivalently, the first Betti number of* Ω_I*). Due to our assumption on* Ω, *it is also the number of "handles" of* Ω_C. *If all the components* $(\Omega_C)_l$ *are simply connected, we have* $g = 0$.

This assumption allows fairly general forms of conductors. For instance, the conducting domain can include finitely many tori which might form together more complicated geometrical figures like the Borromean rings.

The function spaces used in our paper will include complex functions. For instance, $L^p(D)$, $1 \le p < \infty$, is defined as the space of all complex valued functions $v : D \to \mathbb{C}$ such that $|v|^p$ is integrable on $D \subset \mathbb{R}^3$. To distinguish this space from the one with real-valued functions, we introduce

$$L_{\mathbb{R}}^p(D) = \{v : D \to \mathbb{R}, |v|^p \text{ is integrable}\}.$$

The spaces $L^\infty(D)$ (complex) and $L_{\mathbb{R}}^\infty(D)$ (real) are defined accordingly.

Definition 2. *We denote by* $\boldsymbol{\rho}_j$, $j \in \{1, \ldots, g\}$, *a basis of the space of* μ-*harmonic fields*

$$\mathcal{H}_I^\mu = \{\mathbf{v} : \Omega_I \to \mathbb{R}^3 : \operatorname{curl} \mathbf{v} = \mathbf{0} \text{ in } \Omega_I, \operatorname{div}(\mu \mathbf{v}) = 0 \text{ in } \Omega_I, \mu \mathbf{v} \cdot \mathbf{n} = 0 \text{ on } \partial \Omega_I\},$$
(3)

where $\mathbf{n}$ *is the unit outward normal vector on* $\partial \Omega_I$.

The functions $\boldsymbol{\rho}_j$ can be computed once "offline" before the numerical solution of the optimal control problem is started. For associated equations, we refer to [1].

From $(2)_1$ we see that $\operatorname{curl} \mathbf{H} = \mathbf{0}$ holds in Ω_I. Therefore, $\mathbf{H}_{|\Omega_I}$ can be written as $\nabla \psi + \sum_{j=1}^g \alpha_j \boldsymbol{\rho}_j$ (see, e.g., [1, Appen. A.3]). This leads to the weak formulation of our eddy current system: Let $\mathbf{V} = H(\operatorname{curl}; \Omega_C) \times H^1(\Omega_I)/\mathbb{C} \times \mathbb{C}^g$ and define the state space

$$\mathbf{V}_0 = \{(\mathbf{H}, \psi, \boldsymbol{\alpha}) \in \mathbf{V} \text{ that satisfy the interface conditions (4) below}\},$$

where

$$\mathbf{H} \times \mathbf{n} - \nabla \psi \times \mathbf{n} - \sum_{j=1}^g \alpha_j \boldsymbol{\rho}_j \times \mathbf{n} = \mathbf{0} \qquad \text{on } \Gamma. \tag{4}$$

Both spaces $\mathbf{V}$ and $\mathbf{V}_0$ are equipped with the norm

$$\|(\mathbf{H}, \Psi, \boldsymbol{\alpha})\|_{\mathbf{V}} = \left(\|\mathbf{H}\|_{H(\operatorname{curl}; \Omega_C)}^2 + \|\psi\|_{H^1(\Omega_I)/\mathbb{C}}^2 + |\boldsymbol{\alpha}|^2 \right)^{1/2},$$

where $\|\mathbf{H}\|_{H(\operatorname{curl}; \Omega_C)} = \left(\int_{\Omega_C} (\operatorname{curl} \mathbf{H} \cdot \operatorname{curl} \overline{\mathbf{H}} + \mathbf{H} \cdot \overline{\mathbf{H}}) \right)^{1/2}$ and $\|\psi\|_{H^1(\Omega_I)/\mathbb{C}} = \left(\int_{\Omega_I} \nabla \psi \cdot \nabla \overline{\psi} \right)^{1/2}$. We also need the norms $\|\mathbf{Q}\|_{\Omega_C} := \left(\int_{\Omega_C} |\mathbf{Q}(\mathbf{x})|^2 \right)^{\frac{1}{2}}$, $\|\mathbf{Q}\|_{\mu, \Omega_C} := \left(\int_{\Omega_C} \mu(\mathbf{x}) \mathbf{Q}(\mathbf{x}) \cdot \overline{\mathbf{Q}(\mathbf{x})} \right)^{\frac{1}{2}}$, and analogous norms $\|\mathbf{Q}\|_{\sigma, \Omega_C}$ and

$\|\mathbf{Q}\|_{\mu,\Omega_I}$. Further, we introduce a symmetric and positive definite matrix M by

$$M_{nj} = \int_{\Omega_I} \mu\boldsymbol{\rho}_n \cdot \boldsymbol{\rho}_j\,;$$

and the vector norm $|\mathbf{q}|_M = (M\mathbf{q}\cdot\overline{\mathbf{q}})^{\frac{1}{2}}$, where $\mathbf{q} \in \mathbb{C}^g$. Finally, we define an antilinear form $a : \mathbf{V} \times \mathbf{V} \to \mathbb{C}$ by

$$a[\mathbf{u},\mathbf{v}] = \int_{\Omega_C} \sigma^{-1}\operatorname{curl}\mathbf{H}\cdot\operatorname{curl}\overline{\mathbf{W}} + \int_{\Omega_C} i\omega\mu\,\mathbf{H}\cdot\overline{\mathbf{W}} + \int_{\Omega_I} i\omega\mu\,\nabla\psi\cdot\nabla\overline{\eta} + i\omega M\boldsymbol{\alpha}\cdot\overline{\boldsymbol{\beta}},$$

where $\mathbf{u} = (\mathbf{H},\psi,\boldsymbol{\alpha})$ and $\mathbf{v} = (\mathbf{W},\eta,\boldsymbol{\beta})$. The form $a[\cdot,\cdot]$ is continuous and coercive on $\mathbf{V} \times \mathbf{V}$ (see, e.g., [1, p. 37]).

Definition 3. *A triplet* $\mathbf{u} = (\mathbf{H},\psi,\boldsymbol{\alpha}) \in \mathbf{V}_0$ *is said to be a weak solution of the eddy current model associated with* $\mathbf{J}_e \in L^2(\Omega_C)^3$, *if*

$$a[\mathbf{u},\mathbf{v}] = \int_{\Omega_C} \sigma^{-1}\mathbf{J}_e \cdot \operatorname{curl}\overline{\mathbf{W}} \qquad \forall\mathbf{v} := (\mathbf{W},\eta,\boldsymbol{\beta}) \in \mathbf{V}_0. \tag{5}$$

Lemma 1 (Well posedness, [15]). *For all* $\mathbf{J}_e \in L^2(\Omega_C)^3$, *there exists a unique weak solution* $(\mathbf{H},\psi,\boldsymbol{\alpha})$ *of* (7). *Moreover, there is a constant* $c > 0$ *not depending on* $\mathbf{J}_e$ *such that*

$$\|(\mathbf{H},\psi,\boldsymbol{\alpha})\|_\mathbf{V} \le c\,\|\mathbf{J}_e\|_{\Omega_C}. \tag{6}$$

We have shown in [15] that the solution $(\mathbf{H},\psi,\boldsymbol{\alpha}) \in \mathbf{V}_0$ to the variational problem (5) satisfies the following strong eddy current equations, provided that the variational solution is sufficiently smooth:

$$\begin{aligned}
\operatorname{curl}(\sigma^{-1}\operatorname{curl}\mathbf{H}) + i\omega\mu\,\mathbf{H} &= \operatorname{curl}(\sigma^{-1}\mathbf{J}_e) &&\text{in } \Omega_C \\
\mathbf{H}\times\mathbf{n} &= \nabla\psi\times\mathbf{n} + \textstyle\sum_{j=1}^g \alpha_j\boldsymbol{\rho}_j\times\mathbf{n} &&\text{on } \Gamma \\
\mu\mathbf{H}\cdot\mathbf{n} &= \mu\nabla\psi\cdot\mathbf{n} &&\text{on } \Gamma \\
-\operatorname{div}(\mu\,\nabla\psi) &= 0 &&\text{in } \Omega_I \\
\mu\nabla\psi\cdot\mathbf{n}_\Omega &= 0 &&\text{on } \partial\Omega
\end{aligned} \tag{7}$$

with additional geometrical conditions

$$(M\boldsymbol{\alpha})_j = (i\omega)^{-1}\int_\Gamma \sigma^{-1}(\operatorname{curl}\mathbf{H} - \mathbf{J}_e)\cdot(\mathbf{n}\times\boldsymbol{\rho}_j) \quad \forall j \in \{1,\ldots,g\}. \tag{8}$$

3 Optimal Control

3.1 The Optimal Current Problem and Its Well-Posedness

We discuss the following steady state optimal control problem of elliptic type, where the impressed current $\mathbf{J}_e$ is the control. As fixed data, vector functions $\mathbf{H}_d \in \mathbf{L}^2(\Omega)^3$, $\mathbf{E}_d \in \mathbf{L}^2(\Omega_C)^3$ and constants $\nu_C \ge 0$, $\nu_A \ge 0$, $\nu_B \ge 0$, $\nu_E \ge 0$, $\nu \ge 0$ with $\nu_C + \nu_A + \nu_B + \nu_E + \nu > 0$ are given. In Ω_I the reference magnetic field $\mathbf{H}_d$ is split as $\nabla \psi_d + \sum_{j=1}^{g} \alpha_{d,j} \boldsymbol{\rho}_j$. Moreover, a nonempty, bounded, convex and closed set of admissible controls $\mathbf{J}_{ad} \subset L^2(\Omega_C)^3$ is given. Possible choices for $\mathbf{J}_{ad}$ will be specified later.

Thanks to Lemma 1, for each control $\mathbf{J}_e \in \mathbf{J}_{ad}$ there exists a unique weak solution of (7). We express the correspondence of the solution to $\mathbf{J}_e$, by the notation $(\mathbf{H}_{\mathbf{J}_e}, \psi_{\mathbf{J}_e}, \alpha_{\mathbf{J}_e})$ for the solution. Let us now skip the subscript e from the controls and denote them just by $\mathbf{J}$, i.e. $\mathbf{J}$ stands now for the impressed current $\mathbf{J}_e$ and is not the total current. We use the following (reduced) objective functional F,

$$F(\mathbf{J}) = \frac{\nu_C}{2} \|\mathbf{H}_J - \mathbf{H}_d\|_{\mu,\Omega_C}^2 + \frac{\nu_A}{2} \|\nabla \psi_J - \nabla \psi_d\|_{\mu,\Omega_I}^2 + \frac{\nu_B}{2} |\alpha_J - \alpha_d|_M^2$$
$$+ \frac{\nu_E}{2} \|\sigma^{-1}(\operatorname{curl} \mathbf{H}_J - \mathbf{J}) - \mathbf{E}_d\|_{\sigma,\Omega_C}^2 + \frac{\nu}{2} \|\mathbf{J}\|_{\Omega_C}^2. \tag{9}$$

Recalling that the electric field associated with $\mathbf{J}$ is given by $\mathbf{E}_J = \sigma^{-1}(\operatorname{curl} \mathbf{H}_J - \mathbf{J})$, it is easily checked that in F the magnetic energy and the electric energy (per unit time) of $\mathbf{H}$ and $\mathbf{E}$, respectively, appear.

The optimal control problem, written in short form, is

$$\min_{\mathbf{J} \in \mathbf{J}_{ad}} F(\mathbf{J}). \tag{10}$$

A control $\mathbf{J}^* \in \mathbf{J}_{ad}$ is said to be *optimal*, if $F(\mathbf{J}^*) \leq F(\mathbf{J})$ holds for all $\mathbf{J} \in \mathbf{J}_{ad}$.

Theorem 2. *The optimal control problem* (10) *admits at least one optimal control denoted by* $\mathbf{J}^*$. *The optimal control is unique, if* $\nu > 0$.

In view of the continuity of the control-to-state mapping, this is a standard result.

3.2   Necessary Optimality Conditions

The objective functional F is not differentiable, but it is directionally differentiable. This is enough to derive necessary (and by convexity also sufficient) optimality conditions. After quite elementary calculations, the derivative in the direction $\mathbf{J}$ at an arbitrary fixed (not necessarily optimal or admissible) control $\widehat{\mathbf{J}}$ with associated solution $\widehat{\mathbf{H}} := \mathbf{H}_{\widehat{J}}$, $\widehat{\psi} := \psi_{\widehat{J}}$ and $\widehat{\alpha} := \alpha_{\widehat{J}}$ is obtained as

$$F'(\widehat{\mathbf{J}})\, \mathbf{J} = \operatorname{Re} \left\{ \int_{\Omega_C} \nu_C\, \mu(\widehat{\mathbf{H}} - \mathbf{H}_d) \cdot \overline{\mathbf{H}_J} \right.$$
$$+ \int_{\Omega_I} \nu_A\, \mu(\nabla \widehat{\psi} - \nabla \psi_d) \cdot \overline{\nabla \psi_J} + \nu_B\, M(\widehat{\alpha} - \alpha_d) \cdot \overline{\alpha_J}$$
$$+ \int_{\Omega_C} \nu_E (\widehat{\mathbf{E}} - \mathbf{E}_d) \cdot \operatorname{curl} \overline{\mathbf{H}_J}$$
$$\left. - \int_{\Omega_C} \nu_E (\widehat{\mathbf{E}} - \mathbf{E}_d) \cdot \overline{\mathbf{J}} + \nu \int_{\Omega_C} \widehat{\mathbf{J}} \cdot \overline{\mathbf{J}} \right\}. \tag{11}$$

Here, we have inserted the relation $\sigma^{-1}(\operatorname{curl} \widehat{\mathbf{H}} - \widehat{\mathbf{J}}) = \widehat{\mathbf{E}} := \mathbf{E}_{\widehat{J}}$. By an adjoint state, this derivative is transformed to one with explicit appearance of $\mathbf{J}$.

Definition 4 (Adjoint equation). *Let* $\widehat{\mathbf{J}} \in L^2(\Omega_C)^3$ *be a given control with associated states* $\widehat{\mathbf{H}} := \mathbf{H}_{\widehat{\mathbf{J}}}$, $\widehat{\mathbf{E}} := \mathbf{E}_{\widehat{\mathbf{J}}}$, $\widehat{\psi} := \psi_{\widehat{\mathbf{J}}}$, $\widehat{\alpha} := \alpha_{\widehat{\mathbf{J}}}$, *and let* $\mathbf{H}_d \in L^2(\Omega_C)^3$, $\psi_d \in H^1(\Omega_I)/\mathbb{C}$, $\alpha_d \in \mathbb{C}^g$, $\mathbf{E}_d \in L^2(\Omega_C)^3$ *be given as above. The equation for* $(\mathbf{W}, \eta, \boldsymbol{\beta})$,

$$
\int_{\Omega_C} \sigma^{-1} \operatorname{curl} \mathbf{W} \cdot \operatorname{curl} \overline{\mathbf{H}} - i\omega \int_{\Omega_C} \mu \mathbf{W} \cdot \overline{\mathbf{H}} - i\omega \int_{\Omega_I} \mu \nabla \eta \cdot \nabla \overline{\psi} - i\omega M \boldsymbol{\beta} \cdot \overline{\boldsymbol{\alpha}}
$$

$$
= \int_{\Omega_C} \nu_C \, \mu(\widehat{\mathbf{H}} - \mathbf{H}_d) \cdot \overline{\mathbf{H}}
$$

$$
+ \int_{\Omega_I} \nu_A \, \mu(\nabla \widehat{\psi} - \nabla \psi_d) \cdot \nabla \overline{\psi} + \nu_B \, M(\widehat{\alpha} - \alpha_d) \cdot \overline{\boldsymbol{\alpha}}
$$

$$
+ \int_{\Omega_C} \nu_E (\widehat{\mathbf{E}} - \mathbf{E}_d) \cdot \operatorname{curl} \overline{\mathbf{H}} \qquad \forall (\mathbf{H}, \psi, \alpha) \in \mathbf{V}_0
$$

(12)

is said to be the adjoint equation *of equation* (5). *The solution* $(\mathbf{W}_{\widehat{\mathbf{J}}}, \eta_{\widehat{\mathbf{J}}}, \boldsymbol{\beta}_{\widehat{\mathbf{J}}}) \in \mathbf{V}_0$ *is called the* adjoint state *associated with* $\widehat{\mathbf{J}}$.

For the strong form of the adjoint equation, we refer the reader to [15].

For all given $\mathbf{H}_d \in L^2(\Omega_C)^3$, $\psi_d \in H^1(\Omega_I)/\mathbb{C}$, $\alpha_d \in \mathbb{C}^g$, $\mathbf{E}_d \in L^2(\Omega_C)^3$, $\widehat{\mathbf{J}} \in L^2(\Omega_C)^3$, the adjoint equation (12) has a unique solution $(\mathbf{W}_{\widehat{\mathbf{J}}}, \eta_{\widehat{\mathbf{J}}}, \boldsymbol{\beta}_{\widehat{\mathbf{J}}})$. This result follows, analogously to Lemma 1, from the Lemma of Lax and Milgram. By transposition, we can prove the following necessary optimality conditions:

Theorem 3 (Necessary optimality conditions). *Let* $\mathbf{J}^*$ *be an optimal control of problem* (10) *and let* $\mathbf{H}_{J^*}$ *and* $\mathbf{E}_{J^*}$ *be the associated optimal magnetic and electric fields, respectively. Then there exists a unique solution* $(\mathbf{W}_{J^*}, \eta_{J^*}, \boldsymbol{\beta}_{J^*})$ *of the adjoint equation* (12) *such that the variational inequality*

$$
\operatorname{Re} \int_{\Omega_C} \left(\sigma^{-1} \operatorname{curl} \mathbf{W}_{J^*} - \nu_E (\mathbf{E}_{J^*} - \mathbf{E}_d) + \nu \mathbf{J}^* \right) \cdot (\overline{\mathbf{J}} - \overline{\mathbf{J}^*}) \geq 0 \qquad \forall \mathbf{J} \in \mathbf{J}_{ad} \quad (13)
$$

is satisfied.

Proof. The optimal control $\mathbf{J}^*$ must obey the standard variational inequality

$$
F'(\mathbf{J}^*)(\mathbf{J} - \mathbf{J}^*) \geq 0 \quad \forall \mathbf{J} \in \mathbf{J}_{ad}. \tag{14}
$$

We show that this is equivalent to the variational inequality (13). We first consider the expression (11) for $F'(\widehat{\mathbf{J}})$ for the particular choice $\widehat{\mathbf{J}} := \mathbf{J}^*$ and have

$$
F'(\mathbf{J}^*)(\mathbf{J} - \mathbf{J}^*)
$$

$$
= \operatorname{Re} \left[\nu_C \int_{\Omega_C} \mu(\mathbf{H}_{J^*} - \mathbf{H}_d) \cdot \overline{\mathbf{H}_{J-J^*}} \right.
$$

$$
+ \nu_A \int_{\Omega_I} \mu(\nabla \psi_{J^*} - \nabla \psi_d) \cdot \nabla \overline{\psi_{J-J^*}} + \nu_B \, M(\alpha_{J^*} - \alpha_d) \cdot \overline{\alpha_{J-J^*}}
$$

$$
+ \nu_E \int_{\Omega_C} (\mathbf{E}_{J^*} - \mathbf{E}_d) \cdot \operatorname{curl} \overline{\mathbf{H}_{J-J^*}} - \nu_E \int_{\Omega_C} (\mathbf{E}_{J^*} - \mathbf{E}_d) \cdot (\overline{\mathbf{J}} - \overline{\mathbf{J}^*})
$$

$$
\left. + \nu \int_{\Omega_C} \mathbf{J}^* \cdot (\overline{\mathbf{J}} - \overline{\mathbf{J}^*}) \right].
$$

Thanks to a lemma on transposition in [15] that is not deep but a bit technical in the proof, we obtain

$$
\begin{aligned}
F'(\mathbf{J}^*)\,(\mathbf{J} - \mathbf{J}^*) & \\
= \operatorname{Re} & \left[\int_{\Omega_C} \sigma^{-1} \operatorname{curl} \mathbf{W}_{J^*} \cdot (\overline{\mathbf{J}} - \overline{\mathbf{J}^*}) \right. \\
& \left. - \int_{\Omega_C} \nu_E \left(\mathbf{E}_{J^*} - \mathbf{E}_d \right) \cdot (\overline{\mathbf{J}} - \overline{\mathbf{J}^*}) + \int_{\Omega_C} \nu \mathbf{J}^* \cdot (\overline{\mathbf{J}} - \overline{\mathbf{J}^*}) \right] \\
= \operatorname{Re} & \int_{\Omega_C} \left(\sigma^{-1} \operatorname{curl} \mathbf{W}_{J^*} - \nu_E \left(\mathbf{E}_{J^*} - \mathbf{E}_d \right) + \nu \mathbf{J}^* \right) \cdot (\overline{\mathbf{J}} - \overline{\mathbf{J}^*}),
\end{aligned}
\tag{15}
$$

where $\mathbf{W}_{J^*}$ is the first component of the adjoint state associated with $\mathbf{J}^*$. $\quad\square$

Let us define for convenience

$$
\mathbf{D}_{J^*} := \sigma^{-1} \operatorname{curl} \mathbf{W}_{J^*} - \nu_E \left(\mathbf{E}_{J^*} - \mathbf{E}_d \right).
\tag{16}
$$

By this definition, the variational inequality (13) simplifies to

$$
\operatorname{Re} \int_{\Omega_C} \left(\mathbf{D}_{J^*} + \nu \mathbf{J}^* \right) \cdot (\overline{\mathbf{J}} - \overline{\mathbf{J}^*}) \geq 0 \quad \forall \mathbf{J} \in \mathbf{J}_{ad}.
\tag{17}
$$

This is our main necessary condition that will be later used to handle various particular cases for $\mathbf{J}_{ad}$. Though our objective functional F is only directionally differentiable and hence does not have a gradient, we denote for short the direction of steepest ascent of $F'(\widehat{\mathbf{J}})$ as its *reduced gradient*:

$$
\nabla F(\widehat{\mathbf{J}}) := \mathbf{D}_{\widehat{J}} + \nu \widehat{\mathbf{J}}.
\tag{18}
$$

3.3   Modeling the Control and Associated Optimality Conditions

Below, we discuss several types of controls and admissible sets that seem to be useful and establish the associated optimality conditions as conclusions of (17).

Unbounded Complex Control Vectors. If $\nu > 0$, the unbounded control set

$$
\mathbf{J}_{ad} = L^2(\Omega_C)^3
\tag{19}
$$

can be used. Notice that the choice $\nu = 0$ is only useful here, if the desired fields $\mathbf{H}_d$ and $\mathbf{E}_d$ belong to the range of the control-to-state mapping. It follows immediately from the variational inequality (13) that, in the case $\mathbf{J}_{ad} = L^2(\Omega_C)^3$, the equation $\mathbf{D}_{J^*} + \nu \mathbf{J}^* = 0$ is necessary and sufficient for the optimality of $\mathbf{J}^*$, i.e. we have

$$
\mathbf{J}^* = -\frac{1}{\nu} \mathbf{D}_{J^*}.
$$

Complex Control Vectors Bounded by Box Constraints. For all $\nu \geq 0$, the set

$$\mathbf{J}_{ad} = \{\mathbf{J} \in L^2(\Omega_C)^3 : |\operatorname{Re} J_\ell(\mathbf{x})| \leq \operatorname{Re}_{\max}, |\operatorname{Im} J_\ell(\mathbf{x})| \leq \operatorname{Im}_{\max} \atop \text{for } \ell = 1, 2, 3, \text{ and for almost all } \mathbf{x} \in \Omega_C\} \quad (20)$$

might be taken, if positive bounds $\operatorname{Re}_{\max}$ and $\operatorname{Im}_{\max}$ must be imposed on the possible currents. In this case, using the representation (18), the variational inequality (13) can be re-written as

$$\operatorname{Re} \int_{\Omega_C} \nabla F(\mathbf{J}^*) \cdot \overline{\mathbf{J}^*} \leq \operatorname{Re} \int_{\Omega_C} \nabla F(\mathbf{J}^*) \cdot \overline{\mathbf{J}} \quad \forall \mathbf{J} \in \mathbf{J}_{ad}.$$

Expanding the terms under the integral and invoking that $\operatorname{Re} \mathbf{J}$ and $\operatorname{Im} \mathbf{J}$ can be chosen completely independent, we find the following two inequalities:

$$\int_{\Omega_C} \operatorname{Re}(\nabla F(\mathbf{J}^*)) \cdot \operatorname{Re} \mathbf{J}^* \leq \int_{\Omega_C} \operatorname{Re}(\nabla F(\mathbf{J}^*)) \cdot \operatorname{Re} \mathbf{J} \quad \forall \mathbf{J} : |\operatorname{Re} \mathbf{J}(\cdot)| \leq \operatorname{Re}_{\max},$$

$$\int_{\Omega_C} \operatorname{Im}(\nabla F(\mathbf{J}^*)) \cdot \operatorname{Im} \mathbf{J}^* \leq \int_{\Omega_C} \operatorname{Im}(\nabla F(\mathbf{J}^*)) \cdot \operatorname{Im} \mathbf{J} \quad \forall \mathbf{J} : |\operatorname{Im} \mathbf{J}(\cdot)| \leq \operatorname{Im}_{\max}.$$

Here, the inequalities $|\operatorname{Re} \mathbf{J}(\cdot)| \leq \operatorname{Re}_{\max}$ and $|\operatorname{Im} \mathbf{J}(\cdot)| \leq \operatorname{Im}_{\max}$ have to be understood in pointwise and componentwise sense. These inequalities can be discussed further in a pointwise way (for this type of argument, see, e.g., [14, Sect. 2.8]). For instance, the first inequality is equivalent to the condition that

$$\operatorname{Re} \nabla F(\mathbf{J}^*)(\mathbf{x}) \cdot \operatorname{Re} \mathbf{J}^*(\mathbf{x}) \leq \operatorname{Re} \nabla F(\mathbf{J}^*)(\mathbf{x}) \cdot \mathbf{v} \quad \forall \mathbf{v} \in \mathbb{R}^3 : |v_\ell| \leq \operatorname{Re}_{\max}, \ell \in \{1, 2, 3\} \quad (21)$$

holds for almost all $\mathbf{x} \in \Omega_C$. All components of the vector $\mathbf{v} \in \mathbb{R}^3$ can be selected independently. Then the inequality above means for the ℓth component that

$$\min_{v \in \mathbb{R} : |v| \leq \operatorname{Re}_{\max}} \operatorname{Re}(\nabla F(\mathbf{J}^*))_\ell(\mathbf{x}) v = \operatorname{Re}(\nabla F(\mathbf{J}^*))_\ell(\mathbf{x}) \operatorname{Re} J_\ell^*(\mathbf{x}),$$

i.e., that, for a.a. $\mathbf{x} \in \Omega_C$, the minimum at the left-hand side is attained by $\operatorname{Re} J_\ell^*(\mathbf{x})$.

Inserting the concrete expression for the reduced gradient ∇F (see (18)), we find

$$\operatorname{Re} J_\ell^*(\mathbf{x}) = \begin{cases} -\operatorname{Re}_{\max}, & \text{if } \operatorname{Re}(\mathbf{D}_{J^*} + \nu \mathbf{J}^*)_\ell(\mathbf{x}) > 0 \\ \operatorname{Re}_{\max}, & \text{if } \operatorname{Re}(\mathbf{D}_{J^*} + \nu \mathbf{J}^*)_\ell(\mathbf{x}) < 0 \end{cases} \quad (22)$$

for almost all $\mathbf{x} \in \Omega_C$ and all $\ell \in \{1, 2, 3\}$. The formula for the imaginary part is the same with Im substituted for Re. If the Tikhonov regularization parameter is positive, then this is equivalent to the projection formula

$$\operatorname{Re} J_\ell^*(\mathbf{x}) = \mathbb{P}_{[-\operatorname{Re}_{\max}, \operatorname{Re}_{\max}]} \left\{ -\frac{1}{\nu} \operatorname{Re}(\mathbf{D}_{J^*})_\ell(\mathbf{x}) \right\} \quad (23)$$

for almost all $\mathbf{x} \in \Omega_C$ and all $\ell \in \{1, 2, 3\}$. Here, the projection function $\mathbb{P}_{[a,b]} : \mathbb{R} \to [a, b]$ is defined by $\mathbb{P}_{[a,b]}(s) := \max(a, \min(b, s))$.

Example 1 (Optimal control as inverse problem). In electro-encephalography (EEG) or magneto-encephalography (MEG), magnetic or electric fields associated to the electrical activity of the human brain are measured. Then one looks for the electrical currents, located in certain regions of the brain, that generated these fields. Under certain assumptions, this problem can be cast into the form of our optimal control problem, where the desired fields $\mathbf{H}_d$ and $\mathbf{E}_d$ stand for the measurements. Normally, these measurements can be taken only at the boundary Γ of the conductor, say at the surface of the human head Ω_C (which can be assumed to be simply connected). Moreover, they are only given at certain points. Let us assume that these measurements can be interpolated to get a measurement of $\mu \mathbf{H}_d \cdot \mathbf{n}$ on the interface Γ. In view of the interface conditions on Γ, we have then also $\mu \nabla \psi_d \cdot \mathbf{n}$ on Γ. Together with the homogeneous boundary conditions on $\partial \Omega$, we then can determine the harmonic scalar potential ψ_d and hence also $\nabla \psi_d$ in Ω_I that can serve as measurement in Ω_I.

In this inverse problem, one cannot prescribe any particular form or direction of the unknown electrical current $\mathbf{J}_e$. Here the general class $\mathbf{J}_{ad}$ of arbitrary bounded $\mathbf{L}^2$-controls is meaningful indeed. Possible selections of $\mathbf{J}_{ad}$ are the definitions (19) and (20).

Electrical Current in an Induction Coil. Another typical application is the case where the electrical current is prescribed in an induction coil (see, e.g., [13]). A standard induction coil is composed by one wire that is twisted in many windings around the core. Here, the direction of the electrical current in one point is very precisely given by the direction of the wire in that point. The strength j of the current is the only unknown that is to be determined. The control $\mathbf{J}_e$ has the form

$$\mathbf{J}_e(\mathbf{x}) = \frac{N_\star}{Q_{\text{coil}}} \mathbf{e}(\mathbf{x}) j$$

where j is a complex number, the unit vector function $\mathbf{e}$ is the direction of the wire in the point $\mathbf{x}$ of the coil, $N_\star$ is the number of windings and Q_{coil} is the area of the cross section of the coil that is perpendicular to the direction of the windings. Assume for convenience that $N_\star / Q_{\text{coil}} = 1$ to simplify our notation.

Example 2. In [13] the following geometry was chosen for the induction coil, which is topologically equivalent to a torus:

$$\Omega_{\text{coil}} = \{\mathbf{x} \in \mathbb{R}^3 : 0 < r_1 < x_1^2 + x_2^2 < r_2, c_1 < x_3 < c_2\},$$

where $r_2 > r_1 > 0$ and $c_1 < c_2$ are given real numbers. Here the function $\mathbf{e}$ is defined by

$$\mathbf{e}(x_1, x_2, x_3) = \begin{cases} \frac{1}{\sqrt{x_1^2 + x_2^2}} \begin{bmatrix} -x_2 \\ x_1 \\ 0 \end{bmatrix} & \text{in } \Omega_{\text{coil}} \\ 0 & \text{in } \Omega_C \setminus \text{cl}\, \Omega_{\text{coil}}. \end{cases} \qquad (24)$$

In Ω_{coil}, $\mathbf{e}$ is a unit vector.

Notice that in this case the control is just one complex number. Here, the analogue of (20) is

$$\mathbf{J}_{ad} = \{\mathbf{e}(\cdot) j : |\operatorname{Re} j| \leq \operatorname{Re}_{\max} \text{ and } |\operatorname{Im} j| \leq \operatorname{Im}_{\max}\}. \qquad (25)$$

The optimality conditions can be discussed analogously to complex control vector functions, we leave the main steps to the reader. For the real part, we deduce for $\nu \geq 0$

$$\text{Re } j^* = \begin{cases} -\text{Re}_{max}, & \text{if } \int_{\Omega_C} \text{Re} \left(\mathbf{D}_{J^*} \cdot \mathbf{e} + \nu j^* \right) > 0 \\ \text{Re}_{max}, & \text{if } \int_{\Omega_C} \text{Re} \left(\mathbf{D}_{J^*} \cdot \mathbf{e} + \nu j^* \right) < 0. \end{cases} \tag{26}$$

If $\nu > 0$, then we have the projection formula

$$\text{Re } j^* = \mathbb{P}_{[-\text{Re}_{max}, \text{Re}_{max}]} \left\{ -\frac{1}{\nu} \int_{\Omega_C} \text{Re } \mathbf{D}_{J^*} \cdot \mathbf{e} \right\}. \tag{27}$$

Analogous conditions are satisfied by $\text{Im } j^*$ with Im substituted for Re.

Electrical Currents in a Package of Wires. The following situation is somehow intermediate between the two cases mentioned above. Here, the induction coil is composed of a package of single wires that can be controlled separately. Assume that each one of these currents can be controlled independently from the others. The cross section of this package of wires can be viewed as a discrete approximation of a function $j : \Omega_{\text{coil}} \to \mathbb{C}$ that stands for the strength of the current while the direction is still given by a function such as $\mathbf{e}$ above.

Let us consider the geometry of Example 2. Here, the strength j of the current depends only on the radius r and the coordinate x_3, while the direction of the current is given again by $\mathbf{e}$. In terms of cylindrical coordinates, this reads

$$\mathbf{J}_e = \mathbf{e}(r, \varphi, z) \, j(r, z),$$

where $r_1 \leq r \leq r_2$, $0 \leq \varphi < 2\pi$, $c_1 \leq z \leq c_2$. A useful set of admissible control functions might be

$$\mathbf{J}_{ad} = \mathbf{e} \, j_{ad} \tag{28}$$

where we take controls out of the complex space $L^2((r_1, r_2) \times (c_1, c_2))$,

$$j_{ad} = \{ j \in L^2((r_1, r_2) \times (c_1, c_2)) : |\text{Re } j| \leq \text{Re}_{max} \text{ and } |\text{Im } j| \leq \text{Im}_{max} \}$$

and the actual control function would be $j \in L^2((r_1, r_2) \times (c_1, c_2))$. This view is, perhaps, a bit academic but it gives an interpretation on how a controlled distributed current might be generated.

The necessary optimality conditions are analogous to (22) and (23), but $(r_1, r_2) \times (c_1, c_2)$ must be substituted for Ω_C. For instance, the optimal solution obeys, for almost all $(r, z) \in [r_1, r_2] \times [c_1, c_2]$, the projection formula

$$\text{Re } j^*(r, z) = \mathbb{P}_{[-\text{Re}_{max}, \text{Re}_{max}]} \left\{ -\frac{1}{\nu} \int_0^{2\pi} \text{Re } \mathbf{D}_{J^*}(r, z) \cdot \mathbf{e}(r, \varphi, z) \, d\varphi \right\}. \tag{29}$$

Real Current Vectors. A smaller but perhaps more realistic class of controls $\mathbf{J}$ has the particular form

$$\mathbf{J}(\mathbf{x}) = e^{i\phi} J(\mathbf{x}), \tag{30}$$

where J is a real vector function and ϕ is fixed. Here, J varies in the admissible set

$$J_{ad} = \{J \in L_{\mathbb{R}}^2(\Omega_C)^3 : -j_{\max} \leq J_\ell(\mathbf{x}) \leq j_{\max}$$
$$\text{for a.a. } \mathbf{x} \in \Omega_C, \text{ all } \ell \in \{1, 2, 3\}\} \tag{31}$$

with a given bound $j_{\max} > 0$. To cover this ansatz by the control problem (10), we define the functional $f(J) := F(e^{i\phi}J)$ and consider the problem

$$\min_{J \in J_{ad}} f(J). \tag{32}$$

This is nothing more than a particular case of the optimal control problem (10) subject to the particular control set defined by (30) and (31).

The associated optimal control $\mathbf{J}^* = e^{i\phi}J^*$ has to obey the necessary optimality conditions of Theorem 3, in particular (13), i.e.

$$\mathrm{Re} \int_{\Omega_C} (\mathbf{D}_{J^*} + \nu \mathbf{J}^*) \cdot (\overline{\mathbf{J}} - \overline{\mathbf{J}^*}) \geq 0 \quad \forall \mathbf{J} \in \mathbf{J}_{ad}$$

using the notation (16). With the particular ansatz (30), this variational inequality can be further simplified. Finally, inserting the particular form of $\mathbf{J}$, we find

$$\int_{\Omega_C} (D_{J^*} + \nu J^*) \cdot (J - J^*) \geq 0 \quad \forall J \in J_{ad}, \tag{33}$$

with $D_{J^*} := \mathrm{Re}\,(e^{-i\phi}\mathbf{D}_{J^*})$. The further pointwise discussion of (33) is analogous to (22) and (23), where "Re" can be omitted, since all quantities in (33) are real.

References

1. Alonso Rodríguez, A., Valli, A.: Eddy Current Approximation of Maxwell Equations: Theory, Algorithms and Applications. MS&A, vol. 4. Springer-Verlag Italia, Milan (2010)
2. Arnold, L., von Harrach, B.: A unified variational formulation for the parabolic-elliptic eddy current equations (2011, submitted)
3. Bärwolff, G., Hinze, M.: Optimization of semiconductor melts. ZAMM Z. Angew. Math. Mech. **86**(6), 423–437 (2006)
4. Druet, P., Klein, O., Sprekels, J., Tröltzsch, F., Yousept, I.: Optimal control of three-dimensional state-constrained induction heating problems with nonlocal radiation effects. SIAM J. Control Optim. **49**(4), 1707–1736 (2011)
5. Griesse, R., Kunisch, K.: Optimal control for a stationary MHD system in velocity-current formulation. SIAM J. Control Optim. **45**(5), 1822–1845 (2006)
6. Gunzburger, M., Trenchea, C.: Analysis and discretization of an optimal control problem for the time-periodic MHD equations. J. Math. Anal. Appl. **308**(2), 440–466 (2005)
7. Hinze, M.: Control of weakly conductive fluids by near wall Lorentz forces. GAMM-Mitt. **30**(1), 149–158 (2007)
8. Hömberg, D., Sokołowski, J.: Optimal shape design of inductor coils for surface hardening. Numer. Funct. Anal. Optim. **42**, 1087–1117 (2003)

9. Hömberg, D., Volkwein, S.: Control of laser surface hardening by a reduced-order approach using proper orthogonal decomposition. Math. Comput. Model. **38**, 1003–1028 (2003)

10. Kolmbauer, M.: The multiharmonic finite element and boundary element method for simulation and control of eddy current problems. Ph.D. thesis (2012)

11. Kolmbauer, M., Langer, U.: A robust preconditioned MinRes solver for distributed time-periodic eddy current optimal control problems. SIAM J. Sci. Comput. **34**(6), B785–B809 (2012)

12. Nicaise, S., Stingelin, S., Tröltzsch, F.: On two optimal control problems for magnetic fields. Comput. Methods Appl. Math. **14**, 555–573 (2014)

13. Nicaise, S., Stingelin, S., Tröltzsch, F.: Optimal control of magnetic fields in flow measurement. Discret. Contin. Dyn. Syst.-S **8**, 579–605 (2015)

14. Tröltzsch, F.: Optimal Control of Partial Differential Equations: Theory, Methods and Applications, vol. 112. American Mathematical Society, Providence (2010)

15. Tröltzsch, F., Valli, A.: Optimal control of low-frequency electromagnetic fields in multiply connected conductors. Optimization **65**(9), 1651–1673 (2016)

16. Yousept, I., Tröltzsch, F.: PDE-constrained optimization of time-dependent 3D electromagnetic induction heating by alternating voltages. ESAIM M2AN **46**, 709–729 (2012)

Parameter Estimation Algorithms for Kinetic Modeling from Noisy Data

Fabiana Zama[1](✉), Dario Frascari[2], Davide Pinelli[2], and A.E. Molina Bacca[2]

[1] Department of Mathematics, University of Bologna, Bologna, Italy
fabiana.zama@unibo.it
[2] Department of Civil, Chemical,
Environmental and Materials Engineering (DICAM),
University of Bologna, Bologna, Italy

Abstract. The aim of this work is to test the Levemberg Marquardt and BFGS (Broyden Fletcher Goldfarb Shanno) algorithms, implemented by the matlab functions `lsqnonlin` and `fminunc` of the Optimization Toolbox, for modeling the kinetic terms occurring in chemical processes of adsorption. We are interested in tests with noisy data that are obtained by adding Gaussian random noise to the solution of a model with known parameters. While both methods are very precise with noiseless data, by adding noise the quality of the results is greatly worsened. The *semi-convergent* behaviour of the relative error curves is observed for both methods. Therefore a stopping criterion, based on the Discrepancy Principle is proposed and tested. Great improvement is obtained for both methods, making it possible to compute stable solutions also for noisy data.

Keywords: Parameter estimation · Non-linear differential models · Quasi-Newton methods · Discrepancy Principle

1 Introduction

An important topic in many engineering applications is that of estimating parameters of differential models from partial and possibly noisy measurements. For example the removal of pollutants from surface water and groundwater requires the optimization of partial differential models where the dispersion, mass transfer and reaction terms are estimated from data in column reactor experiments [1–3].

We define here the constrained optimization problem connected to the estimation of a parameter, defined by q, in a differential model represented by $c(u, q)$, named state equation, whose solution $u(q)$ is called state variable:

$$\min_q J(u, q) \; s.t. \; c(u, q) = 0 \; \text{ODE-PDE model}$$

By pointing out the implicit dependence of u on the parameter q, the problem is usually presented in its reduced form $\min_q \widehat{J}(q)$ where $\widehat{J}(q) \equiv J(u(q), q)$ represents the fit to the given data.

© IFIP International Federation for Information Processing 2016
Published by Springer International Publishing AG 2016. All Rights Reserved
L. Bociu et al. (Eds.): CSMO 2015, IFIP AICT 494, pp. 517–527, 2016.
DOI: 10.1007/978-3-319-55795-3_49

By defining the observation operator $\mathcal{C}$, that maps the state variable $u \in U$ into the measurements space Y, we obtain the measurements $y \in Y$. The data fidelity term $\widehat{J}(q)$ is defined as: $\widehat{J}(q) \equiv \|F(q) - y\|$ where $F(q) \equiv \mathcal{C}(u(q))$.

Hence the final problem consists in the minimization of the distance between the data and the computed approximation $(F(q))$ measured in a norm dependent on the model of the data noise. The discrete finite dimensional optimization problem is obtained by defining the vector parameter $\mathbf{q} \in \mathbb{R}^P$ and computing the noisy data $\mathbf{y}^\delta \in \mathbb{R}^N$ by sampling y at N points and adding a noise term.

Depending on the type of noise present in the data the discrete minimization problem can be defined in the suitable L_p norm $1 \le p < \infty$. In case of Gaussian random noise the L_2 norm is the optimal choice, obtaining the following nonlinear least squares problem:

$$\min_{\mathbf{q}} \frac{1}{2} \|F(\mathbf{q}) - \mathbf{y}^\delta\|_2^2. \tag{1}$$

Since the present paper focuses on the L_2 norm we define $\| \cdot \| \equiv \| \cdot \|_2$. It is well known that such problems are ill-posed in the sense that noise present in the data leads to poor solutions, hence some form of regularization needs to be introduced. Among the most common ways to regularize problem (1) is the introduction of a suitable regularization constraint, taking into account the smoothness of the solution (see [8] and reference therein). However the application of such methods to the estimation of several parameters of a differential model is quite challenging and has a high computational cost. A more practical way consists in exploiting the possibly semi-convergent behavior of the iterative methods used to solve (1) and compute stable solutions by means of a suitable stopping criterion. The aim of this work is to test the iterative methods implemented by the functions fminunc and lsqnolin of the Matlab Optimization Toolbox and evaluate their efficiency in the solution of problem (1) with noisy data, verifying the *semi-convergence* in presence of medium-high noise. To this purpose we define a test problem where the state equation is a system of two time dependent differential equations, representing the dynamic evolution of the liquid and solid phases of Polyphenolic compounds [4]. The noisy measurements $\mathbf{y}^\delta$ are obtained by adding Gaussian noise to the solution of the state equation with given reference parameters $\mathbf{q}_{true}$. We observe a progressive worsening of the results of both functions while increasing the level of noise in the data. The analysis of the relative errors at each iteration shows that the error curve has the typical *semi-convergent* behaviour: it decreases in the first steps and, after reaching a minimum value, it start to increase, reaching errors possibly higher than those at the initial step. Hence a change in the convergence conditions and tolerances can improve the solutions. Although the semi-convergence of Descent, Gradient and Simultaneous Iterative Reconstruction Technique (SIRT) methods is well understood [9–11], the same does not apply to Gauss Newton, Levemberg Marquardt or quasi Newton Methods such as the BFGS (Broyden Fletcher Goldfarb Shanno). Therefore after the a posteriori verification of such behaviour, we define a stopping rule based on the Morozov's Discrepancy Principle [7], that is proved to be suitable for descent, gradient and SIRT methods. By suitably defining the `options` structure, it is possible to modify the stopping conditions of the matlab optimization functions and evaluate the improvement obtained by computing the solution reached by the application of the stopping rule in case of data with different noise levels. Moreover we can evaluate the efficiency of our stopping rule compared to the optimal solution obtained by minimizing the relative error.

In Sect. 2 we define the state equation and analyze the details of its numerical solution. The optimization methods implemented by the functions `fminunc` and `lsqnolin` are outlined in Sect. 3, together with the proposed stopping rule. Finally in Sect. 4 we report the numerical results and the conclusions.

2   The Adsorption Model

We describe here the differential problem (state equation) used as a test problem. It consists of a system of two time dependent differential equations representing the dynamic evolution of the liquid and solid phases of Polyphenolic compounds [4]. In the hypotheses of not negligible mass transfer and Langmuir adsorption isotherm the liquid phase u is modeled by a convection, diffusion and reaction equation while the solid phase concentration v is characterized by the absence of any dispersion and convection:

$$\begin{cases} \delta_u \frac{\partial u}{\partial t} = -\nu \frac{\partial u}{\partial z} + D\frac{\partial^2 u}{\partial z^2} - R(u,v,\boldsymbol{\theta}) + f_u \\ \delta_v \frac{\partial v}{\partial t} = R(u,v,\boldsymbol{\theta}) + f_v \end{cases} \quad z \in [0,L] \qquad (2)$$

where the adsorption isotherm is given by

$$R(u,v,\boldsymbol{\theta}) = \theta_1 \left(u - \frac{\theta_2 v}{\theta_3 - v} \right)$$

The parameters to be identified are $\boldsymbol{\theta} = (\theta_1, \theta_2, \theta_3)$ while the dispersion coefficient D, the interstitial velocity coefficient ν and the retardation factors δ_u, δ_v are assumed to be known. The spatial domain is given by the height L of the column reactor. Dirichlet boundary conditions are assumed in $z = 0$ while convective flux is assumed in $z = L$ for the liquid phase. Using the Method of Lines, problem (2) is tranformed in the following system of Nonlinear Ordinary Differential Equations:

$$\begin{cases} U'(t) = KU(t) - G(t,\boldsymbol{\theta}) \\ V'(t) = G(t,\boldsymbol{\theta}) \end{cases}, \quad G(t,\boldsymbol{\theta}) \in \mathbb{R}^M, \qquad (3)$$

where M is the number of intervals in the spatial domain, $U(t) = (u_1(t),\dots,u_M(t))$, $V(t) = (v_1(t),\dots,v_M(t))$ are the discrete solutions at time t and G is the discrete isotherm at time t of components $G_i = R(u_i(t),v_i(t),\boldsymbol{\theta})$, $i = 1,\dots,M$. The matrix $K \in \mathbb{R}^{M \times M}$ is the tridiagonal matrix obtained by applying the second order finite differences approximation of the spatial derivatives in (2). It is well known that such systems tend to become very stiff at the increasing of the spatial resolution M, hence a suitable implicit solver is required. In our experiments we use the matlab function `ode15s`, which implements variable order (1–5) method and variable step size, being therefore the most accurate solver available in Matlab ODE-suite for stiff problems.

3   Iterative Regularization

In this section we define the stopping rule applied to the iterative methods used by `fminunc` and `lsqnolin` functions to solve the test problem obtained by the model described in Sect. 2. We start by a brief outline of the iterative numerical methods tested in the numerical experiments.

The first method, Levemberg Marquardt, is specific of the non linear Least squares minimization while the second method, BFGS (Broyden Fletcher Goldfarb Shanno) quasi Newton method, is applied to more general nonlinear minimization problems. (See [5,6] for details.) Both methods compute a sequence of approximate solutions of (1), $\{\mathbf{q}(k)\}$, $k = 0, 1, \ldots$ by the following update relation:

$$\mathbf{q}^{(k+1)} = \mathbf{q}^{(k)} + \alpha_k \mathbf{s}_k \tag{4}$$

where $\alpha_k \in (0, 1]$ is a damping parameter used to guarantee the decrease of the residual norm $\|F(\mathbf{q}^{(k)}) - \mathbf{y}^\delta\|_2$. In the case of Levemberg Marquardt the direction $\mathbf{s}_k$ is computed by solving the linear system obtained by the first order conditions of the linear approximation of the residual at $\mathbf{q}^{(k+1)}$:

$$\left((J_F^{(k)})^t J_F^{(k)} + \lambda_k I\right) \mathbf{s}_k = -(J_F^{(k)})^t (F(\mathbf{q}^{(k)}) - \mathbf{y}^\delta), \lambda_k \geq 0 \tag{5}$$

where $J_F^{(k)}$ is the Jacobian matrix $(J_F^{(k)})_{i,j} = \partial F_i(\mathbf{q}^{(k)})/\partial \mathbf{q}_j$. If the parameter λ_k is zero we have the Gauss Newton Method, otherwise, to overcome possibly singular Jacobians, a diagonal positive matrix is added by means of a small scalar parameter λ_k (Levemberg Marquardt Method). Notice that (5) is the first order condition of the following constrained minimization problem:

$$\min_{\mathbf{s}} \|J_F^{(k)} \mathbf{s} + \mathbf{r}_k\|_2^2, \ s.t. \ \|\mathbf{s}\|_2^2 \leq \Delta_k \tag{6}$$

where $\mathbf{r}_k = \mathbf{y}^\delta - F(\mathbf{q}^{(k)})$ and Δ_k is the level of smoothness required by $\mathbf{s}$ and can be computed by the Trust Region method (see algorithm 4.1 in [6]).

In absence of data noise the following stopping rule is used to stop the iterations:

$$G_F \mathbf{s}^{(k)} < \tau_F \ \text{and} \ \|G_F\|_\infty < 10(\tau_F + \tau_X), \|\mathbf{s}^{(k)}\|_\infty < \tau_X \tag{7}$$

where $G_F = 2(J_F^{(k)})^t (F(\mathbf{q}^{(k)}) - \mathbf{y}^\delta)$ and τ_F, τ_X are tolerance parameters.

The second family of methods is that of Quasi Newton methods where the direction $\mathbf{s}_k$ used in the update step (4) is computed by solving the following linear system:

$$H^{(k)} \mathbf{s}_k = -\nabla_{\mathbf{q}} \widehat{J}(\mathbf{q}^{(k)})$$

where $H^{(k)}$ is the approximate Hessian matrix, whose value is updated by adding a rank one update term. The BFGS method uses the following update term $H^{(k+1)} = H^{(k)} + S(\mathbf{q}^{(k)})$, where

$$S(\mathbf{q}^{(k)}) = \frac{v_k v_k^t}{v_k^t s_k} - \frac{H^{(k)} s_k s_k^t H^{(k)}}{s_k^t H^{(k)} s_k}, \quad v_k = \nabla_\mathbf{q}(\widehat{J}(\mathbf{q}^{(k+1)})) - \nabla_\mathbf{q}(\widehat{J}(\mathbf{q}^{(k)}))$$

The initial Hessian approximation is chosen as $H^{(0)} = \gamma I$ where $\gamma > 0$ is relative to the scaling of the variables. The stopping criterion applied in this case is:

$$\|\nabla \widehat{J}(\mathbf{q}^{(k)})\|_\infty < \tau_F (1 + \|\nabla \widehat{J}(\mathbf{q}^{(0)})\|_\infty), \quad \max_i \left(\frac{|q_i^{(k+1)} - q_i^{(k)}|}{1 + |q_i^{(k)}|} \right) < \tau_X \qquad (8)$$

The main computational effort in each iteration is the computation of the Jacobian matrix. The approach used here is that of finite difference approximation. Although it is not optimal for precision and computational cost, it is simple and readily available in matlab software optimization tools. As reported in Sect. 4, in case of noiseless data, the sequences $\{\mathbf{q}^{(k)}\}$ of both methods converge to good approximate solutions of (1). On the other hand, with noisy data we observe bad results for both algorithms. A thorough analysis reveals that the error curve has the typical *semi-convergent* behaviour: it decreases in the first steps and then start to increase giving a completely wrong solution. An improvement can be obtained by heuristically increasing the tolerances τ_F and τ_X, so as to decrease the number of steps. This strategy would require specific tolerance values for each noise level, which is unknown.

More systematic stopping rules can be obtained by means of the Morozov's Discrepancy Principle (MDP) that proposes to stop (4) at the d-th iteration as soon as the residual norm approximates the data noise:

$$\|F(\mathbf{q}^{(d)}) - \mathbf{y}^\delta\| \simeq \delta, \qquad (9)$$

The main drawback is the need to estimate the noise δ. In order to overcome this difficulty we exploit the decreasing behaviour of the residual norm. We observe that the decrease of the residual norm is fast in the first iterations and tends to become slower and slower as the iterations increase. Hence computing the decrease rate (measured as the difference of the residual norms in two successive steps: d and $d-1$) we stop as soon as it becomes sufficiently small, compared to the initial decrease rate. Therefore we propose the following stopping rule (SR_d)

$$|R_d - R_{d-1}| < \tau |R_1 - R_0|, \quad \tau > 0 \qquad (10)$$

where R_d is the residual norm at d-th step: $R_d = \frac{1}{2}\|F(\mathbf{q}^{(d)}) - \mathbf{y}^\delta\|_2$ and the parameter τ represents the ratio between the change in the residual norm at step d and that at the first step. The optimal parameter should stop the iterations as soon as the noise starts to deteriorate the solution, causing an increase of the relative error. Hence the optimal value depends on the method, the data and the noise as well. Improvement in (10) could be obtained by adding information about the maximum relative change in the solution. However in this application, using only information about the residual norm, it is possible to obtain great improvement in the results. In all our experiments we used the following values for the tolerance τ: 10^{-1} for the Levemberg Marquardt method and 10^{-5} for the BFGS method.

4 Numerical Experiments

The numerical experiments reported in this section are carried out on a test problem obtained by the model equations (2) where the terms f_u and f_v are defined by the known solutions:

$$u(t,z) = e^{(-\pi^2 t)}(\sin(\pi z^2))\cos(0.5\pi z^2), \; v(t,z) = e(-\pi^2 t)(\sin(\pi z))$$

The reference parameter vector $\mathbf{q}_{true}$ has elements $[\theta_1, \theta_2, \theta_3] = [1, 2, 3]$. The retardation factors are $\delta_u = 1$, $\delta_v = 2.2$, and the spatial domain $[0.2, 0.8]$ is discretized using N_z uniform spaced samples. The differential system (3) is solved in the time interval $[0, 0.1]$ by means of the matlab function ode15s with tolerance parameters AbsTol=RelTol=10^{-10}.

The measurements $\mathbf{y} = F(\mathbf{q}_{true})$ are defined on a uniformly spaced grid of $N_t \times N_z$ points on the time space domain and are computed by solving (3) on an oversampled space domain: $\simeq 3N_z$ points. In order to compare the results $(\mathbf{q})$ computed by the different methods, we evaluate the Parameter Relative Error (PRE) and Residual Norm $(ResN)$ defined as follows:

$$PRE = \frac{\|\mathbf{q} - \mathbf{q}_{true}\|}{\|\mathbf{q}\|} \quad ResN = \|F(\mathbf{q}) - \mathbf{y}\| \tag{11}$$

The reported results are computed on a PC Intel(R) equipped with 4 i5 processors 5.8 GB Ram, using Matlab R2010a.

In the first experiment we compare the results obtained by the Levemberg Marquardt and BFGS methods without data noise. The Levemberg Marquardt method with parameter λ (5) is implemented by the matlab function lsqnonlin with the option 'Algorithm',{'levenberg-marquardt',λ}. In this experiment we choose the constant value $\lambda = \varepsilon$ (machine epsilon) throughout all the iterations and we call this method $LM(\varepsilon)$.

The BFGS method is implemented by the matlab function fminunc setting the option {'HessUpdate', 'bfgs'}. The starting value $\mathbf{q}^{(0)}$ is chosen at a relative distance δ_q from $\mathbf{q}_{true}$, i.e. $\mathbf{q}^{(0)} = \mathbf{q}_{true}(1 \pm \delta_q)$. The iterations are stopped using the standard convergence stopping rules (7) and (8) with $\tau_F = \tau_X = 10^{-6}$.

In Table 1 we report the parameter error (PRE) and Residual norm $(NRes)$ obtained by solving the problem with $N_t \times N_z$ samples ranging in the domain $[20, 40] \times [20, 40]$ and with an initial relative distance δ_q given by 25% and 40%. The computational cost is evaluated by the total number of function evaluations fval and by the number of iterations k. We observe that $LM(\varepsilon)$ performs much

Table 1. Test with noiseless measurements. Levemberg Marquardt $LM(\varepsilon)$ with $\lambda = \varepsilon$. BFGS with $H_0 = I$

$\delta_q(\%)$	N_t	N_z	$LM(\varepsilon)$			BFGS		
			PRE	$ResN$	k(fval)	PRE	$RcsN$	k(fval)
25	20	20	1.0455e−12	3.0782e−26	5(24)	4.8683e−03	6.2636e−09	11(56)
	20	30	2.0644e−11	1.9521e−24	5(24)	1.7381e−03	1.1973e−09	9(52)
	30	30	1.3760e−12	4.5015e−25	5(24)	1.7167e−03	1.7694e−09	10(48)
	40	30	3.1294e−10	3.0582e−22	5(24)	4.5629e−05	1.6661e−12	20(92)
	40	40	1.3050e−12	2.1200e−25	5(24)	7.3953e−05	5.9466e−12	20(92)
40	20	20	2.9505e−12	1.8642e−26	7(32)	5.3876e−04	8.7967e−11	22(108)
	20	30	1.4647e−11	5.3618e−25	7(32)	7.5150e−05	2.4042e−12	25(112)
	30	30	4.7820e−12	3.2652e−25	7(32)	5.2751e−05	1.7968e−12	25(112)
	40	30	4.4779e−11	8.5756e−24	7(32)	3.8518e−05	1.2619e−12	25(112)
	40	40	1.3728e−11	1.0526e−24	7(32)	1.3826e−04	2.5034e−11	24(108)

better than BFGS for precision and computational complexity. Furthermore it is more robust in terms of dependence on the initial parameter estimate δ_q. We see that an initial relative error of 25% or 40% doesn't affect much the errors of $LM(\varepsilon)$ while BFGS is more precise for smaller values of δ_q. The behavior of the error curves plotted in Fig. 1 confirms the faster convergence of $LM(\varepsilon)$ compared to BFGS.

In the second experiment we introduce Gaussian random noise of level $\delta \in [10^{-4}, 10^{-1})$ and estimate the parameters starting from the noisy data $\mathbf{y}^\delta$ defined as follows:

$$\mathbf{y}^\delta = \mathbf{y} + \delta\|\mathbf{y}\|\boldsymbol{\eta}, \quad \|\boldsymbol{\eta}\| = 1 \tag{12}$$

We run this analysis for all the cases in Table 1 but, as an example, we report in Table 2 the details of the case $N_t = N_z = 40$ with an initial parameter estimate $\delta_q = 40\%$. The error parameters reported in Table 2 show that both methods are very sensitive to noise, even if BFGS seems to be more stable than $LM(\varepsilon)$ when $\delta \geq 1.e - 2$. Focusing on the BFGS method we observe the *semi-convergence* by

Table 2. Noisy data: $N_t = N_z = 40$, $\delta_q = 40\%$

δ	$LM(\varepsilon)$			BFGS		
	PRE	$ResN$	k (fval)	PRE	$ResN$	k (fval)
1e−4	1.1224e−3	1.1849e−5	7(32)	1.2120e−3	1.7211e−03	21(104)
1e−3	1.3231e−2	1.1852e−3	7(32)	1.3275e−2	1.7229e−02	24(112)
5e−03	1.1362e−01	2.9605e−02	5(24)	1.1415e−01	8.6031e−02	24(108)
1e−2	6.2798e−2	1.1855e−01	6(41)	5.7114e−2	8.6031e−02	24(108)
5e−2	1.2936e+4	2.9593	8(36)	1.7283e+1	8.6013e−01	32(208)

plotting the PRE values at each iteration (see Fig. 2). The PRE curves obtained with noise $\delta = 0.01$ and $\delta = 0.1$ are plotted in Fig. 2, where the iteration at which the relative error is minimum is represented by a red star and the iteration d obtained by rule SR_d (10) by a green circle. By changing the exit condition as in (10), with tolerance $\tau = 10^{-5}$, we obtain a great improvement for the noisy data, as reported in Table 3. The relative errors obtained in column 5 are very close to the minimum value (column 2). The exit condition of the fminunc function is changed by setting the OutputFcn field in the options structure to a user defined mfile.m function that implements the stopping rule (10).

Table 3. BFGS results with best and stopping rule DR_d : $N_t = N_z = 40$, $\delta_q = 40\%$

δ	Best			SR_d		
	PRE	$ResN$	k	PRE	$ResN$	d
5e-03	5.7449e–02	8.6034e–02	11	5.7449e–02	8.6034e–02	10
1e-02	5.7087e–02	1.7215e–01	11	5.7114e–02	1.7215e–01	11
5e-02	7.0872e–02	8.6022e–01	11	7.1133e–02	8.6022e–01	11
1e-01	5.6863e–02	1.7213	5	7.7457e–02	1.7211	9

Concerning the Levemberg Marquardt method, it is possible to improve the results in case of noisy data by using a larger initial value λ_0 and a suitable strategy to update it during the computation. The matlab function lsqnonlin implements the following strategy:

$$\lambda_k = \max(0.1\lambda_{k-1}, \text{eps}) \qquad (13)$$

where eps is the machine epsilon ε. Hence we repeat the experiment with noisy data choosing the default value $\lambda = 0.05$ and refer to this method as $LM(.05)$. Comparing the columns 2 in Tables 2 and 4, we observe an improvement of the errors when $\delta > 1.e - 3$. Also in this case we have *semi-convergence* because the update formula (13) is not optimal. Actually the value of λ_k should be updated by taking into account the constraint in problem (6) (see [3] for details about a possible implementation). By applying the stopping rule (10) with $\tau = 0.1$ we observe a more stable behaviour of the error (Table 4 column 8), which is always better than the standard stopping rule (7) and very close to the minimum value (column 6). The PRE curves obtained with noise $\delta = 0.01$ and $\delta = 0.1$ are plotted in Fig. 3, where the iteration at which the relative error is minimum is represented by a red star and the iteration d obtained by rule SR_d (10) by a green circle.

Finally the global evaluation of the two methods in all the cases reported in Table 1 shows that the Levemberg Marquardt Method is usually slightly better than BFGS in terms of mean PRE (see Fig. 4).

Table 4. Levemberg Marquardt results with $\lambda_0 = 0.05$ case $N_t = N_z = 40$, $\delta_q = 40\%$

δ	$LM(.05)$				min PRE		SR_d	
	PRE	$ResN$	k	fval	PRE	k	PRE	d
1e–3	2.2742e–2	5.3695e–1	6	28	9.8952e–3	5	2.1302e–2	3
5e–3	5.2977e–2	5.4327e–1	6	28	6.8027e–3	4	2.1448e–2	3
1e–2	1.1323e–1	5.6172e–1	6	28	6.7842e–3	4	1.8100e–2	3
5e–2	3.6183e+2	1.0134	11	48	5.6674e–2	3	5.6674e–2	3

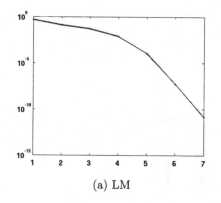
(a) LM

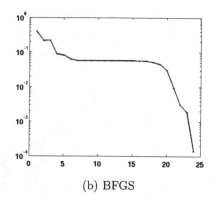
(b) BFGS

Fig. 1. LM and BFGS convergence plots: $\delta_q = 40\%$, $N_t = N_z = 40$

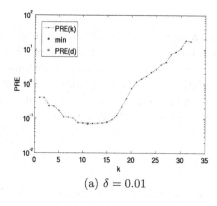
(a) $\delta = 0.01$

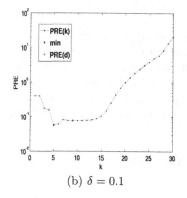
(b) $\delta = 0.1$

Fig. 2. BFGS convergence plots with noisy data, $\delta_q = 40\%$, $N_t = N_z = 40$

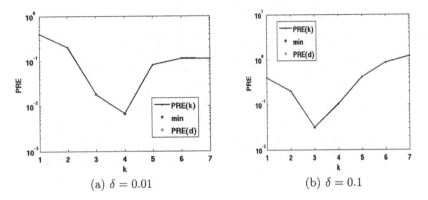

(a) $\delta = 0.01$ (b) $\delta = 0.1$

Fig. 3. Levemberg Marquardt ($LM(.05)$) $\delta_q = 40\%$, $N_t = N_z = 40$ (Color figure online)

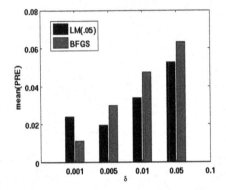

Fig. 4. Mean relative error for all the cases in Table 1 with noise $\delta \in [10^{-3}, 5 \cdot 10^{-2}]$

5   Conclusions

The present work reports tests of the Levemberg Marquardt and BFGS algorithms for modeling the kinetic terms, occurring in chemical processes of adsorption, in the presence of noisy data. The *semi-convergent* behavior of both methods is observed in presence of noise, confirming the need to introduce a suitable stopping criterion. A stopping rule, based on the behavior of the residual norm, is presented and the good performance is reported by the experimental tests.

References

1. Ciavarelli, R., Cappelletti, M., Fedi, S., Pinelli, D., Frascari, D.: Chloroform aerobic cometabolism by butane-growing rhodococcus aetherovorans BCP1 in continuous-flow biofilm reactors. Bioprocess Biosyst. Eng. **35**(5), 667–681 (2012)

2. Frascari, D., Cappelletti, M., Fedi, S., Verboschi, A., Ciavarelli, R., Nocentini, M., Pinelli, D.: Application of the growth substrate pulsed feeding technique to a process of chloroform aerobic cometabolism in a continuous-flow sand-filled reactor. Process Biochem. **47**(11), 1656–1664 (2012)

3. Zama, F., Ciavarelli, R., Frascari, D., Pinelli, D.: Numerical parameters estimation in models of pollutant transport with chemical reaction. In: Hömberg, D., Tröltzsch, F. (eds.) System Modeling and Optimization. volume 391 of IFIP Advances in Information and Communication Technology, pp. 547–556. Springer, Heidelberg (2013)

4. Frascari, D., Bacca, A.E.M., Zama, F., Bertin, L., Fava, F., Pinelli, D.: Olive mill wastewater valorisation through phenolic compounds adsorption in a continuous flow column. Chem. Eng. J. **283**, 293–303 (2016)

5. Fletcher, R.: Practical Methods of Optimization. Wiley, Hoboken (2013)

6. Nocedal, J., Wright, S.J.: Numerical Optimization, 2nd edn. Springer, New York (2006)

7. Scherzer, O.: The use of Morozov's discrepancy principle for Tikhonov regularization for solving nonlinear ill-posed problems. Computing **51**(1), 45–60 (1993)

8. Zama, F.: Parameter identification by iterative constrained regularization. J. Phys: Conf. Ser. **657**(1), 012002 (2015). IOP publishing

9. Elfving, T., Nikazad, T., Hansen, P.C.: Semi-convergence and relaxation parameters for a class of SIRT algorithms. Electron. Trans. Numer. Anal. **37**, 321–336 (2010)

10. Engl, H.W., Hanke, M., Neubauer, A.: Regularization of Inverse Problems. Kluwer, Dordrecht (1996)

11. Hanke, M.: Conjugate Gradient Type Methods for Ill-Posed Problems. Pitman Research Notes in Mathematics. Longman House, Harlow (1995). ISBN 0 582 27370 6

Author Index

Printed in the United States
By Bookmasters